D1559655

# THE GEOMETRY AND TOPOLOGY OF COXETER GROUPS

*London Mathematical Society Monographs Series*

The London Mathematical Society Monographs Series was established in 1968. Since that time it has published outstanding volumes that have been critically acclaimed by the mathematics community. The aim of this series is to publish authoritative accounts of current research in mathematics and high-quality expository works bringing the reader to the frontiers of research. Of particular interest are topics that have developed rapidly in the last ten years but that have reached a certain level of maturity. Clarity of exposition is important and each book should be accessible to those commencing work in its field.

The original series was founded in 1968 by the Society and Academic Press; the second series was launched by the Society and Oxford University Press in 1983. In January 2003, the Society and Princeton University Press united to expand the number of books published annually and to make the series more international in scope.



Vol. 32, *The Geometry and Topology of Coxeter Groups* by Michael W. Davis
Vol. 31, *Analysis of Heat Equations on Domains* by El Maati Ouhabaz

# THE GEOMETRY AND TOPOLOGY OF COXETER GROUPS

Michael W. Davis



PRINCETON UNIVERSITY PRESS
PRINCETON AND OXFORD

Published by Princeton University Press, 41 William Street, Princeton, New Jersey 08540
In the United Kingdom: Princeton University Press, 3 Market Place, Woodstock, Oxfordshire
OX20 1SY



Library of Congress Cataloging-in-Publication Data
Davis, Michael
The geometry and topology of Coxeter groups / Michael W. Davis.
p. cm.
Includes bibliographical references and index.
ISBN-13: 978-0-691-13138-2 (alk. paper)
ISBN-10: 0-691-13138-4
1. Coxeter groups. 2. Geometric group theory. I. Title.
QA183.D38 2007
51s′.2–dc22        2006052879

British Library Cataloging-in-Publication Data is available
This book has been composed in LaTeX
Printed on acid-free paper. ∞
press.princeton.edu

Printed in the United States of America

10 9 8 7 6 5 4 3 2 1

*To Wanda*

# Contents

# Preface

I became interested in the topology of Coxeter groups in 1976 while listening to Wu-chung and Wu-yi Hsiang explain their work [160] on finite groups generated by reflections on acyclic manifolds and homology spheres. A short time later I heard Bill Thurston lecture about reflection groups on hyperbolic 3-manifolds and I began to get an inkling of the possibilities for infinite Coxeter groups. After hearing Thurston's explanation of Andreev's Theorem for a second time in 1980, I began to speculate about the general picture for cocompact reflection groups on contractible manifolds. Vinberg's paper [290] also had a big influence on me at this time. In the fall of 1981 I read Bourbaki's volume on Coxeter groups [29] in connection with a course I was giving at Columbia. I realized that the arguments in [29] were exactly what were needed to prove my speculations. The fact that some of the resulting contractible manifolds were not homeomorphic to Euclidean space came out in the wash. This led to my first paper [71] on the subject. Coxeter groups have remained one of my principal interests.

There are many connections from Coxeter groups to geometry and topology. Two have particularly influenced my work. First, there is a connection with nonpositive curvature. In the mid 1980s, Gromov [146, 147] showed that, in the case of a "right-angled" Coxeter group, the complex $\Sigma$, which I had previously considered, admits a polyhedral metric of nonpositive curvature. Later my student Gabor Moussong proved this result in full generality in [221], removing the right-angled hypothesis. This is the subject of Chapter 12. The other connection has to do with the Euler Characteristic Conjecture (also called the Hopf Conjecture) on the sign of Euler characteristics of even dimensional, closed, aspherical manifolds. When I first heard about this conjecture, my initial reaction was that one should be able to find counterexamples by using Coxeter groups. After some unsuccessful attempts (see [72]), I started to believe there were no such counterexamples. Ruth Charney and I tried, again unsucccessfully, to prove this was the case in [55]. As explained in Appendix J, it is well known that Singer's Conjecture in $L^2$-cohomology implies the Euler Characteristic Conjecture. This led to my paper with Boris Okun [91] on the $L^2$-cohomology of Coxeter groups. Eventually, it also led to my interest in

Dymara's theory of weighted $L^2$-cohomology of Coxeter groups (described in Chapter 20) and to my work with Okun, Dymara and Januszkiewicz [79].

I began working on this book began while teaching a course at Ohio State University during the spring of 2002. I continued writing during the next year on sabbatical at the University of Chicago. My thanks go to Shmuel Weinberger for helping arrange the visit to Chicago. While there, I gave a minicourse on the material in Chapter 6 and Appendices B and C. One of the main reasons for publishing this book here in the *London Mathematical Society Monographs Series* is that in July of 2004 I gave ten lectures on this material for the London Mathematical Society Invited Lecture Series at the University of Southampton. I thank Ian Leary for organizing that conference. Also, in July of 2006 I gave five lectures for a minicourse on "$L^2$-Betti numbers" (from Chapter 20 and Appendix J) at Centre de Recherches mathématiques Université de Montreal.

I owe a great deal to my collaborators Ruth Charney, Jan Dymara, Jean-Claude Hausmann, Tadeusz Januszkiewicz, Ian Leary, John Meier, Gabor Moussong, Boris Okun, and Rick Scott. I learned a lot from them about the topics in this book. I thank them for their ideas and for their work. Large portions of Chapters 15, 16, and 20 come from my collaborations in [80], [55], and [79], respectively. I have also learned from my students who worked on Coxeter groups: Dan Boros, Constantin Gonciulea, Dongwen Qi, and Moussong.

*More acknowledgements*. Most of the figures in this volume were prepared by Sally Hayes. Others were done by Gabor Moussong in connection with our expository paper [90]. The illustration of the pentagonal tessellation of the Poincaré disk in Figure 6.2 was done by Jon McCammond. My thanks go to all three. I thank Angela Barnhill, Ian Leary, and Dongwen Qi for reading earlier versions of the manuscript and finding errors, typographical and otherwise. I am indebted to John Meier and an anonymous "reader" for some helpful suggestions, which I have incorporated into the book. Finally, I acknowledge the partial support I received from the NSF during the preparation of this book.

Columbus, Mike Davis
September, 2006

# THE GEOMETRY AND TOPOLOGY OF COXETER GROUPS

# Chapter One

# INTRODUCTION AND PREVIEW

## 1.1. INTRODUCTION

### Geometric Reflection Groups

Finite groups generated by orthogonal linear reflections on $\mathbb{R}^n$ play a decisive role in

- the classification of Lie groups and Lie algebras;
- the theory of algebraic groups, as well as, the theories of spherical buildings and finite groups of Lie type;
- the classification of regular polytopes (see [69, 74, 201] or Appendix B).

Finite reflection groups also play important roles in many other areas of mathematics, e.g., in the theory of quadratic forms and in singularity theory. We note that a finite reflection group acts isometrically on the unit sphere $\mathbb{S}^{n-1}$ of $\mathbb{R}^n$.

There is a similar theory of discrete groups of isometries generated by affine reflections on Euclidean space $\mathbb{E}^n$. When the action of such a Euclidean reflection group has compact orbit space it is called *cocompact*. The classification of cocompact Euclidean reflection groups is important in Lie theory [29], in the theory of lattices in $\mathbb{R}^n$ and in E. Cartan's theory of symmetric spaces. The classification of these groups and of the finite (spherical) reflection groups can be found in Coxeter's 1934 paper [67]. We give this classification in Table 6.1 of Section 6.9 and its proof in Appendix C.

There are also examples of discrete groups generated by reflections on the other simply connected space of constant curvature, hyperbolic $n$-space, $\mathbb{H}^n$. (See [257, 291] as well as Chapter 6 for the theory of hyperbolic reflection groups.)

The other symmetric spaces do not admit such isometry groups. The reason is that the fixed set of a reflection should be a submanifold of codimension one (because it must separate the space) and the other (irreducible) symmetric spaces do not have codimension-one, totally geodesic subspaces. Hence, they do not admit isometric reflections. Thus, any truly "geometric" reflection group must split as a product of spherical, Euclidean, and hyperbolic ones.

The theory of these geometric reflection groups is the topic of Chapter 6. Suppose $W$ is a reflection group acting on $\mathbb{X}^n = \mathbb{S}^n, \mathbb{E}^n$, or $\mathbb{H}^n$. Let $K$ be the closure of a connected component of the complement of the union of "hyperplanes" which are fixed by some reflection in $W$. There are several common features to all three cases:

- $K$ is geodesically convex polytope in $\mathbb{X}^n$.
- $K$ is a "strict" fundamental domain in the sense that it intersects each orbit in exactly one point (so, $\mathbb{X}^n/W \cong K$).
- If $S$ is the set of reflections across the codimension-one faces of $K$, then each reflection in $W$ is conjugate to an element of $S$ (and hence, $S$ generates $W$).

**Abstract Reflection Groups**

The theory of abstract reflection groups is due to Tits [281]. What is the appropriate notion of an "abstract reflection group"? At first approximation, one might consider pairs $(W, S)$, where $W$ is a group and $S$ is any set of involutions which generates $W$. This is obviously too broad a notion. Nevertheless, it is a step in the right direction. In Chapter 3, we shall call such a pair a "pre-Coxeter system." There are essentially two completely different definitions for a pre-Coxeter system to be an abstract reflection group.

The first focuses on the crucial feature that the fixed point set of a reflection should separate the ambient space. One version is that the fixed point set of each element of $S$ separates the Cayley graph of $(W, S)$ (defined in Section 2.1). In 3.2 we call $(W, S)$ a *reflection system* if it satisfies this condition. Essentially, this is equivalent to any one of several well-known combinatorial conditions, e.g., the Deletion Condition or the Exchange Condition. The second definition is that $(W, S)$ has a presentation of a certain form. Following Tits [281], a pre-Coxeter system with such a presentation is a "Coxeter system" and $W$ a "Coxeter group." Remarkably, these two definitions are equivalent. This was basically proved in [281]. Another proof can be extracted from the first part of Bourbaki [29]. It is also proved as the main result (Theorem 3.3.4) of Chapter 3. The equivalence of these two definitions is the principal mechanism driving the combinatorial theory of Coxeter groups.

The details of the second definition go as follows. For each pair $(s, t) \in S \times S$, let $m_{st}$ denote the order of $st$. The matrix $(m_{st})$ is the *Coxeter matrix* of $(W, S)$; it is a symmetric $S \times S$ matrix with entries in $\mathbb{N} \cup \{\infty\}$, 1's on the diagonal, and each off-diagonal entry $> 1$. Let

$$\mathcal{R} := \{(st)^{m_{st}}\}_{(s,t) \in S \times S}.$$

$(W, S)$ is a *Coxeter system* if $\langle S|\mathcal{R}\rangle$ is a presentation for $W$. It turns out that, given any $S \times S$ matrix $(m_{st})$ as above, the group $W$ defined by the presentation $\langle S|\mathcal{R}\rangle$ gives a Coxeter system $(W, S)$. (This is Corollary 6.12.6 of Chapter 6.)

## Geometrization of Abstract Reflection Groups

Can every Coxeter system $(W, S)$ be realized as a group of automorphisms of an appropriate geometric object? One answer was provided by Tits [281]: for any $(W, S)$, there is a faithful linear representation $W \hookrightarrow GL(N, \mathbb{R})$, with $N = \text{Card}(S)$, so that

- Each element of $S$ is represented by a linear reflection across a codimension-one face of a simplicial cone $C$. (N.B. A "linear reflection" means a linear involution with fixed subspace of codimension one; however, no inner product is assumed and the involution is not required to be orthogonal.)
- If $w \in W$ and $w \neq 1$, then $w(\text{int}(C)) \cap \text{int}(C) = \emptyset$ (here $\text{int}(C)$ denotes the interior of $C$).
- $WC$, the union of $W$-translates of $C$, is a convex cone.
- $W$ acts properly on the interior $\mathcal{I}$ of $WC$.
- Let $C^f := \mathcal{I} \cap C$. Then $C^f$ is the union of all (open) faces of $C$ which have finite stabilizers (including the face $\text{int}(C)$). Moreover, $C^f$ is a strict fundamental domain for $W$ on $\mathcal{I}$.

Proofs of the above facts can be found in Appendix D. Tits' result was extended by Vinberg [290], who showed that for many Coxeter systems there are representations of $W$ on $\mathbb{R}^N$, with $N < \text{Card}(S)$ and $C$ a polyhedral cone which is not simplicial. However, the poset of faces with finite stabilizers is exactly the same in both cases: it is the opposite poset to the poset of subsets of $S$ which generate finite subgroups of $W$. (These are the "spherical subsets" of Definition 7.1.1 in Chapter 7.) The existence of Tits' geometric representation has several important consequences. Here are two:

- Any Coxeter group $W$ is virtually torsion-free.
- $\mathcal{I}$ (the interior of the Tits cone) is a model for $\underline{E}W$, the "universal space for proper $W$-actions" (defined in 2.3).

Tits gave a second geometrization of $(W, S)$: its "Coxeter complex" $\Xi$. This is a certain simplicial complex with $W$-action. There is a simplex $\sigma \subset \Xi$ with $\dim \sigma = \text{Card}(S) - 1$ such that (a) $\sigma$ is a strict fundamental domain and (b) the elements of $S$ act as "reflections" across the codimension-one faces of $\sigma$. When $W$ is finite, $\Xi$ is homeomorphic to unit sphere $\mathbb{S}^{n-1}$ in the canonical representation, triangulated by translates of a fundamental simplex. When $(W, S)$ arises from an irreducible cocompact reflection group on $\mathbb{E}^n$, $\Xi \cong \mathbb{E}^n$. It turns out that $\Xi$ is contractible whenever $W$ is infinite.

The realization of $(W, S)$ as a reflection group on the interior $\mathcal{I}$ of the Tits cone is satisfactory for several reasons; however, it lacks two advantages enjoyed by the geometric examples on spaces of constant curvature:

- The $W$-action on $\mathcal{I}$ is not cocompact (i.e., the strict fundamental domain $C^f$ is not compact).
- There is no natural metric on $\mathcal{I}$ that is preserved by $W$. (However, in [200] McMullen makes effective use of a "Hilbert metric" on $\mathcal{I}$.)

In general, the Coxeter complex also has a serious defect—the isotropy subgroups of the $W$-action need not be finite (so the $W$-action need not be proper). One of the major purposes of this book is to present an alternative geometrization for $(W, S)$ which remedies these difficulties. This alternative is the cell complex $\Sigma$, discusssed below and in greater detail in Chapters 7 and 12 (and many other places throughout the book).

## The Cell Complex $\Sigma$

Given a Coxeter system $(W, S)$, in Chapter 7 we construct a cell complex $\Sigma$ with the following properties:

- The 0-skeleton of $\Sigma$ is $W$.
- The 1-skeleton of $\Sigma$ is Cay$(W, S)$, the Cayley graph of 2.1.
- The 2-skeleton of $\Sigma$ is a Cayley 2-complex (defined in 2.2) associated to the presentation $\langle S|\mathcal{R}\rangle$.
- $\Sigma$ has one $W$-orbit of cells for each spherical subset $T \subset S$. The dimension of a cell in this orbit is Card$(T)$. In particular, if $W$ is finite, $\Sigma$ is a convex polytope.
- $W$ acts properly on $\Sigma$.
- $W$ acts cocompactly on $\Sigma$ and there is a strict fundamental domain $K$.
- $\Sigma$ is a model for $\underline{E}W$. In particular, it is contractible.
- If $(W, S)$ is the Coxeter system underlying a cocompact geometric reflection group on $\mathbb{X}^n = \mathbb{E}^n$ or $\mathbb{H}^n$, then $\Sigma$ is $W$-equivariantly homeomorphic to $\mathbb{X}^n$ and $K$ is isomorphic to the fundamental polytope.

Moreover, the cell structure on $\Sigma$ is dual to the cellulation of $\mathbb{X}^n$ by translates of the fundamental polytope.

- The elements of $S$ act as "reflections" across the "mirrors" of $K$. (In the geometric case where $K$ is a polytope, a mirror is a codimension-one face.)
- $\Sigma$ embeds in $\mathcal{I}$ and there is a $W$-equivariant deformation retraction from $\mathcal{I}$ onto $\Sigma$. So $\Sigma$ is the "cocompact core" of $\mathcal{I}$.
- There is a piecewise Euclidean metric on $\Sigma$ (in which each cell is identified with a convex Euclidean polytope) so that $W$ acts via isometries. This metric is CAT(0) in the sense of Gromov [147]. (This gives an alternative proof that $\Sigma$ is a model for $\underline{E}W$.)

The last property is the topic of Chapter 12 and Appendix I. In the case of "right-angled" Coxeter groups, this CAT(0) property was established by Gromov [147]. ("Right angled" means that $m_{st} = 2$ or $\infty$ whenever $s \neq t$.) Shortly after the appearance of [147], Moussong proved in his Ph.D. thesis [221] that $\Sigma$ is CAT(0) for any Coxeter system. The complexes $\Sigma$ gave one of the first large class of examples of "CAT(0)-polyhedra" and showed that Coxeter groups are examples of "CAT(0)-groups." This is the reason why Coxeter groups are important in geometric group theory. Moussong's result also allowed him to find a simple characterization of when Coxeter groups are word hyperbolic in the sense of [147] (Theorem 12.6.1).

Since $W$ acts simply transitively on the vertex set of $\Sigma$, any two vertices have isomorphic neighborhoods. We can take such a neighborhood to be the cone on a certain simplicial complex $L$, called the "link" of the vertex. (See Appendix A.6.) We also call $L$ the "nerve" of $(W, S)$. It has one simplex for each nonempty spherical subset $T \subset S$. (The dimension of the simplex is Card$(T) - 1$.) If $L$ is homeomorphic to $S^{n-1}$, then $\Sigma$ is an $n$-manifold (Proposition 7.3.7).

There is great freedom of choice for the simplicial complex $L$. As we shall see in Lemma 7.2.2, if $L$ is the barycentric subdivision of any finite polyhedral cell complex, we can find a Coxeter system with nerve $L$. So, the topological type of $L$ is completely arbitrary. This arbitrariness is the source of power for the using Coxeter groups to construct interesting examples in geometric and combinatorial group theory.

### Coxeter Groups as a Source of Examples in Geometric and Combinatorial Group Theory

Here are some of the examples.

- The Eilenberg-Ganea Problem asks if every group $\pi$ of cohomological dimension 2 has a two-dimensional model for its classifying space $B\pi$

(defined in 2.3). It is known that the minimum dimension of a model for $B\pi$ is either 2 or 3. Suppose $L$ is a two-dimensional acyclic complex with $\pi_1(L) \neq 1$. Conjecturally, any torsion-free subgroup of finite index in $W$ should be a counterexample to the Eilenberg-Ganea Problem (see Remark 8.5.7). Although the Eilenberg-Ganea Problem is still open, it is proved in [34] that $W$ is a counterexample to the appropriate version of it for groups with torsion. More precisely, the lowest possible dimension for any $\underline{E}W$ is 3 ($=\dim \Sigma$) while the algebraic version of this dimension is 2.

- Suppose $L$ is a triangulation of the real projective plane. If $\Gamma \subset W$ is a torsion-free subgroup of finite index, then its cohomological dimension over $\mathbb{Z}$ is 3 but over $\mathbb{Q}$ is 2 (see Section 8.5).
- Suppose $L$ is a triangulation of a homology $(n-1)$-sphere, $n \geqslant 4$, with $\pi_1(L) \neq 1$. It is shown in [71] that a slight modification of $\Sigma$ gives a contractible $n$-manifold not homeomorphic to $\mathbb{R}^n$. This gave the first examples of closed apherical manifolds not covered by Euclidean space. Later, it was proved in [83] that by choosing $L$ to be an appropriate "generalized homology sphere," it is not necessary to modify $\Sigma$; it is already a CAT(0)-manifold not homeomorphic to Euclidean space. (Such examples are discussed in Chapter 10.)

### The Reflection Group Trick

This a technique for converting finite aspherical CW complexes into closed aspherical manifolds. The main consequence of the trick is the following.

**Theorem.** (Theorem 11.1). *Suppose $\pi$ is a group so that $B\pi$ is homotopy equivalent to a finite CW complex. Then there is a closed aspherical manifold $M$ which retracts onto $B\pi$.*

This trick yields a much larger class of groups than Coxeter groups. The group that acts on the universal cover of $M$ is a semidirect product $\widetilde{W} \rtimes \pi$, where $\widetilde{W}$ is an (infinitely generated) Coxeter group. In Chapter 11 this trick is used to produce a variety examples. These examples answer in the negative many of questions about aspherical manifolds raised in Wall's list of problems in [293]. By using the above theorem, one can construct examples of closed aspherical manifolds $M$ where $\pi_1(M)$ (a) is not residually finite, (b) contains infinitely divisible abelian subgroups, or (c) has unsolvable word problems. In 11.3, following [81], we use the reflection group trick to produce examples of closed aspherical topological manifolds not homotopy equivalent to closed smooth manifolds. In 11.4 we use the trick to show that if the Borel Conjecture (from surgery theory) holds for all groups $\pi$ which are fundamental groups of closed aspherical manifolds, then it must also hold for any $\pi$ with a finite classifying space. In 11.5 we combine a version of the reflection group trick with the examples of Bestvina and Brady in [24] to show that there are Poincaré duality groups which are not finitely presented. (Hence, there are Poincaré duality groups which do not arise as fundamental groups of closed aspherical manifolds.)

**Buildings**

Tits defined the general notion of a Coxeter system in order to develop the general theory of buildings. Buildings were originally designed to generalize certain incidence geometries associated to classical algebraic groups over finite fields. A building is a combinatorial object. Part of the data needed for its definition is a Coxeter system $(W, S)$. A building of type $(W, S)$ consists of a set $\Phi$ of "chambers" and a collection of equivalence relations indexed by the set $S$. (The equivalence relation corresponding to an element $s \in S$ is called "$s$-adjacency.") Several other conditions (which we will not discuss until 18.1) also must be satisfied. The Coxeter group $W$ is itself a building; a subbuilding of $\Phi$ isomorphic to $W$ is an "apartment." Traditionally (e.g., in [43]), the geometric realization of the building is defined to be a simplicial complex with one top-dimensional simplex for each element of $\Phi$. In this incarnation, the realization of each apartment is a copy of the Coxeter complex $\Xi$. In view of our previous discussion, one might suspect that there is a better definition of the geometric realization of a building where the realization of each chamber is isomorphic to $K$ and the realization of each apartment is isomorphic to $\Sigma$. This is in fact the case: such a definition can be found in [76], as well as in Chapter 18. A corollary to Moussong's result that $\Sigma$ is CAT(0) is that the geometric realization of any building is CAT(0). (See [76] or Section 18.3.)

A basic picture to keep in mind is this: in an apartment exactly two chambers are adjacent along any mirror while in a building there can be more than two. For example, suppose $W$ is the infinite dihedral group. The geometric realization of a building of type $W$ is a tree (without endpoints); the chambers are the edges; an apartment is an embedded copy of the real line.

**(Co)homology**

A recurrent theme in this book will be the calculation of various homology and cohomology groups of $\Sigma$ (and other spaces on which $W$ acts as a reflection group). This theme first occurs in Chapter 8 and later in Chapters 15 and 20 and Appendix J. Usually, we will be concerned only with cellular chains and cochains. Four different types of (co)homology will be considered.

(a) Ordinary homology $H_*(\Sigma)$ and cohomology $H^*(\Sigma)$.

(b) Cohomology with compact supports $H_c^*(\Sigma)$ and homology with infinite chains $H_*^{lf}(\Sigma)$.

(c) Reduced $L^2$-(co)homology $L^2\mathcal{H}^*(\Sigma)$.

(d) Weighted $L^2$-(co)homology $L^2_{\mathbf{q}}\mathcal{H}^*(\Sigma)$.

The main reason for considering ordinary homology groups in (a) is to prove $\Sigma$ is acyclic. Since $\Sigma$ is simply connected, this implies that it is contractible (Theorem 8.2.13).

The reason for considering cohomology with compact supports in (b) is that $H_c^*(\Sigma) \cong H^*(W; \mathbb{Z}W)$. We give a formula for these cohomology groups in Theorem 8.5.1. This has several applications: (1) knowledge of $H_c^1(\Sigma)$ gives the number of ends of $W$ (Theorem 8.7.1), (2) the virtual cohomological dimension of $W$ is $\max\{n | H_c^n(\Sigma) \neq 0\}$ (Corollary 8.5.5), and (3) $W$ is a virtual Poincaré duality group of dimension $n$ if and only if the compactly supported cohomology of $\Sigma$ is the same as that of $\mathbb{R}^n$ (Lemma 10.9.1). (In Chapter 15 we give a different proof of this formula which allows us to describe the $W$-module structure on $H^*(W; \mathbb{Z}W)$.)

When nonzero, reduced $L^2$-cohomology spaces are usually infinite-dimensional Hilbert spaces. A key feature of the $L^2$-theory is that in the presence of a group action it is possible to attach "von Neumann dimensions" to these Hilbert spaces; they are nonnegative real numbers called the "$L^2$-Betti numbers." The reasons for considering $L^2$-cohomology in (c) involve two conjectures about closed aspherical manifolds: the Hopf Conjecture on their Euler characteristics and the Singer Conjecture on their $L^2$-Betti numbers. The Hopf Conjecture (called the "Euler Characteristic Conjecture" in 16.2) asserts that the sign of the Euler characteristic of a closed, aspherical $2k$-manifold $M^{2k}$ is given by $(-1)^k\chi(M^{2k}) \geqslant 0$. This conjecture is implied by the Singer Conjecture (Appendix J.7) which asserts that for an aspherical $M^n$, all the $L^2$-Betti numbers of its universal cover vanish except possibly in the middle dimension. For Coxeter groups, in the case where $\Sigma$ is a $2k$-manifold, the Hopf Conjecture means that the rational Euler characteristic of $W$ satisfies $(-1)^k\chi(W) \geqslant 0$. In the right-angled case this can be interpreted as a conjecture about a certain number associated to any triangulation of a $(2k-1)$-sphere as a "flag complex" (defined in 1.2 as well as Appendix A.3). In this form, the conjecture is known as the Charney-Davis Conjecture (or as the Flag Complex Conjecture). In [91] Okun and I proved the Singer Conjecture in the case where $W$ is right-angled and $\Sigma$ is a manifold of dimension $\leq 4$ (see 20.5). This implies the Flag Complex Conjecture for triangulations of $S^3$ (Corollary 20.5.3).

The fascinating topic (d) of weighted $L^2$-cohomology is the subject of Chapter 20. The weight $\mathbf{q}$ is a certain tuple of positive real numbers. For simplicity, let us assume it is a single real number $q$. One assigns each cell $c$ in $\Sigma$ a weight $\|c\|_q = q^{l(w(c))}$, where $w(c)$ is the shortest $w \in W$ so that $w^{-1}c$ belongs to the fundamental chamber and $l(w(c))$ is its word length. $L^2_q C^*(\Sigma)$ is the Hilbert space of square summable cochains with respect to this new inner product. When $q = 1$, we get the ordinary $L^2$-cochains. The group $W$ no longer acts orthogonally; however, the associated Hecke algebra of weight $q$ is a $*$-algebra of operators. It can be completed to a von Neumann algebra $\mathcal{N}_q$ (see Chapter 19). As before, the "dimensions" of the associated reduced cohomology groups give us $L^2_q$-Betti numbers (usually not rational numbers). It turns out that the "$L^2_q$-Euler characteristic" of $\Sigma$ is $1/W(q)$ where $W(q)$ is the growth series of $W$. $W(q)$ is a rational function of $q$. (These growth series are the subject of Chapter 17.) In 20.7 we give a complete calculation of these $L^2_q$-Betti numbers for $q < \rho$ and $q > \rho^{-1}$, where $\rho$ is the radius of convergence of $W(q)$. When $q$ is the "thickness" (an integer) of a building $\Phi$ of type $(W, S)$ with a chamber transitive automorphism group $G$, the $L^2_q$-Betti numbers are the ordinary $L^2$-Betti numbers (with respect to $G$) of the geometric realization of $\Phi$ (Theorem 20.8.6).

### What Has Been Left Out

A great many topics related to Coxeter groups do not appear in this book, such as the Bruhat order, root systems, Kazhdan–Lusztig polynomials, and the relationship of Coxeter groups to Lie theory. The principal reason for their omission is my ignorance about them.

## 1.2. A PREVIEW OF THE RIGHT-ANGLED CASE

In the right-angled case the construction of $\Sigma$ simplifies considerably. We describe it here. In fact, this case is sufficient for the construction of most examples of interest in geometric group theory.

### Cubes and Cubical Complexes

Let $I := \{1, \dots, n\}$ and $\mathbb{R}^I := \mathbb{R}^n$. The *standard n-dimensional cube* is $[-1, 1]^I := [-1, 1]^n$. It is a convex polytope in $\mathbb{R}^I$. Its vertex set is $\{\pm 1\}^I$. Let $\{e_i\}_{i \in I}$ be the standard basis for $\mathbb{R}^I$. For each subset $J$ of $I$ let $\mathbb{R}^J$ denote the linear subspace spanned by $\{e_i\}_{i \in J}$. (If $J = \emptyset$, then $\mathbb{R}^\emptyset = \{0\}$.) Each face of $[-1, 1]^I$ is a translate of $[-1, 1]^J$ for some $J \subset I$. Such a face is said to be of *type J*.

For each $i \in I$, let $r_i : [-1, 1]^I \to [-1, 1]^I$ denote the orthogonal reflection across the hyperplane $x_i = 0$. The group of symmetries of $[-1, 1]^n$ generated by $\{r_i\}_{i \in I}$ is isomorphic to $(\mathbf{C}_2)^I$, where $\mathbf{C}_2$ denotes the cyclic group of order 2. $(\mathbf{C}_2)^I$ acts simply transitively on the vertex set of $[-1,1]^I$ and transitively on the set of faces of any given type. The stabilizer of a face of type $J$ is the subgroup $(\mathbf{C}_2)^J$ generated by $\{r_i\}_{i \in J}$. Hence, the poset of nonempty faces of $[-1,1]^I$ is isomorphic to the poset of cosets

$$\coprod_{J \subset I} (\mathbf{C}_2)^I / (\mathbf{C}_2)^J.$$

$(\mathbf{C}_2)^I$ acts on $[-1,1]^I$ as a group generated by reflections. A fundamental domain (or "fundamental chamber") is $[0,1]^I$.

A *cubical cell complex* $\Lambda$ is a regular cell complex in which each cell is combinatorially isomorphic to a standard cube. (A precise definition is given in Appendix A.) The *link* of a vertex $v$ in $\Lambda$, denoted $Lk(v, \Lambda)$, is the simplicial complex which realizes the poset of all positive dimensional cells which have $v$ as a vertex. If $v$ is a vertex of $[-1,1]^I$, then $Lk(v, [-1,1]^I)$ is the $(n-1)$-dimensional simplex, $\Delta^{n-1}$.

### The Cubical Complex $P_L$

Given a simplicial complex $L$ with vertex set $I = \{1, \dots, n\}$, we will define a subcomplex $P_L$ of $[-1,1]^I$, with the same vertex set and with the property that the link of each of its vertices is canonically identified with $L$. The construction is similar to the standard way of realizing $L$ as a subcomplex of $\Delta^{n-1}$. Let $\mathcal{S}(L)$ denote the set of all $J \subset I$ such that $J = \text{Vert}(\sigma)$ for some simplex $\sigma$ in $L$ (including the empty simplex). $\mathcal{S}(L)$ is partially ordered by inclusion. Define $P_L$ to be the union of all faces of $[-1,1]^I$ of type $J$ for some $J \in \mathcal{S}(L)$. So, the poset of cells of $P_L$ can be identified with the disjoint union

$$\coprod_{J \in \mathcal{S}(L)} (\mathbf{C}_2)^I / (\mathbf{C}_2)^J.$$

(This construction is also described in [37, 90, 91].)

**Example 1.2.1.** Here are some examples of the construction.

- If $L = \Delta^{n-1}$, then $P_L = [-1,1]^n$.
- If $L = \partial(\Delta^{n-1})$, then $P_L$ is the boundary of an $n$-cube, i.e., $P_L$ is homeomorphic to $S^{n-1}$.
- If $L$ is the disjoint union of $n$ points, then $P_L$ is the 1-skeleton of an $n$-cube.
- If $n = 3$ and $L$ is the disjoint union of a 1-simplex and a 0-simplex, then $P_L$ is the subcomplex of the 3-cube consisting of the top and bottom faces and the 4 vertical edges. (See Figure 1.1.)

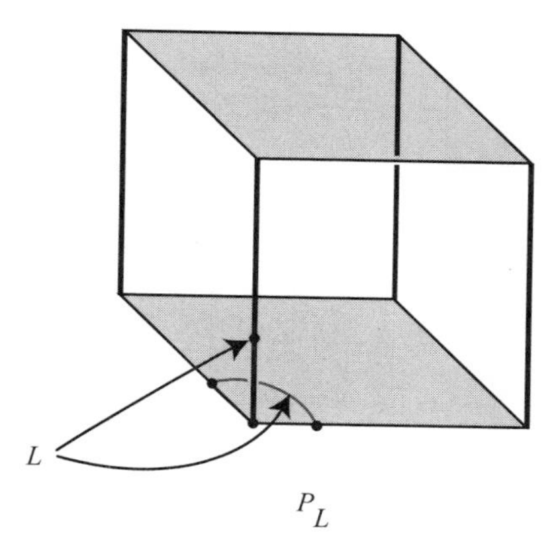


Figure 1.1. $L$ is the union of a 1-simplex and a 0-simplex.

- Suppose $L$ is the join of two simplicial complexes $L_1$ and $L_2$. (See Appendix A.4 for the definition of "join.") Then $P_L = P_{L_1} \times P_{L_2}$.
- So, if $L$ is a 4-gon (the join of $S^0$ with itself), then $P_L$ is the 2-torus $S^1 \times S^1$.
- If $L$ is an $n$-gon (i.e., the triangulation of $S^1$ with $n$ vertices), then $P_L$ is an orientable surface of Euler characteristic $2^{n-2}(4-n)$.

$P_L$ is stable under the $(\mathbf{C}_2)^I$-action on $[-1, 1]^I$. A fundamental chamber $K$ is given by $K := P_L \cap [0, 1]^I$. Note that $K$ is a cone (the cone point being the vertex with all coordinates 1). In fact, $K$ is homeomorphic to the cone on $L$. Since a neighborhood of any vertex in $P_L$ is also homeomorphic to the cone on $L$ we also get the following.

**PROPOSITION 1.2.2.** *If $L$ is homeomorphic to $S^{n-1}$, then $P_L$ is an $n$-manifold.*

*Proof.* The cone on $S^{n-1}$ is homeomorphic to an $n$-disk. □

### The Universal Cover of $P_L$ and the Group $W_L$

Let $\widetilde{P}_L$ be the universal cover of $P_L$. For example, the universal cover of the complex $P_L$ in Figure 1.1 is shown in Figure 1.2. The cubical cell structure on $P_L$ lifts to a cubical structure on $\widetilde{P}_L$. Let $W_L$ denote the group of all lifts of elements of $(\mathbf{C}_2)^I$ to homeomorphisms of $\widetilde{P}_L$ and let $\varphi : W_L \to (\mathbf{C}_2)^I$ be the homomorphism induced by the projection $\widetilde{P}_L \to P_L$. We have a short exact sequence,

$$1 \longrightarrow \pi_1(P_L) \longrightarrow W_L \xrightarrow{\varphi} (\mathbf{C}_2)^I \longrightarrow 1.$$

Since $(\mathbf{C}_2)^I$ acts simply transitively on $\mathrm{Vert}(P_L)$, $W_L$ is simply transitive on $\mathrm{Vert}(\widetilde{P}_L)$. By Theorem 2.1.1 in the next chapter, the 1-skeleton of $\widetilde{P}_L$ is

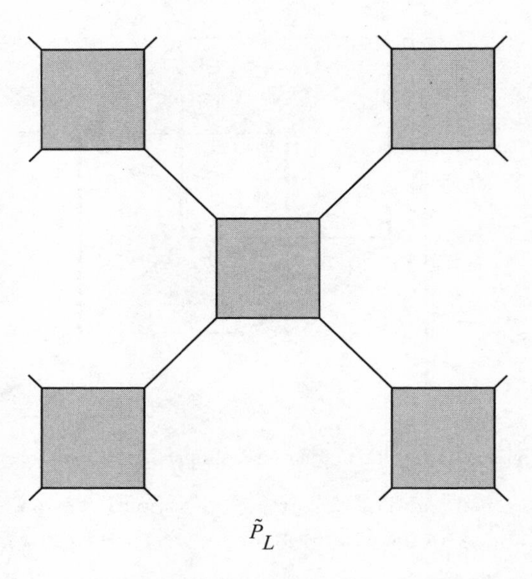


Figure 1.2. The universal cover of $P_L$.

Cay$(W_L, S)$ for some set of generators $S$ and by Proposition 2.2.4, the 2-skeleton of $\widetilde{P}_L$ is a "Cayley 2-complex" associated with some presentation of $W_L$. What is this presentation for $W_L$?

The vertex set of $P_L$ can be identified with $(\mathbf{C}_2)^I$. Fix a vertex $v$ of $P_L$ (corresponding to the identity element in $(\mathbf{C}_2)^I$). Let $\tilde{v}$ be a lift of $v$ in $\widetilde{P}_L$. The 1-cells at $v$ or at $\tilde{v}$ correspond to vertices of $L$, i.e., to elements of $I$. The reflection $r_i$ stabilizes the $i^{\text{th}}$ 1-cell at $v$. Let $s_i$ denote the unique lift of $r_i$ which stabilizes the $i^{\text{th}}$ 1-cell at $\tilde{v}$. Then $S := \{s_i\}_{i \in I}$ is a set of generators for $W_L$. Since $s_i^2$ fixes $\tilde{v}$ and covers the identity on $P_L$, we must have $s_i^2 = 1$. Suppose $\sigma$ is a 1-simplex of $L$ connecting vertices $i$ and $j$. The corresponding 2-cell at $\tilde{v}$ is a square with edges labeled successively by $s_i$, $s_j$, $s_i$, $s_j$. So, as explained in Section 2.2, we get a relation $(s_i s_j)^2 = 1$ for each 1-simplex $\{i, j\}$ of $L$. By Proposition 2.2.4, $W_L$ is the group defined by this presentation, i.e., $(W_L, S)$ is a right-angled Coxeter system, with $S := \{s_1, \dots, s_n\}$. Examining the presentation, we see that the abelianization of $W_L$ is $(\mathbf{C}_2)^I$. Thus, $\pi_1(P_L)$ is the commutator subgroup of $W_L$.

For each subset $J$ of $I$, $W_J$ denotes the subgroup generated by $\{s_i\}_{i \in J}$. If $J \in \mathcal{S}(L)$, then $W_J$ is the stabilizer of the corresponding cell in $\widetilde{P}_L$ which contains $\tilde{v}$ (and so, for $J \in \mathcal{S}(L)$, $W_J \cong (\mathbf{C}_2)^J$). It follows that the poset of cells of $\widetilde{P}_L$ is isomorphic to the poset of cosets,

$$\coprod_{J \in \mathcal{S}(L)} W_L / W_J.$$

### When Is $\widetilde{P}_L$ Contractible?

A simplicial complex $L$ is a *flag complex* if any finite set of vertices, which are pairwise connected by edges, spans a simplex of $L$. (Flag complexes play an important role throughout this book, e.g., in Sections 7.1 and 16.3 and Appendices A.3 and I.6.)

**PROPOSITION 1.2.3.** *The following statements are equivalent.*

*(i) $L$ is a flag complex.*

*(ii) $\widetilde{P}_L$ is contractible.*

*(iii) The natural piecewise Euclidean structure on $\widetilde{P}_L$ is* CAT(0).

*Sketch of Proof.* One shows (ii) $\Longrightarrow$ (i) $\Longrightarrow$ (iii) $\Longrightarrow$ (ii). If $L$ is not a flag complex, then it contains a subcomplex $L'$ isomorphic to $\partial \Delta^n$, for some $n \geqslant 2$, but which is not actually the boundary complex of any simplex in $L$. Each component of the subcomplex of $\widetilde{P}_L$ corresponding to $L'$ is homeomorphic to $S^n$. It is not hard to see that the fundamental class of such a sphere is nontrivial in $H_n(\widetilde{P}_L)$ (cf. Sections 8.1 and 8.2). So, if $L$ is not a flag complex, then $\widetilde{P}_L$ is not contractible, i.e., (ii) $\Longrightarrow$ (i). As we explain in Appendix I.6, a result of Gromov (Lemma I.6.1) states that a simply connected cubical cell complex is CAT(0) if and only if the link of each vertex is a flag complex. So, (i) $\Longrightarrow$ (iii). Since CAT(0) spaces are contractible (Theorem I.2.6 in Appendix I.2), (iii) $\Longrightarrow$ (ii). □

When $L$ is a flag complex, we write $\Sigma_L$ for $\widetilde{P}_L$. It is the cell complex referred to in the previous section.

**Examples 1.2.4.** In the following examples we assume $L$ is a triangulation of an $(n-1)$-manifold as a flag complex. Then $P_L$ is a manifold except possibly at its vertices (a neighborhood of the vertex is homeomorphic to the cone on $L$). If $L$ is the boundary of a manifold $X$, then we can convert $P_L$ into a manifold $M_{(L,X)}$ by removing the interior of each copy of $K$ and replacing it with a copy of the interior of $X$. We can convert $\Sigma_L$ into a manifold $\widehat{\Sigma}_{(L,X)}$ by a similar modification.

A metric sphere in $\Sigma_L$ is homeomorphic to a connected sum of copies of $L$, one copy for each vertex enclosed by the sphere. When $n \geqslant 4$, the fundamental group of such a connected sum is the free product of copies of $\pi_1(L)$ and hence, is not simply connected when $\pi_1(L) \neq 1$. It follows that $\Sigma_L$ is not simply connected at infinity when $\pi_1(L) \neq 1$. (See Example 9.2.7.) As we shall see in 10.3, for each $n \geqslant 4$, there are $(n-1)$-manifolds $L$ with the same homology as $S^{n-1}$ and with $\pi_1(L) \neq 1$ (the so-called "homology spheres"). Any such $L$ bounds a contractible manifold $X$. For such $L$ and $X$, we have that $M_{(L,X)}$ is homotopy equivalent to $P_L$. Its universal cover is $\widehat{\Sigma}_{(L,X)}$, which is contractible. Since $\widehat{\Sigma}_{(L,X)}$ is not simply connected at infinity, it is not homeomorphic to $\mathbb{R}^n$. The $M_{(L,X)}$ were the first examples of closed manifolds with contractible universal cover not homeomorphic to Euclidean space. (See Chapter 10, particularly Section 10.5, for more details.)

Finally, suppose $L = \partial X$, where $X$ is an aspherical manifold with boundary (i.e., the universal cover of $X$ is contractible). It is not hard to see that the closed manifold $M_{(L,X)}$ is also aspherical. This is the "reflection group trick" of Chapter 11.

# Chapter Two

# SOME BASIC NOTIONS IN GEOMETRIC GROUP THEORY

In geometric group theory we study various topological spaces and metric spaces on which a group $G$ acts. The first of these is the group itself with the discrete topology. The next space of interest is the "Cayley graph." It is a certain one dimensional cell complex with a $G$-action. Its definition depends on a choice of a set of generators $S$ for $G$. Cayley graphs for $G$ can be characterized as $G$-actions on connected graphs which are simply transitively on the vertex set (Theorem 2.1.1). Similarly, one can define a "Cayley 2-complex" for $G$ to be any simply connected, two dimensional cell complex with a cellular $G$-action which is simply transitive on its vertex set. To any presentation of $G$ one can associate a two-dimensional cell complex with fundamental group $G$. Its universal cover is a Cayley 2-complex for $G$. Conversely, one can read off from any Cayley 2-complex a presentation for $G$ (Proposition 2.2.4). These one- and two-dimensional complexes are discussed in Sections 2.1 and 2.2, respectively. One can continue attaching cells to the presentation complex, increasing the connectivity of the universal cover ad infinitum. If we add cells to the presentation 2-complex to kill all higher homotopy groups, we obtain a CW complex, $BG$, with fundamental group $G$ and with contractible universal cover. Any such complex is said to be *aspherical*. An aspherical complex is determined up to homotopy equivalence by its fundamental group, i.e., the homotopy type of $BG$ is an invariant of the group $G$. $BG$ is called a *classifying space for* $G$ (or a "$K(G, 1)$-complex"). Its universal cover, $EG$, is a contractible complex on which $G$ acts freely. In 2.3 we discuss aspherical complexes and give examples which are finite complexes or closed manifolds.

## 2.1. CAYLEY GRAPHS AND WORD METRICS

Let $G$ be a group with a set of generators $S$. Suppose the identity element, 1, is not in $S$. Define the *Cayley graph* $\mathrm{Cay}(G, S)$ as follows. The vertex set of $\mathrm{Cay}(G, S)$ is $G$. A two element subset of $G$ spans an edge if and only if it has the form $\{g, gs\}$ for some $g \in G$ and $s \in S$. Label the edge $\{g, gs\}$ by $s$. If the order of $s$ is not 2 (i.e., if $s \neq s^{-1}$), then the edge $\{g, gs\}$ has a direction: its initial vertex is $g$ and its terminal vertex is $gs$. (The labeled graph $\mathrm{Cay}(G, S)$ often will be denoted by $\Omega$ when $(G, S)$ is understood.)

An *edge path* $\gamma$ in $\Omega$ is a finite sequence of vertices $\gamma = (g_0, g_1, \dots, g_k)$ such that any two successive vertices are connected by an edge. Associated to $\gamma$ there is a sequence (or *word*) in $S \cup S^{-1}$, $\mathbf{s} = ((s_1)^{\varepsilon_1}, \dots, (s_k)^{\varepsilon_k})$, where $s_i$ is the label on the edge between $g_{i-1}$ and $g_i$ and $\varepsilon_i \in \{\pm 1\}$ is defined to be $+1$ if the edge is directed from $g_{i-1}$ to $g_i$ (i.e., if $g_i = g_{i-1}s_i$) and to be $-1$ if it is oppositely directed. Given such a word $\mathbf{s}$, define $g(\mathbf{s}) \in G$ by

$$g(\mathbf{s}) = (s_1)^{\varepsilon_1} \cdots (s_k)^{\varepsilon_k}$$

and call it the *value* of the word $\mathbf{s}$. Clearly, $g_k = g_0 g(\mathbf{s})$. This shows there is a one-to-one correspondence between edge paths from $g_0$ to $g_k$ and words $\mathbf{s}$ with $g_k = g_0 g(\mathbf{s})$. Since $S$ generates $G$, $\Omega$ is connected. $G$ acts on $\mathrm{Vert}(\Omega)$ (the vertex set of $\Omega$) by left multiplication and this naturally extends to a simplicial $G$-action on $\Omega$. $G$ is simply transitive on $\mathrm{Vert}(\Omega)$. (Suppose a group $G$ acts on a set $X$. The *isotropy subgroup* at a point $x \in X$ is the subgroup $G_x := \{g \in G \mid gx = x\}$. The $G$-action is *free* if $G_x$ is trivial for all $x \in X$; it is *transitive* if there is only one orbit and it is *simply transitive* if it is both transitive and free.)

Conversely, suppose that $\widetilde{\Omega}$ is a connected simplicial graph and that $G$ acts simply transitively on its 0-skeleton. (A graph is *simplicial* if it has no circuits of length 1 or 2. The *0-skeleton*, $\widetilde{\Omega}^0$, is the union of its vertices.) We can use $\widetilde{\Omega}$ to specify a set of generators $S$ for $G$ by the following procedure. First, choose a base point $v_0 \in \mathrm{Vert}(\widetilde{\Omega})$. Let $\widetilde{S}(v_0)$ denote the set of elements $x \in G$ such that $xv_0$ is adjacent to $v_0$. Noting that $x^{-1}$ takes the edge $\{v_0, xv_0\}$ to $\{x^{-1}v_0, v_0\}$, we see that if $x \in \widetilde{S}(v_0)$, then so is $x^{-1}$. Define $S(v_0)$ to be the set formed by choosing one element from each pair of the form $\{x, x^{-1}\}$. Clearly, $\widetilde{\Omega}$ is $G$-isomorphic to $\mathrm{Cay}(G, S(v_0))$. Explicitly, the isomorphism $\mathrm{Cay}(G, S(v_0)) \to \widetilde{\Omega}$ is induced by the $G$-equivariant isomorphism $g \to gv_0$ of vertex sets. (A map $f : A \to B$ between two $G$-sets is *equivariant* if $f(ga) = gf(a)$, for all $g \in G$.) So, we have proved the following.

**THEOREM 2.1.1.** *Suppose $\widetilde{\Omega}$ is a connected simplicial graph and $G$ is a group of automorphisms of $\widetilde{\Omega}$ which is simply transitive on* $\mathrm{Vert}(\widetilde{\Omega})$. *Let $S(v_0)$ be the set of generators for $G$ constructed above. Then $\widetilde{\Omega}$ is $G$-isomorphic to the Cayley graph,* $\mathrm{Cay}(G, S(v_0))$.

Thus, the study of Cayley graphs for $G$ is the same as the study of $G$-actions on connected, simplicial graphs such that $G$ is simply transitive on the vertex set.

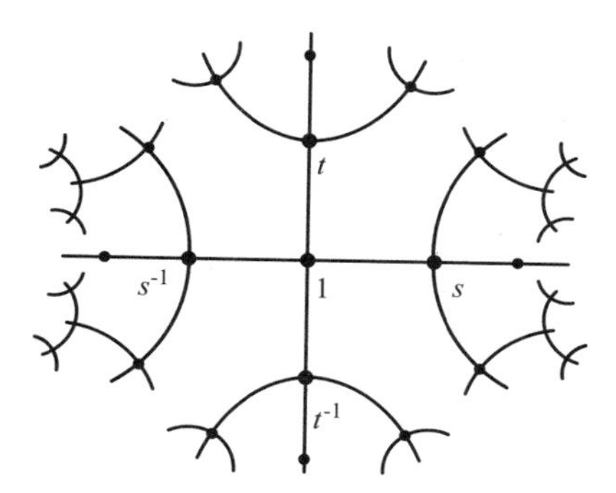


Figure 2.1. Cayley graph of the free group of rank 2.

**Example 2.1.2.** Suppose $S$ is a set and the group in question is $F_S$, the free group on $S$. Then $\text{Cay}(F_S, S)$ is a tree. (Since each element of $F_S$ can be written uniquely as a reduced word in $S \cup S^{-1}$, there is a unique edge path connecting any given element to 1; hence, $\text{Cay}(F_S, S)$ contains no circuits.) See Figure 2.1.

Roughly, any Cayley graph arises as a quotient of the above example. (The reason that this is only roughly true is that there are problems arising from elements of $S$ of order 1 or 2 in $G$.) Given a set of generators $S$ for $G$, we have $G = F_S/N$ for some normal subgroup $N$. Let $\varphi : F_S \to G$ be the projection. Set $T = \text{Cay}(F_S, S)$ and $\Omega = \text{Cay}(G, S)$. The homomorphism $\varphi$, regarded as a map of vertex sets, extends to a $\varphi$-equivariant map $T \to \Omega$. Let $\bar{\varphi} : T/N \to \Omega$ be the induced map. $\bar{\varphi}$ is almost an isomorphism. If $s \in S$ represents $1 \in G$ (i.e., if $s \in N$), then each edge in $T$ labeled by $s$ becomes a loop in $T/N$ and our convention is to omit such loops. If $s$ has order 2 in $G$ (i.e., if $s^2 \in N$), we have edge loops of length 2 in $T/N$ of the form $(g, gs, g)$ and our convention is to collapse such a loop to a single edge in $\Omega$.

## Word Length

We want to define a metric $d : \Omega \times \Omega \to [0, \infty)$. Declare each edge to be isometric to the unit interval. The length of a path in $\Omega$ is then defined in the obvious manner. Set $d(x, y)$ equal to the length of the shortest path from $x$ to $y$. (This procedure works in a much more general context: given any local metric on a path connected space $X$, define the *intrinsic distance* between two points to be the infimum of the set of lengths of paths which connect them. It is easy to see that the triangle inequality is valid, i.e., this procedure defines a metric. For further details, see Appendix I.1 and [37].) $G$ now acts isometrically on $\Omega$. Restricting the metric to the vertex set of $\Omega$, we get the *word metric* $d : G \times G \to \mathbb{N}$ where $\mathbb{N}$ denotes the nonnegative integers.

In other words, $d(h, g)$ is the smallest integer $k$ such that $g = hg(\mathbf{s})$ for a word $\mathbf{s}$ of length $k$ in $S \cup S^{-1}$. The distance from a group element $g$ to the identity element is its *word length* and is denoted $l(g)$.

## 2.2. CAYLEY 2-COMPLEXES

If $G$ acts on a connected, simplicial graph and is simply transitive on its vertex set, then, as in Theorem 2.1.1, the graph is essentially the Cayley graph of $G$ with respect to some set of generators $S$. Moreover, $S$ can be read off by looking at the edges emanating from some base point $v_0$. A *Cayley 2-complex* for $G$ is any simply connected, two-dimensional cell complex such that $G$ is simply transitive on the vertex set. So the 1-skeleton of such a 2-complex is essentially a Cayley graph for $G$. We explain below how one can read off a presentation for $G$ from the set of 2-cells containing a given vertex $v_0$.

### Presentations

Let $S$ be a set. Here, a *word in* $S \cup S^{-1}$ means an element $\mathbf{s}$ in the free group $F_S$ on $S$. In other words, $\mathbf{s} = (s_1)^{\varepsilon_1} \cdots (s_k)^{\varepsilon_k}$, where $s_i \in S$, $\varepsilon_i \in \{\pm 1\}$ and $(s_{i+1})^{\varepsilon_{i+1}} \neq (s_i)^{-\varepsilon_i}$.

**DEFINITION 2.2.1.** A *presentation* $\langle S \mid \mathcal{R} \rangle$ for a group consists of a set $S$ and a set $\mathcal{R}$ of words in $S \cup S^{-1}$. $S$ is the *set of generators*; $\mathcal{R}$ is the *set of relations*. The *group determined by the presentation* is $G := F_S/N(\mathcal{R})$, where $N(\mathcal{R})$ denotes the normal subgroup of $F_S$ generated by $\mathcal{R}$.

Suppose $H$ is a group and $f : S \to H$ a function. If $\mathbf{s} = (s_1)^{\varepsilon_1} \ldots (s_k)^{\varepsilon_k} \in F_S$, then put $f(\mathbf{s}) := f((s_1)^{\varepsilon_1}) \cdots f((s_k)^{\varepsilon_k})$. The group $G$ determined by $\langle S \mid \mathcal{R} \rangle$ satisfies the following universal property: given any group $H$ and any function $f : S \to H$ such that $f(\mathbf{r}) = 1$ for all $\mathbf{r} \in \mathcal{R}$, there is a unique extension of $f$ to a homomorphism $\tilde{f} : G \to H$. Moreover, up to canonical isomorphism, $G$ is characterized by this property.

### The Presentation 2-Complex

Associated with a presentation $\langle S \mid \mathcal{R} \rangle$ for a group $G$, there is a two-dimensional cell complex $X$ with $\pi_1(X) = G$. Its 0-skeleton, $X^0$, consists of a single vertex. Its 1-skeleton, $X^1$, is a bouquet of circles, one for each element of $S$. Each circle is assigned a direction and is labeled by the corresponding element of $S$. For each word $\mathbf{r} = (s_1)^{\varepsilon_1} \cdots (s_k)^{\varepsilon_k}$ in $\mathcal{R}$, take a two-dimensional disk $D_{\mathbf{r}}$, and subdivide its boundary $\partial D_{\mathbf{r}}$ into $k$ intervals. Cyclically label the edges by the $s_i$ which appear in $\mathbf{r}$ and orient them according to the $\varepsilon_i$. These labeled directed edges determine a (cellular) map from $\partial D_{\mathbf{r}}$ to $X^1$. Use it to attach a 2-disk to $X^1$ for each $\mathbf{r} \in \mathcal{R}$. The resulting CW complex $X$ is the *presentation complex*. (See the end of Appendix A.1 for a discussion of CW complexes.) By van Kampen's Theorem, $\pi_1(X) = G$. Its universal cover $\widetilde{X}$ is a Cayley 2-complex for $G$; however, its 1-skeleton need not be $\mathrm{Cay}(G, S)$. The difference is due entirely to the elements in $S$ of order 2. (This is important to us since,we are interested in Coxeter groups, in which case all elements of $S$ have order 2.) Given an element $s$ in $S$ of order 2, there are two edges in $\widetilde{X}$ connecting a vertex $v$ with $vs$, while in $\mathrm{Cay}(G, S)$ there is only one edge. Also $G$ need not act freely on $\mathrm{Cay}(G, S)$ since an edge which is labeled by an element of order 2 has stabilizer a cyclic group of order 2 (which necessarily fixes the midpoint of the edge). Similarly, $G$ need not act freely on a Cayley 2-complex, the stabilizer of a 2-cell can be nontrivial. However, such a 2-cell stabilizer must be finite since it freely permutes the vertices of the 2-cell (in fact, such a stabilizer must be cyclic or dihedral). Associated to a presentation of $G$ there is a Cayley 2-complex with 1-skeleton equal to $\mathrm{Cay}(G, S)$. We describe it below.

### The Cayley 2-Complex of a Presentation

Given a presentation $\langle S \mid \mathcal{R} \rangle$ for $G$, we define a 2-complex $\Lambda$ $(=\mathrm{Cay}(G, \langle S \mid \mathcal{R} \rangle))$ with $G$-action. Let $\mathcal{R}'$ denote the subset of $\mathcal{R}$ consisting of the words which are not of the form $s$ or $s^2$ for some $s \in S$. For each $\mathbf{r} \in \mathcal{R}'$, let $\gamma_{\mathbf{r}}$ be the closed edge path in $\Lambda$ which starts at $v_0$ and which corresponds to the relation $\mathbf{r}$. Let $D_{\mathbf{r}}$ be a copy of the two-dimensional disk. Regard $\gamma_{\mathbf{r}}$ as a map from the circle, $\partial D_{\mathbf{r}}$, to $\Lambda^1$. Call two closed edge paths *equivalent* if one is a reparameterization of the other, i.e., if they differ only by a shift of base point or change of direction. For the remainder of this section, let us agree that a *circuit* in $\Lambda^1$ means an equivalence class of a closed edge path. Let $C_{\mathbf{r}}$ denote the circuit represented by $\gamma_{\mathbf{r}}$. $G$ acts on $\{C_{\mathbf{r}}\}_{\mathbf{r} \in \mathcal{R}'}$. The stabilizer of a circuit can be nontrivial. (If the circuit has length $m$, then its stabilizer is a subgroup of the group of combinatorial symmetries of an $m$-gon.) Let $G_{\mathbf{r}}$ be the stabilizer of $C_{\mathbf{r}}$. $G_{\mathbf{r}}$ acts on $\partial D_{\mathbf{r}}$ in a standard fashion and since $D_{\mathbf{r}}$ is the cone on $\partial D_{\mathbf{r}}$, it also acts on $D_{\mathbf{r}}$. (The "standard action" of a dihedral group on a 2-disk will be discussed in detail in 3.1.)

The 1-skeleton of $\Lambda$ is defined to be the Cayley graph, $\mathrm{Cay}(G, S)$. For each $\mathbf{r} \in \mathcal{R}'$, attach a 2-cell to each circuit in the $G$-orbit of $C_{\mathbf{r}}$. More precisely, equivariantly attach $G \times_{G_{\mathbf{r}}} D_{\mathbf{r}}$ to the $G$-orbit of $C_{\mathbf{r}}$; $\Lambda$ is the resulting 2-complex. ($G \times_H X$ is the *twisted product*, defined as follows: if $H$ acts on $X$, $G \times_H X$ is the quotient space of $G \times X$ via the diagonal action defined by $h \cdot (g, x) := (gh^{-1}, hx)$. The natural left $G$-action on $G \times X$ descends to a left $G$-action on the twisted product.) $G$ acts on $\Lambda$ and is simply transitive on $\mathrm{Vert}(\Lambda)$. We will show in Proposition 2.2.3 below that $\Lambda$ is simply connected. So $\Lambda$ is a Cayley 2-complex.

**Examples 2.2.2.** (i) If $S = \{a, b\}$ and $\mathcal{R} = \{aba^{-1}b^{-1}\}$, then $G = \mathbf{C}_\infty \times \mathbf{C}_\infty$, the product of two infinite cyclic groups. $G$ can be identified with the integer lattice in $\mathbb{R}^2$ and $\mathrm{Cay}(G, S)$ with the grid consisting of the union of all horizontal and vertical lines through points with integral coordinates. The complex $\Lambda$ is the cellulation of $\mathbb{R}^2$ obtained by filling in the squares. In this case, $\Lambda$ is the same as $\widetilde{X}$ (the universal cover of the presentation complex).

(ii) Suppose $S$ is a singleton, say, $S = \{a\}$ and $\mathcal{R} = \{a^m\}$, for some $m > 2$. Then $G$ is cyclic of order $m$, $X$ is the result of gluing a 2-disk onto a circle via a degree-$m$ map $\partial D^2 \to S^1$ and $\widetilde{X}$ consists of $m$-copies of a 2-disk with their boundaries identified. Their common boundary is the single circuit corresponding to the relation $\mathbf{r} = a^m$. On the other hand, the 2-complex $\Lambda$ is the single 2-disk $D_{\mathbf{r}}$ (an $m$-gon) with the cyclic group acting by rotation.

**PROPOSITION 2.2.3.** *$\Lambda$ is simply connected.*

*Proof.* Let $p : \widetilde{\Lambda} \to \Lambda$ be the universal covering. It suffices to show that the $G$-action on $\Lambda$ lifts to a $G$-action on $\widetilde{\Lambda}$. Indeed, suppose the $G$-action lifts. Since $G$ is simply transitive on $\mathrm{Vert}(\Lambda)$, it must also be simply transitive on $\mathrm{Vert}(\widetilde{\Lambda})$. This means that $p$ is a bijection on vertex sets and hence, that the covering map $p : \widetilde{\Lambda} \to \Lambda$ is an isomorphism and therefore $\Lambda$ is simply connected. So, we need to show that we can lift the $G$-action. First we lift the elements in $S$ to $\widetilde{\Lambda}$. Let $v_0 \in \mathrm{Vert}(\Lambda)$ be the vertex corresponding to $1 \in G$. Choose $\tilde{v}_0 \in \mathrm{Vert}(\widetilde{\Lambda}) \in p^{-1}(v_0)$. Given $s \in S$, let $e$ be the edge in $\Lambda$ emanating from $v_0$ which is labeled by $s$ and let $\tilde{e}$ be its lift in $\widetilde{\Lambda}$ with endpoint $\tilde{v}_0$. Let $v_1$ and $\tilde{v}_1$ be the other endpoints of $e$ and $\tilde{e}$, respectively. Any lift of $s$ takes $\tilde{v}_0$ to a lift of $v_1$ and the lift of $s$ is uniquely determined by the choice of lift of $v_1$. Let $\tilde{s} : \widetilde{\Lambda} \to \widetilde{\Lambda}$ be the lift of $s$ which takes $\tilde{v}_0$ to $\tilde{v}_1$. If $s^2 = 1$, then $\tilde{s}^2$ is the identity map on $\widetilde{\Lambda}$ (since it is a lift of the identity and fixes a vertex). Let $\mathbf{r} = (s_1)^{\varepsilon_1} \cdots (s_k)^{\varepsilon_k} \in \mathcal{R}'$ and $D_{\mathbf{r}}$ a corresponding 2-cell in $\Lambda$ which contains $v_0$. Let $\widetilde{D}_{\mathbf{r}}$ be the lift in $\widetilde{\Lambda}$ which contains $\tilde{v}_0$. The corresponding closed edge path $\gamma_{\mathbf{r}}$ starting at $v_0$ lifts to a closed edge path $\tilde{\gamma}_{\mathbf{r}}$ going around $\partial\widetilde{D}_{\mathbf{r}}$ with edge labels corresponding to the word $(\tilde{s}_1)^{\varepsilon_1} \cdots (\tilde{s}_k)^{\varepsilon_k}$, so this element gives the identity map on $\widetilde{\Lambda}$. It follows from the characteristic property of presentations that the function $S \to \mathrm{Aut}(\widetilde{\Lambda})$ defined by $s \to \tilde{s}$ extends to a homomorphism $G \to \mathrm{Aut}(\widetilde{\Lambda})$ giving the desired lift of the $G$-action to $\widetilde{\Lambda}$. □

## Reading off a Presentation

Suppose $\Lambda$ is a Cayley 2-complex for $G$. Choose a base point $v_0 \in \mathrm{Vert}(\Lambda)$. We can read off a presentation $\langle S \mid \mathcal{R} \rangle$ from the set of 1- and 2-cells containing $v_0$, as follows. The set $S$ of generators is chosen by the procedure explained in 2.1: for each edge $e$ emanating from $v_0$, let $s_e \in G$ be the element taking $v_0$ to the other endpoint of $e$, $S$ is the set of all such $s_e$. If $s \in S$ stabilizes its edge, put $s^2$ into $\mathcal{R}$. In other words in $\mathcal{R}$ are given by the procedure indicated earlier. For each 2-cell $c$ containing $v_0$, we get a closed edge path $\gamma_c$ starting at $v_0$ and going around $c$. Cyclically reading the labels on the edges, we get a word $\mathbf{r}_c$ in $S \cup S^{-1}$. The definition of $\mathcal{R}$ is completed by putting all such $\mathbf{r}_c$ into $\mathcal{R}$. The corresponding group element $g(\mathbf{r}_c) \in G$ takes $v_0$ to itself. Since $G$ acts freely on Vert($\Lambda$), $g(\mathbf{r}_c) = 1$. Thus, each $\mathbf{r}_c$ is a relation $\mathbf{r}_c$ in $G$.

**PROPOSITION 2.2.4.** *Suppose $\Lambda$ is a Cayley 2-complex for $G$ and $\langle S \mid \mathcal{R}\rangle$ is the associated presentation. Then $\langle S \mid \mathcal{R}\rangle$ is a presentation for $G$.*

*Proof.* Let $\widetilde{G}$ be the group defined by $\langle S \mid \mathcal{R}\rangle$ . Since each element of $\mathcal{R}$ is a relation in $G$, we have a homomorphism $\rho : \widetilde{G} \to G$ and since $S$ generates $G$, $\rho$ is onto. $\widetilde{G}$ acts on $\Lambda$ via $\rho$. Let $\tilde{g} \in \operatorname{Ker} \rho$. Choose a word $\mathbf{s}$ in $S \cup S^{-1}$ which represents $\tilde{g}$ and let $\gamma$ be the corresponding closed edge path in $\Lambda$ based at $v_0$.

It is a classical result in topology that the fundamental group of a cell complex can be defined combinatorially. (This result is attributed to Tietze in [98, p. 301].) In particular, two closed edge paths are homotopic if and only if one can be obtained from the other by a sequence of moves, each of which replaces a segment in the boundary of a 2-cell by the complementary segment. (See, for example, [10, pp. 131–135].) Since $\Lambda$ is simply connected, $\gamma$ is null homotopic. Since each 2-cell of $\Lambda$ is a translate of a 2-cell corresponding to an element of $\mathcal{R}$, this implies that $\tilde{g}$ $(= \tilde{g}(\mathbf{s}))$ lies in the normal subgroup generated by $\mathcal{R}$, i.e., $\tilde{g} = 1$. So $\operatorname{Ker}\rho$ is trivial and $\rho$ is an isomorphism. □

## 2.3. BACKGROUND ON ASPHERICAL SPACES

A path connected space $X$ is *aspherical* if its homotopy groups, $\pi_i(X)$, vanish for all $i > 1$. So, an aspherical space has at most one nontrivial homotopy group—its fundamental group. A basic result of covering space theory (for example, in [197]) states that if $X$ admits a universal covering space $\widetilde{X}$, then asphericity is equivalent to the condition that $\pi_i(\widetilde{X}) = 0$ for all $i$ (that is to say, $\widetilde{X}$ is *weakly contractible*). If $X$ is homotopy equivalent to a CW complex (and we shall assume this throughout this section), a well-known theorem of J. H. C. Whitehead [301] (see [153, pp. 346–348] for a proof) asserts that if $\widetilde{X}$ is weakly contractible, then it is contractible. So, for spaces homotopy equivalent to CW complexes, the condition that $X$ be aspherical is equivalent to the condition that its universal cover be contractible. (For more on CW complexes, see the end of Appendix A.1.)

For any group $\pi$ there is a standard construction (in fact, several standard constructions) of an aspherical CW complex with fundamental group $\pi$. One such construction starts with the presentation complex defined in the previous section and then attachs cells of dimension $\geqslant 3$ to kill the higher homotopy groups. (See [153, p. 365].) This complex, or any other homotopy equivalent to it, is denoted $B\pi$ and called a *classifying space* for $\pi$. (It is also called an "Eilenberg-MacLane space" for $\pi$ or a "$K(\pi, 1)$-complex.") The universal cover of $B\pi$ is denoted $E\pi$ and called the *universal space* for $\pi$. A classifying space $B\pi$ has the following universal property. Suppose we are given a base point $x_0 \in B\pi$ and an identification of $\pi_1(B\pi, x_0)$ with $\pi$. Let $Y$ be another CW complex with base point $y_0$ and $\varphi : \pi_1(Y, y_0) \to \pi$ a homomorphism. Then there is a map $f : (Y, y_0) \to (B\pi, x_0)$ such that the induced homomorphism on fundamental groups is $\varphi$; moreover, $f$ is unique up to homotopy (relative to the base point). (This is an easy exercise in obstruction theory.) It follows from this universal property that the complex $B\pi$ is unique up to a homotopy equivalence inducing the identity map on $\pi$. In particular, any aspherical CW complex with fundamental group $\pi$ is homotopy equivalent to $B\pi$.

Next we give some examples. (For the remainder of this section, all spaces are path connected.)

### Some Examples of Aspherical Manifolds

*Dimension* 1. The only (connected) closed 1-manifold is the circle $S^1$. Its universal cover is the real line $\mathbb{R}^1$, which is contractible. So $S^1$ is aspherical.

*Dimension* 2. Suppose $X$ is a closed orientable surface of genus $g > 0$. By the Uniformization Theorem of Riemann and Poincaré, $X$ can be given a Riemannian metric so that its universal cover $\widetilde{X}$ is isometrically identified with either the Euclidean plane (if $g = 1$) or the hyperbolic plane (if $g > 1$). Since both planes are contractible, $X$ is aspherical. Similarly, recalling that any closed nonorientable surface $X$ can be written as a connected sum of projective planes, we see that a nonorientable $X$ is aspherical if and only if this connected sum decomposition has more than one term. The two remaining closed surfaces, the 2-sphere and the projective plane are not aspherical since they have $\pi_2 = \mathbb{Z}$. In summary, a closed surface is aspherical if and only if its Euler characteristic is $\leqslant 0$.

*Dimension* 3. Any closed orientable 3-manifold has a unique connected sum decomposition into 3-manifolds which cannot be further decomposed as nontrivial connected sums. Such an indecomposable 3-manifold is said to be *prime*. The 2-spheres along which we take connected sums in this decomposition are nontrivial in $\pi_2$ (provided they are not the boundaries of homotopy balls). Hence, if there are at least two terms in the decomposition which are not homotopy spheres, then the 3-manifold will not be aspherical. (By Perelman's proof [237, 239] of the Poincaré Conjecture, fake homotopy 3-balls or 3-spheres do not exist.) On the other hand, prime 3-manifolds with infinite fundamental group generally are aspherical, the one orientable exception being $S^2 \times S^1$. (This follows from Papakyriakopoulos' Sphere Theorem; see [267] or the original paper [232].)

*Tori.* The $n$-dimensional torus $T^n$ is aspherical since its universal cover is $n$-dimensional Euclidean space $\mathbb{E}^n$. The same is true for all complete Euclidean manifolds (called *flat* manifolds) as well as for all complete affine manifolds.

*Hyperbolic manifolds.* The universal cover of a complete hyperbolic $n$-manifold $X^n$ can be identified with hyperbolic $n$-space $\mathbb{H}^n$. (This is essentially a definition.) In other words, $X^n = \mathbb{H}^n / \Gamma$ where $\Gamma$ is a discrete torsion-free subgroup of $\mathrm{Isom}(\mathbb{H}^n)$ (the isometry group of $\mathbb{H}^n$). Since $\mathbb{H}^n$ is contractible, $X^n$ is aspherical.

We will say more about Euclidean and hyperbolic manifolds in 6.2 and 6.4.

*Lie groups.* Suppose $G$ is a Lie group, $K$ a maximal compact subgroup and $\Gamma$ a torsion-free discrete subgroup of $G$. Then $G/K$ is diffeomorphic to Euclidean space and $\Gamma$ acts freely on $G/K$. It follows that $X = \Gamma\backslash G/K$ is an aspherical manifold (its universal cover is $G/K$). Complete hyperbolic manifolds and other locally symmetric spaces are examples of this type and so are complete affine manifolds. By taking $G$ to be a connected nilpotent or solvable Lie group we get, respectively, nil-manifolds and solv-manifolds.

*Manifolds of nonpositive sectional curvature.* If $X^n$ is a complete Riemannian manifold of nonpositive sectional curvature, then it is aspherical. The reason is the Cartan-Hadamard Theorem which asserts that for any $x \in X^n$, the exponential map, $\exp : T_x X^n \to X^n$, is a covering projection. Hence, the universal cover of $X^n$ is is diffeomorphic to $T_x X^n$ ($\cong \mathbb{R}^n$).

## Some Examples of Finite Aspherical CW Complexes

Here we are concerned with examples where $B\pi$ is a *finite* complex (or at least finite dimensional).

*Dimension* 1. Suppose $X$ is a (connected) graph. Its universal cover $\widetilde{X}$ is a tree which is contractible. Hence, any graph is aspherical. The fundamental group of a graph is a free group.

*Dimension* 2. The presentation 2-complex is sometimes aspherical. For example, a theorem of R. Lyndon [195] asserts that if $\pi$ is a finitely generated 1-relator group and the relation cannot be written as a proper power of another word, then the presentation 2-complex for $\pi$ is aspherical. Another large class of groups for which this holds are groups with presentations as "small cancellation groups." (See R. Strebel's article in [138, pp. 227–273].)

*Nonpositively curved polyhedra.* Our fund of examples of aspherical complexes was greatly increased in 1987 with the appearance of Gromov's landmark paper [147]. He described several different constructions of polyhedra with piecewise Euclidean metrics which were nonpositively curved in the the sense of Aleksandrov. Moreover, he proved such polyhedra were aspherical. He showed some of the main constructions of this book could be explained in terms of nonpositive curvature (see 1.2 and Chapter 12). Gromov developed two other techniques for constructing nonpositively curved polyhedra. These go under the names "branched covers" and "hyperbolization." For example, he showed that a large class of examples of aspherical manifolds can be constructed by taking branched covers of an $n$-torus along a union of totally geodesic codimension-two subtori, [147, pp. 125–126]. The term "hyperbolization" refers to constructions for functorially converting a cell complex into a nonpositively curved polyhedron with the same local structure (but different global topology). (In 12.8 we discuss a technique of "relative hyperbolization" using a version of the reflection group trick.) For more about the branched covering space techniques, see [54]. For expositions of the hyperbolization techniques of [147, pp. 114–117], see [59, 83, 86, 236]. In the intervening years there has been a great deal of work in this area. A lot of it can be found in the book of Bridson and Haefliger [37]. We discuss the general theory of nonpositively curved polyhedra in Appendix I and applications of this theory to the reflection group examples in Chapter 12. For other expositions of the general theory of nonpositively curved polyhedra and spaces, see [1, 14, 45, 78, 90] and Ballman's article [138, pp. 189–201].

*Word hyperbolic groups.* In [147] Gromov considered the notion of what it means for a metric space to be "negatively curved in the large" or "coarsely negatively curved" or in Gromov's terminology "hyperbolic." When applied to the word metric on a group this leads to the notion of a "word hyperbolic group," a notion which had been discovered earlier, independently by Rips and Cooper. For example, the fundamental group of any closed Riemannian manifold of strictly negatve sectional curvature is word hyperbolic. Word hyperbolicity is independent of the choice of generating set. Rips proved that, given a word hyperbolic group $\pi$, there is a contractible simplicial complex $R$ on which $\pi$ acts simplicially with all cell stabilizers finite and with compact quotient. $R$ is called a "Rips complex" for $\pi$. It follows that, when $\pi$ is torsion-free, $R/\pi$ is a finite model for $B\pi$. (For background on word hyperbolic groups, see 12.5, as well as, [37, 144, 147].)

## The Universal Space for Proper $G$-Actions

The action of group of deck transformations on a covering space is proper and free. Conversely, if a group $G$ acts freely and properly on a space $X$, then $X \to X/G$ is a covering projection and $G$ is the group of deck transformations. (The notion of a "proper" action is given in Definition 5.1.5. In the context of cellular actions on CW complexes it means simply that the stabilizer of each cell is finite.) As was first observed by P. A. Smith, a finite cyclic group $\mathbf{C}_m$,

$m > 1$, cannot act freely on a finite-dimensional, contractible CW complex (or even an acyclic one). The reason is that the cohomology of $B\mathbf{C}_m$ is nonzero in arbitrarily high dimensions. It follows that if $G$ has nontrivial torsion, then its classifying space $BG$ cannot be finite dimensional (since $B\mathbf{C}_m$ is a covering space of it). On the other hand, there are many natural examples of groups with torsion acting properly on contractible manifolds or spaces, e.g., discrete groups of isometries of symmetric spaces.

**DEFINITION 2.3.1.** Let $G$ be a discrete group. A CW complex $X$ together with a cellular, proper $G$-action is a *universal space for proper G-actions* if, for each finite subgroup $F$, its fixed point set $X^F$ is contractible.

We note some consequences of this defintion. Since the action is proper, $X^F = \emptyset$ whenever $F$ is infinite. By taking $F$ to be the trivial subgroup, we see that $X$ must be contractible. If $G$ is torsion-free, then $EG$ is a universal space for proper actions.

The notion in Definition 2.3.1 was introduced in [288]. In the same paper it is proved that a universal space for proper $G$-actions always exists and is unique up to $G$-homotopy equivalence. Such a universal space is denoted $\underline{E}G$. Its universal property is the following: given a CW complex $Y$ with a proper, cellular action of $G$, there is a $G$-equivariant map $Y \to \underline{E}G$, unique up to $G$-homotopy. An immediate consequence is the uniqueness of $\underline{E}G$ up to $G$-homotopy equivalence.

# Chapter Three

## COXETER GROUPS

Given a group $W$ and a set $S$ of involutory generators, when does $(W, S)$ deserve to be called an "abstract reflection group" (or a "Coxeter system")? In this chapter we give the two answers alluded to in 1.1. The first is that for each $s \in S$ its fixed set separates $\mathrm{Cay}(W, S)$ (see 3.2). The second is that $W$ has a presentation of a certain form (see 3.3). The main result, Theorem 3.3.4, asserts that these answers are equivalent. Along the way we find three combinatorial conditions (D), (E), and (F) on $(W, S)$, each of which is equivalent to it being a Coxeter system. This line of reasoning culminates with Tits' solution of the word problem for Coxeter groups, which we explain in 3.4.

### 3.1. DIHEDRAL GROUPS

A Coxeter group with one generator is cyclic of order 2. The Coxeter groups with two generators are the dihedral groups. They are key to understanding general Coxeter groups.

**DEFINITION 3.1.1.** A *dihedral group* is a group generated by two elements of order 2.

**Example 3.1.2.** (*Finite dihedral groups*). Given a line $L$ in $\mathbb{R}^2$, let $r_L$ denote orthogonal reflection across $L$. If $L$ and $L'$ are two lines through the origin in $\mathbb{R}^2$ and $\theta$ is the angle between them, then $r_L \circ r_{L'}$ is rotation through $2\theta$. So, if $\theta = \pi/m$, where $m$ is an integer $\geqslant 2$, then $r_L \circ r_{L'}$ is rotation through $2\pi/m$. Consequently, $r_L \circ r_{L'}$ has order $m$. In this case we denote the dihedral subgroup of $O(2)$ generated by $r_L$ and $r_{L'}$ by $\mathbf{D}_m$. (Here $O(2)$ means the group of orthogonal transformations of $\mathbb{R}^2$.) See Figure 3.1. We shall show below that $\mathbf{D}_m$ is finite of order $2m$.

**Example 3.1.3.** (*The infinite dihedral group*). This group is generated by two isometric affine transformations of the real line. Let $r$ and $r'$ denote the reflections about the points 0 and 1, respectively (that is, $r(t) = -t$ and $r'(t) = 2 - t$). Then $r' \circ r$ is translation by 2 (and hence, has infinite order). $\mathbf{D}_\infty$ denotes the subgroup of $\mathrm{Isom}(\mathbb{R})$ generated by $r$ and $r'$. See Figure 3.2.

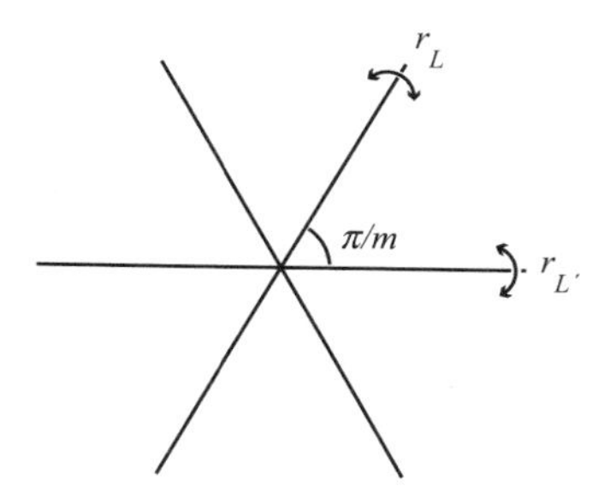


Figure 3.1. The dihedral group $\mathbf{D}_3$.

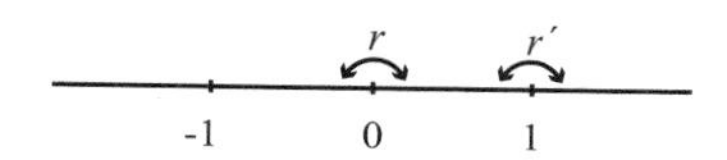


Figure 3.2. The infinite dihedral group $\mathbf{D}_\infty$.

**Example 3.1.4.** For $m$ a positive integer $\geqslant 2$ or for $m = \infty$, let $\mathbf{C}_m$ denote the cyclic group of order $m$ (written multiplicatively). Let $\pi$ be a generator. Regard $\mathbf{C}_2$ as $\{\pm 1\}$. Define an action of $\mathbf{C}_2$ on $\mathbf{C}_m$ by $\varepsilon \cdot x = x^\varepsilon$, where $\varepsilon = \pm 1$. Form the semidirect product $\mathbf{G}_m = \mathbf{C}_m \rtimes \mathbf{C}_2$. In other words, $\mathbf{G}_m$ consists of all pairs $(x, \varepsilon) \in \mathbf{C}_m \times C_2$ and multiplication is defined by

$$(x, \varepsilon) \cdot (x', \varepsilon') = (x^{\varepsilon'} x, \varepsilon\varepsilon').$$

Identify $\pi$ with $(\pi, 1)$ and put $\sigma = (1, -1)$, $\tau = (\pi, -1)$. Thus, $\mathbf{C}_m$ is a normal subgroup of $\mathbf{G}_m$ and $\sigma$, $\tau$ are elements of order 2 which generate $\mathbf{G}_m$. Moreover, the order of $\mathbf{G}_m$ is $2m$ (if $m \neq \infty$) or $\infty$ (if $m = \infty$).

The next lemma shows that a dihedral group is characterized by the order $m$ of the product of the two generators; so, for each $m$ there is exactly one dihedral group up to isomorphism.

**Lemma 3.1.5.** ([29, Prop. 2, p. 2]). *Suppose that $W$ is a dihedral group generated by distinct elements $s$ and $t$.*

*(i) The subgroup $P$ of $W$ generated by $p = st$ is normal and $W$ is the semidirect product, $P \rtimes \mathbf{C}_2$, where $\mathbf{C}_2 = \{1, s\}$. Moreover, $[W : P] = 2$ (where $[W : P]$ denotes the index of $P$ in $W$).*

*(ii) Let $m$ be the order of $p$ and let $\mathbf{G}_m$ be the group defined in Example 3.1.4. Then $m \geqslant 2$ and $\mathbf{G}_m \cong W$ where the isomorphism is defined by $\sigma \to s$, $\tau \to t$.*

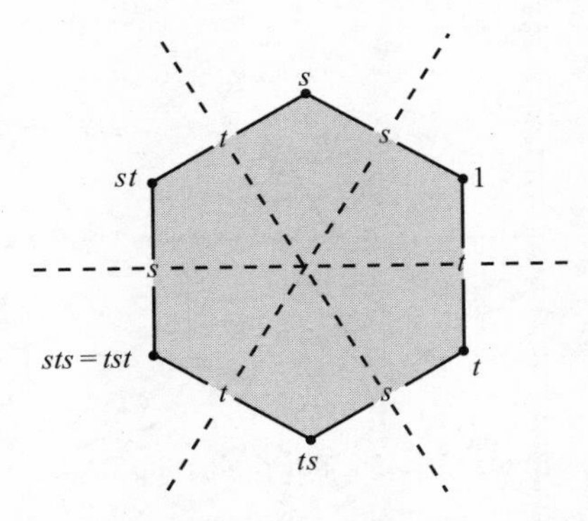


Figure 3.3. The Cayley 2-complex of $\mathbf{D}_3$.

*Proof.* (i) We have $sps^{-1} = ssts = ts = p^{-1}$ and $tpt^{-1} = tstt = ts = p^{-1}$, so $P$ is normal. Since $\mathbf{C}_2P$ contains $s$ and $t$ $(= sp)$, $W = \mathbf{C}_2P = P \cup sP$. So, $[W : P] \leqslant 2$. Suppose $W = P$. Then $W$ is abelian. So, $p^2 = s^2t^2 = 1$ and hence, $W$ is cyclic of order 2, contradicting the hypothesis that it contains at least 3 elements, 1, $s$ and $t$. Therefore, $[W : P] = 2$.

(ii) Since $s \neq t$, we have $p \neq 1$. So, $m \geqslant 2$. Since $\text{Card}(P) = m$ and since $[W : P] = 2$, $\text{Card}(W) = 2m$. There is an isomorphism $\varphi' : \mathbf{C}_m \to P$ sending the generator $\pi$ to $p$ and an isomorphism $\varphi'' : \{\pm 1\} \to \{1, s\}$ sending $-1$ to $s$. These fit together to define an isomorphism $\varphi : \mathbf{G}_m \to W$, where $\mathbf{G}_m = \mathbf{C}_m \rtimes \mathbf{C}_2$. □

**Example 3.1.6.** Let $W$ be the group defined by the presentation $\langle S \mid \mathcal{R} \rangle$ with generating set $S = \{s, t\}$ and with set of relations $\mathcal{R} = \{s^2, t^2, (st)^m\}$. (When $m = \infty$, we omit the relation $(st)^m$.) By the universal property of a group defined by a presentation, there is a surjection $W \to \mathbf{D}_m$ taking $S$ onto the set of two generating reflections for $\mathbf{D}_m$. It follows that $s$ and $t$ are distinct involutions in $W$ and that $st$ has order $m$. Hence, $W$ is a dihedral group and by the previous lemma, $W \to \mathbf{D}_m$ is an isomorphism.

**Example 3.1.7.** (*A Cayley 2-complex for* $\mathbf{D}_m$). Let $\mathbf{D}_m$ be the dihedral group of order $2m$ and $s, t$ its two generators. As in Section 2.2, we have the Cayley 2-complex of the presentation in Example 3.1.6. For $m \neq \infty$, the Cayley 2-complex of $\mathbf{D}_m$ is a $2m$-gon (see Figure 3.3). The Cayley graph of $\mathbf{D}_m$ is the boundary of the polygon (that is, the subdivision of a circle into $2m$ edges). For $m = \infty$, the Cayley 2-complex and the Cayley graph of $\mathbf{D}_\infty$ coincide and they are isomorphic to the the real line subdivided into intervals of length 1. In both cases the edges of the Cayley graph are labeled alternately by the generators $s$ and $t$.

**LEMMA 3.1.8.** *Suppose $\mathbf{D}_m$ is the dihedral group of order $2m$ on generators $s, t$. Then $s$ and $t$ are conjugate in $\mathbf{D}_m$ if and only if $m$ is odd.*

*Proof.* First, suppose $m$ is odd. Put

$$r := \underbrace{st\cdots s}_{m \text{ terms}}.$$

Then $r = r^{-1}$ and

$$rsrt = \underbrace{st\cdots st}_{2m \text{ terms}} = (st)^m = 1.$$

So $r$ conjugates $s$ to $t$.

Next, suppose $m$ is even or $\infty$. Define a function $f : \{s, t\} \to \{\pm 1\}$ by $s \to -1$ and $t \to +1$. Extend this to a function on the set of words in $\{s, t\}$. This function takes the relation $(st)^m$ to $(-1)^m = +1$. Hence, $f$ extends to a homomorphism $\tilde{f}$ from $\mathbf{D}_m$ to the abelian group $\{\pm 1\}$. Since $\tilde{f}(s) \neq \tilde{f}(t)$, $s$ and $t$ are not conjugate. $\square$

There is a more geometric version of the argument in the first paragraph of the above proof. As in Example 3.1.2, represent $\mathbf{D}_m$ as a subgroup of $O(2)$ so that $s$ and $t$ are reflections across lines $L_s$ and $L_t$ making an angle of $\pi/m$. Let $r$ be the element defined above. Since $m$ is odd, $r$ is a reflection. It maps a fundamental sector bounded by the lines $L_s$ and $L_t$ to the antipodal sector bounded by the same lines. Since $r$ is not a rotation, it must interchange these lines, i.e., $r(L_s) = L_t$. Since the reflection across the line $r(L_s)$ is $rsr^{-1}$, we get $rsr = t$.

## Minimal Words in Dihedral Groups

Let $\Omega$ be the Cayley graph of $\mathbf{D}_m$. An edge path (without backtracking) in $\Omega$ starting at 1 corresponds to an alternating word in $\{s, t\}$. Suppose $m \neq \infty$. Since $\Omega$ is a circle with $2m$ vertices, we see that if such a word has length $\leqslant m$, then it corresponds to an edge path of minimal length. Moreover, if its length is $< m$, then it is the unique edge path of minimum length connecting its endpoints. Hence, the maximum length of any element $w \in \mathbf{D}_m$ is $m$ and if $l(w) < m$, it is represented by a unique word of minimum length. There is one element of $\mathbf{D}_m$ of length $m$ corresponding to the vertex antipodal to 1. It is represented by two minimal words: the alternating words $(s, t, \dots)$ and $(t, s, \dots)$ of length $m$.

If $m = \infty$, then $\Omega$ is isomorphic to the real line and there is a unique edge path (without backtracking) between any two vertices. Hence, each element of $\mathbf{D}_\infty$ is represented by a unique minimal word.

## 3.2. REFLECTION SYSTEMS

As we explained in Section 1.1, there are two equivalent notions of an abstract reflection group. The first involves separation properties of a set of generating involutions on the corresponding Cayley graph. We explain this idea below.

**DEFINITION 3.2.1.** A *prereflection system* for a group $W$ consists of a subset $R$ of $W$, an action of $W$ on a connected simplicial graph $\Omega$ and a base point $v_0 \in \mathrm{Vert}(\Omega)$ such that

(a) Each element of $R$ is an involution.

(b) $R$ is closed under conjugation: for any $w \in W$ and $r \in R$, $wrw^{-1} \in R$.

(c) For each edge of $\Omega$, there is a unique element of $R$ which interchanges its endpoints (we say the element of $R$ *flips* the edge) and each $r \in R$ flips at least one edge.

(d) $R$ generates $W$.

An element in $R$ is a *prereflection.*

If $(v_0, v_1, \dots, v_k)$ is an edge path in $\Omega$ and $r_i \in R$ is the unique involution which flips the edge between $v_{i-1}$ and $v_i$, then $r_i v_{i-1} = v_i$ and therefore, $r_k \dots r_1 v_0 = v_k$. So, as a consequence of the connectedness of $\Omega$, we get the following.

**LEMMA 3.2.2.** *Suppose a $W$-action on a connected simplicial graph $\Omega$ is part of a prereflection system. Then $W$ is transitive on* $\mathrm{Vert}(\Omega)$.

Given a base point $v_0 \in \mathrm{Vert}(\Omega)$, let $S$ $(= S(v_0))$ be the set of prereflections that flip an edge containing $v_0$.

**DEFINITION 3.2.3.** Suppose $W$ is a group and $S$ is a set of elements of order two which generate $W$. Then $(W, S)$ is a *pre-Coxeter system.* (The reason for this terminology will become clearer in the next section.)

**Example 3.2.4.** Suppose $(W, S)$ is a pre-Coxeter system and $R$ is the set of all elements of $W$ that are conjugate to an element of $S$. Then $\Omega = \mathrm{Cay}(W, S)$ is a prereflection system. (The base point is the vertex corresponding to the identity element.) In this example $W$ acts freely on $\mathrm{Vert}(\Omega)$ (a requirement which need not be satisfied by a general prereflection system).

Suppose $(\Omega, v_0)$ is a prereflection system for $W$ and $S = S(v_0)$. Note that $R$ is the set of all elements of $W$ which are conjugate to an element of $S$. Indeed, suppose $e$ is an edge of $\Omega$ and that $r$ is the prereflection which flips it. Since $W$ acts transitively on $\mathrm{Vert}(\Omega)$, there is a element $w \in W$ such that $wv_0$ is an endpoint of $e$. Thus, $w^{-1}e = \{v_0, sv_0\}$, for some $s \in S$, and $e = \{wv_0, wsv_0\}$. (As usual, we are identifying an edge with its vertex set.) Since $s$ flips $\{v_0, sv_0\}$, the involution $wsw^{-1}$ flips $e$ and since $r$ is the unique such involution, $r = wsw^{-1}$.

Next suppose $\mathbf{s} = (s_1, \dots, s_k)$ is a word in $S$. For $0 \leqslant i \leqslant k$, define elements $w_i \in W$ by $w_0 = 1$ and $w_i = s_1 \cdots s_i$. For $1 \leqslant i \leqslant k$, define elements $r_i \in R$ by

$$r_i := w_{i-1} s_i w_{i-1}^{-1}. \tag{3.1}$$

Set

$$\Phi(\mathbf{s}) := (r_1, \dots, r_k). \tag{3.2}$$

Note that

$$r_i \cdots r_1 = w_i. \tag{3.3}$$

The word $\mathbf{s} = (s_1, \dots, s_k)$ defines an edge path in $\Omega$ starting at the base point $v_0$. To see this, consider the sequence of vertices $(v_0, \dots, v_k)$, where the vertex $v_i$ is defined by $v_i = w_i v_0$. Since $v_0$ is adjacent to $s_i v_0$, $w_{i-1}v_0$ is adjacent to $w_{i-1}s_i v_0$, i.e., $v_{i-1}$ is adjacent to $v_i$. Thus, $(v_0, \dots, v_k)$ is an edge path. Observe that $r_i$ is the prereflection which flips the edge $\{v_{i-1}, v_i\}$. Conversely, given an edge path $(v_0, \dots, v_k)$ beginning at $v_0$, we can recover the word $\mathbf{s} = (s_1, \dots, s_k)$ by reversing the procedure: if $r_i$ denotes the prereflection which flips $\{v_{i-1}, v_i\}$, set $w_i = r_i \cdots r_1$ and define $s_i$ by $s_i = w_{i-1}^{-1} r_i w_{i-1}$. This establishes a one-to-one correspondence between the set of words in $S$ and the set of edge paths starting at $v_0$.

**Lemma 3.2.5.** *$S$ generates $W$.*

*Proof.* Let $W' = \langle S \rangle$ be the subgroup of $W$ generated by $S$. Since $R$ generates $W$, it suffices to show that $R \subset W'$. Let $r \in R$ and let $e$ be an edge that is flipped by $r$. Choose an edge path starting at $v_0$ so that its last edge is $e$. If $\mathbf{s} = (s_1, \dots, s_k)$ is the corresponding word in $S$ and $\Phi(\mathbf{s}) = (r_1, \dots, r_k)$, then $r = r_k$. Hence, $r = (s_1 \cdots s_{k-1})s_k(s_{k-1} \cdots s_1) \in W'$. □

The choice of the base point $v_0$ in a prereflection system gives us a distinguished set $S$ of generators for $W$. As in Section 2.1, given $w \in W$, $l(w)$ is its word length with respect to $S$. Given a word $\mathbf{s} = (s_1, \dots, s_k)$ in $S$, $w(\mathbf{s}) = s_1 \cdots s_k$ is its value in $W$. We say that $\mathbf{s}$ is a *reduced expression* if it is a word of minimum length for $w(\mathbf{s})$, i.e., if $l(w(\mathbf{s})) = k$.

**Lemma 3.2.6.** *Suppose $\mathbf{s} = (s_1, \dots, s_k)$ is a word in $S$, that $w = s_1 \cdots s_k$ is its value in $W$, and that $\Phi(\mathbf{s}) = (r_1, \dots, r_k)$ is as defined by* (3.2). *Suppose further that $r_i = r_j$ for some $i < j$. If $\mathbf{s}'$ is the subword of $\mathbf{s}$ obtained by deleting the letters $s_i$ and $s_j$, then the edge paths corresponding to $\mathbf{s}$ and $\mathbf{s}'$ have the same endpoints. Moreover, $w = s_1 \cdots \widehat{s_i} \cdots \widehat{s_j} \cdots s_k$.*

*Proof.* Let $(v_0, w_1v_0, \dots, w_kv_0)$ be the edge path corresponding to $\mathbf{s}$. Let $r = r_i = r_j$. Then $r$ interchanges the vertices $w_{i-1}v_0$ and $w_iv_0$ as well as the

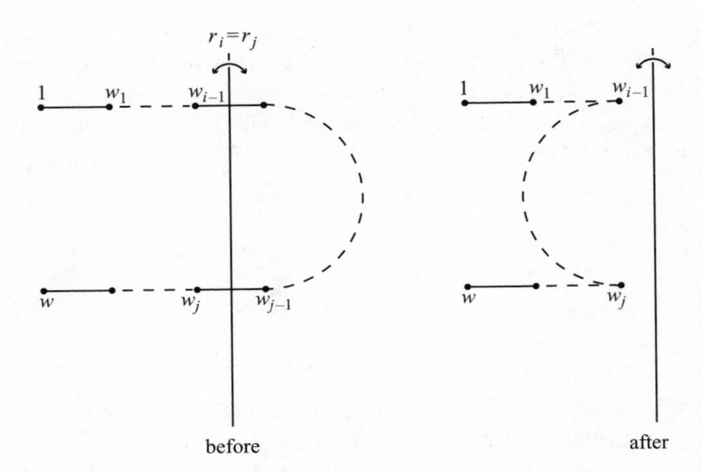


Figure 3.4. The Deletion Condition (D).

vertices $w_{j-1}v_0$ and $w_jv_0$. Hence, it maps the portion of the edge path from $w_iv_0$ to $w_{j-1}v_0$ to an edge path from $w_{i-1}v_0$ to $w_jv_0$. If we replace the piece of the original edge path between these vertices by the transformed piece, we obtain an edge path with two fewer edges and with the same endpoints. (See Figure 3.4.) The corresponding word in $S$ is $\mathbf{s}'$. To prove the last sentence of the lemma we need the following algebraic version of the above argument. The condition $r_i = r_j$ reads, $s_1 \cdots s_{i-1}s_is_{i-1} \cdots s_1 = s_1 \cdots s_{j-1}s_js_{j-1} \cdots s_1$. So, $s_i \cdots s_j = s_{i+1} \cdots s_{j-1}$. Hence, we can replace the subword $(s_i, \dots, s_j)$ of $\mathbf{s}$ by $(s_{i+1}, \dots, s_{j-1})$. □

**COROLLARY 3.2.7.** *Suppose the word $\mathbf{s} = (s_1, \dots, s_k)$ corresponds to an edge path of minimum length from $v_0$ to a vertex $v\ (= v_k)$. Then, in the sequence $\Phi(\mathbf{s}) = (r_1, \dots, r_k)$, no element of $R$ occurs more than once.*

For a given $r \in R$, let $\Omega^r$ denote the set of midpoints of those edges of $\Omega$ which are flipped by $r$. ($\Omega^r$ is contained in the fixed point set of $r$.) Call $\Omega^r$ the *wall* corresponding to $r$. Note that the edge path from $v_0$ to $v_k$ corresponding to a word $\mathbf{s}$ crosses the wall $\Omega^r$ if and only $r$ occurs in the sequence $\Phi(\mathbf{s})$. The argument in the proof of Lemma 3.2.6 shows that, if the edge path crosses $\Omega^r$ more than once, then we can obtain a new edge path, with the same endpoints, which crosses $\Omega^r$ two fewer times.

**LEMMA 3.2.8.** *For each $r \in R$, $\Omega - \Omega^r$ has either one or two connected components. If there are two components, they are interchanged by $r$.*

*Proof.* Since $r$ is conjugate to an element of $S$, we can write $r$ as $wsw^{-1}$ for some $w \in W$ and $s \in S$. Then $w\Omega^s = \Omega^{wsw^{-1}} = \Omega^r$ and $w$ maps $\Omega - \Omega^s$

homeomorphically onto $\Omega - \Omega^r$. So it suffices to prove the lemma for $s$. Let $v$ be a vertex of $\Omega$. We claim that either $v$ or $sv$ lies in the same component of $\Omega - \Omega^s$ as $v_0$. Choose an edge path of minimum length from $v_0$ to $v$ and let $\mathbf{s} = (s_1, \dots, s_k)$ be the corresponding word. If $s$ does not occur in the sequence $\Phi(\mathbf{s}) = (r_1, \dots, r_k)$, then the edge path does not cross $\Omega^s$ and $v$ and $v_0$ lie in the same component. Otherwise, $s$ occurs exactly once in the sequence, say $s = r_i$. Consider the word $\mathbf{s}' = (s, s_1, \dots, s_k)$. It defines an edge path from $v_0$ to $sv$. The corresponding sequence of elements in $R$ is given by $\Phi(\mathbf{s}') = (s, r'_1, \dots, r'_k)$, where $r'_j = sr_js$. It follows that $s$ occurs exactly twice in this sequence: as the first element and as $r'_i$ $(= r_i)$ for some $i$. By Lemma 3.2.6, we can obtain a shorter word $\mathbf{s}'' = (s_1, \dots, \widehat{s_i}, \dots, s_k)$ from $v_0$ to $sv$ by deleting both occurrences. Since this edge path does not cross $\Omega^s$, $v_0$ and $sv$ lie in the same component of $\Omega - \Omega^s$. □

Given a word $\mathbf{s}$ and an element $r \in R$, let $n(r, \mathbf{s})$ denote the number of times $r$ occurs in the sequence $\Phi(\mathbf{s})$, defined by (3.2). In other words, $n(r, \mathbf{s})$ is the number of times the corresponding edge path crosses the wall $\Omega^r$. For the reader who has followed the proofs of Lemmas 3.2.6 and 3.2.8, the following lemma should be clear.

**LEMMA 3.2.9.** *Given a prereflection system $(\Omega, v_0)$ and an element $r \in R$, the following conditions are equivalent.*

(i) *$\Omega^r$ separates $\Omega$.*

(ii) *For any word $\mathbf{s}$ in $S$ corresponding to an edge path from $v_0$ to a vertex $v$, the number $(-1)^{n(r,\mathbf{s})}$ depends only on the endpoint $v$. (In fact, this number is $+1$ if $v_0$ and $v$ lie in the same component of $\Omega - \Omega^r$ and $-1$ if they lie in different components.)*

**DEFINITION 3.2.10.** A prereflection system $(\Omega, v_0)$ for $W$ is a *reflection system* if, for each $s \in S$ $(=S(v_0))$, $\Omega - \Omega^s$ has two components. (It follows that for each $r \in R$, $\Omega^r$ separates $\Omega$.) If $(\Omega, v_0)$ is a reflection system, then the elements of $R$ are called *reflections* and $S$ is the set of *fundamental reflections*.

**DEFINITION 3.2.11.** The closure of a component of $\Omega - \Omega^r$ is a *half-space* bounded by $\Omega^r$. If it contains $v_0$, it is a *positive* half-space.

Suppose $(\Omega, v_0)$ is a reflection system for $W$. The crucial consequence of this hypothesis is that $v_0$ and $v$ lie on the same side of $\Omega^r$ if and only if any edge path from $v_0$ to $v$ crosses $\Omega^r$ an even number of times. The next lemma shows that this implies that $W$ acts freely on Vert($\Omega$) and hence, as was pointed out in 2.1, that $\Omega$ is isomorphic to Cay$(W, S)$. In light of this, we will sometimes call $(W, S)$ a "reflection system" if its Cayley graph is.

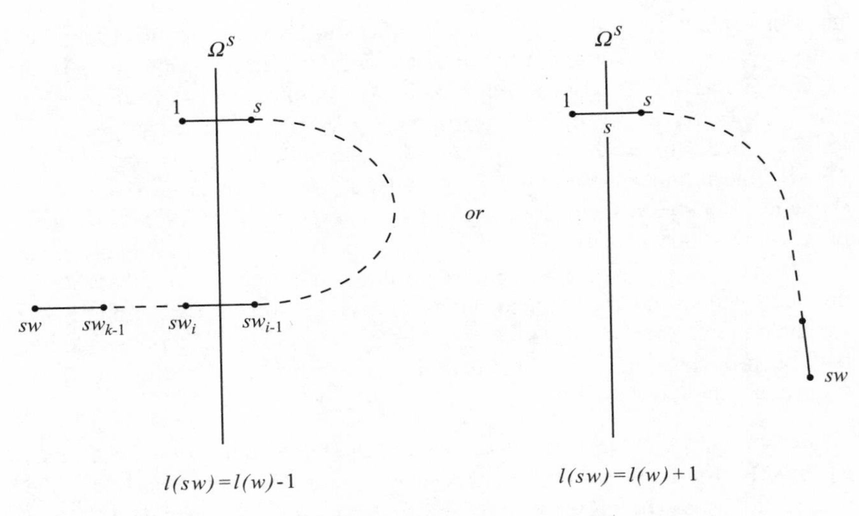


Figure 3.5. The Exchange Condition (E).

**LEMMA 3.2.12.** *Suppose* $(\Omega, v_0)$ *is a reflection system for* $W$. *Then* $W$ *acts freely on* $\mathrm{Vert}(\Omega)$.

*Proof.* Suppose $wv_0 = v_0$ for some $w \neq 1$. By Lemma 3.2.5, we can express $w$ as a word in $S$. Choose such a word, $w = s_1 \cdots s_k$, of minimum length and consider the corresponding edge path from $v_0$ to itself. The first wall crossed by this path is $\Omega^{s_1}$. Since all walls are crossed an even number of times, the path must cross this wall again. But then, by Lemma 3.2.6, we can find a shorter word for $w$, contradicting the assumption that our original word had minimum length. □

**COROLLARY 3.2.13.** *If* $\Omega$ *is a reflection system for* $W$ *with fundamental set of generators* $S$, *then* $\Omega$ *is isomorphic to the Cayley graph* $\mathrm{Cay}(W, S)$.

**LEMMA 3.2.14.** *Suppose Cay* $(W, S)$ *is a reflection system. Let* $\mathbf{s} = (s_1, \dots, s_k)$ *be a word for* $w = w(\mathbf{s})$. *Then* $\mathbf{s}$ *is a reduced expression if and only if the elements of the sequence* $\Phi(\mathbf{s}) = (r_1, \dots, r_k)$ *are all distinct. If this is the case, then* $\{r_1, \dots, r_k\}$ *is the set* $R(1, w)$ *consisting of those elements* $r \in R$ *such that the wall* $\Omega^r$ *separates* $v_0$ *from* $wv_0$.

*Proof.* If $l(w) = k$, then, by Corollary 3.2.7, the elements in $\Phi(\mathbf{s})$ are distinct. For the converse, first note that every element of $R(1, w)$ must occur in the sequence $\Phi(\mathbf{s})$. Hence, $k \geqslant l(w) \geqslant \mathrm{Card}(R(1, w))$. If the elements of $\Phi(\mathbf{s})$ are distinct, then, by Lemma 3.2.9, $\mathrm{Card}(R(1, w)) \geqslant k$ and so, the above inequalities are equalities. In particular, since $l(w) = k$, $\mathbf{s}$ is a reduced expression. □

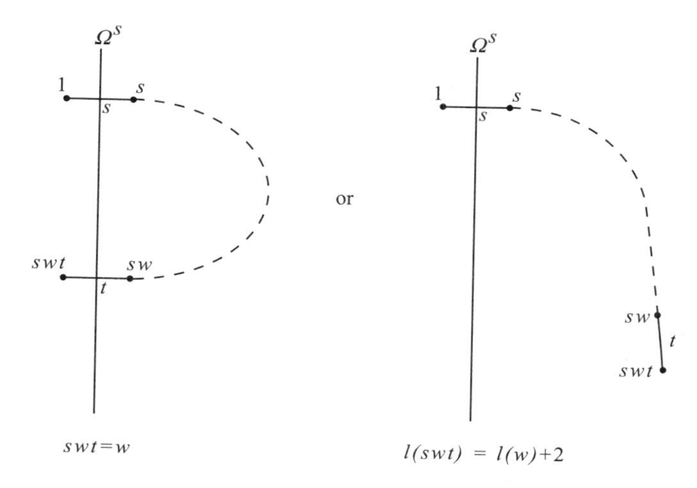


Figure 3.6. The Folding Condition (F).

Given a pre-Coxeter system $(W, S)$, we next discuss three equivalent conditions, (D), (E), and (F), on words in $S$. (Condition (D) is called the "Deletion Condition," (E) is the "Exchange Condition," and (F) is the "Folding Condition.") We will see that each of these conditions is implied by the condition that $\text{Cay}(W, S)$ is a reflection system. Here they are.

- (D) If $\mathbf{s} = (s_1, \dots, s_k)$ is a word in $S$ with $k > l(w(\mathbf{s}))$, then there are indices $i < j$ so that the subword
$$\mathbf{s}' = (s_1, \dots, \widehat{s_i}, \dots, \widehat{s_j}, \dots s_k)$$
is also an expression for $w(\mathbf{s})$.
- (E) Given a reduced expression $\mathbf{s} = (s_1, \dots, s_k)$ for $w \in W$ and an element $s \in S$, either $l(sw) = k + 1$ or else there is an index $i$ such that
$$w = ss_1 \cdots \widehat{s_i} \cdots s_k.$$
- (F) Suppose $w \in W$ and $s, t \in S$ are such that $l(sw) = l(w) + 1$ and $l(wt) = l(w) + 1$. Then either $l(swt) = l(w) + 2$ or $swt = w$.

*Remark.* In regard to condition (E), note that, for a general pre-Coxeter system $(W, S)$, there are only three possibilities:

(a) $l(sw) = l(w) + 1$ (which implies that a reduced expression for $sw$ can be obtained by putting an $s$ in front of any reduced expression for $w$),

(b) $l(sw) = l(w) - 1$ (which implies that $w$ has a reduced expression beginning with $s$), or

(c) $l(sw) = l(w)$.

The meaning of (E) is that possibility (c) does not occur and that in case (b), we can modify an arbitrary reduced expression of $w$ to get one beginning with $s$ by "exchanging" one of its letters for an $s$ in front.

*Remark 3.2.15.* Condition (F) implies that if $w \in W$ and $s, t \in S$ are such that $l(swt) = l(w)$ and $l(sw) = l(wt)$, then $sw = wt$. If $l(sw) = l(wt) = l(w) + 1$, this is immediate from (F). If $l(sw) = l(wt) = l(w) - 1$, put $u := sw$. Then $l(sut) = l(wt) = l(u)$ and $l(ut) = l(swt) = l(w) = l(u) + 1$. So, we can apply (F) to $s, t$ and $u$ to get $su = ut$ and consequently $sw = u = wt$.

**THEOREM 3.2.16.** *Given a pre-Coxeter system* $(W, S)$*, conditions (D), (E), and (F) are equivalent.*

*Proof.* (D) $\Longrightarrow$ (E). This implication is obvious. Suppose $\mathbf{s} = (s_1, \dots, s_k)$ is a reduced expression of $w$ and $s \in S$ is such that $l(sw) \leqslant k$. Since $(s, s_1, \dots, s_k)$ is not reduced, Condition (D) says that we can find a shorter word for $sw$ by deleting two letters. Since $\mathbf{s}$ is reduced, both letters cannot belong to $\mathbf{s}$, so one of the letters must be the initial $s$. Thus, $sw = \hat{s}s_1 \cdots \hat{s}_i \cdots s_k$ and so $w = ss_1 \cdots \hat{s}_i \cdots s_k$. In other words, we have exchanged a letter of $\mathbf{s}$ for an $s$ in front.

(E) $\Longrightarrow$ (F). This implication is also easy. Suppose that $\mathbf{s} = (s_1, \dots, s_k)$ is a reduced expression of $w$, that $s, t \in S$ are such that $l(sw) = k + 1 = l(wt)$ and that $l(swt) < k + 2$. Applying Condition (E) to the word $(s_1, \dots, s_k, t)$ and the element $s$, we see that a letter can be exchanged for an $s$ in front. The exchanged letter cannot be part of $\mathbf{s}$, because this would contradict the assumption that $l(sw) = k + 1$. So, the exchanged letter must be the final $t$. This yields $ss_1 \cdots s_k = s_1 \cdots s_k t$, which can be rewritten as $sw = wt$ or $swt = w$.

(F) $\Longrightarrow$ (D). Suppose the word $\mathbf{s} = (s_1, \dots, s_k)$ is not reduced. Necessarily, $k \geqslant 2$. We must show that we can delete two letters from $\mathbf{s}$ while leaving its value unchanged. The proof is by induction on $k$. We may assume that the words $(s_1, \dots, s_{k-1})$ and $(s_2, \dots, s_k)$ are both reduced (otherwise we are done by induction). Let $w = s_2 \cdots s_{k-1}$. Apply Condition (F) with $s = s_1$ and $t = s_k$. This yields $s_1 w s_k = w$, i.e., $(s_1, \dots, s_k)$ can be shortened by deleting its first and last letters. □

**THEOREM 3.2.17.** *Suppose* Cay$(W, S)$ *is a reflection system. Then (the equivalent) conditions (D), (E) and (F) hold.*

*Proof.* Let $\mathbf{s} = (s_1, \dots, s_k)$ be a word for $w$ $(=w(\mathbf{s}))$ and let $\Phi(\mathbf{s}) = (r_1, \dots, r_k)$. If $\mathbf{s}$ is not reduced, then $r_i = r_j$ for some $i < j$ (Lemma 3.2.14). By Lemma 3.2.6, we can obtain another word $\mathbf{s}'$ for $w$ by deleting $s_i$ and $s_j$, i.e., (D) holds. □

*Remark 3.2.18.* Suppose (D) holds for $(W, S)$. Let $\mathbf{s}$ be a word for $w$. Then the length of $\mathbf{s}$ is congruent to $l(w)$ modulo 2. It follows that the function $\varepsilon : W \to \{\pm 1\}$ defined by $w \to (-1)^{l(w)}$ is a homomorphism.

## 3.3. COXETER SYSTEMS

The second notion of an abstract reflection group involves a presentation of the type explained below.

**DEFINITION 3.3.1.** A *Coxeter matrix* $M = (m_{st})$ on a set $S$ is an $S \times S$ symmetric matrix with entries in $\mathbb{N} \cup \{\infty\}$ such that

$$m_{st} = \begin{cases} 1 & \text{if } s = t; \\ \geqslant 2 & \text{otherwise.} \end{cases}$$

One can associate to $M$ a presentation for a group $\widetilde{W}$ as follows. For each $s \in S$, introduce a symbol $\tilde{s}$. Let $\mathcal{I} = \{(s, t) \in S \times S \mid m_{st} \neq \infty\}$. The set of generators for $\widetilde{W}$ is $\widetilde{S} = \{\tilde{s}\}_{s \in S}$ and the set $\mathcal{R}$ of relations is

$$\mathcal{R} = \{(\tilde{s}\tilde{t})^{m_{st}}\}_{(s,t) \in \mathcal{I}}.$$

Given any pre-Coxeter system $(W, S)$, we have a Coxeter matrix $M$ on $S$, defined by the formula, $m_{st} = m(s, t)$, where $m(s, t)$ denotes the order of $st$.

**DEFINITION 3.3.2.** A pre-Coxeter system $(W, S)$ is a *Coxeter system* if the epimorphism $\widetilde{W} \to W$, defined by $\tilde{s} \to s$, is an isomorphism. If this is the case, then $W$ is a *Coxeter group* and $S$ is a *fundamental set of generators*.

In other words, $(W, S)$ is a Coxeter system if $W$ is isomorphic to the group defined by the presentation associated to its Coxeter matrix. The following basic lemma will be needed in Section 18.1.

**LEMMA 3.3.3.** ([29, p. 5]). *Suppose $(W, S)$ is a Coxeter system. Define an equivalence relation $\sim$ on $S$ by $s \sim s'$ if and only if there is a sequence of elements of $S$, $s = s_0, s_1, \dots, s_n = s'$ such that for any two adjacent elements in the sequence $m(s_i, s_{i+1})$ is an odd integer. Then $s$ and $s'$ are conjugate in $W$ if and only if $s \sim s'$.*

*Proof.* By Lemma 3.1.8, two fundamental generators for the dihedral group $\mathbf{D}_m$ are conjugate if and only if $m$ is odd. So, if $s \sim s'$, then $s$ and $s'$ are conjugate.

Conversely, given $s \in S$, define $f_s : S \to \{\pm 1\}$ by $f_s(t) = -1$ if and only if $s \sim t$. We see that all the defining relations of $W$ are sent to $+1$; so, $f_s$ extends to a homomorphism $\tilde{f}_s : W \to \{\pm 1\}$. Since $\tilde{f}_s$ is a homomorphism, if $s$ is conjugate to $s'$, then $f_s(s') = f_s(s) = -1$, i.e., $s$ and $s'$ are equivalent. $\square$

The next theorem is the main result of this chapter. Its proof will not be completed until the end of the next section.

**THEOREM 3.3.4.** *The following conditions on a pre-Coxeter system $(W,S)$ are equivalent.*

*(i)* $(W,S)$ *is a Coxeter system.*

*(ii)* $\mathrm{Cay}(W,S)$ *is a reflection system.*

*(iii)* $(W,S)$ *satisfies the Exchange Condition (E).*

*Remark.* Suppose $(W,S)$ is a pre-Coxeter system. For each set $\{s,t\}$ of two elements in $S$, let $W_{\{s,t\}}$ denote dihedral subgroup which they generate. Let $\Omega := \mathrm{Cay}(W,S)$ and let $\Lambda$ be the 2-complex formed by gluing a $2m$-gon, $m = m(s,t)$, onto $\Omega$ for each coset of $W_{\{s,t\}}$ in $W$ with $m(s,t) \neq \infty$. How do we interpret Theorem 3.3.4 in light of the results of Sections 2.1 and 2.2? Condition (ii) means that for each $s \in S$ the fixed set of $s$ separates $\Omega$ (and hence, also $\Lambda$) . Condition (i) means that $\Lambda$ is simply connected. The equivalence of these two conditions is by no means geometrically obvious.

As in the previous section, given a word $\mathbf{s}$ in $S$ and an element $r \in R$, let $n(r,\mathbf{s})$ denote the number of times $r$ occurs in the sequence $\Phi(\mathbf{s})$ (in other words, $n(r,\mathbf{s})$ is the number of times the corresponding edge path between 1 and $w(\mathbf{s})$ in the Cayley graph crosses the wall corresponding to $r$). The proof of Theorem 3.3.4 depends on the following.

**LEMMA 3.3.5.** *Suppose $(W,S)$ is a Coxeter system.*

*(i) For any word $\mathbf{s}$ with $w = w(\mathbf{s})$ and any element $r \in R$, the number $(-1)^{n(r,\mathbf{s})}$ depends only on $w$. We denote this number $\eta(r,w) \in \{\pm 1\}$.*

*(ii) There is a homomorphism, $w \to \phi_w$ from $W$ to the group of permutations of the set $R \times \{\pm 1\}$, where the permutation $\phi_w$ is defined by the formula*

$$\phi_w(r,\varepsilon) = (wrw^{-1}, \eta(r,w^{-1})\varepsilon).$$

Before proving this, let us discuss the geometric idea behind it. Suppose $\Omega$ $(:= \mathrm{Cay}(W,S))$. is a reflection system (where the identity element 1 is the base vertex). Then each wall $\Omega^r$ separates $\Omega$ into two half-spaces: the positive one, $\Omega^r(+1)$, which contains the vertex 1 and the negative one, $\Omega^r(-1)$, which does not contain it. The number $(-1)^{n(r,\mathbf{s})}$ is $+1$ if 1 and $w$ lie on the same side of $\Omega^r$ and it is $-1$ if they lie on opposite sides. The set of half-spaces is indexed by $R \times \{\pm 1\}$. The group $W$ acts on the set of half-spaces. How do we describe the induced action on the set of indices? For a given $w \in W$, since the reflection across the wall $w\Omega^r$ is $wrw^{-1}$, $w$ maps the half-space $\Omega^r(+1)$ to $\Omega^{wrw^{-1}}(\varepsilon)$, where $\varepsilon \in \{\pm 1\}$. The only question is to decide the sign of $\varepsilon$. The condition $\varepsilon = +1$ means that $w\Omega^r(+1) = \Omega^{wrw^{-1}}(+1)$ and this holds if and only if $w$ and 1 are on the same side of $\Omega^{wrw^{-1}}$, i.e., if and only if 1 and $w^{-1}$ are on the same side of $\Omega^r$. So, $\varepsilon$ is determined by whether or not 1 and $w^{-1}$ are on the same side of $\Omega^r$: it is $+1$ if they are and $-1$ if they are not.

*Proof of Lemma 3.3.5.* For each $s \in S$ define $\phi_s : R \times \{\pm 1\} \to R \times \{\pm 1\}$ by

$$\phi_s(r, \varepsilon) = (srs, \varepsilon(-1)^{\delta(s,r)}),$$

where $\delta(s, r)$ is the Kronecker delta. It is clear that $(\phi_s)^2$ is the identity map; hence, $\phi_s$ is a bijection. Let $\mathbf{s} = (s_1, \dots, s_k)$. Put $v = s_k \cdots s_1$ and $\phi_{\mathbf{s}} = \phi_{s_k} \circ \cdots \circ \phi_{s_1}$. We claim that

$$\phi_{\mathbf{s}}(r, \varepsilon) = (vrv^{-1}, \varepsilon(-1)^{n(\mathbf{s},r)}).$$

The proof is by induction on $k$. The case $k = 1$ is trivial. Suppose $k > 1$, set $\mathbf{s}' = (s_1, \dots, s_{k-1})$, $u = s_{k-1} \cdots s_1$, and suppose, by induction, that the claim holds for $\mathbf{s}'$. Then

$$\begin{aligned}\phi_{\mathbf{s}}(r, \varepsilon) &= \phi_{s_k}(uru^{-1}, \varepsilon(-1)^{n(\mathbf{s}',r)}) \\ &= (vrv^{-1}, \varepsilon(-1)^{n(\mathbf{s}',r)+\delta(s_k,uru^{-1})}).\end{aligned}$$

Put $\Phi(\mathbf{s}) = (r_1, \dots, r_k)$. Since $r_k = (s_1 \cdots s_{k-1})s_k(s_{k-1} \cdots s_1) = u^{-1}s_k u$, we have $\Phi(\mathbf{s}) = (\Phi(\mathbf{s}'), u^{-1}s_k u)$. Hence, $n(\mathbf{s}, r) = n(\mathbf{s}', r) + \delta(u^{-1}s_k u, r)$, proving the claim.

Next we claim that the map $\mathbf{s} \to \phi_{\mathbf{s}}$ descends to a homomorphism from $W$ to the group of permutations of $R \times \{\pm 1\}$. To check this, we need only show that the map $\mathbf{s} \to \phi_{\mathbf{s}}$ takes each relation to the identity permutation. We have already noted that $(\phi_s)^2 = 1$, so this is true for the relations of the form $s^2 = 1$. The other relations have the form $(st)^m = 1$, where $s, t \in S$ and $m = m(s, t)$ is the order of $st$. To see that $(\phi_s \circ \phi_t)^m = 1$, all we need to check is that if $\mathbf{s} = (s, t, \dots)$ is the alternating word of length $2m$, then $n(\mathbf{s}, r)$ is even for all $r \in R$. If $r$ is not in the dihedral group $\langle s, t\rangle$, then $n(\mathbf{s}, r) = 0$. If $r \in \langle s, t\rangle$, then it follows from our analysis of dihedral groups in Section 3.1 that $n(\mathbf{s}, r) = 2$. This establishes assertion (ii) of the lemma. Since the formula for $\phi_w$ depends only on $w$ (and not on the word representing it), we have also established (i). □

Let $\widehat{R}_w$ denote the set of $r \in R$ such that $\eta(r, w) = -1$. We can now prove the implication (i) $\Longrightarrow$ (ii) of Theorem 3.3.4. We state this as follows.

**PROPOSITION 3.3.6.** *If $(W, S)$ is a Coxeter system, then $\Omega := \mathrm{Cay}(W, S)$ is a reflection system. Moreover, given $r \in R$, the vertices 1 and $w$ lie on opposite sides of $\Omega^r$ if and only if $r \in \widehat{R}_w$ (and so $\widehat{R}_w$ is the set $R(1, w)$ defined in 3.2).*

*Proof.* If $\mathbf{s}$ is a word for $w$, then, by the previous lemma, each element of $\widehat{R}_w$ occurs an odd number of times in $\Phi(\mathbf{s})$. In other words, for each $r \in \widehat{R}_w$, any edge path in $\Omega$ from 1 to $w$ must cross $\Omega^r$ an odd number of times and consequently 1 and $w$ lie on opposite sides. Similarly, if $r \notin \widehat{R}_w$, then there is an edge path connecting 1 to $w$ which does not cross $\Omega^r$. □

## 3.4. THE WORD PROBLEM

Suppose $\langle S \mid \mathcal{R}\rangle$ is a presentation for a group $G$. The *word problem* for $\langle S \mid \mathcal{R}\rangle$ is the following question: given a word $\mathbf{s}$ in $S \cup S^{-1}$, is there an algorithm for determining if its value $g(\mathbf{s})$ is the identity element of $G$?

In this section we give Tits' solution (in [283]) to the word problem for Coxeter groups. Suppose $(W, S)$ is a pre-Coxeter system and $M = (m_{st})$ the associated Coxeter matrix.

**DEFINITION 3.4.1.** An *elementary M-operation* on a word in $S$ is one of the following two types of operations:

(i) Delete a subword of the form $(s, s)$

(ii) Replace an alternating subword of the form $(s, t, \dots)$ of length $m_{st}$ by the alternating word $(t, s, \dots)$, of the same length $m_{st}$.

A word is *M-reduced* if it cannot be shortened by a sequence of elementary M-operations.

**THEOREM 3.4.2.** (Tits [283].) *Suppose $(W, S)$ satisfies the Exchange Condition (or equivalently, Condition (D) or (F)). Then*

*(i) A word $\mathbf{s}$ is a reduced expression if and only if it is M-reduced.*

*(ii) Two reduced expressions $\mathbf{s}$ and $\mathbf{t}$ represent the same element of $W$ if and only if one can be transformed into the other by a sequence of elementary M-operations of type (II).*

*Proof.* We first prove (ii). Suppose that $\mathbf{s} = (s_1, \dots, s_k)$ and $\mathbf{t} = (t_1, \dots, t_k)$ are two reduced expressions for an element $w \in W$. The proof is by induction on $k = l(w)$. If $k = 1$, then the two words are the same and we are done. Suppose $k > 1$. To simplify notation, set $s = s_1$ and $t = t_1$. There are two cases to consider. The first case is where $s = t$. Then $(s_2, \dots, s_k)$ and $(t_2, \dots, t_k)$ are two reduced words for the same element $sw$. By induction, we can transform one into the other by a sequence of type (II) operations. This takes care of the first case. The second case is where $s \neq t$. Put $m = m(s, t)$. We claim that $m$ is finite and that we can find a third reduced expression $\mathbf{u}$ for $w$ which begins with an alternating word $(s, t, \dots)$ of length $m$. Assuming this claim for the moment, let $\mathbf{u}'$ be the word obtained from $\mathbf{u}$ by the type (II) operation which replaces the initial segment of $\mathbf{u}$ by the alternating word $(t, s, \dots)$ beginning with $t$. Then we can transform $\mathbf{s}$ into $\mathbf{t}$ by a sequence of moves indicated schematically as follows:

$$\mathbf{s} \to \mathbf{u} \to \mathbf{u}' \to \mathbf{t},$$

where the first and third arrows represent sequences of moves guaranteed by the first case ($\mathbf{s}$ and $\mathbf{u}$ begin with the same letter as do $\mathbf{u}'$ and $\mathbf{t}$) and the second arrow is the elementary M-operation of type (II).

It remains to prove the claim. Since $l(tw) < l(w)$, the Exchange Condition implies that we can find another reduced expression for $w$ by exchanging a letter of $(s, s_2, \dots, s_k)$, for a $t$ in front. The exchanged letter cannot be the first $s$ (since $t \neq s$). Hence, we obtain a reduced expression beginning with $(t, s)$. This process can be continued. For any integer $q \geqslant 2$, let $\mathbf{s}_q$ be the alternating word in $s$ and $t$ of length $q$ with final letter $s$. Thus, $\mathbf{s}_q$ begins either with $s$ (if $q$ is odd) or $t$ (if $q$ is even). We will show that for any $q \leqslant m$, we can find a reduced word for $w$ that begins with $\mathbf{s}_q$. Suppose, by induction, that we have such a word $\mathbf{s}'$ beginning with $\mathbf{s}_{q-1}$. Let $s'$ be the element of $\{s, t\}$ with which $\mathbf{s}_{q-1}$ does not begin. Since $l(s'w) < l(w)$, the Exchange Condition says that we can find another reduced expression by exchanging a letter of $\mathbf{s}'$ for an $s'$ in front. The exchanged letter cannot belong to the initial segment $\mathbf{s}_{q-1}$, since in the dihedral group of order $2m$, a reduced expression for an element of length $\neq m$ is unique. (See the discussion at the end of 3.1.) So $w$ has a reduced expression beginning with $\mathbf{s}_q$. Since this works for any $q \leqslant m$ and since $q$ is bounded above by $l(w)$, we must have $m < \infty$. Thus, $w$ has a reduced expression beginning with $\mathbf{s}_m$. This reduced expression is either $\mathbf{u}$ (if $m$ is odd) or $\mathbf{u}'$ (if $m$ is even). We can replace $\mathbf{s}_m$ by the other alternating word of length $m$ (also a reduced expression of $w(\mathbf{s}_m)$) to obtain the other one. This proves the claim.

Finally, we prove the first statement in the theorem. One direction is obvious: if $\mathbf{s}$ is reduced, it is M-reduced. So, suppose that $\mathbf{s} = (s_1, \dots, s_k)$ is M-reduced. We will show by induction on $k$ that it is a reduced expression. This is clear for $k = 1$. Suppose $k > 1$. By induction, the word $\mathbf{s}' = (s_2, \dots, s_k)$ is reduced. Suppose $\mathbf{s}$ is not reduced. Set $w = s_1 \cdots s_k$ and $w' = s_2 \cdots s_k$. Since $l(s_1 w') = l(w) \leqslant k - 1$, the Exchange Condition implies that $w'$ has another reduced expression, call it $\mathbf{s}''$, beginning with $s_1$. By statement (ii), $\mathbf{s}'$ can be transformed to $\mathbf{s}''$ by a sequence of M-operations of type (II). Thus, $\mathbf{s}$ can be transformed by a sequence of M-operations to a word beginning with $(s_1, s_1)$, contradicting the assumption that it is M-reduced. So, $\mathbf{s}$ must be a reduced expression. □

We can now complete the proof of the main result of this chapter.

*Proof of Theorem 3.3.4.* We want to show that the following conditions on a pre-Coxeter system $(W, S)$ are equivalent.

(i) $(W, S)$ is a Coxeter system.

(ii) $\mathrm{Cay}(W, S)$ is a reflection system.

(iii) $(W, S)$ satisfies the Exchange Condition.

The implications (i) $\Longrightarrow$ (ii) and (ii) $\Longrightarrow$ (iii) were proved in Proposition 3.3.6 and Theorem 3.2.17, respectively. So it remains to prove that (iii) $\Longrightarrow$ (i). Suppose, as in the beginning of Section 3.3, that $(W, S)$ is a pre-Coxeter system, that $(\widetilde{W}, \widetilde{S})$ is the Coxeter system associated to the Coxeter matrix of $(W, S)$, and that $p : \widetilde{W} \to W$ is the natural surjection. We must show that $p$ is injective. Let $\tilde{w} \in \mathrm{Ker}(p)$ and let $\tilde{\mathbf{s}} = (\tilde{s}_1, \dots, \tilde{s}_k)$ be a reduced expression for $\tilde{w}$. Then $\tilde{\mathbf{s}}$ is M-reduced. Let $\mathbf{s} = (s_1, \dots, s_k)$ be the corresponding word in $S$. Since $(W, S)$ and $(\widetilde{W}, \widetilde{S})$ have the same Coxeter matrices, the notion of M-operations on words in $S$ and $\widetilde{S}$ coincide and so, $\mathbf{s}$ is also M-reduced. But since $\mathbf{s}$ represents the identity element in $W$, it must be the empty word. Consequently, $\tilde{\mathbf{s}}$ is also the empty word, i.e., $\tilde{w} = 1$. □

## 3.5. COXETER DIAGRAMS

There is a well-known method, due to Coxeter, of encoding the information in a Coxeter matrix (see Definition 3.3.1) into a graph with edges labeled by integers $> 3$ or the symbol $\infty$.

**DEFINITION 3.5.1.** Suppose that $M = (m_{ij})$ is a Coxeter matrix on a set $I$. We associate to $M$ a graph $\Gamma$ $(= \Gamma_M)$ called its *Coxeter graph*. The vertex set of $\Gamma$ is $I$. Distinct vertices $i$ and $j$ are connected by an edge if and only if $m_{ij} \geqslant 3$. The edge $\{i, j\}$ is labeled by $m_{ij}$ if $m_{ij} \geqslant 4$. (If $m_{ij} = 3$, the edge is left unlabeled.) The graph $\Gamma$ together with the labeling of its edges is called the *Coxeter diagram* associated to $M$. The vertices of $\Gamma$ are often called the *nodes* of the diagram.

**Example 3.5.2.** The Coxeter diagram of $\mathbf{D}_m$, the dihedral group of order $2m$, is $\circ\overset{m}{\text{——}}\circ$. (If $m = 2$, there is no edge; if $m = 3$, the edge is not labeled.)

Other examples of diagrams of some classical geometric reflection groups can be found in Tables 6.1 and 6.2 of Section 6.9.

The Coxeter diagram obviously carries the same information as its Coxeter matrix. There are other possible ways to encode the same information in a labeled graph. For example, one could connect $s$ and $t$ by an edge labeled by $m_{st}$ whenever $m_{st} < \infty$ and not connect them when $m_{st} = \infty$. In fact, later (in Example 7.1.6) we will want to make use of precisely this labeled graph.

**DEFINITION 3.5.3.** A Coxeter system $(W, S)$ is *irreducible* if its Coxeter graph is connected.

Another way to express the condition in the above definition is that $S$ cannot be partitioned into two nonempty disjoint subsets $S'$ and $S''$ such that each element in $S'$ commutes with each element of $S''$.

Given a subset $T \subset S$, let $\Gamma_T$ denote the full subgraph of $\Gamma$ spanned by $T$. $T$ is an *irreducible component* of $S$ (or simply a *component*) if $\Gamma_T$ is a connected component of $\Gamma$. The main advantage of using the Coxeter diagram is that it makes transparent when $W$ decomposes as a direct product. (See Proposition 4.1.7 in the next chapter.)

## NOTES

This chapter is a reworking of basic material in [29, Chapter IV]. Other references include [43, 163, 248, 298].

**3.1.** Much of the material in this section is taken from [29, Chapter IV, §1.2].

**3.2.** Elsewhere in the literature a "reduced expression" for an element of $W$ is often called a "reduced decomposition."

**3.4.** The proof of Theorem 3.4.2 is taken from [43].

# Chapter Four

# MORE COMBINATORIAL THEORY OF COXETER GROUPS

Except in Section 4.8, $(W, S)$ is a Coxeter system. A *special* subgroup of $W$ is one that is generated by a subset of $S$. In 4.1 we show, among other things, that special subgroups are Coxeter groups (Theorem 4.1.6). In 4.3 we show there is a unique element of minimum length in each coset of a special subgroup. This allows us to discuss, in 4.5, certain "convex subsets" of $W$ such as "half-spaces" and "sectors." In 4.6 we prove that each finite Coxeter group has a unique element of longest length. We use this in proving an important result, Lemma 4.7.2: for any given $w \in W$, the set of letters with which a reduced expression for $w$ can end generates a finite special subgroup. In 4.9 we prove that any subgroup of $W$ that is generated by reflections is itself a Coxeter group. In 4.10 we prove a theorem of Deodhar describing normalizers of special subgroups.

## 4.1. SPECIAL SUBGROUPS IN COXETER GROUPS

We begin with a simple consequence of Theorem 3.4.2: for any $w \in W$, the set of letters that can occur in a reduced expression for it is independent of the choice of reduced expression. We state this as the following.

**PROPOSITION 4.1.1.** *For each $w \in W$, there is a subset $S(w) \subset S$ so that for any reduced expression $(s_1, \dots, s_k)$ for $w$, $S(w) = \{s_1, \dots, s_k\}$.*

*Proof.* The elementary M-operations of type (II) in Definition 3.4.1 do not change the set of letters in a reduced expression; so the proposition follows from Theorem 3.4.2. □

Recall that, for each $T \subset S$, $W_T$ denotes the subgroup generated by $T$. $W_T$ is a *special subgroup* of $W$.

**COROLLARY 4.1.2.** *For each $T \subset S$, $W_T$ consists of those elements $w \in W$ such that $S(w) \subset T$.*

*Proof.* If $(s_1, \dots, s_k)$ is a reduced expression for $w$, then $(s_k, \dots, s_1)$ is a reduced expression for $w^{-1}$. Hence,

$$S(w^{-1}) = S(w). \tag{4.1}$$

Given reduced expressions for $v$ and $w$, we can concatenate them to get a word for $vw$. Although this word might not be reduced, by using the Deletion Condition, we can get a reduced expression by simply deleting letters. It follows that

$$S(vw) \subset S(v) \cup S(w). \tag{4.2}$$

Let $X = \{w \in W \mid S(w) \subset T\}$. Clearly, $X$ is contained in $W_T$. It follows from (4.1) and (4.2) that $X$ is actually a subgroup of $W_T$. Since $T \subset X$ and $W_T$ is the subgroup generated by $T$, $W_T \subset X$. Hence, $W_T = X$. □

**COROLLARY 4.1.3.** *For each $T \subset S$, $W_T \cap S = T$.*

**COROLLARY 4.1.4.** *$S$ is a minimal set of generators for $W$.*

**COROLLARY 4.1.5.** *For each $T \subset S$ and each $w \in W_T$, the length of $w$ with respect to $T$ (denoted $l_T(w)$) is equal to the length of $w$ with respect to $S$ (denoted $l_S(w)$).*

*Proof.* Suppose $(s_1, \dots, s_k)$ is a reduced expression for $w \in W_T$. By Proposition 4.1.1, each $s_i$ lies in $T$. Hence, $l_T(w) = k = l_S(w)$. □

**THEOREM 4.1.6**

*(i) For each $T \subset S$, $(W_T, T)$ is a Coxeter system.*

*(ii) Let $(T_i)_{i \in I}$ be a family of subsets of $S$. If $T = \bigcap_{i\in I} T_i$, then*

$$W_T = \bigcap_{i \in I} W_{T_i}.$$

*(iii) Let $T, T'$ be subsets of $S$ and $w, w'$ elements of $W$. Then $wW_T \subset w'W_{T'}$ (resp. $wW_T = w'W_{T'}$) if and only if $w^{-1}w' \in W_{T'}$ and $T \subset T'$ (resp. $T = T'$).*

*Proof.* (i) $(W_T, T)$ is a pre-Coxeter system. Let $t \in T$ and $w \in W_T$ be such that $l_T(tw) \le l_T(w)$. By the previous corollary, $l_S(tw) \le l_S(w)$. Let $\mathbf{t} = (t_1, \dots, t_k)$, $t_i \in T$, be a reduced expression for $w$. Since $(W, S)$ satisfies the Exchange Condition, a letter of $\mathbf{t}$ can be exchanged for a $t$ in front. Hence, $(W_T, T)$ satisfies the Exchange Condition; so, by Theorem 3.3.4, it is a Coxeter system.

Assertions (ii) and (iii) follow from Corollary 4.1.2. □

The next proposition is an immediate consequence of Corollary 4.1.3 and the fact that $m_{st} = 2$ means $s$ and $t$ commute.

**PROPOSITION 4.1.7.** *Suppose $S$ can be partitioned into two nonempty disjoint subsets $S'$ and $S''$ such that $m_{st} = 2$ for all $s \in S'$ and $t \in S''$. Then $W = W_{S'} \times W_{S''}$.*

## 4.2. REFLECTIONS

As in 3.2, $R$ denotes the set of elements of $W$ which are conjugate to an element of $S$ and as 2.1, $\Omega = \mathrm{Cay}(W, S)$ is the Cayley graph. For each $r \in R$, $\Omega^r$ denotes the fixed set of $r$ on $\Omega$. For each pair $(u, v) \in W \times W$, let $R(u, v)$ denote the set of $r \in R$ such that $\Omega^r$ separates $u$ from $v$.

**LEMMA 4.2.1.** *Suppose $u, v \in W$. Then*

*(i)* $R(u, uv) = uR(1, v)u^{-1}$,

*(ii)* $R(1, uv)$ *is the symmetric difference of* $R(1, u)$ *and* $R(u, uv)$,

*(iii)* $l(u) = \mathrm{Card}(R(1, u))$, *and*

*(iv)* $d(u, v) = \mathrm{Card}(R(u, v))$, *where* $d(,)$ *is the word metric.*

*Proof.* $\Omega^r$ separates $u$ and $uv$, if and only if $u^{-1}\Omega^r$ $(= \Omega^{u^{-1}ru})$ separates 1 and $v$. Hence, $r \in R(u, uv)$ if and only if $u^{-1}ru \in R(1, v)$. This proves (i). If $\Omega^r$ separates 1 from $uv$, then either it separates 1 from $u$ or it separates $u$ from $uv$. If it does both, then $u$ lies on the negative side of $\Omega^r$ and $uv$ lies on the opposite side from $u$, i.e., $uv$ is on the positive side. This proves (ii). Property (iii) follows from Lemma 3.2.14. Since $d(u, v) = d(1, u^{-1}v) = l(u^{-1}v)$, (iv) follows from (ii) and (iii). □

**LEMMA 4.2.2.** *Given a reflection $r \in R$ and an element $w \in W$, $r \in R(1, w)$ if and only if $l(w) > l(rw)$.*

*Proof.* Suppose $\mathbf{s} = (s_1, \dots, s_k)$ is a reduced expression for $w$. The corresponding edge path from 1 to $w$ is $(1, w_1, \dots, w_k = w)$. Let $\Phi(\mathbf{s}) = (r_1, \dots, r_k)$ be as defined by (2). First, suppose $r \in R(1, w)$. Then the edge path crosses $\Omega^r$, so $r = r_i$ for some index $i$. Now reflect the portion of the edge path from $w_i$ to $w = w_k$ about $\Omega^r$ to obtain an edge path from $rw_i = w_{i-1}$ to $rw$ and adjoin this new path to the original path from 1 to $w_{i-1}$. This gives an edge path $(1, \dots, w_{i-1}, rw_{i+1}, \dots, rw_k)$ of length $k - 1$ from 1 to $rw$. (The corresponding word is $(s_1, \dots, \widehat{s_i}, \dots, s_k)$.) Hence, $l(rw) < l(w)$. The other case is where $r \notin R(1, w)$, so that 1 and $w$ are in the same half-space bounded by $\Omega^r$. Since $r$ interchanges the half-spaces, $rw$ belongs to the negative half-space, i.e., $r \in R(1, rw)$. The first case applied to $rw$ gives $l(w) < l(rw)$. □

The next lemma shows that the two possible definitions of a "reflection" in a special subgroup coincide.

**LEMMA 4.2.3.** *Suppose $T$ is a subset of $S$. If $r \in R \cap W_T$, then there is an element $w \in W_T$ such that $w^{-1}rw \in T$.*

*Proof.* Suppose $\mathbf{s} = (s_1, \ldots, s_k)$ is a reduced expression for $r$. For $1 \leqslant i \leqslant k$, put $w_i := s_1 \cdots s_i$. As in formula (1) of Section 3.2, put $r_i := w_{i-1}s_iw_{i-1}^{-1}$. By Corollary 4.1.2, each $s_i$ belongs to $T$. Since $l(r) > 0 = l(r^2)$, Lemma 4.2.2 implies $r \in R(1, r)$. So, the gallery from 1 to $r$ must cross the wall corresponding to $r$, i.e., $r_i = r$ for some $i$. Hence, for $w = w_{i-1}$, $w^{-1}rw = s_i \in T$. □

## 4.3. THE SHORTEST ELEMENT IN A SPECIAL COSET

**LEMMA 4.3.1.** *Suppose $T$ and $T'$ are subsets of $S$ and that $w$ is an element of minimum length in the double coset $W_TwW_{T'}$. Then any element $w'$ in this double coset can be written in the form $w' = awa'$, where $a \in W_T$, $a' \in W_{T'}$, and $l(w') = l(a) + l(w) + l(a')$. In particular, there is a unique element of minimum length in each such double coset.*

*Proof.* Write $w'$ as $bwb'$, where $b \in W_T$, $b' \in W_{T'}$. Let $\mathbf{s}$, $\mathbf{u}$, and $\mathbf{s}'$ be reduced expressions for $b$, $w$, and $b'$, respectively. The concatenation of these words, denoted $\mathbf{sus}'$, is a word representing $w'$. It might not be reduced. If not, then, by the Deletion Condition, we can shorten it by deleting two letters at a time. Neither of the deleted letters can be in $\mathbf{u}$, since $w$ is the shortest element in its double coset. Hence, one of the deleted letters must occur in $\mathbf{s}$ and the other in $\mathbf{s}'$. After carrying out this process as far as possible, we get a reduced expression for $w'$ of the form $\mathbf{tut}'$ where $\mathbf{t}$ and $\mathbf{t}'$ are words obtained by deleting letters from $\mathbf{s}$ and $\mathbf{s}'$, respectively. Setting $a = w(\mathbf{t})$ and $a' = w(\mathbf{t}')$, we get $w' = awa'$ with $l(w') = l(a) + l(w) + l(a')$. If $w'$ were another element of minimum length in $W_TwW'_T$, then writing $w' = awa'$ as above, we get $l(w) = l(w') = l(a) + l(w) + l(a')$. Consequently, $l(a) = 0 = l(a')$, $a = 1 = a'$, and $w' = w$. □

**DEFINITION 4.3.2.** Suppose $T$, $T'$ are subsets of $S$. An element $w \in W$ is $(T, T')$-reduced if it is the shortest element in its double coset $W_TwW'_T$.

**LEMMA 4.3.3.** ([29, Ex. 3, pp. 31–32]). *Suppose $T \subset S$.*

*(i) Any element $w' \in W$ can be written uniquely in the form $w' = aw$, where $a \in W_T$ and $w$ is $(T, \emptyset)$-reduced.*

*(ii) An element $w$ is $(T, \emptyset)$-reduced if and only if for each $t \in T$, $l(tw) = l(w) + 1$.*

*Similar statements hold for left cosets of $W_T$ and $(\emptyset, T)$-reduced elements.*

*Proof.* By Lemma 4.3.1, $w'$ can be written as $aw$ and since $w$ is the unique element of minimum length in $W_T w'$, this expression is unique. This proves (i). To prove (ii), first note that if $w$ is $(T, \emptyset)$-reduced, then $l(tw) = l(w) + l(t) = l(w) + 1$, for all $t \in T$. Conversely, suppose $w$ satisfies the condition in (ii). By (i), we can write $w$ as $au$, where $a \in W_T$ and $u$ is $(T, \emptyset)$-reduced. Since $l(w) = l(a) + l(u)$ and $l(tw) = l(ta) + l(u)$, the condition in (ii) gives $l(ta) = l(a) + 1$ for all $t \in T$. But this is only possible if $a = 1$. (If $a \neq 1$, then the first letter of any reduced expression for $a$ must be an element $T$.) Hence, $w = u$ is $(T, \emptyset)$-reduced. □

## 4.4. ANOTHER CHARACTERIZATION OF COXETER GROUPS

Suppose $(W, S)$ is a Coxeter system. For each $s \in S$, put

$$A_s := \{w \in W \mid l(sw) > l(w). \tag{4.3}$$

$A_s$ is a *fundamental half-space* of $W$. A translate $wA_s$ of $A_s$ is also a *half-space*; $sA_s$ is the *opposite half-space* to $A_s$. Note that $A_s$ is the set of elements of $W$ which are $(\{s\}, \emptyset)$-reduced.

*Remark 4.4.1.* Suppose $r \in R$. Then $r = wsw^{-1}$ for some $s \in S$. Note that

(i) $wA_s$ is the vertex set of the half-space of the Cayley graph $\Omega$ bounded by $\Omega^r$ and containing the vertex $w$.

(ii) $rwA_s = wsA_s$ is the vertex set of the opposite half-space.

(iii) $wA_s$ is the vertex set of the positive side of $\Omega^r$ if and only if $\eta(r, w) = +1$. If this is the case, then $wA_s = \{u \in W \mid l(u) < l(ru)\}$ (cf. Lemma 4.2.2). (The number $\eta(r, w) \in \{\pm 1\}$ was defined in Lemma 3.3.5.)

(iv) Elements $u$ and $ru$ in $W$ are connected by an edge if and only if $r = utu^{-1}$ for some $t \in S$.

A *gallery* in a pre-Coxeter system $(W, S)$ means an edge path in the Cayley graph of $(W, S)$, that is, a sequence $\gamma = (w_0, w_1, \dots, w_k)$ of adjacent elements. The elements $w_0$ and $w_k$ are, respectively, the *initial* and *final* elements of the $\gamma$. The other $w_i$ are its *intermediate* elements. The *type* of $\gamma$ is the word $\mathbf{s} = (s_1, \dots, s_k)$, defined by $w_i = w_{i-1}s_i$. $\gamma$ is a *minimal gallery* (or a *geodesic gallery*) if it corresponds to a geodesic edge path in $\Omega$, i.e., if its type $\mathbf{s}$ is a reduced expression.

Given a gallery $\alpha = (w_0, \dots, w_k)$ from $u$ $(= w_0)$ to $v$ $(= w_k)$ and another one $\beta = (w'_0, \dots, w'_l)$ from $v$ to $w$, we can *concatenate* them to form a gallery $\alpha\beta := (w_0, \dots, w_k, w'_1, \dots, w'_l)$ from $u$ to $w$. The following result is another useful characterization of Coxeter systems.

**PROPOSITION 4.4.2.** ([29, Prop. 6, p. 11]). *Suppose $(W, S)$ is a pre-Coxeter system and that $\{P_s\}_{s \in S}$ is a family of subsets of $W$ satisfying the following conditions:*

*(A)* $1 \in P_s$ *for all* $s \in S$.

*(B)* $P_s \cap sP_s = \emptyset$ *for all* $s \in S$.

*(C) Suppose* $w \in W$ *and* $s, t \in S$. *If* $w \in P_s$ *and* $wt \notin P_s$, *then* $sw = wt$.

*Then* $(W, S)$ *is a Coxeter system and* $P_s = A_s$.

Note that when $P_s = A_s$, condition (C) is equivalent to the Folding Condition (F) of 3.2.

*Proof.* Suppose $s \in S$ and $w \in W$. There are two possibilities:

(a) $w \notin P_s$. Let $s_1 \cdots s_k$ be a reduced expression for $w$ and let $(w_0, w_1, \dots, w_k)$ be the corresponding gallery where $w_0 := 1$ and $w_i := s_1 \cdots s_i = w_{i-1}s_i$, for $1 \leqslant i \leqslant k$. Since $w_0 \in P_s$ (by (A)) and $w_k \notin P_s$, there is an integer $i$ between 1 and $k$ such that $w_{i-1} \in P_s$ and $w_i \notin P_s$. Apply (C) with $w = w_{i-1}$ and $t = s_i$ to get $sw_{i-1} = w_{i-1}s_i$. This gives the formula from the Exchange Condition: $sw = w_1 \cdots w_{i-1} w_{i+1} \cdots w_k$ and $l(sw) < l(w)$.

(b) $w \in P_s$. Let $w' = sw$. By (B), $w' \notin P_s$; so, by (a), $l(sw) = l(w') > l(sw') = l(w)$.

Hence, the Exchange Condition holds. So, by Theorem 3.3.4, $(W, S)$ is a Coxeter system. □

## 4.5. CONVEX SUBSETS OF $W$

**DEFINITION 4.5.1.** Given a subset $T$ of $S$, let

$$A_T := \{w \in W \mid w \text{ is } (T, \emptyset)\text{-reduced}\}.$$

$A_T$ is called the *fundamental* $T$*-sector*. For a given $v \in W$, the subset $vA_T$ is the $T$*-sector based at* $v$. As in (4.3), if $T$ is a singleton $\{s\}$, write $A_s$ instead of $A_{\{s\}}$.

By Lemma 4.3.3 (i), $A_T$ is a set of representatives for the set of right cosets $W_T\backslash W$. Similarly, the set $B_T$ of $(\emptyset, T)$-reduced elements is a set of representatives for the left cosets $W/W_T$. By Lemma 4.3.3 (ii), $B_T = (A_T)^{-1}$. It follows from the same lemma that the sector $A_T$ is the intersection of the half-spaces $A_s$, $s \in T$.

**DEFINITION 4.5.2.** By Lemma 4.3.3, any $w \in W$ can be written uniquely in the form $w = au$, with $a \in W_T$ and $u \in A_T$. The *fundamental* $T$*-retraction* is the map $p_T : W \to A_T$ defined by $w \to u$. The *retraction* onto the $T$-sector based at $v$ is the map $vp_Tv^{-1} : W \to vA_T$ which sends $w$ to $vp_T(v^{-1}w)$.

Given elements $u, w \in W$ and a reflection $r$ such that $r \notin R(u, w)$, a half-space bounded by the wall $\Omega^r$ either contains both $u$ and $w$ or neither of them. Let $D(u, w) \subset W$ denote the intersection of all half-spaces which contain both $u$ and $w$. An important property of $D(u, w)$ is described in the next lemma.

**LEMMA 4.5.3.** *Suppose $u, w \in W$ and $v \in D(u, w)$. If $\alpha$ is a minimal gallery from $u$ to $v$ and $\beta$ is a minimal gallery from $v$ to $w$, then $\alpha\beta$ is a minimal gallery from $u$ to $w$. It follows that $D(u, w)$ is the set of all elements of $W$ that can occur as the intermediate elements of some minimal gallery from $u$ to $w$.*

*Proof.* If $r \in R(u, w)$, then exactly one of the following possibilities holds: $\Omega^r$ separates $u$ and $v$ or $\Omega^r$ separates $v$ and $w$. In other words, $R(u, w)$ is the disjoint union of $R(u, v)$ and $R(v, w)$. The lemma follows. □

**DEFINITION 4.5.4.** A subset $U$ of $W$ is *connected* if any two elements of $U$ can be connected by a gallery in $U$ (i.e., the intermediate elements of the gallery lie in $U$).

**DEFINITION 4.5.5.** A subset $U$ of $W$ is *convex* if, for any minimal gallery $\gamma$ with endpoints in $U$, the intermediate elements of $\gamma$ lie in $U$. The subset $U$ is *starlike* (with respect to the element 1) if $1 \in U$ and if, for any minimal gallery with initial element 1 and with final element in $U$, the intermediate elements are also in $U$. An element $u \in U$ is *extreme* if no minimal gallery in $U$ with initial element 1 can contain $u$ as an intermediate element. (In other words, if $u$ is an element of such a gallery, then it must be the final element.)

**Example 4.5.6.** Suppose $S$ is finite. Order the elements of $W$ as $w_1, \dots, w_n \dots,$ in such a fashion that $l(w_k) \leqslant l(w_{k+1})$ for all $k \geqslant 1$. For $n \geqslant 0$, set

$$U_n := \{w_1, \dots, w_n\}.$$

Then $1 \in U_n$ (since $w_1 = 1$) and each $U_n$ is starlike with respect to 1.

**LEMMA 4.5.7.** *A subset of $W$ is convex if and only if it is an intersection of half-spaces.*

*Proof.* It follows from the discussion in 3.2 that any intersection of half-spaces is convex. Conversely, suppose $U$ is a convex subset of $W$. Let $D$ denote the intersection of all half-spaces that contain $U$. We claim $D = U$. Suppose to the contrary that $v \in D - U$. Take a gallery of minimum length from $v$ to a chamber in $U$. The first wall $\Omega^r$ that it crosses must separate $v$ from $U$, a contradiction. □

**Example 4.5.8.** Any sector is a convex subset of $W$.

For each $T \subset S$, put

$$\Omega(T) := \Omega - \bigcup_{r \in R \cap W_T} \Omega^r. \tag{4.4}$$

Example 4.5.8 can then be restated as follows.

**LEMMA 4.5.9.** *$A_T$ is the set of vertices in the connected component of $\Omega(T)$ that contains the identity vertex.*

### Parabolic Subgroups

A subgroup of $W$ is *parabolic* if it is conjugate to some special subgroup.

**PROPOSITION 4.5.10.** *Suppose special subgroups $W_T$ and $W_U$ are conjugate. Then $T$ and $U$ are conjugate subsets of $S$.*

*Proof.* Suppose $wW_Tw^{-1} = W_U$. By Lemma 4.2.3, conjugation by $w$ maps $R \cap W_T$ to $R \cap W_U$; hence, it takes the connected components of $\Omega(T)$ to the connected components of $\Omega(U)$. In particular, $w$ maps the component of $\Omega(T)$ containing 1 to a component of $\Omega(U)$, i.e., $wA_T = uA_U$ for some $u \in W_U$. So $v := uw^{-1}$ takes $A_T$ to $A_U$.

Suppose $\Omega(U)_o$ is a component of $\Omega(U)$. Call an edge of $\Omega$ a *boundary edge* of $\Omega(U)_o$ if one of its endpoints lies in $\Omega(U)_o$ and the other does not. We will also say that it is a *boundary edge* of the sector $uA_U$, where $uA_U := \text{Vert}(\Omega(U)_o)$. The element $v$ takes the boundary edges of $A_T$ to the boundary edges of $A_U$. It is not hard to see that, in the case of $A_T$, the midpoints of these boundary edges are indexed by elements of $T$ and in the case of $A_U$ by elements of $U$. Hence, $vTv^{-1} = U$. $\square$

**Exercise 4.5.11.** Prove the assertion in the previous proof: the midpoints of the boundary edges of $A_T$ are indexed by elements of $T$.

Let $G := wW_Tw^{-1}$ be a parabolic subgroup. Its *rank* is defined by

$$\text{rk}(G) := \text{Card}(T). \tag{4.5}$$

This is well defined by Proposition 4.5.10.

## 4.6. THE ELEMENT OF LONGEST LENGTH

**LEMMA 4.6.1.** ([29, Ex. 22, p. 40]). *Let $(W, S)$ be a Coxeter system. The following conditions on an element $w_0 \in W$ are equivalent.*

(a) *For each $u \in W$, $l(w_0) = l(u) + l(u^{-1}w_0)$.*

(b) *For each $r \in R$, $l(w_0) > l(rw_0)$.*

*Moreover, $w_0$ exists if and only if $W$ is finite. If $w_0$ satisfies either (a) or (b), then*

*(i)* *$w_0$ is unique,*

*(ii)* $l(w_0) = \text{Card}(R)$,

*(iii)* *$w_0$ is an involution, and*

*(iv)* $w_0 S w_0 = S$.

The element $w_0$ is called the *element of longest length.*

*Proof.* Condition (a) obviously implies (b). By Lemma 4.2.2, condition (b) means that $R(1, w_0) = R$. (So condition (b) entails $\text{Card}(R) < \infty$ and since $S \subset R$, $\text{Card}(S) < \infty$.) It follows from Lemma 4.2.1 (ii) that $R$ is the disjoint union of $R(1, u)$ and $R(u, w_0)$ and from the other assertions of the same lemma that $l(w_0) = \text{Card}(R(1, u)) + \text{Card}(R(u, w_0)) = l(u) + l(u^{-1}w_0)$. Hence, (b) implies (a).

Since the length of any element in $W$ is bounded by $l(w_0)$, if $w_0$ exists, $W$ is finite. Conversely, suppose $W$ is finite and $v$ is an element of maximum length. Since $w \to (-1)^{l(w)}$ is a homomorphism from $W$ to $\{\pm 1\}$ and since, for any $r \in R$, $(-1)^{l(r)} = -1$,

$$l(rv) \equiv l(r) + l(v) \equiv 1 + l(v) \mod 2,$$

so $rv$ cannot have the same length as $v$. Hence, $l(rv) < l(v)$, i.e., $v$ satisfies (b). So, if $W$ is finite, $w_0 := v$ exists.

Suppose that $w_0$ and $v$ are two elements satisfying (a). Then

$$l(w_0) = l(v) + l(v^{-1}w_0),$$

$$l(v) = l(w_0) + l(w_0^{-1}v).$$

Because any element and its inverse have the same length, $l(v^{-1}w_0) = l(w_0^{-1}v)$. The above equations then imply that $l(v^{-1}w_0) = 0$, i.e., $w_0 = v$. This proves (i). Property (ii) follows from (b) and Lemma 4.2.1 (iii). If $W$ is finite, then, since $w_0^{-1}$ is also an element of maximum length, it follows from the previous argument and (i) that $w_0^{-1} = w_0$. This proves (iii). If $s \in S$, then, since $l(w_0)$ is maximum, $l(sw_0) = l(w_0) - 1$. Rewriting (a) as, $l(u^{-1}w_0) = l(w_0) - l(u)$, and applying it to the case $u = sw_0$, we obtain $l(w_0 s w_0) = l(w_0) - l(sw_0) = 1$. Since $w_0 s w_0$ has length 1, it belongs to $S$, proving (iv). □

**LEMMA 4.6.2.** *Suppose there is an element $w_0 \in W$ so that $l(sw_0) < l(w_0)$ for all $s \in S$. Then $W$ is finite and $w_0$ is the element of longest length.*

*Proof.* Given a (possibly infinite) sequence $\mathbf{s} = (s_1, \dots, s_i, \dots)$ of elements in $S$, let $\mathbf{s}_i = (s_i, \dots, s_1)$ be an initial subsequence written in reverse order.

Assume each $\mathbf{s}_i$ is a reduced expression. Suppose $l(sw_0) < l(w_0)$ for all $s \in S$. We claim that it is a consequence of the Exchange Condition that $w_0$ has a reduced expression beginning with $\mathbf{s}_i$. The argument is similar to the proof of Theorem 3.4.2. It is clear for $s_1$. Suppose by induction that $w_0$ has a reduced expression beginning with $\mathbf{s}_{i-1}$. Since $l(s_i w_0) < l(w_0)$, the Exchange Condition implies that we can find another reduced expression by exchanging a letter for an $s_i$ in front. The exchanged letter cannot be in the initial string because otherwise we could cancel the final parts to obtain, $s_i \cdots s_{j+1} = s_{i-1} \cdots s_j$ for some $j < i$, contradicting the assumption that $\mathbf{s}_i$ is a reduced expression. This proves the claim. Hence, condition (a) of Lemma 4.6.1 holds; so, $W$ finite and $w_0$ is the element of longest length. □

## 4.7. THE LETTERS WITH WHICH A REDUCED EXPRESSION CAN END

**DEFINITION 4.7.1.** Given $w \in W$, define a subset $\mathrm{In}(w)$ of $S$ by

$$\mathrm{In}(w) := \{s \in S \mid l(ws) < l(w)\}.$$

Also, set

$$\mathrm{Out}(w) := S - \mathrm{In}(w).$$

By the Exchange Condition (applied to the element $w^{-1}$ and the letter $s$), $\mathrm{In}(w)$ is the set of letters with which a reduced expression for $w$ can end.

The following lemma will play a key role in Chapters 7, 8, 9, 10, and 15.

**LEMMA 4.7.2.** *For any $w \in W$, the special subgroup $W_{\mathrm{In}(w)}$ is finite.*

*Proof.* By Lemma 4.3.1, $w$ can be written uniquely in the form $w = aw_0$, where $a$ is $(\emptyset, \mathrm{In}(w))$-reduced and where for any $u \in W_{\mathrm{In}(w)}$, $l(au) = l(a) + l(u)$. For any $s \in \mathrm{In}(w)$,

$$l(w) = l(a) + l(w_0),$$

$$l(ws) = l(aw_0 s) = l(a) + l(w_0 s).$$

Since $l(ws) = l(w) - 1$, the above equations give $l(w_0 s) = l(w_0) - 1$ for all $s \in \mathrm{In}(w)$. By Lemma 4.6.2 (applied to $w_0^{-1}$), this implies that $W_{\mathrm{In}(w)}$ is finite and that $w_0$ is its element of longest length. □

**LEMMA 4.7.3**

*(i) $w$ is the longest element in $wW_{\mathrm{In}(w)}$. In other words, for any $u \in W_{\mathrm{In}(w)}$, $l(wu) = l(w) - l(u)$.*

*(ii) $w$ is the shortest element in $wW_{\mathrm{Out}(w)}$. In other words, for any $u \in W_{\mathrm{Out}(w)}$, $l(wu) = l(w) + l(u)$.*

*Proof.* The first assertion is implied by the proof of the previous lemma. For the second, note that it follows from the definition of $\mathrm{Out}(w)$ that for any $s \in \mathrm{Out}(w)$, $l(ws) > l(w)$. Hence, by Lemma 4.3.3 (ii), $w$ is $(\emptyset, \mathrm{Out}(w))$-reduced. □

**Definition 4.7.4.** Given a subset $T$ of $S$, define a subset of $W^T$ of $W$ by

$$W^T := \{w \in W \mid \mathrm{In}(w) = T\}.$$

For example, $W^{\emptyset} = \{1\}$. Lemma 4.7.2 shows that if $W^T$ is nonempty, then $W_T$ is finite. In Chapters 8 and 10 we will need the following.

**Lemma 4.7.5.** ([77, Lemma 1.10, p. 301]). *Suppose $W^T$ is a singleton. Then $W_T$ is finite and $W$ splits as a direct product*

$$W = W_T \times W_{S-T}.$$

To prove this we need the following.

**Lemma 4.7.6.** ([77, Lemma 1.9, p. 300]). *Suppose $T \subset S$ is such that $W_T$ is finite and that $w_T \in W_T$ is its element of longest length. Let $s \in S - T$, Then $sw_T = w_T s$ if and only if $m_{st} = 2$ for all $t \in T$.*

*Proof.* If $m_{st} = 2$, then $s$ and $t$ commute. Hence, if $m_{st} = 2$ for all $t \in T$, then $s$ and $w_T$ commute.

Conversely, suppose $s$ and $w_T$ commute. Since $l(w_T s) = l(w_T) + 1$, $s \in \mathrm{In}(w_T s)$. Since $w_T s = s w_T$, $T \subset \mathrm{In}(w_T s)$. Therefore, $\mathrm{In}(w_T s) = T \cup \{s\}$. We want to show that $m_{st} = 2$ for all $t \in T$. Suppose, to the contrary, that $m_{st} > 2$ for some $t \in T$. Consider the dihedral subgroup $W_{\{s,t\}}$. Since $\{s,t\} \subset \mathrm{In}(w_T s)$, $W_{\{s,t\}} \subset W_{\mathrm{In}(w_T s)}$. By Lemma 4.6.1 (i), $l(w_T s u) = l(w_T s) - l(u)$ for any $u \in W_{\{s,t\}}$. Since $m_{st} > 2$, the element $u := sts$ has length 3. So, $l((w_T s)(sts)) = l(w_T s) - 3$. On the other hand, $l((w_T s)(sts)) = l(w_T ts) = l(w_T t) + 1 = l(w_T) = l(w_T s) - 1$; a contradiction. □

*Proof of Lemma 4.7.5.* By Lemma 4.7.2, $W_T$ is finite. Let $w_T$ be its element of longest length. By Lemma 4.6.1 (ii), $w_T \in W^T$ and since $W^T$ is a singleton, this singleton is $\{w_T\}$. We will show that $w_T$ commutes with each element $s \in S - T$. Let $s \in S - T$. Clearly, $T \subset \mathrm{In}(sw_T) \subset \{s\} \cup T$ and since $sw_T \notin W^T$, we must have $\mathrm{In}(sw_T) = \{s\} \cup T$. Thus, $l(s(w_T s)) = l((sw_T)s) = l(sw_T) - 1 = l((sw_T)^{-1}) - 1 = l(w_T s) - 1$. In other words, $w_T s$ does not lie in the half-space $A_s$. Since $w_T \in A_s$, the Folding Condition (F) implies that $sw_T = w_T s$. Then, by Lemma 4.7.6, $m_{st} = 2$ for all $t \in T$. If $T \neq \emptyset$, this means that $(W, S)$ is reducible and that $W$ splits as $W = W_T \times W_{S-T}$. □

## 4.8. A LEMMA OF TITS

In this section $(W, S)$ is required only to be a pre-Coxeter system. We will state and prove a lemma of Tits [282] which will be needed in the next section and again in Appendix D.

Suppose a group $G$ acts on a set $\Pi$. A subset $B$ of $\Pi$ is *prefundamental* for $G$, if for any $g \in G$, $gB \cap B \neq \emptyset \implies g = 1$.

As usual, for each $s \in S$, $W_{\{s\}}$ is the subgroup of order two generated by $s$ and for each pair of distinct elements $s, t \in S$, $W_{\{s,t\}}$ is the dihedral subgroup generated by $\{s, t\}$. Suppose $W$ acts on a set $\Pi$ and that for each $s \in S$ we are given a subset $B_s$ of $\Pi$ which is prefundamental for $W_{\{s\}}$. Finally, suppose that $B := \bigcap_{s\in S} B_s$ is nonempty. Consider the following property of $(W, S, \{B_s\}_{s\in S})$:

(P) For any $w \in W$ and $s \in S$, either $wB \subset B_s$ or $wB \subset sB_s$. Moreover, in the second case, $l(sw) = l(w) - 1$.

($l(w)$ denotes word length with respect to $S$.)

*Remark 4.8.1.* Property (P) implies that $B$ is prefundamental for $W$. (Proof: Suppose, to the contrary, that $wB \cap B \neq \emptyset$ for some $w \neq 1$. Then for some $s \in S$, $w = sw'$ with $l(w') = l(w) - 1$. Since $wB \not\subset sB_s$, $w'B = swB \not\subset B_s$. So, by (P) applied to $w'$, $w'B \subset sB_s$ and $l(w) = l(sw') = l(w') - 1$, a contradiction.)

*Remark 4.8.2.* Property (P) is not enough to guarantee that a pre-Coxeter system $(W, S)$ is a Coxeter system. For example, if $\Pi = W$, $B = \{1\}$ and $B_s = \{w \in W \mid l(sw) < l(w)\}$, then (P) holds whenever $l(sw) = l(w) \pm 1$. Moreover, this condition holds whenever the function $\varepsilon : W \to \{\pm 1\}$ defined by $\varepsilon(w) := (-1)^{l(w)}$ is a homomorphism or equivalently, if every element in the kernel of $\widetilde{W} \to W$ has even length (where $\widetilde{W}$ is the Coxeter group associated to the Coxeter matrix of $(W, S)$). Nontrivial examples of this are easily constructed. For example, let $\widetilde{W}$ be a finite Coxeter group of rank $> 2$ in which the element $w_0$ of longest length is central and of even length (e.g., when the Coxeter diagram of $\widetilde{W}$ is $\mathbf{B}_n$ from Table 6.1 in Section 6.9, with $n$ even). Putting $W := \widetilde{W}/\langle w_0\rangle$, we have that $\varepsilon : W \to \{\pm 1\}$ is a homomorphism.

To guarantee $(W, S)$ is a Coxeter system we must add something like Condition (C) in Proposition 4.4.2, say,

(C′) Suppose $w \in W$ and $s, t \in S$. If $wB \subset B_s$ and $wtB \not\subset B_s$, then $sw = wt$.

If Property (P) and Condition (C′) both hold, then it follows from Proposition 4.4.2 and Remark 4.8.1 that $(W, S)$ is a Coxeter system.

**Lemma 4.8.3.** (Tits [282, Lemma 1]). *With notation as above, suppose that, for each pair of distinct elements $s, t \in S$, $(W_{\{s,t\}}, \{s, t\}, \{B_s, B_t\})$ satisfies (P). Then $(W, S, \{B_s\}_{s\in S})$ satisfies (P).*

Let $(\mathrm{P}_n)$ denote the statement of property (P) for all $w$ with $l(w) = n$. Let $(\mathrm{Q}_n)$ denote the following property:

$(\mathrm{Q}_n)$ For any $w \in W$, with $l(w) = n$, and any two distinct elements $s, t \in S$, there exists $u \in W_{\{s,t\}}$ such that $wB \subset u(B_s \cap B_t)$ and $l(w) = l'(u) + l(u^{-1}w)$.

($l'(u)$ denotes word length with respect to $\{s, t\}$.)

We shall prove Lemma 4.8.3 by establishing (P) for $(W, S, \{B_s\}_{s\in S})$ by using the inductive scheme

$$((\mathrm{P}_n) \text{ and } (\mathrm{Q}_n)) \implies (\mathrm{P}_{n+1}) \quad \text{and} \quad ((\mathrm{P}_{n+1}) \text{ and } (\mathrm{Q}_n)) \implies (\mathrm{Q}_{n+1}).$$

$(\mathrm{P}_0)$ and $(\mathrm{Q}_0)$ hold trivially.

*Proof that* $((\mathrm{P}_n)$ *and* $(\mathrm{Q}_n)) \implies (\mathrm{P}_{n+1})$. Suppose $l(w) = n + 1$ and $s \in S$. Then there exists $t \in S$ with $w = tw'$ and $l(w') = n$. If $s = t$, then $(\mathrm{P}_n)$ applied to $w'$ shows that $w'B \subset B_s$; hence, $wB \subset sB_s$ and $l(sw) = l(w') = l(w) - 1$. So, suppose $s \neq t$. Applying $(\mathrm{Q}_n)$ to $w'$, there exists $u \in W_{\{s,t\}}$ such that $w'B \subset u(B_s \cap B_t)$ and $l(u^{-1}w') = l(w') - l'(u)$. Put $v = tu \in W_{\{s,t\}}$. By hypothesis, (P) holds for $(W_{\{s,t\}}, \{s, t\}, \{B_s, B_t\})$. Hence, either

(a) $tu(B_s \cap B_t) \subset B_s$, in which case, $wB \subset B_s$ or

(b) $tu(B_s \cap B_t) \subset sB_s$, in which case, $wB \subset sB_s$.

Moreover, in case (b), $l'(stu) = l'(tu) - 1$ and hence,

$$\begin{aligned} l(sw) = l(stw') = l((stu)(u^{-1}w')) &\leqslant l'(stu) + l(u^{-1}w') \\ = l'(tu) + l(u^{-1}w') - 1 &= l'(tu) + l(w') - l'(u) - 1 \leqslant l(w) - 1. \end{aligned}$$

But this implies $l(sw) = l(w) - 1$. □

*Proof that* $((\mathrm{P}_{n+1})$ *and* $(\mathrm{Q}_n)) \implies (\mathrm{Q}_{n+1})$. Suppose $l(w) = n + 1$ and $s, t$ are distinct elements of $S$. If $wB \subset (B_s \cap B_t)$, then $(\mathrm{Q}_{n+1})$ holds with $u = 1$. To fix ideas, suppose $s$ is the given element of $\{s, t\}$ and that $wB \not\subset B_s$. By $(\mathrm{P}_{n+1})$, $wB \subset sB_s$ and $l(sw) = n$. Applying $(\mathrm{Q}_n)$ to $sw$, there exist $v \in W_{\{s,t\}}$ such that $swB \subset v(B_s \cap B_t)$ and $l(sw) = l'(v) + l(v^{-1}sw)$. Hence, $wB \subset sv(B_s \cap B_t)$ and

$$\begin{aligned} l(w) = 1 + l(sw) &= 1 + l'(v) + l(v^{-1}sw) \\ &\geqslant l'(sv) + l((sv)^{-1}w) \geqslant l(w). \end{aligned}$$

So the inequalities must be equalities and $(\mathrm{Q}_{n+1})$ holds with $u = sv$. □

## 4.9. SUBGROUPS GENERATED BY REFLECTIONS

As usual, $(W, S)$ is a Coxeter system, $\Omega := \mathrm{Cay}(W, S)$, and $R$ is the set of reflections in $W$. Suppose we are given a subset $X$ of $R$. Let $W' := \langle X \rangle$ be the subgroup generated by $X$. We want to prove that $W'$ is a Coxeter group and determine a fundamental set of generators for it. Put $R' := R \cap W'$ and

$$Z(X) := \Omega - \bigcup_{r \in R'} \Omega^r. \tag{4.6}$$

$W'$ acts on $Z(X)$. Let us call the vertex set of a path component of $Z(X)$ a *chamber for $W'$ on $\Omega$*. Such a chamber is a subset of $W$. Given a chamber $D$ and a reflection $r \in R'$, one of the two components of $\Omega - \Omega^r$ contains $D$. Denote the vertex set of this component by $A_r(D)$. It is a half-space of $W$. Since $D$ is the intersection of the $A_r(D)$, it is a convex subset of $W$ (Definition 4.5.5).

**DEFINITION 4.9.1.** Suppose $D$ is a chamber for $W'$ on $\Omega$. An element $r \in R'$ is a boundary reflection of $D$ if there is a vertex $w \in D$ such that $\{w, rw\}$ is a boundary edge (i.e., an edge of $\Omega$ such that $rw \notin D$). Let $S'(D)$ denote the set of boundary reflections for $D$. Chambers $D_1, D_2$ are called adjacent if $D_2 = rD_1$ for some boundary reflection $r$ of $D_1$.

Clearly,

$$D = \bigcap_{r \in S'(D)} A_r(D). \tag{4.7}$$

The proof of Lemma 3.2.5 shows that $S'(D)$ generates $W'$. Our goal in this section is to prove the following.

**THEOREM 4.9.2.** (Compare [94, 107].) *Property (P) of the previous section holds for* $(W', S'(D), \{A_s(D)\}_{s \in S'(D)})$. *Hence,* $(W', S'(D))$ *is a Coxeter system.*

($S'(D)$ can be infinite.)

To simplify notation, write $A_s$ instead of $A_s(D)$.

**COMPLEMENT 4.9.3.** Suppose we are given two half-spaces $A_s, A_t$ of $W$ corresponding to a random choice of two distinct reflections $s, t \in R$. Let $W'$ be the dihedral subgroup generated by $\{s, t\}$ and $m$ $(:= m(s,t))$ the order of $st$. We want to analyze when $A_s \cap A_t$ is prefundamental for $W'$. We discuss the cases $m = \infty$ and $m \neq \infty$ separately.

*Case 1.* $m = \infty$. In this case one can show that the picture for the possible intersections of $A_s$ and $A_t$ is exactly the same as for the standard action of the infinite dihedral group $\mathbf{D}_\infty$ on the real line (Example 3.1.3). In other words, there are three possibilities:

(a) $A_s \cap A_t = \emptyset$,

(b) $A_s \subset A_t$ or $A_t \subset A_s$,

(c) neither (a) nor (b).

(P) fails if (a) or (b) holds. On the other hand, if (c) is true, then (P) holds. (The proofs of the two previous statements are left as exercises for the reader; to prove the second, one argues by induction on the word length of $v \in W'$.) Hence, in case (c), $A_s \cap A_t$ is prefundamental for $W'$ and the vertex set of any component of $Z(\{s,t\})$ has the form $v(A_s \cap A_t)$ for some $v \in W'$.

*Case 2.* $m \neq \infty$. Once again one can show that the possible patterns of intersection are the same as in the picture of the standard orthogonal linear action of the dihedral group $\mathbf{D}_m$ on $\mathbb{R}^2$ (Example 3.1.2). Thus, if $A_s \cap A_t$ is prefundamental for $W'$, then $A_s \cap A_t$ is a component of $Z(\{s,t\})$ and (P) holds. Suppose it is not prefundamental. Then we can find a half-space $A_r \subset W$, corresponding to a reflection $r \in W'$ ($r$ is conjugate to $s$ or $t$), so that $A_s \cap A_r$ is prefundamental for $W'$ and $A_s \cap A_r \subset A_s \cap A_t$. Thus, $\Omega^r$ separates $A_s \cap A_t$ and so (P) fails for either $s$ or $t$. Moreover, there is another chamber for $W'$ of the form $u(A_s \cap A_r)$, $u \in W'$, so that $u(A_s \cap A_t) \subset A_s \cap A_t)$ and so that $t$ is a boundary reflection for it.

*Proof of Theorem 4.9.2.* According to Tits' Lemma 4.8.3 we need to show that for any two elements $s, t \in S'(D)$, given $s' \in \{s,t\}$ and $v \in W'_{\{s,t\}}$, either (a) $v(A_s \cap A_t) \subset A_{s'}$ or (b) $v(A_s \cap A_t) \subset s'A_{s'}$ and $l'(s'v) = l'(v) - 1$. (As before, $l'(\ )$ is the word length in the dihedral group $W'_{\{s,t\}}$.) To fix ideas, suppose $s' = s$.

First suppose $m(s,t) = \infty$. Since $A_s \cap A_t \supset D \neq \emptyset$ and since $s$ and $t$ are boundary reflections for $D$, possibilities (a) and (b) in Case 1 of Complement 4.9.3 are excluded. So, we are left with possibility (c) and the discussion in Complement 4.9.3 shows that the assertion in the previous paragraph is true.

So, suppose $m(s,t) \neq \infty$. Since $s$ is a boundary reflection for $D$, there is $w \in D$ such that $sw \notin A_s$ and $\{w, sw\}$ is an edge of $\Omega$. By the discussion in Case 2 of Complement 4.9.3, there is a half-space $A_r$, for some reflection $r \in W'_{\{s,t\}}$, such that $A_s \cap A_r \subset A_s \cap A_t$ and $A_s \cap A_r$ is prefundamental for $W'_{\{s,t\}}$. If $A_r = A_t$, we are done. So, suppose not. First we claim that the element $w$ lies in $A_s \cap A_r$. Indeed, it cannot lie in any other nontrivial translate $u(A_s \cap A_r) \subset A_s \cap A_t$ because such a translate does not have $s$ as a boundary reflection. Since $w \in A_s \cap A_r$ and since $D$ is convex, $D \subset A_s \cap A_r$. Similarly, since $t$ is a boundary reflection of $D$, there is a $w' \in D$ such that $tw \notin A_t$ and $\{w, tw\}$ is an edge of $\Omega$. Arguing as before, we see that $w' \in u(A_s \cap A_r)$ where, as in Case 2 of Complement 4.9.3, $u \in W'_{\{s,t\}}$ is such that $t$ is a boundary reflection of $u(A_s \cap A_r)$. Hence, $D \subset u(A_s \cap A_r)$. This implies $D$ also lies in $rA_r$, a contradiction. □

We continue to suppose that $D$ is a chamber for $W'$ on $\Omega$. Let $D'$ be an adjacent chamber to $D$. Recall that this means that $D' = sD$ for some boundary reflection $s \in S'(D)$. We claim the boundary reflection $s$ is determined by $\{D, D'\}$. Indeed, suppose that $D' = tD$ for some other element $t \in S'(D)$ distinct from $s$. Then $stD = D$, contradicting the fact from the proof of Theorem 4.9.2 that $A_s(D) \cap A_t(D)$ is prefundamental for the dihedral group $W'_{\{s,t\}}$. Thus, $s = t$.

Define a graph $\Omega'$ by declaring its vertex set to be $\pi_0(Z(X))$ and connecting two vertices by an edge if and only if the corresponding chambers are adjacent. Since each $s \in S'(D)$ separates $\Omega'$, $(\Omega', [D])$ is a reflection system in the sense of Definition 3.2.10. (Here [D] denotes the vertex of $\Omega'$ corresponding to $D$.) Theorem 3.3.4 then provides a different proof that $(W', S'(D))$ is a Coxeter system. Moreover, $\Omega' \cong \mathrm{Cay}(W', S'(D))$.

The techniques in the proof of Theorem 4.9.2 can also be used to answer the following.

**Question.** Suppose $D$ is an arbitrary convex set in $W$, $S'(D)$ is the corresponding set of boundary reflections and $W' := \langle S'(D) \rangle$. When is $(W', S'(D))$ a Coxeter system?

By Remark 4.8.1 (ii) and Tits' Lemma, $(W', S'(D))$ is a Coxeter system if and only if Property (P) holds for $(W'_{\{s,t\}}, \{s, t\}, \{A_s(D), A_t\})$ for every pair of distinct elements $s, t \in S'(D)$. The argument in the proof of Theorem 4.9.2 (using Case 1 of Complement 4.9.3) shows that if $m(s, t) = \infty$, then Property (P) automatically holds. So, we have proved the following.

**THEOREM 4.9.4.** *Suppose $D$ is a convex set in $W$. Let $S'(D)$ be the corresponding set of boundary reflections and $W' := \langle S'(D) \rangle$. Then $(W', S'(D))$ is a Coxeter system if and only if Property (P) holds for $(W'_{\{s,t\}}, \{s, t\}, \{A_s(D), A_t(D)\})$ for all $s \neq t$ in $S'(D)$ with $m(s, t) \neq \infty$.*

## 4.10. NORMALIZERS OF SPECIAL SUBGROUPS

In this section we explain a result of Deodhar which gives a description of the normalizers of special subgroups. It will be needed later in Section 12.7.

Given $T \subset S$, let $A_T$ be the fundamental $T$-sector in $W$ (Definition 4.5.1) and let $N(W_T)$ be the normalizer of the special subgroup $W_T$. Define a subgroup

$$G_T := \{w \in W \mid wA_T = A_T\} \tag{4.8}$$

**LEMMA 4.10.1**

*(i)* $G_T$ *normalizes* $W_T$.

*(ii) Each* $w \in G_T$ *is* $(T, T)$*-reduced (Definition 4.3.2).*

*(iii)* $W_T \cap G_T = \{1\}$.

*(iv) For any* $w \in N(W_T)$*, there exists a unique element* $u \in W_T$ *such that* $wA_T = uA_T$.

*Proof.* (i) Since $G_T$ maps $A_T$ to itself, it must map the set of boundary reflections to itself. That is to say, each element of $G_T$ conjugates $T$ to itself.

(ii) By definition, $A_T$ is the set of $(T, \emptyset)$-reduced elements in $W$. Since $1 \in A_T$, $w = w \cdot 1$ is in $A_T$ for any $w \in G_T$. So, $w$ is $(T, \emptyset)$-reduced. Similarly, $w^{-1} \in A_T$; so, $w$ is $(\emptyset, T)$-reduced.

(iii) This follows from (ii).

(iv) For any $w \in W$, $wA_T$ is a sector for $wW_Tw^{-1}$. Since $w \in N(W_T)$, $wA_T$ is a sector for $wW_Tw^{-1} = W_T$. But any such sector has the form $uA_T$ for some unique $u \in W_T$. □

**PROPOSITION 4.10.2.** ([38, Prop. 2.1] or [181, Prop. 3.1.9]).

$$N(W_T) = W_T \rtimes G_T.$$

*Proof.* We define a projection $p : N(W_T) \to G_T$. By Lemma 4.10.1 (iv), for any $w \in N(W_T)$, there is a unique $u \in W_T$ with $wA_T = uA_T$. Put $p(w) := u^{-1}w$. One checks easily that this is a well-defined homomorphism with kernel $W_T$. □

**DEFINITION 4.10.3.** Suppose $T \subset S$ and $s \in S - T$. Let $U$ be the irreducible component of $T \cup \{s\}$ containing $s$ (i.e., $\Gamma_U$, the Coxeter diagram of $W_U$, is the component of $\Gamma_{T \cup \{s\}}$ containing $s$). If $W_U$ is finite, define

$$\nu(T, s) := w_{U-\{s\}} w_U,$$

where $w_{U-\{s\}}$ and $w_U$ are the elements of longest length in $W_{U-\{s\}}$ and $W_U$, respectively. When $W_U$ is infinite, $\nu(T, s)$ is not defined.

**LEMMA 4.10.4.** *Suppose* $\nu(T, s)$ *is defined. Then* $\nu(T, s)^{-1} A_T = A_{T'}$ *for some* $T' = (T \cup \{s\}) - \{t\}$, $t \in U$.

*Proof.* Conjugation by $w_{U-\{s\}}$ stabilizes $T$, permuting the elements of $U-\{s\}$ and fixing the elements of $T-U$. Moreover, for each $t \in U-\{s\}$, $A_t$ and $w_{U-\{s\}}A_t$ are opposite half-spaces. Similarly, conjugation by $w_U$ stabilizes $T \cup \{s\}$, permuting $U$ and fixing $T-U$. Put $t := w_U s w_U$. (Recall that elements of longest length are involutions.) Since conjugation by $w_U$ stabilizes $U$, $t \in U$. Setting $\nu = \nu(T, s)$, we have $\nu^{-1}T\nu = w_U(w_{U-\{s\}}Tw_{U-\{s\}})w_U = w_U T w_U = T'$, where $T' := (T \cup \{s\}) \cup \{t\}$. Moreover, for any $t \in T$, $\nu^{-1}$ maps $A_t$ to a positive half-space indexed by an element of $T'$. Hence, $\nu^{-1}$ takes $A_T$ to $A_{T'}$. □

Define a directed labeled graph $\Upsilon$: its vertices are the subsets of $S$; there is an edge from $T$ to $T'$ labeled $s$ whenever $\nu(T, s)$ is defined and $\nu(T, s)^{-1}A_T = A_{T'}$. Write $T \xrightarrow{s} T'$ to mean there is an edge from $T$ to $T'$ labeled $s$. The proof of the next proposition is left as an exercise for the reader.

**PROPOSITION 4.10.5.** (See [181, Prop. 3.1.2]). *If $T \xrightarrow{s} T'$ is an edge of $\Upsilon$, then so is $T' \xrightarrow{t} T$, where either $\{t\} = T - T'$ or $T = T'$ and $t = s$.*

The next theorem, due to Deodhar (and Howlett [158, Lemma 4] in the case where $W$ is finite), can be used to give generators and relations for the subgroup $G_T$ of $N(W_T)$ (as in [38]). For a proof we refer the reader to [93, Prop. 5.5] or [38, Prop. 2.3].

**THEOREM 4.10.6.** (Deodhar [93]). *Let $T, T' \subset S$ and $w \in W$ be such that $w^{-1}A_T = A_{T'}$. Then there is a directed path*

$$T = T_0 \xrightarrow{s_0} T_1 \xrightarrow{s_1} \cdots \xrightarrow{s_k} T_{k+1} = T'$$

*in $\Upsilon$ such that $w = \nu(T_0, s_0) \cdots \nu(T_k, s_k)$. Moreover,*

$$l(w) = \sum_{i=0}^{k} l(\nu(T_i, s_i)).$$

## NOTES

**4.1.** The results in this section are taken from [29, pp. 12–13]. Paris [233] has proved a converse to Proposition 4.1.7.

**4.4.** In much of the literature, e.g., in [51, 248, 298], the subspaces $wA_s$ are called "roots" rather than "half-spaces" (as in Definition 4.5.1). (In standard terminology a root is a certain vector in the "canonical representation" of Appendix D which defines a linear reflection and a linear half-space in its dual representation space. These linear half-spaces carry the same combinatorial information as do the half-spaces in the sense of 4.5.1.)

**4.7.** In [26, p. 17] the subset $\mathrm{In}(w)$ of $S$ is denoted $D_R(w)$ and called the *right descent set* of $w$. These subsets play an important role in [26]. In the same book the subset $W^T$ of $W$ is denoted $\mathcal{D}_T$ and called a set of *descent classes*.

**4.8.** The proof of Tits' Lemma 4.8.3 can also be found in [29, pp. 98–99]. In [282], Tits actually proves a more general lemma. Instead of considering groups generated by involutions, he deals with a group $G$ which is generated by a family of subgroups $\{G_i\}$ (rather than just groups of order 2).

**4.10.** The proofs in [38, 94, 181] of the results in this section all make use of the "canonical representation" of $W$ and the notions of "roots" and "root systems" as defined in Appendix D.3. Here we have, instead, framed things in terms of half-spaces and sectors.

# Chapter Five

## THE BASIC CONSTRUCTION

Given a Coxeter system $(W, S)$, a space $X$ and a family of subspaces $(X_s)_{s\in S}$, there is a classical construction of a space $\mathcal{U}(W, X)$ with $W$-action. $\mathcal{U}(W, X)$ is constructed by pasting together copies of $X$, one for each element of $W$. The purpose of this chapter is to give the details of this construction. More generally, in 5.1 we describe the construction for an arbitrary group $G$ together with a family of subgroups. This greater generality will not be needed until Chapter 18 when it will be used in the discussion of geometric realizations of buildings.

### 5.1. THE SPACE $\mathcal{U}$

A *mirror structure* on a space $X$ consists of an index set $S$ and a family of closed subspaces $(X_s)_{s\in S}$. The subspaces $X_s$ are the *mirrors* of $X$. The space $X$ together with a mirror structure is a *mirrored space over* $S$. $X$ is a *mirrored CW complex* if it is a CW complex and the $X_s$ are subcomplexes. (The notion of a "CW complex" is explained at the end of Appendix A.1.) When $S$ is infinite we also will assume that $(X_s)_{s\in S}$ is a locally finite family, i.e., that each point has a neighborhood which intersects only finitely many of the $X_s$. For each $x \in X$, put

$$S(x) := \{s \in S \mid x \in X_s\}. \tag{5.1}$$

For each nonempty subset $T$ of $S$, let $X_T$ (resp. $X^T$) denote the intersection (resp. union) of the mirrors indexed by $T$, that is,

$$X_T := \bigcap_{t\in T} X_t \tag{5.2}$$

and

$$X^T := \bigcup_{t\in T} X_t. \tag{5.3}$$

Also, for $T = \emptyset$, put $X_\emptyset = X$ and $X^\emptyset = \emptyset$. We shall sometimes call $X_T$ a *coface* of $X$. (The reason for this terminology will become clearer in Chapter 7.)

**DEFINITION 5.1.1.** Suppose $(X_s)_{s\in S}$ is a mirror structure on $X$. Associated to the mirror structure there is an abstract simplicial complex $N(X)$, called its *nerve*, defined as follows. The vertex set of $N(X)$ is the set of $s \in S$ such that $X_s \neq \emptyset$. A nonempty subset $T$ of $S$ is a simplex of $N(X)$ if and only if $X_T \neq \emptyset$. (The notion of an "abstract simplicial complex" is explained in Appendix A.2.)

A *family of groups* over a set $S$ consists of a group $G$, a subgroup $B \subset G$ and a family $(G_s)_{s\in S}$ of subgroups of such that each $G_s$ contains $B$. Put $G_\emptyset := B$ and for each nonempty subset $T$ of $S$, let $G_T$ denote the subgroup of $G$ generated by $\{G_s\}_{s\in T}$. Also, suppose $G$ is a topological group and $B$ is an open subgroup so that $G/B$ has the discrete topology. (Except in Chapter 18, $B$ will be the trivial subgroup so that $G$ will be discrete.)

**Example 5.1.2.** Suppose $(W, S)$ is a pre-Coxeter system (i.e., as in Definition 3.2.3 $W$ is a group and $S$ is a set of involutions which generate $W$). For each $s \in S$, let $W_s$ denote the subgroup of order two generated by $s$. Then $(W_s)_{s\in S}$ is a family of subgroups as above. (Here $B = W_\emptyset := \{1\}$.)

Suppose $X$ is a mirrored space over $S$ and $(G_s)_{s\in S}$ is a family of subgroups of $G$ over $S$. Define an equivalence relation $\sim$ on $G \times X$ by $(h, x) \sim (g, y)$ if and only if $x = y$ and $h^{-1}g \in G_{S(x)}$ (where $S(x)$ is defined by (5.1)). Give $G/B \times X$ the product topology and let $\mathcal{U}(G, X)$ denote the quotient space:

$$\mathcal{U}(G,X) := (G/B \times X)/\sim. \tag{5.4}$$

The image of $(gB, x)$ in $\mathcal{U}(G, X)$ is denoted $[g, x]$.

The natural $G$-action on $G/B \times X$ is compatible with the equivalence relation; hence, it descends to an action on $\mathcal{U}(G, X)$. The map $i : X \to \mathcal{U}(G, X)$ defined by $x \to [1, x]$ is an embedding. We identify $X$ with its image under $i$ and call it the *fundamental chamber*. For any $g \in G$, the image of $gB \times X$ in $\mathcal{U}(G, X)$ is denoted $gX$ and is called a *chamber* of $\mathcal{U}(G, X)$. Since the stabilizer of $X$, the set of chambers is identified with $G/B$. The orbit space of the $G$-action on $G/B \times X$ is $X$ and the orbit projection is projection onto the second factor, $G/B \times X \to X$. This descends to a projection map $p : \mathcal{U}(G, X) \to X$. Since $p \circ i = id_X$, $p$ is a retraction. It is easy to see that $p$ is an open mapping. Since the orbit relation on $G \times X$ is coarser than the equivalence relation $\sim$, $p$ induces a continuous bijection $\bar{p} : \mathcal{U}(G, X)/G \to X$. Since $p$ is open so is $\bar{p}$. So, $\bar{p}$ is a homeomorphism.

**DEFINITION 5.1.3.** Suppose $G$ acts on a space $Y$. A closed subset $C \subset Y$ is a *fundamental domain* for $G$ on $Y$ if each $G$-orbit intersects $C$ and if for each point $x$ in the interior of $C$, $Gx \cap C = \{x\}$. $C$ is a *strict fundamental domain* if it intersects each $G$-orbit in exactly one point.

The preceding discussion shows that $X$ is a strict fundamental domain for $G$ on $\mathcal{U}(G,X)$.

**Notation**

For any coset $a = gB \in G/B$, let $aX := gX$. For any subset $A$ of $G/B$, define a subspace $AX$ of $\mathcal{U}(W,X)$ to be the corresponding union of chambers:

$$AX := \bigcup_{a \in A} aX.$$

Given a point $x \in X$, set

$$C_x := \bigcup_{s \notin S(x)} X_s \quad \text{and} \quad V_x := X - C_x. \tag{5.5}$$

Since $(X_s)_{s \in S}$ is a locally finite family of closed subspaces of $X$, each $C_x$ is closed and each $V_x$ is open. Consequently, $G_{S(x)} V_x$ is an open neighborhood of $[1,x]$ in $\mathcal{U}(G,X)$.

**Lemma 5.1.4.** *$\mathcal{U}(G,X)$ is connected if the following three conditions hold:*

*(a) The family of subgroups $(G_s)_{s \in S}$ generates $G$.*

*(b) $X$ is connected.*

*(c) $X_s \neq \emptyset$ for all $s \in S$.*

*Conversely, if $\mathcal{U}(G,X)$ is connected, then (a) and (b) hold.*

*Proof.* Suppose (a), (b), and (c) hold. Since $\mathcal{U}(G,X)$ has the quotient topology, a subset of $\mathcal{U}(G,X)$ is open (resp. closed) if and only if its intersection with each chamber $gX$ is open (resp. closed). Since $X$ is connected, it follows that any subset of $\mathcal{U}(G,X)$ which is both open and closed is a union of chambers, i.e., it has the form $AX$ for some $A \subset G/B$. Suppose we can find a proper nonempty subset $A$ of $G/B$ so that $AX$ is open and closed in $\mathcal{U}(G,X)$. Let $H$ be the inverse image of $A$ in $G$. If $X_s \neq \emptyset$ and $x \in X_s$, then for any $g_s \in G_s$ and $hB \in A$, any open neighborhood of $[hg_s, x]$ must intersect $hX$ as well as $hg_sX$. Thus, $HG_s \subset H$. It follows that $H$ must be the subgroup $\widehat{G}$ of $G$ generated by the $G_s$, $s \in S$. So, if $\widehat{G} = G$ (condition (a)), then $AX = \mathcal{U}(G,X)$, i.e., $\mathcal{U}(G,X)$ is connected.

Conversely, suppose $\mathcal{U}(G,X)$ is connected. Since the orbit map $p : \mathcal{U}(G,X) \to X$ is a retraction, $X$ is also connected, i.e., (b) holds. Since $\widehat{G}$ contains all the isotropy subgroups $G_{S(x)}$, $x \in X$, it follows that $\widehat{G}X$ is open in $\mathcal{U}(G,X)$. It is clearly closed. Hence, $\widehat{G} = G$, i.e., (a) holds. □

**Definition 5.1.5.** Suppose $G$ is discrete. A $G$-action on a Hausdorff space $Y$ is *proper* (or "properly discontinuous") if the following three conditions hold.

(i) $Y/G$ is Hausdorff.

(ii) For each $y \in Y$, the isotropy subgroup $G_y$ is finite. ($G_y = \{g \in G \mid gy = y\}$.)

(iii) Each $y \in Y$ has a $G_y$-stable neighborhood $U_y$ such that $gU_y \cap U_y = \emptyset$ for all $g \in G - G_y$.

**DEFINITION 5.1.6.** A mirror structure on $X$ is *G-finite* (with respect to a family of subgroups for $G$) if $X_T = \emptyset$ for any subset $T \subset S$ such that $G_T/B$ is infinite.

**LEMMA 5.1.7.** *Suppose G is discrete. The G-action on $\mathcal{U}(G,X)$ is proper if and only if the following two conditions hold.*

*(a) X is Hausdorff.*

*(b) The mirror structure is G-finite.*

*Proof.* If $G$ acts properly on $\mathcal{U}(G,X)$, then (a) and (b) follow from conditions (i) and (ii) of Definition 5.1.5. Conversely, conditions (a) and (b) obviously imply conditions (i) and (ii) of Definition 5.1.5. It suffices to establish condition (iii) at $[1,x] \in \mathcal{U}(G,X)$ for an arbitrary $x \in X$. Let $V_x$ be the open neighborhood defined by (5.5). Then $U_x := G_{S(x)}V_x$ is an open $G_{S(x)}$-stable neighborhood of $[1,x]$ in $\mathcal{U}(G,X)$ and clearly $gU_x \cap U_x = \emptyset$ for all $g \in G - G_{S(x)}$. □

## 5.2. THE CASE OF A PRE-COXETER SYSTEM

Suppose $(W,S)$ is a pre-Coxeter system. As in Example 5.1.2, this gives a family of subgroups indexed by $S$.

**DEFINITION 5.2.1.** Given a space $Z$, the *cone* on $Z$, denoted by $\mathrm{Cone}(Z)$, is the space formed from $Z \times [0,1]$ by identifying $Z \times 0$ to a point. (This definition is repeated in Appendix A.4.)

**Example 5.2.2.** Suppose $X = \mathrm{Cone}(S)$, the cone on the set of generators $S$. Define a mirror structure on $\mathrm{Cone}(S)$ by putting $\mathrm{Cone}(S)_s$ equal to the point $(s,1)$. Then $\mathcal{U}(W,\mathrm{Cone}(S))$ is the Cayley graph of $(W,S)$. (The segment $s \times [0,1]$ in $\mathrm{Cone}(S)$ is a "half edge.") The inverse image of this segment in $\mathcal{U}(W,\mathrm{Cone}(S))$ is the union of all edges labeled $s$ in the Cayley graph.

Definition 4.5.4 works for pre-Coxeter systems as well as Coxeter systems: a subset $A$ of $W$ is *connected* if it is the vertex set of a connected subgraph of the Cayley graph $\mathrm{Cay}(W,S)$.

**LEMMA 5.2.3.** *Suppose $X$ is connected (resp. path connected) and $X_s \neq \emptyset$ for each $s \in S$. Given a subset $A$ of $W$, $AX$ is connected (resp. path connected) if and only if $A$ is connected.*

*Proof.* As in the proof of Lemma 5.1.4 a subset of $AX$ that is both open and closed must be of the form $BX$ for some $B \subset A$. Let $B$ be a proper nonempty subset of $A$ such that $BX$ is open and closed in $AX$. Let $\check{B} = A - B$. Suppose that $A$ is connected. Without loss of generality, we can suppose there are elements $b \in B$ and $\check{b} \in \check{B}$ which are connected by an edge in the Cayley graph. If $s$ is the label on the edge, then $bX_s = \check{b}X_s$ lies in $BX \cap \check{B}X$. Since $X_s \neq \emptyset$, $BX$ and $\check{B}X$ cannot be disjoint. So $AX$ is connected. Conversely, if $AX$ is connected, then the above analysis shows that $A$ cannot be partitioned into disjoint subsets $B$ and $\check{B}$ with the property that no element of $B$ can be connected by an edge to an element of $\check{B}$. But this means that $A$ is connected. When "connected" is replaced by "path connected" the argument is even easier and is left for the reader. □

The special case $A = W$ of the above lemma is the following.

**COROLLARY 5.2.4.** *$\mathcal{U}(W, X)$ is connected (resp. path connected) if the following two conditions hold:*

*(a) $X$ is connected (resp. path connected) and*

*(b) $X_s$ is nonempty for each $s \in S$.*

*Remark.* If $(W, S)$ is only required to be a pre-Coxeter system, then it is *not* true that condition (b) is necessary for $\mathcal{U}(W, X)$ to be path connected. For example, suppose $W = \mathbf{C}_2 \times \mathbf{C}_2$ and $S = \{s, t, st\}$ is the set of its nontrivial elements. Then $\text{Cay}(W, S)$ is the 1-skeleton of a tetrahedron. If we remove the interiors of the two edges labeled $st$, we get a $\mathcal{U}(W, X)$, where $X$ is the cone on $\{s, t\}$. $\mathcal{U}(W, X)$ is connected (it is a square) and $X_{st} = \emptyset$. On the other hand, in Section 8.2 we will show that when $(W, S)$ is a Coxeter system, conditions (a) and (b) are necessary for $\mathcal{U}(W, X)$ to be path connected (Corollary 8.2.3).

To simplify notation we often shorten $\mathcal{U}(W, X)$ to $\mathcal{U}$ when the mirrored space $X$ is understood.

The next lemma gives a universal property of the construction. Its proof is immediate.

**LEMMA 5.2.5.** (Vinberg [290]). *Suppose that $Y$ is any space with $W$-action. For each $s \in S$, let $Y^s$ denote the fixed point set of $s$ on $Y$. Let $f : X \to Y$ be any map such that $f(X_s) \subset Y^s$. Then there is a unique extension of $f$ to a $W$-equivariant map $\tilde{f} : \mathcal{U} \to Y$, given by the formula $\tilde{f}([w, x]) = wf(x)$.*

(In this lemma as well as in the rest of the book a *map* between topological spaces means a continuous map.)

**Reflection Groups**

In the following chapters we shall often be proving results of the following nature: if a group $W$ acts on a space $Y$ and if $W$ is generated by some sort of "reflections," then $W$ is a Coxeter group and its action on $Y$ arises by applying our basic construction to some fundamental domain $X$. In light of this and the results of Chapter 3, we make the following definition.

**DEFINITION 5.2.6.** The action of a discrete group $\widehat{W}$ on a space $Y$ is a *reflection group* if there is a Coxeter system $(W, S)$ and a subspace $X \subset Y$ such that

(a) $\widehat{W} = W$.

(b) If a mirror structure on $X$ is defined by setting $X_s$ equal to the intersection of $X$ with the fixed set of $s$ on $Y$, then the map $\mathcal{U}(W, X) \to Y$, induced by the inclusion of $X$ in $Y$, is a homeomorphism.

The isotropy subgroup at a point $[w, x]$ in $\mathcal{U}(W, X)$ is $wW_{S(x)}w^{-1}$. Suppose $(W, S)$ is a Coxeter system. Since $W_{S(x)}$ is a special subgroup, this means that each isotropy subgroup is a parabolic subgroup (defined in 4.5).

**Example 5.2.7.** (*The Coxeter complex.*) Suppose $(W, S)$ is a Coxeter system, that $\Delta$ is a simplex of dimension $\mathrm{Card}(S) - 1$ and that the codimension-one faces of $\Delta$ are indexed by the elements of $S$. The set of codimension-one faces $\{\Delta_s\}_{s\in S}$ is a mirror structure on $\Delta$. The space $\mathcal{U}(W, \Delta)$ is naturally a simplicial complex, called the *Coxeter complex* of $(W, S)$. We shall see in Chapter 6 that, when $W$ is finite, $\mathcal{U}(W, \Delta)$ is homeomorphic to a sphere. (See, in particular, Theorems 6.4.3 and 6.12.9 and Lemma 6.3.3.) On the other hand, if $W$ is infinite, then it is proved in [255] that $\mathcal{U}(W, \Delta)$ is contractible. (This also follows from the results in Chapters 8 and 9.) Since $\Delta_T$ is a nonempty simplex for each proper subset of $S$, each proper parabolic subgroup occurs as an isotropy subgroup at some point of $\mathcal{U}(W, \Delta)$.

## 5.3. SECTORS IN $\mathcal{U}$

Throughout this section, $(W, S)$ is a Coxeter system, $X$ is a mirrored space over $S$, and $\mathcal{U} = \mathcal{U}(W, X)$.

As in Proposition 4.1.1, for any $w \in W$, $S(w)$ denotes the set of letters in $S$ which occur in any reduced expression for $w$. The next lemma shows that any two chambers of $\mathcal{U}$ intersect in a common coface (possibly an empty coface).

**LEMMA 5.3.1.** *For any $w \in W$, $X \cap wX = X_{S(w)}$. Hence, for any two elements of $v, w \in W$, $vX \cap wX = vX_{S(v^{-1}w)} = wX_{S(v^{-1}w)}$.*

*Proof.* It suffices to prove the first sentence (since it implies the second). Suppose $x \in X \cap wX$. Then $w \in W_{S(x)}$ and, by Corollary 4.1.2, $S(w) \subset S(x)$, i.e., $x \in X_{S(w)}$. Thus, $X \cap wX \subset X_{S(w)}$. Conversely, suppose $x \in X_{S(w)}$. Let $(s_1, \dots, s_k)$ be a reduced expression for $w$. So $S(w) = \{s_1, \dots, s_k\}$ and $x \in X_{s_i}$ for each $s_i$. Thus, each $s_i$ fixes $x$ and consequently so does $w$. Therefore, $x \in X \cap wX$, i.e., $X_{S(w)} \subset X \cap wX$. □

**DEFINITION 5.3.2.** A *sector* (or *half-space*) of $\mathcal{U}(W, X)$ is a union of chambers of the form $AX$ where $A \subset W$ is a sector (or half-space) of $W$.

For each $r \in R$, the fixed point set of $r$ on $\mathcal{U}$ is denoted $\mathcal{U}^r$ and called a *wall* of $\mathcal{U}$. If $r = wsw^{-1}$, $s \in S$, then $wA_s$ and $rwA_s$ are the two half-spaces into which $r$ separates $W$. (Recall $A_s := A_{\{s\}}$.) If $uX$ and $ruX$ are adjacent chambers in these half-spaces, then, by Remark 4.4.1 (iv), $r = utu^{-1}$ for some $t \in S$. It follows that $wA_sX \cap rwA_sX = \mathcal{U}^r$.

The next three results concern the following situation: a subset $T$ of $S$ is given, $A_T$ is the fundamental $T$-sector in $W$ (cf. Definition 4.5.1) and $\widehat{X} = A_TX$. Define a mirror structure (over $T$) on $\widehat{X}$ by putting $\widehat{X}_t = \mathcal{U}^t \cap \widehat{X}$. As before, for $\hat{x} \in \widehat{X}$, set $T(\hat{x}) = \{t \in T \mid \hat{x} \in \widehat{X}_t\}$.

**LEMMA 5.3.3**

*(i)* $W_T\widehat{X} = \mathcal{U}$.

*(ii) If $\hat{x}, \hat{y} \in \widehat{X}$ are such that $w\hat{x} = \hat{y}$ for some $w \in W_T$, then $\hat{x} = \hat{y}$ and $\hat{x} \in W_{T(\hat{x})}$.*

*Thus, $\widehat{X}$ is a strict fundamental domain for the $W_T$-action on $\mathcal{U}$ (that is, it intersects each $W_T$-orbit in exactly one point).*

*Proof.* Since $A_T$ is the set of $(T, \emptyset)$-reduced elements, it is a set of coset representatives for $W_T\backslash W$. This proves (i). The proof of (ii) is by induction on $l(w)$. Suppose $l(w) = 1$. This means $w = t$ for some $t \in T$. If $t\hat{x} = \hat{y}$, then $\hat{y}$ lies in $\widehat{X} \cap t\widehat{X} \subset \mathcal{U}^t$, so $t$ fixes $\hat{y}$, and consequently $\hat{x} = \hat{y}$. Next, suppose $l(w) > 1$ and $(t_1, \dots, t_k)$ is a reduced expression for $w$. Put $w' = t_1w = t_2 \cdots t_k$. Apply $t_1$ to the equation $w\hat{x} = \hat{y}$, to obtain

$$w'\hat{x} = t_1\hat{y}. \tag{5.6}$$

Since $l(t_1w') > l(w')$, $w'\widehat{X}$ is on the positive side of $\mathcal{U}^{t_1}$. Since $t_1\widehat{X}$ is on the negative side, (5.6) implies that $w'\hat{x} = t_1\hat{y} \in \widehat{X} \cap t_1\widehat{X} \subset \mathcal{U}^{t_1}$, so $t_1$ fixes $\hat{y}$ and (5.6) can be rewritten as $w'\hat{x} = \hat{y}$. By induction, $\hat{x} = \hat{y}$ and $w' \in W_{T(\hat{x})}$. So, $w = t_1w'$ is also in $W_{T(\hat{x})}$ and the lemma is proved. □

Applying the basic construction to the mirror structure on $\widehat{X}$, we obtain $\widehat{\mathcal{U}} = \mathcal{U}(W_T, \widehat{X})$. Letting $\mathcal{U}$ stand for $\mathcal{U}(W, X)$, we have, by Lemma 5.2.5, that the inclusion $\iota : \widehat{X} \to \mathcal{U}$ induces a $W_T$-equivariant map $\tilde{\iota} : \widehat{\mathcal{U}} \to \mathcal{U}$.

**COROLLARY 5.3.4.** *$\tilde{\iota} : \widehat{\mathcal{U}} \to \mathcal{U}$ is a homeomorphism.*

*Proof.* By Lemma 5.3.3, $\tilde{\iota}$ is a continuous bijection. $\widehat{\mathcal{U}}$ has the quotient topology from $W_T \times A_T \times X$ and $\mathcal{U}$ has the quotient topology from $W \times X$. It follows that $\tilde{\iota}$ is an open map and hence, a homeomorphism. □

**COROLLARY 5.3.5.** *Let $p_T : W \to A_T$ be the fundamental retraction, defined in 4.5.2, which sends $w$ to its $(T, \emptyset)$-reduced representative in $W_T w$. Then $p_T$ descends to a retraction of spaces, $\mathcal{U} \to A_T X$. (This map will also be denoted by $p_T$.)*

*Proof.* We have the orbit projection $\widehat{\mathcal{U}} \to \widehat{X} = A_T X$ defined by $[u, \hat{x}] \to \hat{x}$. Conjugating by the $W_T$-equivariant homeomorphism $\tilde{\iota}$, we get a retraction $\mathcal{U} \to A_T X$ defined by $[w, x] \to [p_T(w), x]$. □

We will need the following lemma in Chapter 13.

**LEMMA 5.3.6.** *The intersection of two parabolic subgroups of $W$ is a parabolic subgroup.*

*Proof.* Given subsets $T, U$ of $S$ and an element $w \in W$, it suffices to prove that $wW_Tw^{-1} \cap W_U$ is a parabolic subgroup. Consider the action of $W$ on the Coxeter complex $\mathcal{U}(W, \Delta)$ of Example 5.2.7. We can assume that $T$ is a proper subset of $S$ (otherwise there is nothing to prove). Hence, there is a point $x \in \mathcal{U}(W, \Delta)$ with isotropy subgroup $wW_Tw^{-1}$. By Corollary 5.3.4, we can identify the $W_U$-spaces $\mathcal{U}(W, \Delta)$ and $\widehat{\mathcal{U}}$ $(:= \mathcal{U}(W_U, A_U\Delta))$. As we observed in the paragraph preceding Example 5.2.7, any isotropy subgroup of the $W_U$-action is a parabolic subgroup of $W_U$ (*afortiori* a parabolic subgroup of $W$). But the isotropy subgroup at $x$ is $wW_Tw^{-1} \cap W_U$. □

*Remark.* The above proof shows that the parabolic subgroup $wW_Tw^{-1} \cap W_U$ can be explicitly described as follows. By Lemma 4.3.3, we can write $w$ uniquely in the form $w = ua$ where $u \in W_U$ and $a \in A_U$. Set $U' := aTa^{-1} \cap U$. The intersection in question is then $uW_{U'}u^{-1}$.

## NOTES

The basic construction gives a case where a $G$-space can be recovered from its orbit space and isotropy subgroup data. However, it ignores all the aspects of covering space theory that can enter into such an undertaking.

For reflection groups the basic construction is classical. In this case, the construction was emphasized in [180] and by Vinberg in [290]. Its importance was also emphasized in [71]. The construction was discussed by Tits in [285] in the full generality of 5.1. Tits' motivation was to use it to define the geometric realization of buildings with chamber-transitive automorphism groups. The construction was also used in [57] to define "buildinglike" complexes for Artin groups. In [37, pp. 381–387], Bridson and Haefliger also call this "the basic construction." (Actually, their definitions are slightly more general than those in 5.1. They replace the mirrored space $X$ and its the nerve, $N(X)$, by a "stratified space" over a poset $\mathcal{P}$. The family of subgroups $(G_s)_{s\in S}$ is replaced by a "simple complex of groups" over $\mathcal{P}$. (This last notion is explained in Appendix E.2.) Bridson and Haefliger show that any action of a discrete group $G$ on a space $Y$ with a strict fundamental domain can be recovered from their more general version of the basic construction.

**5.1.** Definition 5.1.1 (the "nerve of a mirror structure") will be used in Section 7.1.

# Chapter Six

## GEOMETRIC REFLECTION GROUPS

This chapter deals with the classical theory of reflection groups on the three geometries of constant curvature: the $n$-sphere, Euclidean $n$-space, and hyperbolic $n$-space. Let $\mathbb{X}^n$ stand for one of these. The main result, Theorem 6.4.3, states that, if $P^n$ is a convex polytope in $\mathbb{X}^n$ with all dihedral angles integral submultiples of $\pi$, then the group $W$ generated by the isometric reflections across the codimension one faces of $P^n$ is (1) a Coxeter group and (2) a discrete subgroup of isometries of $\mathbb{X}^n$; moreover, $\mathbb{X}^n \cong \mathcal{U}(W, P^n)$. In the spherical case, $P^n$ is a spherical simplex and $W$ is finite (Lemma 6.3.3). Conversely, every Coxeter system $(W, S)$ with $W$ finite, essentially has a unique representation of this form (Theorem 6.12.9). In the Euclidean case, $P^n$ is a product of simplices (Lemma 6.3.10). In Section 6.5 on "polygon groups," we discuss the two-dimensional examples. In 6.8, and 6.9 and Appendix C, we determine all such geometric reflection groups in the case where the polytope is a simplex, getting the standard lists of possibilities for spherical and Euclidean reflection groups (Table 6.1). In the case of hyperbolic $n$-space, these simplicial examples exist only in dimensions $\leqslant 4$: there are an infinite number of hyperbolic triangle groups, nine hyperbolic tetrahedral groups and five more with fundamental chamber a hyperbolic 4-simplex (Table 6.2). Section 6.10 deals with reflection groups on hyperbolic 3-space, in which case is a beautiful result of Andreev (Theorem 6.10.2), giving a complete classification. It says that the combinatorial type of $P^3$ can be that of any simple polytope and that any assignment of dihedral angles to the edges is possible, subject to a few obviously necessary inequalities. These inequalities have a topological interpretation: they mean that any quotient of $\mathcal{U}(W, P^3)$ by a torsion-free subgroup of finite index is an aspherical and atoroidal 3-manifold (see Remark 6.10.4). Andreev's Theorem provided the first major evidence for Thurston's Geometrization Conjecture. In 6.11, we state Vinberg's Theorem (6.11.8) that cocompact hyperbolic reflection groups do not exist in dimensions $\geqslant 30$. In 6.12, we discuss the canonical linear representation of any Coxeter system. Its definition is motivated by the geometric examples. The dual of this representation is even more important. We call it the "geometric representation" and discuss it further in Appendix D.

## 6.1. LINEAR REFLECTIONS

**DEFINITION 6.1.1.** A *linear reflection* on a vector space $V$ is a linear automorphism $r : V \to V$ such that $r^2 = 1$ and such the fixed subspace of $r$ is a hyperplane.

Suppose $r : V \to V$ is a linear reflection. Let $H$ be the fixed hyperplane. Choose a $(-1)$-eigenvector $e$ (the $(-1)$-eigenspace of $r$ is one dimensional) and a linear form $\alpha$ on $V$ with kernel $H$. Then $r$ can be defined by the following formula:

$$r(v) = v - \frac{2\alpha(v)}{\alpha(e)}e. \tag{6.1}$$

Conversely, if $\alpha$ is a linear form on $V$ and $e \in V$ is any vector such that $\alpha(e) \neq 0$, then the linear transformation $r$ defined by (6.1) is a reflection.

## 6.2. SPACES OF CONSTANT CURVATURE

In each dimension $n \geqslant 2$, there are three simply connected, complete Riemannian manifolds of constant sectional curvature: $\mathbb{S}^n$ (the $n$-sphere), $\mathbb{E}^n$ (Euclidean $n$-space), and $\mathbb{H}^n$ (hyperbolic $n$-space). $\mathbb{S}^n$ is positively curved, $\mathbb{E}^n$ is flat (curvature 0), and $\mathbb{H}^n$ is negatively curved. By scaling the metric appropriately we may assume that $\mathbb{S}^n$ has curvature $+1$ and that $\mathbb{H}^n$ has curvature $-1$.

One of the main features of these spaces, important in the study of reflection groups, is that each admits many totally geodesic codimension one subspaces which we shall call "hyperplanes." Each such hyperplane separates the ambient space into two geodesically convex "half-spaces" and for each such hyperplane there is an isometric "reflection" which fixes the hyperplane and interchanges the the two half-spaces.

The standard models for these spaces are very well known (see, for example, [2], [37], [279], or [304]). We shall briefly review them below. Since the hyperbolic case is less well-known than the spherical and Euclidean cases, we discuss it in slightly more detail.

### Euclidean $n$-Space $\mathbb{E}^n$

$\mathbb{R}^n$ denotes the vector space of all real-valued functions on $\{1, \dots, n\}$. Write $x = (x_1, \dots, x_n) \in \mathbb{R}^n$ for the function that sends $i$ to $x_i$. Let $e_i : \{1, \dots, n\} \to \mathbb{R}$ denote the Kronecker delta function $\delta_{ij}$. Then $\{e_i\}_{1 \leqslant i \leqslant n}$ is the *standard basis* for $\mathbb{R}^n$. The *standard inner product* $\langle\, ,\rangle$ on $\mathbb{R}^n$ is defined by

$$\langle x, y\rangle = \sum_{i=1}^{n} x_i y_i. \tag{6.2}$$

This inner product is part of the structure of $\mathbb{R}^n$. $\mathbb{E}^n$ is the affine space associated to $\mathbb{R}^n$. (Affine spaces are discussed in Appendix A.1.) Thus, $\mathbb{R}^n$ acts simply transitively on $\mathbb{E}^n$. The action of a vector $v \in \mathbb{R}^n$ on $\mathbb{E}^n$ is *translation* by $v$. So, $\mathbb{E}^n$ is essentially the same as $\mathbb{R}^n$ except that there is no longer a distinguished point of $\mathbb{E}^n$ (as the origin is a distinguished vector in $\mathbb{R}^n$). By definition, the *tangent space* of $\mathbb{E}^n$ (at any point $x \in \mathbb{E}^n$), is $\mathbb{R}^n$ equipped with its standard inner product given by (6.2). So, $\mathbb{E}^n$ is a Riemannian manifold. It has constant sectional curvature 0.

By a *hyperplane* or *half-space* in $\mathbb{E}^n$ we mean, respectively, an affine hyperplane or an affine half-space. The "official" definition of a *convex polytope* in $\mathbb{E}^n$ is given as Definition A.1.1 of Appendix A.1. Briefly, it is a compact intersection of a finite number of half-spaces.

Suppose $H$ is a hyperplane in $\mathbb{E}^n$, $x_0 \in H$ and $u \in \mathbb{R}^n$ is a unit vector orthogonal to $H$. Then the orthogonal reflection $r_H$ across $H$ is given by the formula

$$r_H(x) = x - 2\langle u, x - x_0\rangle u, \tag{6.3}$$

where, of course, $x - x_0$ means the vector in $\mathbb{R}^n$ which translates the point $x_0 \in \mathbb{E}^n$ to $x \in \mathbb{E}^n$ (It is easy to see that the affine map $r_H : \mathbb{E}^n \to \mathbb{E}^n$ depends only on the hyperplane $H$ and not on the choices of $x_0$ and $u$.)

The group of affine automorphisms of $\mathbb{E}^n$ is $\mathbb{R}^n \rtimes GL(n)$. (In the semidirect product the action of $GL(n)$ on $\mathbb{R}^n$ is via the standard representation.) Here $\mathbb{R}^n$ is the group of translations of $\mathbb{E}^n$. The component in $GL(n)$ is the *linear part* of the automorphism. The subgroup $\mathbb{R}^n \rtimes O(n)$ is the group of isometries of $\mathbb{E}^n$. ($O(n)$ is the group of linear isometries of $\mathbb{R}^n$.) For example, the linear part of the reflection $r_H$ defined by (6.3) is the linear reflection $v \to v - 2\langle u, v\rangle u$.

We can identify $\mathbb{E}^n$ with an affine hyperplane in $\mathbb{R}^{n+1}$; for the sake of definiteness, say with the affine hyperplane $x_{n+1} = 1$. We can think of this as identifying $\mathbb{E}^n$ with the complement of the "hyperplane at infinity" (defined by $x_{n+1} = 0$) in real projective space $\mathbb{R}P^n$. The group of affine automorphisms of $\mathbb{E}^n$ is then identified with a subgroup of $PGL(n+1, \mathbb{R})$, the group of projective transformations of $\mathbb{R}P^n$. ($PGL(n+1, \mathbb{R})$ is the quotient of $GL(n+1, \mathbb{R})$ by the group of scalar multiples of the identity matrix.) The subgroup which fixes the hyperplane at infinity can be identified with the image of the set of matrices of the form

$$\begin{pmatrix} a_{11} & \dots & a_{1n} & b_1 \\ \vdots & \ddots & \vdots & \vdots \\ a_{n1} & \dots & a_{nn} & b_n \\ 0 & \dots & 0 & 1 \end{pmatrix}.$$

Multiplication by such a matrix is the affine transformation $x \to Ax + b$ where we are identifying $A \in GL(n, \mathbb{R})$, $x \in \mathbb{E}^n$ and $b \in \mathbb{R}^n$ with the matrices:

$$A = \begin{pmatrix} a_{11} & \dots & a_{1n} & 0 \\ \vdots & \ddots & \vdots & \vdots \\ a_{n1} & \dots & a_{nn} & 0 \\ 0 & \dots & 0 & 1 \end{pmatrix}, \quad x = \begin{pmatrix} x_1 \\ \vdots \\ x_n \\ 1 \end{pmatrix}, \quad b = \begin{pmatrix} b_1 \\ \vdots \\ b_n \\ 1 \end{pmatrix}.$$

This is an isometry of $\mathbb{E}^n$ if and only if $A \in O(n)$.

### The *n*-Sphere

$\mathbb{S}^n$ is the quadratic hypersurface in $\mathbb{R}^{n+1}$ defined by $\langle x, x \rangle = 1$. The tangent space of $\mathbb{S}^n$ at a point $x \in \mathbb{S}^n$ is denoted $T_x\mathbb{S}^n$. It is naturally identified with $x^\perp$, the subspace of $\mathbb{R}^{n+1}$ orthogonal to $x$. The inner product on $\mathbb{R}^{n+1}$ induces an inner product on $T_x\mathbb{S}^n$ and hence, a Riemannian metric on $\mathbb{S}^n$. It turns out that this metric has constant sectional curvature 1.

A *hyperplane* in $\mathbb{S}^n$ is its intersection with a linear hyperplane in $\mathbb{R}^{n+1}$. A *half-space* in $\mathbb{S}^n$ is its intersection with a linear half-space of $\mathbb{R}^{n+1}$. In other words, a hyperplane in $\mathbb{S}^n$ is a great subsphere and a half-space in $\mathbb{S}^n$ is a hemisphere.

In Appendix A.1 we define a *convex polyhedral cone* in $\mathbb{R}^{n+1}$ to be the intersection of a finite number of linear half-spaces (Definition A.1.8). The polyhedral cone is *essential* if it contains no line. A *convex polytope* in $\mathbb{S}^n$ is the intersection of $\mathbb{S}^n$ with an essential convex polyhedral cone in $\mathbb{R}^{n+1}$.

Suppose $H$ is a hyperplane in $\mathbb{S}^n$ and $u$ is a unit vector in $\mathbb{R}^{n+1}$ orthogonal to $H$. The *spherical reflection* $r_H$ of $\mathbb{S}^n$ across $H$ is given by the formula

$$r_H(x) = x - 2\langle u, x \rangle u. \tag{6.4}$$

$r_H$ is an isometry of $\mathbb{S}^n$. Its fixed set is $H$. It follows that $H$ is a totally geodesic submanifold of codimension one in $\mathbb{S}^n$.

The group $O(n+1)$ of linear automorphisms of $\mathbb{R}^{n+1}$ which preserve the standard inner product defined by (6.2), stabilizes $\mathbb{S}^n$ and acts on it via isometries. In fact, it is the full group of isometries of $\mathbb{S}^n$.

### Hyperbolic *n*-Space $\mathbb{H}^n$

A symmetric bilinear form on an $(n+k)$-dimensional real vector space is *type* $(n, k)$ if it has $n$ positive eigenvalues and $k$ negative eigenvalues. A positive semidefinite form on a $(n+1)$-dimensional vector space with precisely one eigenvalue 0 is said to be of *corank* 1.

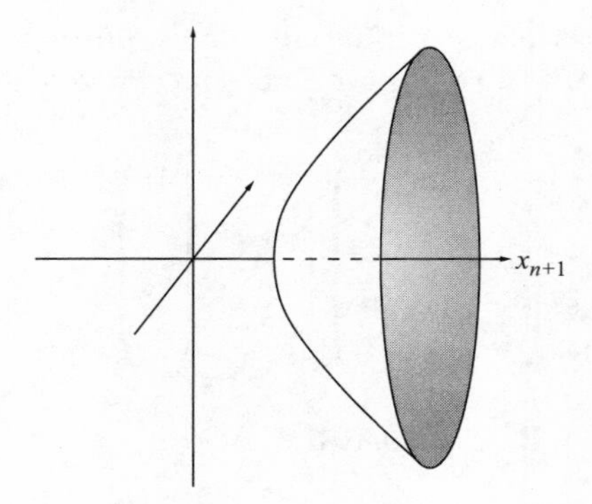


Figure 6.1. The hyperbolic plane in $\mathbb{R}^{2,1}$.

Let $\mathbb{R}^{n,1}$ denote an $(n+1)$-dimensional real vector space equipped with the symmetric bilinear form $\langle\ ,\ \rangle$ of type $(n, 1)$ defined by

$$\langle x, y\rangle = x_1y_1 + \cdots + x_ny_n - x_{n+1}y_{n+1}, \tag{6.5}$$

where $x = (x_1, \ldots, x_{n+1})$, $y = (y_1, \ldots, y_{n+1})$. $\mathbb{R}^{n,1}$ is *Minkowski space*. A $(k+1)$-dimensional subspace $V$ of $\mathbb{R}^{n,1}$ is *spacelike*, *timelike*, or *lightlike* as the restriction of the bilinear form to $V$ is, respectively, positive definite, type $(k, 1)$, or positive semidefinite of corank 1.

The quadratic hypersurface defined by $\langle x, x\rangle = -1$ is a hyperboloid of two sheets. The sheet defined by $x_{n+1} > 0$ is (the quadratic form model of) *hyperbolic n-space* $\mathbb{H}^n$.

Given $x \in \mathbb{H}^n$, $T_x\mathbb{H}^n$, the tangent space at $x$, is naturally identified with $x^\perp$, the subspace of $\mathbb{R}^{n,1}$ orthogonal to $x$. Since $x$ is timelike, $x^\perp$ is spacelike, i.e., the restriction of the bilinear form to $T_x\mathbb{H}^n$ is positive definite. Thus, $\mathbb{H}^n$ is naturally a Riemannian manifold. It turns out that it has constant sectional curvature $-1$.

A nonzero linear subspace of $\mathbb{R}^{n,1}$ has nonempty intersection with $\mathbb{H}^n$ if and only if it is timelike. A *hyperplane* in $\mathbb{H}^n$ is the intersection of $\mathbb{H}^n$ with a timelike linear hyperplane of $\mathbb{R}^{n,1}$. It is a submanifold of $\mathbb{H}^n$ isometric to $\mathbb{H}^{n-1}$. Similarly, a half-space of $\mathbb{H}^n$ is the intersection of $\mathbb{H}^n$ and a linear half-space of $\mathbb{R}^{n,1}$ bounded by a timelike hyperplane.

The *positive light cone* in Minkowski space is the set of $x \in \mathbb{R}^{n,1}$ such that $\langle x, x\rangle \geqslant 0$ and $x_{n+1} \geqslant 0$. A *convex polytope* in $\mathbb{H}^n$ is its intersection with a convex polyhedral cone $C \subset \mathbb{R}^{n,1}$ such that $C - 0$ is contained in the interior of the positive light cone.

Suppose $H$ is a hyperplane in $\mathbb{H}^n$ and $u$ is a unit spacelike vector orthogonal to $H$. The *hyperbolic reflection* $r_H$ of $\mathbb{H}^n$ across $H$ is the restriction to $\mathbb{H}^n$ of the orthogonal reflection of $\mathbb{R}^{n,1}$ defined by

$$r_H(x) = x - 2\langle u, x\rangle u \; . \tag{6.6}$$

This is an isometry of $\mathbb{H}^n$. Its fixed point set is $H$; so, $H$ is a totally geodesic submanifold of codimension one in $\mathbb{H}^n$. Similar considerations show that if $V$ is any timelike linear subspace of dimension $k+1$, then $V \cap \mathbb{H}^n$ is a totally geodesic subspace isometric to $\mathbb{H}^k$.

The group $O(n,1)$ of isometries of the bilinear form on $\mathbb{R}^{n,1}$ has four connected components. There is a subgroup of index two, $O_+(n,1)$, which preserves the sheets of the hyperboloid. This subgroup acts on $\mathbb{H}^n$ as a group of isometries. It is easily seen that $O_+(n,1)$ is transitive on $\mathbb{H}^n$ and that its isotropy subgroup at $x$ is the full group of isometries of the restriction of the bilinear form to $T_x\mathbb{H}^n$, i.e., it is isomorphic to $O(n)$. It follows that $O_+(n,1)$ is the full group of isometries of $\mathbb{H}^n$.

The model of hyperbolic $n$-space discussed above is usually called the *quadratic form model*. There are several other models of $\mathbb{H}^n$ (see [279]). One of the most useful is the *Poincaré disk model*. In dimension 2, points of $\mathbb{H}^2$ correspond to points in the interior of the unit disk and geodesics are circular arcs which meet the boundary orthogonally. A picture of a tessellation of $\mathbb{H}^2$ in the Poincaré disk model can be found as Figure 6.2 of Section 6.5.

### The Notation $\mathbb{X}^n$

In what follows we will use $\mathbb{X}^n$ to stand for $\mathbb{E}^n$, $\mathbb{S}^n$, or $\mathbb{H}^n$ and $\mathrm{Isom}(\mathbb{X}^n)$ for its isometry group. A reflection across a hyperplane is called a *geometric reflection*. We leave as an exercise for the reader to prove the following well-known facts.

- Every element of $\mathrm{Isom}(\mathbb{X}^n)$ can be written as a product of reflections.
- $\mathrm{Isom}(\mathbb{X}^n)$ acts transitively on $\mathbb{X}^n$.
- For each $x \in \mathbb{X}^n$, the isotropy subgroup of $\mathrm{Isom}(\mathbb{X}^n)$ at $x$ acts on $T_x\mathbb{X}^n$ and this gives an identification of the isotropy subgroup with the group of linear isometries of $T_x\mathbb{X}^n$, i.e., with $O(n)$.

For use in Section 6.4, we want to define the notion of a "polyhedral cone in $\mathbb{X}^n$." Basically, this should mean the image of a linear polyhedral cone in $T_x\mathbb{X}^n$ under the exponential map $\exp : T_x\mathbb{X}^n \to \mathbb{X}^n$. Given $x \in \mathbb{X}^n$ and $r \in (0, \infty)$, the *open polyhedral cone of radius* $r$ at a *vertex* $x$ is the intersection of an open ball of radius $r$ about $x$ with a finite number of half-spaces in $\mathbb{X}^n$ such that each bounding hyperplane passes through $x$. If $\mathbb{X}^n = \mathbb{S}^n$, we further require that $r < \pi$. The *tangent cone* of such a polyhedral cone $C^n$ is the linear polyhedral cone $C_x \subset T_x\mathbb{X}^n$ consisting of all tangent vectors that point into $C^n$. The *link* of $x$ in $C^n$ is the $(n-1)$-dimensional spherical polytope formed by intersecting $C_x$ and the unit sphere in $T_x\mathbb{X}^n$. $C^n$ is a *simplicial cone* if this link is a simplex.

## 6.3. POLYTOPES WITH NONOBTUSE DIHEDRAL ANGLES

Suppose $H_1$ and $H_2$ are hyperplanes in $\mathbb{X}^n$ bounding half-spaces $E_1$ and $E_2$ and $E_1 \cap E_2 \neq \emptyset$. Let $u_1$ and $u_2$ be the inward-pointing unit normals at a point $x \in H_1 \cap H_2$. Then $\theta := \cos^{-1}\langle u_1, u_2\rangle$ is the *exterior dihedral angle* along $H_1 \cap H_2$. The supplementary angle $\pi - \theta$ is the *dihedral angle*.

**DEFINITION 6.3.1.** Suppose that $\{E_1, \dots, E_k\}$ is a family of half-spaces in $\mathbb{X}^n$ with nonempty intersection and that $H_1, \dots, H_k$ are the bounding hyperplanes. The family $\{E_1, \dots, E_k\}$ has *nonobtuse dihedral angles* if for every two distinct indices $i$ and $j$ either (a) $H_i \cap H_j = \emptyset$ or (b) $H_i \cap H_j \neq \emptyset$ and the dihedral angle along $H_i \cap H_j$ is $\leqslant \pi/2$. If all the dihedral angles are $\pi/2$, then the family is *right angled.* (In the spherical case condition (a) is vacuous. In the Euclidean case, the hyperplanes $H_i$ and $H_j$ do not intersect if and only if they are parallel.)

Let $P^n \subset \mathbb{X}^n$ be a convex polytope and $F_1, \dots, F_k$ its codimension-one faces. For $1 \leqslant i \leqslant k$, let $H_i$ be the hyperplane determined by $F_i$ and $E_i$ the half-space bounded by $H_i$ which contains $P^n$. The polytope $P^n$ has *nonobtuse dihedral angles* if the family $\{E_1, \dots, E_k\}$ has this property. Similarly, $P^n$ is *right angled* if $\{E_1, \dots, E_k\}$ is right angled.

*Remark.* The previous definition does not only refer to the dihedral angles along the codimension-two faces of $P^n$. It does not rule out the possibility that the subspace $H_i \cap H_j$ is nonempty but only contains a face of $P^n$ of codimension $>2$ (possibly the empty face). In fact, this situation does not occur. This is a consequence of the following result of Andreev (which we shall not prove here).

**PROPOSITION 6.3.2.** (Andreev [9].) *Suppose that $P^n \subset \mathbb{X}^n$ is a convex polytope and that the dihedral angles along all codimension-two faces of $P^n$ are $\leqslant \pi/2$. Then the hyperplanes of any two nonadjacent codimension-one faces do not intersect. Hence, $P^n$ has nonobtuse dihedral angles.*

**LEMMA 6.3.3.** *Suppose $P^n \subset \mathbb{S}^n$ is a convex spherical polytope with nonobtuse dihedral angles. Then $P^n$ is a spherical simplex.*

**COROLLARY 6.3.4.** *Suppose $C^{n+1} \subset \mathbb{R}^{n+1}$ is an essential polyhedral cone with nonobtuse dihedral angles. Then $C^{n+1}$ is a simplicial cone.*

Lemma 6.3.3 is a consequence of the following result from linear algebra.

**LEMMA 6.3.5.** ([29, p. 82].) *Suppose that $B$ is a positive semidefinite symmetric bilinear form on $\mathbb{R}^n$. Let $u_1, \dots, u_k$ be vectors in $\mathbb{R}^n$ such that $B(u_i, u_j) \leqslant 0$ for $i \neq j$.*

(i) *Suppose $v = \sum c_i u_i$, $c_i \in \mathbb{R}$, is a linear combination of the $u_i$ such that $B(v, v) = 0$. Then*

$$B\left(\sum |c_i| u_i, \sum |c_i| u_i\right) = 0.$$

(ii) *If $B$ is nondegenerate and if there is a linear form $f$ such that $f(u_i) > 0$ for all $i$, then the vectors $u_1, \dots, u_k$ are linearly independent.*

*Proof.* Since $B(u_i, u_j) \leqslant 0$ for all $i \neq j$, we get

$$B\left(\sum |c_i| u_i, \sum_i |c_i| u_i\right) \leqslant B\left(\sum c_i u_i, \sum c_i u_i\right);$$

hence, (i).

Suppose $B$ is nondegenerate and $\sum c_i u_i = 0$ is a linear relation. Using (i) (and the nondegeneracy of $B$), we get $\sum |c_i| u_i = 0$. So, if $f$ is any linear form on $\mathbb{R}^n$, then $\sum |c_i| f(u_i) = 0$. If, in addition, $f(u_i) > 0$, then each $c_i = 0$, proving (ii). □

*Proof of Lemma 6.3.3.* Suppose $P^n \subset \mathbb{S}^n$ is a convex polytope and $F_1, \dots, F_k$ its codimension-one faces. Let $u_i \in \mathbb{R}^{n+1}$ be the inward-pointing unit vector normal to $F_i$. Choose a point $x$ in the interior of $P^n$ and consider the linear form $f$ on $\mathbb{R}^{n+1}$ defined by $f(v) = \langle v, x \rangle$. Since $x$ lies on the positive side of the half-space determined by $u_i$, $f(u_i) = \langle u_i, x \rangle > 0$ for $i = 1, \dots, k$. So, by Lemma 6.3.5, the vectors $u_1, \dots, u_k$ are linearly independent. Consequently, $k = n + 1$, $\{u_i\}_{1 \leqslant i \leqslant n+1}$ is a basis for $\mathbb{R}^{n+1}$ and $P^n$ is a spherical simplex. □

Later we will need Lemma 6.3.7 below. Its proof is along the same lines as that of Lemma 6.3.5. Before stating it we first recall a definition from matrix theory.

**DEFINITION 6.3.6.** An $n \times n$ matrix $(a_{ij})$ is *decomposable* if there is a nontrivial partition of the index set as $\{1, \dots, n\} = I \cup J$, so that $a_{ij} = a_{ji} = 0$ whenever $i \in I, j \in J$. Otherwise, it is *indecomposable*.

**LEMMA 6.3.7.** ([29, p.83].) *Suppose that $A = (a_{ij})$ is an indecomposable, symmetric, positive semidefinite $n \times n$ matrix and that $a_{ij} \leqslant 0$ for all $i \neq j$.*

*(i) If A is degenerate, then its corank is 1 (i.e., its kernel is dimension 1). Moreover, its kernel is spanned by a vector with all coordinates* $>0$.

*(ii) The smallest eigenvalue of A has multiplicity 1 and this eigenvalue has an eigenvector with all coordinates* $>0$. *(This statement is true independently of whether A is degenerate or nondegenerate.)*

*Proof.* (i) Associated to $A$ there is a quadratic form $q$ on $\mathbb{R}^n$ defined by $q(x) = x^t Ax$. Let $N$ be the null space of $q$ ($= \operatorname{Ker} A$) and suppose $\sum c_j e_j$ is a nonzero vector in $N$. By Lemma 6.3.5 (i), $\sum |c_j| e_j$ is also in $N$, i.e.,

$$\sum_j a_{ij} |c_j| = 0. \tag{6.7}$$

Now let $I$ denote the set of $j \in \{1, \dots, n\}$ such that $c_j \neq 0$. Suppose $I$ is a proper subset of $\{1, \dots, n\}$ and fix an index $i$ not in $I$. If $j \in I$, then $a_{ij}|c_j| \leqslant 0$ (since $j \neq i$). If $j \notin I$, then $a_{ij}|c_j| = 0$ (since $|c_j| = 0$). So, each term in the sum on the left-hand side of (6.7) is nonpositive. Therefore, all terms are 0. In particular, $a_{ij} = 0$ for all $i \notin I$ and $j \in I$. Thus, $A$ is decomposable whenever $I \neq \{1, \dots, n\}$. Since, by hypothesis, $A$ is indecomposable $I = \{1, \dots, n\}$. In other words, all coordinates of a nonzero vector in $N$ are nonzero. This is obviously impossible if $\dim N \geqslant 2$. So $\dim N = 1$. Moreover, if $v$ is a nonzero vector in $N$, then we can assume all its coordinates are $> 0$ (since we can replace $v = \sum c_j e_j$ by $\sum |c_j| e_j$).

(ii) Let $\lambda \geqslant 0$ be the smallest eigenvalue of $A$. The preceding argument applies to the matrix $A - \lambda I$. □

**DEFINITION 6.3.8.** An $n$-dimensional polytope $P^n$ is *simple* if the link of each vertex is an $(n-1)$-simplex. (See Appendix A.6 for the definition of "link.") Equivalently, $P^n$ is simple if exactly $n$ codimension-one faces meet at each vertex.

For example, a cube is simple; an octahedron is not. The next result is also an immediate corollary of Lemma 6.3.3.

**PROPOSITION 6.3.9.** *Suppose* $P^n$ *is a convex polytope with nonobtuse dihedral angles in a space of constant curvature* $\mathbb{X}^n$. *Then* $P^n$ *is simple.*

A convex subset $C \subset \mathbb{E}^n$ is *reducible* if it is congruent to one of the form $C' \times C''$, where $C' \subset \mathbb{E}^m$, $C'' \subset \mathbb{E}^{n-m}$ and neither $C'$ nor $C''$ is a point. $C$ is *irreducible* if it is not reducible.

Suppose a convex subset $C \subset \mathbb{E}^n$ has a finite number of supporting hyperplanes $H_i$, $i \in I$. Then $C$ is reducible if and only if there is a nontrivial partition of the index set, $I = I' \cup I''$, so that $H_i$ and $H_j$ intersect orthogonally whenever $i \in I'$ and $j \in I''$.

**LEMMA 6.3.10.** *Suppose $P^n \subset \mathbb{E}^n$ is an irreducible convex polytope with nonobtuse dihedral angles. Then $P^n$ is an n-simplex.*

*Proof.* Let $u_1, \dots, u_k$ be the inward-pointing unit vectors normal to the codimension-one faces of $P^n$ and let $U := (\langle u_i, u_j \rangle)$ be the matrix of inner products of the $u_i$. It is a $k \times k$ positive semidefinite symmetric matrix (called the "Gram matrix" in Section 6.8). Since $P^n$ is irreducible, $U$ is indecomposable. By Lemma 6.3.7 there are only two possibilities. The first is that $U$ is positive definite, in which case $k = n$, $\{u_1, \dots, u_n\}$ is a basis for $\mathbb{R}^n$ and $P^n$ is a simplicial cone. This case is impossible since, by definition, a polytope is compact. The second possibility is that $k = n + 1$ and $U$ has corank 1. After picking an origin for $\mathbb{E}^n$ (i.e., an identification with $\mathbb{R}^n$), this means that $P^n$ is defined by inequalities: $\langle u_i, x \rangle \geqslant c_i$, for some real numbers $c_i$, $i = 1, \dots, n + 1$. That is to say, $P^n$ is a simplex. □

**COROLLARY 6.3.11.** (Coxeter [67].) *Suppose a convex set $P^n \subset \mathbb{E}^n$ is the intersection of a finite number of half-spaces and that the dihedral angle along any codimension-two face of $P^n$ is nonobtuse. Then $P^n$ is congruent to a product of convex sets of the following three types: (i) a simplex, (ii) a simplicial cone, (iii) a Euclidean subspace. In particular, if $P^n$ is compact (so that it is a polytope), then it is a product of simplices.*

## 6.4. THE DEVELOPING MAP

### Geometric Structures on Manifolds

As before, $\mathbb{X}^n$ stands for $\mathbb{S}^n$, $\mathbb{E}^n$, or $\mathbb{H}^n$ and Isom($\mathbb{X}^n$) is its group of isometries. An *$\mathbb{X}^n$-structure* on a manifold $M^n$ is an atlas of charts $\{\psi_i : U_i \to \mathbb{X}^n\}_{i \in I}$, where $\{U_i\}$ is an open cover of $M^n$, each $\psi_i$ is a homeomorphism onto its image and each overlap map

$$\psi_i \psi_j^{-1} : \psi_j(U_i \cap U_j) \to \psi_i(U_i \cap U_j)$$

is the restriction of an isometry in Isom($\mathbb{X}^n$). The Riemannian metric on $\mathbb{X}^n$ induces one on $M^n$ so that each chart $\psi_i$ is an isometry onto its image. An $\mathbb{X}^n$-structure on $M^n$ is essentially the same thing as a Riemannian metric of constant sectional curvature $+1$, 0 or $-1$ as $\mathbb{X}^n$ is, respectively, $\mathbb{S}^n$, $\mathbb{E}^n$, or $\mathbb{H}^n$.

As explained in [279, §3.3 and §3.4], an $\mathbb{X}^n$-structure on $M^n$ induces one on its universal cover $\widetilde{M}^n$. Furthermore, there is a "developing map" $D : \widetilde{M}^n \to \mathbb{X}^n$. (The condition that is needed for the developing map to be defined is that Isom($\mathbb{X}^n$) acts transitively and real-analytically on the manifold $\mathbb{X}^n$. This is certainly true in our case, when $\mathbb{X}^n$ is a space of constant curvature.) The $\mathbb{X}^n$-manifold $M^n$ is *complete* if $D : \widetilde{M}^n \to \mathbb{X}^n$ is a homeomorphism.

Suppose $P^n \subset \mathbb{X}^n$ is a convex polytope with codimension-one faces $(F_i)_{i \in I}$. For each $i \in I$, let $r_i$ denote the isometric reflection of $\mathbb{X}^n$ across the hyperplane supported by $F_i$. Let $\overline{W}$ be the subgroup of Isom($\mathbb{X}^n$) generated by $\{r_i\}_{i \in I}$. We want to determine when $P^n$ is a fundamental domain for the $\overline{W}$-action on $\mathbb{X}^n$ and when $\overline{W}$ is a discrete subgroup of Isom($\mathbb{X}^n$). One obvious condition must hold: if the hyperplanes supported by $F_i$ and $F_j$ intersect, then the subgroup $\overline{W}_{ij}$ generated by $r_i$ and $r_j$ must be finite and the sector bounded by these hyperplanes which contains $P^n$ must be a fundamental domain for the $\overline{W}_{ij}$-action. As in Example 3.1.2, the condition that the sector be a fundamental domain implies that the dihedral angle between these hyperplanes is an integral submultiple of $\pi$. Since this forces all the dihedral angles of $P^n$ to be nonobtuse (because $\pi/m \leqslant \pi/2$ if $m \geqslant 2$), the polytope $P^n$ must be simple (Proposition 6.3.9). If $P^n$ is simple, and if, for $i \neq j$, the intersection $F_i \cap F_j$ is nonempty, then this intersection is a face $F_{ij}$ of codimension two. So suppose $P^n$ is simple and that whenever $F_i \cap F_j \neq \emptyset$, the dihedral angle along $F_{ij}$ is of the form $\pi/m_{ij}$, for some integer $m_{ij} \geqslant 2$. If $F_i \cap F_j = \emptyset$, set $m_{ij} = \infty$. (Also, set $m_{ij} = 1$ when $i = j$.)

The matrix $(m_{ij})$ is the Coxeter matrix of the pre-Coxeter system $(\overline{W}, \{r_i\}_{i \in I})$ (cf. Definition 3.3.1). Let $(W, S)$ be the corresponding Coxeter system, with generating set $S = \{s_i\}_{i \in I}$. In view of Example 3.1.2, the order of $r_i r_j$ is $m_{ij}$; so the function $s_i \to r_i$ extends to a homomorphism $\phi : W \to \overline{W}$. There is a *tautological mirror structure* on $P^n$: the mirror corresponding to $i$ is $F_i$. So, the basic construction of 5.1 gives the space $\mathcal{U}(W, P^n)$. As in Lemma 5.2.5, the inclusion $\iota : P^n \to \mathbb{X}^n$ induces a $\phi$-equivariant map $\tilde{\iota} : \mathcal{U}(W, P^n) \to \mathbb{X}^n$, defined by $[w, x] \to \phi(w)x$. We want to prove that $\tilde{\iota}$ is a homeomorphism.

**Example 6.4.1.** (*One-dimensional Euclidean space.*) Let $I$ be an interval in $\mathbb{E}^1$ and let $r_1$ and $r_2$ be the reflections across its endpoints. Set $m_{12} = \infty$. The group $\overline{W}$ generated by $r_1$ and $r_2$ is the infinite dihedral group $\mathbf{D}_\infty$ of Example 3.1.3. $W$ is also infinite dihedral and $\phi : W \to \overline{W}$ is an isomorphism. The map $\tilde{\iota} : \mathcal{U}(W, I) \to \mathbb{E}^1$ is obviously a homeomorphism.

**Example 6.4.2.** (*One-dimensional sphere.*) Suppose $I$ is a circular arc of length $\pi/m$ and that $r_1$ and $r_2$ are the orthogonal reflections of $\mathbb{S}^1$ across the endpoints of $I$. This example is somewhat exceptional in that although the endpoints of $I$ do not intersect, we nevertheless want to define $m_{12}$ to be $m$ (instead of $\infty$). The group $\overline{W}$ generated by $r_1$ and $r_2$ is the dihedral group $\mathbf{D}_m$ of Example 3.1.2. By Lemma 3.1.5, $\phi : W \to \overline{W}$ is an isomorphism and $\tilde{\iota} : \mathcal{U}(W, I) \to \mathbb{S}^1$ is a homeomorphism.

The next result is the main theorem in this chapter.

**THEOREM 6.4.3.** *Suppose $P^n$ is a simple convex polytope in $\mathbb{X}^n$, $n \geqslant 2$, with dihedral angles of the form $\pi/m_{ij}$, and with $m_{ij} \in \mathbb{N}$ whenever $F_i \cap F_j \neq \emptyset$. If $F_i \cap F_j = \emptyset$, put $m_{ij} = \infty$. Let $W$ be the Coxeter group defined by the Coxeter matrix $(m_{ij})$. Then the natural map $\tilde{\iota} : \mathcal{U}(W, P^n) \to \mathbb{X}^n$ is a homeomorphism. This implies*

*(i)* $\phi : W \to \overline{W}$ *is an isomorphism.*

*(ii)* $\overline{W}$ *acts properly on* $\mathbb{X}^n$.

*(iii)* $P^n$ *is a strict fundamental domain for the* $\overline{W}$*-action on* $\mathbb{X}^n$ *(as in Definition 5.1.3).*

*Proof.* The proof is by induction on the dimension $n$. Let $(s_n)$ stand for the statement of the theorem in the case where $\mathbb{X}^n = \mathbb{S}^n$ and $P^n$ is a spherical simplex $\sigma^n$. Let $(c_n)$ stand for the statement of the theorem with $P^n$ replaced by an open simplicial cone $C^n$ (of some radius) in $\mathbb{X}^n$ and with $\mathbb{X}^n$ replaced by the open ball (of the same radius) in $\mathbb{X}^n$ about the vertex of $C^n$. Finally, $(t_n)$ stands for the theorem. The structure of the argument is

$$(s_n) \implies (c_{n+1}) \implies (t_{n+1}).$$

Since, by Example 6.4.2, $(s_1)$ and $(c_2)$ hold, establishing these implications will constitute a proof.

$(s_n) \implies (c_{n+1})$. Suppose that $C^{n+1} \subset \mathbb{X}^{n+1}$ is a simplicial cone of radius $r$ with nonobtuse dihedral angles of the form $\pi/m_{ij}$. Let $\sigma^n$ be its intersection with $\mathbb{S}^n$. The Coxeter group $W$ associated to $C^{n+1}$ is the same as the one assoicated to $\sigma^n$. Suppose that $(s_n)$ is true. Then $\phi : W \to \overline{W}$ is an isomorphism and $\mathcal{U}(W, \sigma^n)$ is homeomorphic to $\mathbb{S}^n$. Since $\mathbb{S}^n$ is compact, so is $\mathcal{U}(W, \sigma^n)$. Hence, $W$ is finite. We know that: (1) $C^{n+1}$ is the cone on $\sigma^n$, (2) $\mathcal{U}(W, C^{n+1})$ is the cone on $\mathcal{U}(W, \sigma^n)$ and (3) an open ball in $\mathbb{X}^{n+1}$ is the cone on $\mathbb{S}^n$. It follows that $\tilde{\iota}$ takes $\mathcal{U}(W, C^{n+1})$ homeomorphically onto the open ball. The same argument works even if the polyhedral cone is not essential, i.e., if $C^{n+1}$ stands for the intersection of $k$ half-spaces, $k < n+1$, in general position in $\mathbb{X}^{n+1}$ with a ball about some point in the intersection of supporting hyperplanes.

$(c_{n+1}) \implies (t_{n+1})$. Let $W$, $\overline{W}$, and $P^{n+1}$ be as in the statement of the theorem. We first show that $(c_{n+1})$ implies $\mathcal{U}(W, P^{n+1})$ has an $\mathbb{X}^{n+1}$-structure so that $\tilde{\iota} : \mathcal{U}(W, P^{n+1}) \to \mathbb{X}^{n+1}$ is a local isometry. Given $x \in P^{n+1}$, let $S(x)$ denote the set of reflections $s_i$ across the codimension one faces $F_i$ which contain $x$. By Theorem 4.1.6 (i), $(W_{S(x)}, S(x))$ is a Coxeter system. Let $r_x$ denote the distance to the nearest face of $P^{n+1}$ which does not contain $x$ and let $C_x$ (resp. $B_x$) be an open conical neighborhood (resp. ball) of radius $r_x$ about $x$ in $P^{n+1}$ (resp. $\mathbb{X}^{n+1}$). An open neighborhood of $[1, x]$ in $\mathcal{U}(W, P^{n+1})$ has the form $\mathcal{U}(W_{S(x)}, C_x)$. By $(c_{n+1})$, $\tilde{\iota}$ maps this neighborhood homeomorphically onto $B_x$. By equivariance, it maps the $w$-translate of such a neighborhood homeomorphically onto $\phi(w)(B_x)$. This defines an atlas on $\mathcal{U}(W, P^{n+1})$. The atlas gives $\mathcal{U}(W, P^{n+1})$, first of all, the structure of a smooth manifold and secondly, an $\mathbb{X}^{n+1}$-structure.

The $\mathbb{X}^{n+1}$-structure induces a Riemannian metric (of constant curvature) on $M^{n+1} := \mathcal{U}(W, P^{n+1})$. Since $W$ acts isometrically on $M^{n+1}$ and since the quotient space $M^{n+1}/W$ $(= P^{n+1})$ is compact, a standard argument shows that $M^{n+1}$ is metrically complete. It is well known (cf. [279, Prop. 3.4.15]) that this is equivalent to the condition that the $\mathbb{X}^{n+1}$-structure is complete. In other words, the developing map $D : \widetilde{M}^{n+1} \to \mathbb{X}^{n+1}$ is a covering projection. By Lemma 5.2.3, $\mathcal{U}(W, P^{n+1})$ is connected. Since the developing map is locally given by $\tilde{\iota} : \mathcal{U}(W, P^{n+1}) \to \mathbb{X}^{n+1}$ and since $\tilde{\iota}$ is globally defined, $\tilde{\iota}$ must be covered by $D$, i.e., $\tilde{\iota}$ is also a covering projection. Since $\mathbb{X}^{n+1}$ is simply connected, $\widetilde{M}^{n+1} = M^{n+1}$ $(= \mathcal{U}(W, P^{n+1}))$, $D = \tilde{\iota}$ and $\tilde{\iota}$ is a homeomorphism. □

**DEFINITION 6.4.4.** A *geometric reflection group* is the action of a group $\overline{W}$ on $\mathbb{X}^n$, which, as in Theorem 6.4.3, is generated by the reflections across the faces of a simple convex polytope with dihedral angles integral submultiples of $\pi$. The reflection group is *spherical*, *Euclidean*, or *hyperbolic* as $\mathbb{X}^n$ is, respectively, $\mathbb{S}^n$, $\mathbb{E}^n$, or $\mathbb{H}^n$.

*Remarks 6.4.5*

(a) Theorem 6.4.3 (ii) implies that $\overline{W}$ is a discrete subgroup of $\text{Isom}(\mathbb{X}^n)$.

(b) Part (iii) of the same theorem implies that, $\mathbb{X}^n$ is tiled by congruent copies of $P^n$.

(c) Since the orbit space of a reflection is a half-space, each point in the relative interior of a codimension-one face of the convex polytope $P^n$ has a neighborhood which is isomorphic to an open subset of the orbit space of a reflection. The hypothesis of Theorem 6.4.3 is that the dihedral angles of $P^n$ are integral submultiples of $\pi$, in other words, each point in the relative interior of a codimension-two face has a neighborhood isomorphic to an open subset of the orbit space of a linear action of a finite dihedral group on $\mathbb{R}^n$. (See Examples 3.1.2 and 6.4.2.) The conclusion of the theorem implies that, globally, $P^n$ is the orbit space of a group action on a smooth manifold. In particular, $P^n$ can be given the structure of a smooth "orbifold." This means that it is locally isomorphic to an orbit space of a finite group on $\mathbb{R}^n$. (See [279] and [149].) In other words, the hypothesis is that $P^n$ is an orbifold at points in the complement of the strata of codimension $\geqslant 3$ and the conclusion is that it is everywhere an orbifold.

(d) The previous remark is a special case of the important general principle "codimension-two conditions suffice," explained in [280, Theorem 4.1, p. 532]. Although we will not define all the relevant terms, this principle states that if $X^n$ is a $\mathbb{X}^n$-cone manifold and if $X^n$ is an orbifold at points in the complement of the strata of codimension $\geqslant 3$, then it is an orbifold. The proof of this principle is similar to that of Theorem 6.4.3—it is proved by induction on dimension.

(e) In the statement of Theorem 6.4.3 we can regard $P^n$ as an orbifold, $\mathcal{U}(W, P^n)$ as an "orbifoldal covering space" and $\tilde{\iota} : \mathcal{U}(W, P^n) \to \mathbb{X}^n$ as the developing map. Except in the case $\mathbb{X}^n = \mathbb{S}^1$ of Example 6.4.2, $\mathcal{U}(W, P^n)$ $(= \mathbb{X}^n)$ is simply connected, i.e., it is the universal orbifoldal covering of $P^n$. (This follows from the proof of Theorem 6.4.3. It also follows from Theorem 9.1.3 in Chapter 9.)

(f) There are a couple of remarkable features to the proof of Theorem 6.4.3. First of all, we have managed to prove that $\overline{W}$ is a Coxeter group, seemingly without using any of the separation properties developed in Chapters 3 and 4. (The one exception is the use of Theorem 4.1.6 in showing that $W_{S(x)}$ is a Coxeter group but, in fact, the use of this theorem could have been avoided.) Second, we have proved that the space $\mathcal{U}(W, P^n)$ is a manifold. *A priori*, neither fact is obvious. The second fact followed once we had used the inductive argument to show $\mathcal{U}(W, P^n) \to \mathbb{X}^n$ was a covering projection. As for the first fact, the point is that if we replace $W$ by a quotient $W'$, then the existence of a homomorphism $W' \to \overline{W}$ is no longer guaranteed. In fact, if the kernel of $W \to W'$ is nontrivial, no such homomorphism can exist (because either $\mathcal{U}(W', P^n)$ is not locally isometric to $\mathbb{X}^n$ or because it is not simply connected).

## 6.5. POLYGON GROUPS

The exterior angles in a Euclidean polygon sum to $2\pi$. So, if $P^2$ is an $m$-gon in $\mathbb{E}^2$ with interior angles $\alpha_1, \ldots, \alpha_m$, then $\sum(\pi - \alpha_i) = 2\pi$ or equivalently;

$$\sum_{i=1}^{m} \alpha_i = (m-2)\pi. \tag{6.8}$$

More generally, if $\mathbb{X}^2_\varepsilon$ stands for $\mathbb{S}^2$, $\mathbb{E}^2$, or $\mathbb{H}^2$ as $\varepsilon = 1, 0$, or $-1$, respectively, and $P^2$ is an $m$-gon in $\mathbb{X}^2_\varepsilon$, then the Gauss–Bonnet Theorem asserts that

$$\varepsilon \operatorname{Area}(P^2) + \sum(\pi - \alpha_i) = 2\pi. \tag{6.9}$$

Hence, $\sum \alpha_i$ is $>$, $=$, or $<$ $(m-2)\pi$, as $\mathbb{X}^2_\varepsilon$ is, respectively, $\mathbb{S}^2$, $\mathbb{E}^2$, or $\mathbb{H}^2$.

In the spherical case, the $\alpha_i$ satisfy some additional inequalities, namely, any exterior angle must be less than the sum of all the others. (To see this, consider the "polar dual" of $P^2 \subset \mathbb{S}^2$. It is the boundary of another spherical polygon with vertices the inward-pointing unit vectors normal to the edges of $P^2$. The triangle inequalities on the vertices in the resulting metric on $S^1$ are the inequalities in question.) Conversely, given a combinatorial $m$-gon and any assignment of angles $\alpha_i \in (0, \pi)$ to its vertices, we can realize it as a convex polygon in some $\mathbb{X}^n_\varepsilon$ with the prescribed interior angles, where $\varepsilon$ is determined by the sum of the $\alpha_i$ as above and where, in the spherical case when $\varepsilon = 1$, the exterior angles $\pi - \alpha_i$ satisfy the additional inequalities mentioned above.

In this section we consider convex $m$-gons $P^2 \subset \mathbb{X}^n_\varepsilon$ with vertices $v_1, \dots, v_m$ where the angle $\alpha_i$ at $v_i$ is of the form $\alpha_i = \pi/m_i$, for an integer $m_i \geqslant 2$.

**Example 6.5.1.** (*Spherical triangle groups.*) Suppose $P^2$ is a spherical polygon. Since each $\alpha_i \leqslant \pi/2$, the condition $\sum \alpha_i > (m-2)\pi$ forces $m < 4$, i.e., $P^2$ must be a triangle. (This also follows from Lemma 6.3.3.) What are the possibilities for $\alpha_i$? The inequality $\pi/m_1 + \pi/m_2 + \pi/m_3 > \pi$ can be rewritten as

$$\frac{1}{m_1} + \frac{1}{m_2} + \frac{1}{m_3} > 1. \tag{6.10}$$

Supposing that $m_1 \leqslant m_2 \leqslant m_3$, it is easy to see that the only triples $(m_1, m_2, m_3)$ of integers $\geqslant 2$ satisfying (6.10) are $(2, 3, 3)$, $(2, 3, 4)$, $(2, 3, 5)$, and $(2, 2, n)$, for $n$ an integer $\geqslant 2$. By Theorem 6.4.3, each such triple corresponds to a spherical reflection group. The Coxeter groups corresponding to the first three triples are the symmetry groups of the Platonic solids. (See Appendix B.) The triple $(2, 3, 3)$ gives the symmetry group of the tetrahedron, $(2, 3, 4)$ the symmetry group of the cube (or octahedron), and $(2, 3, 5)$ the symmetry group of the dodecahedron (or icosahedron). The Coxeter system corresponding to $(2, 2, n)$ is reducible; its diagram is o  o$\overset{n}{\text{——}}$o; the group is $\mathbf{C}_2 \times \mathbf{D}_n$.

**Example 6.5.2.** (*Two-dimensional Euclidean groups.*) $\sum \alpha_i = (m-2)\pi$ forces $m \leqslant 4$. If $m = 4$, there is only one possibility for the $m_i$, namely, $m_1 = m_2 = m_3 = m_4 = 2$, in which case, $P^2$ is a rectangle and we get a standard rectangular tiling of $\mathbb{E}^2$. The corresponding Coxeter group is $\mathbf{D}_\infty \times \mathbf{D}_\infty$. If $m = 3$, the relevant equation is

$$\frac{1}{m_1} + \frac{1}{m_2} + \frac{1}{m_3} = 1. \tag{6.11}$$

There are only three triples $(m_1, m_2, m_3)$ of integers $\geqslant 2$ with $m_1 \leqslant m_2 \leqslant m_3$ that satisfy this equation: $(2, 3, 6)$, $(2, 4, 4)$, and $(3, 3, 3)$. The corresponding reflection groups are the *Euclidean triangle groups*.

Figure 6.2. Tessellation by right-angled pentagons.

**Example 6.5.3.** (*Hyperbolic polygon groups.*) Given any assignment of angles of the form $\pi/m_i$ to the vertices of a combinatorial $m$-gon so that

$$\sum_{i=1}^{m} \frac{1}{m_i} < m - 2,$$

we can find a convex realization of it in $\mathbb{H}^2$. By Theorem 6.4.3, this yields a corresponding reflection group on $\mathbb{H}^2$. (See [278, Chapter 5: orbifolds].) For example, given a right-angled pentagon in $\mathbb{H}^2$, one gets a group generated by the reflections across its edges. The corresponding tessellation of $\mathbb{H}^2$ in the Poincaré disk model is illustrated in Figure 6.2. The dual tessellation by squares is marked with dotted lines. Since a hyperbolic (or spherical) triangle is determined, up to congruence, by its angles, when $m = 3$ we get a discrete subgroup of Isom($\mathbb{H}^2$), well defined up to conjugation. If $m > 3$, there is a continuous family (moduli) of hyperbolic polygons with the same angles and hence, a moduli space of representations of the Coxeter group.

The conclusion to be drawn from these examples is that any assignment of angles of the form $\pi/m_i$ to the vertices of an $m$-gon can be realized by a convex polygon in a two-dimensional space $\mathbb{X}^2$ of constant curvature; moreover, apart from a few exceptional cases, $\mathbb{X}^2 = \mathbb{H}^2$.

## 6.6. FINITE LINEAR GROUPS GENERATED BY REFLECTIONS

We have postponed our discussion of finite groups generated by linear reflections long enough. This classical situation provides much of the motivation for the previous material and terminology in Chapters 3, 4, and 5. In short, the main result of this section is that every finite linear group that is generated by reflections is isomorphic to the orthogonal linear group associated to a spherical reflection group (defined in the previous section). We start with the following elementary lemma.

**LEMMA 6.6.1.** *Any representation of a finite group $G$ on a finite-dimensional real vector space $V$ admits a $G$-invariant, positive definite, symmetric bilinear form (i.e., a $G$-invariant inner product).*

*Proof.* The space of positive definite bilinear forms on a vector space is convex. This means that any convex linear combination of positive definite forms is again positive definite. So, we can choose a positive definite form on $V$ and average it over $G$ to obtain a $G$-invariant one. □

So, given a finite group $G$ of linear automorphisms of an $n$-dimensional real vector space $V$, we can choose a $G$-invariant inner product. $V$ is then isometric to $\mathbb{R}^n$ by an isometry which takes $G$ to a subgroup of $O(n)$. Hence, we can assume that any finite linear group is a subgroup of $O(n)$.

**COROLLARY 6.6.2.** *Any representation of a finite group $G$ on a finite-dimensional real vector space $V$ is semisimple (i.e., any $G$-stable subspace of $V$ is a direct summand).*

*Proof.* Suppose $U \subset V$ is a $G$-stable subspace. Then its orthogonal complement $U^{\perp}$ is also $G$-stable and $V = U \oplus U^{\perp}$. □

Now suppose $W$ is a finite group generated by linear reflections on a finite dimensional vector space $V$. The $W$-action is *essential* if no nonzero vector is fixed by $W$. As above, we can assume that $V = \mathbb{R}^n$ and that the reflections are orthogonal. If $V^W$ denotes the subspace fixed by $W$ and $V'$ its orthogonal complement, then $W$ acts as a group generated by reflections on $V'$ and the action is essential. Let $R$ denote the set of all reflections in $W$. For each $r \in R$, let $H_r$ be the hy.perplane fixed by $r$. Such a fixed hyperplane is a *wall* of the $W$-action. Since the $W$-action is orthogonal, $r$ is uniquely determined by its fixed hyperplane $H_r$. We note that if $r \in R$ and $w \in W$, then $wrw^{-1}$ is also a reflection with fixed hyperplane $H_{wrw^{-1}} = wH_r$. In particular, this shows that the union of the reflecting hyperplanes is a $W$-stable subset of $\mathbb{R}^n$. Consider the set of points in $\mathbb{R}^n$ which do not lie in any reflecting hyperplane of $W$:

$$\mathbb{R}^n_{(1)} := \mathbb{R}^n - \bigcup_{r \in R} H_r \,.$$

$\mathbb{R}^n_{(1)}$ is the complement in $\mathbb{R}^n$ of the "strata" of codimension $\geqslant 1$. An *open chamber* for $W$ is a connected component of this complement. A *chamber* is the closure of an open chamber. Such a chamber is clearly a polyhedral cone. A *mirror* of a chamber $C$ is a codimension one face of $C$. A hyperplane spanned by a mirror of $C$ a *wall* of $C$. Two chambers are *adjacent* if they intersect in a common codimension one face. A *gallery* is a finite sequence of adjacent chambers. Since $\mathbb{R}^n_{(1)}$ is a $W$-stable subset, $W$ acts on the set $\mathcal{C}$ of chambers in $\mathbb{R}^n$. The action clearly preserves adjacency. As we shall see below, all this terminology is consistent with that introduced earlier in 3.2, 4.5, and 5.1.

Fix a chamber $C$ and call it the *fundamental chamber*. Let $S$ be the subset of $R$ consisting of those reflections $s$ such that $H_s$ is a wall of $C$. In the next theorem we list the basic properties of finite linear groups generated by reflections.

**THEOREM 6.6.3.** *Let $W$ be a finite group generated by linear reflections on a finite-dimensional vector space $V$. Then, with the notation above:*

*(i)* *$(W, S)$ is a Coxeter system.*

*(ii)* *$W$ acts simply transitively on the set $\mathcal{C}$ of chambers.*

*(iii)* *If the $W$-action is essential, then $C$ is a simplicial cone (so $\operatorname{Card} S = n$).*

*(iv)* *The natural map $\mathcal{U}(W, C) \to \mathbb{R}^n$, induced by the inclusion $C \hookrightarrow \mathbb{R}^n$, is a homeomorphism. (Here $C$ has the tautological mirror structure indexed by $S$ and $\mathcal{U}(W, C)$ is the basic construction of Chapter 5.)*

To prove this we need to introduce some notation and establish a sequence of lemmas.

Let $\Omega$ be the simplicial graph with vertex set $\mathcal{C}$ and with two vertices $C, C' \in \mathcal{C}$ connected by an edge if and only if the chambers are adjacent. A gallery of chambers gives an edge path in $\Omega$. (This agrees with the use of the term "gallery" in 4.5.)

**LEMMA 6.6.4.** *$\Omega$ is connected.*

*Proof.* The point is that the complement of a union of subspaces of codimension $\geqslant 2$ in $\mathbb{R}^n$ is connected. Suppose $\mathbb{R}^n_{(2)}$ denotes the complement of the union of all subspaces of the form $H_s \cap H_{s'}$, with $s \neq s'$. In other words, $\mathbb{R}^n_{(2)}$ is the complement of the strata of codimension $\geqslant 2$. Given $C, C' \in \mathcal{C}$, choose a piecewise linear path from a point in the interior of $C$ to a point in the interior of $C'$. If the path is in general position with respect to fixed hyperplanes, then it misses the strata of codimension $\geqslant 2$, i.e., its image lies in $\mathbb{R}^n_{(2)}$. Any path in general position crosses a sequence of adjacent chambers. So, it defines a gallery, that is, an edge path in $\Omega$. Since any (piecewise linear or smooth) path can be approximated by one in general position, we get an edge path in $\Omega$ from $C$ to $C'$. □

**LEMMA 6.6.5.** *The set of reflections $R \subset W$ and the graph $\Omega$ (together with the basepoint $C$) are the data for a prereflection system for $W$ (Definition 3.2.1).*

*Proof.* $\Omega$ is connected by the previous lemma. As we previously remarked, each reflection in $W$ is determined by its fixed hyperplane. It follows that if $C', C''$ are two adjacent vertices in $\Omega$, then there is a unique $r \in R$ which interchanges them. □

**COROLLARY 6.6.6.** *With hypotheses as above, the following statements are true.*

(i) *$W$ acts transitively on $\mathcal{C}$.*

(ii) *Every element in $R$ is conjugate to one in $S$.*

(iii) *$S$ generates $W$.*

*Proof.* Statement (i) follows from Lemma 6.6.4. Statements (ii) and (iii) follow from Lemma 6.6.5 and the properties of prereflection systems developed in 3.2, e.g., in Lemma 3.2.5. □

**LEMMA 6.6.7**

(i) *$(\Omega, C)$ is a reflection system in the sense of Definition 3.2.10.*

(ii) *$(W, S)$ is a Coxeter system.*

*Proof.* (i) To show that the prereflection system $(\Omega, C)$ is a reflection system we must show that for each $r \in R$, the fixed set $\Omega^r$ separates $\Omega$. The hyperplane $H_r$ fixed by a reflection $r$ separates $\mathbb{R}^n$ into two components. It follows that $\Omega - \Omega^r$ has two components. (Any edge path in $\Omega$ can be lifted to a path in $\mathbb{R}^n$ between the chambers corresponding to its endpoints; these chambers lie on opposite sides of $H_r$ if and only if the lifted path crosses $H_r$ an odd number of times or equivalently, if the edge path in $\Omega$ crosses $\Omega^r$ an odd number of times.)

(ii) By Lemma 3.2.12, $W$ acts freely on $\text{Vert}(\Omega)$ and hence, by Theorem 2.1.1, $\Omega$ is isomorphic to $\text{Cay}(W, S)$. By Theorem 3.3.4, (i) $\Longrightarrow$ (ii). □

For each $s \in S$, let $C_s$ be the mirror corresponding to $s$ (i.e., $C_s = C \cap H_s$). As in formula (5.1) of Section 5.1, put $S(x) := \{s \in S \mid x \in C_s\}$.

**LEMMA 6.6.8.** (Compare [29, p. 80].) *Suppose $x, y \in C$ and $w \in W$ are such that $wx = y$. Then $x = y$ and $w \in W_{S(x)}$.*

*Proof.* The proof is by induction on the length $k$ of $w$. The case $k = 0$ is trivial. If $k \geqslant 1$, then there is a wall $H$ of $C$ such that $w = s_H w'$, where $s_H \in S$ is orthogonal reflection across $H$ and where $l(w') = k - 1$. Since $l(s_H w) < l(w)$, $C$ and $wC$ lie on opposite sides of $H$ (Lemma 4.2.2). Hence, $y \in H$. Therefore, $y = s_H y = s_H(wx) = w'x$. So, by the inductive hypothesis, $x = y$ and $w' \in W_{S(x)}$. Since we also have $s_H \in S(x)$, this gives $w \in W_{S(x)}$. □

**COROLLARY 6.6.9.** *$C$ is a strict fundamental domain for $W$ on $\mathbb{R}^n$ (Definition 5.1.3). In other words, $C$ intersects each $W$-orbit in exactly one point.*

*Proof.* By Corollary 6.6.6 (i), $WC = \mathbb{R}^n$, i.e., $C$ intersects each $W$-orbit on $\mathbb{R}^n$. By the previous lemma, such an intersection is a singleton. □

*Proof of Theorem 6.6.3.* Statement (i) is Lemma 6.6.7 (ii). By Corollary 6.6.6 and Lemma 3.2.12, $W$ acts simply transitively on $\mathcal{C}$, which is statement (ii).

(iii) Suppose $H$ and $H'$ are walls of $C$ and $s, s' \in S$ are the corresponding reflections. The dihedral group $\langle s, s' \rangle$ acts on the 2-dimensional Euclidean space $E = (H \cap H')^{\perp}$ and the image of $C$ in $E$ under orthogonal projection is a fundamental sector. It follows that the dihedral angle of $C$ along $H_s \cap H_{s'}$ is $\pi/m_{ss'}$, which is $\leqslant \pi/2$. So, by Corollary 6.3.4, $C$ is a simplicial cone.

(iv) Let $f : \mathcal{U}(W, C) \to \mathbb{R}^n$ be the natural $W$-equivariant map defined by $[w, x] \to wx$. Since $WC = \mathbb{R}^n$, $f$ is onto. It is injective by Lemma 6.6.8. It is immediate from the definitions in Chapter 5 that $f$ is an open map; hence, a homeomorphism. □

**COMPLEMENT 6.6.10.** *(On fundamental domains.)* Suppose $G$ is a discrete group of isometries of a metric space $Y$ and $x_0$ is a point in $Y$. The *Dirichlet domain* centered at $x_0$ is the set of points $x \in Y$ such that the distance from $x_0$ to $x$ is no greater than the distance from $x_0$ to any other point in the $G$-orbit of $x$. Obviously, a Dirichlet domain is a fundamental domain for $G$ on $Y$ (as in Definition 5.1.3); however, there is no reason for it to be a strict fundamental domain. The Dirichlet domain depends on the choice of $x_0$. In general, any movement of the point $x_0$ can produce a nontrivial variation of the Dirichlet domain.

Suppose, as above, $C$ is a chamber for a finite reflection group $W$ on $\mathbb{R}^n$ and that $x_0$ is a point in the interior of $C$. Since $C$ is contained in the positive half-space bounded by $H_s$ for any $s \in S$, a simple geometric argument shows

$$d(sx, x_0) \geqslant d(x, x_0).$$

For any $w \in W$, one can then argue by induction on $l(w)$ that

$$d(wx, x_0) \geqslant d(x, x_0);$$

hence, $C$ is the Dirichlet domain centered at $x_0$. This shows that these fundamental domains have the following properties which are special to reflection groups:

- Different choices for $x_0$ in the interior of $C$ give the same Dirichlet domain, namely, $C$ itself.
- Chambers are intrinsically defined (a chamber is the closure of a component of the hyperplane complement in $\mathbb{R}^n$).
- Each chamber is a strict fundamental domain for $W$ on $\mathbb{R}^n$ (and each such chamber is homeomorphic to the orbit space, $\mathbb{R}^n/W$).

## 6.7. EXAMPLES OF FINITE REFLECTION GROUPS

Notation in this section comes from Table 6.1 in Section 6.9, where the diagrams of all irreducible, spherical reflection groups are listed.

**Example 6.7.1.** (*The symmetric group, type* $\mathbf{A}_{n-1}$.) The group $S_n$ of all permutations of $\{1,\dots,n\}$ is the *symmetric group on n letters*. It acts orthogonally on $\mathbb{R}^n$ by permutation of coordinates. The transposition $(ij)$ acts as orthogonal reflection across the linear hyperplane $\widehat{H}_{ij}$ defined by $x_i = x_j$. The (nonessential) polyhedral cone $\widehat{C}$ in $\mathbb{R}^n$ defined by $x_1 \geqslant \cdots \geqslant x_n$ is clearly a fundamental domain for $S_n$ on $\mathbb{R}^n$. Let $D$ be the diagonal line, $x_1 = \cdots = x_n$, and $V$ the orthogonal hyperplane defined by $\sum x_i = 0$. Since $S_n$ fixes $D$, it acts on the orthogonal complement, $D^\perp = V$. Let $p : \mathbb{R}^n \to V$ be the orthogonal projection. The reflecting hyperplanes in $V$ are denoted by $H_{ij} = p(\widehat{H}_{ij})$. The simplicial cone $C = p(\widehat{C}) \subset V$ is a fundamental domain. The walls of $C$ are the hyperplanes $H_{i,i+1}$, $i = 1,\dots,n-1$. Let $s_i$ be the reflection across $H_{i,i+1}$ and set $S = \{s_1,\dots,s_{n-1}\}$. By Theorem 6.6.3, $(S_n, S)$ is a Coxeter system.

What is its associated Coxeter matrix? The product of the transpositions (12) and (23) is the 3-cycle (123). It follows that $s_i s_{i+1}$ has order 3. On the other hand, if $|i-j| > 1$, then $s_i$ and $s_j$ correspond to the commuting transpositions $(i, i+1)$ and $(j, j+1)$ and so, $s_i s_j$ has order 2. Thus,

$$m_{ij} = \begin{cases} 1 & \text{if } i = j, \\ 2 & \text{if } |i-j| > 1, \\ 3 & \text{if } |i-j| = 1. \end{cases}$$

Hence, the diagram of $S_n$ is $\mathbf{A}_{n-1}$: o—o · · · · · · o—o .

**Example 6.7.2.** (*The n-octahedral group, type* $\mathbf{B}_n$.) Let $V = \mathbb{R}^n$. By a *simple sign change* we will mean an orthogonal reflection of $\mathbb{R}^n$ which, for some $i$, sends $e_i$ to $-e_i$ and which fixes the other $e_j$. (This is an orthogonal reflection across the coordinate hyperplane $x_i = 0$.) The *n-octahedral* group $G_n$ is the subgroup of $O(n)$ generated by all sign changes and permutations of coordinates. (This group is also discussed in Example A.1.7 of Appendix A.) Since the simple sign changes and transpositions act as reflections on $\mathbb{R}^n$, $G_n$ is generated by reflections. It is not hard to see that $G_n$ is the semidirect product

$$G_n = (\mathbf{C}_2)^n \rtimes S_n,$$

where $S_n$ acts on $(\mathbf{C}_2)^n$ by permuting the factors. So, $G_n$ is a group of order $2^n n!$. A fundamental domain for the $G_n$-action is the simplicial cone $C$ defined by $x_1 \geqslant \cdots \geqslant x_n \geqslant 0$. For $1 \leqslant i \leqslant n-1$, let $s_i$ denote orthogonal reflection across the hyperplane $x_i = x_{i+1}$ and let $s_n$ be reflection across $x_n = 0$. Put $S := \{s_1, \dots, s_n\}$. By Theorem 6.6.3, $(G_n, S)$ is a Coxeter system. We note that $\{s_1, \dots, s_{n-1}\}$ is the usual set of generators for $S_n$. So, the Coxeter subdiagram determined by $\{s_1, \dots, s_{n-1}\}$ is $\mathbf{A}_{n-1}$. It remains to determine the order of $s_i s_n$.

The inward-pointing unit normal vector $u_i$ to the wall $x_i = x_{i+1}$ is given by $u_i = \frac{1}{\sqrt{2}}(e_i - e_{i+1})$. The inward-pointing unit normal to $x_n = 0$ is $e_n$. Hence, $\langle e_n, u_i \rangle = 0$ for $i \leqslant n-2$, while $\langle e_n, u_{n-1} \rangle = \frac{-1}{\sqrt{2}} = -\cos(\pi/4)$. It follows that $s_n$ commutes with $s_i$ for $i \leqslant n-2$ and that the order of $s_n s_{n-1}$ is 4. So its Coxeter matrix is given by

$$m_{ij} = \begin{cases} 1 & \text{if } i = j, \\ 2 & \text{if } |i-j| > 1, \\ 3 & \text{if } |i-j| = 1 \text{ and } i,j \neq n, \\ 4 & \text{if } \{i,j\} = \{n-1, n\}, \end{cases}$$

and its Coxeter diagram is type $\mathbf{B}_n$: o——o · · · · · · o—4—o .

**Example 6.7.3.** (*The group of type* $\mathbf{D}_n$.) Let $G_n$ be as in the previous example and let $H_n \subset G_n$ be the subgroup of index two consisting of all permutations of coordinates and even sign changes. Thus,

$$H_n = (\mathbf{C}_2)^{n-1} \rtimes S_n,$$

where $(\mathbf{C}_2)^{n-1}$ denotes the diagonal matrices with $\pm 1$ as diagonal entries and with an even number of $-1$'s. This group is generated by the reflections corresponding to the transpositions (i.e., the reflections $e_i - e_j \to e_j - e_i, i \neq j$) and the reflections $e_i + e_j \to -(e_i + e_j)$. The second type of reflection has the effect of transposing $x_i$ and $x_j$ and multiplying both by $-1$. A fundamental simplicial cone $C$ is defined by the inequalities

$$x_1 \geqslant \cdots \geqslant x_{n-1} \geqslant x_n \quad \text{and} \quad x_{n-1} \geqslant -x_n.$$

Then $S = \{s_1, \ldots, s_n\}$, where the first $n-1$ reflections are exactly the same as those for $\mathbf{A}_{n-1}$ (or for $\mathbf{B}_n$), that is, for $i = 1, \ldots, n-1$, $s_i$ is reflection across $x_i = x_{i+1}$. However, $s_n$ is reflection across $x_{n-1} = -x_n$. The first $n-1$ generators give a subgroup of type $\mathbf{A}_{n-1}$. So it remains to see how $s_n$ interacts with the other fundamental generators. The inward-pointing unit normals for the walls corresponding to $s_{n-2}, s_{n-1}$, and $s_n$ are, respectively,

$$u_{n-2} = \frac{1}{\sqrt{2}}(e_{n-2} - e_{n-1}),$$

$$u_{n-1} = \frac{1}{\sqrt{2}}(e_{n-1} - e_n),$$

$$u_n = \frac{1}{\sqrt{2}}(e_{n-1} + e_n).$$

Calculation gives $\langle u_{n-1}, u_n \rangle = 0$ and $\langle u_{n-2}, u_n \rangle = -1/2$. It is also clear that $\langle u_i, u_n \rangle = 0$ for $i < n-2$. Hence,

$$m_{ni} = \begin{cases} 2 & \text{if } i \neq n-2, n, \\ 3 & \text{if } i = n-3. \end{cases}$$

and the Coxeter diagram of $H_n$ is type $\mathbf{D}_n$. (See Table 6.1 of Section 6.9.)

**Regular Polytopes**

A classical source of examples of spherical reflection groups comes from the theory of regular polytopes. In the next few paragraphs, we briefly sketch this connection. More details will be given in Appendix B.

By a *chamber* for a convex polytope $P^n$ we mean a maximal chain of proper faces of $P$, $F_0 < F_1 \cdots < F_{n-1}$ where $\dim F_i = i$. Two chambers are *adjacent* if the chains are the same except for one face. They are *i-adjacent* if the face in which they differ is of dimension $i$. As explained in Appendix A.3, a chamber in the above sense corresponds to a top-dimensional simplex in the barycentric subdivision of $\partial P$. (The barycenters of the $F_i$ span the $(n-1)$-simplex.) Moreover, two chambers are adjacent in the above sense if and only if the corresponding simplices are adjacent in the usual sense that they share a codimension-one face.

Given a polytope $P^n \subset \mathbb{E}^n$, let $\mathrm{Isom}(P)$ denote its symmetry group. $\mathrm{Isom}(P)$ is finite (since an element $g \in \mathrm{Isom}(P)$ is determined by its action on $\mathrm{Vert}(P)$). The polytope $P^n$ is *regular* if $\mathrm{Isom}(P)$ acts transitively on the set of chambers of $P^n$. (As shown in Lemma B.1.3 of Appendix B, this action on the set of chambers is necessarily free.)

Suppose $P$ is regular. Taking its center as the origin, we get an identification of $\mathbb{E}^n$ with the vector space $\mathbb{R}^n$ and $\mathrm{Isom}(P)$ with a subgroup of $O(n)$. We can radially project $\partial P$ onto $\mathbb{S}^{n-1}$, obtaining a cellulation of $\mathbb{S}^{n-1}$ by spherical polytopes. Fix a chamber for $P$ and let $\sigma$ denote the image in $\mathbb{S}^{n-1}$ of the corresponding simplex of the barycentric subdivision of $\partial P$. For $0 \leqslant i \leqslant n-1$, let $\sigma_i$ be the codimension-one face of $\sigma$ which is opposite to the vertex corresponding to the barycenter of the face of dimension $i$ in the chain. Let $s_i$ be the unique element of Isom($P$) which maps the given chamber to the $i$-adjacent one. Since $s_i$ fixes the barycenters of the faces in the chain which are not of dimension $i$, it must fix $\sigma_i$. Hence, it must be reflection across $\sigma_i$. It follows that Isom($P$) is a group generated by spherical reflections on $\mathbb{S}^{n-1}$, that $\sigma$ is a fundamental simplex and that $S = \{s_0, \dots, s_{n-1}\}$ is the corresponding fundamental set of generators.

As is explained in Appendix B, the Coxeter diagram of (Isom($P$), $S$) is a straight line. More precisely, if $|i-j| \geqslant 2$, then $s_i$ and $s_j$ commute, while if $|i-j| = 1$ then $m_{ij}$, the order of $s_i s_j$, is $\geqslant 3$. Let $m_i$ denote the order of $s_i s_{i+1}$. The $n$-tuple $(m_1, \dots, m_{n-1})$ is called the *Schläfli symbol* of $P$. (A different but equivalent definition of the Schläfli symbol is given in B.2.)

Two polytopes $P$ and $Q$ are *dual* if their face posets are anti-isomorphic. Thus, dual polytopes have isomorphic barycentric subdivisions. (This is discussed more fully in Appendix A.) In Appendix B we will see that regular polytopes $P$ and $Q$ are dual if and only if the Schläfli symbol of $P$ is the Schläfli symbol of $Q$ in reverse order. This is, of course, consistent with the obvious fact that dual regular polytopes have isomorphic symmetry groups.

In each dimension $n > 1$, we always have at least three regular polytopes:

- The $n$-simplex $\Delta^n$ of Example A.1.4 with Schläfli symbol $(3, \dots, 3)$. Its symmetry group is the symmetric group $S_{n+1}$ with diagram of type $\mathbf{A}_n$. (See Example 6.7.1 above.) It is self-dual.
- The $n$-cube $\square^n$ of Example A.1.5 with Schläfli symbol $(4, 3, \dots, 3)$.
- The $n$-octahedron of Example A.1.7 with Schläfli symbol $(3, \dots, 3, 4)$. The $n$-octahedron and the $n$-cube are dual polytopes and therefore, have the same symmetry group (the $n$-octahedral group) with diagram of type $\mathbf{B}_n$. (The two-dimensional octahedron and the two-dimensional cube are both squares.)

It turns out the only further examples of regular polytopes occur in dimensions 2, 3, and 4. In dimension 2 we have the regular polygons. The Schläfli symbol of an $m$-gon is $(m)$; its symmetry group is the dihedral group $\mathbf{D}_m$. (See Example 3.1.2.) In dimension 3 we have the dodecahedron and the icosahedron with Schläfli symbols $(5, 3)$ and $(3, 5)$, respectively. They are dual to one another. Their Coxeter diagram is type $\mathbf{H}_3$: $\circ\overset{5}{\text{—}}\circ\text{—}\circ$ . The corresponding symmetry group was mentioned previously in Example 6.5.1. In dimension 4 there are three exotic polytopes and since two of them are dual, two new diagrams, types $\mathbf{F}_4$: o—o$\overset{4}{—}$o—o and $\mathbf{H}_4$: o$\overset{5}{—}$o—o—o. These four-dimensional polytopes are called "the 24-cell," "the 120-cell," and "the 600-cell," and they have Schläfli symbols $(3,4,3)$, $(5,3,3)$, and $(3,3,5)$, respectively. (See [69] for a detailed discussion of these polytopes.) The 24-cell has 24 octahedra as its three-dimensional faces. It is self-dual. The 120-cell and the 600-cell are dual to one another. The 120-cell has 120 dodecahedra as its three-dimensional faces; the 600-cell has 600 tetrahedra as its three-dimensional faces.

## 6.8. GEOMETRIC SIMPLICES: THE GRAM MATRIX AND THE COSINE MATRIX

### Spherical Simplices

Suppose $\sigma \subset \mathbb{S}^n \subset \mathbb{R}^{n+1}$ is a spherical $n$-simplex. It has $n+1$ faces of codimension one. Index them by $\{0, 1, \dots, n\}$. So $\{\sigma_i\}_{0 \leqslant i \leqslant n}$ is the set of codimension-one faces of $\sigma$. Let $u_i \in \mathbb{R}^{n+1}$ be the inward-pointing unit vector normal to $\sigma_i$. The condition that the $u_i$ is *inward pointing* means that $\sigma$ and $u_i$ lie on the same side of the hyperplane $\langle u_i, x\rangle = 0$. Thus, $\sigma$ is the locus of points $x \in \mathbb{S}^n$ satisfying

$$\langle u_i, x\rangle \geqslant 0 \quad \text{for} \quad 0 \leqslant i \leqslant n.$$

The *Gram matrix* of $\sigma$ is the $(n+1) \times (n+1)$ matrix $c_{ij}(\sigma)$ defined by

$$c_{ij}(\sigma) := \langle u_i, u_j\rangle, \quad 0 \leqslant i, j \leqslant n. \tag{6.12}$$

In other words, if $U$ is the $(n+1) \times (n+1)$ matrix with column vectors $u_0, \dots, u_n$, then $(c_{ij}(\sigma))$ is the matrix $U^t U$. ($U^t$ is the transpose of $U$.) It follows that $(c_{ij}(\sigma))$ is symmetric and positive definite and all diagonal entries are 1. (The reason it is positive definite is that $U$ is nonsingular. This, in turn, is because $\{u_0, \dots, u_n\}$ is a basis for $\mathbb{R}^{n+1}$.) For $i \neq j$, the dihedral angle $\theta_{ij}$ between $\sigma_i$ and $\sigma_j$ is given by

$$\theta_{ij} = \pi - \cos^{-1}(\langle u_i, u_j\rangle)\ (= \cos^{-1}(-\langle u_i, u_j\rangle)). \tag{6.13}$$

(By convention, when $i = j$, put $\theta_{ii} = \pi$.)

**LEMMA 6.8.1.** *A spherical simplex is determined, up to isometry, by its Gram matrix (or, in view of (6.13), by its dihedral angles).*

*Proof.* Suppose $\sigma$ and $\sigma'$ are spherical $n$-simplices with the same Gram matrix. In other words, if $u_i$ (resp. $u_i'$) is the inward-pointing unit normal vector to $\sigma_i$ (resp. $\sigma_i'$), then

$$\langle u_i, u_j\rangle = \langle u_i', u_j'\rangle. \tag{6.14}$$

Since $\{u_0, \dots, u_n\}$ and $\{u'_0, \dots, u'_n\}$ are bases for $\mathbb{R}^{n+1}$, there is a unique linear automorphism $g$ of $\mathbb{R}^{n+1}$ determined by $gu_i = u'_i$, for $0 \leqslant i \leqslant n$. This means that $g$ takes the set of defining half-spaces for $\sigma$ to the set of defining half-spaces for $\sigma'$. By (6.14), $\langle gu_i, gu_j\rangle = \langle u'_i, u'_j\rangle = \langle u_i, u_j\rangle$. Since the $u_i$ form a basis of $\mathbb{R}^{n+1}$, $g$ is an isometry. □

Next, we address the question of finding a spherical simplex with prescribed dihedral angles. Suppose that for each unordered pair of distinct integers $i, j \in \{0, \dots, n\}$ we are given numbers $\theta_{ij} \in (0, \pi)$ (with $\theta_{ij} = \theta_{ji}$). When is there a spherical simplex $\sigma$ such that the dihedral angle along $\sigma_i \cap \sigma_j$ is $\theta_{ij}$? If such a $\sigma$ exists, we will say that $\sigma$ has *dihedral angles prescribed by* $(\theta_{ij})$. Its Gram matrix would be the matrix $(c_{ij}(\theta))$ defined by

$$c_{ij}(\theta) = \begin{cases} 1 & \text{if } i = j, \\ -\cos\theta_{ij} & \text{if } i \neq j. \end{cases} \tag{6.15}$$

**PROPOSITION 6.8.2.** *Suppose that, as above, we are given numbers $\theta_{ij} \in (0, \pi)$, for $0 \leqslant i, j \leqslant n$ and $i \neq j$. Then there is a spherical simplex $\sigma$ with dihedral angles prescribed by $(\theta_{ij})$ if and only if the matrix $(c_{ij}(\theta))$, defined by (6.15), is positive definite.*

*Proof.* Given any positive definite symmetric matrix $A$, we can find a nonsingular matrix $U$ such that $U^tU = A$. (For example, we could take $U$ to be the square root of $A$.) Supposing $A = (c_{ij}(\theta))$ to be positive definite, we apply this observation to get a matrix $U$. Let $u_0, \dots, u_n$ be its column vectors. Since the diagonal entries of $A$ are all 1, the $u_i$ are unit vectors. The half-spaces $u_i \geqslant 0$ have nonempty intersection (as does any collection of half-spaces defined by linearly independent $u_i$) and hence, these half-spaces define a spherical simplex $\sigma$ with Gram matrix $A$. The implication in the other direction (that a Gram matrix of a spherical simplex is positive definite) was explained in the first paragraph of this section. □

### Hyperbolic Simplices

The discussion of Gram matrices of hyperbolic simplices is similar to the spherical case. Suppose $\sigma \subset \mathbb{H}^n$ is a hyperbolic $n$-simplex and that $u_0, \dots, u_n$ are its unit inward-pointing normals. (The $u_i$ are unit spacelike vectors in $\mathbb{R}^{n,1}$.) The Gram matrix $(c_{ij}(\sigma))$ is defined exactly as before by (6.12). Let $J$ be the $(n+1) \times (n+1)$ diagonal matrix with diagonal entries $(1, \dots, 1, -1)$ and let $U$ be the $(n+1) \times (n+1)$ matrix with column vectors $u_0, \dots, u_n$. Then $c_{ij}(\sigma)$ is the matrix $U^tJU$. It follows that it is symmetric, nondegenerate of type $(n, 1)$ and that all its diagonal entries are 1. (As before, it is type $(n, 1)$ because the $u_i$ form a basis for $\mathbb{R}^{n,1}$.) We omit the proof of the next lemma since it is virtually identical to that of Lemma 6.8.1.

**LEMMA 6.8.3.** *A hyperbolic simplex is determined, up to isometry, by its Gram matrix (or, in view of (6.13), by its dihedral angles).*

There is one further condition that the Gram matrix of a hyperbolic simplex always satisfies. We discuss it below. The $k^{\text{th}}$*-principal submatrix* of a square matrix $A$ is the matrix obtained by deleting the $k^{\text{th}}$ row and $k^{\text{th}}$ column of $A$. Suppose that $A = (\langle u_i, u_j \rangle)$ is the Gram matrix of a hyperbolic simplex $\sigma$ and let $A_{(k)}$ be the $k^{\text{th}}$-principal submatrix. Let $v_k \in \mathbb{H}^n$ denote the vertex opposite to $\sigma_k$ and let $L_k$ be the line through the origin in $\mathbb{R}^{n,1}$ determined by $v_k$, i.e.,

$$L_k := \bigcap_{i \neq k} u_i^{\perp}. \tag{6.16}$$

Since $L_k$ is timelike, its orthogonal complement $L_k^{\perp}$ is spacelike. Since $\{u_i\}_{i \neq k}$ spans $L_k^{\perp}$, the matrix of $A_{(k)}$ of inner products of the $\{u_i\}_{i \neq k}$ is positive definite. A slightly different way to say this is that $A_{(k)}$ is the Gram matrix for the spherical $(n-1)$-simplex $\text{Lk}(v_k, \sigma)$. ( The definition of $\text{Lk}(v_k, \sigma)$, the "link" of $v_k$ in $\sigma$, is given at the end of 6.2 and in more detail in Appendix A.6.) Thus, every principal submatrix of the Gram matrix of a hyperbolic simplex is positive definite.

As in the spherical case, suppose that for each pair of distinct integers $i, j \in \{0, \dots, n\}$ we are given numbers $\theta_{ij} \in (0, \pi)$. When is there a hyperbolic simplex $\sigma$ such that the dihedral angle along $\sigma_i \cap \sigma_j$ is $\theta_{ij}$? If such a simplex exists, its Gram matrix would be the matrix $(c_{ij}(\theta))$ defined by (6.15).

**PROPOSITION 6.8.4.** *Suppose, as above, we are given numbers $\theta_{ij} \in (0, \pi)$, for $0 \leqslant i, j \leqslant n$ and $i \neq j$ and let $(c_{ij}(\theta))$ be the matrix defined by (6.15). Then there is a hyperbolic simplex $\sigma$ with dihedral angles prescribed by $(\theta_{ij})$ if and only if*

- *the matrix $(c_{ij}(\theta))$ is type $(n, 1)$ and*
- *each principal submatrix of $(c_{ij}(\theta))$ is positive definite.*

*Proof.* Since any two nondegenerate bilinear forms of the same type are equivalent over $\mathbb{R}$, given any nonsingular symmetric matrix $A$ of type $(n, 1)$, there is a nonsingular matrix $U$ with $U^t J U = A$. Apply this observation to the matrix $A = (c_{ij}(\theta))$ to get $U$. Let $u_0, \dots, u_n$ be its column vectors. Since $\langle u_i, u_i \rangle = c_{ii}(\theta) = 1$, each $u_i$ is spacelike. It follows that the simplicial cone in $\mathbb{R}^{n,1}$ defined by the inequalities $\langle u_i, x \rangle \geqslant 0$, for $0 \leqslant i \leqslant n$, intersects either the positive or the negative light cone. If it intersects the negative light cone, then replace each $u_i$ by $-u_i$. Since the principal submatrix $A_{(k)}$ is positive definite, the hyperplane spanned by $\{u_i\}_{i \neq k}$ is spacelike. Hence, its orthogonal complement $L_k$, defined by (6.16), is timelike and the intersection $R_k$ of $L_k$ with the positive light cone is an extremal ray of the simplicial cone. Since each extremal ray of the simplicial cone lies inside the positive light cone, the entire simplicial cone lies inside the positive light cone and hence, its intersection with $\mathbb{H}^n$ is a hyperbolic simplex. □

### Euclidean Simplices

The analysis of Gram matrices of Euclidean simplices is a bit more complicated than in the spherical and hyperbolic cases.

Suppose $\sigma$ is a Euclidean $n$-simplex in $\mathbb{E}^n$, that $\sigma_0, \dots, \sigma_n$ are its codimension-one faces, that $u_i \in \mathbb{R}^n$ is the inward-pointing unit vector normal to $\sigma_i$, that $v_i$ is the vertex of $\sigma$ opposite to $\sigma_i$ and that $H_i$ is the affine hyperplane spanned by $\sigma_i$. After choosing a basepoint, we can identify $\mathbb{E}^n$ with $\mathbb{R}^n$.

**LEMMA 6.8.5.** *The vectors $u_0, \dots, u_n$ determine $\sigma$ up to translation and homothety.*

*Proof.* After changing $\sigma$ by a translation, we may assume that the vertex $v_0$ is the origin in $\mathbb{R}^n$. Then $\sigma$ is defined by the inequalities

$$\langle u_i, x\rangle \geqslant 0, \quad i = 1, \dots, n, \tag{6.17}$$

$$\langle u_0, x\rangle \geqslant -d, \tag{6.18}$$

where $d$ is the distance from the hyperplane $H_0$ to the origin. After scaling by $1/d$, we can put $\sigma$ in a "standard form" where $d = 1$. □

Since the $u_i$ span $\mathbb{R}^n$ and since there are $n+1$ such vectors, there is a nontrivial linear relation of the form

$$c_0u_0 + \cdots + c_nu_n = 0, \tag{6.19}$$

and this relation is unique up to multiplication by a scalar.

**LEMMA 6.8.6.** *The coefficients $c_i$ in the relation (6.19) are all nonzero and all have the same sign (which we could take to be positive).*

*Proof.* Suppose $\sigma$ is defined by the inequalities in (6.17) and (6.18). For $1 \leqslant i,j \leqslant n$ and $i \neq j$, $\langle u_j, v_i\rangle = 0$ and $\langle u_0, v_i\rangle = -d$. Take the inner product of both sides of the relation (6.19) with $v_i$ to obtain

$$-c_0d + c_i\langle u_i, v_i\rangle = 0 \tag{6.20}$$

or

$$\frac{\langle u_i, v_i\rangle}{d} = \frac{c_0}{c_i}.$$

Since $\langle u_i, v_i\rangle > 0$, $c_0$ and $c_i$ are both nonzero and both have the same sign. □

*Remark.* We can take $c_i$ to be the $(n-1)$-dimensional volume of the face $\sigma_i$.

**LEMMA 6.8.7.** *Suppose $\{u_0, \ldots, u_n\}$ is a set of $n+1$ unit vectors spanning $\mathbb{R}^n$. Then $\{u_0, \ldots, u_n\}$ is the set of inward-pointing unit normal vectors to a Euclidean simplex if and only if the nontrivial linear relation $c_0 u_0 + \cdots + c_n u_n$ satisfies the conclusion of Lemma 6.8.6 (that is, the coefficients $c_i$ can all be taken to be positive).*

*Proof.* Suppose $\{u_0, \ldots, u_n\}$ satisfies the conclusion of Lemma 6.8.6. Fix a positive number $d$, let $C$ be the convex cone in $\mathbb{R}^n$ defined by the inequalities in (6.17), and let $\sigma$ be the convex subset of $C$ defined by adjoining the inequality (6.18). Since the coefficients $c_i$ are all nonzero, the vectors $u_1, \ldots, u_n$ are linearly independent. So $C$ is a simplicial cone. We must show that $\sigma$ is a simplex. Let $R_i$ be the extremal ray of $C$ opposite the face defined by $\langle u_i, x \rangle = 0$ and let $L_i$ be the line spanned by $R_i$. Let $v_i$ be the unique point on $L_i$ satisfying $\langle u_0, v_i \rangle = -d$. Then $\sigma$ is a simplex if and only if each $v_i \in R_i$. In other words, we must show that $\langle u_i, v_i \rangle > 0$, for all $0 \leqslant i \leqslant n$. By (6.20),

$$\langle u_i, v_i \rangle = \frac{c_0 d}{c_i}.$$

So if $c_0$ and $c_i$ have the same sign, $\langle u_i, v_i \rangle$ is positive. □

The Gram matrix of $\sigma$ is defined, just as before, by (6.12): it is the matrix $(\langle u_i, u_j \rangle)_{0 \leqslant i,j \leqslant n}$. In other words, it is the matrix $A = U^t U$ where $U$ is the $n \times (n+1)$ matrix with the $u_i$ as column vectors. Since the vectors $u_i$ span $\mathbb{R}^n$, $A$ has rank $n$. Therefore, $A$ is positive semidefinite of corank one. A final observation is that the one-dimensional null space of $A$ is spanned by the column vector

$$v = \begin{pmatrix} c_0 \\ \vdots \\ c_n \end{pmatrix}$$

where the $c_i$ are the coefficients in the relation (6.19). The reason is that (6.19) can be rewritten as the equation, $Uv = 0$; hence, $v^t A v = v^t U^t U v = 0$.

We can now give the necessary and sufficient conditions for the existence of a Euclidean simplex with prescribed dihedral angles.

**PROPOSITION 6.8.8.** *As before, suppose we are given numbers $\theta_{ij} \in (0, \pi)$, for $0 \leqslant i, j \leqslant n$ and $i \neq j$. Let $(c_{ij}(\theta))$ be the matrix defined by (6.15). Then there is a Euclidean simplex $\sigma$ with dihedral angles along its codimension two faces $\sigma_i \cap \sigma_j$ as prescribed by the $\theta_{ij}$ if and only if*

(a) *the matrix $(c_{ij}(\theta))$ is positive semidefinite of corank 1 and*

(b) *the null space of $(c_{ij}(\theta))$ is spanned by a vector $v$ with positive coordinates $c_0, \ldots, c_n$.*

*Proof.* Let $A$ be the positive semidefinite matrix $(c_{ij}(\theta))$ and let $U$ be its square root. Then $A$ and $U$ have the same null space, namely, the line spanned by $v$. Let $u_0, \dots, u_n$ be the column vectors of $U$. We regard $U$ as a linear transformation $\mathbb{R}^{n+1} \to \mathbb{R}^{n+1}$ taking $e_i$ to $u_i$. Since the kernel of $U$ is the line spanned by $v$, the image of $U$ is the hyperplane $H$ orthogonal to $v$. So, the $u_i$ are unit vectors in the Euclidean space $H$ satisfying the linear relation (6.19) with positive coefficients $c_i$. Hence, by Lemma 6.8.7, there is a simplex $\sigma$ in $H$ with unit inward-pointing normal vectors $u_0, \dots, u_n$. □

*Remarks 6.8.9*

(i) If the $\theta_{ij}$ are all nonobtuse, then, by Lemma 6.3.7, condition (b) of Proposition 6.8.8 is automatic.

(ii) Any principal submatrix of a positive semidefinite matrix $A$ is either positive semidefinite or positive definite. If condition (b) holds for a positive semidefinite matrix $A$, then each principal submatrix is automatically positive definite. For example, suppose to the contrary that $v_1$ is a nonzero vector in the nullspace of $A_{(k)}$ and $w \in \mathbb{R}^{n+1}$ is the vector obtained from $v_1$ by inserting a 0 as the $k^{\text{th}}$ coordinate, then $w^tAw = v_1^tA_{(k)}v_1 = 0$. So, $w$ is in the null space of $A$. Since the $k^{\text{th}}$ coordinate of $w$ is 0, this contradicts (b).

**Exercise 6.8.10.** Suppose $n = 2$, that we are given numbers, $\theta_{01}, \theta_{02}, \theta_{12} \in (0, \pi)$, that $a = \theta_{01} + \theta_{02} + \theta_{12}$ is the angle sum and that $A = (c_{ij}(\theta))$ is the $3 \times 3$ matrix defined by (6.15). Show that $\det A$ is $> 0$, $= 0$ or $< 0$ as $a$ is $> \pi$, $= \pi$ or $< \pi$, respectively. Deduce that in these three cases we have, respectively, a spherical, Euclidean or hyperbolic triangle with angles as prescribed by the $\theta_{ij}$.

### The Cosine Matrix

Suppose $M = (m_{ij})$ is a Coxeter matrix on a set $I$.

**DEFINITION 6.8.11.** The *cosine matrix* associated to a Coxeter matrix $M$ is the $I \times I$ matrix $(c_{ij})$ defined by

$$c_{ij} = -\cos(\pi/m_{ij}). \tag{6.21}$$

When $m_{ij} = \infty$ we interpret $\pi/\infty$ to be 0 and $-\cos(\pi/\infty) = -\cos(0) = -1$. (Note all diagonal entries of $(c_{ij})$ are $-\cos(\pi/1) = 1$.)

Combining Theorem 6.4.3, Propositions 6.8.2, 6.8.4, 6.8.8, and Remark 6.8.9, we have the following.

**THEOREM 6.8.12.** *Let $M = (m_{ij})$ be a Coxeter matrix over $I$, $W$ the associated Coxeter group, and $C = (c_{ij})$ the associated cosine matrix. Suppose that no $m_{ij}$ is $\infty$. Then*

*(i) $W$ can be represented as a spherical reflection group generated by the reflections across the faces of a spherical simplex (with Gram matrix $C$) if and only if $C$ is positive definite.*

*(ii) Suppose, in addition, that $M$ is irreducible. Then $W$ can be represented as a Euclidean reflection group generated by the reflections across the faces of a Euclidean simplex (with Gram matrix $C$) if and only if $C$ is positive semidefinite of corank 1.*

*(iii) $W$ can be represented as a hyperbolic reflection group generated by the reflections across the faces of a hyperbolic simplex (with Gram matrix $C$) if and only if $C$ is nondegenerate of type $(n, 1)$ and each principal submatrix is positive definite.*

## 6.9. SIMPLICIAL COXETER GROUPS: LANNÉR'S THEOREM

By Theorem 6.8.12, if the cosine matrix of a Coxeter diagram is positive definite, the corresponding Coxeter group is finite (since it is a discrete subgroup of isometries of a sphere). We shall see in Section 6.12 (Theorem 6.12.9), that, conversely, if a Coxeter group is finite, then the cosine matrix of its Coxeter matrix is positive definite.

Let $\Delta^n$ stand for the $n$-simplex. Index its codimension-one faces by the set $I = \{0, 1, \dots, n\}$. The set $\{\Delta_i\}_{i \in I}$ of its codimension-one faces is a mirror structure as in 5.1. In this section we describe results of Lannér [184] classifying those Coxeter groups $W$ which can act as proper reflection groups on a simply connected space with fundamental chamber a simplex. In other words, Lannér determined which $W$ act properly on $\mathcal{U}(W, \Delta^n)$. Such a $W$ is a *simplicial Coxeter group*. As we shall see in Proposition 6.9.1, it turns out that any such $W$ is a geometric reflection group generated by reflections across the faces of an $n$-simplex in either $\mathbb{S}^n$, $\mathbb{E}^n$, or $\mathbb{H}^n$.

Suppose $W$ is the Coxeter group defined by an $I \times I$ Coxeter matrix $(m_{ij})$ and $C$ is the associated cosine matrix. By Lemma 5.1.7, $W$ acts properly on $\mathcal{U}(W, \Delta^n)$ if and only if the mirror structure on $\Delta^n$ is $W$-finite (Definition 5.1.6). By the remarks in the first paragraph of this section, this is the case if and only if each principal submatrix of $C$ is positive definite. So, the problem is reduced to finding all Coxeter diagrams $\Gamma$ with the property that every proper subdiagram is positive definite. We first note that if $\Gamma$ has this property and it is not connected, then each of its connected components is positive definite. It follows that there are only three possibilities depending on the determinant of the cosine matrix $C$:

(i) If $\det C > 0$, then $C$ is positive definite.

(ii) If $\det C = 0$, then $\Gamma$ is connected and $C$ is positive semidefinite of corank 1.

(iii) If $\det C < 0$, then $\Gamma$ is connected and $C$ is type $(n, 1)$.

Theorem 6.8.12 implies the first sentence of the following theorem; the second sentence will be proved in Appendix C.

**THEOREM 6.9.1.** (Lannér [184].) *Any simplicial Coxeter group can be represented as a geometric reflection group with fundamental chamber an $n$-simplex in either $\mathbb{S}^n$, $\mathbb{E}^n$, or $\mathbb{H}^n$. The (irreducible) spherical and Euclidean diagrams are given in Table 6.1, the hyperbolic diagrams in Table 6.2.*

## 6.10. THREE-DIMENSIONAL HYPERBOLIC REFLECTION GROUPS: ANDREEV'S THEOREM

Let us review the situation so far. A geometric reflection group on $\mathbb{S}^n$, $\mathbb{E}^n$, or $\mathbb{H}^n$ is determined by its fundamental chamber. This chamber is a convex polytope with dihedral angles integral submultiples of $\pi$ and any such polytope gives a reflection group (Theorem 6.4.3). In the spherical case the fundamental polytope must be a simplex and in the Euclidean case it must be a a product of simplices (Corollary 6.3.11). In the hyperbolic case all we know so far is that the polytope must be simple. In the converse direction, if the fundamental chamber is a simplex (i.e., if the reflection group is simplicial), then we have a complete classification in all three cases (Theorem 6.9.1). There is nothing more to said in the spherical and Euclidean cases. In the hyperbolic case we know what happens in dimension 2: the fundamental polygon can be an $m$-gon for any $m \geqslant 3$ and almost any assignment of angles can be realized by a hyperbolic polygon (there are a few exceptions when $m = 3$ or 4). What happens in dimension $n \geqslant 3$?

In dimension 3, there is a beautiful theorem due to Andreev [7], which gives a complete answer. Roughly speaking, it says that, given a simple polytope $P^3$, for it to be the fundamental polytope of a hyperbolic reflection group, (a) there is no restriction on its combinatorial type and (b) subject to the condition that the mirror structure be $W$-finite, almost any assignment of dihedral angles to the edges of $P^3$ can be realized (provided a few simple inequalities hold). Moreover, in contrast to the picture in dimension 2, the three-dimensional hyperbolic polytope is uniquely determined, up to isometry, by its dihedral angles—the moduli space is a point.

This situation reflects the nature of the relationship between geometry and topology in dimensions 2 and 3. A closed 2-manifold admits a hyperbolic

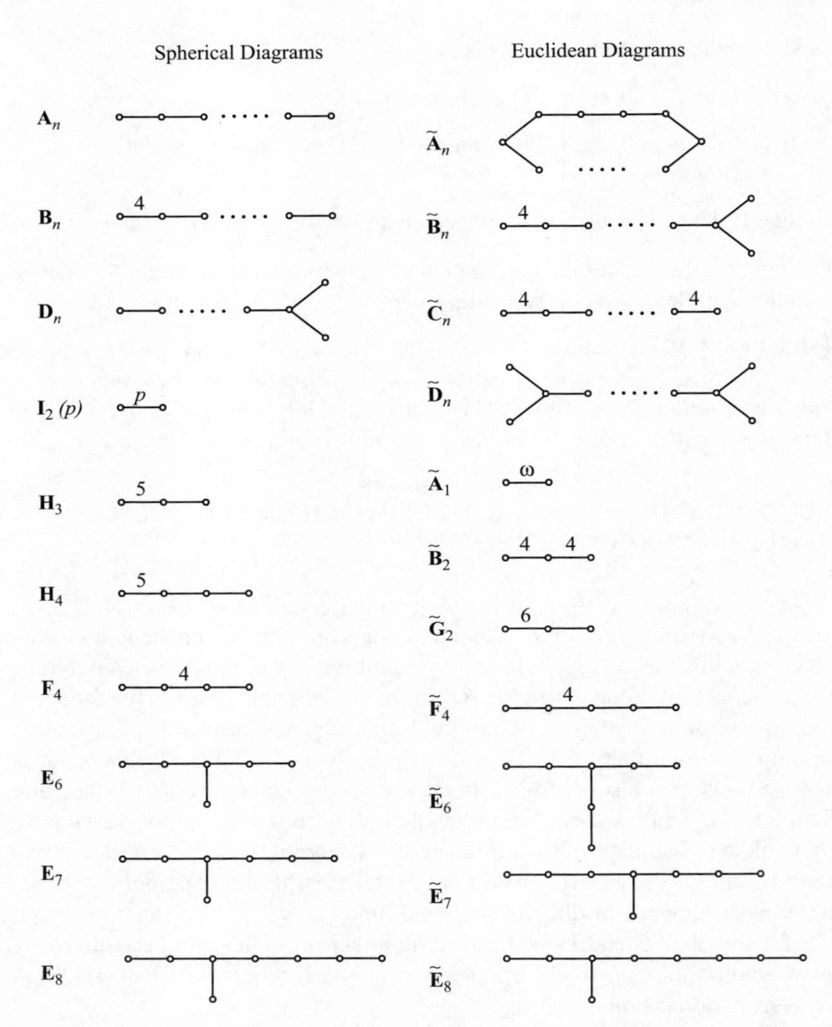


**Table 6.1.** Irreducible spherical and Euclidean diagrams.

structure if and only if its Euler characteristic is $< 0$ and there is a moduli space of such hyperbolic structures. In dimension 3 we have the famous Geometrization Conjecture of Thurston [276] (now a theorem of Perelman [237, 239, 238]). Roughly, it says that a closed 3-manifold $M^3$ admits a hyperbolic structure if and only if it satisfies certain simple topological conditions. Moreover, in contrast to the situation in dimension 2, the hyperbolic structure on $M^3$ is uniquely determined, up to isometry, by $\pi_1(M^3)$. (This is a consequence of the Mostow Rigidity Theorem [220], which is not true in dimension 2.)

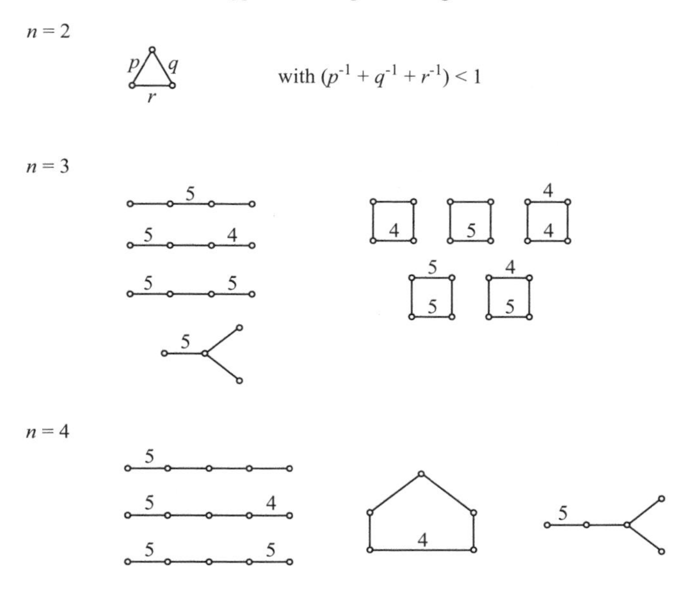


**Table 6.2.** Hyperbolic diagrams.

As one might expect, Andreev's inequalities on the dihedral angles of $P^3$ precisely correspond to the topological restrictions on $M^3$ in Thurston's Geometrization Conjecture. Thus, Andreev's Theorem is a special case of (an orbifoldal version of) the Geometrization Conjecture.

**CONJECTURE 6.10.1.** (Thurston's Geometrization Conjecture.) *A closed 3-manifold $M^3$ admits a hyperbolic structure if and only if it satisfies the following two conditions:*

*(a) Every embedded 2-sphere bounds a 3-ball in $M^3$.*

*(b) There is no incompressible torus in $M^3$ (i.e., $M^3$ is "atoroidal").*

(An *incompressible torus* is an embedded $\mathbb{T}^2$ in $M^3$ such that the inclusion induces an injection $\pi_1(\mathbb{T}^2) \hookrightarrow \pi_1(M^3)$.)

Conditions (a) and (b) imply conditions (a)′ and (b)′ below on the algebraic topology of $M^3$. (In fact, since the Poincaré Conjecture is true, (a) and (a)′ are equivalent.)

(a)′ $\pi_2(M^3) = 0$.

(b)′ $\mathbb{Z} \times \mathbb{Z} \not\subset \pi_1(M^3)$.

(When the fundamental group of $M^3$ is infinite, (a)′ means that $M^3$ is aspherical.)

Here is the statement of Andreev's Theorem.

**THEOREM 6.10.2.** (Andreev [7].) *Suppose $P^3$ is (the combinatorial type of) a simple polytope, different from a tetrahedron. Let $E$ be its edge set and $\theta : E \to (0, \pi/2]$ any function. Then $(P^3, \theta)$ can be realized as a convex polytope in $\mathbb{H}^3$ with dihedral angles as prescribed by $\theta$ if and only if the following conditions hold:*

- *(i) At each vertex, the angles at the three edges $e_1, e_2, e_3$ which meet there satisfy $\theta(e_1) + \theta(e_2) + \theta(e_3) > \pi$.*
- *(ii) If three faces intersect pairwise but do not have a common vertex, then the angles at the three edges of intersection satisfy $\theta(e_1) + \theta(e_2) + \theta(e_3) < \pi$.*
- *(iii) Four faces cannot intersect cyclically with all four angles $= \pi/2$ unless two of the opposite faces also intersect.*
- *(iv) If $P^3$ is a triangular prism, then the angles along the base and top cannot all be $\pi/2$.*

*Moreover, when $(P^3, \theta)$ is realizable, it is unique up to an isometry of $\mathbb{H}^3$.*

**COROLLARY 6.10.3.** *Suppose that $P^3$ is (the combinatorial type of) a simple polytope, different from a tetrahedron, that $\{F_i\}_{i \in I}$ is its set of codimension-one faces and that $e_{ij}$ denotes the edge $F_i \cap F_j$ (when $F_i \cap F_j \neq \emptyset$). Given an angle assignment $\theta : E \to (0, \pi/2]$, with $\theta(e_{ij}) = \pi/m_{ij}$ and $m_{ij}$ an integer $\geqslant 2$, then $(P^3, \theta)$ is the fundamental polytope of a hyperbolic reflection group $W \subset$ Isom$(\mathbb{H}^3)$ if and only if the $\theta(e_{ij})$ satisfy Andreev's Conditions in Theorem 6.10.2. Moreover, $W$ is unique up to conjugation in* Isom$(\mathbb{H}^3)$.

*Remark 6.10.4.* In the case where all the dihedral angles are integral submultiples of $\pi$, the conditions in Andreev's Theorem have interesting topological interpretations. Condition (i) means that the link of each vertex in $P^3$ can be given the structure of a spherical triangle. In view of Example 6.5.1, this means that the group generated by the reflections across the three faces, which meet at a given vertex, is finite. In other words, Condition (i) is the condition that $P^3$ can be given the structure of a three-dimensional orbifold (i.e., it has local models of the form $\mathbb{R}^3/G$, where $G$ is a finite group acting on $\mathbb{R}^3$. (See Remarks 6.4.5 (c).)

Condition (ii) is the main condition of Theorem 6.10.2. It means that $P^3$ has no spherical suborbifold which does not bound (the quotient of) a 3-ball and that it has no suborbifold corresponding to a Euclidean triangle group (Example 6.5.2). To be specific, suppose we have three edges as in

Condition (ii). Then we can find an embedding $\Delta \hookrightarrow P^3$ of a triangle $\Delta$ so that the edges of $\Delta$ are mapped to the three faces of $P^3$ and the vertices to the edges $e_1, e_2, e_3$. Let $W'$ denote the special subgroup generated by the three faces of $P^3$. If $\theta(e_1) + \theta(e_2) + \theta(e_3) > \pi$, then $W'$ is finite and $\mathcal{U}(W', \Delta) = S^2$. Moreover, since $\Delta$ divides $P^3$ into two pieces, neither of which is a conical neighborhood of a vertex, $\mathcal{U}(W', \Delta)$ divides $\mathcal{U}(W, P^3)$ into two pieces neither of which is a 3-ball. So, $\pi_2(\mathcal{U}(W, P^3)) = H_2(\mathcal{U}(W, P^3)) \neq 0$, i.e., $P^3$ is not aspherical as an orbifold. Similarly, if $\theta(e_1) + \theta(e_2) + \theta(e_3) = \pi$, then $W'$ is a Euclidean triangle group and $\mathbb{Z} \times \mathbb{Z} \subset W'$; so $W$ cannot be a discrete group of isometries of $\mathbb{H}^3$. Condition (iii) is similar. It just means that the two-dimensional Euclidean 4-gon group with all angles $\pi/2$ (that is, $\mathbf{D}_\infty \times D_\infty$) is not a special subgroup of $W$. If Condition (iv) fails, then $W = W' \times D_\infty$, where $W'$ is a triangle group. If $W'$ is infinite, this means that $W$ contains a subgroup $\cong \mathbb{Z} \times \mathbb{Z}$, which is incompatible with $\mathcal{U}(W, P^3)$ being $\mathbb{H}^3$. (If $W'$ is a hyperbolic triangle group, then $\mathcal{U}(W, P^3)$ looks like $\mathbb{H}^2 \times \mathbb{R}$.)

The last sentence of Theorem 6.10.2 (uniqueness) follows from the following rigidity result.

**PROPOSITION 6.10.5.** *A simple convex polytope $P^3 \subset \mathbb{H}^3$ is determined up to isometry by its dihedral angles.*

The first step in proving this is to note that the angles of each two-dimensional face of $P^3$ are determined by its dihedral angles. Indeed, since $P$ is simple, the link of each vertex is a spherical triangle. Each angle of such a spherical triangle is the corresponding dihedral angle of $P$. A spherical triangle is determined by its angles. (This is not true for spherical $m$-gons, if $m > 3$.) So, the edge lengths of such a spherical triangle are determined by the dihedral angles. But these edge lengths are the face angles at the given vertex (which has the spherical triangle as its link).

The second step is to show that the two-dimensional faces of $P^3$ are determined, up to congruence, by the face angles. Of course, it is not true that a hyperbolic $m$-gon is determined by its angles if $m > 3$; however, we have the following lemma of Cauchy (who proved it in the spherical case). A proof which works equally well in the hyperbolic case can be found in [20, Theorem 18.7.6, p. 298, vol. II].

**LEMMA 6.10.6.** (Cauchy's Geometric Lemma.) *Suppose $F$ and $F'$ are two hyperbolic $m$-gons with the same angles. Label each edge of $F$ either $+, 0$, or $-$ as its length is $>, =$, or $<$ the length of the corresponding edge of $F'$. Then either all edges of $F$ are labeled by 0 (and $F$ is congruent to $F'$) or there are at least four changes of sign as we go around $\partial F$.*

A second, purely topological, lemma of Cauchy asserts that it is impossible to have a cellulation of $S^2$ by polygons with such a (nontrivial) labeling of edges. (These two lemmas are the key steps in Cauchy's proof of his famous Rigidity Theorem for convex polytopes in $\mathbb{E}^3$.) A proof of this second lemma of Cauchy can be found in [20, Section 12.8, pp. 52–54, vol. II]. It is a simple counting argument using the fact that the Euler characteristic of $S^2$ is 2.

**LEMMA 6.10.7.** (Cauchy's Topological Lemma.) *Suppose we are given a cellulation of $\mathbb{S}^2$, combinatorially equivalent to the boundary complex of a three-dimensional polytope and that each edge of this cellulation is labeled by an element of $\{+, 0, -\}$ in such a fashion that for any 2-cell either all its edges are labeled 0 or there are at least four sign changes. Then, in fact, all edges of the cellulation are labeled 0.*

*Proof of Proposition 6.10.5.* The dihedral angles of $P^3$ determine the angles in the link of each vertex. Since $P^3$ is simple, each such link is a spherical triangle and hence, is determined by up to congruence by its angles. The edge lengths of the spherical triangles are angles of the faces of $P^3$. So Lemmas 6.10.6 and 6.10.7 show that the dihedral angles of $P^3$ determine its faces. But $P^3$ is obviously determined, up to an element of Isom($\mathbb{H}^3$), by the knowledge of the isometry types of its faces together with its dihedral angles. □

*Remarks 6.10.8*

(i) The above argument shows that a convex polytope in $\mathbb{H}^3$ is determined by its face angles (whether or not it is simple). See [246, Theorem 4.1].

(ii) A simple argument by induction on dimension shows that Proposition 6.10.5 remains valid in all dimensions $\geqslant 3$.

### A Dimension Count

Next, we do a simple dimension count which indicates why Andreev's Theorem is true. The point is that the dimension of the space of isometry classes of three-dimensional, hyperbolic, simple polytopes of a given combinatorial type is the same as the dimension of the Euclidean space of all real-valued functions on the edge set.

**DEFINITION 6.10.9.** The *de Sitter sphere* $\mathbb{S}^{n-1,1}$ is the $n$-dimensional hypersurface in $\mathbb{R}^{n,1}$ consisting of all spacelike vectors of length 1:

$$\mathbb{S}^{n-1,1} = \{u \in \mathbb{R}^{n,1} \mid \langle u, u \rangle = 1\}.$$

Suppose $P$ is a simple polytope in $\mathbb{H}^3$ and $\mathcal{F}$ is its set of (two-dimensional) faces. For each $F \in \mathcal{F}$ let $u_F \in \mathbb{S}^{2,1}$ be the inward-pointing unit vector normal to $F$. The $(u_F)_{F\in\mathcal{F}}$ determine $P$ (since $P$ is the intersection of the half-spaces determined by the $u_F$). The assumption that $P$ is simple means that the hyperbolic hyperplanes normal to the $u_F$ intersect in general position. So a slight perturbation of the $u_F$ will not change the combinatorial type of $P$. That is to say, the set of $\mathcal{F}$-tuples $(u_F)$ which define a polytope combinatorially equivalent to $P$ is an open subset $Y$ of $(\mathbb{S}^{2,1})^{\mathcal{F}}$. If we change $P$ by an element $g \in \mathrm{Isom}(\mathbb{H}^3) = O_+(3,1)$, then $u_F$ is changed to $gu_F$. Since the set of unit normals to the faces of $P$ spans $\mathbb{R}^{3,1}$, the group $O_+(3,1)$ acts freely on $Y$. We can identify the space of isometry classes of hyperbolic polytopes combinatorially equivalent to $P$ with the smooth manifold $Y/O_+(3,1)$. If $f = \mathrm{Card}(\mathcal{F})$ denotes the number of faces of $P$, we see that the dimension of this manifold is $f\dim(\mathbb{S}^{2,1}) - \dim(O_+(3,1)) = 3f-6$.

Let $C(P)$ denote the manifold of isometry classes of hyperbolic polytopes which (a) are combinatorially equivalent to $P$ and (b) have all dihedral angles $< \pi/2$. As explained above, we have

$$\dim C(P) = 3f - 6. \tag{6.22}$$

Let $\tilde{C}(P)$ denote the slightly larger space where the dihedral angles are allowed to take $\pi/2$ as a value. Let

$$\Theta : \tilde{C}(P) \to (0,\pi/2]^E$$

be the map which sends a hyperbolic polytope $Q$ to the $E$-tuple $(\theta(e))_{e\in E}$, where $E$ is the edge set of $P$ and $\theta(e)$ is the dihedral angle of $Q$ along $e$. Let $\tilde{V} \subset (0,\pi/2]^E$ be the open convex subset defined by Andreev's inequalities in Theorem 6.10.2 and let $V = \tilde{V} \cap (0,\pi/2)^E$. The previous explanation of the necessity of Andreev's inequalities shows that the image of $\Theta$ is contained in $\tilde{V}$. Andreev's Theorem is equivalent to the statement that $\Theta : \tilde{C}(P) \to \tilde{V}$ is a homeomorphism. Proposition 6.10.5 shows it is injective.

Let $f$, $e$, and $v$ denote, respectively, the number of faces, edges, and vertices of $P$. By (6.22), $C(P)$ is a manifold of dimension $3f-6$ while $V$ is a manifold of dimension $e$ (it is a open convex subset of $e$-dimensional Euclidean space). Since $\chi(\mathbb{S}^2) = 2$, $f - e + v = 2$; hence, $3f - 6 = 3e - 3v$. Since three edges meet at each vertex, $3v = 2e$ and therefore $3f - 6 = 3e - 3v = e$. So the manifolds $C(P)$ and $V$ have the same dimension.

Since $\Theta$ is a continuous injection, it follows from invariance of domain that its image is an open subset $U$ of $V$. (Alternatively, one could write down a formula for $\Theta$ in local coordinates, see that it is differentiable with nonsingular derivative at each point and conclude by the Inverse Function Theorem that it is a local diffeomorphism onto an open subset of $V$.)

To complete the proof of Theorem 6.10.2 it remains to show $U = V$. This amounts to showing that $U \neq \emptyset$ and that $\partial U = \partial V$. (Here $\partial U$ means the frontier of $U$, i.e., $\partial U := \overline{U} - U$.) We will not prove either of these facts, both of which require geometric arguments, but instead refer the reader to [7] or [278].

*Remark 6.10.10.* (*The dimension count when* $n > 3$.) The above dimension count does not work in dimensions >3. As before, $P^n$ is a simple $n$-polytope, $f_i$ is the number of its $i$-dimensional faces, and $C(P)$ is the space of isometry classes of convex hyperbolic polytopes of the same combinatorial type as $P$. Calculating as before, we have $\dim C(P) = nf_{n-1} - \dim O_+(n,1)$, where $\dim O_+(n,1) = \binom{n+1}{2}$. By Remarks 6.10.8 (ii), the map $\theta : C(P) \to [0,\pi)^{f_{n-2}}$ which assigns to each polytope its dihedral angles is an injection for $n \geqslant 3$. Hence, $\dim C(P)$ is $\leqslant$ the dimension of the range. That is to say,

$$nf_{n-1} - \binom{n+1}{2} \leqslant f_{n-2}. \tag{6.23}$$

This inequality is precisely the Lower Bound Theorem from combinatorics, proved by Barnette [16]. Moreover, as shown in [39, §19], for $n > 3$, equality holds in (6.23) if and only if $P^n$ is a "truncation polytope," i.e., if it is obtained from an $n$-simplex by successively truncating vertices. For example, if we truncate a vertex of a simplex, we obtain a simplicial prism (see Example 6.11.2 below). Since the dimension count works for truncation polytopes, it might be a reasonable project to determine all possible hyperbolic reflection groups with fundamental domain a truncation polytope. This would almost certainly add to the store of examples discussed in the next section. (This remark is taken from [58, Section 6].)

## 6.11. HIGHER-DIMENSIONAL HYPERBOLIC REFLECTION GROUPS: VINBERG'S THEOREM

The main result of this section is that cocompact hyperbolic reflection groups do not exist in arbitrarily high dimensions (Theorem 6.11.8). In the right-angled case they do not exist in dimensions >4. We give the proof as Corollary 6.11.7 below. In small dimensions ⩾4 no general picture of hyperbolic reflection group has emerged. As of 1985 the highest dimension of any known cocompact example was known to exist was 7 [291, p. 64].

### Some Isolated Examples

The four-dimensional simplicial examples are listed in Table 6.2. The nonsimplicial examples are surveyed in Vinberg's paper [291]. We discuss some of these in the next three examples.

**Example 6.11.1.** (*The right-angled* 120-*cell.*) Consider the four-dimensional simplicial hyperbolic group $W$ with diagram, o—5—o—o—o—4—o. Number the nodes of this diagram from left to right as $s_0, s_1, \dots, s_4$. Let $v_i$ denote the vertex of the fundamental simplex opposite to the face fixed by $s_i$. The isotropy subgroup at $v_4$ has diagram $\mathbf{H}_4$. It is the symmetry group of a regular 120-cell. (See Appendix B.2.) It follows that the union of the translates of the fundamental simplex under the isotropy subgroup at $v_4$ is a regular 120-cell $P^4$ in $\mathbb{H}^4$. Since the label on the furthest edge on the right of this diagram is 4, each dihedral angle is $\pi/2$. By Theorem 6.4.3, we get a hyperbolic reflection group $W_{P^4}$ with fundamental chamber $P^4$. It is a subgroup of index $(120)^2$ in $W$. (This is the order of the Coxeter group of type $\mathbf{H}_4$, the isotropy subgroup at $v_4$.)

**Example 6.11.2.** (*Simplicial prisms.*) Suppose $(W, S)$ is a Coxeter system with all $m_{st} \neq \infty$, with cosine matrix of type $(n,1)$, with one principal minor of type $(n-1,1)$ and with all other principal minors positive definite. We can realize $W$ as a reflection group on $\mathbb{H}^n$ across the mirrors of an $n$-simplex with one missing vertex (which is "outside the sphere at infinity"). Suppose $u_0, \dots, u_n$ are inward-pointing unit normals to the mirrors such that $\mathrm{Span}(u_1, \dots, u_n)$ is timelike (i.e., the $0^{\text{th}}$-principal minor is type ($n$-1, 1)). Let $v$ be the inward-pointing unit vector orthogonal to $\mathrm{Span}(u_1, \dots, u_n)$. Then $u_0, \dots, u_n$, $v$ are the inward-pointing normal vectors to a hyperbolic polytope combinatorially equivalent to the product of an ($n$-1)-simplex and an interval. (We have introduced a new face by "truncating" the missing vertex.) The "top" and "bottom" faces of this prism do not intersect; they have unit normals $u_0$ and $v$, respectively. The new bottom face (orthogonal to $v$) intersects all the other mirrors, except the top one, orthogonally. What is an example of such a Coxeter system with the properties in the first sentence of this paragraph? For $n = 5$, the diagram $\circ\overset{5}{\text{---}}\circ\text{---}\circ\text{---}\circ\text{---}\circ\overset{4}{\text{---}}\circ$ does the job. Other examples of such diagrams can be found in [291, p. 61]. All such examples are classified in [169].

**Example 6.11.3.** (*Doubling along a face.*) Suppose the polytope $P^n$ is a fundamental chamber for a hyperbolic reflection group $W$ on $\mathbb{H}^n$. Let $F$ be a codimension-one face such that the dihedral angle between it and all other faces which it intersects is $\pi/2$. Let $s$ be the reflection across $F$. Put $D(P) := P \cup s(P)$. $D(P)$ is the convex polytope formed by gluing together two copies of $P$ along $F$. All its dihedral angles are identified with dihedral angles in $P$; hence, they are all integral submultiples of $\pi$. It follows that the reflectons across the faces of $D(P)$ generate a Coxeter group $W'$ which has index 2 in $W$. For example, if $P^2$ is a right-angled $m$-gon in $\mathbb{H}^2$, then $D(P^2)$ is a right-angled $(2m-4)$-gon. As another example, if $P^4$ is the regular 120-cell of Example 6.11.1, then $D(P^4)$ is another right-angled polytope in $H^4$ with 226 three-dimensional faces. If $P$ has two or more disjoint faces with this property than we can obtain an infinite number of different examples by iterating this process. This method can be used to produce infinite families of cocompact examples in dimensions $\leqslant 6$ (see [4, 291]).

### The *h*-Polynomial of a Simple Polytope

Given an $n$-dimensional simple polytope $P^n$, denote by $f_i(P)$, or simply $f_i$, the number of $i$-faces of $P$. The $(n+1)$-tuple $(f_0, \dots, f_n)$ is the *f-vector* of $P$. Define a degree $n$-polynomial $h(t)$, called the *h-polynomial* of $P$ by

$$h(t) = \sum_{i=0}^{n} h_i t^i := \sum_{i=0}^{n} f_i (t-1)^i. \tag{6.24}$$

The $(n+1)$-tuple of coefficients $(h_0, \dots, h_n)$ is the *h-vector*. The $h$-vector and the $f$-vector obviously carry the same information; however, certain relations which the $f_i$ satisfy are simpler when expressed in terms of the $h_i$. The most important of these relations are stated in the following lemma.

**LEMMA 6.11.4.** (See [39]). *Suppose* $(h_0, \dots, h_n)$ *is the h-vector of a simple polytope* $P^n$. *Then for* $0 \leqslant i \leqslant n$,

*(i)* $h_i = h_{n-i}$ *and*

*(ii)* $h_i \geqslant 0$.

The demonstration of the above relations in [39] goes as follows. Suppose $P^n \subset \mathbb{R}^n$. Choose a linear function $\lambda : \mathbb{R}^n \to \mathbb{R}$ which is "generic" with respect to $P^n$ in the sense that no edge of $P^n$ is contained in a level set of $\lambda$. Then, on each edge of $P^n$, $\lambda$ is either monotone increasing or monotone decreasing. This defines an orientation on each edge which we indicate by an arrow. Since $P^n$ is simple exactly $n$ edges are incident to any given vertex $v$. Define the *index* of $\lambda$ at $v$ to be the number of edges incident to $v$ where the arrow points towards $v$. Let $g_i$ be the number of vertices of index $i$. It is proved in [39, §18] that $g_i = h_i$. In particular, the integer $g_i$ is independent of the choice of the linear function $\lambda$. Given this, formulas (i) and (ii) in Lemma 6.11.4 are now obvious. If we replace $\lambda$ by $-\lambda$, then the index at any given vertex is changed from $i$ to $n - i$; hence, formula (i). By definition each $g_i$ is a nonnegative integer; hence, (ii).

*Remark.* We will return to the $h$-polynomial in 17.4. There, we will use a slightly different definition than (6.24). Instead, we will define $h(t)$ in terms of the $f$-vector of the simplicial complex dual to $\partial P$. This leads to formula (17.16), which is equivalent to (6.24).

Next we estimate the average number $A_2$ of vertices in a 2-dimensional face of $P^n$. Since $P^n$ is simple, each vertex is contained in exactly $\binom{n}{2}$ two-dimensional faces. So the total number of pairs of the form $(F, v)$, where

$F$ is a 2-face and $v$ is a vertex of $F$, is $f_0\binom{n}{2}$. So $A_2$ is this number divided by the total number of 2-faces, i.e.,

$$A_2 = \frac{f_0}{f_2}\binom{n}{2}. \tag{6.25}$$

From (17.16) we get

$$f_i = \sum_{p=0}^{n} \binom{p}{i} h_p. \tag{6.26}$$

Set $m = [\frac{n}{2}]$. Taking into account formula (ii) of Lemma 6.11.4, we only need to know the $h_p$ for $p$ at or below the middle dimension. So, set

$$\hat{h}_p = \begin{cases} h_p & \text{if } p < \dfrac{n}{2}, \\ \dfrac{1}{2}h_p & \text{if } p = \dfrac{n}{2}. \end{cases}$$

Then (6.26) can be rewritten as

$$f_i = \sum_{p=0}^{m} \left[\binom{n-p}{i} + \binom{p}{i}\right] \hat{h}_p. \tag{6.27}$$

In particular, for $i = 0$ and $i = 2$, we have

$$f_0 = 2\sum_{p=0}^{m} \hat{h}_p \tag{6.28}$$

and

$$f_2 = \sum_{p=0}^{m} \left[\binom{n-p}{2} + \binom{p}{2}\right] \hat{h}_p. \tag{6.29}$$

Next, we claim that the coefficients in (6.29) are decreasing. Indeed,

$$\begin{aligned} &\left[\binom{n-(p-1)}{2} + \binom{p-1}{2}\right] - \left[\binom{n-p}{2} + \binom{p}{2}\right] \\ &\qquad = \left[\binom{n-(p-1)}{2} - \binom{n-p}{2}\right] - \left[\binom{p}{2} - \binom{p-1}{2}\right] \\ &\qquad = (n-p) - (p-1) = n - 2p + 1 > 0, \end{aligned}$$

where the final inequality is because $p \leqslant m$. So the smallest coefficient in (6.29) occurs for $p = m$. Thus,

$$f_2 > \left[\binom{n-m}{2} + \binom{m}{2}\right] \sum_{p=0}^{m} \hat{h}_p. \tag{6.30}$$

Substituting (6.28) and (6.30) into (6.25), we get

$$A_2 < \frac{2\binom{n}{2}}{\left[\binom{n-m}{2} + \binom{m}{2}\right]}. \tag{6.31}$$

**LEMMA 6.11.5.** *The right-hand side of (6.31) is*

$$\begin{cases} \dfrac{4(n-1)}{n-2}, & \text{if } n \text{ is even,} \\ \dfrac{4n}{n-1}, & \text{if } n \text{ is odd.} \end{cases}$$

*Hence, if $n > 4$, then $A_2 < 5$.*

*Proof.* If $n = 2m$, then the right hand side of (6.31) is

$$\frac{2\binom{n}{2}}{2\binom{m}{2}} = \frac{2m(2m-1)}{m(m-1)} = \frac{4(2m-1)}{2(m-1)} = \frac{4(n-1)}{n-2}.$$

If $n = 2m + 1$, it is

$$\frac{2\binom{n}{2}}{\left[\binom{m+1}{2} + \binom{m}{2}\right]} = \frac{2(2m+1)(2m)}{(m+1)m + m(m-1)}$$

$$= \frac{4m(2m+1)}{2m^2} = \frac{4(2m+1)}{2m} = \frac{4n}{n-1}.$$

To prove the last sentence of the lemma, note that in both cases these are strictly decreasing functions of $n$ (for $n > 2$). Moreover, when $n = 5$, $\frac{4n}{n-1} = 5$, and when $n = 6$, $\frac{4(n-1)}{n-2} = 5$. □

The idea of using the $h$-polynomial to make this estimate (and similar ones) is due to Nikhulin.

**COROLLARY 6.11.6.** *For $n > 4$, any simple polytope $P^n$ must have a two-dimensional face which is either a triangle or a quadrilateral.*

### Hyperbolic Reflection Groups Do not Exist in Arbitrarily High Dimensions

**COROLLARY 6.11.7.** (Vinberg.) *There do not exist right-angled convex (compact) polytopes in hyperbolic space of dimension* $>4$.

*Proof.* If $P^n \subset \mathbb{H}^n$ is right angled, then at any vertex the set of inward-pointing tangent vectors is isometric to the simplicial cone in $\mathbb{R}^n$ defined by $x_i \geqslant 0$, for $1 \leqslant i \leqslant n$. (In other words, the picture looks like the arrangement of coordinate hyperplanes in $\mathbb{R}^n$.) It follows that the angles of any 2-face are all $\pi/2$. But right-angled $m$-gons in the hyperbolic plane exist if and only if $m > 4$ (see Example 6.5.3). □

This corollary shows that there are no (cocompact) right-angled hyperbolic reflections groups in dimensions $> 4$.

*Remark.* On the other hand, both the dodecahedron and the 120-cell obviously have $A_2 = 5$. Moreover, both of these polytopes have right-angled hyperbolic versions. Thus, neither Corollary 6.11.6 nor 6.11.7 holds for $n = 3$ or 4.

Using a slightly more complicated version of this argument, Vinberg proved the following.

**THEOREM 6.11.8.** (Vinberg [291].) *Suppose* $P^n \subset \mathbb{H}^n$ *is a convex (compact) polytope with all dihedral angles integral submultiples of* $\pi$. *Then* $n < 30$. *In other words, hyperbolic reflection groups do not exist in dimensions* $\geqslant 30$.

## 6.12. THE CANONICAL REPRESENTATION

**DEFINITION 6.12.1.** Suppose $V$ is a vector space over a field $k$. A linear endomorphism $r : V \to V$ is a *pseudoreflection* if $1 - r$ is of rank 1.

Of course, any reflection on a real vector space is a pseudoreflection. For another example, if $k = \mathbb{C}$, $V = \mathbb{C}^n$ and $\lambda \neq 1$ is a complex number, then the endomorphism $r : \mathbb{C}^n \to \mathbb{C}^n$ defined by $r(e_1) = \lambda e_1$ and $r(e_i) = e_i$, for $i > 1$, is a pseudoreflection.

**LEMMA 6.12.2.** ([29, p. 70].) *Let* $\rho$ *be an irreducible linear representation of a group* $G$ *on a finite-dimensional vector space* $V$. *Suppose there is an element* $r \in G$ *such that* $\rho(r)$ *is a pseudoreflection.*

(i) *If* $u : V \to V$ *is a linear endomorphism which commutes with* $\rho(G)$, *then* $u$ *is a homothety (i.e.,* $u = c \cdot 1$, *for some constant* $c$ *in the field* $k$*).*

(ii) *Any nonzero $G$-invariant bilinear form on $V$ is nondegenerate. Moreover, any such form is either symmetric or skew-symmetric.*

(iii) *Any two such $G$-invariant bilinear forms are proportional.*

*Proof.* Let $D = \mathrm{Im}(1 - \rho(r))$. Then $\dim(D) = 1$. If $x = (1 - \rho(r))(y) \in D$, then, since $u$ commutes with $1 - \rho(r)$, $u(x) = (1 - \rho(r))(u(y)) \in D$. Thus, $u(D) \subset D$. Since any endomorphism of a one-dimensional vector space is a homothety, $u|_D = (c \cdot 1)|_D$, for some scalar $c$. Since $\rho$ is irreducible and since $u - c \cdot 1$ commutes with $\rho(G)$, the endomorphism $u - c \cdot 1$ is either an automorphism or the zero map. Since $D$ is contained in its kernel, it is the zero map, i.e., $u = c \cdot 1$. This proves (i).

Suppose $B$ is a nonzero $G$-invariant bilinear form. Then

$$N := \{x \in V \mid B(x, y) = 0, \text{ for all } y \in V\}$$

is a proper $\rho(G)$-stable subspace; hence, 0. Similarly,

$$N' := \{y \in V \mid B(x, y) = 0, \text{ for all } x \in V\} = 0.$$

This proves the first sentence of (ii). Let $B'$ be another nonzero $G$-invariant form. Since $V$ is finite dimensional, the nondegenerate form $B$ induces an isomorphism from $V$ to its dual space. It follows that any bilinear form $B'$ can be written as $B'(x, y) = B(u(x), y)$ for some linear endomorphism $u$. Since $B$ and $B'$ are both $G$-invariant, $u$ commutes with $\rho(G)$. By (i), $u = c \cdot 1$. Hence, $B'(x, y) = B(cx, y) = cB(x, y)$. This proves (iii). To prove the second sentence of (ii), we apply (iii) to the case where $B'$ is defined by $B'(x, y) = B(y, x)$. This yields $B(y, x) = cB(x, y) = c^2B(y, x)$. So $c^2 = 1$, $c = \pm 1$ and $B$ is either symmetric or skew-symmetric. □

As in Definition 3.3.1, let $M = (m_{ij})$ be a Coxeter matrix on a set $I$. For each $i \in I$, introduce a symbol $s_i$, set $S = \{s_i\}_{i \in I}$, and let $W$ be the group defined by the presentation $\langle S \mid \mathcal{R} \rangle$ associated to $M$. We will see below in Corollary 6.12.6 that $(W, S)$ is, in fact, a Coxeter system. (This is not a tautology.)

As in 6.8, the cosine matrix $(c_{ij})$ associated to $M$ is defined by (6.21). Let $\mathbb{R}^I$ be a real vector space of dimension $\mathrm{Card}(I)$ with basis $\{e_i\}_{i \in I}$. Let $B_M( , )$ be the symmetric bilinear form on $\mathbb{R}^I$ associated to the cosine matrix:

$$B_M(e_i, e_j) = c_{ij}. \tag{6.32}$$

We are going to define a linear representation $\rho : W \to GL(\mathbb{R}^I)$ so that the form $B_M$ will be $\rho(W)$-invariant. In other words, the image of $\rho$ will lie in $O(B_M)$, the isometry group of the form.

For each $i \in I$, let $H_i$ be the hyperplane in $\mathbb{R}^I$ defined by

$$H_i := \{x \mid B_M(e_i, x) = 0\}$$

and $\rho_i : \mathbb{R}^I \to \mathbb{R}^I$ the linear reflection

$$\rho_i(x) := x - 2B_M(e_i, x)e_i \; . \tag{6.33}$$

**LEMMA 6.12.3.** *The order of $\rho_i \rho_j$ is $m_{ij}$.*

*Proof.* Let $W_{ij} = \langle \rho_i, \rho_j \rangle$ be the dihedral group generated by $\rho_i$ and $\rho_j$. The subspace $E_{ij}$ of $\mathbb{R}^I$ spanned by $e_i$ and $e_j$ is $W_{ij}$-stable. To simplify notation, set $m = m_{ij}$. The restriction of the bilinear form $B_M$ to $E_{ij}$ is given by the matrix

$$\begin{pmatrix} 1 & -\cos\dfrac{\pi}{m} \\ -\cos\dfrac{\pi}{m} & 1 \end{pmatrix}$$

which is positive definite for $m \neq \infty$. For $m = \infty$,we get

$$\begin{pmatrix} 1 & -1 \\ -1 & 1 \end{pmatrix}$$

which is positive semidefinite with kernel the line spanned by the vector $e_i + e_j$. We consider the two cases separately.

Suppose $m = \infty$. Let $u = e_i + e_j$. Then

$$(\rho_i \rho_j)(e_i) = \rho_i(e_i + 2e_j) = 3e_i + 2e_j = 2u + e_i.$$

Hence, for all $n \in \mathbb{Z}$,

$$(\rho_i \rho_j)^n(e_i) = 2nu + e_i.$$

which shows $\rho_i \rho_j$ has infinite order.

Suppose $m \neq \infty$. Since $B_M$ is positive definite on $E_{ij}$, we can identify $E_{ij}$ with $\mathbb{R}^2$. Let $L_i$ (resp. $L_j$) be the line in $E_{ij}$ orthogonal to $e_i$ (resp. $e_j$). The restriction of $\rho_i$ (resp. $\rho_j$) to $E_{ij}$ is orthogonal reflection across $L_i$ (resp. $L_j$), the lines $L_i$ and $L_j$ make an angle of $\pi/m$ and the action of $W_{ij}$ on $E_{ij}$ is equivalent to the standard action, defined in Example 3.1.2, of the dihedral group $\mathbf{D}_m$ on $\mathbb{R}^2$. In particular, the restriction of $(\rho_i \rho_j)$ to $E_{ij}$ is a rotation through $2\pi/m$. Since $B_M$ is positive definite on $E_{ij}$, $\mathbb{R}^I$ decomposes as a $W_{ij}$ representation as a direct sum of $E_{ij}$ and its orthogonal complement (with respect to $B_M$). Moreover, $W_{ij}$ fixes this orthogonal complement. So, the order of $\rho_i \rho_j$ is $m$. □

**COROLLARY 6.12.4.** *The map $S \to GL(\mathbb{R}^I)$ defined by $s_i \to \rho_i$ extends to a homomorphism $\rho : W \to GL(\mathbb{R}^I)$.*

*Proof.* We must check that the relations in $\mathcal{R}$ are sent to the identity element of $GL(\mathbb{R}^I)$. The relation $s_i^2$ goes to $\rho_i^2$ which equals 1, since $\rho_i$ is a reflection. The relation $(s_i s_j)^{m_{ij}}$ is sent to $(\rho_i \rho_j)^{m_{ij}}$, which is also equal to 1, by Lemma 6.12.3. □

**DEFINITION 6.12.5.** The homomorphism $\rho : W \to GL(\mathbb{R}^I)$ of Corollary 6.12.4 is called the *canonical representation*.

The next result shows that there is a Coxeter system for any given Coxeter matrix $M$.

**COROLLARY 6.12.6.** *Suppose $M$ is a Coxeter matrix and $W$ is the group with generating set $S$ defined by the presentation associated to $M$.*

(i) *For each $i \in I$, $s_i$ has order* 2.

(ii) *The $s_i$, $i \in I$, are distinct.*

(iii) *$s_i s_j$ has order $m_{ij}$.*

*Hence, $(W, S)$ is a Coxeter system.*

*Proof.* The representation $\rho$ takes the $s_i$ to the distinct reflections $\rho_i$ on $\mathbb{R}^I$. This proves (i) and (ii). By Lemma 6.12.3, the order of $\rho_i\rho_j$ is $m_{ij}$. Hence, this is also the order of $s_i s_j$, proving (iii). It follows that $(W, S)$ is a pre-Coxeter system and that $M$ is its associated matrix (defined in 3.3). So $(W, S)$ is a Coxeter system. □

**PROPOSITION 6.12.7.** ([29, Prop. 7, p. 102].) *Suppose the Coxeter system $(W, S)$ is irreducible (Definition 3.5.3). Consider the canonical representation on $E = \mathbb{R}^I$. Let $E^0$ be the kernel of the bilinear form $B_M$ (i.e., $E^0 = \{x \in E \mid B_M(x, y) = 0 \quad \forall y \in E\}$). Then $W$ acts trivially on $E^0$ and every proper $W$-stable subspace of $E$ is contained in $E^0$.*

*Proof.* It follows from equation (6.33) that each generator $\rho_i$ acts trivially on $E^0$; hence, $W$ fixes $E^0$. Let $E'$ be a $W$-stable subspace of $E$. Suppose that some basis vector $e_i$ lies in $E'$. If $j \in I$ is another index such that $m_{ij} \neq 2$, then from (6.33) we see that the coefficient of $e_j$ in $\rho_j(e_i)$ is nonzero; hence, $e_j$ is also in $E'$. Since $(W, S)$ is irreducible, it follows that each basis vector must lie in $E'$. The conclusion is that if any basis vector lies in $E'$, then they all do. So if $E'$ is a proper subspace it cannot contain any $e_i$. Now consider $E$ as a $\langle\rho_i\rangle$-representation, where $\langle\rho_i\rangle$ denotes the cyclic group of order 2 generated by $\rho_i$. It decomposes as the direct sum of the $\pm 1$ eigenspaces of $\rho_i$, i.e., as the direct sum of line generated by $e_i$ and the hyperplane $H_i$ which is orthogonal to $e_i$ (with respect to $B_M$). Since $e_i \notin E'$, $E' \subset H_i$ for all $i \in I$. That is to say, $E' \subset \cap H_i = E^0$. □

**COROLLARY 6.12.8.** *Suppose $(W, S)$ is irreducible.*

(i) *If $B_M$ is nondegenerate, then the canonical representation on $E$ is irreducible (i.e., there is no nontrivial proper $W$-stable subspace of $E$).*

(ii) *If $B_M$ is degenerate, then the canonical representation is not semisimple (i.e., $E$ has a nontrivial $W$-stable subspace which is not a direct summand).*

*Proof.* Statement (i) follows immediately from Proposition 6.12.7, since if $B_M$ is nondegenerate, then $E^0 = 0$. As for (ii), if $B_M$ is degenerate, then $E^0 \neq 0$ and by Proposition 6.12.7, $E$ does not contain any $W$-stable complementary subspace. □

### A Finiteness Criterion

The following classical result gives a necessary and sufficient condition for a Coxeter group to be finite. It is important for the classification results in Appendix C.

**THEOREM 6.12.9.** *Suppose $M = (m_{ij})$ is a Coxeter matrix on a set $I$, that $(c_{ij})$ is its associated cosine matrix defined by (6.21) and that $(W, S)$ is its associated Coxeter system. Then the following statements are equivalent:*

(i) *$W$ is a reflection group on $\mathbb{S}^n$, $n = \mathrm{Card}(I) - 1$, so that the elements of $S$ are represented as the reflections across the codimension-one faces of a spherical simplex $\sigma$.*

(ii) *$(c_{ij})$ is positive definite.*

(iii) *$W$ is finite.*

*Proof.* We first establish the equivalence of (i) and (ii). By Theorem 6.4.3, (i) is equivalent to the existence of a spherical simplex with dihedral angles of the form $\pi/m_{ij}$ and by Proposition 6.8.2, this is equivalent to the positive definiteness of $(c_{ij})$. Hence, (i) $\Leftrightarrow$ (ii). Obviously, (i) $\Rightarrow$ (iii). It remains to show that (iii) $\Rightarrow$ (ii). So, suppose $W$ is finite. Without loss of generality we can assume that $(W, S)$ is irreducible. (For when $(W, S)$ is reducible, its cosine matrix is positive definite if and only if the cosine matrix of each factor is positive definite.) Consider the canonical representation of $W$ on $E$. Since $W$ is finite, Corollary 6.6.2 implies that $E$ is semisimple. So by Corollary 6.12.8, the bilinear form $B_M$ is nondegenerate and the representation $E$ is irreducible. By Lemma 6.6.1, $E$ admits a $W$-invariant inner product, let us call it $B'$. By Lemma 6.12.2, $B_M$ is proportional to $B'$. So, $B_M$ is either positive definite or negative definite. Since $B_M(e_i, e_i) = c_{ii} = 1$, the possibility of being negative definite is excluded. □

### The Dual of the Canonical Representation (or the Geometric Representation)

In Appendix D we will establish the following result of Tits.

**THEOREM 6.12.10.** ([29, Corollary 2, p. 97].) $\rho^* : W \to GL((\mathbb{R}^I)^*)$, *the dual of the canonical representation, is faithful.*

**COROLLARY 6.12.11.** $\rho$ *is faithful.*

**COROLLARY 6.12.12.** *Every finitely generated Coxeter group is virtually torsion-free (i.e., contains a torsion-free subgroup of finite index).*

*Proof.* Selberg's Lemma (see [254]) asserts that every finitely generated subgroup of $GL(n, \mathbb{R})$ is virtually torsion-free. □

## NOTES

The treatment in 6.4 is based on [279]. Section 6.10 is largely based on an unpublished manuscript [278] of Thurston continuing [279]. Sections 6.3 and 6.12 come from [29]. Sections 6.9 and 6.11 are based on their treatments in [291].

**6.9.** In [29, Ex. 12c), p. 141] a hyperbolic simplicial Coxeter group is called a Coxeter group of "compact hyperbolic type." If its fundamental simplex is not required to be compact but is still required to have finite volume (possibly with ideal vertices), then it is said to be of "hyperbolic type." (A list of the diagrams of the hyperbolic groups of with a fundamental simplex of finite volume can be found in [163, §6.9].) Similar terminology is still widely used, particularly by people working on the theory of buildings. This terminology is misleading. It gives the false impression that the fundamental chamber of a hyperbolic reflection group is always a simplex, completely ignoring the polygonal reflection groups of Example 6.5.3 as well as the 3-dimensional examples of 6.10. The situation is particularly egregious when a result such as Theorem 6.8.12 is presented in [180] as giving a "classification of hyberbolic reflection groups."

**6.10.** By the early 1970's the following general picture of closed 3-manifolds had emerged. Kneser had proved in 1928 that every closed 3-manifold $M^3$ had unique decomposition as a connected sum: $M^3 = M_1 \# \cdots \# M_k$ such that each of summands $M_i$ could not be further decomposed. Such a summand is called *prime*. Moreover, apart from a few exceptional cases (only $S^2 \times S^1$ in the orientable case), every prime $M^3$ was either aspherical or had finite fundamental group. Using Waldhausen's ideas, Jaco and Shalen and independently, Johannson, proved that every prime (or irreducible) 3-manifold can be canonically decomposed along incompressible tori into pieces $N_1, \ldots N_m$ such that each $N_j$ is a manifold with boundary, each component of the boundary is a torus and each $N_j$ is either Seifert fibered (i.e., essentailly an $S^1$-bundle) or *simple*, which means that it is atoroidal (i.e., any incompressible torus is parallel to the boundary). This is called the *JSJ decomposition* of the 3-manifold. Around 1976 Thurston conjectured that the interior of each simple piece in this decomposition could be given the structure of a hyperbolic manifold of finite volume. In particular, a closed 3-manifold is aspherical and atoroidal if and only if it admits a hyperbolic structure. The Geometrization Conjecture was a further generalization. First, Thurston showed that there are exactly eight 3-dimensional geometries in the following sense: a Lie group $G$ and a compact subgroup $K$ such that (a) $G/K$ is a simply connected 3-manifold, (b) $G$ admits a discrete subgroup $\Gamma$ such that $\Gamma\backslash G/K$ has finite volume, and (c) $G$ is maximal with respect to these properties. Three of these geometries yield the constant curvature spaces, $\mathbb{S}^3$, $\mathbb{E}^3$, and $\mathbb{H}^3$. The other five all have Siefert fibered structures. The full statement of the Geometrization Conjecture (6.10.1) is that the interior of each piece in the JSJ decomposition has a geometric structure. In particular, this means that if the fundamental group of $M^3$ is finite then it is covered by $S^3$. So, the Geometrization Conjecture implies the Poincaré Conjecture. (See 10.3.) The Geometrization Conjecture now should be called Perelman's Theorem.

The Mostow Rigidity Theorem [220] was a milestone of twentieth-century mathematics. Suppose $Y$ is a locally symmetric, closed Riemannian manifold with no local factor locally isometric to either (a) Euclidean space, (b) a compact symmetric space, or (c) the hyperbolic plane. Then $Y$ is determined up to isometry (and homotheties of the local factors) by its fundamental group. More precisely, the Rigidity Theorem states that if $Y'$ is another such locally symmetric space, then any isomorphism $\pi_1(Y) \to \pi_1(Y')$ is induced by an isometry $Y \to Y'$. The theorem is definitely false in the presence of local $\mathbb{E}^n$ or $\mathbb{H}^2$ factors. Indeed, the $n$-torus has many different flat metrics (a moduli space of them) and the same is true for hyperbolic metrics on any closed surface of genus $> 1$. Here is another way to state the same result. Suppose $G$ is a semisimple Lie group of noncompact type, $K$ a maximal compact subgoup and $\Gamma \subset G$ a torsion-free uniform lattice. (This means that $\Gamma$ is a cocompact discrete subgroup of $G$.) Suppose further that $(\Gamma, G, K)$ has no factor isometric to one of the form $(\Gamma', G', K')$, where $G' = O_+(2,1)$, $K = O(2)$ and $\Gamma'$ is a surface group. (However, factors where $\Gamma'$ is an irreducible lattice in a product of more than one copy of $O_+(2,1)$ are allowed.) The theorem says that if $(\Gamma_1, G_1, K_1)$ and $(\Gamma_2, G_2, K_2)$ are two such (where the $G_i$ are normalized to be centerless), then any isomorphism $\Gamma_1 \to \Gamma_2$ extends to an isomorphism $G_1 \to G_2$. For example, the Rigidity Theorem means that if two closed hyperbolic $n$-manifolds, $n \geqslant 3$, have isomorphic fundamental groups, then they are isometric. The Rigidity Theorem is valid for locally symmetric orbifolds (i.e., the lattice $\Gamma$ does not have to be torsion-free). It is also valid in the cofinite volume case as well (i.e., for non-uniform lattices). If the rank is $\geqslant 2$, this follows from Margulis' "Super Rigidity Theorem." In the rank one case (e.g. for hyperbolic manifolds) it was proved by Prasad, provided one assumes that the isomorphism $\Gamma_1 \to \Gamma_2$ "preserves the peripheral structure" (i.e., maps the fundamental group of the cusps to themselves). This last point is important to the Geometrization Conjecture. It means that the hyperbolic pieces in the JSJ decomposition have a unique hyperbolic structure. So, the Geometrization Conjecture means that geometric pieces in the decomposition are unique up to isometry.

**6.11.** We have avoided discussing hyperbolic reflection groups of finite covolume in 6.9, 6.10, and 6.11. In fact, there are some beautiful examples of these. One should have the following picture in mind. The fundamental polytope $P^n$ has finite volume and one or more ideal vertices lying on the sphere at infinity. Its combinatorial type is the same as that of a compact polytope. The special subgroup generated by the reflecting hyperplanes through an ideal vertex is a Euclidean reflection group. (This is because a "horosphere" at an ideal point is isometric to Euclidean space.) For example, in the three-dimensional case the isotropy subgroup at an ideal vertex can be a Euclidean triangle group (Example 6.5.2) or $\mathbf{D}_\infty \times \mathbf{D}_\infty$. This last case shows that the three-dimensional polytope need not be simple; the link of an ideal vertex can be a square. More generally, the link of an ideal vertex in $P^n$ can be a product simplices (Corollary 6.3.11 and Theorem 6.8.12 (ii)). Using Tables 6.1 and 6.2, it is easy to find all hyperbolic Coxeter groups with fundamental polytope a simplex with one or more ideal vertices (one allows some of the principal minors to be semidefinite rather than positive definite). The diagrams can be found in the exercises of [29, Ch.V, §4, Ex. 15–17]. There are 11 such three-dimensional examples and 3 more in dimension 9 (all coming from the Euclidean reflection group $\widetilde{\mathbf{E}}_8$. Interesting nonsimplicial examples include the regular right-angled three-dimensional octahedron with all its vertices ideal and the regular right-angled 24-cell with all its vertices ideal. In dimension 3, there is also a complete classification, once again due to Andreev [8]. Just as in Theorem 6.10.2 it says an assignment of dihedral angles is possible if and only if the obvious necessary conditions are satisfied (the previous conditions plus the condition that special subgroups which are Euclidean reflection groups can occur only at the ideal vertices.) Using the construction in Example 6.11.3, Allcock [4] has shown that there exist infinite families of such ideal polytopes in all dimensions through 19. On the other hand, Prokhorov [241] has proved there are no such examples with cofinite volume in dimensions >992.

**6.12.** Serre [255, p. 99] says that Selberg's Lemma (used in the proof of Corollary 6.12.12) follows from a classical argument due to Minkowski [217].

# Chapter Seven

# THE COMPLEX $\Sigma$

As we mentioned in 1.1, associated to any Coxeter system $(W, S)$ there is a cell complex $\Sigma$ equipped with a proper, cocompact $W$-action. It is the natural geometric object associated to $(W, S)$. In this chapter we define $\Sigma$ and describe some of its basic properties. In Section 8.2 of the next chapter we will prove that $\Sigma$ is contractible and in 12.3, that its natural piecewise Euclidean metric is CAT(0).

Several other important notions are explained in this chapter. In 7.1 we define the nerve $L$ of $(W, S)$: it is a certain simplicial complex with vertex set $S$. In 7.3 we define a *Coxeter polytope*: it is the convex hull of a generic orbit of a finite Coxeter group acting on $\mathbb{R}^n$. $\Sigma$ can be cellulated so that each cell is a Coxeter polytope and so that there is one $W$-orbit of $n$-cells for each $(n-1)$-simplex in $L$ and so that the link of each 0-cell is $L$. In 7.5 we determine the fixed point set of any finite special subgroup on $\Sigma$.

## 7.1. THE NERVE OF A COXETER SYSTEM

**Definition 7.1.1.** A subset $T$ of $S$ is *spherical* if $W_T$ is a finite subgroup of $W$. If this is the case, we say that the special subgroup $W_T$ is *spherical*.

By Theorem 4.1.6, for any subset $T$ of $S$, $(W_T, T)$ is a Coxeter system. By Theorem 6.12.9, if $T$ is spherical, then $W_T$ can be represented as a geometric reflection group on the unit sphere in $\mathbb{R}^T$. (This is the explanation for the "spherical" terminology.) Denote by $\mathcal{S}(W, S)$ (or simply $\mathcal{S}$) the set of all spherical subsets of $S$. $\mathcal{S}$ is partialy ordered by inclusion. Let $\mathcal{S}^{(k)}$ denote the set of spherical subsets of cardinality $k$.

The poset $\mathcal{S}_{>\emptyset}$ of all nonempty spherical subsets is an abstract simplicial complex. (This just means that if $T \in \mathcal{S}_{>\emptyset}$ and $\emptyset \neq T' \subset T$, then $T' \in \mathcal{S}_{>\emptyset}$; cf. Definition A.2.6.) We will also denote this simplicial complex $L(W, S)$ (or simply $L$) and call it the *nerve* of $(W, S)$. In other words, the vertex set of $L$ is $S$ and a nonempty set $T$ of vertices spans a simplex $\sigma_T$ if and only if $T$ is spherical. Thus, $\mathcal{S}^{(k)}$ is the set of $(k-1)$-simplices in $L$. (See Appendix A.2 for a further discussion of posets and simplicial complexes.)

**Example 7.1.2.** If $W$ is finite, then $\mathcal{S}$ is the power set of $S$ and $L$ is the full simplex on $S$. (Compare Example A.2.4.)

**Example 7.1.3.** If $W$ is the infinite dihedral group $\mathbf{D}_\infty$, then $L(W, S)$ consists of two points (i.e., $L = S^0$).

**Example 7.1.4.** (*The nerve of a geometric reflection group*). As in Chapter 6, suppose $W$ is a geometric reflection group (cf. Definition 6.4.4) generated by the set $S$ of reflections across the codimension-one faces of a simple convex polytope $P^n$ in $\mathbb{E}^n$ or $\mathbb{H}^n$. Then $(W, S)$ is a Coxeter system and $S$ is naturally bijective with the set of codimension-one faces of $P$. Since $P$ is simple, the boundary of its dual polytope $\partial P^*$ is a simplicial complex (whose vertex set can be identified with $S$). It turns out that *the nerve $L$ of the Coxeter system can be identified with $\partial P^*$*. To understand this, consider the poset $\mathcal{F}(P)$ of nonempty faces of $P$ as in Example A.2.3. For each $F \in \mathcal{F}(P)$, let $S(F) \subset S$ be the set of reflections across codimension one faces which contain $F$. Since $W$ acts properly, each isotropy subgroup is finite, i.e., each $S(F)$ is a spherical subset of $S$. Thus, $\mathcal{F}(P)^{op} \subset \mathcal{S}$ (where $\mathcal{F}(P)^{op}$ denotes the dual poset of $\mathcal{F}(P)$). (A discussion of dual polytopes is given in Appendix A.2.) So, $\mathcal{F}(\partial P^*) = \mathcal{F}(\partial P)^{op}$ is a subset of $\mathcal{S}_{>\emptyset}$, i.e., $\partial P^*$ is a subcomplex $L$. (See Lemma 7.1.9 below.) The opposite inclusion $L \subset \partial P^*$ follows from [290]. We shall give a different argument in the next chapter (Corollary 8.2.10).

**Example 7.1.5.** Suppose $(W, S)$ decomposes as

$$(W, S) = (W_1 \times W_2, S_1 \cup S_2)$$

where the elements of $S_1$ commute with those of $S_2$. A subset $T = T_1 \cup T_2$, with $T_i \subset S_i$, is spherical if and only if $T_1$ and $T_2$ are both spherical. Hence,

$$\mathcal{S}(W, S) \cong \mathcal{S}(W_1, S_1) \times \mathcal{S}(W_2, S_2) .$$

Similarly, the simplicial complex $L(W, S)$ decomposes as the join

$$L(W, S) = L(W_1, S_1) * L(W_2, S_2) .$$

(See Appendix A.4 for the definition of the *join* of two simplicial complexes.)

**Example 7.1.6.** (*The Coxeter system associated to a labeled simplicial graph*). Suppose $\Upsilon$ is a simplicial graph with vertex set $S$ and that the edges of $\Upsilon$ are labeled by integers $\geqslant 2$. This gives the data for a Coxeter matrix

$(m_{st})$ over $S$:

$$m_{st} = \begin{cases} 1 & \text{if } s = t, \\ \text{the label on } \{s,t\} & \text{if } \{s,t\} \text{ is an edge,} \\ \infty & \text{otherwise.} \end{cases}$$

As in the beginning of Section 3.3, associated to the matrix $(m_{st})$ there is a Coxeter system $(W, S)$. Note that the 1-skeleton of $L(W, S)$ is $\Upsilon$.

**Example 7.1.7.** (*The right-angled Coxeter system associated to a graph.*) Suppose, as above, that $\Upsilon$ is a simplicial graph. Put the label 2 on each edge by default. Then there is an associated right-angled Coxeter system $(W, S)$ such that the 1-skeleton of $L(W, S)$ is $\Upsilon$. As in Example A.3.8, $L(W, S)$ is the flag complex determined by $\Upsilon$. (The official definition of a "flag complex" is given as Definition A.3.5. We repeat it here. A simplicial complex $L$ is a *flag complex* if a finite, nonempty set of vertices $T$ spans a simplex of $L$ if and only if any two elements of $T$ are connected by an edge.)

**Lemma 7.1.8.** *If $(W, S)$ is right angled, then $L(W, S)$ is a flag complex.*

*Proof.* Suppose $T$ is a subset of $S$ such that any two elements of $T$ are connected by an edge in $L$. Since $(W, S)$ is right angled, this implies that $W_T \cong (\mathbf{C}_2)^{\mathrm{Card}(T)}$. Since this group is finite, $T$ is spherical, i.e., $\sigma_T \in L$. So $L$ is a flag complex. □

## The Poset of Spherical Cosets

A *spherical coset* is a coset of a spherical special subgroup in $W$. The set of all spherical cosets is denoted $W\mathcal{S}$, i.e.,

$$W\mathcal{S} = \bigcup_{T \in \mathcal{S}} W/W_T. \tag{7.1}$$

It is partially ordered by inclusion.

By Theorem 4.1.6 (iii), $wW_T = w'W_{T'}$ if and only if $T = T'$ and $w^{-1}w' \in W_{T'}$. It follows that the union in (7.1) is a disjoint union. It also follows that there is well-defined projection map $W\mathcal{S} \to \mathcal{S}$ given by $wW_T \to T$. A section $\mathcal{S} \hookrightarrow W\mathcal{S}$ of the projection is defined by $T \to W_T$. $W$ acts naturally on the poset $W\mathcal{S}$ and the quotient poset is $\mathcal{S}$.

## The Nerve of a Mirrored Space

After recalling the definitions of the "nerve of a mirrored space" in Definition 5.1.1 and of "$W$-finite" in Definition 5.1.6, the following lemma is immediate.

**LEMMA 7.1.9.** *A mirror structure* $(X_s)_{s \in S}$ *on a space* $X$ *is* $W$*-finite if and only if* $N(X) \subset L(W, S)$.

## 7.2. GEOMETRIC REALIZATIONS

As we will explain in more detail in Appendix A.2, associated to any poset $\mathcal{P}$ there is an abstract simplicial complex Flag($\mathcal{P}$) consisting of all finite chains in $\mathcal{P}$. (A *chain* is a nonempty totally ordered subset of $\mathcal{P}$.) The topological space corresponding to Flag($\mathcal{P}$) is denoted $|\mathcal{P}|$ and called the *geometric realization* of $\mathcal{P}$. Since a simplex in Flag($\mathcal{P}$) is a chain in $\mathcal{P}$, the vertex set of any simplex is totally ordered. In particular, each simplex in Flag($\mathcal{P}$) has a minimum vertex as well as a maximum one.

*Remark 7.2.1.* If $\mathcal{P}$ is the poset of cells in a convex cell complex $\Lambda$, then the simplicial complex Flag($\mathcal{P}$) is the barycentric subdivision of $\Lambda$. (See Example A.4.6.)

The next lemma is one of the underpinnings for most of the examples constructed in Chapters 8, 10 and 11. It implies that the condition of being the nerve of Coxeter system imposes no conditions on the topology of a simplicial complex: it can be any polyhedron.

**LEMMA 7.2.2.** ([71, Lemma 11.3, p. 313].) *Suppose* $\Lambda$ *is a convex cell complex. Then there is a right-angled Coxeter system* $(W, S)$ *with nerve the barycentric subdivision of* $\Lambda$. *(Barycentric subdivisions are defined in Appendix A.3.)*

*Proof.* Let $b\Lambda$ be the barycentric subdivision of $\Lambda$ and let $S = \text{Vert}(b\Lambda)$. As in Example 7.1.7, the 1-skeleton of $b\Lambda$ gives the data for a presentation of a right-angled Coxeter system $(W, S)$. Flag complexes are determined by their 1-skeletons (Appendix A.3). The proof of Lemma 7.1.8 shows that $L(W, S)$ is flag complex determined by the 1-skeleton of $b\Lambda$. Since $b\Lambda$ is a flag complex (associated to the poset of cells of $\Lambda$), it coincides with $L(W, S)$. □

Here we are dealing with two posets, $\mathcal{S}$ and $W\mathcal{S}$. The geometric realization of $\mathcal{S}$ is denoted $K(W, S)$ (or simply $K$). The geometric realization of $W\mathcal{S}$ is denoted $\Sigma(W, S)$ (or simply $\Sigma$). The simplicial complex $\Sigma$ is one of the fundamental objects of study in this book.

The projection $W\mathcal{S} \to \mathcal{S}$ induces a simplicial projection $\pi : \Sigma \to K$. Similarly, the inclusion $\mathcal{S} \hookrightarrow W\mathcal{S}$ induces an inclusion $\iota : K \hookrightarrow \Sigma$. We identify $K$ with its image under $\iota$. The $W$-action on $W\mathcal{S}$ induces a $W$-action on $\Sigma$. $K$, as well as any one of its translates by an element of $W$, is called a *chamber* of $\Sigma$.

**LEMMA 7.2.3.** *Any simplex of* $\Sigma$ *is a translate of a simplex of* $K$.

*Proof.* By Theorem 4.1.6 (iii), $wW_T \subset w'W_{T'}$ if and only if $T \subset T'$ and $wW_{T'} = w'W_{T'}$. It follows that if $w_0W_{T_0} \subset \cdots \subset w_kW_{T_k}$, then $T_0 \subset \cdots \subset T_k$ and that $w_iW_{T_i} = w_0W_{T_i}$ for all $0 \leqslant i \leqslant k$. In other words, the corresponding simplex in $\Sigma$ is the translate by $w_0$ of the simplex in $K$ corresponding to $T_0 \subset \cdots \subset T_k$. □

$K$ is the cone on the barycentric subdivision of $L$. (By Remark 7.2.1, the simplicial complex $\mathrm{Flag}(\mathcal{S}_{>\emptyset})$ is the barycentric subdivision of $L$; the empty set provides the cone point.) For each $s \in S$ let

$$K_s := |\mathcal{S}_{\geqslant \{s\}}|,$$

i.e., $K_s$ is the union of the (closed) simplices in $K$ with minimum vertex $\{s\}$. (In other words, $K_s$ is the closed star of the vertex corresponding to $s$ in the barycentric subdivision of $L$.) The family $(K_s)_{s\in S}$ is a mirror structure on $K$. ("Mirror structures" are defined at the beginning of 5.1.) The isotropy subgroup at an interior point of a simplex $(T_0 \subset \cdots \subset T_k)$ is $W_{T_0}$. Thus, $\iota(K_s)$ is contained in the fixed set of $s$ on $\Sigma$. By Lemma 5.2.5, $\iota : K \to \Sigma$ induces a $W$-equivariant map $\tilde{\iota} : \mathcal{U}(W,K) \to \Sigma$. By Lemma 7.2.3, $\tilde{\iota}$ is a bijection. It is clear from the definitions that $\mathcal{U}(W,K)$ and $\Sigma$ have the same topology. So, we have proved the following theorem.

**THEOREM 7.2.4.** *$\tilde{\iota} : \mathcal{U}(W,K) \to \Sigma$ is a $W$-equivariant homeomorphism.*

For use in Section 8.1 we record the following.

**LEMMA 7.2.5**

*(i) $K$ is contractible.*

*(ii) For each spherical subset $T$, the coface $K_T$ ($= \bigcap_{s\in T} K_s$) is contractible.*

*(iii) For each nonempty spherical subset $T$, the union of mirrors $K^T$ is contractible. (Recall $K^T = \bigcup_{s\in T} K_s$.)*

*Proof.* $K = |\mathcal{S}|$ is contractible because it is a cone. (The simplest explanation for the fact that it is a cone is that the poset $\mathcal{S}$ has a minimum element, namely $\emptyset$.) Similarly, $K_T = |\mathcal{S}_{\geqslant T}|$ is a cone since $T$ is the minimum element of $\mathcal{S}_{\geqslant T}$. This takes care of assertions (i) and (ii).

There are two ways to prove assertion (iii). The first is to say that, by assertion (ii), $K^T$ is a union of a finite number of contractible pieces, namely, the $K_s$, $s \in T$, and that the intersection of any subcollection of the $K_s$ is contractible. It follows from van Kampen's Theorem that $K^T$ is simply connected and by the Acyclic Covering Lemma (Lemma E.3.3 of Appendix E.3) that $K^T$ is acyclic. So, by the Hurewicz Theorem, $\pi_i(K^T)$ is trivial for all $i$ (that is to say, $K^T$ is weakly contractible). Since $K^T$ is a CW complex, a well-known theorem of J.H.C. Whitehead implies that it is contractible.

The second method for proving (iii) is more direct. $|\mathcal{S}_{>\emptyset}|$ is the barycentric subdivision $bL$ of $L$. For any $T \in \mathcal{S}_{>\emptyset}$, $\sigma_T$ is a simplex of $L$, so its barycentric subdivision $b\sigma_T$ is a subcomplex of $bL$. $K^T$ is the union of the closed stars in $bL$ of the vertices $s$, $s \in T$. In other words, $K^T$ is the first derived neighborhood of $b\sigma_T$ in $bL$. We claim that there is a simplicial deformation retraction $r : K^T \to b\sigma_T$. Since $b\sigma_T$ is contractible, this gives the result. The retraction is defined as follows. A vertex of $K^T$ lies in $K_s$ for some $s \in T$. A vertex of $K_s$ corresponds to some spherical subset $T'$, with $s \in T'$. Hence, if $T'$ corresponds to a vertex of $K^T$, we can consider the spherical subset $T' \cap T$. This intersection is nonempty since $s \in T' \cap T$ for some $s \in T'$. Hence, $T' \cap T$ corresponds to a vertex of $b\sigma_T$ and the map $T' \to T' \cap T$ defines the simplicial retraction $r : K^T \to b\sigma_T$. The map $r$ collapses a simplex of the form $\{T'_0, \dots, T'_k\}$ to its face $\{T'_0 \cap T, \dots, T'_k \cap T\}$. In the geometric realization of this simplex we have the line segment from any point $x$ to its image under $r$. Therefore, the straight line homotopy $h_t(x) = (1-t)x + tr(x)$ is a well-defined homotopy from $id$ to $r$. □

## 7.3. A CELL STRUCTURE ON Σ

For any poset $\mathcal{P}$ there are two decompositions of $|\mathcal{P}|$ into closed subspaces. Both decompositions are indexed by $\mathcal{P}$. For the first decomposition take the geometric realizations of the subposets $\mathcal{P}_{\leqslant p}$, $p \in \mathcal{P}$. $|\mathcal{P}_{\leqslant p}|$ is a *face* of $|\mathcal{P}|$. For the second decomposition take the geometric realizations of the subposets $\mathcal{P}_{\geqslant p}$, $p \in \mathcal{P}$. $|\mathcal{P}_{\geqslant p}|$ is a *coface* of $|\mathcal{P}|$. (All this is explained further in Appendix A.5.)

For example, if $\mathcal{P}$ is the poset of cells in a regular cell complex $\Lambda$, then the faces of $|\mathcal{P}|$ are the cells of $\Lambda$. A cell $p$ is the union of all simplices in the barycentric subdivision with maximum vertex $p$. If $\mathcal{P}$ is the poset of cells in a PL cellulation of a manifold, then the cofaces are the dual cells.

The principal result of this section is that in the case of the poset $W\mathcal{S}$, the faces are cells. We begin by considering the case where the Coxeter group is finite.

### Coxeter Polytopes

In this subsection we suppose $W$ is finite. As in 6.12, consider the canonical representation of $W$ on $\mathbb{R}^S$ ($\mathbb{R}^S \cong \mathbb{R}^n$, $n = \text{Card}(S)$). Choose a point $x$ in the interior of the fundamental chamber; such an $x$ is called a *generic point* ($x$ is determined by specifying its distance to each of the bounding hyperplanes, i.e., by specifying an element of $(0, \infty)^S$.)

**DEFINITION 7.3.1.** The *Coxeter polytope* (or the *Coxeter cell*) associated to $(W, S)$ is the convex polytope $C_W$ (or simply $C$) defined as the convex hull of $Wx$ (a generic $W$-orbit).

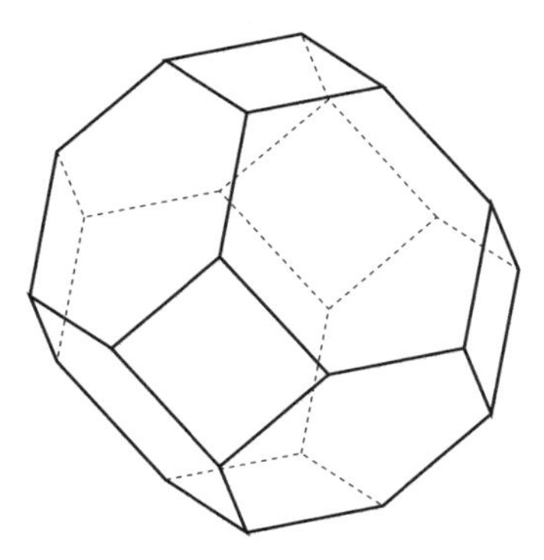

Figure 7.1. Permutohedron.

**Examples 7.3.2**

(i) If $W = \mathbf{C}_2$, the cyclic group of order 2, then $C$ is the interval $[-x, x]$.

(ii) If $W = \mathbf{D}_m$ is the dihedral group of order $2m$, then $C$ is a $2m$-gon. (It is a regular $2m$-gon if $x$ is equidistant from the two rays which bound the fundamental sector containing $x$.)

(iii) If $(W, S)$ is reducible and decomposes as $W = W_1 \times W_2$, then $C_W$ decomposes as $C_W = C_{W_1} \times C_{W_2}$. In particular, if $W = (\mathbf{C}_2)^n$, then $C$ is a product of intervals (and if $x$ is equidistant from the bounding hyperplanes, then $C$ is a regular $n$-cube).

(iv) If $W$ is the symmetric group on $n + 1$ letters (the group associated to the Coxeter diagram $\mathbf{A}_n$), then $C$ is the $n$-dimensional *permutahedron*. The picture for $n = 3$ is given in Figure 7.1 above.

**LEMMA 7.3.3.** ([56, Lemma 2.1.3, p. 117].) *Suppose $W$ is finite and that $C$ is its associated Coxeter polytope. Let $\mathcal{F}(C)$ be its face poset. Then the correspondence $w \to wx$ induces an isomorphism of posets, $W\mathcal{S} \cong \mathcal{F}(C)$. (In other words, a subset of $W$ corresponds to the vertex set of a face of $C$ if and only if it is a coset of a special subgroup of $W$.)*

*Proof.* As in Definition A.1.8, let $K \subset \mathbb{R}^n$ be a fundamental simplicial cone for $W$ on $\mathbb{R}^n$ and let $x \in \text{int}(K)$. For each $s \in S$, let $e_s$ be the vector on the extremal ray of $K$ which is opposite to the face fixed by $s$. Let $\varphi$ be the linear form on $\mathbb{R}^n$ defined by $v \to \langle v, e_s \rangle$. Put $c = \langle x, e_s \rangle$. We claim that $c$ is the maximum value of $\varphi$ on $K$. By Complements 6.6.10, $K$ is a Dirichlet domain for $W$ on $\mathbb{R}^n$, i.e., any point of $K$ is at least as close to $x$ as it is to any other point in the orbit of $x$. In particular, $|wx - e_s|^2 \geqslant |x - e_s|^2$ for all $w \in W$. This implies $\langle wx, e_s \rangle \leqslant \langle x, e_s \rangle$ for all $w \in W$; hence, the maximum value of $\varphi$ on $C$ is attained at $x$. Let $T = S - \{s\}$. The affine hyperlane $\varphi(v) = c$ contains the orbit of $x$ under $W_T$ and is spanned by this orbit. It follows that $\varphi(v) = c$ is a supporting hyperplane of $C$ and that the convex hull of $W_T x$ is a codimension one face of $C$. Letting $s$ vary over $S$ we obtain all the supporting hyperplanes of $C$ containing the vertex $x$. Replacing $x$ by $wx$ and $e_s$ by $we_s$, we get a description of all the supporting hyperplanes of $C$. The lemma follows. □

Lemma 7.3.3 means that when $W$ is finite we can identify the simplicial complex $\Sigma(W, S)$ with the barycentric subdivision of the associated Coxeter polytope, that is, $\Sigma(W, S)$ is topologically a cell. (It follows from Lemma 7.3.3 that the combinatorial type of $C$ does not depend on the choice of the generic point $x$.)

### The General Case

We return to the case where $W$ can be infinite. Since the poset $(W\mathcal{S})_{\leqslant wW_T}$ is isomorphic to $W_T(\mathcal{S}_{\leqslant T})$, the face corresponding to $\mathcal{S}_{\leqslant wW_T}$ is isomorphic to the barycentric subdivision of a Coxeter cell of type $W_T$. So we can put a cell structure on $\Sigma$, coarser than its simplicial structure, by identifying each such barycentric subdivision with the corresponding Coxeter cell.

Let us review what this means. The 0-cells (or vertices) of $\Sigma$ correspond to the cosets of $W_\emptyset$, i.e., they correspond to the elements of $W$. A subset of $W$ corresponds to the vertex set of a cell of $\Sigma$ if and only if it is a spherical coset of the form $wW_T$. The cell is a Coxeter polytope of type $W_T$; its dimension is Card($T$). With this cell structure, the 0-skeleton of $\Sigma$ is $W$, its 1-skeleton is the Cayley graph of $(W, S)$ discussed in Section 2.1, and its 2-skeleton is Cay$(W, \langle S \mid \mathcal{R} \rangle)$ (the Cayley 2-complex of $(W, S)$ discussed in Section 2.2). We summarize the preceding discussion in the following proposition.

**PROPOSITION 7.3.4.** *There is a natural cell structure on $\Sigma$ so that*

- *its vertex set is $W$, its 1-skeleton is the Cayley graph,* Cay$(W, S)$, *and its 2-skeleton is a Cayley 2-complex,*
- *each cell is a Coxeter polytope,*
- *the link of each vertex is isomorphic to $L(W, S)$,*
- *a subset of $W$ is the vertex set of a cell if and only if it is a spherical coset,*
- *the poset of cells in $\Sigma$ is $W\mathcal{S}$.*

*Remark.* There are many Coxeter systems $(W, S)$. Almost none of them arise as geometric reflection groups on $\mathbb{E}^n$ or $\mathbb{H}^n$. To convince oneself that there are are many Coxeter systems, consider Lemma 7.2.2: the simplicial complex $L(W, S)$ can have arbitrary topological type. To see that there are very few geometric reflection groups, consider the classification of Euclidean reflection groups (Theorem C.1.3 of Appendix C) and Vinberg's Theorem (Theorem 6.11.8) as well as Corollary 6.11.7 of Section 6.11. On the other hand, the cell complex $\Sigma$ is always defined. For many reasons, it should be considered a satisfactory replacement for the constant curvature space $\mathbb{X}^n = \mathbb{E}^n$ or $\mathbb{H}^n$. For example, $W$ acts on it properly and cocompactly as a reflection group. In Section 8.2 (Theorem 8.2.13), we will prove that $\Sigma$ is contractible. Of course, the fundamental chamber $K$ need not be a convex polytope (or even a topological disk) and $\Sigma$ need not be a manifold. A feature of $\mathbb{X}^n$ which is preserved in $\Sigma$ is the cellullation of $\mathbb{X}^n$ by dual cells to the cofaces of chambers. In $\mathbb{X}^n$ such dual cells are Coxeter polytopes and they give the cell structure on $\Sigma$. (See Example 7.4.3 below.) As we shall see in Chapter 12, there is a natural piecewise Euclidean metric on $\Sigma$ and it is nonpositively curved in the sense of Aleksandrov [1] and Gromov [147].

**LEMMA 7.3.5.** *$\Sigma$ is simply connected.*

*Proof.* This follows from two standard facts:

- The fundamental group of any cell complex is isomorphic to the fundamental group of its 2-skeleton (since any map of a 1-sphere or 2-disk to the cell complex can be homotoped into its 2-skeleton).
- The Cayley 2-complex of a presentation, $\mathrm{Cay}(G, \langle S \mid \mathcal{R}\rangle)$ of any presentation is simply connected (by Proposition 2.2.3).

Since $\mathrm{Cay}(W, \langle S \mid \mathcal{R}\rangle)$ is the 2-skeleton of $\Sigma$, $\pi_1(\Sigma) = 1$. □

**Examples 7.3.6.** If $(W, S)$ is right angled, then each cell of $\Sigma$ is a cube and $\Sigma$ can be identified with the cubical complex $\widetilde{P}_L$ of Section 1.2. By Lemma 7.2.2, $L$ can have the topological type of an arbitrary polyhedron. Hence, for any cell complex $\Lambda$, there is a right-angled $(W, S)$ such that the link $L$ of each vertex of $\Sigma(W, S)$ is isomorphic to the barycentric subdivision of $\Lambda$.

The following proposition is a corollary of assertion in Proposition 7.3.4 that the link of every vertex of $\Sigma$ is isomorphic to $L$.

**PROPOSITION 7.3.7.** *Suppose $L$ is a triangulation of $S^{n-1}$, then $\Sigma$ is a topological $n$-manifold.*

*Remark.* In Corollary 7.5.3 as well as in Theorem 10.6.1 of Section 10.6, we will see that if, in addition, $L$ is a PL manifold, then

(i) $\Sigma$ is a PL manifold and

(ii) for each $T \in \mathcal{S}$, the fixed set of $W_T$ on $\Sigma$, denoted by $\Sigma^{W_T}$, is a PL submanifold of codimension $\mathrm{Card}(T)$.

Next consider the cofaces of $K$. Since $\mathcal{S} = \mathcal{S}_{\geqslant \emptyset}$, $K = |\mathcal{S}|$ is itself a coface (the maximum coface). The other cofaces are intersections of mirrors:

$$|\mathcal{S}_{\geqslant T}| = K_T = \bigcap_{s \in S} K_s.$$

(The use "face" and "coface" here, agrees with the terminology in Section 5.1.) The cofaces of $\Sigma$ are translates of the cofaces of $K$. In particular, maximal cofaces of $\Sigma$ are chambers.

## 7.4. EXAMPLES

**Example 7.4.1.** ($\dim \Sigma = 1$.) Suppose Card($S$) $= k$ and that $m_{st} = \infty$ for all $s \neq t$. Then the nerve $L$ is the disjoint union of $k$ points, the chamber $K$ is the the cone on $L$, each Coxeter cell is an interval, $W$ is the free product of $k$ copies of $\mathbf{C}_2$ and $\Sigma$ is a regular $k$-valent tree. (Its edges are the Coxeter 1-cells.)

**Example 7.4.2.** ($\dim \Sigma = 2$.) Suppose $\Upsilon$ is a simplicial graph with vertex set $S$ and as in Example 7.1.6, suppose each edge is labeled by an integer $\geqslant 2$. Let $(W, S)$ be the associated Coxeter system and let $L$ be its nerve. When is $L = \Upsilon$? A necessary and sufficient condition is that there are no spherical subsets of cardinality 3. It follows from Example 6.5.1, equation (6.10), that this is equivalent to the following condition on the 3-circuits in $\Upsilon$: if $m_1, m_2$, and $m_3$ are the labels on the edges of any 3-circuit in $\Upsilon$, then

$$\frac{1}{m_1} + \frac{1}{m_2} + \frac{1}{m_3} \leqslant 1.$$

(This condition holds vacuously if $\Upsilon$ does not contain any 3-circuits.) If the condition holds, then $\dim \Sigma = 2$ and the link of each vertex is $\Upsilon$. There is one orbit of Coxeter 2-cells for each edge of $\Upsilon$ and such a cell is a $2m_e$-gon where $m_e$ is the label on the edge. In particular, if all edges are labeled by the same integer $m$, then all 2-cells are $2m$-gons. Moreover, the above condition is nonvacuous only for $m = 2$ (in which case it means that there are no 3-circuits). Such 2-complexes (where all 2-cells are isomorphic to the same polygon and where the links of any two vertices are isomorphic) can be thought of as generalizations of the Platonic solids or of the regular tessellations of $\mathbb{E}^2$ and $\mathbb{H}^2$ by polygons. There is a fairly extensive literature on such complexes; for example, [14, 18, 150, 152, 275] and [37, pp. 393–396]. We will return to the discussion of these examples in Section 12.6. (See, in particular, Proposition 12.6.4 and Corollary 12.6.5.)

**Example 7.4.3.** (*Geometric reflection groups.*) As in Chapter 6, suppose $(W, S)$ arises as a group generated by reflections across the faces of a convex polytope $P^n$ in $\mathbb{E}^n$ or $\mathbb{H}^n$. As explained in Example 7.1.4, $L(W,S)$ can be identified the boundary complex of the dual polytope to $P^n$ and $K \cong P^n$. For example, if $P^n$ is a cube in $\mathbb{E}^n$, then $L$ is the boundary complex of an $n$-octahedron. (The relevant definitions can be found in Appendix A; the "$n$-cube" and "$n$-octahedron" are defined in Examples A.1.5 and A.1.7, respectively, and the "dual polytope" in Appendix A.2.) The cell structure on $\Sigma$ given in 7.3 is dual to the tessellation of $\mathbb{E}^n$ or $\mathbb{H}^n$ by copies of $P^n$. For example, suppose $W$ is the group generated by reflections across the edges of a right-angled pentagon in $\mathbb{H}^2$. Then $\Sigma$ is the dual complex to the tessellation of $\mathbb{H}^2$ by pentagons. It is pictured in Figure 6.2 of Section 6.5.

**Example 7.4.4.** (*Products.*) Suppose, as in Example 7.1.5, that $W$ decomposes as $(W,S) = (W_1 \times W_2, S_1 \cup S_2)$. Then $\mathcal{S} = \mathcal{S}(W_1,S_1) \times \mathcal{S}(W_2,S_2)$. By Examples 7.3.2 (iii), $\Sigma = \Sigma(W_1,S_1) \times \Sigma(W_2,S_2)$. (Here $\mathcal{S} = \mathcal{S}(W,S)$ and $\Sigma = \Sigma(W,S)$.)

**Example 7.4.5.** (*Octahedral links in the right-angled case.*) As in Examples 7.1.4 and 7.4.1, if $L = S^0$, then $W = \mathbf{D}_\infty$ and $\Sigma$ is the real line cellulated by intervals of the same length. By the previous example, if $W$ is the $n$-fold product $\mathbf{D}_\infty \times \cdots \times \mathbf{D}_\infty$, then $L$ is the $n$-fold join $S^0 * \cdots * S^0$ (i.e., $L$ is the boundary complex of an $n$-octahedron) and $\Sigma = \mathbb{R} \times \cdots \mathbb{R} = \mathbb{R}^n$, cellulated by $n$-cubes.

## 7.5. FIXED POSETS AND FIXED SUBSPACES

Given a spherical special subgroup $W_T$, we analyze the fixed set of its action on the poset $W\mathcal{S}$. Of course, this will also provide information about its fixed point set on the cell complex $\Sigma$. Given $T \in \mathcal{S}$, let $F(T)$ denote the fixed poset of $W_T$ on $W\mathcal{S}$, i.e.,

$$\begin{aligned} F(T) &:= \{wW_U \mid W_T w W_U = wW_U\}. \\ &= \{wW_U \mid W_T \subset wW_U w^{-1}\}. \end{aligned}$$

$F(T)$ indexes the set of Coxeter cells in $\Sigma$ stabilized by $W_T$. Denote the minimal elements of $F(T)$ by $F_{\min}(T)$, i.e., $F_{\min}(T) := \{wW_U \mid W_T = wW_U w^{-1}\}$.

**LEMMA 7.5.1.** *Suppose $T \in \mathcal{S}$ and $\alpha := wW_U \in F(T)$. Then*

*(i) $F(T)_{\leqslant \alpha}$ is isomorphic to the poset of faces of a convex cell of dimension* $\mathrm{Card}(U) - \mathrm{Card}(T)$.

*(ii) $F(T)_{>\alpha}$ is isomorphic to $\mathcal{S}_{>U}$.*

*Proof.* When $T = \emptyset$, $F(\emptyset) = W\mathcal{S}$. By Theorem 4.1.6, $(W\mathcal{S})_{\leqslant\alpha} \cong W_U(\mathcal{S}_{\leqslant U})$ and $(W\mathcal{S})_{>\alpha} \cong \mathcal{S}_{>U}$. By Lemma 7.3.3, $W_U(\mathcal{S}_{\leqslant U})$ is isomorphic to the poset of faces for a Coxeter polytope of type $W_U$. So, (i) and (ii) hold for $T = \emptyset$.

In the general case, set $H := w^{-1}W_T w$. $H$ is a subgroup of $W_U$. It fixes a linear subspace of $\mathbb{R}^U$ of codimension Card$(T)$. The intersection of this subspace with the Coxeter polytope for $W_U$ is a convex polytope of dimension Card$(U)$ − Card$(T)$. This proves (i). If $\alpha$ is fixed by $W_T$, then so is any larger spherical coset. Hence, $F(T)_{>\alpha} = W\mathcal{S}_{>\alpha} \cong \mathcal{S}_{>U}$, so (ii) holds. □

It follows that $F(T)$ is an "abstract convex cell complex" as in Definition A.2.12. Fix$(W_T, \Sigma)$ denotes the corresponding cell complex. For $\alpha \in F(T)$, let $c_\alpha$ be the corresponding cell in Fix$(W_T, \Sigma)$. Note that if $\alpha \in F_{\min}(T)$, then $c_\alpha$ is a vertex in this cell structure on Fix$(W_T, \Sigma)$.

**COROLLARY 7.5.2.** *For any $T \in \mathcal{S}$ and $\alpha = wW_U \in F_{\min}(T)$, the link of $c_\alpha$ in* Fix$(W_T, \Sigma)$ *is isomorphic to* Lk$(\sigma_U, L)$ *(where $\sigma_U$ denotes the simplex of $L$ corresponding to $U$).*

*Proof.* For both links the abstract simplicial complex is $\mathcal{S}_{>U}$. □

**COROLLARY 7.5.3.** *Suppose $T \in \mathcal{S}^{(k)}$.*

(i) *If $\Sigma$ is a pseudomanifold of dimension $n$, then* Fix$(W_T, \Sigma)$ *is a pseudomanifold of dimension $(n-k)$.*

(ii) *If $\Sigma$ is a homology $n$-manifold, then* Fix$(W_T, \Sigma)$ *is a homology $(n-k)$-manifold.*

(iii) *If $\Sigma$ is a PL $n$-manifold, then* Fix$(W_T, \Sigma)$ *is a PL $(n-k)$-manifold.*

The notion of a "pseudomanifold" is defined in 13.3, "homology manifolds" and "PL manifolds" in 10.4; see Definitions 10.4.1 and 10.4.8, respectively. The point of the corollary is that if $\Lambda$ is a cell complex which has one of these properties, then the link of any cell in $\Lambda$ also has this property. (For further discussion, see Theorem 10.6.1.)

**LEMMA 7.5.4.** *Suppose $x \in$* Fix$(W_T, \Sigma)$*. Then there is a neighborhood of $x$ in $\Sigma$ which is $W_T$-equivariantly homeomorphic to one of the form $F \times V$, where $F$ is a neighborhood of $x$ in* Fix$(W_T, \Sigma)$ *and $V$ is a neighborhood of the origin in the canonical representation of $W_T$ on $\mathbb{R}^T$. Moreover, if $x$ belongs to the relative interior of the Coxeter cell corresponding to an element $wW_U \in F(T)$, then $F$ is homeomorphic to a neighborhood of the cone point in* Cone$(S^{m-1} *$ Lk$(\sigma_U, L))$, *where $m =$* Card$(U)$ − Card$(T)$ *and $\sigma_U$ is the simplex of $L$ corresponding to $U$.*

*Proof.* Suppose $x \in$ int$(C)$, where $C$ is the Coxeter polytope corresponding to $wW_U \in F(T)$. $W_T$ acts on $C$ via $W_T \subset wW_Uw^{-1}$. If $m =$ Card$(U)$ − Card$(T)$, then $\text{Fix}(W^T, \text{int}(C)) \cong \mathbb{R}^m$. So a neighborhood of $x \in \text{Fix}(W_T, \Sigma)$ is homeomorphic to $\text{Cone}(S^{m-1} * \text{Lk}(\sigma_U, L))$. Moreover, $\text{int}(C)$ is $W_T$-equivariantly homeomorphic to $\mathbb{R}^{m-1} \times \mathbb{R}^T$. The lemma follows. □

## NOTES

**7.1, 7.2.** As we explained in 1.1, a major theme of this book is that Coxeter groups are a good source of examples in geometric group theory. The basic method underlying the construction of such examples is that of Examples 7.1.6, 7.1.7, and Lemma 7.2.2. One starts with simplicial complex $L$. Possibly after replacing $L$ by its barycentric subdivision, we can assume that $L$ is a flag complex (Lemma 7.2.2). As in Example 7.1.7, associated to the 1-skeleton $L^1$ there is a right-angled Coxeter group with nerve $L$. Various topological properties of $L$ are then reflected in $\Sigma$ (as well as in $W$).

**7.3.** In 7.2 we defined $\Sigma$ to be a certain simplicial complex (the geometric realization of the poset of spherical cosets). This simplicial complex was first described in [71]. The cellulation in 7.3 of $\Sigma$ by Coxeter polytopes was first explained in Moussong's thesis [221], where it was used to prove that $\Sigma$ is CAT(0). (See Chapter 12.) For these reasons, $\Sigma$ is sometimes called the "Davis-Moussong complex." More recently, this has been shortened to the "Davis complex."

**7.5.** The material in this section will be used in Chapter 13.

*Chapter Eight*

# THE ALGEBRAIC TOPOLOGY OF $\mathcal{U}$ AND OF $\Sigma$

As in Chapter 5, $(W, S)$ is a Coxeter system, $X$ is a mirrored space over $S$, and $\mathcal{U}$ $(= \mathcal{U}(W, X))$ is the result of applying the basic construction to these data. The two main results of this chapter are formulas for the homology of $\mathcal{U}$ (Theorem 8.1.2) and for its cohomology with compact supports (Theorem 8.3.1). The two formulas are similar in appearance and in their proofs. (Of course, they give completely different answers whenever $W$ is infinite.) We will give alternative proofs in Chapter 15.

The most important applications of these formulas are to the case where $\mathcal{U}$ is the complex $\Sigma$ of Chapter 7. In 8.2 we give necessary and sufficient conditions on $X$ for $\mathcal{U}$ to be acyclic. Using this, we prove, in Theorem 8.2.13, that $\Sigma$ is contractible. Our formula for cohomology with compact supports gives a calculation of $H_c^*(\Sigma)$ in terms of the cohomology of the "punctured nerves" $L - \sigma_T$ (Corollary 8.5.3). Here $L$ is the nerve of $(W, S)$ and $\sigma_T$ ranges over the simplices of $L$ (including the empty simplex). Since $H_c^*(\Sigma) = H^*(W; \mathbb{Z}W)$, this gives a calculation of the cohomology of $W$ with coefficients in its group ring. We give several applications of this in 8.5. First of all, in Corollary 8.5.5, we get a formula for the virtual cohomological dimension (= vcd) of $W$. By definition, this is the cohomological dimension of a torsion-free, finite index subgroup $\Gamma \subset W$. Appropriate choices of $L$ yield examples of $\Gamma$ which have different cohomological dimension over $\mathbb{Z}$ than over $\mathbb{Q}$, as well as, examples which show that cohomological dimension is not additive for direct products. If $\text{vcd}(W) = 1$, then $W$ is virtually free. In 8.8, we use our calculation of $H_c^*(W; \mathbb{Z}W)$ to characterize when $W$ is virtually free: it is if and only if its nerve $L$ decomposes as a "tree of simplices." Further applications of these formulas will be given in later chapters. For example, in 10.9, we will use the formula for $H_c^*(W; \mathbb{Z}W)$ to characterize when $W$ is a virtual Poincaré duality group. In the last section of this chapter, we use similar techniques to get a formula for the cohomology with compact supports of the fixed point set of any spherical special subgroup on $\Sigma$. This gives a calculation of the cohomology of the normalizer of such a special subgroup with coefficients in its group ring. We will need this computation in Chapter 13.

Throughout this chapter, except in 8.4, we assume $X$ is a mirrored CW complex (i.e., $X$ is a CW complex and the $X_s$ are subcomplexes). Then $\mathcal{U}$ is also a CW complex. Also, except in 8.4, $C_*(Y)$ will denote the cellular chain complex of a CW-complex $Y$ and $H_*(Y)$ its homology; similarly, $C_c^*(Y)$

will mean the finitely supported cellular cochains and $H_c^*(Y)$ the cohomology of this complex. $C_k(Y)$ can be identified with the free abelian group on the set of $k$-cells of $Y$ and $C_c^k(Y)$ with the set of finitely supported functions on this set. The reason for the assumption that $X$ is a mirrored CW complex is that the proofs of our formulas are technically much easier with cellular chains and cochains. We indicate how to deal with a general $X$ in 8.4.

## 8.1. THE HOMOLOGY OF $\mathcal{U}$

### Symmetrization and Alternation

For each spherical subset $T$ of $S$, define elements $\tilde{a}_T$ and $\tilde{h}_T$ in the group ring $\mathbb{Z}W_T \subset \mathbb{Z}W$ by

$$\tilde{a}_T := \sum_{w \in W_T} w \quad \text{and} \quad \tilde{h}_T := \sum_{w \in W_T} (-1)^{l(w)} w \tag{8.1}$$

called, respectively, *symmetrization* and *alternation* with respect to $T$. (In this section, where we calculate homology, we will only use alternation; symmetrization is more useful for calculations in cohomology.) Our first goal is to show that $\tilde{h}_T$ induces a homomorphism $H_*(X, X^T) \to H_*(W_T X)$, where, as usual,

$$X^T := \bigcup_{s \in T} X_s \quad \text{and} \quad W_T X := \bigcup_{u \in W_T} uX.$$

Multiplication by $\tilde{h}_T$ defines a homomorphism $\tilde{h}_T : C_m(X) \to C_m(W_T X)$. We claim $\tilde{h}_T$ vanishes on the subgroup $C_m(X^T)$. To see this, suppose $e$ is an $m$-cell in $X^T$. Then $e$ lies in $X_s$ for some $s \in T$. For any $u \in W_T$, the coset $uW_{\{s\}}$ contains two elements $u$ and $us$. Both elements act the same way on $e$. However, in $\tilde{h}_T$ the coefficient of one is $+1$ and of the other $-1$. So, $\tilde{h}_T e = 0$ and hence $\tilde{h}_T$ vanishes on $C_m(X^T)$. Therefore, it induces a chain map $\phi : C_*(X, X^T) \to C_*(W_T X)$ and a corresponding map on homology $\phi_* : H_*(X, X^T) \to H_*(W_T X)$. (Recall $C_*(X, X^T) := C_*(X)/C_*(X^T)$.)

Similarly, for any $w \in W$, we have a map

$$\rho_w : H_*(X, X^T) \to H_*(wW_T X), \tag{8.2}$$

defined as the composition of $\phi_*$ and the map induced by translation by $w$.

### An Increasing Union of Chambers

As in Example 4.5.6, order the elements of $W$: $w_1, \dots, w_n, \dots$, so that $l(w_k) \leqslant l(w_{k+1})$. For $n \geqslant 1$, set

$$U_n = \{w_1, \dots, w_n\}$$

and

$$P_n = U_n X := \bigcup_{w \in U_n} w_n X.$$

Recall Definition 4.7.1: $\mathrm{In}(w) = \{s \in S \mid l(ws) < l(w)\}$. So $\mathrm{In}(w)$ indexes the set of mirrors of the chamber $wX$ with the property that the adjacent chamber across the mirror is closer to the base chamber $X$ than is $wX$. The next lemma explains how the $n^{\text{th}}$ chamber intersects the union of the previous ones. It says that this intersection has the form $X^T$, where $T$ is the spherical subset $\mathrm{In}(w_n)$. This is the key to all the calculations in this chapter and the next.

**LEMMA 8.1.1.** (Compare [255, Lemma 4, p. 108] and [71, Lemma 8.2].) *For each $n \geqslant 2$,*

$$w_n X \cap P_{n-1} = w_n X^{\mathrm{In}(w_n)}.$$

*Proof.* To simplify notation, set $w = w_n$. If $s \in \mathrm{In}(w)$, then $ws \in U_{n-1}$ so $wX_s = wsX_s \subset wsX \subset P_{n-1}$. Hence, $wX^{\mathrm{In}(w)} \subset wX \cap P_{n-1}$. For the opposite inclusion, suppose $v = w_i$ for some $i < n$ and consider the intersection, $wX \cap vX$. By Lemma 5.3.1, this intersection has the form $wX_T$, where $T = S(v^{-1}w)$ is the set of letters appearing in a minimal gallery from $v$ to $w$. By Lemma 4.3.1, there is a unique element of minimum length in $wW_T$. This element is not $w$ since $l(v) \leq l(w)$. By Lemma 4.3.3, this implies that $l(ws) < l(w)$ for some $s \in T$. In other words, $T \cap \mathrm{In}(w) \neq \emptyset$. Hence, $X_T \subset X_s$ for some $s \in \mathrm{In}(w)$ and since $X_s \subset X^{\mathrm{In}(w)}$, we get $wX \cap P_{n-1} \subset wX^{\mathrm{In}(w)}$. □

The main result of this section is the following.

**THEOREM 8.1.2.** ([73, Theorem A].)

$$H_*(\mathcal{U}) \cong \bigoplus_{w \in W} H_*(X, X^{\mathrm{In}(w)}).$$

This follows almost immediately from the next lemma.

**LEMMA 8.1.3**

$$H_*(P_n) \cong \bigoplus_{i=1}^{i=n} H_*(X, X^{\mathrm{In}(w_i)}).$$

*Proof.* We will use the exact sequence in homology of the pair $(P_n, P_{n-1})$, $n > 1$. There is an excision

$$H_*(P_n, P_{n-1}) \cong H_*(w_n X, w_n X^{\mathrm{In}(w_n)}), \tag{8.3}$$

where we have excised the open subset $P_n - w_n X$. Simplifying notation as before, set $w = w_n$ and $T = \mathrm{In}(w)$. The right-hand side of (8.3) is isomorphic to $H_*(X, X^T)$. So, the exact sequence of the pair $(P_n, P_{n-1})$ can be rewritten as

$$\longrightarrow H_*(P_{n-1}) \longrightarrow H_*(P_n) \xrightarrow{f} H_*(X, X^T) \longrightarrow \tag{8.4}$$

where $f$ is the composition of the map to $H_*(P_n, P_{n-1})$, the excision (8.3), and the isomorphism induced by translation by $w^{-1}$. We want to define a map $g$ which splits $f$. Basically, $g$ is the map $\rho_w$ defined in (8.2). However, the range of $\rho_w$ is the homology of $wW_TX$ while we want it to land in the homology of $P_n$. So, we need to verify that $wW_TX \subset P_n$. This follows from two facts: (1) $T := \mathrm{In}(w_n)$ is spherical (Lemma 4.7.2) and (2) $w = w_n$ is the element of longest length in the coset $wW_T$ (Lemma 4.7.3 (i)). Therefore, the index of any element of $wW_T$ is $\leqslant n$. So, define $g$ to be the composition of $\rho_w$ with the map on homology induced by the inclusion $wW_TX \hookrightarrow P_n$. It is then easy to see that $f \circ g$ is the identity map of $H_*(X, X^T)$. Hence, the exact sequence (8.4) is split short exact. Thus,

$$H_*(P_n) \cong H_*(P_{n-1}) \oplus H_*(X, X^{\mathrm{In}(w_n)}). \tag{8.5}$$

When $n = 1$, we have $w_1 = 1$, $\mathrm{In}(w_1) = \emptyset$ and $X^{\mathrm{In}(w_1)} = \emptyset$; hence, (8.5) becomes

$$H_*(P_1) = H_*(X, X^{\mathrm{In}(w_1)}) = H_*(X, \emptyset) = H_*(X).$$

Combining this with the equations in (8.5), we get the formula of the lemma. □

*Proof of Theorem 8.1.2.* Since $\mathcal{U}$ is the increasing union of the $P_n$, $H_*(\mathcal{U})$ is the direct limit of the $H_*(P_n)$. So, Lemma 8.1.3 implies the theorem. □

There is also a relative version of Theorem 8.1.2. Suppose $Y$ is a subcomplex of $X$. Give $Y$ the induced mirror structure and write $\mathcal{U}(X)$ and $\mathcal{U}(Y)$ for $\mathcal{U}(W, X)$ and $\mathcal{U}(W, Y)$, respectively.

**THEOREM 8.1.4**

$$H_*(\mathcal{U}(X), \mathcal{U}(Y)) \cong \bigoplus_{w \in W} H_*(X, X^{\mathrm{In}(w)} \cup Y).$$

*Proof.* The proof is similar to that of Theorem 8.1.2. Write $P_n(X)$ and $P_n(Y)$ for the union of the first $n$ chambers in $\mathcal{U}(X)$ and $\mathcal{U}(Y)$, respectively. Consider the exact sequence of the triple $(P_n(X), P_{n-1}(X) \cup wY, P_n(Y))$, where $w := w_n$. We have excisions $H_*(P_{n-1}(X) \cup wY, P_n(Y)) \cong H_*(P_{n-1}(X), P_{n-1}(Y))$ and $H_*(P_n(X), P_{n-1}(X) \cup wY) \cong H_*(X, X^T \cup Y)$, where $T := \mathrm{In}(w)$. This gives a decomposition of $H_*(P_n(X), P_n(Y))$ analogous to the one in Lemma 8.1.3, which immediately implies the result. □

**COROLLARY 8.1.5.** *Suppose that the homology of* $(X, Y)$ *vanishes in all dimensions. Then*

$$H_*(\mathcal{U}(X), \mathcal{U}(Y)) \cong \bigoplus_{w \in W} H_{*-1}(X^{\mathrm{In}(w)}, Y^{\mathrm{In}(w)}).$$

*Proof.* By the previous theorem, $H_*(\mathcal{U}(X), \mathcal{U}(Y))$ is a direct sum of terms of the form $H_*(X, X^T \cup Y)$, where $T = \mathrm{In}(w)$ for some $w \in W$. Since $H_*(X, Y) = 0$, the exact sequence of the triple $(X, X^T \cup Y, Y)$ gives $H_*(X, X^T \cup Y) \cong H_{*-1}(X^T \cup Y, Y)$. This last term excises to $H_{*-1}(X^T, Y^T)$. □

Recall Definition 4.7.4: for each spherical subset $T \in \mathcal{S}$,

$$W^T = \{w \in W \mid \mathrm{In}(w) = T\}.$$

In other words, $W^T$ is the set of elements of $W$ which can end precisely with the letters in $T$. Let $\mathbb{Z}W^T$ denote the free abelian group on $W^T$.

Obviously, $W$ is the disjoint union of the $W^T$, $T \in \mathcal{S}$. So, collecting the terms on the right-hand side of the formula in Theorem 8.1.2, one can rewrite it as follows.

**THEOREM 8.1.6.** ([73, Theorem A′].)

$$H_*(\mathcal{U}) \cong \bigoplus_{T \in \mathcal{S}} H_*(X, X^T) \otimes \mathbb{Z}W^T.$$

One problem with Theorems 8.1.2 and 8.1.4 is that they do not give an algebraic description of the action of $W$ on $H_*(\mathcal{U})$. We will say more about this in 15.3 (in particular, Theorem 15.3.4).

## 8.2. ACYCLICITY CONDITIONS

$\overline{H}_*(Y)$ denotes the reduced homology of a space $Y$.

*Convention.* The reduced homology of the empty set is defined to be zero in dimensions $\geqslant 0$ and to be $\mathbb{Z}$ in dimension $-1$.

$X$ is *acyclic* if $\overline{H}_i(X) = 0$ for all $i$. Similarly, $X$ is *m-acyclic* ($m$ an integer) if $\overline{H}_i(X) = 0$ for $-1 \leqslant i \leqslant m$. (N.B. The empty set is *not m-acyclic* for $m \geqslant -1$.) A pair $(X, Y)$ is *m-acyclic* if $H_i(X, Y) = 0$ for $0 \leqslant i \leqslant m$. The following is a simple corollary of Theorem 8.1.2.

**COROLLARY 8.2.1.** *$\mathcal{U}$ is acyclic if and only if $X$ is acyclic and $X^T$ is acyclic for each nonempty spherical subset $T$.*

*Proof.* By Lemma 4.7.2, for each $w \in W$, $\mathrm{In}(w)$ is a spherical subset of $S$. Moreover, $\mathrm{In}(w) = \emptyset$ only when $w = 1$. Conversely, any spherical subset $T$

occurs as $\mathrm{In}(w)$ for some $w$ (for example, for $w$ the element of longest length in $W_T$). By the formula in Theorem 8.1.2, $\overline{H}_i(\mathcal{U}) = 0$ if and only if $\overline{H}_i(X) = 0$ and $H_i(X, X^T) = 0$ for all $T \in \mathcal{S}_{>\emptyset}$. When $X$ is acyclic, it follows from the exact sequence of the pair $(X, X^T)$ that the acyclicity of $(X, X^T)$ is equivalent to the acyclicity of $X^T$. □

Essentially the same argument gives the following result.

**COROLLARY 8.2.2.** (Compare [71, Theorem 10.1].) *$\mathcal{U}$ is $m$-acyclic if and only if $(X, X^T)$ is $m$-acyclic for each spherical subset $T$.*

The special case $m = 0$ gives the following.

**COROLLARY 8.2.3.** (Compare Corollary 5.2.4.) *$\mathcal{U}$ is path connected if and only if the following two conditions hold:*

*(a) $X$ is path connected and*

*(b) each $X_s$ is nonempty.*

*Proof.* In Corollary 5.2.4 we proved that (a) and (b) imply that $\mathcal{U}$ is path connected. Conversely, suppose $\mathcal{U}$ is path connected, i.e., 0-acyclic. By the previous corollary, $X$ is 0-acyclic as is $(X, X_s)$ for each $s \in S$. The exact sequence of the pair shows that $X_s$ is $(-1)$-acyclic, i.e., nonempty. □

**DEFINITION 8.2.4.** A space $Y$ *has the same homology as* $S^n$ if its reduced homology is concentrated in dimension $n$ and is isomorphic to $\mathbb{Z}$ in that dimension, i.e., if

$$\overline{H}_i(Y) \cong \begin{cases} 0 & \text{if } i \neq n, \\ \mathbb{Z} & \text{if } i = n. \end{cases}$$

(A space has the same homology as $S^{-1}$ if and only if it is empty.) Similarly, a pair of spaces $(Y, Y')$ *has the same homology as* $(D^n, S^{n-1})$ if $H_*(Y, Y')$ is concentrated in dimension $n$ and is isomorphic to $\mathbb{Z}$ in that dimension.

Here is a related corollary to Theorem 8.1.2 for spaces with the same homology as $S^n$.

**COROLLARY 8.2.5.** *$\mathcal{U}$ has the same homology as $S^n$ if and only if there is a spherical subset $T \in \mathcal{S}$ satisfying the following three conditions:*

*(a) $W$ decomposes as $W = W_T \times W_{S-T}$.*

*(b) For all $T' \in \mathcal{S}$ with $T' \neq T$, $(X, X^{T'})$ is acyclic.*

*(c) $(X, X^T)$ has the same homology as $(D^n, S^{n-1})$. (When $T = \emptyset$ we interpret this to mean that $X$ has the same homology as $S^n$.)*

*Moreover, the spherical subset $T$ is unique.*

*Proof.* First suppose $\mathcal{U}$ has the same homology as $S^n$. Then, in Theorem 8.1.2, the sum on the right-hand side of the formula for $H_n(\mathcal{U})$ can have only one nonzero term and it must have the form $H_n(X, X^{\mathrm{In}(w)}) \cong \mathbb{Z}$, for a unique element $w$. Put $T = \mathrm{In}(w)$. Since $w$ is unique, Lemma 4.7.5 implies that $W = W_T \times W_{S-T}$. Conditions (b) and (c) are immediate. The argument in the converse direction is routine. □

*Remark 8.2.6.* (*Finitely generated homology.*) Corollary 8.2.5 can be generalized to the case where $H_*(\mathcal{U})$ is finitely generated. First note that any Coxeter group decomposes as $W_F \times W_{S-F}$, where $F$ is the union of the spherical components in the diagram of $W$ (so $W_F$ is finite). Suppose $H_*(\mathcal{U})$ is finitely generated. By the formula in Theorem 8.1.6, $H_*(X, X^T)$ can be nonzero only for those $T \in \mathcal{S}$ such that $W^T$ a finite set. It follows from [26, Exercises 17 and 23 (c), pp. 58–59] that if $W^T$ is finite, then $T \subset F$. (This is a generalization of Lemma 4.7.5.) Hence, for every $T \in \mathcal{S}$ such that $H_*(X, X^T) \neq 0$, we have $T \subset F$. $X$ has a mirror structure over $F$ by restriction to the mirrors indexed by $F$. Put $\mathcal{U}' = \mathcal{U}(W_F, X)$. By our formula, the inclusion $\mathcal{U}' \hookrightarrow \mathcal{U}$ induces an isomorphism on homology. If $X$ is a finite complex, then so is $\mathcal{U}'$. In other words, when $X$ is a finite complex and $H_*(\mathcal{U})$ is finitely generated, the homology is carried by a finite complex stable under the action of a finite factor. (In fact, as in Remark 8.2.17 below, one sees that $\mathcal{U}$ "homologically resembles" the product action on $\mathcal{U}' \times \Sigma''$, where $\Sigma'' := \Sigma(W_{S-F}, S - F)$.)

Next we want to translate the condition in Corollary 8.2.2 into one that is more useful. This condition is expressed in terms of the cofaces $X_T$ of $X$.

**THEOREM 8.2.7.** (Compare [71, Theorem 10.1].) *$\mathcal{U}$ is m-acyclic if and only if the following two conditions hold:*

*(a) $X$ is m-acyclic.*

*(b) For each $k \leqslant m + 1$ and each spherical subset $T \in \mathcal{S}^{(k)}$, the coface $X_T$ is $(m - k)$-acyclic.*

Before beginning the proof, we state three corollaries to Theorem 8.2.7. The first is the following alternative version of Corollary 8.2.1.

**COROLLARY 8.2.8.** *$\mathcal{U}$ is acyclic if and only if $X_T$ is acyclic for each $T \in \mathcal{S}$. (Recall that $X_\emptyset = X$.)*

According to Lemma 7.1.9, a mirror structure on $X$ is $W$-finite if and only if its nerve $N(X)$ is a subcomplex of $L(W, S)$. In the case where $\mathcal{U}$ is acyclic, Corollary 8.2.8 gives the opposite inclusion. (The point being that the empty set is not acyclic.) We restate this as follows.

**COROLLARY 8.2.9.** *If $\mathcal{U}(W,X)$ is acyclic, then $L(W,S) \subset N(X)$. If, in addition, the $W$-action on $\mathcal{U}(W,X)$ is proper, then $N(X) = L(W,S)$.*

A special case of this is the following result discussed earlier in Example 7.1.4.

**COROLLARY 8.2.10.** *Suppose $W$ is a geometric reflection group (Definition 6.4.4) generated by the set $S$ of reflections across the codimension-one faces of a convex polytope $P^n$ in $\mathbb{E}^n$ or $\mathbb{H}^n$. Let $P^*$ be the dual polytope. Then $L(W,S) = \partial P^*$.*

Before attempting the proof of Theorem 8.2.7, consider the special case $m = 1$ where the proof is easier to understand. (The case $m = 0$ was already taken care of in Corollary 8.2.3.)

**PROPOSITION 8.2.11.** *$\mathcal{U}$ is 1-acyclic if and only if the following three conditions hold:*

*(a) $X$ is 1-acyclic.*

*(b) For each $s \in S$, the mirror $X_s$ is nonempty and path connected.*

*(c) For each $\{s,t\} \in \mathcal{S}^{(2)}$, $X_s \cap X_t$ is nonempty.*

*Proof.* First suppose that $\mathcal{U}$ is 1-acyclic. Theorem 8.1.2 gives

(a)′ $X$ is 1-acyclic.

(b)′ For each $s \in S$, $(X, X_s)$ is 1-acyclic.

(c)′ For each $\{s,t\} \in \mathcal{S}^{(2)}$, $(X, X^{\{s,t\}})$ is 1-acyclic.

Together with the exact sequence of the pair $(X, X_s)$ conditions (a)′ and (b)′ imply that $H_0(X_s) = \mathbb{Z}$, i.e., that $X_s$ is nonempty and path connected. Similarly, conditions (a)′ and (c)′ imply that $H_0(X^{\{s,t\}}) = \mathbb{Z}$. Now apply the Mayer–Vietoris sequence to $X^{\{s,t\}} = X_s \cup X_t$:

$$H_0(X_{\{s,t\}}) \longrightarrow H_0(X_s) \oplus H_0(X_t) \longrightarrow H_0(X^{\{s,t\}})$$

(The intersection of $X_s$ and $X_t$ is $X_{\{s,t\}}$.) Since we already know $H_0(X_s) \oplus H_0(X_t) = \mathbb{Z} \oplus \mathbb{Z}$ and $H_0(X^{\{s,t\}}) = \mathbb{Z}$, we must have $H_0(X_{\{s,t\}}) \neq 0$, i.e., $X_s \cap X_t \neq \emptyset$. Thus, if $\mathcal{U}$ is 1-acyclic, conditions (a), (b), and (c) hold.

Conversely, suppose the three conditions hold. Then for each nonempty spherical subset $T$, $X^T$ is nonempty and path connected. (The reason is that it is a union of path connected subspaces $X_s$, $s \in T$, any two of which have nonempty intersection.) Thus, $H_0(X^T) = \mathbb{Z}$. Since $X$ is 1-acyclic (condition (a)), the exact sequence of the pair shows that $(X, X^T)$ is 1-acylic. So, $\mathcal{U}$ is 1-acyclic by Corollary 8.2.2. □

Theorem 8.2.7 follows from Corollary 8.2.2 and the next lemma, the proof of which is based on a version of the Acyclic Covering Lemma from Appendix E.3.

**LEMMA 8.2.12.** *The following two conditions on the mirrored CW complex $X$ are equivalent.*

*(a)* $(X, X^T)$ *is $m$-acyclic for all* $T \in \mathcal{S}$.

*(b)* $X$ *is $m$-acyclic and* $X_T$ *is* $(m - \text{Card}(T))$*-acyclic for all spherical subsets $T$ with* $\text{Card}(T) \leqslant m + 1$.

*Proof.* Condition (a) is equivalent to the condition that $X$ is $m$-acyclic and $X^T$ is $(m-1)$-acyclic for all $T \in \mathcal{S}$.

(b) $\Longrightarrow$ (a). Suppose (b) holds. Let $T \in \mathcal{S}_{>\emptyset}$. Consider the cover of $X^T$ by the mirrors, $\{X_s\}_{s\in T}$. For any $U \subset T$ with $\text{Card}(U) \leqslant m+1$, the intersection $\bigcap_{s\in U} X_s$ is $(m - \text{Card}(U))$-acyclic. In particular, such an intersection is nonempty. So, the nerve of the cover has $m$-skeleton equal to the $m$-skeleton of the simplex on $T$. Lemma E.3.4 implies that in dimensions $\leqslant m-1$, the reduced homology of $X^T$ is isomorphic to the reduced homology of this nerve, i.e., it is 0. So, $X^T$ is $(m-1)$-acyclic.

(a) $\Longrightarrow$ (b). Suppose (b) does not hold. Let $U \in \mathcal{S}_{>\emptyset}$ be a counterexample to (b) with minimum number of elements. So, the reduced homology of $X_U$ is nonzero in some dimension $i$ with $i \leqslant m - \text{Card}(U)$. Let $i$ be the minimum such dimension. (If $X_U = \emptyset$, then $\overline{H}_{-1}(X_U) \neq 0$.) By Lemma E.3.5 applied to the covering $\{X_s\}_{s\in U}$ of $X^U$, $\overline{H}_{i+\text{Card}(U)-1}(X^U) \neq 0$. So, $X^U$ is not $(m-1)$-acyclic. □

## Σ is Contractible

We are now in position to prove one of our main results.

**THEOREM 8.2.13.** $\Sigma$ *is contractible.*

*Proof.* By Theorem 7.2.4, $\Sigma = \mathcal{U}(W, K)$. By Lemma 7.2.5, each $K_T$ is acyclic (in fact, each $K_T$ is a cone and hence is contractible). So, by Corollary 8.2.8, $\Sigma$ is acyclic. By Lemma 7.3.5, $\Sigma$ is simply connected. So, by the Hurewicz Theorem, it is weakly contractible. Since $\Sigma$ is a cell complex, this implies that it is contractible. □

*Remark 8.2.14.* Here is a slightly different version of the proof. $\Sigma$ is an increasing union of contractible fundamental domains. Ordering them as before, $P_n$ is obtained from $P_{n-1}$ by gluing on a copy of $K$ along a subspace of the form $K^T$, $T \in \mathcal{S}$. Each $K^T$ is contractible, since it is a union of contractible subcomplexes $(\{K_s\}_{s\in T})$ and the intersection of any subcollection is contractible. So, $P_n$ is formed from $P_{n-1}$ by gluing on a contractible space along a contractible subspace. Hence, each $P^n$ is contractible and therefore, $\Sigma$ is contractible.

### Homology of a Sphere

Just as Corollary 8.2.2 implies Theorem 8.2.7, there is a version of Corollary 8.2.5 which can be stated in terms of the cofaces of $X$. We deal first with the case where $W$ is finite.

**LEMMA 8.2.15.** *Suppose $(W,S)$ is spherical, that $\mathcal{U}(W,X)$ has the same homology as $S^n$ and that $(X,X^S)$ has the same homology as $(D^n,S^{n-1})$. Then*

*(i) For any proper subset $U$ of $S$, $X_U$ is acyclic.*

*(ii)* $\mathrm{Card}(S) \leqslant n+1$ *and $X_S$ has the same homology as $S^{n-\mathrm{Card}(S)}$. (When* $\mathrm{Card}(S) = n+1$ *this means $X_S = \emptyset$.)*

*Proof.* Since $(X,X^S)$ has the same homology as $(D^n,S^{n-1})$, the unique spherical subset $T$ of Corollary 8.2.5 is $S$ itself. By the same corollary, $X$ is acyclic, $X^U$ is acyclic for all $U \subsetneq S$, and $X^S$ has the same homology as $S^{n-1}$. Given $U \subsetneq S$, consider the covering of $X^U$ by the mirrors indexed by $U$. Since $X^U$ is acyclic, induction and the Acyclic Covering Lemma imply, as before, that $X_{T'}$ is acyclic (and nonempty) for all $T' \subset U$. In particular, $X_U$ is acyclic. Next consider the covering of $X^S$ by the mirrors of $X$. Since $X_U$ is nonempty for each proper subset $U$ of $S$, the nerve of this covering contains $\partial\Delta$ where $\Delta$ is the full simplex on $S$. Since $X^S$ has the homology of $S^{n-1}$, Lemma E.3.5 implies (ii). □

The general version of Corollary 8.2.5 is the following.

**THEOREM 8.2.16.** *Suppose $\mathcal{U}$ has the same homology as $S^n$ and $T$ is the unique spherical subset satisfying the three conditions of Corollary 8.2.5. Let $U$ be a subset of $T$ and $V$ a spherical subset of $S-T$. Then*

*(i) If $U$ is a proper subset of $T$, then $X_{U\cup V}$ is acyclic.*

*(ii)* $\mathrm{Card}(T) \leqslant n+1$ *and $X_{T\cup V}$ has the same homology as $S^{n-\mathrm{Card}(T)}$. (When* $\mathrm{Card}(T) = n+1$*, this means $X_T = \emptyset$.)*

*Proof.* Put $Y := \mathcal{U}(W_{S-T},X)$. $Y$ has a natural mirror structure over $T$, defined by $Y_t := \mathcal{U}(W_{S-T},X_t)$. For any $U \subset T$, we have $Y^U = \mathcal{U}(W_{S-T},X^U)$. Theorem 8.1.4 applied to $(Y,Y^U)$ yields

$$H_*(Y,Y^U) \cong \bigoplus_{w\in W_{S-T}} H_*(X, X^{\mathrm{In}(w)} \cup X^U).$$

If $U \neq T$, every term on the right-hand side vanishes. Hence, for $U \neq T$, $(Y,Y^U)$ is acyclic.

Consider the $W_T$-space $\mathcal{U}(W_T, Y)$. By Corollary 5.3.4, it can be identified with $\mathcal{U}$ $(= \mathcal{U}(W, X))$. So, $\mathcal{U}(W_T, Y)$ has the same homology of $S^n$. By Corollary 8.2.5, there is a unique subset $U$ of $T$ such that $(Y, Y^U)$ has the same homology as $(D^n, S^{n-1})$. By the previous paragraph, the only possibility is $U = T$. Hence, we can apply the previous lemma to $\mathcal{U}(W_T, Y)$ to conclude

(i)$'$ If $U \neq T$, then $Y_U$ is acyclic.

(ii)$'$ Card$(T) \leqslant n + 1$ and $Y_T$ has the same homology as $S^{n-\text{Card}(T)}$.

We also have $Y_U = \mathcal{U}(W_{S-T}, X_U)$. So, when $U \neq T$, Corollary 8.2.8 states that $(X_U)_V$ $(= X_{U \cup V})$ is acyclic, i.e., (i) holds. Since $(X, X^T)$ has the same homology as $(D^n, S^{n-1})$, statement (i) and the Acyclic Covering Lemma imply that $X_T$, as well as, each $X_{T \cup V}$ have the same homology as $S^{n-\text{Card}(T)}$. □

*Remark 8.2.17.* What does a geometric action of a finite Coxeter group $W_T$ on the unit sphere $\mathbb{S}^n$ in $\mathbb{R}^{n+1}$ look like? Put $m = \text{Card}(T) - 1$. If $n = m$, the linear action on $\mathbb{R}^{n+1}$ is essential and the fundamental chamber $K' \subset \mathbb{S}^n$ is a simplex $\Delta^m$. If $n > m$, then the fixed subsphere of $W_T$ has dimension $n - m - 1$ and $K'$ is the join, $S^{n-m-1} * \Delta^m$. (See Appendix A.4 for the definition of a "join.") In other words, $K'$ is a suspended simplex (an $(n-m)$-fold suspension). Suppose $\Sigma'' = \Sigma(W_{S-T}, S - T)$ and $K'' = K(W_{S-T}, S - T)$. The meaning of Theorem 8.2.16 is that if $\mathcal{U}$ has the same homology as $S^n$, then $W$ splits as a product $W_T \times W_{S-T}$ and the action "homologically resembles" the product action of $W_T \times W_{S-T}$ on $\mathbb{S}^n \times \Sigma''$ in that the cofaces of $X$ have the same homology as those of $K' \times K''$.

## 8.3. COHOMOLOGY WITH COMPACT SUPPORTS

In this section $X$ is a finite CW complex and its mirror structure is $W$-finite (Definition 5.1.6). Recall Definition 4.7.1: Out$(w)$ is the complement of In$(w)$ in $S$. Out$(w)$ indexes the set of mirrors of $wX$ such that the adjacent chamber across the mirror is farther away from the base chamber $X$ than is $wX$. Our goal is to prove the following theorem of [77]. (Our proof is from [89].)

**THEOREM 8.3.1**

$$H^*_c(\mathcal{U}) \cong \bigoplus_{w \in W} H^*(X, X^{\text{Out}(w)}).$$

A corollary of this and the Universal Coefficient Theorem is the corresponding result for locally finite homology. (See Appendix G.2.)

**THEOREM 8.3.2**

$$H^{lf}_*(\mathcal{U}) \cong \prod_{w \in W} H_*(X, X^{\text{Out}(w)}).$$

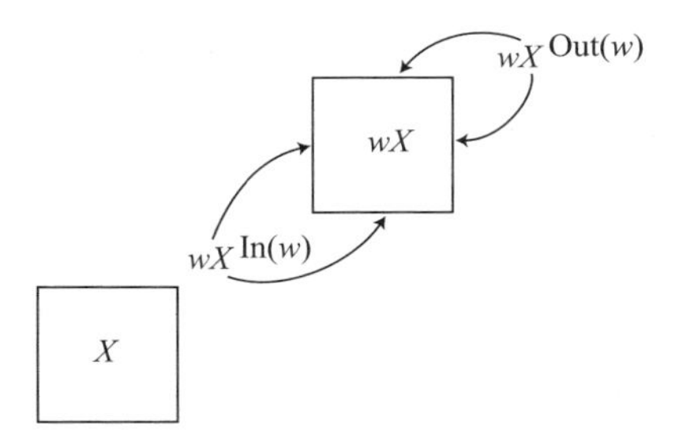


Figure 8.1. $wX^{\mathrm{In}(w)}$ and $wX^{\mathrm{Out}(w)}$.

Let us start the proof of Theorem 8.3.1. It is fairly similar to the proof of Theorem 8.1.2. In outline it goes as follows. For any subset $U$ of $W$, let $\check{U}$ denote the complementary subset, $\check{U} = W - U$. As in 8.1, order the elements of $W$ compatibly with word length and set $U_n = \{w_1, \dots, w_n\}$. Then $U_1X \subset \cdots \subset U_nX \subset \cdots$ is an exhaustive sequence of compact subsets of $\mathcal{U}$ and $\check{U}_1X \supset \cdots \supset \check{U}_nX \supset \cdots$ is essentially the inverse sequence of their complements in $\mathcal{U}$. Hence,

$$H_c^*(\mathcal{U}) = \varinjlim H^*(\mathcal{U}, \check{U}_nX). \tag{8.6}$$

(In 8.1 we wrote $P_n$ instead of $U_nX$.) Consider the exact sequence in cohomology of the triple $(\mathcal{U}, \check{U}_{n-1}X, \check{U}_nX)$:

$$\to H^*(\mathcal{U}, \check{U}_{n-1}X) \to H^*(\mathcal{U}, \check{U}_nX) \to H^*(\check{U}_{n-1}X, \check{U}_nX) \to .$$

As in equation (8.3), we have an excision:

$$H^*(\check{U}_{n-1}X, \check{U}_nX) \cong H^*(w_nX, w_nX^{\mathrm{Out}(w_n)}). \tag{8.7}$$

We will show that the exact sequence of the triple splits and hence, that

$$H^*(\mathcal{U}, \check{U}_nX) \cong H^*(\mathcal{U}, \check{U}_{n-1}X) \oplus H^*(X, X^{\mathrm{Out}(w_n)}). \tag{8.8}$$

From this we conclude that

$$H^*(\mathcal{U}, \check{U}_nX) \cong \bigoplus_{i=1}^{i=n} H^*(X, X^{\mathrm{Out}(w_i)}), \tag{8.9}$$

which implies the theorem.

Here are the details. First, recall some definitions from 4.5. In Definition 4.5.1, for each $T \subset S$, we defined the notion of the "fundamental $T$-sector" $A_T \subset W$ and a retraction $p_T : W \to A_T$. ($A_T$ is the set of $(T, \emptyset)$-reduced elements and $p_T$ sends $u$ to the shortest element in the coset $W_Tu$.) We also have the notion of a "starlike" subset of $W$ and an "extreme element" in such a subset (Definition 4.5.5).

For each $w \in W$, put $T = \text{In}(w)$ and define $p_w : W \to wA_T$ by

$$p_w(u) = wp_T(w^{-1}u). \tag{8.10}$$

**LEMMA 8.3.3**

*(i) For each $w \in W$, $p_w^{-1}(w) = wW_T$.*

*(ii) Suppose that $U \subset W$ is starlike (with respect to 1) and that $w$ is an extreme element of $U$. Then $p_w(\check{U}) = w(A_T - 1)$.*

(Here $\check{U} = W - U$ and $T = \text{In}(w)$.)

*Proof* Assertion (i) follows from the fact that $p_T^{-1}(1) = W_T$. By Lemma 4.7.3 (i), $w$ is the unique element of longest length in $wW_T$ and by Lemma 4.6.1, given any minimal gallery $\gamma$ in $W_T$ connecting 1 to the element $w_T$ of longest length, there is a minimal gallery from 1 to $w$ with terminal segment $\gamma$. Hence, $wW_T \subset U$. So, by (i), $p_w^{-1}(w) \cap \check{U} = \emptyset$, which is equivalent to (ii). □

In Corollary 5.3.5 we showed that $p_T : W \to A_T$ induces a retraction $\mathcal{U} \to A_TX$ of spaces, also denoted $p_T$. Furthermore, $p_T$ can be identified with the orbit projection $\mathcal{U} \to \mathcal{U}/W_T$. Similarly, $p_w : W \to wA_T$ induces a topological retraction $p_w : \mathcal{U} \to wA_TX$ defined by $p_w = w \circ p_T \circ w^{-1}$. (Here $w$ stands for the map $\mathcal{U} \to \mathcal{U}$ given by translation by $w$.) For each $T \subset S$, the inclusion of pairs $(X, X^{S-T}) \to (A_TX, (A_T - 1)X)$ is an excision,

$$H^*(A_TX, (A_T - 1)X) \cong H^*(X, X^{S-T}) \tag{8.11}$$

(where we have excised $A_TX - X$).

For the remainder of the proof of Theorem 8.3.1, we take the hypotheses of Lemma 8.3.3 (ii): $U \subset W$ is a starlike subset, $w \in U$ is an extreme element, and $T = \text{In}(w)$. Set $V = U - w$. Consider the sequence of the triple $(\mathcal{U}, \check{V}X, \check{U}X)$:

$$\longrightarrow H^*(\mathcal{U}, \check{V}X) \longrightarrow H^*(\mathcal{U}, \check{U}X) \longrightarrow H^*(\check{V}X, \check{U}X) \cdots .$$

The inclusion $(wX, wX^{S-T}) \to (\check{V}X, \check{U}X)$ is an excision,

$$H^*(\check{V}X, \check{U}X) \cong H^*(wX, wX^{S-T}). \tag{8.12}$$

Hence, the exact sequence of the triple can be rewritten as

$$\longrightarrow H^*(\mathcal{U}, \check{V}X) \longrightarrow H^*(\mathcal{U}, \check{U}X) \xrightarrow{f} H^*(wX, wX^{S-T}) \cdots$$

where $f$ is the composition of the map from the sequence of the triple with the excision (8.12).

Next, we want to define a splitting of $f$, i.e., a map $g : H^*(wX, wX^{S-T}) \to H^*(\mathcal{U}, \check{U}X)$ so that $f \circ g = id$. Let

$$q_w : C_*(\mathcal{U}) \to C_*(wA_TX, w(A_T - 1)X) \tag{8.13}$$

be the composition of the maps induced by $p_w$ with the inclusion $(wA_TX, \emptyset) \to (wA_TX, w(A_T - 1)X)$. By Lemma 8.3.3 (ii), $q_w$ takes $C_*(\check{U}X)$ to 0. Hence, we get an induced map $\pi_w : C^*(wA_TX, w(A_T - 1)X) \to C^*(\mathcal{U}, \check{U}X)$. Define $g$ to be the composition of the excision (8.11) with the map on cohomology induced by $\pi_w$. It is routine to unwind these definitions and check that $f \circ g = id$. Therefore,

$$H^*(\mathcal{U}, \check{U}X) \cong H^*(\mathcal{U}, \check{V}X) \oplus H^*(wX, wX^{S-T}), \tag{8.14}$$

which is the same as (8.8) in the case $U = U_n$, $w = w_n$, and $V = U_{n-1}$.

*Completion of the proof of Theorem 8.3.1.* We have proved (8.14) and therefore, (8.9). Since $H_c^*(\mathcal{U}) = \varinjlim H^*(\mathcal{U}, \check{U}_nX)$, the theorem follows. □

*Remark 8.3.4* In (8.13) we defined $q_w : C_*(\mathcal{U}) \to C_*(wA_TX, w(A_T - 1)X)$. Composing this with translation by $w^{-1}$ and the excision (8.11), we get a chain map, $C_*(\mathcal{U}) \to C_*(X, X^{\mathrm{Out}(w)})$ and its dual

$$\lambda_w : C^*(X, X^{\mathrm{Out}(w)}) \to C^*(\mathcal{U}). \tag{8.15}$$

As in the sentence following (8.13), the image of $\lambda_w$ lies in $C^*(wW_TX)$ and hence in the finitely supported cochains $C_c^*(\mathcal{U})$. The proof shows that the isomorphism in Theorem 8.3.1 is induced by the map of cochain complexes

$$\oplus\lambda_w : C_c^*(\mathcal{U}) \to \bigoplus_{w\in W} C^*(X, X^{\mathrm{Out}(w)}).$$

Just as in Theorem 8.1.6, we can rewrite Theorem 8.3.1 in terms of the free abelian groups $\mathbb{Z}W^T$, as follows.

**THEOREM 8.3.5**

$$H_c^*(\mathcal{U}) \cong \bigoplus_{T\in\mathcal{S}} \mathbb{Z}W^T \otimes H^*(X, X^{S-T}).$$

There is a relative version of Theorem 8.3.1. As in Theorem 8.1.4, suppose $Y$ is a subcomplex of $X$; give $Y$ the induced mirror structure and write $\mathcal{U}(X)$ and $\mathcal{U}(Y)$ for $\mathcal{U}(W, X)$ and $\mathcal{U}(W, Y)$, respectively.

**THEOREM 8.3.6.** (Compare Theorem 8.1.4.)

$$H_c^*(\mathcal{U}(X), \mathcal{U}(Y)) \cong \bigoplus_{w\in W} H^*(X, X^{\mathrm{Out}(w)} \cup Y).$$

*Proof.* As in (8.6),

$$H_c^*(\mathcal{U}(X), \mathcal{U}(Y)) = \varinjlim H^*(\mathcal{U}(X), \mathcal{U}(Y) \cup \check{U}_n X).$$

We use the exact sequence of the triple $(\mathcal{U}(X), \mathcal{U}(Y) \cup \check{U}_{n-1}X, \mathcal{U}(Y) \cup \check{U}_n X)$ in cohomology together with the excision

$$H^*(\mathcal{U}(Y) \cup \check{U}_{n-1}X, \mathcal{U}(Y) \cup \check{U}_n X) \cong H^*(X, Y \cup X^{\mathrm{Out}(w_n)})$$

to get, as in (8.9),

$$H^*(\mathcal{U}(X), \mathcal{U}(Y) \cup \check{U}_n X) = \bigoplus_{i=1}^{i=n} H^*(X, Y \cup X^{\mathrm{Out}(w_i)}).$$

The theorem follows. □

The proof of the following corollary is left as an exercise for the reader.

**COROLLARY 8.3.7.** (Compare Corollary 8.1.5.) *Suppose* $(X, Y)$ *is acyclic. Then*

$$H_c^*(\mathcal{U}(X), \mathcal{U}(Y)) \cong \bigoplus_{w \in W} H^{*-1}(X^{\mathrm{Out}(w)}, Y^{\mathrm{Out}(w)})$$

$$\cong \bigoplus_{T \in \mathcal{S}} \mathbb{Z}W^T \otimes H^{*-1}(X^{S-T}, Y^{S-T}).$$

## 8.4. THE CASE WHERE *X* IS A GENERAL SPACE

In this section we weaken the assumption that $X$ is a mirrored CW complex and only assume that it is a mirrored space. Do the formulas in 8.1 and 8.3 remain valid for singular homology and for compactly supported singular cohomology? The crucial ingredient needed to make the earlier arguments is that excision holds for certain subspaces in $X$. In order to give a condition which guarantees this, we define the notion of a "mirrored subspace" of $X$. It plays the role of a subcomplex. Recall from 5.1 that a *coface* of $X$ is any intersection of the form $X_T = \bigcup_{t \in T} X_t$. A *mirrored subspace* of $X$ is any union of cofaces.

A pair of spaces $(A, B)$, with $B$ a closed in $A$, is a *collared pair* if there is an open neighborhood of $B$ in $A$ which deformation retracts onto $B$. For example, if $B$ is any subcomplex of a CW complex $A$, then $(A, B)$ is collared. If $(A, B)$ is collared and $C \subset B$, then the inclusion $(A - C, B - C) \hookrightarrow (A, B)$ is an excision, i.e., $H_*(A - C, B - C) \cong H_*(A, B)$ (Proof: replace $B$ by a neighborhood $U$ which deformation retracts onto it and then excise $C$.) A mirrored space $X$ satisfies the *Collared Condition* if any pair $(A, B)$ of mirrored subspaces in $X$ is collared. In practice this will be used as follows: if $V \subset B$ is such that $A - V$ is a mirrored subspace of $X$, then $(A - V, B - V) \hookrightarrow (A, B)$ is an excision.

In this section only, chains, cochains, homology, and cohomology of a space will mean the singular versions. Assume for the remainder of the section that the mirrored space $X$ satisfies the Collared Condition. We will show this is enough to guarantee that the singular versions of the previous formulas in homology and cohomology hold.

The first place we encounter a problem is in the definition of the alternation map at the chain level. Given $T \in \mathcal{S}$, the element $\tilde{h}_T \in \mathbb{Z}W_T$ from (8.1) induces a map of singular chains $C_*(X) \to C_*(\mathcal{U})$; however, it is not true that this map vanishes on $C_*(X^T)$. The problem is that although $\tilde{h}_T$ does vanish on any singular simplex whose image is contained in $X_t$, $t \in T$, this does not imply that it vanishes on any singular simplex whose image lies in $X^T$ (the union of the $X_t$). However, the next lemma shows that alternation does work the level of homology.

**LEMMA 8.4.1.** *Given $T \in \mathcal{S}$, multiplication by $\tilde{h}_T \in \mathbb{Z}W_T$ induces a well-defined map $H_*(X, X^T) \to H_*(W_T X)$.*

*Proof.* The proof uses some standard ideas in singular homology theory. (For example, the same arguments are used in the proof of Mayer-Vietoris sequence.) For each $t \in T$, let $U_t$ be an open neighborhood of $X_t$ in $X^T$ which deformation retracts back onto it. Consider the open cover $\{U_t\}_{t\in T}$ of $X^T$. Let $C'_k(X^T)$ denote the subgroup of $C_k(X^T)$ generated by those singular simplices $\sigma : \Delta^k \to X^T$ whose image is contained in some $U_t$. It is standard (cf. [153, pp. 119–124]) that the inclusion $C'_*(X^T) \hookrightarrow C_*(X^T)$ is a chain homotopy equivalence. Let $C''_k(X^T) \subset C'_k(X^T)$ be the subgroup generated by those $\sigma : \Delta^k \to X^T$ whose image is contained in some $X_t$. Since $U_t$ is homotopy equivalent to $X_t$, $C''_*(X^T)$ is chain homotopy equivalent to $C'_*(X^T)$. So, $C_*(X, X^T) := C_*(X)/C_*(X^T)$ is chain homotopy equivalent to $C_*(X)/C''_*(X^T)$. The alternation map vanishes on $C''_*(X^T)$ (by the argument in the beginning of 8.1); hence, the lemma. □

The proof of Theorem 8.1.2 proceeds as before. The Collared Condition gives us the excision (8.3): $H_*(P_n, P_{n-1}) \cong H_*(w_n X, w_n X^{\mathrm{In}(w_n)})$. The results of 8.2 also remain valid. The reason is that the Collared Condition is enough to insure that Mayer-Vietoris sequences and spectral sequences hold for mirrored subspaces of $X$.

What about the results in 8.3 on compactly supported cohomology? Assume, as before, that $X$ is compact with a $W$-finite mirror structure (so that $\mathcal{U}$ is locally compact). The cohomology of $\mathcal{U}$ with compact supports is defined by

$$H^*_c(\mathcal{U}) := \varinjlim H^*(\mathcal{U}, \mathcal{U} - C),$$

where the direct limit is taken over some cofinal system of compact subsets $C \subset \mathcal{U}$. (See (G.8) in Appendix G.2.) Again we have (8.6): $H_c^*(\mathcal{U}) := \varinjlim H^*(\mathcal{U}, \check{U}_n X)$, where the $\check{U}_n := W - U_n$ are the subsets of $W$ defined in 8.3. Since the Collared Condition allows our arguments with excision to work, Theorems 8.3.1 and 8.3.5 hold for $\mathcal{U}$.

### The Local Topology of $\mathcal{U}$

Here are some exercises for the reader.

**Exercise 8.4.2.** Prove:

(i) $\mathcal{U}$ is locally path connected if and only if, $X$ is locally path connected.

(ii) $\mathcal{U}$ is locally $m$-acyclic if and only if for each $k \leqslant m$ and $T \in \mathcal{S}^{(k)}$, $X_T$ is locally $(m-k)$-acyclic.

(iii) $\mathcal{U}$ is locally simply connected if and only if $X$ is locally simply connected and each mirror $X_s$ is locally path connected.

A path connected and locally path connected space $Y$ admits a universal covering space if and only if it is semilocally simply connected [197]. Recall this means that for every point $y \in Y$ there is a neighborhood $U$ so that the homomorphism $\pi_1(U, y) \to \pi_1(Y, y)$, induced by the inclusion, is the zero map.

**Exercise 8.4.3.** Assume $X$ is path connected and locally path connected and that each mirror $X_s$ is nonempty. Then $\mathcal{U}$ is semilocally simply connected if and only if $X$ is semilocally simply connected.

## 8.5. COHOMOLOGY WITH GROUP RING COEFFICIENTS

In this section we use Theorem 8.3.1 to compute the cohomology of $W$ with coefficients in the group ring $\mathbb{Z}W$. We then give a number of applications of this to computations of the virtual cohomological dimension of $W$.

### The Cohomology with Compact Supports of $\Sigma$

With notation as in 7.2, we have $\Sigma = \mathcal{U}(W, K)$. The next result is a special case of Theorem 8.3.1.

**THEOREM 8.5.1.** ([77, Theorem A].)

$$H_c^*(\Sigma) \cong \bigoplus_{w \in W} H^*(K, K^{\mathrm{Out}(w)}) = \bigoplus_{T \in \mathcal{S}} \mathbb{Z}W^T \otimes H^*(K, K^{S-T}).$$

As explained in Appendix F.2 (Lemma F.2.2), $H_c^*(W;\mathbb{Z}W)\cong H_c^*(\Sigma)$, giving the following.

**COROLLARY 8.5.2.** ([77, Cor. 4.4, p. 305].)

$$H^*(W;\mathbb{Z}W)\cong\bigoplus_{w\in W}H^*(K,K^{\mathrm{Out}(w)})=\bigoplus_{T\in\mathcal{S}}\mathbb{Z}W^T\otimes H^*(K,K^{S-T}).$$

Let $L$ be the nerve of $(W,S)$. For each $T\in\mathcal{S}$, $\sigma_T$ denotes the corresponding closed simplex in $L$. (When $T=\emptyset$, $\sigma_T$ is the empty simplex.) The complement $L-\sigma_T$ is called a *punctured nerve*. It is proved in Appendix A.5 that $L-\sigma_T$ deformation retracts onto $K^{S-T}$ (Lemma A.5.5). Hence, Corollary 8.5.2 can be rewritten as follows.

**COROLLARY 8.5.3**

$$H_c^i(\Sigma)=H^i(W;\mathbb{Z}W)\cong\bigoplus_{T\in\mathcal{S}}\mathbb{Z}W^T\otimes\overline{H}^{i-1}(L-\sigma_T).$$

**COROLLARY 8.5.4.** *Given a positive integer m, the following two conditions are equivalent:*

(i) *$H_c^i(\Sigma)$ vanishes for all $i\leqslant m$.*

(ii) *$\overline{H}^{i-1}(L-\sigma_T)$ vanishes for all $T\in\mathcal{S}$ and all $i\leqslant m$.*

## Virtual Cohomological Dimension of W

One definition of the *cohomological dimension* of a group $\Gamma$ is

$$\operatorname{cd}\Gamma:=\sup\{n\mid H^n(\Gamma;M)\neq 0\text{ for some }\mathbb{Z}\Gamma\text{-module }M\}.\tag{8.16}$$

(See Appendix F.3 or [42, p. 185].) The notion of a group being of "type *FP*" is defined in Appendix F.4. It is slightly weaker than the requirement that $\Gamma$ have a finite model for its classifying space. If $\Gamma$ is type *FP* (e.g., if $\Gamma$ is a torsion-free subgroup of finite index in a finitely generated Coxeter group), then there is a better formula for $\operatorname{cd}\Gamma$ (cf. Proposition F.4.1 or [42, Prop. 6.7, p. 202]):

$$\operatorname{cd}\Gamma=\max\{n\mid H^n(\Gamma;\mathbb{Z}\Gamma)\neq 0\}.\tag{8.17}$$

Recall that a group *virtually* has some property if a subgroup of finite index has the property. For example, a finite group is virtually trivial. Since Coxeter groups have faithful linear representations, they are virtually torsion-free (Corollary D.1.4 of Appendix D.1).

If $G$ has nontrivial torsion, then $\operatorname{cd}G=\infty$. For a virtually torsion-free group $G$, a more useful notion is its *virtual cohomological dimension*, denoted vcd $G$: it is the cohomological dimension of any torsion-free subgroup of finite index. (More explanation is given in Appendix F.3.)

By (8.17), $\text{vcd}\, W = \max\{n \mid H^n(\Gamma; \mathbb{Z}\Gamma) \neq 0\}$ where $\Gamma$ is any torsion-free finite index subgroup of $W$. Since $H^*(\Gamma; \mathbb{Z}\Gamma) = H^*_c(\Sigma) = H^*(W; \mathbb{Z}W)$, we can replace $H^n(\Gamma; \mathbb{Z}\Gamma)$ by $H^n(W; \mathbb{Z}W)$. So the next result follows from Corollary 8.5.2.

**COROLLARY 8.5.5**

$$\text{vcd}\, W = \max\{n \mid \overline{H}^{n-1}(L - \sigma_T) \neq 0, \textit{ for some } T \in \mathcal{S}\}.$$

**COROLLARY 8.5.6.** *$W$ is virtually free if and only if for all $T \in \mathcal{S}$, $H^n(L - \sigma_T)$ vanishes for all $n \geqslant 1$.*

*Proof.* Stallings [266] proved that a virtually torsion-free group (such as $W$) is virtually free if and only if its virtual cohomological dimension is 1 (Theorem F.3.3). □

We completely analyze virtually free Coxeter groups in 8.8. In particular, in that section we describe all simplicial complexes $L$ satisfying the condition of Corollary 8.5.6 and, in Proposition 8.8.5, we will give a direct proof of the corollary, without invoking Stallings' Theorem.

## vcd$W$ = 2: The Eilenberg-Ganea Problem

The *geometric dimension* of a group $\Gamma$, denoted by $\text{gd}\,\Gamma$, is the smallest possible dimension of a CW model for $B\Gamma$. The Eilenberg-Ganea Problem asks if there is a group $\Gamma$ with $\text{cd}\,\Gamma = 2$ and $\text{gd}\,\Gamma = 3$. (As explained in Theorem F.3.2, whenever $\text{cd}\,\Gamma \neq 2$, the cohomological and geometric dimensions are equal.) As Bestvina pointed out in [22], torsion-free, finite-index subgroups of certain Coxeter groups are good candidates for counterexamples. We now explain why.

Let $L$ be a two-dimensional, finite simplicial complex which is acyclic but not simply connected. (There are many examples of such by Remark 8.5.7 (i), below.) Assume $(W, S)$ is a Coxeter system with $L(W, S) = L$. (By Lemma 7.2.2, we can achieve this by replacing $L$ by its barycentric subdivision and taking the associated right-angled Coxeter system.) Let $\Gamma$ be any torsion-free subgroup of finite index in $W$. Since $\dim L = 2$, $\dim \Sigma = 3$. Hence, $\text{gd}\,\Gamma \leqslant 3$. Recall that the fundamental chamber $K$ for $W$ on $\Sigma$ is the geometric realization of $\mathcal{S}$. Let $\partial K$ be the geometric realization of $\mathcal{S}_{>\emptyset}$. ($\partial K$ is the barycentric subdivision of $L$.) $\mathcal{U}(W, \partial K)$ is the singular set of $\Sigma$ (i.e., it is the set of points in $\Sigma$ with nontrivial isotropy subgroup). By Corollary 8.2.8, $\mathcal{U}(W, \partial K)$ is acyclic. (Indeed, for a nonempty spherical subset $T$ we have $(\partial K)_T = K_T$, which is a cone, while for $T = \emptyset$, we have $(\partial K)_\emptyset = \partial K$, which is acyclic.) Since $\Gamma$ acts freely on $\mathcal{U}(W, \partial K)$ and $\dim \mathcal{U}(W, \partial K) = 2$, $\operatorname{cd} \Gamma \leqslant 2$. It turns out that $\operatorname{cd} \Gamma = 2$ and hence, $\operatorname{vcd} W = 2$. (This will follow from the discussion in 8.8, below; in fact, by Proposition 8.8.5 if $\Gamma$ is free, then $L$ is in the class $\mathcal{G}'$ of simplicial complexes defined in 8.8 and such complexes are contractible.) It seems quite likely that any such $\Gamma$ associated to a nonsimply connected, acyclic 2 complex $L$ actually is a counterexample to the Eilenberg-Ganea Problem, that is, $\operatorname{gd} \Gamma = 3$ and $\operatorname{cd} \Gamma = 2$ (see Remark 8.5.7 (iv)). What is needed is a suitable invariant to show that $\operatorname{gd} \Gamma > 2$.

*Remark 8.5.7*

(i) (*The 2-skeleton of Poincaré's homology sphere.*) Here is a specific construction of an acyclic 2-complex $L$. *Poincaré's homology 3-sphere* is the 3-manifold formed by identifying opposite faces of a dodecahedron (with a $\pi/10$ twist). Its fundamental group is the binary icoahedral group (a perfect group of order 120). If we remove the interior of the dodecahedron, we get an acyclic 2-complex made from six pentagons. Take its barycentric subdivision to get a simplicial complex $L$. Since $L$ is a flag complex, there is an associated right-angled Coxeter group $W$ with nerve $L$ (Lemma 7.2.2).

(ii) Note that if the complex $L$ is contractible, then $\mathcal{U}(W, \partial K)$ is contractible, so $\operatorname{gd} \Gamma = \operatorname{cd} \Gamma$ and we will not get a counterexample.

(iii) The construction of right-angled Coxeter groups with nonsimply connected acyclic nerves works in any dimension $\geqslant 2$; however, in higher dimensions it does not provide counterexamples to the Eilenberg-Ganea Theorem (in Appendix F.3). The reason is that if $L$ is any acyclic complex, then it is always possible to attach 2- and 3-cells to it to obtain an contractible complex $L'$. We can assume these cells are attached via piecewise linear maps and that $L'$ is triangulated as a flag complex with $L$ a full subcomplex. If $\Gamma'$ is a torsion-free subgroup of the Coxeter group $W'$ corresponding to $L'$ and $\Gamma = \Gamma' \cap W$, then for $\dim L \geqslant 3$, $\operatorname{gd} \Gamma \leqslant \operatorname{gd} \Gamma' \leqslant \dim L' = \dim L$, in compliance with the Eilenberg-Ganea Theorem.

(iv) On the other hand, if $\dim L = 2$ and $\pi_1(L) \neq 1$, then it is conjectured that $L$ can never be embedded in a contractible 2-complex. If such an embedding into a finite contractible 2-complex $L'$ were possible, we could attach 1- and 2-cells to $L$ to obtain $L'$. Since $L$ and $L'$ both have Euler characteristic 1, the number of 1-cells attached is equal to the number of 2-cells. But the Kervaire Conjecture asserts that it is impossible to kill any nontrivial group $G$ by adding the same number of generators as relations. Moreover, the Kervaire Conjecture has been proved for many classes of groups. These are the intuitive reasons in [22] for believing that the above constructions actually give counterexamples to the Eilenberg-Ganea Problem.

(v) In [24] Bestvina and Brady describe an even more convincing candidate for a counterexample to the Eilenberg-Ganea Problem. This time the group $H$ is one of the "Bestvina-Brady groups" described in Section 11.6. It is subgroup of a "right-angled Artin group" instead of a right-angled Coxeter group. The Artin group is associated to the same flag triangulation $L$ of the 2-skeleton of Poincaré's homology sphere discussed in (i). It is shown in [24] that $\operatorname{cd} H = 2$ and either $\operatorname{gd} H = 3$ or else Whitehead's Conjecture on aspherical 2-complexes is false. (Whitehead's Conjecture asserts that any connected subcomplex of an aspherical 2-complex is aspherical.)

(vi) In [34] Brady, Leary, and Nucinkis define the appropriate notions of "cohomological dimension" and "geometric dimension" for groups $G$ with torsion. In this context the *geometric dimension* is the smallest dimension of any CW model for $\underline{E}G$, the universal space for proper $G$-actions (Definition 2.3.1). The cohomological version of this definition involves resolutions of the constant object $\mathbb{Z}$ by direct sums of modules of the form $\mathbb{Z}(G/F)$ where $F$ ranges over the finite subgroups of $G$. Denote these notions by $\operatorname{gd}'(\ )$ and $\operatorname{cd}'(\ )$. There is a generalization of the Eilenberg-Ganea Theorem: the cohomological and geometric dimensions coincide except in the case where the geometric dimension is 3 and the cohomological version is 2. The main result of [34] is that the Coxeter group examples discussed above are counterexamples to this version of the Eilenberg-Ganea Problem for groups with torsion: $\operatorname{cd}'(W) = 2$ and $\operatorname{gd}'(W) = 3$.

**More Examples Concerning vcd($W$)**

We continue to use the method, explained in the Notes to Chapter 7, for producing examples of right-angled Coxeter systems with various nerves. The first is a variant of an example of Bestvina-Mess [25].

**Example 8.5.8.** (*Different dimension over $\mathbb{Z}$ than $\mathbb{Q}$.*) Suppose $L$ is a triangulation of $\mathbb{R}P^2$ as a flag complex and $(W, S)$ is the right-angled Coxeter system with nerve $L$. What is the cohomology of the punctured nerves $L - \sigma_T$? When $T \neq \emptyset$, $L - \sigma_T$ is the complement of a simplex in $\mathbb{R}P^2$, i.e., it is an open Möbius band. On the other hand, for $T = \emptyset$, $L$ is $\mathbb{R}P^2$. Corollary 8.5.3 yields

$$H^i(W; \mathbb{Z}W) \cong \begin{cases} 0 & \text{if } i = 0, 1, \\ \mathbb{Z}^\infty & \text{if } i = 2, \\ \mathbb{Z}/2 & \text{if } i = 3. \end{cases}$$

With rational coeffcients, we have

$$H^i(W;\mathbb{Q}W) \cong \begin{cases} 0 & \text{if } i \neq 2, \\ \mathbb{Q}^\infty & \text{if } i = 2. \end{cases}$$

(Here $\mathbb{Z}^\infty$ means a free abelian group of countably infinite rank and $\mathbb{Q}^\infty$ means a rational vector space of countably infinite dimension.) From these formulas and (8.17) (or from Corollary 8.5.5), we get $\mathrm{vcd}_{\mathbb{Z}} W = 3$ while $\mathrm{vcd}_{\mathbb{Q}} W = 2$.

We have the following example of Dranishnikov [104].

**Example 8.5.9.** (*Nonadditivity of cohomological dimension.*) $\mathbb{R}P^2$ is a Moore space, i.e., its reduced homology is concentrated in a single dimension (in this case, dimension 1) and in that dimension it is isomorphic to a given abelian group (in this case, $\mathbb{Z}/2$). We have a similar construction for any abelian group. Let $L''$ be the result of attaching a 2-disk to $S^1$ by a PL map $\partial D^2 \to S^1$ of degree 3. Then $L''$ is a Moore space with homology $\cong \mathbb{Z}/3$ in dimension 1. Assume $L''$ is triangulated as a flag complex; $W''$ is the associated right-angled Coxeter group and $\Sigma''$ the associated complex. Exactly as in the previous example, we calculate

$$H^i_c(\Sigma'') = H^i(W'';\mathbb{Z}W'') \cong \begin{cases} 0 & \text{if } i = 0, 1, \\ \mathbb{Z}^\infty & \text{if } i = 2, \\ \mathbb{Z}/3 & \text{if } i = 3. \end{cases}$$

Let $L'$ $(= \mathbb{R}P^2)$ be as in the previous example and $W'$ and $\Sigma'$ the associated Coxeter group and cell complex. Put $W := W' \times W''$ and $\Sigma := \Sigma' \times \Sigma''$. $\Sigma$ is six dimensional. On the other hand, by the Künneth Formula, $H^6_c(\Sigma) \cong H^3_c(\Sigma') \otimes H^3_c(\Sigma_2) = \mathbb{Z}/2 \otimes \mathbb{Z}/3 = 0$, while $H^5_c(\Sigma) \neq 0$. Hence, by (8.17)

$$\mathrm{vcd}(W' \times W'') = 5 \neq 6 = \mathrm{vcd}(W') + \mathrm{vcd}(W'').$$

(This answers Problem C2 in Wall's Problem List [293, p. 376] in the negative.)

## 8.6. BACKGROUND ON THE ENDS OF A GROUP

Suppose $G$ is a group with a finite set of generators $S$. As in 2.1, let $\Omega$ be its Cayley graph. In Appendix G.3 we explain the notion of the set of "ends" of a noncompact space. In the case of a locally finite graph, such as $\Omega$, the set of ends can be described as follows. Let $\mathcal{C}$ be the poset of finite subgraphs of $\Omega$ ordered by inclusion. This gives an inverse system, $\{\Omega - C\}_{C\in\mathcal{C}}$. So, we can take path components and then form the inverse limit

$$\mathrm{Ends}(\Omega) := \varprojlim \pi_0(\Omega - C),$$

and define

$$\mathrm{Ends}(G) := \mathrm{Ends}(\Omega). \tag{8.18}$$

By [37, Prop. 8.29, p. 145], Ends($G$) is well defined up to a canonical homeomorphism. (The proposition in [37] states that any quasi-isometry $X_1 \to X_2$ of geodesic spaces induces a homeomorphism $\mathrm{Ends}(X_1) \to \mathrm{Ends}(X_2)$. We can apply this because different generating sets for $G$ give quasi-isometric Cayley graphs. The definition of "quasi-isometry" can be found in 12.5.) A proof of the following famous result of H. Hopf can be found in [37, pp. 146–147].

**THEOREM 8.6.1.** (Hopf [157].) *Suppose $G$ is a finitely generated group. Let $e(G)$ denote the number of its ends.*

*(i) $G$ has either 0, 1, 2, or infinitely many ends.*

*(ii) $e(G) = 0$ if and only if $G$ is finite.*

*(iii) $e(G) = 2$ if and only if $G$ has an infinite cyclic subgroup of finite index (i.e., if $G$ is virtually infinitely cyclic).*

*(iv) If $e(G)$ is infinite, then it is uncountable. Moreover, each point of* Ends($G$) *is an accumulation point.*

A good method for calculating the number of ends of a group is to invoke the following proposition (basically a restatement of Proposition G.3.3).

**PROPOSITION 8.6.2.** *Suppose an infinite discrete group $G$ acts properly and cocompactly on a simply connected CW complex $X$. Then*

*(i) $G$ is one-ended if and only if $H^1_c(X) = 0$.*

*(ii) $G$ has two ends if and only if $H^1_c(X) \cong \mathbb{Z}$.*

*(iii) $G$ has infinitely many ends if and only if the rank of $H^1_c(X)$ is infinite.*

*Proof.* The proof uses results from Appendix G.2. Since $\Omega$ is quasi-isometric to $X$, $e(G)$ is the rank of $H^e_0(X)$, the homology of $X$ at infinity in dimension 0. By the Universal Coefficient Theorem, when finite, $H^e_0(X)$ and $H^0_e(X)$ have the same rank and if the rank of one is infinite so is the other. (However, when the rank of $H^0_e(X)$ is countable, $H^e_0(X)$ has uncountable rank.) By Proposition G.2.1 we have an exact sequence:

$$\mathbb{Z} \longrightarrow H^0_e(X) \longrightarrow H^1_c(X) \longrightarrow H^1(X) = 0;$$

hence, the result. (See also Remark G.3.4.) □

When $G$ in the previous proposition is finitely presented, we can take $X$ to be a Cayley 2-complex.

The next result is a famous theorem of Stallings [266]. It characterizes groups with infinitely many ends in a fashion analogous to parts (ii) and (iii) of Theorem 8.6.1. A proof of Stallings' Theorem can be found in [95]. The notions of "amalgamated product" and "HNN construction" are explained in Appendix E.1 (Examples E.1.1 and E.1.2, respectively).

**THEOREM 8.6.3.** (Stallings [266].) *Suppose G is a finitely generated group with more than one end. Then G splits as an amalgamated product or HNN construction over a finite subgroup.*

**COMPLEMENT 8.6.4.** For a group with more than one end, the question arises: does the process of splitting over finite subgroups terminate after a finite number of steps (i.e., is the group "accessible") or not (is it "inaccessible")? If the group is accessible, it can be written as the fundamental group of a finite graph of groups where each edge group is finite and each vertex group is one ended. (Graphs of groups are explained in Appendix E.1.) The accessibility question was answered by Dunwoody. In [105] he showed that every finitely presented group is accessible and then in [106] he gave examples of finitely generated groups which were inaccessible. (See [95].)

## 8.7. THE ENDS OF $W$

An immediate corollary of Proposition 8.6.2 is the following.

**THEOREM 8.7.1.** *Let $(W, S)$ be a Coxeter system. Then*

*(i) $W$ is one ended if and only if $H^1_c(\Sigma) = 0$,*

*(ii) $W$ has two ends if and only if $H^1_c(\Sigma) \cong \mathbb{Z}$, and*

*(iii) $W$ has infinitely many ends if and only if the rank of $H^1_c(\Sigma)$ is infinite.*

Applying Theorem 8.5.1 and the homotopy equivalence $K^{S-T} \sim L - \sigma_T$ of Lemma A.5.5, the above result yields the next three theorems, which give concrete conditions for deciding the number of ends of $W$.

**THEOREM 8.7.2.** *$W$ is one ended if and only if, for each $T \in \mathcal{S}$, the punctured nerve $L - \sigma_T$ is connected.*

Conversely, given a spherical subset $T$ such that $L - \sigma_T$ is disconnected we get an explicit splitting of $W$ over finite subgroups as follows. Suppose $l_1, \dots, l_k$ are the components of $L - \sigma_T$. For $1 \leqslant i \leqslant k$ let $L_i$ denote the closure of $l_i$, $S_i$ the vertex set of $L_i$ and $T_i = S_i \cap T$. Define a graph of groups as follows. The graph $\Gamma$ is the cone on $\{1, \dots, k\}$; it has edges $e_1, \dots, e_k$ where $e_i$ denotes the cone on the vertex $i$. The group associated to the vertex $i$ is $W_{S_i}$, the group associated to the cone point is $W_T$ and the group associated to $e_i$ is $W_{T_i}$. The fundamental group of this graph of groups is clearly $W$.

$W$ has two ends if and only if there is a unique $T \in \mathcal{S}$ such that $L - \sigma_T$ has exactly two components and a unique $w \in W$ such that $T = \mathrm{In}(w)$. The following theorem is a special case of Theorem 10.9.2 in Chapter 10. Although it follows fairly easily from the above remarks and Lemma 4.7.5, we postpone the details of the proof until we get to Theorem 10.9.2.

**THEOREM 8.7.3.** *$W$ is two ended if and only if $(W, S)$ decomposes as*

$$(W, S) = (W_0 \times W_1, S_0 \cup S_1)$$

*where $W_1$ is finite and $W_0$ is the infinite dihedral group.*

In the above case, $L(W, S)$ is the suspension of the simplex on $S_1$. The two suspension points are the two elements of $S_0$.

Combining Theorems 8.7.2 and 8.7.3 we get the following.

**THEOREM 8.7.4.** *$W$ has infinitely many ends if and only if it is not as in Theorem 8.7.3 and there is at least one $T \in \mathcal{S}$ so that the punctured nerve $L - \sigma_T$ is disconnected.*

## 8.8. SPLITTINGS OF COXETER GROUPS

As usual, $(W, S)$ is a Coxeter system and $L$ is its nerve. Suppose $L_0$ is a full subcomplex of $L$. (This means that if $T$ is any simplex of $L$ such that its vertices are in $L_0$, then $T \subset L_0$.) Set $S_0 = \mathrm{Vert}(L_0)$ and $W_0 = W_{S_0}$.

Suppose that $L_0$ disconnects $L$, in other words, that $L - L_0$ has more than one component. (We allow the possibility that $L_0$ might be empty.) Then we can write $L - L_0$ as a disjoint union, $L - L_0 = l_1 \cup l_2$ where for $i = 1, 2$, $l_i$ is a nonempty union of components of $L - L_0$. For $i = 1, 2$, let $S_i$ denote the union of $S_0$ with the vertices in $l_i$, let $L_i$ be the full subcomplex of $L$ spanned by $S_i$ and let $W_i = W_{S_i}$. Clearly, $L_i$ is the nerve of $(W_i, S_i)$, $S_1 \cap S_2 = S_0$ and $L_1 \cap L_2 = L_0$. Moreover, if a vertex $s_1 \in S_1$ is connected by an edge to a vertex $s_2 \in S_2$, then either $s_1$ or $s_2$ lies $S_0$. From this and the standard presentation of a Coxeter group, we deduce the following.

**PROPOSITION 8.8.1.** *With notation as above, $W$ is the amalgamated product of $W_1$ and $W_2$ along $W_0$.*

If $W$ has more than one end, then, by Theorem 8.7.2, we can choose $L_0$ to be a (possibly empty) simplex $\sigma_T$ of $L$. Then $W_0 = W_T$ is finite and we obtain a splitting of $W$ as an amalgamated product over a finite group. (The existence of such a splitting is predicted by Stalling's Theorem 8.6.3.) This process can be continued. We can ask if $W_1$ or $W_2$ has more than one end. If, say, $W_1$ has more than one end, then we can decompose $L_1$ as a union of two pieces along a simplex and decompose $W_1$ as an amalgamated product along another spherical special subgroup. Since the number of generators of one of the components decreases at each step, this process terminates. So we have proved the following proposition, which we will need in 9.2.

**PROPOSITION 8.8.2.** *Any Coxeter system decomposes as a tree of groups, where each vertex group is a* 0- *or* 1-*ended special subgroup and each edge group is a finite special subgroup.*

(The notion of a graph of spaces is defined in Appendix E.1. A *tree of groups* (or *spaces*) is the special case where the underlying graph is a tree. To say that a group decomposes as a tree of groups just means that it can be written as an iterated amalgamated product of vertex groups, amalgamated along the edge groups.) The proof of the proposition shows that any finite simplicial complex $L$ can be decomposed as a tree of subcomplexes where the subcomplex associated to each edge is a simplex (or empty) and for any vertex $v$ the associated subcomplex $L_v$ satisfies the condition of Theorem 8.7.2, i.e., for any simplex $\sigma$ in $L_v$, $L_v - \sigma$ is connected (or empty).

### Virtually Free Coxeter Groups

Following [141], define $\mathcal{G}$ to be the smallest class of Coxeter groups such that

- each spherical Coxeter group is in $\mathcal{G}$, and
- if $W_1, W_2 \in \mathcal{G}$ and if $W_0$ is a common spherical special subgroup of both, then $W_1 *_{W_0} W_2 \in \mathcal{G}$.

Equivalently, $W$ is in $\mathcal{G}$ if and only if it has a tree of groups decomposition where each vertex group is a spherical special subgroup.

Our previous discussion suggests the following: define $\mathcal{G}'$ to be the smallest class of finite simplicial complexes such that

- for any $n \in \mathbb{N}$, each $n$-simplex is in $\mathcal{G}'$, and
- if $L_1, L_2 \in \mathcal{G}'$ and if $L_0$ is a subcomplex of both which is either empty or a simplex, then the simplicial complex formed by gluing $L_1$ to $L_2$ along the simplex $L_0$ is in $\mathcal{G}'$.

Equivalently, $L$ is in $\mathcal{G}'$ if and only if it can be decomposed as a "tree of simplices," i.e., a tree of subcomplexes where the subcomplex associated to each vertex is a simplex and the subcomplex associated to an edge is a common face (possibly empty) of the simplices associated to its endpoints.

**Example 8.8.3.** Any finite tree is in $\mathcal{G}'$. We can think of each component of a simplicial complex in $\mathcal{G}'$ as being a "thick tree."

**LEMMA 8.8.4.** *Let $L$ be a finite simplicial complex such that for all simplices $\sigma \subsetneq L$ (including $\sigma = \emptyset$), $L - \sigma$ is acyclic. Then $L$ is a simplex.*

*Proof.* For each $v \in \text{Vert}(L)$, let $\text{Lk}(v)$ denote its link in $L$ (defined in Appendix A.6). We claim that $\text{Lk}(v)$ also satisfies the hypothesis of the lemma. To see this, first note the excision $H_*(L, L - v) \cong H_*(\text{Star}(v), \text{Lk}(v))$ where $\text{Star}(v)$ denotes the union of all closed simplices with $v$ as a vertex. We can assume $L \neq v$ (otherwise $\text{Lk}(v) = \emptyset$). Since $L$ and $L - v$ are both acyclic and $\text{Star}(v)$ is a cone, it follows from the exact sequence of $(L, L - v)$ that $\text{Lk}(v)$ is acyclic. Let $\sigma$ be a simplex of $\text{Lk}(v)$. Then $\sigma$ corresponds to a simplex $\sigma'$ of $L$ containing $v$. ($\sigma'$ is the join of $v$ and $\sigma$.) We have another excision, $H_*(L - v, L - \sigma') \cong \overline{H}_{*-1}(\text{Lk}(v) - \sigma)$. By induction on the number of vertices, $\text{Lk}(v)$ is a simplex $\sigma$. We can identify $\sigma$ with a simplex of $L$, namely the face opposite to $v$ in $\sigma'$. If $L = \sigma'$, we are done. Otherwise, $L - \sigma$ has a least two components, contradicting the hypothesis that it is acyclic. □

For any Coxeter system $(W, S)$, it is obvious that $W \in \mathcal{G}$ if and only if the nerve of $(W, S)$ is in $\mathcal{G}'$.

**PROPOSITION 8.8.5.** *The following conditions on a Coxeter group $W$ are equivalent:*

*(a)* $L(W, S) \in \mathcal{G}'$.

*(b)* $W \in \mathcal{G}$.

*(c)* *$W$ is virtually free.*

*(d)* *For all $T \in \mathcal{S}$, $H^n(L - \sigma_T)$ vanishes for all $n \geqslant 1$.*

*Proof.* We pointed out before that (a) $\Longrightarrow$ (b). Obviously, (b) $\Longrightarrow$ (c). Also, if $W$ is virtually free, then $\text{vcd}\, W = 1$; so (c) $\Longrightarrow$ (d) by Corollary 8.5.6. Suppose that $W$ is infinite and (d) holds. Since $L$ is not a simplex, there is a $T \in \mathcal{S}$ such that $\overline{H}^0(L - \sigma_T) \neq 0$ (by Lemma 8.8.4). As in Proposition 8.8.1, this gives a splitting of $W$ as an amalgamated product of two special subgroups over the finite special subgroup $W_T$. It follows (by induction on $\text{Card}(S)$) that $W \in \mathcal{G}$ and $L(W, S) \in \mathcal{G}'$. So, (d) $\Longrightarrow$ (a). □

In 14.2 we will give another condition equivalent to the conditions in Proposition 8.8.5. (The condition is that $W$ does not contain subgroup isomorphic to the fundamental group of a closed orientable surface of genus $> 0$.)

## 8.9. COHOMOLOGY OF NORMALIZERS OF SPHERICAL SPECIAL SUBGROUPS

For any $T \in \mathcal{S}$, the vertex set of $\mathrm{Lk}(\sigma_T, L)$ can be identified with

$$\{U \in \mathcal{S}_{>T} \mid \mathrm{Card}(U - T) = 1\}.$$

Hence, the vertices of $\mathrm{Lk}(\sigma_T, L)$ are indexed by the set

$$V(T) := \{a \in S - T \mid T \cup \{a\} \in \mathcal{S}\}. \tag{8.19}$$

For each $a \in V(T)$, put $\partial_a K_T := K_{T\cup\{a\}}$ and for any $A \subset V(T)$, put

$$\partial_A K_T := \bigcup_{a\in A} \partial_a K_T. \tag{8.20}$$

For any $T \in \mathcal{S}$, in 7.5, we defined $F_{\min}(T) := \{wW_{T'} \mid W_T = wW_{T'}w^{-1}\}$. Let $\mathcal{F}(T)$ be the set of $w \in W$ such that for some $T' \in \mathcal{S}$ the coset $wW_{T'} \in F_{\min}(T)$ and $w$ is the longest element in $wW_{T'}$. (See 4.6 for facts about the longest element.) For each $w \in \mathcal{F}(T)$, let $T_w$ denote the corresponding element $T'$ of $\mathcal{S}$, i.e., $T_w = T' = w^{-1}Tw$.

As in 8.3, assume $X$ is a finite complex with a $W$-finite mirror structure. For each $w \in \mathcal{F}(T)$, set $A(w,T) := V(T_w) - \mathrm{In}(w)$ and

$$\delta_w X_T := \partial_{A(w,T)} X_{T_w}, \tag{8.21}$$

where $V(T_w)$ is defined by (8.19), $\partial_{A(w,T)}$ by (8.20) and $\mathrm{In}(w)$ by Definition 4.7.1. We have the following generalization of Theorem 8.3.1.

**THEOREM 8.9.1.** ([60, p. 456].) *For each $T \in \mathcal{S}$,*

$$H^*_c(\mathrm{Fix}(W_T, \mathcal{U})) \cong \bigoplus_{w\in\mathcal{F}(T)} H^*(X_{T_w}, \delta_w X_{T_w}).$$

*Remarks*

(i) In the above formula, $T_w$ ranges over all subsets $T'$ of $S$ which are conjugate to $T$.

(ii) The $wX_{T_w}$, with $w \in \mathcal{F}(T)$, are cofaces of codimension 0 in $\mathrm{Fix}(W_T, \Sigma)$.

(iii) If we order the elements of $\mathcal{F}(T)$ by word length, then $\delta_w X_{T_w}$ is the union of cofaces of the form $\partial_a X_{T_w}$ such that the adjacent coface to $wX_{T_w}$ across $w\partial_a X_{T_w}$ is further from the base ($= X_T$) than is $wX_{T_w}$.

*Sketch of proof of Theorem 8.9.1.* In Remark 8.3.4 we defined, for each $w \in W$, a map of cochain complexes $\lambda_w : C^*(X, X^{\mathrm{Out}(w)}) \to C^*_c(\mathcal{U})$. If $w \in \mathcal{F}(T)$, then $\lambda_w$ restricts to a map

$$\lambda^T_w : C^*(X_{T_w}, \delta_w X_{T_w}) \to C^*_c(\mathrm{Fix}(W_T, \mathcal{U})),$$

and hence, we get

$$\lambda^T := \oplus \lambda^T_w : \bigoplus_{w \in \mathcal{F}(T)} C^*(X_{T_w}, \delta_w X_{T_w}) \to C^*_c(\mathrm{Fix}(W_T, \mathcal{U})).$$

An argument similar to the proof of Theorem 8.3.1 shows that $\lambda^T$ induces an isomorphism in cohomology. □

Applying Theorem 8.9.1 to the case $\mathcal{U} = \Sigma$, we get the following.

**COROLLARY 8.9.2.** *For each $T \in \mathcal{S}$,*

$$H^*_c(\mathrm{Fix}(W_T, \Sigma)) \cong \bigoplus_{w \in \mathcal{F}(T)} H^*(K_{T_w}, \delta_w K_{T_w}).$$

$N(W_T)$, the normalizer of $W_T$ in $W$, acts on $\mathrm{Fix}(W_T, \Sigma)$ properly and cocompactly. We will prove later (in Theorem 12.3.4 (ii)) that $\mathrm{Fix}(W_T, \Sigma)$ is contractible. This is a consequence of Moussong's Theorem (12.3.3), which states that the natural piecewise Euclidean metric on $\Sigma$ is CAT(0). Hence, we have the following generalization of Corollary 8.5.2.

**COROLLARY 8.9.3.** *For each $T \in \mathcal{S}$,*

$$H^*(N(W_T); \mathbb{Z}N(W_T)) \cong \bigoplus_{w \in \mathcal{F}(T)} H^*(K_{T_w}, \delta_w K_{T_w}).$$

## NOTES

**8.2.** The argument in Remark 8.2.14 is often used to prove some space with group action is contractible. One shows there is a contractible fundamental domain and that its translates can be ordered so that the union of the first $n$ of them is formed by gluing on the $n^{\text{th}}$ to the previous ones along a contractible subspace.

**8.4.** In order to make our arguments with excision work, we assumed that the mirrored space $X$ satisfied the "Collared Condition." The following weaker assumption would suffice: for any two mirrored subspaces $X_1, X_2$ of $X$, the triple $(X_1 \cup X_2, X_1, X_2)$ should be an "exact triad." (See [142, p. 98].) This means that $(X_1, X_1 \cap X_2) \hookrightarrow (X_1 \cup X_2, X_2)$ is an excision (and similarly with the roles of $X_1$ and $X_2$ switched).

**8.5.** The results of this section are from [77]. The complex in Example 8.5.8 will be discussed again in Example 12.4.1 of Section 12.4, where we consider its "visual boundary." For more discussion of the vcd of a Coxeter group, see [22, 97].

**8.6.** The discussion here is taken from Bridson and Haefliger [37]. The definition of a "quasi-isometry" can be found there or in Definition 12.5.1.

**8.9.** This section is taken from [60]. One could give a direct proof of the fact that $\mathrm{Fix}(W_T, \Sigma)$ is acyclic for each $T \in \mathcal{S}$, without invoking Moussong's Theorem. The argument is along the lines laid down in 8.1 and 8.2 (e.g., in the proof of Theorem 8.1.2).

# Chapter Nine

# THE FUNDAMENTAL GROUP AND THE FUNDAMENTAL GROUP AT INFINITY

Section 9.1 deals with $\pi_1(\mathcal{U})$. The universal cover of $\mathcal{U}$ can be constructed as a space of the form $\widetilde{\mathcal{U}} = \mathcal{U}(\widetilde{W}, \widetilde{X})$, where $\widetilde{X}$ is the universal cover of the fundamental chamber $X$ and the Coxeter group $\widetilde{W}$ has a fundamental generator for each component of the inverse image in $\widetilde{X}$ of each mirror of $X$. This gives a short exact sequence computing $\pi_1(\mathcal{U})$ in terms of $\pi_1(X)$ and $\widetilde{W}$ (Theorem 9.1.5).

The main result of 9.2 gives a necessary and sufficient condition for $\Sigma$ to be simply connected at infinity (Theorem 9.2.2). The condition is phrased in terms of the nerve $L$ and its "punctured" versions, $L - \sigma_T$. The proof of the theorem is based on some general results from Appendix G.4 on semistability and the fundamental group at infinity.

## 9.1. THE FUNDAMENTAL GROUP OF $\mathcal{U}$

### When Is $\mathcal{U}$ Simply Connected?

We pointed out in the beginning of 5.1 that the projection map $p : \mathcal{U} \to X$ is a retraction. This implies the following.

**LEMMA 9.1.1.** *For each nonnegative integer $i$, $p_* : \pi_i(\mathcal{U}) \to \pi_i(X)$ is onto.*

**COROLLARY 9.1.2.** *If $\mathcal{U}$ is simply connected, then so is $X$.*

We assume for the rest of this chapter that $X$ is a connected, mirrored CW complex. One of the main results in this section is the following.

**THEOREM 9.1.3.** (Compare Proposition 8.2.11.) *$\mathcal{U}$ is simply connected if and only if the following three conditions hold:*

(a) *$X$ is simply connected.*

(b) *For each $s \in S$, $X_s$ is nonempty and path connected.*

(c) *For each spherical subset $\{s, t\} \in \mathcal{S}^{(2)}$, $X_s \cap X_t$ is nonempty.*

*Proof.* Suppose $\mathcal{U}$ is simply connected. By Corollary 9.1.2, (a) holds. Since $H_1(\mathcal{U})$ is the abelianization of $\pi_1(\mathcal{U})$, $H_1(\mathcal{U}) = 0$. Hence, Proposition 8.2.11 implies that (b) and (c) hold.

Conversely, suppose (a), (b), and (c) hold. Then, as in the proof of Proposition 8.2.11, for each nonempty spherical subset $T$, $X^T$ is nonempty and path connected. As in 8.1, $\mathcal{U}$ is an increasing union of chambers, $P_1 \subset \cdots \subset P_n \subset \cdots$, where $P_1 = X$ and $P_n$ is obtained from $P_{n-1}$ by gluing on a copy of $X$ along a subspace of the form $X^T$, $T \in \mathcal{S}_{>\emptyset}$. We claim that each $P_n$ is simply connected. For $n = 1$, this is condition (a). Assume by induction that $P_{n-1}$ is simply connected. We plan to apply the Seifert–van Kampen Theorem ([253, 168] or [197]) to $P_n = P_{n-1} \cup X$. Since $X$ is simply connected ((a) again) and since their intersection $P_{n-1} \cap X = X^T$ is path connected, the Seifert–van Kampen Theorem implies that $P_n$ is simply connected. Since $\mathcal{U}$ is the increasing union of the $P_n$, $\pi_1(\mathcal{U})$ is trivial. □

There is an alternative argument using covering space theory for the second implication in the proof of the previous theorem.

*Alternative proof of Theorem 9.1.3.* Suppose $\mathcal{U}$ admits a universal covering space $q : \widetilde{\mathcal{U}} \to \mathcal{U}$. We will show that if (a), (b), and (c) hold, then $q$ is a homeomorphism and hence $\mathcal{U}$ is simply connected. As usual, identify $X$ with a chamber of $\mathcal{U}$. Since $X$ is simply connected, $q$ maps each path component of $q^{-1}(X)$ homeomorphically onto $X$. Choose such a component and call it $Y$. Define $c : X \to \widetilde{\mathcal{U}}$ to be the inverse of the homeomorphism $q|_Y$. Next we lift the $W$-action to $\widetilde{\mathcal{U}}$. First we lift the generators. Each $s \in S$ has a unique lift $\tilde{s} : \widetilde{\mathcal{U}} \to \widetilde{\mathcal{U}}$ fixing a given basepoint in $c(X_s)$. Since $\tilde{s}^2$ covers the identity on $\mathcal{U}$ and fixes a basepoint, it is the identity on $\widetilde{\mathcal{U}}$, i.e., $\tilde{s}$ is an involution. Since, by (b), $X_s$ is path connected, $\tilde{s}$ must fix all of $c(X_s)$. So, the definition of $\tilde{s}$ is independent of the choice of basepoint in $c(X_s)$. Suppose $\{s, t\} \in \mathcal{S}^{(2)}$ (in other words, $m(s,t) \neq \infty$). Since, by (c), $X_{\{s,t\}}$ $(= X_s \cap X_t)$ is nonempty, we can choose the basepoint in $X_{\{s,t\}}$. Then $\tilde{s}\tilde{t}$ is the unique lift of $st$ fixing the basepoint. For $m = m(s,t)$, $(\tilde{s}\tilde{t})^m$ is the lift of the identity which fixes the basepoint; hence, $(\tilde{s}\tilde{t})^m$ is the identity on $\widetilde{\mathcal{U}}$. This shows that the $W$-action on $\mathcal{U}$ lifts to a $W$-action on $\widetilde{\mathcal{U}}$ (so that the projection map $q$ is $W$-equivariant). By the universal property of $\mathcal{U}$ in Lemma 5.2.5, $c : X \to \widetilde{\mathcal{U}}$ extends to a $W$-equivariant map $\tilde{c} : \mathcal{U} \to \widetilde{\mathcal{U}}$. Since $c$ is a section of $q|_Y$, $\tilde{c}$ is a section of $q$. Therefore, $q : \widetilde{\mathcal{U}} \to \mathcal{U}$ is the trivial covering and consequently, $\mathcal{U}$ is simply connected. □

Combining Theorem 9.1.3 and Corollary 8.2.2, we get the next result.

**THEOREM 9.1.4.** ([71, Cor. 10.3].) *Suppose $X$ is a mirrored CW complex. Then $\mathcal{U}$ is contractible if and only if the following two conditions hold:*

*(a) $X$ is contractible.*

*(b) For each $T \in \mathcal{S}_{>\emptyset}$, $X_T$ is acyclic.*

**The Universal Cover of $\mathcal{U}$**

Let $\pi : \widetilde{X} \to X$ be the universal cover. Define a mirror structure on $\widetilde{X}$ as follows. Let $\widetilde{S}$ denote the disjoint union over all $s \in S$ of the path components of $\pi^{-1}(X_s)$, that is,

$$\widetilde{S} = \bigcup_{s \in S} \pi_0(\pi^{-1}(X_s)). \tag{9.1}$$

If $a \in \widetilde{S}$, then it is a component of $\pi^{-1}(X_s)$ for some $s \in S$. Let $f : \widetilde{S} \to S$ be the function $a \to s$. This gives a tautological mirror structure on $\widetilde{X}$ over $\widetilde{S}$, namely, $\widetilde{X}_a := a$. If $(m_{st})$ is the Coxeter matrix for $(W, S)$, define a Coxeter matrix $(\tilde{m}_{ab})$ on $\widetilde{S}$ by

$$\tilde{m}_{ab} = \begin{cases} 1 & \text{if } a = b, \\ m_{f(a)f(b)} & \text{if } a \cap b \neq \emptyset, \\ \infty & \text{otherwise.} \end{cases} \tag{9.2}$$

Let $(\widetilde{W}, \widetilde{S})$ be the associated Coxeter system. ($\widetilde{S}$ can be infinite.) The function $f : \widetilde{S} \to S$ induces a homomorphism $\varphi : \widetilde{W} \to W$. Put $\widetilde{\mathcal{U}} = \mathcal{U}(\widetilde{W}, \widetilde{X})$ and define $\overline{\pi} : \widetilde{\mathcal{U}} \to \mathcal{U}$ by $[\tilde{w}, \tilde{x}] \to [\varphi(\tilde{w}), f(\tilde{x})]$. The map $\overline{\pi}$ is $\varphi$-equivariant and clearly is a covering projection. By Theorem 9.1.3, $\widetilde{\mathcal{U}}$ is the universal cover of $\mathcal{U}$.

Next, we want to describe the group of all lifts of the $W$-action on $\mathcal{U}$ to $\widetilde{\mathcal{U}}$. Let $H = \pi_1(X)$, regarded as the group of deck transformation of $\widetilde{X}$. $H$ acts on $\widetilde{S}$ and leaves the Coxeter matrix invariant. So there is an induced action of $H$ on $\widetilde{W}$ by automorphisms. With respect to this action, form the semidirect product $\widetilde{W} \rtimes H$. The formula

$$(\tilde{w}, h) \cdot [\tilde{v}, \tilde{x}] = [\tilde{w}h(\tilde{v}), hx], \tag{9.3}$$

gives a well-defined action of $\widetilde{W} \rtimes H$ on $\widetilde{\mathcal{U}}$ (see Proposition 9.1.7 below). Clearly, $\widetilde{W} \rtimes H$ is the group of all lifts of the $W$-action to $\widetilde{\mathcal{U}}$. So, we have the following generalization of Theorem 9.1.3, a computation of $\pi_1(\mathcal{U})$.

**THEOREM 9.1.5.** *With notation as above, the following sequence is short exact:*

$$1 \to \pi_1(\mathcal{U}) \to \widetilde{W} \rtimes H \to W \to 1.$$

**The Semidirect Product Construction**

It is worth abstracting the above construction, since it will be needed again in Chapter 11.

**DEFINITION 9.1.6.** Suppose $W$ is a Coxeter system. An automorphism $\varphi$ of $W$ is a *diagram automorphism* of $(W, S)$ if $\varphi(S) = S$.

If $\varphi$ is a diagram automorphism, then it preserves the associated Coxeter matrix:

$$m_{\varphi(s)\varphi(t)} = m_{st}. \tag{9.4}$$

Since $S$ generates $W$, $\varphi$ is determined by $\varphi|_S$. Conversely, any bijection $\varphi : S \to S$ satisfying (9.4) extends to an automorphism of $W$ (since $W$ is the group determined by the presentation associated to $(m_{st})$). Suppose

- $H$ is a group of diagram automorphisms of $(W, S)$,
- $X$ is a mirrored space over $S$, and
- $H$ acts on $X$ compatibly with its action on $S$, i.e., $hX_s = X_{hs}$ for all $h \in H$.

Put $G := W \rtimes H$ and $\mathcal{U} := \mathcal{U}(W, X)$. $G$ acts on $\mathcal{U}$ as follows: given $g = (w, h) \in G$ and $[u, x] \in \mathcal{U}$,

$$g \cdot [u, x] := [wh(u), hx]. \tag{9.5}$$

**PROPOSITION 9.1.7.** *The expression for $g \cdot [u, x]$ in (9.5) is well defined and gives an action of $W \rtimes H$ on $\mathcal{U}(W, X)$.*

*Proof.* The equivalence relation in the definition of $\mathcal{U}(W, X)$ is such that $[uv, x] = [u, x]$ for all $v \in W_{S(x)}$. So, to prove formula (9.5) is well defined, we must show that when we replace $[u, x]$ by $[uv, x]$ on the left-hand side, the right-hand side remains unchanged. By the compatibility of the $H$-actions on $S$ and $X$, $S(hx) = h(S(x))$. Hence, $g \cdot [uv, x] = [wh(uv), hx] = [wh(u)h(v), hx] = [wh(u), hx] = g \cdot [u, x]$, where the next to last equality holds since $h(v) \in W_{S(hx)}$.

Suppose $g_1 = (w_1, h_1)$ and $g_2 = (w_2, h_2)$. By definition of multiplication in the semidirect product, $g_1g_2 = (w_1h_1(w_2), h_1h_2)$. To prove that (9.5) defines an action, we must show $g_1 \cdot (g_2 \cdot [u, x]) = (g_1g_2) \cdot [u, x]$. Indeed,

$$\begin{aligned} g_1 \cdot (g_2 \cdot [u, x]) &= g_1 \cdot [w_2h_2(u), h_2x] = [w_1h_1(w_2h_2(u)), h_1h_2x] \\ &= [w_1h_1(w_2)(h_1h_2)(u), h_1h_2x] = (g_1g_2) \cdot [u, x]. \end{aligned}$$ □

**Example 9.1.8.** (*The case $\mathcal{U} = \Sigma$.*) If $H$ is a group of diagram automorphisms of $(W, S)$, then the $H$-action on $S$ induces a $H$-action on the poset of spherical subsets $\mathcal{S}$ and its geometric realization $K$. Hence, $G$ acts on $\Sigma = \mathcal{U}(W, S)$.

We record some properties of the semidirect product construction.

**PROPOSITION 9.1.9**

*(i)* $\mathcal{U}/G \cong X/H$.

*(ii) Given* $x \in X$, *the isotropy subgroup* $H_x$ *permutes* $S(x)$ *(defined by (5.1)). So,* $H_x$ *is a group of diagram automorphisms of* $(W_{S(x)}, S(x))$. *The isotropy subgroup of* $G$ *at the point* $[1,x] \in \mathcal{U}$ *is* $W_{S(x)} \rtimes H_x$.

*(iii) If* $W$ *acts properly on* $\mathcal{U}$ *and* $H$ *acts properly on* $X$, *then* $G$ *acts properly on* $\mathcal{U}$.

(For the meaning of "proper action," see Definition 5.1.5.)

*Proof.* (i) $\mathcal{U}/W \cong X$; hence, $\mathcal{U}/G \cong X/H$.

(ii) The first two sentences of (ii) are obvious. If $(w,h)$ fixes $[1,x]$, then $(w,h) \cdot [1,x] = [w,hx] = [1,x]$. So, $h \in H_x$ and $w \in W_{S(x)}$.

(iii) Conditions for the $W$-action on $\mathcal{U}$ to be proper are given in Lemma 5.1.7. So, (iii) follows from (ii). □

## 9.2. WHEN IS Σ SIMPLY CONNECTED AT INFINITY?

A *neighborhood of infinity* in a noncompact space $Y$ is the complement of a compact subspace. A one-ended space $Y$ is *simply connected at infinity* if for every neighborhood of infinity $U$ there is a smaller neighborhood of infinity $V$ so that every loop in $V$ is null-homotopic in $U$. The notion of being simply connected at an end is defined analogously.

Appendix G.4 explains notions related to the "fundamental group at infinity" of a noncompact space $Y$. Given a proper ray $r : [0,\infty) \to Y$ and an exhaustion $C_1 \subset C_2 \cdots$ of $Y$ by compact subsets, we get an inverse sequence $\{\pi_1(Y - C_i, y_i)\}$ where the basepoints $y_i$ are chosen on the ray $r$. The ray determines an end $e$ of $Y$. $Y$ is *semistable at* $e$ if the inverse sequence $\{\pi_1(Y - C_i)\}$ satisfies the Mittag-Leffler Condition; it is *semistable* if it is semistable at each end. (Semistability does not depend on the choices of ray or compact subsets.) If $Y$ is semistable at $e$, the inverse limit,

$$\pi_1^e(Y) := \varprojlim \pi_1(Y - C_i),$$

is well defined up to isomorphism and it vanishes if and only if $Y$ is simply connected at $e$ (Proposition G.4.3).

As usual, $(W,S)$ is a Coxeter system, $L$ is its nerve and $\Sigma$ is the associated complex. For each spherical subset $T$, $\sigma_T$ is the corresponding (closed) simplex in $L$. The goal of this section is to prove the next two theorems.

**THEOREM 9.2.1.** ([207, Theorem 1.1].) *$\Sigma$ is semistable at each of its ends.*

Recall Theorem 8.7.2 states that $\Sigma$ is one ended if and only if for each $T \in \mathcal{S}$, $L - \sigma_T$ is connected. Similarly, we have the following.

**THEOREM 9.2.2.** ([89]). *$\Sigma$ is one ended and simply connected at infinity if and only if for each $T \in \mathcal{S}$, $L - \sigma_T$ is simply connected.*

**LEMMA 9.2.3.** *For any $T \in \mathcal{S}_{>\emptyset}$,*

*(i) $K^T$ is contractible, and*

*(ii) $K$ deformation retracts onto $K^T$.*

*Proof.* Statement (i) is Lemma 7.2.5. Since $K^T$ is a subcomplex of $K$ and since both $K$ and $K^T$ are contractible, (ii) follows. □

As in Chapter 8, given a subset $U \subset W$, put

$$UK := \bigcup_{w \in U} wK. \tag{9.6}$$

For any $T \subset S$, let $A_T \subset W$ be the fundamental $T$-sector (Definition 4.5.1).

**LEMMA 9.2.4.** *For any $T \subset S$, $(A_T - 1)K$ deformation retracts onto $K^{S-T}$.*

(By Lemma A.5.5 in Appendix A.5, $K^{S-T}$ and $L - \sigma_T$ are homotopy equivalent.)

*Proof.* For the notion of a "starlike" subset of $W$, see Definition 4.5.5. Put $V_1 := \{1\} \cup S - T$. Then $V_1$ is starlike (with respect to 1). Extend this to an increasing sequence of finite, starlike subsets $V_1 \subset V_2 \subset \cdots$ that exhaust $A_T$ so that each $V_n - V_{n-1}$ consists of a single extreme element $w_n$. (One way to do this is to order the elements of $A_T$ by word length, $w_2, w_3, \dots$, starting with the elements of length 2.) As in Lemma 8.1.1, $w_{n+1}K \cap V_nK = w_{n+1}K^{\mathrm{In}(w_{n+1})} \cong K^{\mathrm{In}(w_{n+1})}$. By Lemmas 4.7.2 and 9.2.3, $V_{n+1}K$ deformation retracts onto $V_nK$ for any $n \geqslant 1$. Hence, $A_TK$ deformation retracts onto $V_1K$ and $(A_T - 1)K$ deformation retracts onto $(V_1 - 1)K$. Using Lemma 9.2.3 again, we see that the last space deformation retracts onto $K^{S-T}$. The above argument works even when $T = \emptyset$. We conclude that $(W - 1)K$ deformation retracts onto $K^S$ $(= \partial K)$. □

As in Chapter 8, for any subset $U$ of $W$, $\check{U}$ denotes its complement.

**LEMMA 9.2.5.** ([89, Lemma 3.4].) *Suppose $U \subset W$ is starlike (with respect to a base element $w_0$), that $w$ is an extreme element of $U$, and that $V = U - w$.*

*Set* $T := \mathrm{In}(w_0^{-1}w)$. *Then*

(i) $\check{V}K$ *is homotopy-equivalent to* $\check{U}K$ *with* $wK^{S-T}$ *coned off.*

(ii) $\check{U}K$ *retracts onto* $wK^{S-T}$.

*Suppose further that* $W$ *is infinite. Then*

(a) $\pi_0(\check{U}K) \to \pi_0(\check{V}K)$ *is surjective.*

(b) *If we choose basepoints* $\{p_i\}_{1 \leqslant i \leqslant k}$ *in each component of* $wK^{S-T}$, *then the* $p_i$*'s lie in distinct components of* $\check{U}K$ *and* $\pi_1(\check{V}K)$ *is the quotient of the free product* $\pi_1(\check{U}K, p_1) * \cdots * \pi_1(\check{U}K, p_k)$.

*Proof.* Without loss of generality we can assume $w_0 = 1$.

(i) $\check{V}K = \check{U}K \cup wK$ and $\check{U}K \cap wK = wK^{S-T}$. Since $K$ is contractible, (i) follows.

(ii) As in Definition 4.5.2, we have the fundamental $T$-retraction $p_T : W \to A_T$ (sending $u$ to the shortest element in $uW_T$) and the retraction $p_w : W \to wA_T$ onto the $T$-sector based at $w$ (where, as in (8.10), $p_w(u) = wp_T(w^{-1}u)$). In Corollary 5.3.5, we showed that $p_w$ induces a retraction $\overline{p}_w : \Sigma \to wA_TK$ of topological spaces. By Lemma 8.3.3 (ii), $p_w(\check{U}) = w(A_T - 1)$. Hence, $\overline{p}_w(\check{U}K) = w(A_T - 1)K$. By Lemma 9.2.4, $w(A_T - 1)K$ deformation retracts onto $wK^{S-T}$, proving (ii).

Statement (a) follows from (i) and the fact that $K^{S-T} \neq \emptyset$ (since $W$ is infinite).

To prove (b), let $\Upsilon_1, \dots, \Upsilon_k$ be the path components of $\check{U}K$. Then

$$K^{S-T} = \bigcup_{i=1}^{i=k} \Lambda_i,$$

where the $\Lambda_i := w^{-1}(wK^{S-T} \cap \Upsilon_i)$ are disjoint subcomplexes of $K^{S-T}$. Hence, $\pi_1(\check{V}K)$ is the free product of the $\pi_1(\check{U}K, p_i)$ $(= \pi_1(\Upsilon_i))$ with the $\pi_1(\Lambda_i)$ killed off. □

As in Example 4.5.6, order the elements of $W$: $w_1, w_2, \dots,$ so that $l(w_n) \geqslant l(w_{n-1})$. Put $U_n := \{w_1, \dots, w_n\}$. Then $U_1K \subset U_2K \subset \cdots$ is an exhaustive sequence of finite subcomplexes of $\Sigma$. By Lemma 9.2.5 (a), each of the bonds $\pi_0(\check{U}_nK) \to \pi_0(\check{U}_{n-1}K)$ is surjective.

We turn to the issue of semistability. Let $e$ be an end of $\Sigma$. By Definition G.4.1 in Appendix G.4, the cell complex $\Sigma$ is semistable at $e$ if and only if $\{\pi_1(\check{U}_nK, r(n))\}$ is a semistable inverse sequence of groups (Definition G.1.1). This means that for $m > n$ the images of the $\{\pi_1(\check{U}_mK, r(m))\}$ in $\pi_1(\check{U}_nK, r(n))$

must stabilize. (Here $r : [0, \infty) \to \Sigma$ is an appropriately chosen proper ray whose equivalence class is $e$.)

*Proof of Theorem 9.2.1.* First suppose $\Sigma$ is one ended. Each of the $\check{U}_n K$ is path connected and by Lemma 9.2.5 (b) (applied to $U = U_n$, $V = U_{n-1}$) each of the bonds $\pi_1(\check{U}_n K) \to \pi_1(\check{U}_{n-1} K)$ is an epimorphism. Hence, the inverse sequence of fundamental groups is semistable.

In the general case we use a result of Mihalik [206, Theorem 1.1] which states that an amalgamated product of two finitely presented, one ended groups along a finite subgroup is semistable if and only if both factors in the amalgamation are semistable. By Proposition 8.8.2, $W$ can be decomposed as a tree of groups where the vertex groups are 0- or 1-ended special subgroups and the edge groups are finite special subgroups. So, the general case follows from the one ended case. □

*Proof of Theorem 9.2.2.* By Lemma A.5.5, for a given $T \in \mathcal{S}$, $L - \sigma_T$ is simply connected if and only if $K^{S-T}$ is simply connected. Suppose first $K^{S-T}$ is simply connected for all $T \in \mathcal{S}$. By Lemma 9.2.4, $\check{U}_1 K$ deformation retracts onto $K^S$, so $\pi_1(\check{U}_1 K)$ is trivial. By Lemma 9.2.5 (i), $\check{U}_n K$ is homotopy equivalent to $\check{U}_{n+1} K$ with a copy of $K^{S-T}$ coned off, where $T = \mathrm{In}(w_{n+1})$. So, $\pi_1(\check{U}_{n+1} K) \cong \pi_1(\check{U}_n K)$, for $n \geqslant 1$. Therefore, $\pi_1(\check{U}_n K)$ is trivial for all $n$ and consequently, $\pi_1^\infty(\Sigma) = \varprojlim \pi_1(\check{U}_n K)$ is also trivial. Since $\Sigma$ is semistable, this implies it is simply connected at infinity (Proposition G.4.3).

Conversely, suppose $\Sigma$ is simply connected at infinity. Since $\Sigma$ is one ended, each $K^{S-T}$ is path connected (Theorem 8.7.2). By Lemma 9.2.5 (b), each of the bonds $\pi_1(\check{U}_{n+1} K) \to \pi_1(\check{U}_n K)$ is onto. Since the inverse limit is trivial, this means that each $\pi_1(\check{U}_n K)$ is trivial. By Lemma 9.2.5 (ii), $\check{U}_n K$ retracts onto $K^{S-\mathrm{In}(w_n)}$; so this space is also simply connected. As $w_n$ varies, $\mathrm{In}(w_n)$ varies over all $T \in \mathcal{S}$. So, $K^{S-T}$ is simply connected for all $T \in \mathcal{S}$. □

Given any starlike subset $U$ of $W$, put

$$\partial(UK) := UK \cap \check{U}K. \tag{9.7}$$

The proof of the next lemma is similar to that of Lemma 9.2.4.

**LEMMA 9.2.6.** *$\check{U}K$ is homotopy equivalent to $\partial(UK)$.*

*Proof.* Let $V_1$ be the union of $U$ with all the elements of $\check{U}$ which are adjacent to some element of $U$. As in Lemma 9.2.4, extend $V_1$ to an exhaustive sequence of starlike subsets, $U \subset V_1 \subset V_2 \subset \cdots$ so that for each $n \geqslant 2$, $V_n - V_{n-1}$ consists of a single extreme element $v_n$. As before, for $n \geqslant 2$, $V_n$ deformation retracts onto $V_{n-1}$ and $V_1$ deformation retracts onto $\partial(UK)$. □

*Remark.* Consider the problem of computing the groups $\pi_1(\check{U}_n K)$ for an arbitrary one-ended Coxeter group. By Lemma 9.2.6, $\check{U}_n K$ deformation retracts onto $\partial(\check{U}_n K) := \partial(U_n K)$. $\partial(U_n K)$ is formed from $\partial(U_{n-1}K)$ by removing a copy of $\mathrm{int}(K^T)$ from both $\partial(U_n K)$ and $\partial K$ and then gluing the pieces together along $\partial K^T$ $(:= K^T \cap K^{S-T})$ (where $T := \mathrm{In}(w_n)$). So, $\pi_1(\partial U_n K)$ is the amalgamated product of $\pi_1(\partial(U_{n-1}K - \mathrm{int}(K^T))$ and $\pi_1(K^{S-T})$ amalgamated along $\pi_1(\partial K^T)$. The problem is that this last group is not easy to control. $\partial K^T$ may not be simply connected (or for that matter, even connected). Hence, in the general case it does not seem possible to write down a good formula for the inverse limit, $\pi_1^\infty(\Sigma) := \varprojlim(\pi_1(\partial(U_n K))$. However, in the next example we discuss a special case when it is possible.

**Example 9.2.7.** (*The nerve is a PL manifold.*) Suppose the nerve $L$ of $W$ is a PL $m$-manifold, $m \geqslant 3$. (The definition of "PL manifold" is given in 10.4.8 of the next chapter.) As explained in 10.6, if $L$ is a PL $m$-manifold, then, for each $s \in S$, the mirror $K_s$ is PL homeomorphic to an $m$-disk. Similarly, for each nonempty spherical subset $T$, $K^T$ is an $m$-disk. So each $K^{S-T}$ is the complement of the interior of an $m$-disk in $L$. Moreover, $\partial(U_n K)$ is formed by removing an open $m$-disk and gluing on a copy of $K^{S-T}$. Since $\partial(U_1 K) = \partial K = L$, $\partial(U_n K)$ is a PL $m$-manifold homeomorphic to the connected sum of $\partial(U_{n-1}K)$ and $L$. Thus, $\partial(U_n K)$ is the $n$-fold connected sum:

$$\partial(U_n K) = L\sharp \cdots \sharp L$$

(where $\sharp$ denotes connected sum). Since $m \geqslant 3$, the fundamental group of a connected sum is the free product of the fundamental groups of its factors. So

$$\pi_1(\check{U}_n K) = \pi_1(\partial(U_n K) = \underbrace{\pi_1(L) * \cdots * \pi_1(L)}_{n}.$$

Thus, $\pi_1^\infty(\Sigma) = \varprojlim G_n$, where $G_n$ is the free product of $n$ copies of $\pi_1(L)$. (In other words, $\pi_1^\infty(\Sigma)$ is the "projective free product" of an infinite number of copies of $\pi_1(L)$.) We will return to this example in 10.5 in the case where the PL manifold $L$ is a nonsimply connected homology sphere.

Given a CW complex $Y$, we define, in Appendix G.2, its chain and cochain groups at infinity, as well as, their respective homology groups $H_*^e(Y)$ and $H_e^*(Y)$ (formulas (G.2) and (G.4)). Also, in Definition G.2.4, we explain what it means for $Y$ to be *homologically semistable* (given an exhaustion of $Y$ by compact subsets, $C_1 \subset C_2 \subset \cdots$, the corresponding inverse sequence of homology groups $\{H_*(Y - C_n)\}$ is semistable). We point out in Proposition G.2.5 that if $Y$ is homologically semistable, then $H_*^e(Y) \cong \varprojlim H_*(Y - C_n)$.

**DEFINITION 9.2.8.** Let $m$ be a nonnegative integer. $Y$ is *$m$-acyclic at infinity* if it is one ended and $H_i^e(Y)$ vanishes for $i > 0$.

The notion of being $m$-connected at infinity (generalizing the idea of being simply connected at infinity) is defined at the end of Appendix G.4. If $Y$ is semistable and homologically semistable, then it follows from the Hurewicz Theorem that when $m \geqslant 1$, $Y$ is $m$-connected at infinity if and only if it is both simply connected at infinity and $m$-acyclic at infinity. Here are two more corollaries of Lemma 9.2.5.

**THEOREM 9.2.9.** *$\Sigma$ is homologically semistable.*

*Proof.* This also follows from the proof of Theorem 8.3.1 in the previous chapter. □

**THEOREM 9.2.10.** (Compare Corollary 8.5.4.) *Suppose $W$ is infinite and $m$ is an integer $\geqslant 0$. Then*

*(i) $\Sigma$ is $m$-acyclic at infinity if and only if $L - \sigma_T$ is $m$-acyclic for all $T \in \mathcal{S}$.*

*(ii) $\Sigma$ is $m$-connected at infinity if and only if $L - \sigma_T$ is $m$-connected for all $T \in \mathcal{S}$.*

*Proof.* Starting with our usual exhaustive sequence $U_1 \subset U_2 \subset \cdots$ of starlike subsets of $W$, Lemma 9.2.5 gives

$$\overline{H}_*(\check{U}_n K) \cong \bigoplus_{i=1}^{i=n} \overline{H}_*(K^{\mathrm{Out}(w_i)}).$$

Also, $K^{\mathrm{Out}(w_i)} = K^{S-T}$, where $T = \mathrm{In}(w_i)$. By Lemma A.5.5, $K^{S-T}$ and $L - \sigma_T$ are homotopy equivalent. Hence, $\check{U}_n K$ is $m$-acyclic for all $n$ if and only if $L - \sigma_T$ is $m$-acyclic for all $T \in \mathcal{S}$. This proves (i). Similarly, (ii). □

## NOTES

**9.1.** For Theorem 9.1.3 to hold, it is enough to assume that $X$ is a mirrored space satisfying the Collared Condition of 8.4. For Theorem 9.1.5 to hold, we need only assume that, in addition, $X$ is locally path connected and semilocally simply connected. The group $\widetilde{W} \rtimes \pi_1(X)$ in Theorem 9.1.5 can be thought of as the "orbihedral fundamental group" of $X$ or equivalently, as the "fundamental group of the scwol" associated to $X$. (See Appendix E.2.)

**9.2.** It would be nice to have a direct proof of the semistability of $W$ (Theorem 9.2.1) without invoking Mihalik's result that an amalgamated product of two finitely presented one-ended groups along a finite subgroup is semistable if and only if each of the factors is semistable.

# *Chapter Ten*

## ACTIONS ON MANIFOLDS

The classical examples of geometric reflection groups in Chapter 6 act on simply connected manifolds of constant curvature. This chapter concerns actions of groups generated by reflections on manifolds without geometric assumptions. In several respects such actions resemble the geometric examples. The first result, Theorem 10.1.13, is that the group $W$ is a Coxeter group and the manifold $M$ has the form $M = \mathcal{U}(W, C)$ for some fundamental chamber $C$. This is easiest to understand in the case where the $W$-action is smooth (or, at least, locally linear). In this case $C$ resembles a simple polytope in that it is a manifold with corners as are its mirrors and as are all intersections of its mirrors. As we shall see in 10.4, the same result is nearly true without the local linearity hypothesis, where "nearly" refers to the fact that it is necessary to replace the phrase "manifold with corners" by "homology manifold with corners." If $M$ is a contractible manifold and the action is cocompact, the resemblance of $C$ to a simple polytope is even more striking: $C$ is contractible and each of its cofaces is acyclic (Theorem 9.1.4). In this case the nerve $L(W, S)$ must be a "generalized homology sphere" (Definition 10.4.5).

In 10.5, by choosing $L$ to be a nonsimply connected PL homology sphere of dimension $(n - 1)$, we get examples of aspherical manifolds whose universal covers are not homeomorphic to $\mathbb{E}^n$, the reason being that their universal covers are not simply connected at infinity (by Example 9.2.7). In Example 10.5.3 we improve this to get examples where the contractible cell complex $\Sigma$ is a topological manifold not homeomorphic to $\mathbb{E}^n$. (This answers Gromov's question if every CAT(0) manifold must be homeomorphic to $\mathbb{E}^n$.)

In Theorem 10.6.1 we give necessary and sufficient conditions for $\Sigma$ to be a PL manifold (resp., topological manifold or homology manifold). In particular, $\Sigma$ is a homology $n$-manifold if and only if $L$ is a generalized homology $(n - 1)$-sphere. A Coxeter system whose nerve is a generalized homology $(n - 1)$-sphere is said to be *type* $\mathrm{HM}^n$. In 10.9 we characterize when $W$ is a virtual Poincaré duality group: it is if and only if it decomposes as $W_0 \times W_1$ with $W_0$ spherical and $W_1$ of type $\mathrm{HM}^n$.

## 10.1. REFLECTION GROUPS ON MANIFOLDS

### Locally Linear Reflections

Suppose $G$ is a discrete group of diffeomorphisms acting properly on a smooth manifold $M$. Properness of the action implies that for any $x \in M$, the isotropy subgroup $G_x$ is finite. Since $G_x$ is a group of diffeomorphisms, it acts linearly on the tangent space $T_xM$. If we choose a $G_x$-invariant Riemannian metric on $M$, then the exponential map provides a $G_x$-equivariant diffeomorphism from a neighborhood of the origin in $T_xM$ onto a $G_x$-stable open neighborhood $U_x$ of $x$ in $M$. As in Definition 5.1.5, properness of the action also means that we can choose $U_x$ small enough so that $gU_x \cap U_x = \emptyset$ for all $g \in G - G_x$.

**DEFINITION 10.1.1.** A proper action of a discrete group $G$ on a manifold $M^n$ is *locally linear* if each $x \in M$ has a $G_x$-stable open neighborhood $U_x$, as above, which is $G_x$-equivariantly homeomorphic to a neighborhood of the origin in some linear representation of $G_x$ on $\mathbb{R}^n$.

Thus, a smooth proper action of a discrete group $G$ on a smooth manifold $M$ is locally linear. (For this reason locally linear actions are called "locally smooth" in [36].)

It is immediate from Definition 10.1.1 that if $G$ acts properly and locally linearly on a manifold $M$ and $H \subset G$ is a finite subgroup, then its fixed set $\text{Fix}(H, M)$ is a locally flat submanifold of $M$; the point being that the fixed set of a linear action of $H$ on $\mathbb{R}^n$ is a linear subspace. (To say $N^k \subset M^n$ is a *locally flat* submanifold means that $(M^n, N^k)$ is locally homeomorphic to $(\mathbb{R}^n, \mathbb{R}^k)$.) In general, the fixed set of a finite group of homeomorphisms on a manifold need not be a submanifold.

**DEFINITION 10.1.2.** An involution $r$ on a connected manifold $M$ is a *reflection* if its fixed set $M_r$ separates $M$.

**LEMMA 10.1.3.** *Suppose $r$ is a locally linear reflection on a connected manifold $M$. Then the fixed submanifold $M_r$ is codimension one in $M$. Moreover, it separates $M$ into exactly two components, which are interchanged by $r$.*

*Proof.* Since $M_r$ separates $M$, at least one point in $M_r$ must have an $\langle r\rangle$-stable linear neighborhood which is separated by $M_r$. The only linear involutions on $\mathbb{R}^n$ with the property that their fixed subspace separates $\mathbb{R}^n$ are the linear reflections. Hence, for at least one component of $M_r$, $r$ is locally modeled on a linear reflection at each point in the component. Such a component is a locally flat submanifold of codimension one in $M$. Let $M'_r$ denote the union of the codimension one components of $M_r$. Let $\overline{M} := M/\langle r\rangle$ be the orbit space. We can identify $M_r$ with its image $\overline{M}_r$ in $\overline{M}$. Since the orbit space of a linear reflection on $\mathbb{R}^n$ is a half-space, $\overline{M}$ is a manifold with boundary near each point of $\overline{M}'_r$ (the boundary being $\overline{M}'_r$). Since $\overline{M}'_r$ has a collared neighborhood in $\overline{M}$, the spaces $\overline{M}$ and $\overline{M} - \overline{M}'_r$ are homotopy equivalent. Since $M$ is connected, so is its continuous image $\overline{M}$ and therefore, also $\overline{M} - \overline{M}'_r$. The other components of $M_r$ have codimension $> 1$ and so cannot separate $\overline{M}$ even locally. Hence, $\overline{M} - \overline{M}_r$ is also connected. Since $\overline{M} - \overline{M}_r$ is connected and since $M - M_r \to \overline{M} - M_r$ is a principal $\mathbf{C}_2$-bundle, its total space has either one or two components. Since $M_r$ separates, $M - M_r$ has two components, the $\mathbf{C}_2$-bundle is trivial and the components of $M - M_r$ are interchanged by $r$. It follows that all components of $M_r$ are codimension one, that $\overline{M}$ is a manifold with boundary and that $M$ is the double of $\overline{M}$ along its boundary. □

### Nonexamples

One might be tempted to define a locally linear reflection on a manifold $M$ simply as an involution with fixed point set a (locally flat) submanifold of codimension one and then hope that the separation property would follow automatically. The following simple examples show this is not the case. First consider the involution $r$ on $T^2$ ($= S^1 \times S^1$) defined by $r(x, y) := (y, x)$. The fixed set is the diagonal circle, which does not separate. The orbit space $T^2/\langle r \rangle$ is a Möbius band. Next consider the case where $M = \mathbb{R}P^2$ and $r : \mathbb{R}P^2 \to \mathbb{R}P^2$ is induced by a linear involution $\tilde{r}$ on $\mathbb{S}^2$. The fixed set of $r$ is the disjoint of a copy of $\mathbb{R}P^1$ (the image of the fixed set of $\tilde{r}$) and a point (the image of the line in $\mathbb{R}^3$ normal to the fixed subspace). This time the fixed set has components of different dimensions. Again, it doesn't separate.

### The Basic Theorem for Groups Generated by Reflections

In the next lemma, as well as in Theorem 10.1.5 and Lemma 10.1.6 below, we do not assume the actions are locally linear.

**LEMMA 10.1.4.** *Suppose $r, s$ are distinct reflections on a connected manifold $M$ and that the dihedral group $\langle r, s \rangle$ acts properly on $M$. Then*

*(i) $M_r$ is nowhere dense in $M$.*

*(ii) $M_r \cap M_s$ is nowhere dense in $M_r$.*

When the $\langle r, s \rangle$-action is locally linear, the lemma is obvious. We postpone the proof of the general case until 10.4. The proof is based on a basic result in the cohomological theory of compact transformation groups known as the "Local Smith Theorem" ("Smith" refers to P. A. Smith, one of the founders of the theory.)

Suppose $W$ acts properly on a connected manifold $M$ and that it is generated by reflections. Let $R$ denote the set of all reflections in $W$. The fixed set, $M_r$, of an element $r \in R$ is the *wall* associated to $r$. (By Lemma 10.1.4 (ii), a reflection is determined by its wall. This fact is used implicitly in the proof of Theorem 10.1.5.) We proceed as in 6.6. Let

$$M_{(1)} := M - \bigcup_{r \in R} M_r$$

denote the complement of the union of all walls. A connected component of $M_{(1)}$ is an *open chamber*. A *chamber* is the closure of an open chamber. A *wall of a chamber* $C$ is a wall $M_r$ so that there is at least one point $x \in C \cap M_r$ such that $x$ belongs to no other wall. If this is the case, $C_r := C \cap M_r$ is a *mirror* of $C$. Two chambers are *adjacent* if they intersect in a common mirror. Fix a chamber $C$ and call it the *fundamental chamber*. Let $S$ be the set of reflections $s$ such that $M_s$ is a wall of $C$. Give $C$ its *tautological mirror structure*, i.e., the mirror structure $(C_s)_{s \in S}$.

In the next theorem we prove that a group generated by reflections on a manifold is a reflection group in the sense of Definition 5.2.6.

**THEOREM 10.1.5.** *Suppose that $W$ acts properly as a group generated by reflections on a connected manifold $M$. With the notation above, the following statements are true.*

(a) *$(W, S)$ is a Coxeter system.*

(b) *The natural $W$-equivariant map $\mathcal{U}(W, C) \to M$, induced by the inclusion $C \hookrightarrow M$, is a homeomorphism.*

The proof is entirely similar to that of Theorem 6.6.3. Define $\Omega$ to be the simplicial graph with vertex set equal to the set $\mathcal{C}$ of chambers and with two vertices $C$, $C' \in \mathcal{C}$ connected by an edge if and only if they are adjacent. As in Lemma 6.6.4, $\Omega$ is connected and as in Lemmas 6.6.5 and 6.6.7, $(\Omega, C)$ is a reflection system. Hence, $(W, S)$ is a Coxeter system, establishing statement (a) of the theorem. It also follows that two adjacent chambers are separated by the wall corresponding to their common mirror (a fact used implicitly in the proof of Lemma 10.1.6 below). To establish (b) we need the analog of Lemma 6.6.8. As in Sections 5.1 (formula (5.1)) and 6.6, $S(x) := \{s \in S \mid x \in C_s\}$. The proof of 6.6.8 goes through to give the following.

**LEMMA 10.1.6.** (Compare [29, p.80].) *Suppose $x, y \in C$ and $w \in W$ are such that $wx = y$. Then $x = y$ and $w \in W_{S(x)}$.*

The proof of Theorem 10.1.5 (b) is then exactly the same as that of Theorem 6.6.3 (iv): the natural map $\mathcal{U}(W, C) \to M$ is a continuous surjection and is easily seen to be open; by Lemma 10.1.6, it is injective; hence, a homeomorphism.

### More nonexamples

Suppose, as in Example 6.5.1, that $W$ is a spherical triangle group corresponding to a triple $(m_1, m_2, m_3)$. It is instructive to consider the $W$-action on $\mathbb{R}P^2$ in order to understand how the conclusion of Theorem 10.1.5 can fail to hold. Let $a := -id$ denote the antipodal map on $\mathbb{S}^2$. There are two cases to consider depending on whether or not $a$ belongs to $W$. The first case, when $a \notin W$, occurs only for the triples $(2, 2, n)$, with $n$ odd, and $(2, 3, 3)$. In this case, the orbit space is a spherical triangle but not with the predicted angles (they are $(2, 2, 2n)$- and $(2, 3, 4)$-triangles instead) and the three generators are not the predicted ones. In the second case when $a \in W$, the action is not effective; instead, we get an effective action of $W/\langle a \rangle$. In neither case do we get a Coxeter system and $\mathbb{R}P^2$ cannot be recovered from the basic construction.

### Manifolds with Corners

A second countable Hausdorff space $C$ is a *smooth n-manifold with corners* if it is differentiably modeled on the standard $n$-dimensional simplicial cone $\mathbb{R}^n_+ := [0, \infty)^n$. This means that $C$ is equipped with a maximal atlas of local charts from open subsets of $C$ to open subsets of $\mathbb{R}^n_+$ so that the overlap maps are diffeomorphisms. (A homeomorphism between two subsets of $\mathbb{R}^n$ is a *diffeomorphism* if it extends to a local diffeomorphism on some open neighborhood.) Given a point $x \in C$ and a chart $\varphi$ defined on a neighborhood of $x$, the number $c(x)$ of coordinates of $\varphi(x)$ which are equal to 0 is independent of the chart $\varphi$. For $0 \leqslant k \leqslant n$, a connected component of $c^{-1}(k)$ is a *stratum* of $C$ of *codimension k*. The closure of a stratum is a *closed stratum*.

While a local diffeomorphism of $\mathbb{R}^n_+$ must take strata to strata and preserve the codimension of a stratum, the same cannot be said of a local homeomorphism. So, if we want to define a topological version of the above, we must add a condition. Thus, $C$ is a *topological n-manifold with corners* if it is locally modeled on $\mathbb{R}^n_+$ and if each overlap map is a strata-preserving homeomorphism (between two open subsets of $\mathbb{R}^n_+$). For each $x \in C$, let $\Upsilon(x)$ denote the set of closed codimension one strata which contain $x$. The manifold with corners $C$ is *nice* if $\mathrm{Card}(\Upsilon(x)) = 2$ for all $x$ with $c(x) = 2$. For example, the manifold with corners structure on $D^2$ pictured in Figure 10.1 is not nice. If $C$ is a nice manifold with corners, then, by an easy argument, $\mathrm{Card}(\Upsilon(x)) = c(x)$ for all $x \in C$. Moreover, any closed stratum of codimension $k$ in $C$ is also a nice manifold with corners.

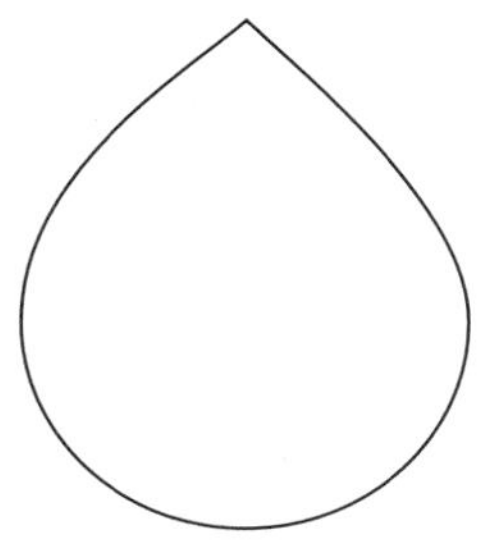

Figure 10.1. A manifold with corners which is not nice.

**Example 10.1.7.** A simplical cone is a nice manifold with corners, so is any simple polytope. If a polytope is not simple, then its natural differentiable structure is not that of a manifold with corners.

**DEFINITION 10.1.8.** A *mirrored manifold with corners* is a nice manifold with corners $C$ together with a mirror structure $(C_s)_{s \in S}$ on $C$ indexed by some set $S$ such that

- each mirror is a disjoint union of closed codimension one strata of $C$ and
- each closed codimension one stratum is contained in exactly one mirror.

**PROPOSITION 10.1.9.** *Suppose $W$ acts properly and locally linearly on a connected manifold $M$ and that $W$ is generated by reflections. Let $C$ be a fundamental chamber endowed with its tautological mirror structure $(C_s)_{s \in S}$. Then $C$ is a mirrored manifold with corners. If the action is smooth, then $C$ is a smooth manifold with corners.*

*Proof.* The point is that the $W$-action is locally modeled on finite linear reflection groups and by Theorem 6.6.3 (iii), a fundamental chamber for such a linear action is a simplicial cone. So the proposition follows from Example 10.1.7 □

Conversely, we have the following result.

**PROPOSITION 10.1.10.** *Suppose $(W, S)$ is a Coxeter system and $C$ is a mirrored manifold with corners with $W$-finite mirror structure $(C_s)_{s \in S}$. Then $\mathcal{U}(W, C)$ is a manifold and $W$ acts properly and locally linearly on it as a group generated by reflections.*

*Proof.* Given $x \in C$, let $S(x) = \{s \in S \mid x \in C_s\}$. Since the mirror structure is $W$-finite (Definition 5.1.6), $S(x)$ is spherical. We can find a contractible neighborhood of $x$ in $C$ of the form $U \times C_x$ where $U$ is an open subset in the relative interior of $C_{S(x)}$ and $C_x$ is homeorphic to a simplicial cone in $\mathbb{R}^{S(x)}$. By Theorem 6.12.9, $\mathcal{U}(W_{S(x)}, C_x) \cong \mathbb{R}^{S(x)}$. Hence, there is a $W_{S(x)}$-stable neighborhood of $[1, x]$ in $\mathcal{U}(W, C)$ of the form $U \times \mathbb{R}^{S(x)}$. The proposition follows. □

*Remark 10.1.11.* If $C$ is a smooth manifold with corners, then it is shown in [71, p. 322] that $C$ can be given the structure of a "smooth orbifold," meaning that $\mathcal{U}(W, M)$ has the structure of a smooth manifold with a smooth $W$-action.

**Reflection Groups on Acyclic Manifolds and Homology Spheres**

Here we combine results from this section with those of 8.2. We continue to assume that the $W$-action on $M^n$ is proper, locally linear, and generated by reflections, that $C$ is a fundamental chamber, and that $S$ is the set of fundamental reflections with respect to $C$. Then $(W, S)$ is a Coxeter system, $C$ is a mirrored manifold with corners, and $M^n \cong \mathcal{U}(W, C)$. The next result is an immediate corollary of Lemma 8.2.15.

**PROPOSITION 10.1.12.** (W. C. and W. Y. Hsiang [160, Chapter I]). *Suppose $W$ is finite and $M^n$ is a closed manifold with the same homology as $S^n$ (i.e., $M^n$ is a "homology sphere"). Put $m =$* Card$(S) - 1$. *Then*

*(i) $n \geqslant m$.*

*(ii) $C_S$ is a homology sphere of dimension $n - m - 1$. (When $n = m$, this means $C_S = \emptyset$.)*

*(iii) For each $T \subsetneq S$, $C_T$ is an compact, acyclic manifold with corners of codimension* Card$(T)$ *in $C$.*

(Homology spheres are discussed in the beginning of 10.3.)

The proposition means that the $W$-action on $M^n$ resembles the linear action on $S^n$ in that the cofaces of $C$ have the same homology as those of an $(n - m)$-fold suspension of an $m$-simplex. (See Remark 8.2.17.) Similarly, if $M^n$ is acyclic and $W$ is finite, the action resembles the linear action on $R^n$ in that the cofaces of its fundamental chamber have the same homology as those of a simplicial cone. (This also was proved in [160].) When $W$ is infinite and acts properly on an acyclic $M^n$, we have the following corollary to Corollary 8.2.8

**THEOREM 10.1.13.** ([71, Theorem 10.1].) *Suppose $W$ acts cocompactly on an acyclic manifold $M^n$. Then $C$ resembles an $n$-dimensional simple polytope in the following sense: for each $T \in \mathcal{S}$, $C_T$ is a compact, acyclic manifold with corners of codimension* Card$(T)$ *in $C$.*

Since acyclic implies nonempty, this means that $N(C) = L(W, S)$, where $N(C)$ denotes the the nerve of the mirror structure on $C$ (Definition 5.1.1). Thus, $L(W, S)$ plays the role of the boundary complex of the dual polytope.

## 10.2. THE TANGENT BUNDLE

In this section we describe a relationship between the equivariant tangent bundle of a manifold with a reflection group action and the tangent bundle of its fundamental chamber.

To avoid the technicalities of discussing tangent "bundles" of topological or PL manifolds, let us suppose in this section that $M$ is smooth and that the reflection group $W$ acts smoothly and properly on it. By Theorem 10.1.5, $M \cong \mathcal{U}(W, C)$. By Proposition 10.1.9, the fundamental chamber $C$ is a smooth manifold with corners. Let $p : M \to C$ be the orbit map. $TM$ and $TC$ denote the tangent bundles of $M$ and $C$, respectively.

**THEOREM 10.2.1.** ([75, Prop. 1.4, p. 110].) *The tangent bundle $TM$ is stably $W$-equivariantly isomorphic to the pullback $p^*(TC)$.*

(Recall that two vector bundles with the same base space are *stably isomorphic* if they become isomorphic after adding a trivial bundle to each. Two $W$-vector bundles are *$W$-equivariantly stably isomorphic*, if, up to addition of trivial bundles, there is a $W$-equivariant bundle isomorphism between them.)

To prove Theorem 10.2.1 we first need to explain how $M$ is a pullback of the geometric representation of Appendix D. As in 6.12, we have the canonical representation of $W$ on $\mathbb{R}^S$. Its dual on $E = (\mathbb{R}^S)^*$ is the *geometric representation*. In Appendix D we show

- $W$ is generated by the reflections across the faces of a simplicial cone $D$.
- $WD$, the union of translates of $D$, is a convex cone.
- $W$ acts properly on the interior $\mathcal{I}$ of this cone.
- Let $D^f = \mathcal{I} \cap D$. Then $D^f$ is the union of all faces of $D$ with finite stabilizers.
- $D^f$ is a strict fundamental domain for $W$ on $\mathcal{I}$ and $\mathcal{I} = \mathcal{U}(W, D^f)$.

We can find a strata-preserving map $\varphi : C \to D^f$. The reason is that each stratum of $D^f$ is a face of a simplicial cone and hence, is contractible. Also, we can obviously assume that $\varphi$ is smooth. Suppose $T \in \mathcal{S}^{(k)}$ is a spherical subset of cardinality $k$. Since $C$ and $D^f$ are manifolds with corners, the codimension-$k$ strata $C_T$ and $D^f_T$ have tubular neighborhoods of the form $C_T \times \mathbb{R}^k_+$ and $D^f_T \times \mathbb{R}^k_+$, respectively. Clearly, we can also assume that $\varphi$ is linear on such a tubular neighborhood of $C_T$, i.e., the restriction of $\varphi$ to the tubular neighborhood has the form

$$\varphi_T \times id : C_T \times \mathbb{R}^k_+ \to D^f_T \times \mathbb{R}^k_+,$$

where $\varphi_T$ denotes the restriction of $\varphi$ to $C_T$.

By the universal property of the basic construction (Lemma 5.2.5), the map $\varphi$ extends to a $W$-equivariant map $\tilde{\varphi} : M = \mathcal{U}(W, C) \to \mathcal{I} \subset E$. Let $p : M \to C$ and $q : \mathcal{I} \to D^f$ denote the orbit maps. Then $M$ can be identified as the following subset of $C \times \mathcal{I}$: $M \cong \{(c, u) \in C \times \mathcal{I} \mid \varphi(c) = q(u)\}$, i.e., $M$ is the fiber product of $C$ and $\mathcal{I}$. We have the pullback square:

$$\begin{array}{ccc} M & \xrightarrow{\tilde{\varphi}} & \mathcal{I} \\ {\scriptstyle p}\downarrow & & \downarrow{\scriptstyle q} \\ C & \xrightarrow{\varphi} & D^f. \end{array}$$

**LEMMA 10.2.2.** *$M$ is embedded in $C \times \mathcal{I}$ with trivial normal bundle.*

*Proof.* Let $\psi := \varphi - q : C \times \mathcal{I} \to E$. As we have just seen, $M = \psi^{-1}(0)$. The differential of $p$ takes the tangent space of $p^{-1}(C_T)$ isomorphically onto the tangent space of $C_T$ and the differential of $\varphi$ takes the normal space to $p^{-1}(C_T)$ isomorphically onto the normal space to $q^{-1}(D^f_T)$. It follows that 0 is a regular value of $\psi$ and that $d\psi$ maps the normal space to $M$ isomorphically onto $E$. □

*Proof.* Since $E$, with its linear $W$-action, is $W$-equivariantly contractible, the inclusion $M \hookrightarrow C \times E$ is $W$-homotopic to $p$. Hence,

$$T(C \times \mathcal{I})|_M \cong p^*(TC) \times E. \tag{10.1}$$

The left-hand side of equation (10.1) is isomorphic to the sum of $TM$ and the normal bundle of $M$, which, by Lemma 10.2.2, is trivial. □

**COROLLARY 10.2.3.** *$TM$ is stably $W$-equivariantly trivial if and only if $TC$ is trivial.*

*Proof.* Since the boundary of $C$ is nonempty, $TC$ is stably trivial if and only if it is trivial. (Here we are assuming $W$ is nontrivial.) If $TM$ is stably trivial, then so is its restiction to the fundamental chamber $C$ and $TM|_C = TC$. Conversely, by Therem 10.2.1, if $TC$ is stably trivial, then $TM$ is stably $W$-equivariantly trivial. □

*Remark.* A manifold with trivial (resp., stably trivial) tangent bundle is *parallelizable* (resp., *stably parallelizable*). Closed manifolds which are stably parallelizable need not be parallelizable. The primary obstruction for parallelizability is the Euler class of the tangent bundle: a closed manifold with nonzero Euler characteristic is not parallelizable. The word "stably" cannot be omitted from Corollary 10.2.3. Indeed, in the case where $W$ is a nontrivial finite geometric reflection group on $\mathbb{S}^n$, $C$ is a disk and $TC$ is trivial while $\mathbb{S}^n$ is not parallelizable for $n \neq 1, 3$ or 7. In fact, for $n$ even, $\chi(\mathbb{S}^n) = 2$ (not 0).

**COROLLARY 10.2.4.** *Suppose $\Gamma \subset W$ is a torsion-free subgroup of finite index, that $M' = M/\Gamma$, and that $r : M' \to C$ is the retraction induced by the orbit map. Then $TM'$ is stably isomorphic to $r^*(TC)$.*

## 10.3. BACKGROUND ON CONTRACTIBLE MANIFOLDS

### Homology Spheres

A closed $m$-manifold is a *homology sphere* if it has the same homology as $S^m$. For $m \leqslant 2$, the only $m$-dimensional homology sphere is $S^m$. In 1900 Poincaré conjectured that the same was true for $m = 3$. He soon found a counterexample. His new homology sphere was $S^3/G$, where $S^3$ is identified with the group of quaternions of norm 1 and $G \subset S^3$ is the binary icosahedral group (a finite subgroup of order 120.) Slightly earlier Poincaré had discovered the concept of the fundamental group and proved it was a topological invariant. Since $\pi_1(M^3) = G$ while $\pi_1(S^3) = 1$, $M^3 \neq S^3$. $M^3$ is now called *Poincaré's homology sphere*. (Another description of $M^3$ was given in Remark 8.5.7 (i).) After discovering this counterexample, Poincaré modified his conjecture to the version, discussed below, which remained open until this century. Later it was shown that many other groups $G$ occur as fundamental groups of homology 3-spheres. Over the last twenty years, homology 3-spheres have become objects of intense study in 3-manifold theory (e.g., Casson invariants, Vassiliev invariants, Floer homology).

For $m \geqslant 3$, the following three conditions on a group $G$ are necessary for it to be the fundamental group of a homology $m$-sphere:

(a) $G$ is finitely presented,

(b) $H_1(G) = 0$, and

(c) $H_2(G) = 0$.

Condition (a) is necessary because, by a result of Kirby and Siebenmann [177], any closed manifold has the homotopy type of a finite complex; hence, its fundamental group has a finite presentation. The reason for (b) is that for any space $X$, $H_1(X)$ is the abelianization of $\pi_1(X)$; hence, the fundamental group of a homology $m$-sphere, $m \geqslant 2$, must be a perfect group. The reason for (c) is Hopf's Theorem [156] (see Theorem F.1.4 in Appendix F) that for any space $X$ with fundamental group $G$, the canonical map $X \to BG$ induces a surjection $H_2(X) \to H_2(G)$.

Conversely, Kervaire [174] proved that when $m \geqslant 5$, conditions (a), (b) and (c) are sufficient for a group $G$ to occur as the fundamental group of a homology $m$-sphere.

*Remark.* It is known that in dimensions 3 and 4 these conditions are not sufficient. The argument in dimension 4 uses $L^2$-homology. As Shmuel Weinberger explained to me, when $G$ is infinite and the dimension is 4, there is an additional restriction on $G$:

$$2 \geqslant L^2b^2(EG; G) - 2L^2b^1(EG; G). \tag{10.2}$$

(Here $L^2b^i(Y; G)$ denotes the $i^{\text{th}}$ $L^2$-Betti number of $Y$, defined, in Appendix J.5, as the "von Neumann dimension" of $L^2\mathcal{H}_i(Y)$, the reduced $L^2$-homology group of $Y$.) The proof of (10.2) is based on the following two facts.

(i) There is an analogous result to Hopf's Theorem but with $L^2$-homology: the natural map $X \to BG$ induces an isomorphism $L^2\mathcal{H}_1(\widetilde{X}) \to L^2\mathcal{H}_1(EG)$ and a (weak) surjection $L^2\mathcal{H}_2(\widetilde{X}) \to L^2\mathcal{H}_2(EG)$.

(ii) Atiyah's Formula (Theorem J.5.3 of Appendix J.5) states that the alternating sum of the $L^2$-Betti numbers of $\widetilde{X}$ is equal to the ordinary Euler characteristic of $X$ (which is 2).

By Poincaré duality and (i), $L^2b^3(\widetilde{X}; G) = L^2b^1(\widetilde{X}; G) = L^2b^1(EG; G)$ and $L^2b^2(\widetilde{X}; G) \geqslant L^2b^2(EG; G)$. Using (ii) we get (10.2). Furthermore, it is known that there are groups $G$ which satisfy Kervaire's conditions (a), (b), and (c) above but which fail to satisfy the above inequality, [154].

In dimension 3, Thurston's Geometrization Conjecture (proved by Perelman [237, 239, 238]) imposes severe restrictions on $G$. For example, if $G$ is infinite freely indecomposable, and does not contain $\mathbb{Z} \times \mathbb{Z}$, then it must be a discrete subgroup of $\mathrm{Isom}(\mathbb{H}^3)$.

If $C^n$ is a compact, contractible manifold with boundary, one can ask if $C^n$ must be homeomorphic to the $n$-disk. By Poincaré duality, $(C^n, \partial C)$ has the same homology as $(D^n, \partial D)$. It then follows from the exact sequence of the pair $(C^n, \partial C)$ that $\partial C$ has the same homology as $\partial D = S^{n-1}$, i.e., $\partial C$ is a homology

$(n-1)$-sphere. To be explicit:

$$H_i(C^n, \partial C) \cong \begin{cases} \mathbb{Z} & \text{if } i = n, \\ 0 & \text{otherwise;} \end{cases}$$

$$H_i(\partial C) \cong \begin{cases} \mathbb{Z} & \text{if } i = 0, n-1, \\ 0 & \text{otherwise.} \end{cases}$$

However, for $n \geqslant 4$, there is no reason for $\partial C$ to be simply connected. For $n \geqslant 5$, the first example with $\partial C$ not simply connected was constructed by Newman, [226], by manipulating Poincaré's homology sphere. In his example $\pi_1(\partial C)$ is the binary icosahedral group. We give the details in Example 10.3.2 below. A few years later Mazur [198] gave a four-dimensional example with $\pi_1(\partial C)$ nontrivial (and infinite). In fact, we now know the following.

**Theorem 10.3.1.** *Every homology sphere bounds a contractible topological manifold.*

When the dimension of the contractible manifold is $\geqslant 5$, Theorem 10.3.1 is a consequence of surgery theory (see [159, Theorem 5.6]). So, for $n \geqslant 5$, the theorem holds not only for topological manifolds but in the PL category as well. (For $n \geqslant 5$, it also holds in the smooth category provided we are willing to modify the homology sphere by taking connected sum with an exotic sphere.) For $n = 4$, the fact that any homology 3-sphere bounds a contractible 4-manifold is due to Freedman [131, Theorem 1.4$'$, p. 367]. The four-dimensional result is true only in the topological category (in dimension 4, in the PL and smooth categories, surgery theory does not work). For example, Poincaré's homology 3-sphere does not bound a contractible 4-manifold with a PL structure. (This follows from (a) Rohlin's Theorem [247], which asserts that the signature of a closed PL manifold with vanishing first and second Stiefel-Whitney classes must be divisible by 16 and (b) the fact that Poincaré's homology 3-sphere bounds a parallelizable 4-manifold of signature 8, the "$\mathbf{E}_8$-plumbing.")

**Example 10.3.2.** (*Newman's example*, [226].) Let $A^3$ be the complement of a small open ball in Poincaré's homology 3-sphere. $A^3$ is an acyclic 3-manifold with fundamental group the binary icosahedral group $G$. $A^3$ deformation retracts onto the 2-skeleton $L$ of Poincare's homology 3-sphere. Embed $L$ as a subcomplex of some PL triangulation of $S^n$, $n \geqslant 5$. (In this range of dimensions we can embed any finite 2-complex as a subcomplex of the sphere.) Let $R$ be a regular neighborhood of $L$ in $S^n$ (i.e., after replacing the triangulation by its barycentric subdivision take $R$ to be the union of all closed simplices which intersect $L$). More explicitly, we could take $R$ to be a tubular neighborhood of $A^3$ in $S^n$. Since $A^3$ is acyclic, any vector bundle over it is trivial.

(Proof: Since $A$ is acyclic every homomorphism $\pi_1(A) \to \mathbb{Z}/2$ is trivial; so any vector bundle over $A$ is orientable. A principal $G$-bundle, $P \to A$, with $G = GL(n, \mathbb{R})_0$ is trivial if and only if it admits a section. Since $G$ is a simple space, $\pi_1(A)$ acts trivially on its homotopy groups. The obstructions to a section lie in groups of the form $H^i(A; \pi_{i-1}(G))$, all of which vanish.) So, $R = A^3 \times D^{n-3}$. Therefore, $\partial R = (A^3 \times S^{n-4}) \cup (S^2 \times D^{n-3})$. By van Kampen's Theorem, $\pi_1(\partial R) = G = \pi_1(R)$. Finally, let $C^n$ denote the complement of the interior of $R$ in $S^n$. By Alexander duality, $C^n$ is acyclic. Since the codimension of $L$ in $S^n$ is at least 3, $\pi_1(C^n) = 1$. (Any loop in $S^n - L$ is null homotopic in $S^n$; if the null homotopy is in general position with respect to $L$, then it misses $L$.) Hence, $C^n$ is a contractible $n$-manifold with boundary. Moreover, since the fundamental group of its boundary is $G$, $C^n$ is not homeomorphic to a disk.

### The Generalized Poincaré Conjecture

Note that a homology sphere $M^n$, $n \geqslant 2$, is simply connected if and only if it is homotopy equivalent to $S^n$. (Proof: Any closed orientable manifold $M^n$ admits a degree one map onto $S^n$; if $M^n$ is a homology sphere this map induces an isomorphism on homology; if, in addition, $M^n$ is simply connected, then, by Whitehead's Theorem, this map is a homotopy equivalence.) Soon after discovering his homology 3-sphere, Poincaré conjectured that any simply connected, closed 3-manifold is homeomorphic to $S^3$. (If $M^3$ is simply connected, then it is orientable and $H_1(M^3) = 0$. By Poincaré duality it is a homology sphere and hence, as noted above, a homotopy sphere.) These remarks led others to observe that the correct generalization of the Poincaré Conjecture to higher dimensions is the statement that if a closed manifold $M^n$ is homotopy equivalent to $S^n$, then it is homeomorphic to it. In other words, for $n \geqslant 2$, a simply connected homology $n$-sphere should be homeomorphic to the standard $n$-sphere. Poincaré's original conjecture has been proved by G. Perelman [237, 239, 238] by using ideas of R. Hamilton on the Ricci flow. (Also, see [49, 179, 219] for expositions of Perelman's work.) Its generalization to higher dimensions was proved much earlier, for $n \geqslant 5$, by Smale [259] (with contributions by others, [227, 264, 265]), and for $n = 4$, by Freedman [131].

**THEOREM 10.3.3.** (The Generalized Poincaré Conjecture of Smale [259], Freedman [131] and Perelman [237, 239, 238].)

*(i) If $M^{n-1}$ is a simply connected homology sphere, then $M^{n-1}$ is homeomorphic to $S^{n-1}$.*

*(ii) If $C^n$ is a compact, contractible manifold with boundary, and if $\partial C$ is simply connected, then $C^n$ is homeomorphic to $D^n$.*

What Smale actually proved in [259] was the smooth h-Cobordism Theorem in dimensions $\geqslant 6$. A cobordism $W^n$ between two manifolds $M_0^{n-1}$ and $M_1^{n-1}$ is an *h-cobordism* if the inclusion of each end $M_i^{n-1} \hookrightarrow W^n$ is a homotopy equivalence. The h-Cobordism Theorem asserts that any simply connected h-Cobordism $W^n$ is isomorphic (relative to $M_0$) to the cylinder $M_0 \times [0, 1]$. (Here "isomorphism" means either "homeomorphism," "PL homeomorphism," or "diffeomorphism" depending on the category of manifold under discussion.) The h-Cobordism Theorem immediately implies Theorem 10.3.3 (ii). Indeed, if we remove an open disk from the interior of $C^n$, what remains is an h-cobordism from $S^{n-1}$ to $\partial C$. Hence, it is isomorphic to $S^{n-1} \times [0, 1]$ and so, $C^n$ is isomorphic to $D^n$. The h-Cobordism Theorem also implies statement (i) of the theorem. For, if we remove an open $(n-1)$-disk from $M^{n-1}$, what remains is a contractible $(n-1)$-manifold with boundary $S^{n-2}$. Hence, it is isomorphic to $D^{n-1}$. So, $M^{n-1}$ is the union of two $(n-1)$-disks glued along their boundaries and therefore, is homeomorphic to $S^{n-1}$. (Notice that we cannot conclude that they are diffeomorphic.) So, as a corollary to the smooth h-Cobordism Theorem in dimensions $\geqslant 6$, Smale obtained statement (ii) of Theorem 10.3.3 for smooth $C^n$, $n \geqslant 6$, with the stronger conclusion that $C^n$ is diffeomorphic to $D^n$, as well as, statement (i) for $M^{n-1}$ smooth and $n - 1 \geqslant 6$. To get statement (i) for $n - 1 = 5$ he reasoned as follows. As was mentioned previously, by surgery theory, $M^5$ bounds a smooth contractible 6-manifold $C^6$ (it might be necessary to first alter $M^5$ by taking connected sum with an exotic sphere, but Kervaire and Milnor [175] had previously shown that there are no exotic 5-spheres). So, $C^6 \cong D^6$ and hence, $M^5 \cong S^5$. A short time later Stallings [264] showed that these results also held for PL (= piecewise linear) manifolds. Subsequently, Newman [227] extended these results to the topological category. In 1982 Freedman [131, Theorems 1.3 and 1.6] proved the five-dimensional h-Cobordism Theorem as well as Theorem 10.3.3 for $n = 5$, but only in the topological category. It is now known that the five-dimensional h-Cobordism Theorem is false for general simply connected manifolds in the PL and smooth categories; however, what happens with the four-dimensional version of Theorem 10.3.3 in these categories remains unknown. For two other short discussion of the Poincaré Conjecture in dimension 3, as well as in higher dimensions; see [211, 214].

**Open Contractible Manifolds**

Suppose $N^n$ is an open contractible $n$-manifold ("open" means that $N^n$ is a noncompact manifold without boundary). One might ask if all such $N^n$ are homeomorphic to $\mathbb{R}^n$. For $n \geqslant 3$ an obvious necessary condition is that $N^n$ be simply connected at infinity. It turns out that for $n \geqslant 4$ this condition is also sufficient. We state this as the following theorem.

Figure 10.2. Whitehead's example.

**THEOREM 10.3.4.** (Stallings [265], Freedman [131].) *For $n \geqslant 4$, an open contractible $n$-manifold is homeomorphic to $\mathbb{R}^n$ if and only if it is simply connected at infinity.*

Note that this implies the Generalized Poincaré Conjecture (Theorem 10.3.3 (i)).

For $n \geqslant 5$, Theorem 10.3.4 was proved by Stallings in [265]. His proof was distinctly different from Smale's proof of the h-Cobordism Theorem using handle cancellation. Instead, Stalling's proof introduced the new technique of engulfing. (A simple exposition of the ideas of Stalling's proof can be found in [127].) In dimension 4, the theorem is due to Freedman in the paper [131, Corollary 1.2, p. 366] cited earlier.

In view of our previous discussion of homology spheres, we see there are many open contractible manifolds not homeomorphic to Euclidean space, at least in dimensions $\geqslant 4$. Indeed, if $C^n$ is a compact contractible manifold with nonsimply connected boundary, then its interior, $\dot{C}$, is not simply connected at infinity and hence, is exotic. $\dot{C}$ is semistable (actually, stable) and $\pi_1^\infty(\dot{C}) \cong \pi_1(\partial C)$. The picture can be more complicated: $N^n$ need not be homeomorphic to the interior of any compact manifold. In dimension 3, there is the following well-known example of Whitehead [300] of an open contractible 3-manifold which is not simply connected at infinity (and therefore, not homeorphic to the interior of any compact 3-manifold).

**Example 10.3.5.** (*Whitehead's example*, [300].) In Figure 10.2 there is a picture of two solid tori $T_1$ and $T_0'$ in $S^3$. Let $T_0$ be the complement of the interior of $T_0'$. Since $T_0'$ is unknotted, $T_0$ is also a solid torus with $T_0 \supset T_1$ and $\pi_1(T_0 - T_1)$ nonabelian. Choose a self-homeomorphism $f$ of $S^3$ which takes $T_0$ onto $T_1$ and inductively define a decreasing sequence of solid tori

$$T_0 \supset T_1 \supset \cdots T_n \supset \cdots$$

by $T_{n+1} = f(T_n)$. The intersection

$$Z = \bigcap_{i=0}^{\infty} T_i$$

is called the *Whitehead continuum.*) Then $S^3 - Z$ is an increasing union of compact subsets:

$$S^3 - \dot{T}_0 \subset S^3 - \dot{T}_1 \subset \cdots S^3 - \dot{T}_n \subset \cdots.$$

Each $T_i$ is unknotted in $S^3$. So each $S^3 - T_i$ has the homotopy type of $S^1$. It is easy to see that the inclusion $S^3 - T_i \subset S^3 - T_{i+1}$ induces the trivial map on the fundamental group. Thus, $S^3 - Z$ is homotopy equivalent to the infinite mapping telescope of the degree 0 map of $S^1$ to itself. So $S^3 - Z$ is contractible. Since there is a homeomorphism which takes $T_i - Z$ onto $T_{i+1} - Z$, the groups $\pi_1(T_{i+1} - Z)$ and $\pi_1(T_i - Z)$ are isomorphic. It is not hard to see that the inclusion $T_{i+1} - Z \to T_i - Z$ induces a monomorphism $\pi_1(T_{i+1} - Z) \to \pi_1(T_i - Z)$ which is not onto. Since $S^3 - \dot{T}_i$ is a cofinal sequence of compact sets, $S^3 - Z$ is not semistable in the sense of Definition G.4.1 of Appendix G.4. In particular, $S^3 - Z$ is not simply connected at infinity.

*Remark.* In 1934, a year before the above example appeared, Whitehead [299] had published a "proof" of the three-dimensional Poincaré Conjecture; the above example was a counterexample to the statement in [299] that any open contractible 3-manifold is homeomorphic to $\mathbb{R}^3$.

In 10.5 we will construct some higher-dimensional examples, which, although semistable, do not have a finitely generated fundamental group at infinity and hence are not interiors of compact manifolds.

## 10.4. BACKGROUND ON HOMOLOGY MANIFOLDS

**Definition 10.4.1.** A space $X$ is a *homology n-manifold* if it has the same local homology groups as $\mathbb{R}^n$, i.e., if for each $x \in X$,

$$H_i(X, X - x) = \begin{cases} \mathbb{Z} & \text{if } i = n, \\ 0 & \text{otherwise.} \end{cases}$$

$X$ is *orientable* if there is a class $[X] \in H_n^{lf}(X)$ which restricts to a generator of $H_n(X, X - x)$ for each $x \in X$. (The superscript in $H_*^{lf}(\ )$ stands for "locally finite." When $X$ is a CW complex $H_*^{lf}(X)$ simply means cellular homology with infinite chains.)

The usual argument for manifolds (e.g., in [142]) shows that orientable homology manifolds satisfy Poincaré duality.

**Definition 10.4.2.** (Compare [244].) A pair $(X, \partial X)$, with $\partial X$ a closed subspace of $X$, is a *homology n-manifold with boundary* if it has the same local homology groups as does a manifold with boundary, i.e., if the following conditions hold:

- $X - \partial X$ is a homology $n$-manifold,
- $\partial X$ is a homology $(n-1)$-manifold, and
- for each $x \in \partial X$, the local homology groups $H_*(X, X - x)$ all vanish.

### The Local Smith Theorem and Newman's Theorem

If a group $G$ acts on a space $X$, then Fix $(G, X)$ means the fixed point set $\{x \in X \mid gx = x\}$. The next theorem is due to P. A. Smith [260, Theorem 1]. In its statement the phrase "$\mathbb{Z}/p$-homology manifold" means that the formula in Definition 10.4.1 holds with coefficients in $\mathbb{Z}/p$ and with $\mathbb{Z}$ replaced by $\mathbb{Z}/p$. A proof can be found in [27, pp. 74–80].

**THEOREM 10.4.3.** (The Local Smith Theorem.) *Let $p$ be a prime and $P$ a finite $p$-group acting effectively on a $\mathbb{Z}/p$-homology manifold $M$. Then* $\text{Fix}(P, M)$ *is a $\mathbb{Z}/p$-homology manifold. Moreover, if $p$ is odd, then* $\text{Fix}(P, M)$ *has even codimension in $M$.*

In other words, actions of $p$-groups on topological manifolds (or on homology manifolds) have a certain amount of local linearity built in: the fixed sets are submanifolds, up to homology with coefficients in $\mathbb{Z}/p$.

*Remark.* (*Cohomology manifolds.*) To simplify the exposition, we are stating Theorem 10.4.3 incorrectly. Actually the statements in both [260] and [27] are in terms of "cohomology manifolds" rather than homology manifolds (where the definition of a "cohomology manifold" is the obvious one). Under mild hypotheses on the local topology, these two notions coincide.

A corollary of Theorem 10.4.3 is the following result of M.H.A. Newman, [225] (see [36, pp. 154–158] for a proof).

**THEOREM 10.4.4.** (Newman's Theorem). *Suppose a finite group $G$ acts effectively on a homology manifold $M$. Then* $\text{Fix}(G, M)$ *is nowhere dense in $M$.*

(Of course, the Local Smith Theorem and Newman's Theorem are obvious for locally linear actions of finite groups on manifolds.)

### Polyhedral Homology Manifolds

**DEFINITION 10.4.5.** A space $X$ is a *generalized homology $n$-sphere* (for short, a "GHS$^n$" or a "GHS") if it is a homology $n$-manifold with the same homology as $S^n$. A pair $(X, \partial X)$ is a *generalized homology $n$-disk* (for short, a "GHD$^n$" or a "GHD") if it is a homology $n$-manifold with boundary and if it has the same homology as $(D^n, S^{n-1})$.

*Remark on terminology.* There is a conflict between the definition of a "homology sphere" in the beginning of 10.3 and that of a "homology manifold" in Definition 10.4.1. A "homology sphere" is a manifold while a homology manifold need not be. So, in Definition 10.4.5 we use the term "generalized homology sphere" as being less absurd than "homology homology sphere." A justification for this terminology is that homology manifolds are sometimes called "generalized manifolds."

The definition of a "convex cell complex" is given in Appendix A.1 (Definition A.1.9); the definition of a "link" of a cell in a cell complex is given in Appendix A.6. A convex cell complex which is also a homology manifold is a *polyhedral homology manifold.*

**LEMMA 10.4.6.** *The following conditions on an n-dimensional convex cell complex $\Lambda$ are equivalent.*

(a) *$\Lambda$ is a homology n-manifold.*

(b) *For each cell $\sigma$ of $\Lambda$,* $\mathrm{Lk}(\sigma, \Lambda)$ *is a* $\mathrm{GHS}^{n-\dim\sigma-1}$.

(c) *For each vertex v of $\Lambda$,* $\mathrm{Lk}(v, \Lambda)$ *is a* $\mathrm{GHS}^{n-1}$.

*Proof.* It is clear that a space is a homology $n$-manifold if and only if its product with an open interval is a homology $(n+1)$-manifold. By Proposition A.6.2, if a point $x$ in $\Lambda$ lies in the interior of some $k$-cell $\sigma$, then a neighborhood of $x$ is homeomorphic to the cone on the $k$-fold suspension of $\mathrm{Lk}(\sigma, \Lambda)$. It follows that (a) $\iff$ (b). The implication (b) $\implies$ (c) is trivial. If $\sigma$ is a cell of $\Lambda$ and $v$ is a vertex of $\sigma$, then, by Lemma A.6.3, $\mathrm{Lk}(\overline{\sigma}, \mathrm{Lk}(v, \Lambda)) = \mathrm{Lk}(\sigma, \Lambda)$, where $\overline{\sigma}$ is the cell of $\mathrm{Lk}(v, \Lambda)$ corresponding to $\sigma$. So, (c) $\implies$ (b). □

The next proposition describes one way in which polyhedral generalized homology spheres enter the theory of reflection groups on manifolds.

**PROPOSITION 10.4.7.** *Suppose W is a proper, cocompact, locally linear reflection group on an acyclic n-manifold $M^n$. Let C be a fundamental chamber and S the set of fundamental reflections with respect to C. Then $L(W, S)$ is a $GHS^{n-1}$.*

In 10.7 we will remove the assumption of local linearity.

*Proof.* By Corollary 8.2.9, $L(W, S)$ $(= L)$ is the nerve of the mirror structure on $C$ (Definition 5.1.1). By Theorem 10.1.13, for each $T \in \mathcal{S}$, $C_T$ is a compact acyclic manifold with boundary. Its dimension is $n - \mathrm{Card}(T)$. So, $\partial C_T$ is a homology sphere of dimension $n - \mathrm{Card}(T) - 1$. Since $L$ is the nerve of the covering of $\partial C$ by the closed codimension one strata of $C$, the Acyclic Covering Lemma (E.3.3) implies that it has has the same homology as $S^{n-1}$. Similarly, for each $T \in \mathcal{S}_{>\emptyset}$, $\mathrm{Lk}(\sigma_T, L)$ is the nerve of the covering of $\partial C_T$ by the codimension one strata of $C_T$. So, $\mathrm{Lk}(\sigma_T, L)$ has the same homology as $S^{n-\mathrm{Card}(T)-1}$. This completes the proof. □

**DEFINITION 10.4.8.** A convex cell complex $\Lambda$ is a *PL n-manifold* if the link of each vertex is PL homeomorphic to $S^{n-1}$.

**Example 10.4.9.** (*Suspensions and double suspensions of homology spheres.*) The prototypical example of a polyhedron which is a homology manifold but not a manifold is the suspension of a nonsimply connected homology sphere. Suppose $M^{n-1}$, $n \geqslant 4$, is a homology sphere with $\pi_1(M) \neq 1$. Its suspension $SM$ is a GHS$^n$. However, it is not a manifold. The reason is that for $n \geqslant 3$, every point $x \in \mathbb{R}^n$ has an arbitrarily small neighborhood $U$ such that $U - x$ is simply connected, while a suspension point in $SM$ does not have such simply connected deleted neighborhoods.

On the other hand, if we suspend again, $S^2M$ does have this property. $S^2M$ is certainly not a PL manifold, since the link of a 1-cell in the suspension circle is $M^{n-1}$ and $M^{n-1} \not\cong S^{n-1}$. An amazing theorem, due in many cases to Edwards and in complete generality to Cannon [47], asserts that $S^2M$ is always a topological manifold. This solved the "Double Suspension Problem," Milnor had listed as one of the seven hardest and most important problems in topology at the 1963 Seattle Conference on topology. (Actually Milnor gave this problem only for the double suspension of a homology 3-sphere. Another problem on the list was the Poincaré Conjecture in dimensions 3 and 4. See [185, Problems 23–29, p.579] for the complete list of problems.) Since $S^2M$ is homotopy equivalent to $S^{n+1}$, the Generalized Poincaré Conjecture, Theorem 10.3.3, implies that $S^2M$ is, in fact, homeomorphic to $S^{n+1}$.

The following is the definitive result for polyhedral homology manifolds. It implies the Double Suspension Theorem.

**THEOREM 10.4.10.** (Edwards [116], Freedman [131].) *A polyhedral homology n-manifold X, $n \geqslant 3$, is a topological manifold if and only if the link of each of its vertices is simply connected.*

For $n \leqslant 2$, every GHS$^n$ is homeomorphic to $S^n$. Consequently, for $n \leqslant 3$, every polyhedral homology $n$-manifold is a manifold (in fact, a PL manifold). In [116] Edwards stated the above theorem for $n \geqslant 5$ and calls it the "Polyhedral Manifold Characterization Theorem." For $n = 4$, this comes down to the question of whether or not the cone on a fake 3-sphere is necessarily a 4-manifold. Independent of the question of whether or not fake 3-spheres exist,

Freedman showed that this was indeed the case: he showed that the product of a homotopy sphere with $\mathbb{R}$ was homeomorphic to $S^3 \times \mathbb{R}$, [131, Corollary 1.3′, p. 371], and hence, that the open cone on such a homotopy sphere is an open 4-disk.

## 10.5. ASPHERICAL MANIFOLDS NOT COVERED BY EUCLIDEAN SPACE

In this section we describe the main examples of [71] to prove the following.

**THEOREM 10.5.1.** ([71].) *In each dimension $n \geqslant 4$, there are closed, aspherical manifolds $M^n$ with universal cover $\widetilde{M}^n$ not homeomorphic to Euclidean space.*

In view of Theorem 10.3.4 of Stallings and Freedman, the only way to distinguish $\widetilde{M}^n$ from $\mathbb{R}^n$ is to show that it is not simply connected at infinity.

Suppose a finite simplicial complex $L$ is a PL $(n-1)$-manifold and a homology sphere. Assume that there is a Coxeter system $(W, S)$ with $L(W, S) = L$. (By Lemma 7.2.2, we can always achieve this, without changing the topological type of $L$, by replacing $L$ by its barycentric subdivision and then taking the right-angled Coxeter system associated to its 1-skeleton.) Let $\Sigma$ $(= \Sigma(W, S))$ be the cell complex constructed in Chapter 7. It is a contractible, homology $n$-manifold. Moreover, $\Sigma$ is a PL manifold except possibly at its vertices. The vertices are PL singularities if and only if $L$ is not PL homeomorphic to $S^{n-1}$.

We wish to modify $\Sigma$ to a topological manifold. By Theorem 7.2.4, we know that $\Sigma = \mathcal{U}(W, K) = (W \times K)/\sim$, where the fundamental chamber $K$ is the geometric realization $|\mathcal{S}|$ of $\mathcal{S}$. It follows that $K$ is the cone on $\partial K$ $(= |\mathcal{S}_{>\emptyset}|)$. Also, $\partial K$ is homeomorphic to $L$ (with the cell structure dual to its simplicial structure). By Theorem 10.3.1, the homology sphere $\partial K$ is the boundary of a contractible $n$-manifold $C$. (If $n > 4$, we can take $C$ to be a PL manifold.) The idea is to hollow out each copy of $K$ in $\mathcal{U}(W, K)$ and replace it by a copy of $C$. The details go as follows. Since $\partial C$ is identified with $\partial K$, we can use the decomposition of $\partial K$ into mirrors to get a mirror structure on $C$. In other words, for each $s \in S$, set

$$C_s := K_s \subset \partial K.$$

Then $\mathcal{U}(W, C)$ is a contractible $n$-manifold with $W$-action. (It is contractible because it is homotopy equivalent to $\mathcal{U}(W, K)$; it is a manifold because $\mathcal{U}(W, K)$ is a manifold except at the cone points and we have desingularized them.)

Suppose $n \geqslant 4$ and $L$ is not simply connected. By Theorem 9.2.2, $\pi_1^\infty(\Sigma)$ is not trivial. The obvious map $C \to K$, extending the identity on the boundary, induces a $W$-equivariant proper homotopy equivalence $\mathcal{U}(W,K) \to \mathcal{U}(W,C)$. Since such a proper homotopy equivalence induces an isomorphism of fundamental groups at infinity, we see that $\mathcal{U}(W,C)$ is also not simply connected at infinity and hence, not homeomorphic to $\mathbb{R}^n$.

*Remark 10.5.2.* As in Chapters 8 and 9, let $U_n = \{w_1, \dots, w_n\}$ be an exhaustive sequence of starlike subsets of $W$. As explained in Example 9.2.7, $\partial(U_nC) = \partial(U_nK)$ is the connected sum of $n$ copies of $L$. Since $n \geqslant 3$, $\pi_1(\partial(U_nC))$ is the free product of $n$ copies of $\pi_1(L)$. Since $\check{U}_nC$ deformation retracts onto $\partial(U_nC)$, $\pi_1(\check{U}_nC)$ is also equal to this free product. The bond $\pi_1(\partial(U_{n+1}C)) \to \pi_1(\partial(U_nC))$ is the natural projection which kills the last factor. Hence, as in Example 9.2.7, $\pi_1^\infty(\mathcal{U}(W,M)) = \varprojlim(\pi_1(L) * \cdots * \pi_1(L))$ and when $\pi_1(L)$ is nontrivial, so is the inverse limit.

To actually get an aspherical $n$-manifold $M$ as in Theorem 10.5.1 we argue as follows. As explained in Corollary D.1.4 of Appendix D.1, the fact that $W$ has a faithful linear representation implies that it is virtually torsion-free. So, we can choose a torsion-free subgroup $\Gamma$ of finite index in $W$. Since $W$ acts properly on $\mathcal{U}(W,C)$, all its isotropy subgroups are finite and therefore, $\Gamma$ acts freely on $\mathcal{U}(W,C)$. So, $\mathcal{U}(W,C) \to \mathcal{U}(W,C)/\Gamma$ is a covering projection. Therefore, $M := \mathcal{U}(W,C)/\Gamma$ is a manifold. It is compact since $\Gamma$ has finite index in $W$. Since $\widetilde{M} := \mathcal{U}(W,C)$ is simply connected, it is the universal cover of $M$ and since it is not simply connected at infinity, it is not homeomorphic to $\mathbb{R}^n$. This completes the proof of Theorem 10.5.1.

**Example 10.5.3.** ([83, Remark 5b.2, p.384].) Next we give examples, for each $n \geqslant 5$, where the cell complex $\Sigma$ is a contractible topological $n$-manifold not homeomorphic to Euclidean $n$-space. Suppose $A$ is an $(n-1)$-dimensional PL manifold with boundary such that

(a) $A$ is acyclic,

(b) $\pi_1(A) \neq 1$, and

(c) $\pi_1(\partial A) \to \pi_1(A)$ is onto.

For $n-1 \geqslant 4$, such $A$ exist. Indeed, if $C^3$ is denotes the complement of an open 3-ball in Poincaré's homology 3-sphere (cf. Section 10.3), then one could take $A = C^3 \times D^k$, for any $k > 0$.

Triangulate $A$ as a flag complex so that $\partial A$ is a full subcomplex and put

$$L := A \cup \mathrm{Cone}(\partial A).$$

Let $S := \mathrm{Vert}(L)$. $(W,S)$ is the associated right-angled Coxeter system and $\Sigma$ is the associated contractible cell complex. Let $t \in S$ be the cone point of $\mathrm{Cone}(\partial A)$. Since $A$ is an acyclic manifold with boundary, $L$ is a $\mathrm{GHS}^{n-1}$. Hence, $\Sigma$ is a homology $n$-manifold. By (c) and the Seifert–van Kampen

Theorem, $L$ is simply connected. So, by Edwards' Polyhedral Manifold Characterization Theorem (10.4.10), $\Sigma$ is a topological manifold. If we delete the vertex $t$ from $L$, what remains is homotopy equivalent to $A$, which is not simply connected. So by Theorem 9.2.2, $\pi_1^\infty(\Sigma)$ is not trivial and therefore, $\Sigma$ cannot be homeomorphic to Euclidean space.

As we shall see in Chapter 12, the natural piecewise Euclidean metric on $\Sigma$ is always nonpositively curved. So, the above construction provides an answer to a question of Gromov [144, p. 187]: for each $n \geqslant 5$ there is a closed $n$-manifold $M^n$ with a nonpositively curved piecewise Euclidean metric such that its universal cover $\widetilde{M}^n$ is not homeomorphic to Euclidean space.

## 10.6. WHEN IS $\Sigma$ A MANIFOLD?

**THEOREM 10.6.1**

*(i)* $\Sigma$ *is a homology n-manifold if and only if L is a generalized homology* $(n-1)$*-sphere.*

*(ii)* $\Sigma$ *is an n-manifold if and only if L is a generalized homology* $(n-1)$*-sphere and is simply connected* $(n-1 \neq 0, 1)$*.*

*(iii)* $\Sigma$ *is a PL n-manifold if and only if L is a PL triangulation of* $S^{n-1}$.

*Proof.* The link of any vertex in $\Sigma$ is isomorphic to $L$. So (i) and (iii) follow immediately from the definitions. (See Lemma 10.4.6.) Assertion (ii) follows from Theorem 10.4.10 (Edwards' Polyhedral Manifold Characterization Theorem). □

**DEFINITION 10.6.2.** A Coxeter system is *type* $\mathrm{HM}^n$ (or simply, type HM) if its nerve is a $\mathrm{GHS}^{n-1}$. ("HM" stands for "homology manifold.")

## 10.7. REFLECTION GROUPS ON HOMOLOGY MANIFOLDS

This section concerns reflection groups on topological manifolds without the assumption of local linearity. It turns out that our conclusions are substantially unchanged if we only assume that the ambient space is a homology manifold. We begin with a version of Lemma 10.1.3 when the reflection is not required to be locally linear.

**LEMMA 10.7.1.** *Suppose r is a reflection on a connected homology manifold M. Then the fixed set* $M_r$ *is a homology manifold of codimension one in M.* $M_r$ *separates M into exactly two components ("half-spaces") and each such half-space is a homology manifold with boundary (the boundary being* $M_r$*).*

*Proof.* Suppose $\dim M = n$. By the Local Smith Theorem (Theorem 10.4.3), $M_r$ is a homology manifold over $\mathbb{Z}/2$ of dimension $\leqslant n-1$. Arguing as in the proof of Lemma 10.1.3, we prove successively that

- at least one component of $M_r$ has dimension $n-1$,
- the orbit space $\overline{M}$ is connected,
- $\overline{M} - \overline{M}_r$ is connected,
- $M - M_r$ has exactly two components, and
- each component of $M_r$ has dimension $n-1$.

Moreover, if $X$ denotes the closure of a component of $M - M_r$, then $M \cong \mathcal{U}(\mathbf{C}_2, X)$, where $\mathbf{C}_2$ is the cyclic group of order two generated by $r$.

Put $\partial X := M_r$. We must show that $(X, \partial X)$ satisfies the conditions in Definition 10.4.2. First, $X - \partial X$ is a homology $n$-manifold, since it is an open subset of $M$. Let $x \in \partial X$. Define

$$Z := X \cup \mathrm{Cone}(X - x),$$

$$Z_r := M_r \cup \mathrm{Cone}(M_r - x) = \partial X \cup \mathrm{Cone}(\partial X - x),$$

$$\mathcal{U}(Z) := M \cup \mathrm{Cone}(M - x).$$

(In fact, $M \cup \mathrm{Cone}(M - x) \cong \mathcal{U}(\mathbf{C}_2, Z)$, where $Z$ has the single mirror $Z_r$.) Since $M$ is a homology $n$-manifold, $\mathcal{U}(Z)$ has the same homology as $S^n$. By Corollary 8.2.5, there are only two possibilities, either

(i) $Z$ has the same homology as $S^n$ and $H_*(Z, Z_r)$ vanishes in all dimensions or

(ii) $Z$ is acyclic and $Z_r$ has the same homology as $S^{n-1}$.

Since $M_r$ is a homology $(n-1)$-manifold over $\mathbb{Z}/2$, the first possibility is excluded. So the second holds. But this translates to the conditions that the local homology groups $H_*(X, X - x)$ all vanish and that $\partial X$ has the same local homology groups over $\mathbb{Z}$ as does $\mathbb{R}^{n-1}$. So the conditions in Definition 10.4.2 hold. □

We are now in position to complete the proof of Lemma 10.1.4.

*Proof of Lemma 10.1.4.* This lemma concerns the action of a dihedral group $\langle r, s\rangle$ on a manifold (or homology manifold) $M$ as a group generated by reflections. The first assertion of the lemma is that the wall $M_r$ is nowhere dense in $M$. Of course, this follows the previous lemma. (In fact, we do not yet need this lemma—the first assertion follows from Newman's Theorem or from the Local Smith Theorem.) The second assertion is that $M_r \cap M_s$ (the fixed set of $\langle r, s\rangle$) is nowhere dense in $M_r$. Supposing, as we may, that $M_r \cap M_s \neq \emptyset$, the dihedral group $\langle r, s \rangle$ is finite. Let $m$ be the order of $rs$. Choose a prime $p$ dividing $m$ and let $\mathbf{C}_p$ be the cyclic subgroup generated by $(rs)^{m/p}$. We have $M_r \cap M_s \subset \mathrm{Fix}(\mathbf{C}_p, M)$. By the Local Smith Theorem, $\mathrm{Fix}(\mathbf{C}_p, M)$ is a homology manifold over $\mathbb{Z}/p$. Moreover, if $p$ is odd, its codimension in $M$ must be even and hence, it must have positive codimension in $M_r$ (which, by the previous lemma, is codimension one). If $p = 2$, then $\mathbf{C}_2$ acts on the (integral) homology manifold $M_r$ and so its fixed set, $M_r \cap M_s$, must have positive codimension in $M_r$ by the Local Smith Theorem. □

For the remainder of this section, $W$ is a group generated by reflections acting properly on a connected homology manifold $M$. Since we have finished proving Lemma 10.1.4, we have also completed the proof of Theorem 10.1.5. So, we know $(W, S)$ is a Coxeter system and $M = \mathcal{U}(W, C)$ for some strict fundamental domain $C$, where $S$ is the set of reflections across the walls of $C$. Next, we investigate the local structure of $C$ and prove an analog of Proposition 10.1.9.

**Definition 10.7.2.** Suppose we have a space $C$ and a filtration:

$$C = C_n \supset C_{n-1} \supset \cdots C_0 \supset C_{-1} = \emptyset,$$

so that for $0 \leqslant k \leqslant n$, $C_k - C_{k-1}$ is a homology $k$-manifold. A connected component of $C_k - C_{k-1}$ is a *k-dimensional stratum*. The closure of such a stratum is a *closed stratum*. The filtered space $C$ is a *homology n-manifold with corners* if each closed stratum is a homology manifold with boundary, i.e., whenever $X$ is a $k$-dimensional stratum and $x \in \overline{X}$,

$$H_i(\overline{X}, \overline{X} - x) = \begin{cases} \mathbb{Z} & \text{if } i = k \text{ and } x \in X, \\ 0 & \text{otherwise.} \end{cases}$$

(Actually, to allow the type of picture in Figure 10.1 we should only require that the above holds for any "local stratum," i.e., in the above we should replace $X$ by any connected component of $U \cap X$ for any open neighborhood $U$ of $x$.) As before, a homology manifold with corners is *nice* if each codimension-$p$ stratum is contained in the closure of exactly $p$ codimension-one strata (Definition 10.1.8).

It follows from this definition that any homology $n$-manifold with corners $C$ is a homology manifold with boundary, where $\partial C := C_{n-1}$ is the union of all strata of positive codimension. Also, if $C$ is nice, then any one of its closed $k$-dimensional strata is a nice homology $k$-manifold with corners. A *mirrored homology manifold with corners* is defined as in 10.1.8.

**Proposition 10.7.3.** *Suppose $W$ is a finite reflection group on a connected homology n-manifold $M = \mathcal{U}(W, C)$, with $C$ a fundamental chamber, $S$ is the* *set of reflections across walls of $C$, and $C$ is equipped with its tautological mirror structure $(C_s)_{s\in S}$. Then*

(i) $\mathrm{Fix}(W, M)$ *is a homology manifold of dimension $n - |S|$ and*

(ii) *$C$ is a mirrored homology n-manifold with corners.*

(Here $|S| = \mathrm{Card}(S)$.)

*Proof.* The proof is by induction on $|S|$. The case $|S| = 1$ was taken care of in Lemma 10.7.1. Suppose, by induction, that $|S| > 1$ and the proposition is true for all special subgroups $W_T$ with $T$ a proper subset of $S$. Given $x \in C$, let $S(x) = \{s \in S \mid x \in C_s\}$. If $S(x) \neq S$, then it follows from the induction hypothesis that for any subset $U$ of $S(x)$ the local homology groups $H_*(C_U, C_U - x)$ are correct. (This is because $C$ is a neighborhood of $x$ in a fundamental domain for $W_{S(x)}$ on $M$.) So, we are reduced to the case $S(x) = S$. As in 10.7.1, put

$$Z := C \cup \mathrm{Cone}(C - x)$$

and

$$\mathcal{U}(Z) := \mathcal{U}(W, C) = M \cup \mathrm{Cone}(M - x).$$

Since $M$ is a homology $n$-manifold, $\mathcal{U}(Z)$ has the same homology as $S^n$. By Corollary 8.2.5, there is a unique subset $T \subset S$ such that $(Z, Z^T)$ has the same homology as $(D^n, S^{n-1})$ (or the same homology as $S^n$ if $T = \emptyset$). Moreover,

- $W = W_T \times W_{S-T}$ and
- $H_*(Z, Z^U) = 0$ for all $U \neq T$.

*Claim.* $T = S$. Suppose to the contrary that $T \neq S$. We first dispose of the case $T = \emptyset$. In this case, $Z$ has the homology of $S^n$ and $H_*(Z, Z^U) = 0$ for all $U \neq \emptyset$. Then $Z^U$ has the homology of $S^n$ and in particular, $Z_s$ has the same homology as $S^n$ for each $s \in S$. But this implies that $(C_s, C_s - x)$ has the same homology as $(D^n, S^{n-1})$. However, by Lemma 10.7.1, $M_s$ is a homology manifold of dimension $n - 1$. Therefore, $H_n(C_s) = 0 = H_{n-1}(C_s - x)$ and consequently, $H_n(C_s, C_s - x) = 0$. This contradiction shows that $T = \emptyset$ is impossible. The argument when $T$ is a nonempty proper subset of $S$ is similar. Since $T \neq \emptyset$, $Z$ is acyclic. Hence, $Z^T$ has the homology of $S^{n-1}$ and if $s \in S - T$, $Z_s$ and $Z^{T\cup\{s\}}$ $(= Z^T \cup Z_s)$ are both acyclic. From the Mayer-Vietoris sequence, we see that $Z^T \cap Z_s$ has the same homology as $S^{n-1}$. Therefore, $(Z_s, Z^T \cap Z_s)$ has the same homology as $(D^n, S^{n-1})$ and hence, $W_T Z_s = \mathcal{U}(W_T, Z_s)$ has the same homology as $S^n$. So, $W_T C_s$ is a homology $n$-manifold at $x$. Since $W_T$ and $s$ commute, $W_T C_s \subset M_s$. But this contradicts the fact that the homology manifold $M_s$ has dimension $n - 1$. So, we have proved the claim.

If $T = S$, then, arguing as before, we get that $Z^S$ has the same homology as $S^{n-1}$ and by Theorem 8.2.16, $Z_S$ has the same homology as $S^{n-|S|}$ and for any

$U \subsetneq S$, $Z_U$ is acyclic. Hence, $\mathrm{Fix}(W_S, M)$ $(= C_S)$ is a homology manifold of codimension $|S|$ in $M$ and $H_*(C_U, C_U - x) = 0$. □

There is no problem removing the finiteness hypothesis from Proposition 10.7.3 to get the following.

**THEOREM 10.7.4.** *Suppose that $W$ is a proper reflection group on a connected homology $n$-manifold $M = \mathcal{U}(W, C)$, where $C$ is a fundamental chamber with its tautological mirror structure $(C_s)_{s \in S}$. Then $C$ is a mirrored homology $n$-manifold with corners.*

*Proof.* We must show that for each $x \in C$, the pair $(C_{S(x)}, C_{S(x)} - x)$ has the same homology as $(D^{n-|S(x)|}, S^{n-|S(x)|-1})$ and that for each proper subset $U$ of $S(x)$, $H_*(C_U, C_U - x) = 0$. Since $C$ is neighborhood of $x$ in a fundamental domain for the $W_{S(x)}$-action on $M$ and since, by Proposition 10.7.3, the corresponding statements are true for this action, they also hold for $C$. □

We have the following converse to Theorem 10.7.4. Its proof is left as an exercise for the reader.

**PROPOSITION 10.7.5.** *Suppose $(W, S)$ is a Coxeter system and $C$ is a mirrored homology manifold with corners with $W$-finite mirror structure $(C_s)_{s \in S}$. Then $\mathcal{U}(W, C)$ is a homology manifold.*

**DEFINITION 10.7.6.** A mirrored homology $n$-manifold with corners $C$ is a *generalized simple polytope* (or a *generalized polytope* for short) if $(C, \partial C)$ is a generalized homology $n$-disk and if for each $\sigma_T \in N(C)$, $(C_T, \partial C_T)$ is a generalized homology disk of dimension $n - |T|$.

The proof of Proposition 10.4.7 now goes through to give the following.

**THEOREM 10.7.7.** *Suppose an infinite, discrete group $W$ acts properly and cocompactly on a acyclic homology $n$-manifold $M$ and that $W$ is generated by reflections. Let $C$ be a fundamental chamber endowed with its tautogical mirror structure $(C_s)_{s \in S}$. Then $C$ is a generalized polytope and $L(W, S)$ is a $GHS^{n-1}$ (i.e., $(W, S)$ is type* $\mathrm{HM}^n$ *as in Definition 10.6.2).*

In 10.9 we will see that the hypothesis of this theorem can be weakened even further (Theorem 10.9.2).

## 10.8. GENERALIZED HOMOLOGY SPHERES AND POLYTOPES

Given a simplicial complex $L$ with vertex set $S$, let $\mathcal{S}(L)$ denote the poset of vertex sets of simplices in $L$, including the empty simplex. (The notation is the same as in Appendix A.2, Example A.2.3.) As in Definition A.3.4 or in 7.2, $K := |\mathcal{S}(L)|$ is the geometric realization of $\mathcal{S}(L)$. For each $T \in \mathcal{S}(L)$, $K_T := |\mathcal{S}(L)_{\geqslant T}|$ is the corresponding coface and $\partial K_T := |\mathcal{S}(L)_{>T}|$. ($K_T$ is sometimes called the "dual cone" to the simplex $\sigma_T$.) $K$ is the *dual complex* to $L$. Also, for each $s \in S$, put $K_s := K_{\{s\}}$. As in 5.1 and 7.2, for each $T \subset S$, put

$$K^T := \bigcup_{s \in T} K_s.$$

In Lemma A.5.5, we showed that for each $T \in \mathcal{S}(L)$, $L - \sigma_T$ is homotopy equivalent to $K^{S-T}$.

Suppose $K$ is the dual complex of a simplicial complex $L$. $K$ is a *polyhedral generalized polytope* if for each $T \in \mathcal{S}(L)$, $(K_T, \partial K_T)$ is a $\mathrm{GHD}^{n-|T|}$. As in Theorem 10.7.7, note that $L$ is the nerve of the covering of $\partial K$ by the $K_s$ and that $\mathrm{Lk}(\sigma_T, L)$ is the nerve of the covering of $\partial K_T$ by the codimension-one strata of $K_T$. By the Acyclic Covering Lemma (E.3.3 in Appendix E.3), for any $T \in \mathcal{S}(L)$, $\mathrm{Lk}(\sigma_T, L)$ and $\partial K_T$ have the same homology. (For $T = \emptyset$, this means that $L$ and $\partial K$ have the same homology.) Thus, $\mathrm{Lk}(\sigma_T, L)$ has the same homology as $S^{n-|T|-1}$ if and only if $\partial K_T$ does. It follows that $K$ is a generalized $n$-polytope if and only if $L$ is a $\mathrm{GHS}^{n-1}$.

*Remark.* Of course, the model to keep in mind is where $L$ is the boundary complex of a simplicial polytope, $K$ is the dual polytope, and the $K_T$, $T \in \mathcal{S}(L)$, are the faces of $K$.

A simplicial complex $L$ has the *punctured n-sphere property in homology* (for short, $L$ is a $\mathrm{PHS}^n$) if the following two conditions hold:

(a) $L$ has the same homology as $S^n$ and

(b) for each nonempty simplex $\sigma$ of $L$, $L - \sigma$ is acyclic.

In terms of the dual complex $K$, these two conditions translate as follows:

(a)′ $(K, \partial K)$ has the same homology as $(D^{n+1}, S^n)$ and

(b)′ for each $T \in \mathcal{S}(L)_{>\emptyset}$, $H_*(K, K^{S-T})$ vanishes in all dimensions.

**LEMMA 10.8.1.** ([77, Lemma 5.1, p.306].) *Suppose L is a simplicial complex and $K = |\mathcal{S}(L)|$ is its dual complex. Then the following three conditions are equivalent:*

*(i) L is a generalized homology $(n-1)$-sphere.*

*(ii) K is a generalized n-polytope.*

*(iii) L has the punctured $(n-1)$-sphere property in homology.*

*Proof.* We have already observed that (i) and (ii) are equivalent.
For $s \in S$, set

$$\overline{L} := \mathrm{Lk}(s, L), \quad \overline{K} := |\mathcal{S}(\overline{L})| = K_s, \quad \text{and} \quad \overline{S} := S - s.$$

The vertex set of $\overline{L}$ is $\overline{S}$ and $T \subset S - s$ is in $\mathcal{S}(\overline{L})$ if and only if $T \cup \{s\}$ is a simplex of $L$. □

*Claim.* If $L$ is a $\mathrm{PHS}^{n-1}$, then $\overline{L}$ is a $\mathrm{PHS}^{n-2}$.

*Proof of Claim.* For any $T \in \mathcal{S}(\overline{L})$,

$$(\overline{K}, \overline{K}^{\overline{S}-T}) = (K_s, K_s \cap K^{S-(T\cup\{s\})}).$$

By excision,

$$H_*(K_s, K_s \cap K^{S-(T\cup\{s\})}) \cong H_*(K^{S-T}, K^{S-(T\cup\{s\})}),$$

where we have excised the complement of $K_s$ in $K^{S-T}$. By the punctured sphere property, $K^{S-(T\cup\{s\})}$ is acyclic and for $T \neq \emptyset$, $K^{S-T}$ is also acyclic. So, for $T \in \mathcal{S}(\overline{L})_{>\emptyset}$, $H_*(\overline{K}, \overline{K}^{\overline{S}})$ vanishes in all dimensions. When $T = \emptyset$, $H_*(\overline{K}, \partial\overline{K}) = H_*(K^S, K^{S-\{s\}})$, which is concentrated in dimension $n - 1$ and is isomorphic to $\mathbb{Z}$ in that dimension.

We continue with the proof of the lemma by showing (iii) $\Longrightarrow$ (ii) by induction on $n$. The case $n = 0$ (i.e., $L = \emptyset$) is trivial. So, suppose $L$ is a $\mathrm{PHS}^{n-1}$, $n \geqslant 1$. It suffices to show, for each $T \in \mathcal{S}(L)$, that $(K_T, \partial K_T)$ has the same homology as $(D^{n-|T|}, S^{n-|T|-1})$ . (Indeed, if we prove this, then $\mathrm{Lk}(\sigma_T, L)$ has the same homology as $S^{n-|T|-1}$ for each $T \in \mathcal{S}(L)$ and hence $K_T$ ($= \mathrm{Cone}(\mathrm{Lk}(\sigma_T, L))$) is a homology manifold with boundary.) The case $T = \emptyset$ is part of the definition of the punctured $(n-1)$-sphere property (the second version). So suppose $T \in \mathcal{S}(L)_{>\emptyset}$. Choose $s \in T$ and let $T' := T - s$ (so that $T = T' \cup \{s\}$). Let $\overline{L} = \mathrm{Lk}(s, L)$ and $\overline{K} = K_s$ be as above. According to the claim $\overline{L}$ is a $PHS^{n-2}$. By the inductive hypothesis, (iii) holds for $\overline{L}$ and $\overline{K}$. On the other hand, we have natural identifications $\mathrm{Lk}(\sigma_T, L) = \mathrm{Lk}(\sigma_{T'}, \overline{L})$ and $(K_T, \partial K_T) = (\overline{K}_{T'}, \partial\overline{K}_{T'})$. Since $(\overline{K}_{T'}, \partial\overline{K}_{T'})$ has the same homology as $(D^{n-|T'|-1}, S^{n-|T'|-2})$, $(K_T, \partial K_T)$ has the same homology as $(D^{n-|T|}, S^{n-|T|-1})$.

Finally, we prove (i) $\Longrightarrow$ (iii). Suppose $L$ is a $\mathrm{GHS}^{n-1}$ and $T \in \mathcal{S}(L)$. If $T = \emptyset$, then $\partial K$ ($= L$) has the same homology as $S^{n-1}$ and so $(K, \partial K)$ has the same homology as $(D^n, S^{n-1})$. For $T \neq \emptyset$, $K^T$ is a regular neighborhood of the simplex $\sigma_T$ in the barycentric subdivision of $L$; hence, $K^T$ is contractible. Since $L$ is a polyhedral homology $(n-1)$-manifold, $(K^T, \partial(K^T))$ is a homology $(n-1)$-manifold with boundary, where $\partial(K^T) = K^T \cap K^{S-T}$. By Poincaré duality, $(K^T, \partial(K^T))$ has the same homology as $(D^{n-1}, S^{n-2})$. Hence,

$$H_*(K^S, K^{S-T}) \cong H_*(K^T, K^T \cap K^{S-T}) \cong H_*(D^{n-1}, S^{n-2}),$$

which is the condition we needed to check to verify (iii). □

**DEFINITION 10.8.2.** Suppose $K$ is a polyhedral generalized polytope dual to a simplicial complex $L$ which is a $\mathrm{GHS}^{n-1}$. A *resolution* of $K$ is a mirrored manifold with corners $C$ such that $N(C) = L$ and such that for each $T \in \mathcal{S}(L)$, the stratum $C_T$ is contractible.

**THEOREM 10.8.3.** ([71, Theorem 12.2].) *Any polyhedral generalized polytope K admits a resolution.*

This is basically a consequence of the fact that any homology sphere bounds a contractible manifold (Theorem 10.3.1).

*Proof.* Suppose $\dim K = n$. Let $K^k$ denote the union of strata of $K$ of dimension $\leqslant k$. These strata are indexed by

$$\mathcal{S}(k) := \{T \in \mathcal{S}(L) \mid n - |T| \leqslant k\}.$$

We are going to inductively define a sequence of spaces $C^0 \subset \cdots \subset C^n = C$ so that the strata of $C^k$ are indexed by $\mathcal{S}(k)$ and so that each stratum $C_T$ is a contractible, $(n - |T|)$ dimensional manifold with corners. Since every generalized homology sphere of dimension $\leqslant 2$ is a sphere, each stratum of $K$ of dimension $\leqslant 3$ is a cell. So, we can set $C^3 := K^3$. Suppose $C^{k-1}$ has been defined and $T \in \mathcal{S}(k) - \mathcal{S}(k-1)$. Put

$$\partial C_T := \bigcup_{U \supsetneq T} C_U.$$

□

*Claim.* $\partial C_T$ is a homology sphere.

*Proof of Claim.* To prove this we must show $\partial C_T$ is a manifold with the same homology as a sphere. The reason it is a manifold is that the link of each stratum in $\partial C_T$ is the boundary of a simplex. More precisely, the link of $C_U$ in $\partial C_T$ is $\mathrm{Lk}(\sigma_T, \partial\sigma_U)$. To see that it has the same homology as a sphere, consider the covering of $\partial C_T$ by its top dimensional strata. The nerve of this covering is $\mathrm{Lk}(\sigma_T, L)$ which has the same homology as $S^{n-|T|-1}$ (since $L$ is a $\mathrm{GHS}^{n-1}$). By the Acyclic Covering Lemma (in Appendix E.3), $\partial C_T$ also has the same homology as $S^{n-|T|-1}$, proving the claim. By Theorem 10.3.1, $\partial C_T$ bounds a contractible manifold, which is our $C_T$. □

*Remark.* The same argument shows that if a simplicial complex $L$ is a polyhedral homology manifold, then there is a manifold $M$ $(= \partial C)$ and a cell-like map $M \to L$, which is a "resolution" in the classical sense explained in the Notes to 10.4.

**COROLLARY 10.8.4.** ([71, Theorem 15.5].) *Suppose* $(W, S)$ *is type* $\mathrm{HM}^n$. *Then* $W$ *acts as a proper, cocompact, locally linear reflection group on a contractible n-manifold* $M^n$.

*Proof.* $L(W, S)$ is a $\mathrm{GHS}^{n-1}$ and $K$ is a generalized polytope. By Theorem 10.8.3 $K$ has a resolution $C$. Put $M^n = \mathcal{U}(W, C)$. By Proposition 10.1.10, $M^n$ is a manifold and the $W$-action is locally linear. By Theorem 9.1.4, $M^n$ is contractible. □

This leaves open the question of whether we can choose $C$ to be a smooth manifold with corners so that the action on $M^n$ is smooth (Remark 10.1.11). To achieve this it might be necessary to weaken Definition 10.8.2 by allowing cofaces of $C$ to be acyclic rather than contractible. Call the manifold with corners $C$ an *acyclic resolution* of the polyhedral generalized polytope $K$ if for each $T \in \mathcal{S}(L)$, the stratum $C_T$ is acylic. The following theorem and its corollary are proved in [71] (and we will not repeat the proofs here).

**THEOREM 10.8.5.** ([71, Theorem 17.1].) *Suppose* $L$ *is a* $\mathrm{GHS}^{n-1}$ *with dual polyhedral generalized polytope* $K$. *Then* $K$ *admits an acyclic resolution by a smooth manifold with corners* $C$. *Furthermore, we may take each coface of* $C$ *of dimension* $\neq 3, 4$ *to be contractible. If, for every simplex* $\sigma$ *of codimension* 4 *in* $L$, *the homology* 3*-sphere* $\mathrm{Lk}(\sigma, L)$ *smoothly bounds a contractible* 4*-manifold, then we can take* $C$ *to be a resolution.*

**COROLLARY 10.8.6.** ([71, Cor. 17.2].) *Suppose* $(W, S)$ *is type* $\mathrm{HM}^n$. *Then* $W$ *acts as a proper,cocompact, smooth reflection group on a smooth acyclic manifold* $M^n$. *Moreover, we can choose* $M^n$ *to be contractible unless* $n = 4$ *and the homology* 3*-sphere* $L(W, S)$ *does not smoothly bound a contractible manifold.*

## 10.9. VIRTUAL POINCARÉ DUALITY GROUPS

We recall some definitions from Appendix F. Details can be found in [42]. A group $\Gamma$ is *type FP* if $\mathbb{Z}$ (regarded as the trivial $\mathbb{Z}\Gamma$-module) has a finite resolution by finitely generated projective $\mathbb{Z}\Gamma$-modules. (If $\Gamma$ has a classifying space $B\Gamma$ which is a finite CW complex, then the augmented cellular chain complex for $E\Gamma$ is a resolution of $\mathbb{Z}$ by finitely generated free $\mathbb{Z}\Gamma$-modules.) A group of type FP is torsion-free and finitely generated; however, it need not be finitely presented (As we shall see in 11.6, Bestvina and Brady [24] have provided examples of groups of type *FP* which are not finitely presented.)

$\Gamma$ is an $n$-dimensional *Poincaré duality group* (or a PD$^n$-*group* for short) if it is type *FP* and

$$H^i(\Gamma; \mathbb{Z}\Gamma) \cong \begin{cases} 0 & \text{if } i \neq n, \\ \mathbb{Z} & \text{if } i = n. \end{cases}$$

Equivalently, $\Gamma$ is PD$^n$ if it is type *FP* and if $B\Gamma$ satisfies Poincaré duality with arbitrary local coefficients (see Appendix F.5 and [42, p. 222]). Of course, the fundamental group of any closed aspherical $n$-manifold $M^n$ is a PD$^n$-group. $\Gamma$ is a *virtual* PD$^n$-*group* if it possesses a (necessarily torsion-free) subgroup of finite index which is a PD$^n$-group.

**LEMMA 10.9.1.** *Let $(W, S)$ be a Coxeter system. Then $W$ is a virtual* PD$^n$ *group if and only if*

$$H^i_c(\Sigma) \cong \begin{cases} 0 & \text{if } i \neq n, \\ \mathbb{Z} & \text{if } i = n. \end{cases}$$

*Proof.* As in Lemma F.2.2, if $\Gamma$ is a discrete group acting properly and cocompactly on a contractible CW complex $Y$, then $H^*(\Gamma, \mathbb{Z}\Gamma) \cong H^*_c(Y)$. As in Corollary D.1.4, the existence of a faithful linear representation and Selberg's Lemma imply that $W$ contains a torsion-free subgroup $\Gamma$ of finite index. Since $W$ acts properly on $\Sigma$, $\Gamma$ acts freely. Since $\Sigma$ is contractible, $\Sigma/\Gamma$ is a finite model for $B\Gamma$; so, $\Gamma$ is type *FP* (in fact, type *F*). Therefore, $\Gamma$ is a PD$^n$-group (and $W$ is a virtual PD$^n$-group) if and only $H^i_c(\Sigma)$ is given by the formula in the lemma. □

*Remark.* By Farrell's Theorem (F.5.2) and Lemma F.2.2, $W$ is a virtual Poincaré duality group if and only if $H^*_c(\Sigma)$ is finitely generated.

The next theorem is one of the main results of [77].

**THEOREM 10.9.2.** ([77, Theorem B].) *Let $(W, S)$ be a Coxeter system. Then $W$ is a virtual* PD$^n$ *group if and only if it decomposes as*

$$(W, S) = (W_{S_0} \times W_{S_1}, S_0 \cup S_1),$$

*where $W_{S_1}$ is finite and $(W_{S_0}, S_0)$ is type* HM$^n$.

*Proof.* First, suppose $W$ decomposes as in the theorem with $W_{S_1}$ is finite and $(W_{S_0}, S_0)$ of type HM$^n$. For $i = 0, 1$, set

$$W_i := W_{S_i}, \quad L_i := L(W_i, S_i), \quad K_i := K(W_i, S_i), \quad \Sigma_i := \Sigma(W_i, S_i).$$

Since $\Sigma_0$ is a contractible homology manifold of dimension $n$, $W_0$ is a virtual PD$^n$-group and and therefore, so is $W$ (since $W_1$ is finite).

Conversely, suppose $W$ is a virtual $\mathrm{PD}^n$-group. By Lemma 10.9.1, this implies $H_c^*(\Sigma)$ is concentrated in dimension $n$ and is isomorphic to $\mathbb{Z}$ in that dimension. By Theorem 8.5.1, $H_c^*(\Sigma)$ is a direct sum of terms of the form $H^*(K, K^{\mathrm{Out}(w)})$. Hence, only one of these summands can be nonzero (and it must be isomorphic to $\mathbb{Z}$ in dimension $n$). Therefore, there is a unique spherical subset $S_1 \in \mathcal{S}$ such that

(a) If $T \in \mathcal{S}$ and $T \neq S_1$, then $H^*(K, K^{S-T}) = 0$ in all dimensions.

(b) $(K, K^{S-S_1})$ has the same homology as $(D^n, S^{n-1})$.

(c) $W^{S_1}$ is a singleton.

(Here $W^{S_1}$ is as in Definition 4.7.4: it is the set of $w$ in $W$ such that $\mathrm{In}(w) = S_1$.) According to Lemma 4.7.5, (c) implies a decomposition as in the theorem: $(W, S) = (W_0 \times W_1, S_0 \cup S_1)$. It follows that $K = K_0 \times K_1$, $\Sigma = \Sigma_0 \times \Sigma_1$ and $L = L_0 * L_1$ (the join of $L_0$ and $L_1$ defined in A.4.3). Since $S_1$ is spherical, $L_1$ is a simplex, $\Sigma_1$ is a Coxeter cell and $K_1$ is a Coxeter block (Definition 7.3.1). Thus,

- $(K_0, \partial K_0)$ has the same homology as $(D^n, S^{n-1})$ and
- if $T \in \mathcal{S}(W_0, S_0)_{>\emptyset}$, then $H^*(K_0, (K_0)^{S_0-T})$ vanishes in all dimensions.

That is to say, $L_0$ has the punctured $(n-1)$-sphere property in homology. By Lemma 10.8.1, $L_0$ is a $\mathrm{GHS}^{n-1}$, i.e., $(W_0, S_0)$ is type $\mathrm{HM}^n$. □

Theorem 10.9.2 has an important corollary: the notion of being type $\mathrm{HM}^n$ depends only on the Coxeter group $W$ and not on the choice of a fundamental set of generators $S$. We state this as follows.

**COROLLARY 10.9.3.** ([60, Theorem 2.3].) *Suppose $S$ and $S'$ are two sets of fundamental generators for $W$. If $(W, S)$ is type $\mathrm{HM}^n$, then so is $(W, S')$.*

*Proof.* We claim $(W, S)$ is type $\mathrm{HM}^n$ if and only if the following two conditions hold:

(a) $W$ is a virtual $\mathrm{PD}^n$-group.

(b) $W$ has no nontrivial finite subgroup as a direct factor.

Indeed, Theorem 10.9.2 states that if (a) holds and the diagram of $(W, S)$ has no nontrivial spherical component, then it is type $\mathrm{HM}^n$. Conversely, suppose $(W, S)$ is type $\mathrm{HM}^n$ and that a finite subgroup $F$ is a direct factor. As we shall see in Chapter 12 (Theorem 12.3.4 (ii)), since $F$ is finite, $\mathrm{Fix}(F, \Sigma) = \mathrm{Fix}(G, \Sigma)$ for some spherical parabolic subgroup $G$. Since $F$ is normal we can assume that $G$ is a spherical special subgroup $W_T$. By Corollary 7.5.3 (ii), $\mathrm{Fix}(W_T, \Sigma)$ is a homology $(n-k)$-manifold, where $k = \mathrm{Card}(T)$. Let $N(F)$

denote the normalizer of $F$ in $W$. Since $N(F) = W$, we have $\mathrm{vcd}(W) = \mathrm{vcd}(N(F)) = \dim \mathrm{Fix}(W_T, \Sigma) = n - k$. So, $k = 0$ and $F$ is trivial.

Since conditions (a) and (b) are independent of the fundamental set of generators, the corollary follows. □

In Chapter 13 we shall prove another version of this corollary for groups of "type PM$^n$" (Theorem 13.3.10).

Using these ideas, we get the following manifold version of Theorem 8.2.16.

**PROPOSITION 10.9.4.** *Suppose $W$ is a proper, cocompact reflection group on an $N$-manifold $M$ with fundamental chamber $C$ and fundamental set of generators $S$. Suppose $M$ has the same homology as $S^n$. Let $W = W_T \times W_{S-T}$ be the decomposition of Corollary 8.2.5. Then $W_{S-T}$ is type* $\mathrm{HM}^{N-n}$.

*Proof.* As in Remark 8.2.17, let $K'$ be the fundamental chamber for the linear $W_T$-action on $S^n$, let $K'' = K(W_{S-T}, S - T)$ and $\Sigma'' = \Sigma(W_{S-T}, S - T)$. By Remark 8.2.17, $K'$ is the suspended simplex $S^{n-m-1} * \Delta^m$, where $m = \mathrm{Card}(T) - 1$ and $C$ homologically resembles $K' \times K''$. This means there is a coface-preserving map $C \to K' \times K''$ inducing a $W$-equivariant map $M \to S^n \times \Sigma''$ which is an isomorphism on homology. This implies that $M$ and $S^n \times \Sigma''$ have isomorphic compactly supported cohomology. By Poincaré duality (applied to $M^n$), this is the same as $H^*_c(S^n \times \mathbb{R}^{N-n})$. So, by the Künneth Theorem, $H^*_c(\Sigma)$ is isomorphic to $H^*_c(\mathbb{R}^{N-n})$. By Theorem 10.9.2 this means $W_{S-T}$ decomposes as the product of a spherical Coxeter group and a Coxeter group of type $\mathrm{HM}^{N-n}$. Since $\dim \Sigma'' = \dim K'' = N - n$, the spherical factor is trivial. □

The proof shows $M$ homologically resembles the product action on the product of $S^n$ with a contractible manifold in that the chamber $C$ homologically resembles the product of the suspended simplex $K'$ with the generalized polytope $K''$.

*Remark 10.9.5.* Let $R$ be a commutative ring. Definitions 10.4.1 and 10.4.5 can be extended in an obvious fashion to define, respectively, a *homology manifold over* $R$ and a *generalized homology n-sphere over* $R$ (a $\mathrm{GHS}^n_R$, for short). A homology manifold over $R$ satisfies Poincaré duality with coefficients in $R$. Similarly, the notion of a $\mathrm{PD}^n_R$*-group* makes sense. (See Remark F.5.5.) The proof of Theorem 10.9.2 works over any principal ideal domain $R$. (The reason we need to assume $R$ is a principal ideal domain is that we want to be able to conclude from the assumption $\overline{H}^*(\Sigma; R) \cong R$ that there is only one nonzero term in the sum of Theorem 8.5.1.)

**Example 10.9.6.** (*Poincaré duality groups over R.*) There are groups of type $FP$ which satisfy Poincaré duality over some ring $R$ but not over $\mathbb{Z}$. For example, suppose $R = \mathbb{Z}[\frac{1}{m}]$. For each $n > 3$ we can find an $L$ which is a generalized homology $(n-1)$-sphere over $R$ but not over $\mathbb{Z}$. For $n$ even, take $L$ to be a lens space $S^{n-1}/\mathbf{C}_m$, while for $n = 2k+1$, take $L^{2k}$ to be the suspension of a $(2k-1)$-dimensional lens space. Proceed as in Lemma 7.2.2. Triangulate $L$ as a flag complex and let $W$ be the right-angled Coxeter group associated to its 1-skeleton. Then $W$ is a virtual $\mathrm{PD}^n_R$-group. However, since $H^*(L)$ has nonzero $m$-torsion, it follows from Corollary 8.5.2 that $H^*(W;\mathbb{Z}W)$ has nonzero $m$-torsion. Hence, $W$ is not a virtual $\mathrm{PD}^n$-group over $\mathbb{Z}$.

Next we use Theorem 10.9.2 to prove that if a Coxeter group acts effectively, properly and cocompactly on an acyclic manifold, then it acts as a group generated by reflections.

**PROPOSITION 10.9.7.** ([77, Cor. 5.6]). *Suppose a Coxeter group $W$ acts effectively, properly and cocompactly on a $\mathbb{Z}/2$ acyclic $n$-manifold $M$. Then $W$ acts on $M$ as group generated by reflections. (Here "reflection" is used as in Definition 10.1.2.) In particular, for any fundamental set of generators $S$ and any $s \in S$, $M_s$ is a $\mathbb{Z}/2$-acyclic, $\mathbb{Z}/2$-homology manifold of codimension 1 in $M$.*

*Proof.* By the Local Smith Theorem (10.4.3), $M_s$ $(= \mathrm{Fix}(\langle s\rangle, M))$ is a $\mathbb{Z}/2$-homology manifold and by Smith theory, it is $\mathbb{Z}/2$-acyclic. Since $H^*_c(M;\mathbb{Z}/2)$ is the same as $H^*_c(\mathbb{R}^n;\mathbb{Z}/2)$, it follows from Lemma F.2.2 that $W$ is a virtual Poincaré duality group over $\mathbb{Z}/2$. By Remark 10.9.5, Theorem 10.9.2 goes through when $\mathbb{Z}$ is replaced by $\mathbb{Z}/2$. So, if $S$ is any fundamental set of generators for $W$, then $S = S_0 \cup S_1$ and $W = W_0 \times W_1$, where $W_1$ is finite and $L(W_0, S_0)$ is a $\mathrm{GHS}^{n-1}$ over $\mathbb{Z}/2$. For any $s \in S$, its centralizer, $N_W(s)$ acts properly and cocompactly on $M_s$. Since $M_s$ is $\mathbb{Z}/2$-acyclic, $\dim M_s = \mathrm{vcd}_{\mathbb{Z}/2}(N_W(s))$ (where the virtual cohomological dimension over $\mathbb{Z}/2$ is defined as in Remark F.4.2). If $s \in S_1$, this virtual cohomological dimension is $n$; hence, $M_s = M$. Since the $W$-action is effective, this cannot happen. So, $S_1 = \emptyset$ and $W = W_0$. If $s \in S_0$, then $\dim M_s = \mathrm{vcd}_{\mathbb{Z}/2}(N_W(s)) = \dim \Sigma_s = n-1$. Then, by Alexander duality, $M - M_s$ has two components, i.e., $s$ is a reflection. □

**COROLLARY 10.9.8.** *Suppose $M$ is a symmetric space of noncompact type and $W$ is a discrete cocompact group of isometries of $M$. Then $M$ must be a product of a Euclidean space and real hyperbolic spaces.*

*Proof.* Since $W$ acts on $M$ by isometries, for each $s \in S$, $\mathrm{Fix}(\langle s\rangle, M)$ is a totally geodesic submanifold of $M$. By the previous proposition, this submanifold must be of codimension one. But symmetric spaces of noncompact type do not contain totally geodesic submanifolds of codimension unless each irreducible factor is $\mathbb{E}^1$ or a real hyperbolic space $\mathbb{H}^k$. □

## NOTES

In this chapter we have discussed some of the most famous results in topology: the Generalized Poincaré Conjecture, the h-Cobordism Theorem, the Double Suspension Theorem, and the Polyhedral Manifold Characterization Theorem. We have also explained some of topology's most famous examples: nonsimply connected homology spheres and Whitehead's example of a contractible 3-manifold which is not simply connected at infinity.

**10.1.** There is an extensive literature on the cohomology theory of finite transformation groups, e.g., see [27, 36].

The motivation of the Hsiangs [160] for proving Proposition 10.1.12 is interesting. They were studying smooth actions of compact connected Lie groups on homotopy spheres and on contractible manifolds. They wanted to use finite reflection groups similarly to the way they are used in Lie group theory. The theorems they were aiming for said that, under mild hypotheses, the smooth Lie group action would be "modeled on" a linear representation. Typically their arguments went as follows. Suppose a Lie group $G$ acts on a homotopy sphere $M$. Let $T \subset G$ be a maximal torus. By Smith theory, the fixed set $M^T$ is a homology sphere. The Weyl group $W$ acts on it. If we somehow know that $W$ acts by reflections on $M^T$, then we can apply Proposition 10.1.12 to conclude that all proper special subgroups of $W$ have fixed points. From this one can often draw conclusions about the original $G$-action.

**10.3.** Figure 10.2 and much of the discussion in Example 10.3.5 on Whitehead's example are taken from [215].

**10.4.** A compact metrizable space is an *ANR* if it is locally contractible. When is an ANR homology $n$-manifold a topological manifold? When $n \geqslant 5$, a complete answer was given by Edwards and Quinn. An ANR homology manifold $X^n$ is *resolvable* if it is the image of a cell-like map from an $n$-manifold. $X^n$ has the *disjoint 2-disk property* if given embeddings $f_i : D^2 \to X^n$, $i = 1, 2$, $f_2$ can be approximated by another embedding with image disjoint from $f_1(D^2)$. (When $X^n$ is a polyhedron this property fails whenever the link of a vertex is a homology sphere with nontrivial $\pi_1$.) Generalizing Theorem 10.4.10, Edwards showed that $X^n$ is a topological manifold if and only if it is resolvable and has the disjoint 2-disk property. Quinn showed that there is a single obstruction in $1 + 8\mathbb{Z}$ for $X^n$ to admit a resolution. (In fact, he incorrectly believed that he could show this obstruction was always trivial.) Finally, it was proved in [44] that for each $n \geqslant 5$, there are examples, satisfying the disjoint 2-disk property, where the obstruction to resolvability is nonzero.

**10.5.** For several years it was conjectured that the universal cover of any closed aspherical manifold must be homeomorphic to Euclidean space. (See, for example, [167, 170] and [293, Problem F16, p. 388].) Some positive results in this direction had been achieved in [188, 164]. This speculation was put to rest by Theorem 10.5.1, which was proved in [71] by the method of 10.5.

In [71, Theorem 16.1] we claimed that when $L$ is a GHS, $\Sigma$ is simply connected at infinity if and only if $L$ is simply connected. Certainly, if $\pi_1^\infty(\Sigma)$ is trivial, then so is $\pi_1(L)$. However, the converse is patently false. We gave a counterexample in Example 10.5.3. The correct result is Theorem 9.2.2: $\Sigma$ is simply connected at infinity if and only if $L$, as well as, all "punctured nerves" (of the form $L - \sigma_T$) are simply connected.

Another version of Example 10.5.3 is explained in [5]. Suppose $L^{n-1}$ is a non-simply-connected PL homology sphere, $n \geqslant 5$. One can then find a codimension one homology sphere $L' \subset L$ whose fundamental group maps onto $\pi_1(L)$. $L - L'$ has two components. Let $L_1$, $L_2$ be their closures. For $i = 1, 2$, put $\widehat{L_i} := L_i \cup \text{Cone}\,(L')$. Triangulate $L$ as a flag complex so that $L'$ is a full subcomplex. $W$ is the associated right-angled Coxeter group. One can construct a CAT(0) cubical complex $\widehat{K}$ with $\partial\widehat{K} \cong L$ and with two interior vertices $v_1$, $v_2$ such that $\text{Lk}(v_i, \widehat{K}) = \widehat{L_i}$. Since $\widehat{L_i}$ is simply connected, it follows from Theorem 10.4.10 that $\widehat{K}$ is a topological manifold with boundary and hence, $\mathcal{U}(W, \widehat{K})$ is a manifold, as well as a CAT(0)-space.

# Chapter Eleven

## THE REFLECTION GROUP TRICK

The term "reflection group trick" refers to a method for converting a finite aspherical CW complex into a closed aspherical manifold which retracts back onto it. The upshot is that closed aspherical manifolds are at least as complicated as finite aspherical complexes. The basic technique underlying the trick is the semidirect product construction of 9.1.

Following [81], in 11.3 the trick is used to produce examples of topological closed aspherical manifolds which are not homotopy equivalent to smooth closed manifolds. Another consequence is discussed in 11.4: if the Borel Conjecture is true for all closed aspherical manifolds, then its relative version for aspherical manifolds with boundary is also true. In 11.6 we discuss the Bestvina-Brady examples of groups of type $FL$ which are not finitely presented. Then we show how the reflection group trick can be used to promote these to examples of Poincaré duality groups which are not finitely presented.

### 11.1. THE FIRST VERSION OF THE TRICK

A group $\pi$ is *type F* if its classifying space $B\pi$ has the homotopy type of a finite CW complex. (See Appendix F.4 for more about groups of type $F$.) Given a group $\pi$ of type $F$, the reflection group trick is a technique for constructing a closed aspherical manifold $M$ such that $\pi_1(M)$ retracts onto $\pi$. In a nutshell the construction goes as follows. Assume, as we may, that $B\pi$ is a finite polyhedron. "Thicken" $B\pi$ to $X$, a compact PL manifold with boundary. (For example, if $n > 2(\dim B\pi)$, we can piecewise linearly embed $B\pi$ in some triangulation of $\mathbb{E}^n$ and then take $X$ to be a regular neighborhood of $B\pi$ in $\mathbb{E}^n$.) $X$ is homotopy equivalent to $B\pi$ (since it collapses onto it). Let $(W, S)$ be a Coxeter system whose nerve $L$ is a triangulation of $\partial X$. (As in Lemma 7.2.2, we can take $L$ to be any triangulation of $\partial X$ as a flag complex and then, as in Example 7.1.7, take $(W, S)$ to be the right-angled Coxeter system associated to the 1-skeleton of $L$.) Let $\mathcal{S}$ be the set of spherical subsets of $S$. For each $s \in S$, $X_s$ is the geometric realization of $\mathcal{S}_{\geqslant\{s\}}$ (regarded as a subset of $\partial X$). In other words, $X_s$ is the closed star of the vertex $s$ of $L$ in the barycentric subdivision of $L$. $(X_s)_{s\in S}$ is a $W$-finite mirror structure on $X$. As usual, $\mathcal{U} := \mathcal{U}(W, X)$.

Since $L$ is a PL triangulation of $\partial X$, this mirror structure gives $X$ the structure of a mirrored manifold with corners (in the sense of Definition 10.1.8). By Proposition 10.1.10, this implies $\mathcal{U}$ is a manifold with a proper $W$-action. By Corollary 7.5.3 (iii) and its proof, the action is locally linear.

Let $\Gamma \subset W$ be any torsion-free subgroup of finite index. (For example, in the right-angled case we could take $\Gamma$ to be the commutator subgroup of $W$.) Define $M$ (or $M(\Gamma, X)$) to be the quotient space

$$M := \mathcal{U}/\Gamma. \tag{1}$$

Since $\Gamma$ acts properly and freely on $\mathcal{U}$, the quotient map $\mathcal{U} \to M$ is a covering projection. Hence, $M$ is a manifold.

**THEOREM 11.1.1.** *Suppose $M$ is defined by* (1). *Then*

*(i) $M$ is a closed aspherical manifold and*

*(ii) $M$ retracts onto $B\pi$.*

To prove this we need the following.

**LEMMA 11.1.2.** *$\mathcal{U}$ is aspherical.*

*Proof.* We construct the universal cover of $\widetilde{\mathcal{U}}$ of $\mathcal{U}$ as in 9.1. Let $\widetilde{X}$ be the universal cover of $X$, let $\widetilde{L} = \widetilde{\partial X}$ be the inverse image of $\partial X$ in $\widetilde{X}$ and let $\widetilde{S} = \text{Vert}(\widetilde{L})$ be the inverse image of $S$. As in 9.1, give $\widetilde{X}$ the induced mirror structure indexed by $\widetilde{S}$ and let $(\widetilde{W}, \widetilde{S})$ be the (infinitely generated) Coxeter system defined by formula (2) in Section 9.1. We must show $\widetilde{\mathcal{U}}$ is contractible. By Theorem 9.1.4, this is equivalent to showing that

(a) $\widetilde{X}$ is contractible and

(b) for each $\widetilde{T} \in \mathcal{S}(\widetilde{W}, \widetilde{S})_{>\emptyset}$, the coface $\widetilde{X}_{\widetilde{T}}$ is acyclic.

Condition (a) holds since $X$ is aspherical. Condition (b) holds since for any $\widetilde{T} \neq \emptyset$, $\widetilde{X}_{\widetilde{T}}$ is homeomorphic to the disk $X_T$ (where $T$ denotes the image of $\widetilde{T}$). □

*Proof of Theorem 11.1.1.* $M$ is aspherical since it is covered by $\mathcal{U}$ (which is aspherical by the previous lemma). Since $X$ is a strict fundamental domain for the $W$-action, it can be identified with a subspace of $M$. Since $[W : \Gamma] < \infty$, $M$ is tiled by finitely many copies of $X$. Since $X$ $(= \mathcal{U}/W)$ is compact, so is $M$. The orbit map $\mathcal{U} \to X$ induces a retraction $M \to X$. Since $X$ collapses onto $B\pi$, (ii) holds. □

There is a more explicit description of $\pi_1(M)$. As in 9.1, $\widetilde{W} \rtimes \pi$ is the group of all lifts of the $W$-action to $\widetilde{\mathcal{U}}$. Let $q : \widetilde{W} \to W$ be the natural projection and put $\widetilde{\Gamma} := q^{-1}(\Gamma)$. Since $\Gamma$ is torsion-free, so is $\widetilde{\Gamma}$. For any $w \in W$, $q^{-1}(w)$ is a $\pi$-stable subset of $\widetilde{W}$. Hence, $\widetilde{\Gamma}$ is $\pi$-stable. $\pi_1(M)$ is the group of deck transformations of $\widetilde{\mathcal{U}} \to M$, i.e., it consists of the elements of $\widetilde{W} \rtimes \pi$ which act on $\mathcal{U}$ as an element of $\Gamma$. That is to say, $\pi_1(M) = \widetilde{\Gamma} \rtimes \pi$ and the retraction $\pi_1(M) \to \pi$ is the canonical homomorphism from the semidirect product onto $\pi$.

In the next proposition, $(X, \partial X)$ is not required to be a manifold with boundary (however, $\partial X$ is always required to be homeomorphic to the simplicial complex $L$).

**PROPOSITION 11.1.3.** (Compare Theorem 10.6.1.) *Suppose $M = \mathcal{U}/\Gamma$ as in* (1). *The following statements hold.*

*(i) If $(X, \partial X)$ is a Poincaré pair and $\partial X$ is a homology manifold, then $M$ is a Poincaré space. (For the notions of a "Poincaré pair" and "homology manifold," see Definitions H.1.3 and 10.4.1, respectively.)*

*(ii) If $(X, \partial X)$ is a homology manifold with boundary, then $M$ is a homology manifold.*

*(iii) If $(X, \partial X)$ is a topological (resp. PL) manifold with boundary and if $L$ is a triangulation of $\partial X$ as a PL manifold, then $M$ is a topological (resp. PL) manifold.*

*(iv) If $(X, \partial X)$ is a smooth manifold with boundary and if $L$ is a smooth triangulation of $\partial X$, then $M$ is a smooth manifold.*

*Proof.* If $(X, \partial X)$ is a homology manifold with boundary, then the dual complex of $L$ $(= \partial X)$ gives $X$ the structure of a mirrored homology manifold with corners (as defined in 10.7.2). The proof is similar to that of Lemma 10.8.1. If $L$ is a PL triangulation of $\partial X$, then $X$ has the structure of a manifold with corners. Similarly, if $L$ is a smooth triangulation, $X$ is a smooth manifold with corners. Assertions (iii) and (iv) now follow from Proposition 10.1.10 (although PL manifolds are not mentioned in the statement of Proposition 10.1.10). Assertion (ii) follows from Proposition 10.7.5.

To prove (i), write $\mathcal{U}$ as an increasing union of chambers as in 8.1:

$$X = P_1 \subset \cdots \subset P_n \subset \cdots,$$

where the intersection of the $n^{\text{th}}$ chamber $w_n X$ with $P_{n-1}$ is a generalized homology disk of codimension 0 in $w_n \partial X$. (This uses the fact that $\partial X$ is a homology manifold.) It follows that $(P_n, \partial P_n)$ is a Poincaré pair (and that $\partial P_n$ is a homology manifold). This implies that $\mathcal{U}$ is a Poincaré space and therefore, so is $M$ (since $\mathcal{U}$ is a covering of it). $\square$

## 11.2. EXAMPLES OF FUNDAMENTAL GROUPS OF CLOSED ASPHERICAL MANIFOLDS

We begin with two examples of G. Mess [204], based on Baumslag-Solitar groups.

Given a pair of integers $(p, q)$, the *Baumslag-Solitar group* $BS(p, q)$ is the 1-relator group defined by the presentation:

$$BS(p,q) := \langle a, b \mid ab^p a^{-1} = b^q \rangle. \tag{2}$$

If $p$ and $q$ are relatively prime, then the word $ab^p a^{-1} b^{-q}$ is not a proper power. So it follows from Lyndon's Theorem on 1-relator groups ([195]) that the presentation complex for $BS(p, q)$ is aspherical. In other words, $BS(p, q)$ is type $F$ and of geometric dimension 2. (This was explained earlier in 2.3. For the definitons of "geometric dimension" and "type $F$," see Appendix F.)

Of course, every two-dimensional polyhedron can be embedded in $\mathbb{E}^5$. Although it is not true that every 2-complex can be embedded in $\mathbb{E}^4$, every finite two-dimensional CW complex can be thickened to a compact 4-manifold. (Proof: start with a boundary connected sum of $S^1 \times D^3$ s as a thickened 1-skeleton, then add a 2-handle for each 2-cell.) Thus, for each Baumslag-Solitar group $\pi = BS(p, q)$ with $p, q$ relatively prime, $B\pi$ can be thickened to a compact aspherical $n$-manifold with boundary for any $n \geqslant 4$.

A group $\pi$ is *residually finite* if given any two elements $g_1, g_2 \in \pi$ there is a homomorphism $\varphi$ to some finite group $F$ such that $\varphi(g_1) \neq \varphi(g_2)$.

**Example 11.2.1.** (*Not residually finite*; Mess [204].) The Baumslag-Solitar group $\pi = BS(2, 3)$ is not residually finite. Neither is any other group (such as $\widetilde{\Gamma} \rtimes \pi$) which retracts onto $\pi$. Hence, for each $n \geqslant 4$, there is a closed aspherical $n$-manifold whose fundamental group is not residually finite. (This example answered Problem F.2 in Wall's Problem List [293, p. 386] in the negative.)

**Example 11.2.2.** (*Infinitely divisible abelian subgroups*; Mess [204].) This time take $\pi = BS(1, 2)$. The centralizer of $b$ in this group is isomorphic to a copy of the dyadic rationals. Hence, for each $n \geqslant 4$, there is a closed aspherical $n$-manifold whose fundamental group contains an infinitely divisible abelian group.

**Example 11.2.3.** (*Unsolvable word probem*, Weinberger [297, p. 106].) There are examples of finitely presented groups $\pi$ with unsolvable word problem such that $B\pi$ is a finite 2-complex. (See [209, Theorem 4.12].) Since any group which retracts onto such a group also has unsolvable word problem, the reflection group trick gives the following: for each $n \geqslant 4$, there is a closed aspherical $n$-manifold whose fundamental group has unsolvable word problem.

## 11.3. NONSMOOTHABLE ASPHERICAL MANIFOLDS

The first examples of aspherical manifolds which could not be smoothed were constructed in [81] by using the reflection group trick (answering part of Problem G2 in Wall's Problem List [293, p.391] in the negative).

As in 11.1, suppose $(X, \partial X)$ is a compact aspherical $n$-manifold with boundary, that $\partial X$ is triangulable, that $L$ is a triangulation of it as a flag complex and that $M = \mathcal{U}/\Gamma$ is the result of applying the reflection group trick to these data.

As explained in the last paragraph of Appendix H.2, associated to any topological manifold with boundary (indeed, to any Poincaré pair), say $(Y, \partial Y)$, there is a stable spherical fibration $\nu(Y)$ over $Y$, called its "Spivak normal fibration." If $Y$ is a smooth manifold, then the sphere bundle associated to its stable normal bundle is $\nu(Y)$. So, there is an obstruction for $(Y, \partial Y)$ to be homotopy equivalent to a smooth manifold: its Spivak normal fibration must reduce to a linear vector bundle. In more concrete terms, this comes down to the question of whether the classifying map $c_{\nu(Y)} : Y \to BG$ lifts to $BO$. (As explained in Appendix H.2, $BG$ is the classifying space for stable spherical fibrations and $BO$ is the classifying space for stable vector bundles.)

**PROPOSITION 11.3.1.** *The Spivak normal fibration of $X$ reduces to a linear vector bundle if and only if the Spivak normal fibration of $M$ reduces to a linear vector bundle. Hence, if the Spivak normal fibration of $X$ does not reduce to a linear vector bundle, $M$ is not homotopy equivalent to a smooth manifold.*

*Proof.* Since $X$ is a submanifold of codimension 0 in $M$, $\nu(M)|_X = \nu(X)$. So, if $c_{\nu(M)} : M \to BG$ lifts to $BO$, then so does $c_{\nu(X)} : X \to BG$.

To prove the converse, we first note that the proof of Theorem 10.2.1 shows that $\nu(M) \cong p^*(\nu(X))$, where $p : M \to X$ is the retraction. Hence, $c_{\nu(M)} = c_{\nu(X)} \circ p$. So, if $\tilde{c} : X \to BO$ is a lift of $c_{\nu(X)}$, then $\tilde{c} \circ p$ lifts $c_{\nu(M)}$. $\square$

**THEOREM 11.3.2.** ([81].) *In each dimension $\geqslant 13$, there are closed aspherical manifolds not homotopy equivalent smooth manifolds.*

*Proof.* The proof uses Proposition 11.3.1. Let $B\widetilde{PL}(i)$ be the classifying space for PL block bundles of rank $i$. Then

$$BPL \sim B\widetilde{PL} := \lim_{i \to \infty} B\widetilde{PL}(i).$$

Construct $X$ as follows. First, find an element $a \in \pi_k(BPL)$ so that its image $\alpha \in \pi_k(BG)$ does not lie in the image of $\pi_k(BO) \to \pi_k(BG)$. Let $a_r : S^k \to B\widetilde{PL}(r)$ represent $a$. Take a degree one map $T^k \to S^k$ and compose it with $a_r$ to get $b_r : T^k \to B\widetilde{PL}(r)$. We can also compose it with $\alpha$ to get $\beta : T^k \to BG$. By the classification of abstract regular PL neighborhoods ([250, Cor 4.7]), there is a compact PL manifold $X$ of dimension $k + r$ such that

(i) $X$ collapses onto $T^k$ (so $X$ is aspherical and $\pi \cong \mathbb{Z}^k$), and

(ii) $b_r$ classifies the normal block bundle of $T^k$ in $X$ (so the stable tangent bundle of $X$ is classified by the composition of $X \to T^k$ with $b_r$).

Since $T^k$ is parallelizable, it follows from (ii) that the composition of $X \to T^k$ with $\beta$ classifies the inverse of the Spivak normal fibration $\nu(X)$. We claim that $c_{\nu(X)} : X \to BG$ cannot lift to $BO$. Suppose, to the contrary, it does. Since $BO$ and $BG$ are infinite loop spaces, $BO \sim \Omega(\Omega^{-1}BO)$ and $BG \sim \Omega(\Omega^{-1}BG)$. Since $c_{\nu(X)}$ lifts to $BO$, its adjoint, $ad(c_{\nu(X)}) : ST^k \to \Omega^{-1}BG$, lifts to $\Omega^{-1}BO$. (Here $ST^k$ is the suspension of $T^k$.) It is well known that $ST^k$ is homotopy equivalent to $S^{k+1} \vee A$, where $A$ is a wedge of spheres of dimension $\leqslant k$; moreover, $ad(c_{\nu(X)})|_{S^{k+1}}$ is $ad(-\alpha)$. But this contradicts the assumption that $\alpha$ does not lift to $BO$.

It remains to show that there exists such an element $a \in \pi_k(BPL)$. As shown in [81, p. 140]), there is such an element in $\pi_9(BPL)$ coming from an $a_4 \in \pi_9(B\widetilde{PL}(4))$. □

## 11.4. THE BOREL CONJECTURE AND THE PD$^n$-GROUP CONJECTURE

The most famous problem concerning aspherical manifolds is the Borel Conjecture. It asserts that a homotopy equivalence between closed aspherical manifolds is homotopic to a homeomorphism. A relative version is that if $f : (X, \partial X) \to (X', \partial X')$ is a homotopy equivalence of pairs between manifolds with boundary with $X$ and $X'$ aspherical and $f|_{\partial X}$ a homeomorphism, then $f$ is homotopic rel $\partial X$ to a homeomorphism.

The Borel Conjecture is closely related to the PD$^n$-Group Conjecture. This asserts that if $\pi$ is a finitely presented Poincaré duality group, then it is the fundamental group of a closed aspherical manifold. (We will see in Theorem 11.6.4 that without the hypothesis of finite presentation, the PD$^n$-Group Conjecture is false.) A relative version is that if $\pi$ is a finitely presented group of type $FP$ and if $(X, \partial X)$ is a Poincaré pair with $X$ homotopy equivalent to $B\pi$ and $\partial X$ a manifold, then $(X, \partial X)$ is homotopy equivalent rel $\partial X$ to a compact manifold with boundary. (For the definition of what it means for a group to be of type $F$ or $FP$, see Appendix F.4; for the definition of "Poincaré duality group," see Appendix F.5.1.) The Borel Conjecture and the PD$^n$-Group Conjecture are discussed in greater detail in Appendix H.1. In this section we use the reflection group trick to show that for either of the above conjectures, the absolute version implies the relative version. More precisely, we prove the following.

**THEOREM 11.4.1**

*(i)* *Suppose $(X, \partial X)$ is a counterexample to the relative version of the* $\mathrm{PD}^n$*-Group Conjecture in dimension $n \geqslant 5$ and $L$ is a flag triangulation of $\partial X$. Then $M := \mathcal{U}(W, X)/\Gamma$ is an aspherical Poincaré duality space and $\pi_1(M)$ is a counterexample to the* $\mathrm{PD}^n$*-Group Conjecture.*

*(ii)* *Suppose $f : (X, \partial X) \to (X', \partial X')$ is a counterexample to the relative version of the Borel Conjecture in dimension $n \geqslant 5$ and $L$ is a flag triangulation of $\partial X$ $(= \partial X')$. Let $M := \mathcal{U}(W, X)/\Gamma$ and $M' := \mathcal{U}(W, X')/\Gamma$. Then the homotopy equivalence $\tilde{f} : M \to M'$, induced by $f$, is a counterexample to the Borel Conjecture.*

As explained in Appendix H, in dimensions $\geqslant 5$, the relative versions of the Borel Conjecture and the $\mathrm{PD}^n$-Group Conjecture for a given group $\pi$ are basically equivalent to three algebraic conjectures. The first two conjectures are in algebraic K-theory: the Reduced Projective Class Group Conjecture (Conjecture H.1.8) and the Whitehead Group Conjecture (Conjecture H.1.9). The Reduced Projective Class Group Conjecture implies that if $\pi$ is type *FP*, then it is automatically type *F*. The Whitehead Group Conjecture implies that, in dimensions $\geqslant 5$, any h-cobordism between manifolds with fundamental group $\pi$ is a cylinder. The third conjecture is the Assembly Map Conjecture in L-theory (Conjecture H.3.2). Our argument for Theorem 11.4.1 consists of proving the corresponding results for these algebraic conjectures. As usual, $\pi = \pi_1(X)$, $M = \mathcal{U}/\Gamma$ and $\tilde{\pi} := \pi_1(M) = \pi \rtimes \Gamma$.

**THEOREM 11.4.2.** *Suppose a group $\pi$ of type $F$ is a counterexample to the Whitehead Group Conjecture, i.e., $Wh(\pi) \neq 0$. Then $\tilde{\pi}$ is the fundamental group of a closed aspherical manifold $M$ and $Wh(\tilde{\pi}) \neq 0$.*

*Proof.* Since $\pi$ is a retract of $\tilde{\pi}$, $Wh(\pi)$ is a direct summand of $Wh(\tilde{\pi})$. □

In the case of the Assembly Map Conjecture, we have the following.

**THEOREM 11.4.3.** *With notation as in Appendix H.3, suppose $(X, \partial X)$ is a counterexample to the Assembly Map Conjecture, i.e., suppose the assembly map $A : H_*(X; \mathbb{L}) \to L_*(\pi)$ fails to be an isomorphism. Then $A : H_*(M; \mathbb{L}) \to L_*(\tilde{\pi})$ fails to be an isomorphism.*

*Proof.* As explained in Appendix H.3, after stabilizing by taking a product with $D^4$, the structure set $\mathcal{S}_*(X)$ is an abelian group; moreover, it is functorial in $X$. The Assembly Map Conjecture is equivalent to the statement that $\mathcal{S}_*(X)$ is trivial. Since $X$ is a retract of $M$, $\mathcal{S}_*(X)$ is a direct summand of $\mathcal{S}_*(M)$. This proves the theorem. A more pedestrian version of this argument is to consider the commutative diagram

$$\begin{array}{ccc} H_*(X;\mathbb{L}) & \xrightarrow{A} & L_*(\pi) \\ \downarrow & & \downarrow \\ H_*(M;\mathbb{L}) & \xrightarrow{A} & L_*(\tilde{\pi}). \end{array}$$

$H_*(X;\mathbb{L})$ is a direct summand of $H_*(M;\mathbb{L})$ and $L_*(\pi)$ is a direct summand of $L_*(\tilde{\pi})$. So, if $A : H_*(X;\mathbb{L}) \to L_*(\pi)$ fails to be an isomorphism, then so does $A : H_*(M;\mathbb{L}) \to L_*(\tilde{\pi})$.) □

**Question 11.4.4.** Suppose $\pi$ is *FP* and not $F$. (Conjecturally no such $\pi$ exists.) Is it possible to find a triangulable manifold $\partial B$ and a map $i : \partial B \to B\pi$ (which we may assume is an embedding) so that $(B\pi, \partial B)$ is a Poincaré pair (defined in Appendix H.1)?

If we can find such a Poincaré pair, then we can apply the reflection group trick (at least its second version, defined in the next section) to obtain a space $Y = \mathcal{U}(W, B\pi)/\Gamma$ which is an aspherical Poincaré space (see Proposition 11.5.3 below). Since $\pi$ is not type $F$ neither is any group, such as $\pi_1(Y)$, which retracts onto it. Hence, if there is such a pair $(B\pi, \partial B)$, then there is also a Poincaré duality group not of type $F$.

*Proof of Theorem 11.4.1.* (i) Suppose $(X, \partial X)$ is an $n$-dimensional Poincaré pair with $\partial X = L$ a manifold triangulated as a flag complex, with $X$ finitely dominated and homotopy equivalent to $B\pi$, where $\pi$ is a group of type *FP*. By Proposition 11.1.3 (i), $M$ is Poincaré complex; moreover, $M$ is also finitely dominated and hence, $\tilde{\pi}$ is a finitely presented PD$^n$-group. There are two algebraic ways in which $(X, \partial X)$ could provide a counterexample to the relative version of the PD$^n$-group Conjecture, either

- Wall's finiteness obstruction $\sigma(X) \in \widetilde{K}_0(\mathbb{Z}\pi)$ could be nonzero or
- $\mathcal{S}_{n-1}(X)$ could be nonzero.

(For an explanation of these two conditions, see Appendices F.4 and H.3.) Moreover, for $n \geqslant 5$, these are the only two ways in which the PD$^n$-Group Conjecture can fail. The failure of either condition for $X$ causes the corresponding failure in $M$. In the first case, since $\sigma(M) \in K_0(\mathbb{Z}\tilde{\pi})$ restricts to $\sigma(X)$, we see that if $\sigma(X) \neq 0$, then $\sigma(M) \neq 0$. Similarly, in the second case, Theorem 11.4.3 shows that if $\mathcal{S}_{n-1}(X) \neq 0$, then $\mathcal{S}_{n-1}(M) \neq 0$.

(ii) Suppose $(X, \partial X)$ is a compact aspherical $n$-manifold with boundary and $L$ is a PL triangulation of $\partial X$ as a flag complex. By Proposition 11.1.3 (iii), $M$ is a closed aspherical manifold. There are two algebraic ways in which $(X, \partial X)$ could fail the relative version of the Borel Conjecture, either:

- $Wh(\pi) \neq 0$ (and hence, there is a nontrivial h-cobordism of $X$ rel boundary) or
- $\mathcal{S}_n(X) \neq 0$.

Moreover, for $n \geqslant 5$, these are the only the only two ways in which the Borel Conjecture can fail. In the first case, by Theorem 11.4.2 (ii), $Wh(\tilde{\pi}) \neq 0$. In the second case, by Theorem 11.4.3, $\mathcal{S}_n(M) \neq 0$. □

## 11.5. THE SECOND VERSION OF THE TRICK

By weakening the conditions in 11.1, we get a more general version of the reflection group trick. As data, suppose we are given

- a space $X$, a group $\pi$, and an epimorphism $\psi : \pi_1(X) \to \pi$,
- a Coxeter system $(W, S)$ with nerve $L$, and
- an embedding $L \hookrightarrow X$, identifying $L$ with a closed subpace of $X$.

As before, this gives $X$ a mirror structure: $X_s := |\mathcal{S}_{\geqslant\{s\}}|$.

We want to define a covering space $\tilde{p} : \widetilde{\mathcal{U}} \to \mathcal{U}$ with group of deck transformations $\pi$. Let $p : \widetilde{X} \to X$ be the covering space corresponding to $\operatorname{Ker} \psi$. Let $\widetilde{L} := p^{-1}(L)$ with the induced structure of a simplicial complex. Let $\widetilde{S} = p^{-1}(S)$. As in formula (2) of Section 9.1, define a Coxeter matrix $m_{\tilde{s}\tilde{t}}$ on $\widetilde{S}$ by

$$m_{\tilde{s}\tilde{t}} = \begin{cases} 1 & \text{if } \tilde{s} = \tilde{t}, \\ m_{st} & \text{if } \tilde{s} \text{ and } \tilde{t} \text{ span an edge in } \widetilde{L}, \\ \infty & \text{otherwise.} \end{cases}$$

where $\tilde{s}, \tilde{t} \in \widetilde{S}$ are vertices lying over $s, t \in S$. Let $(\widetilde{W}, \widetilde{S})$ be the corresponding Coxeter system. If $\widetilde{T}$ is a spherical subset of $\widetilde{S}$, then its image $T$ is a spherical subset of $S$. Since $T$ spans the simplex $\sigma_T$ in $L$, $p^{-1}(\sigma_T)$ contains a simplex spanned by $\widetilde{T}$. It follows that $\widetilde{L}$ is the nerve of $(\widetilde{W}, \widetilde{S})$. As before, $\widetilde{L}$ gives a mirror structure on $\widetilde{X}$ over $\widetilde{S}$. Put $\widetilde{\mathcal{U}} := \mathcal{U}(\widetilde{W}, \widetilde{X})$ and let $\tilde{p} : \widetilde{\mathcal{U}} \to \mathcal{U}$ be the natural projection. The group $\pi$ acts on $\widetilde{X}$ via deck transformations and on $\widetilde{W}$ through diagram automorphisms (Definition 9.1.6). So, the semidirect product construction of 9.1 gives an action of the group $G := \widetilde{W} \rtimes \pi$ on $\widetilde{\mathcal{U}}$. The subgroup $\pi$ acts freely on $\widetilde{\mathcal{U}}$ and the orbit space is identified with $\mathcal{U}$ via $\tilde{p}$. (So, $\widetilde{\mathcal{U}}$ is the covering space of $\mathcal{U}$ associated to the kernel of $\psi \circ r_* : \pi_1(\mathcal{U}) \to \pi$, where $r : \mathcal{U} \to X$ is the natural retraction.) Just as in 9.1, $G$ is the group of all lifts of the $W$-action on $\mathcal{U}$ to $\widetilde{\mathcal{U}}$. Let $q : G \to W$ be the projection. Given a subgroup $\Gamma \subset W$, put $\widetilde{\Gamma} := q^{-1}(\Gamma)$. Clearly, $\widetilde{\Gamma} = \Gamma' \rtimes \pi$, where $\Gamma'$ is the inverse image of $\Gamma$ in $\widetilde{W}$.

**LEMMA 11.5.1.** *Suppose $S$ is finite and $\pi$ is torsion-free. Then $G$ is virtually torsion-free.*

*Proof.* By Corollary 6.12.12, $W$ has a finite-index, torsion-free subgroup $\Gamma$. Let $\Gamma'$ be its inverse image in $\widetilde{W}$. Then $\Gamma'$ is finite index in $\widetilde{W}$. We claim it is also torsion-free. This is because (a) by construction, $\widetilde{W} \to W$ maps each spherical special subgroup of $\widetilde{W}$ isomorphically onto a spherical special subgroup of $W$ and (b) by Corollary D.2.9 (or Theorem 12.3.4 (i)), each finite subgroup of $\widetilde{W}$ is contained in some conjugate of a spherical special subgroup; hence, $\widetilde{W} \to W$ maps each finite subgroup injectively. Since $\Gamma'$ and $\pi$ are both torsion-free, so is their semidirect product $\widetilde{\Gamma} = \Gamma' \rtimes \pi$. □

**PROPOSITION 11.5.2.** *$\widetilde{\mathcal{U}}$ is acyclic if and only if $\widetilde{X}$ is acyclic. Moreover, if this is the case and if $X$ is a finite CW complex, then $G$ is type VFL.*

(See Appendix F.4 for the definition of "type *VFL*.")

*Proof.* By construction, for each nonempty spherical subset $T \subset \widetilde{S}$, $\widetilde{X}_T$ is a cone and hence, is acyclic. So, the first sentence follows from Corollary 8.2.7. The second sentence is just the observation that if $\widetilde{\Gamma}$ is a torsion-free subgroup of finite index, then the chain complex $C_*(\widetilde{\mathcal{U}})$ is a finite resolution of $\mathbb{Z}$ by finitely generated free $\mathbb{Z}\widetilde{\Gamma}$-modules. □

**PROPOSITION 11.5.3.** (Compare Lemma F.5.3). *Suppose $(X, L)$ is a compact $n$-manifold with boundary and $\widetilde{X}$ is acyclic. Then $G$ is a virtual Poincaré duality group of dimension $n$.*

*Proof.* Let $\widetilde{\Gamma}$ be a torsion-free subgroup of finite index in $G$, as above. By Proposition 11.5.2, $\widetilde{\Gamma}$ is type *FL* (and so *a fortiori* type *FP*) and $\widetilde{\mathcal{U}}$ is acyclic. By Proposition 10.7.5, $\widetilde{\mathcal{U}}$ is a homology manifold. (If $(X, L)$ is a PL manifold with boundary, then, by Proposition 10.1.10, $\widetilde{\mathcal{U}}$ is actually a manifold.) Hence,

$$H^i_c(\widetilde{\mathcal{U}}) \cong \begin{cases} 0 & \text{if } i \neq n, \\ \mathbb{Z} & \text{if } i = n. \end{cases}$$

Since $H^*(\widetilde{\Gamma}; \mathbb{Z}\widetilde{\Gamma}) \cong H^*_c(\widetilde{\mathcal{U}})$ (Lemma F.2.2), $\widetilde{\Gamma}$ is a PD$^n$-group. □

*Remark.* Proposition 11.5.3 illustrates why we were interested in weakening the conditions from 11.1. We want to replace the condition that $X$ is aspherical by the condition that it has an acyclic covering space $\widetilde{X}$. In the case where $\psi : \pi_1(X) \to \pi$ is the identity map, $\widetilde{X}$ is the universal cover and we are back to the original situation where $X$ is aspherical.

## 11.6. THE BESTVINA-BRADY EXAMPLES

In [24] Bestvina and Brady constructed the first examples of groups of type $FL$ which are not finitely presented (and hence, not type $F$).

### Right-Angled Artin Groups

As usual, $L$ is a finite flag complex with vertex set $S$ and, as in Example 7.1.7, $(W, S)$ is the right-angled Coxeter system associated to its 1-skeleton. For each $s \in S$ introduce a new symbol $x_s$ and let $A_L$ be the group defined by the presentation $\langle \{x_s\}_{s \in S} \mid \mathcal{R} \rangle$, where the set of relations $\mathcal{R}$ consists of all commutators of the form $[x_s, x_t]$ with $\{s, t\} \in \mathrm{Edge}(L)$. In other words, $A_L$ has a generator for each vertex of $L$; two such generators commute if and only if the corresponding vertices are connected by an edge and there are no other relations. $A_L$ is called the *right-angled Artin group* associated to $L$.

**Example 11.6.1**

(i) If $L$ is the union of $n$ points, $A_L$ is the free group of rank $n$.
(ii) If $L$ is an $n$-simplex, $A_L$ is free abelian of rank $n + 1$.

The classifying space of $A_L$ has a simple description, similar to the description of the complex $P_L$ in 1.2. Denote the product of $S$ copies of the circle by $T^S$ (the "$S$-torus"). If $U \subset S$, then $T^U$ is naturally a subspace (in fact, a subgroup) of $T^S$. Define a subcomplex $T_L$ of $T^S$ by

$$T_L := \bigcup_{U \in \mathcal{S}(L)} T^U. \tag{3}$$

It is easy to see $T_L$ is aspherical, i.e., a model for $BA_L$ ([176]). The fact that its universal cover $\widetilde{T}_L$ is contractible can also be explained with the theory of nonpositively curved spaces discussed in Appendix I. $\widetilde{T}_L$ is naturally a cubical cell complex with a piecewise Euclidean metric and it follows from Gromov's Lemma in Appendix I.6 that it is a CAT(0)-space and hence, is contractible. (See [56].)

### Bestvina-Brady Groups

Let $\varphi : A_L \to \mathbb{Z}$ be the homomorphism which sends each generator $x_s$ to 1. The *Bestvina-Brady group* $B_L$ associated to $L$ is defined to be $\operatorname{Ker}\varphi$. Using the cubical structure on $\widetilde{T}_L$ one can construct a $\varphi$-equivariant, piecewise linear "Morse function" $f : \widetilde{T}_L \to \mathbb{R}$ which takes the vertices of $\widetilde{T}_L$ to $\mathbb{Z}$. (The vertices of $\widetilde{T}_L$ play the role of critical points.) Each level set $f^{-1}(t)$, $t \in \mathbb{R}$, is $B_L$-stable. The main idea of [24] is to analyze how the topology of a level set changes as it moves across a vertex. The effect is to cone off a copy of $L$. Hence, if $L$ is acyclic, the homology of all level sets are the same and $f^{-1}(t)$ is acyclic for each $t \in \mathbb{R}$. By using the CAT(0) structure on $\widetilde{T}_L$, Bestvina and Brady are also able to analyze the effect of moving $t$ on the fundamental group of a level set $f^{-1}(t)$. For example, they show $f^{-1}(t)$ is simply connected if and only if $L$ is simply connected. If $L$ is contractible, $f^{-1}(t)/B_L$ is a model for $BB_L$. For some fixed generic $t$ (i.e., $t \notin \mathbb{Z}$), put

$$\widetilde{Y}_L := f^{-1}(t) \quad \text{and} \quad Y_L := f^{-1}(t)/B_L. \tag{4}$$

Since $f$ is a piecewise linear map, $T_L$ can be triangulated so that $Y_L$ is a finite subcomplex. Bestvina and Brady prove the following.

**THEOREM 11.6.2.** ([24].) *$B_L$ is*

*(i) finitely generated if and only if $L$ is connected,*

*(ii) finitely presented if and only if $L$ is simply connected,*

*(iii) type FL (or type FP) if and only if $L$ is acyclic,*

*(iv) type F if and only if is $L$ is contractible.*

In particular, taking $L$ to be a flag complex which is acyclic but not simply connected, they get the following.

**COROLLARY 11.6.3.** ([24]). *There are groups of type FL which are not finitely presented.*

These examples answered Problem F10 in Wall's Problem List [293, p. 387] in the negative. In the same problem Wall asked "Does this at least hold for Poincaré duality groups?"

### Poincaré Duality Groups That Are not Finitely Presented

The next theorem shows how we can use the second version of the reflection group trick to promote the Bestvina-Brady examples to a negative answer to the second part of Wall's problem.

**THEOREM 11.6.4.** ([77, Theorem C].) *In each dimension $\geqslant 4$, there are examples of Poincaré duality groups which are not finitely presented.*

*Proof.* Suppose $L$ is a acyclic two-dimensional flag complex which is not simply connected and that $Y_L$ is as in (4) (i.e., $Y_L$ is one of the Bestvina-Brady examples). $Y_L$ is a finite 2-complex and there is an epimorphism $\varphi : \pi_1(Y_L) \to B_L$. We want to apply the second version of the reflection group trick from 11.5. Thicken $Y_L$ to $X$, a compact PL $n$-manifold with boundary ($n$ can be any integer $\geqslant 4$). Let $r : X \to Y_L$ be the collapse and let $\psi = \varphi \circ r_* : \pi_1(X) \to B_L$. Here we run into a problem with notation: the letter $L$ which we want to use for a triangulation of the boundary of a manifold is already in use for a different flag complex, so we will have to use the letter $\Lambda$ instead. Let $\Lambda$ be a triangulation of $\partial X$ as a flag complex and $(W_\Lambda, S_\Lambda)$ be the associated right-angled Coxeter system. This is the necessary data for the second version of the reflection group trick (where $\pi = B_L$ and $(W, S) = (W_\Lambda, S_\Lambda)$). By Proposition 10.1.10, $\mathcal{U}(W_\Lambda, X)$ is a manifold with a proper, locally linear $W_\Lambda$-action. $\widetilde{X}$ is acyclic (since $\widetilde{Y}_L$ is acyclic and $\widetilde{X}$ is homotopy equivalent to $\widetilde{Y}_L$). By Proposition 11.5.3, $G = \widetilde{W}_\Lambda \rtimes B_L$ is a virtual PD$^n$-group with a finite index, torsion-free subgroup $\widetilde{\Gamma} = \Gamma' \rtimes B_L$, which retracts onto $B_L$. Since $B_L$ is not finitely presented, neither is any group which retracts onto it ([292, Lemma 1.3]). □

Kirby and Siebenmann [177] proved that any compact topological manifold is homotopy equivalent to a finite CW complex. In particular, the fundamental group of a compact manifold must be finitely presented. So, Theorem 11.6.4 has the following corollary.

**COROLLARY 11.6.5.** *For each $n \geqslant 4$, there are* PD$^n$*-groups which are not fundamental groups of closed aspherical manifolds.*

### The Leary-Nucinkis Examples

By throwing finite group actions into the mix, Leary-Nucinkis proved the following.

**THEOREM 11.6.6.** (Leary-Nucinkis [187].) *There are examples of groups $G$ possessing a torsion-free subgroup $\Gamma$ of finite index such that*

*(i) $\Gamma$ is type $F$.*

*(ii) There is no CW model for $\underline{E}G$ with a finite number of $G$-orbits of cells.*

Let $Q$ be a group of simplicial automorphisms of the finite flag complex $L$. We can assume the $Q$-action is *admissible*, i.e., for any simplex $\sigma$ in $L$, its setwise stabilizer is equal to its pointwise stabilizer. The $Q$-action on $S$ (= Vert($L$)) induces a $Q$-action on $A_L$ through automorphisms and the subgroup $B_L$ is $Q$-stable. Hence, we can form the semidirect products

$$A_L \rtimes Q \quad \text{and} \quad B_L \rtimes Q$$

The homomorphism $\varphi : A_L \to \mathbb{Z}$ extends to a homomorphism (also denoted $\varphi$) from $A_L \rtimes Q$ to $\mathbb{Z}$ by sending $Q$ to 0. Let $T_L$ be the cubical complex defined by (3). The $Q$-action on $L$ induces an isometric action on $T_L$. The group of all lifts of this action to $\widetilde{T}_L$ is $A_L \rtimes Q$. Fix $t \in \mathbb{R}$. The level set $f^{-1}(t)$ is $B_L$-stable and $Q$ acts on $Y_L$ $(= f^{-1}(t)/B_L)$. (When $L$ is contractible, the homotopy type of $f^{-1}(t)$ does not change as $t$ varies in $\mathbb{R}$.)

As explained in [187], in the above construction, the process of taking fixed sets of $Q$ works as one would predict:

$$(T_L)^Q = T_{L^Q} \quad \text{and} \quad (Y_L)^Q = Y_{L^Q}.$$

It is proved in [70] that the subgroup of $A_L$ fixed by $Q$ is $A_{L^Q}$ and in [187] that the subgroup of $B_L$ fixed by $Q$ is $B_{L^Q}$. Regarding $Q$ as a subgroup of $B_L \rtimes Q$ and letting $V_Q$ stand for $N_{B_L \rtimes Q}(Q)/Q$, we have $V_Q = B_{L^Q}$ ([187]).

By Theorem 11.6.2 (iv), $B_L$ is type $F$ if and only if $L$ is contractible. The main idea of [187] is that if $L$ is contractible and the $Q$-action is such that $L^Q$ is not contractible, then $G := B_L \rtimes Q$ cannot have a finite model for $\underline{E}G$. The reason is that if $\underline{E}G$ has a finite model, then so does $\underline{E}V_Q$, namely, the fixed set of $Q$ on $\underline{E}G$. But if $L^Q$ is not contractible, then the torsion-free group $V_Q$ $(= B_{L^Q})$ is not type $F$. Since there are many examples of such $Q$-actions on contractible, finite simplicial complexes $L$, we get Theorem 11.6.6 with $G = B_L \rtimes Q$ and $\Gamma = B_L$. For examples of such $Q$-actions on contractible, finite complexes with noncontractible fixed sets, see [36].

## 11.7. THE EQUIVARIANT REFLECTION GROUP TRICK

Suppose $G$ is a discrete, cocompact group of isometries of a Riemannian manifold $M$ of nonpositive sectional curvature. For any finite subgroup $H \subset G$, the fixed set $M^H$ is a totally geodesic, Riemannian submanifold; hence, a convex subset; hence, contractible. So, $M$ is a finite model for $\underline{E}G$. (By saying that the model is "finite" we mean that it can be cellulated so that there are only finitely many $G$-orbits of cells.) The group $V_H := N_H(G)/H$ is a cocompact isometry group of $M^H$. Since $M^H$ is a contractible manifold, this implies any torsion-free subgroup of finite index in $V_H$ is a Poincaré duality group. (See 10.9 and Appendix F.5.) Such examples might lead one to speculate that if $G$ is a virtual Poincaré duality with a finite model for $\underline{E}G$, then, for any finite subgroup $H \subset G$, $V_H$ is also a virtual Poincaré duality group. We will see in Theorem 11.7.3 below that such speculations are false.

The "equivariant reflection group trick" is not so different from the ordinary reflection group trick. Again, it is basically the semidirect product construction of 9.1. The main difference is that we will not assume that the group acts freely on $L$.

Suppose, as before, we are given as input, a CW pair $(X, \partial X)$ and a triangulation of $\partial X$ by a flag complex $L$ with vertex set $S$. From this we get a right-angled Coxeter system $(W, S)$ and a mirror structure $(X_s)_{s \in S}$. Assume that a finite group $Q$ acts on $(X, \partial X)$ and that the action on $\partial X$ is induced by a simplicial action on $L$. Further assume that the action of $Q$ on $L$ is admissible (as defined in the previous section). As in 9.1, $G := W \rtimes Q$ acts on $\mathcal{U}(W, X)$.

For $H$ a subgroup of $Q$, let $S^H$ denote the vertex set of $L^H$ and $W_{S^H}$ the special subgroup generated by $S^H$. $N_Q(H)/H$ acts on $S^H$ and hence, on $(W_{S^H}, S^H)$ through diagram automorphisms. Let

$$V_H := W_{S^H} \rtimes N_Q(H)/H. \tag{5}$$

In the next proposition we list two properties of the equivariant reflection group trick. Its proof is left as an exercise for the reader.

**PROPOSITION 11.7.1.** *Suppose, as above, $G = W \rtimes Q$ and $H$ is a subgroup of $Q$. Put $\mathcal{U} = \mathcal{U}(W, X)$ and let $\mathcal{U}^H$ be the fixed set of $H$ on $\mathcal{U}$. Then*

*(i) $\mathcal{U}^H = \mathcal{U}(W_{S^H}, X^H)$ and*

*(ii) $N_G(H)/H = V_H$, where $V_H$ is defined by* (5).

If $X = \text{Cone}(\partial X)$, then $\mathcal{U}(W, X)$ is our favorite cubical complex $\Sigma$. In the next chapter we will see that the natural piecewise Euclidean metric on $\Sigma$ is CAT(0). Since $G$ acts via isometries, it follows from Corollary I.2.12 in Appendix I.2 that $\Sigma$ is a model for $\underline{E}G$. If $L$ ($\cong \partial X$) is a triangulation of $S^{n-1}$, then $\Sigma$ is a manifold. Since $W_{S^H}$ has finite index in $V_H$, we see that $V^H$ is a virtual Poincaré duality group if and only if $W_{S^H}$ is and by Theorem 10.9.2, this is the case if and only if $L^H$ is a generalized homology sphere.

There are many classical examples of actions of finite groups on spheres which are not equivalent to linear actions (e.g., see [36]). In particular, the fixed set of a finite group $H$ on $S^{n-1}$ need not be a GHS. The idea of [88] is that these exotic actions on spheres can be promoted to give VPD$^n$ groups $G$ with finite models for $\underline{E}G$ such that the normalizers of certain finite subgroups are not VPD$^m$ for any $m$. Hence, there can be no cocompact action of any such $G$ on a contractible manifold so that the fixed sets of finite subgroups are contracible submanifolds. Before stating this theorem, we give a concrete example of an exotic action on a sphere.

**Example 11.7.2.** (*Actions on Brieskorn spheres.*) Let $N(a_1, \dots, a_n)$ denote the link of the origin in the complex hypersurface

$$(z_1)^{a_1} + \cdots + (z_n)^{a_n} = 0.$$

It is a smooth manifold of dimension $2n - 3$, called a "Brieskorn manifold." As Brieskorn showed, it is often homeomorphic to $S^{2n-3}$; see [213]. For example, $N(3, 2, 2, 2)$ is diffeomorphic to $S^5$. If $\zeta$ denotes a sixth root of unity, define an action of $\mathbf{C}_6$ on $N(3, 2, 2, 2)$ by

$$\zeta \cdot (z_1, z_2, z_3, z_4) = (\zeta^2 z_1, \zeta^3 z_2, z_3, z_4).$$

The fixed set of $\mathbf{C}_3$ on $N(3, 2, 2, 2)$ is $N(2, 2, 2)$ which is $\mathbb{R}P^3$, the fixed set of $\mathbf{C}_2$ is $N(3, 2, 2)$ which is a 3-dimensional lens space with fundamental group $\mathbb{Z}/3$ and the fixed set of $\mathbf{C}_6$ is $N(2, 2)$ which is the disjoint union of two circles.

**THEOREM 11.7.3.** ([88], [125]). *For each $n \geqslant 6$, there are examples of* VPD$^n$ *groups $G$ such that*

(i) *$\underline{E}G$ has a finite model (by a* CAT(0)*-manifold).*

(ii) *There is a finite subgroup $H \subset G$ so that $N_G(H)/H$ is not a virtual Poincaré duality group.*

*Proof.* Equivariantly triangulate $N(3, 2, 2, 2)$ $(\cong S^5)$ as a flag complex and form the $CAT(0)$ 6-manifold $\Sigma$ as above. For any nontrivial subgroup $H$ of $\mathbf{C}_6$, $L^H$ is not a GHS; hence, by Theorem 10.9.2, $V_H$ is not a virtual Poincaré duality group. In fact, we can use the results of 8.9, to calculate the cohomology of $V_H$ with group ring coefficients for any subgroup $H \subset \mathbf{C}_6$. For $H = \mathbf{C}_3$ or $\mathbf{C}_2$, $\mathrm{vcd}(V_H) = 4$ and $H^4(V_H; \mathbb{Z}V_H) = \mathbb{Z}$. However, $H^3(V_H; \mathbb{Z}V_H)$ is either an infinite sum of $\mathbb{Z}/2$'s (when $H = \mathbf{C}_3$) or of $\mathbb{Z}/3$'s (when $H = \mathbf{C}_2$). For $H = \mathbf{C}_6$, the group $V_H$ has cohomological dimension 2 and its second cohomology group is free abelian of infinite rank. By suspending $L$ or by taking the product of $G$ with $\mathbb{Z}^k$, we can promote such examples to any dimension $\geqslant 6$ □

*Remark.* A VPD$^n$ group $G$ need not have any finite model for $\underline{E}G$. This is proved in [88] by using the equivariant reflection group trick to promote the Leary-Nucinkis examples to cocompact actions on contractible manifolds.

## NOTES

Thirty years ago there were many techniques for producing finite aspherical complexes (e.g., the theory of small cancellation groups) and relatively few examples of closed aspherical manifolds (principally, locally symmetric spaces). The reflection group trick changed our perception of the possibilities.

The reflection group trick was motivated by an idea of Thurston. In the mid-1970s he discussed the following version of it in dimension 3. Suppose $(X, \partial X)$ is a compact 3-manifold with boundary and that the interior of $X$ can be given the structure of a complete hyperbolic manifold. Thurston showed that by taking a suitable triangulation $L$ of $\partial X$ and proceeding as in 11.1 one can give $X$ the structure of a closed hyperbolic orbifold. In other words, the universal cover of $\mathcal{U}(W, X)$ is identified with $\mathbb{H}^3$ and $G$ with a discrete group of isometries of $\mathbb{H}^3$. This result was a generalization of Andreev's Theorem (6.10.2) and a special case of Thurston's Geometrization Conjecture (6.10.1). In fact, this version of the reflection group trick is a key step in Thurston's proof of the Geometrization Conjecture for Haken manifolds [277].

In its present form the reflection group trick was first described in [71, Remark 15.9]. It is also discussed in [146, p. 390]. In the next chapter, in 12.8, we will explain a version of the reflection group trick which takes nonpositive curvature into account. This version will allow us to start with a nonpositively curved, piecewise Euclidean polyhedron and then find a closed, nonpositively curved, piecewise Euclidean manifold which retracts onto it.

**11.3.** The results of this section were improved in [83]. It is proved there that for each $n \geqslant 4$, there are closed aspherical topological $n$-manifolds which are not homotopy equivalent to any PL manifold. The idea is to apply Gromov's hyperbolization technique to the $\mathbf{E}_8$-homology 4-manifold, obtaining an aspherical homology 4-manifold of signature 8. By [131] this can be resolved to a topological 4-manifold, which, by Rohlin's Theorem, cannot be given a PL structure. By taking the product of this example with a $k$-torus we also obtain examples in all dimensions $>4$.

**11.5.** Suppose $X$ is a finite CW complex, $\widetilde{X}$ is acyclic and $\pi$ is nontrivial. In [77, Theorem 6.5 (i)] I calculated the compactly supported cohomology of $\widetilde{\mathcal{U}}$ and proved the following: $\widetilde{W} \rtimes \pi$ is a virtual PD$^n$ group if and only if (a) $L$ is an $(n-1)$-dimensional homology manifold and (b) $(X, L)$ is an $n$-dimensional Poincaré pair over $\pi$ ([77, Theorem 6.10, p.313]).

**11.6.** To any Coxeter system $(W, S)$ one can associate an *Artin group* as follows. For each $s \in S$ introduce a symbol $x_s$ and let $= A_{(W,S)}$ be the group with generating set $\{x_s\}_{s \in S}$ and relations of the form:

$$\underbrace{x_s x_t \cdots}_{m_{st}} = \underbrace{x_t x_s \cdots}_{m_{st}}$$

for any two distinct elements $s,t \in S$ with $m_{st} \neq \infty$. In other words, the alternating word of length $m_{st}$ in $x_s$ and $x_t$ is equal to the alternating word of the same length in the other order. Since these relations hold in the Coxeter group, we have an epimorphism $A_{(W,S)} \to W$ defined by $x_s \to s$. When $W$ is the symmetric group on $n$ letters, $A_{(W,S)}$ is the *braid group on n strands*. The justification for the name "Artin group" is that E. Artin initiated the study of braid groups. The theory of Artin groups is closely tied to the theory of Coxeter groups and could have been dealt with extensively in this book; however, we instead confine ourselves to a few remarks here. Suppose $W$ is finite. Complexifying the geometric representation we get a complex representation on $\mathbb{C}^n$. Let $M$ be the complement of the union of reflecting hyperplanes in $\mathbb{C}^n$. $W$ acts freely on $M$. Deligne [92] proved that $M/W$ is the classifying space for $BA_{(W,S)}$.

It is conjectured that a similar result holds when $W$ is infinite. This is known to be true in many cases (see [57, 289]) but is not known in general. The problem is that the complex hyperplane complement is not known to be aspherical. It is also known that the quotient of the complex hyperplane complement is homotopy equivalent to a finite cell complex $\Lambda$ called the "Salvetti complex" (see [56]). In many ways, the universal cover of $\Lambda$ is quite similar to $\Sigma$. For example, it has a cellulation by Coxeter polytopes which projects to the cellulation on $\Sigma$. However, in general, it is still not known if $BA_{(W,S)}$ is homotopy equivalent to a finite CW complex or even to a finite dimensional one. In the right-angled case, everything works (in fact, the universal cover of $\Lambda$ is CAT(0)). Right-angled Artin groups are sometimes called "graph groups." More information about Artin groups can be found in [56, 57, 70, 78, 96].

Corollary 11.6.5 should not be regarded as a counterexample to the PD$^n$-Group Conjecture (H.1.2) and in fact, it is not (because in our statement of the conjecture we explicitly assume the group to be finitely presented). The point is that the condition of finite presentability is not the issue in surgery theory with which the conjecture is usually associated. Our discussion of the Leary–Nucinkis examples is taken from [187] and [88].

**11.7.** The discussion in this section comes from [88]. In [125] Farrell and Lafont give a proof of Theorem 11.7.3 in which $G$ is word hyperbolic. (See 12.5 for a discussion of word hyperbolic groups.) Their construction uses the strict hyperbolization procedure of [59].

# Chapter Twelve

## $\Sigma$ IS CAT(0): THEOREMS OF GROMOV AND MOUSSONG

One of the main results in this book is that the natural piecewise Euclidean metric on $\Sigma(W,S)$ is nonpositively curved. Although this result will be explained here, a general discussion of nonpositively curved spaces and many of the details of our arguments will be postponed until Appendix I. The "natural piecewise Euclidean metric on $\Sigma$" means, roughly, that each cell of $\Sigma$ is identified with a convex polytope in Euclidean space. This is explained in 12.1. The notion of "nonpositive curvature" is defined via comparison triangles. It means that any sufficiently small geodesic triangle in $\Sigma$ is "thinner" than a comparison triangle in $\mathbb{E}^2$ or in more precise language, that it satisfies the "CAT(0)-inequality." The question of whether a polyhedral metric on a cell complex such as $\Sigma$ is locally CAT(0) comes down to the question of showing that the link $L$ of each vertex, with its natural piecewise spherical structure, is CAT(1) (i.e., triangles in $L$ are thinner than comparison triangles in $\mathbb{S}^2$). In the right-angled case the fact that $L$ is CAT(1) is due to Gromov [147]. It is proved in Appendix I.6 as "Gromov's Lemma." The same result holds for the nerve of any Coxeter system (cf. Appendix I.7 and Section 12.3). This was proved by Moussong in [221].

Since $\Sigma$ is simply connected, the fact that it is nonpostively curved (i.e., is locally CAT(0)) implies that the CAT(0)-inequality holds globally for all triangles (i.e., $\Sigma$ is a "CAT(0)-space"). It follows that there is a unique geodesic between any two points in $\Sigma$ and that $\Sigma$ is a finite model for $\underline{E}W$).

Moussong also determined exactly when $\Sigma$ admits a piecewise hyperbolic metric which is CAT($-1$). $\Sigma$ has such a metric if and only if certain obviously necessary conditions hold. (These conditions are analogous to Andreev's Conditions in Theorem 6.10.2.) From this, we show in Corollary 12.6.3 that the following conditions on $(W,S)$ are equivalent:

- $W$ is "word hyperbolic" in the sense of [147].
- $\mathbb{Z}\times\mathbb{Z} \not\subset W$.
- $\Sigma(W,S)$ admits a piecewise hyperbolic, CAT($-1$)-metric.

Moussong's result precisely determines when every nontrivial free abelian subgroup of $W$ has rank 1. Krammer [181] generalized this by determining all free abelian subgroups of $W$. His result (Theorem 12.7.4) is that the only such subgroups are the "obvious ones."

In 12.8 we discuss a variant of the reflection group trick which preserves nonpositive curvature. It is a simultaneous generalization of the results of 11.1 and 12.2.

## 12.1. A PIECEWISE EUCLIDEAN CELL STRUCTURE ON $\Sigma$

In Appendix I.3 we define the notion of an $\mathbb{X}_0$-*polyhedral complex* or a *piecewise Euclidean cell complex*. It is explained there how such a structure on a cell complex $\Lambda$ induces a length metric on it, called a *piecewise Euclidean metric*. (A metric is a *length metric* if the distance between any two points is the infimum of the lengths of all paths between them.) Under mild conditions (e.g., finitely many shapes of cells), $\Lambda$ is a "geodesic space" (defined in Appendix I.1).

In 7.3 we defined a cellulation of $\Sigma$ by Coxeter polytopes. Each Coxeter polytope is naturally realized as polytope in some Euclidean space. The realization is unique once we specify edge lengths and $\Sigma$ gets the structure of a piecewise Euclidean cell complex.

### The Euclidean Structure on a Coxeter Polytope

In this paragraph, $W$ is finite. $S$ is a fundamental set of generators. As explained in Sections 6.8 and 6.12, there is a canonical representation of $W$ on the Euclidean space $\mathbb{R}^S$ (of real-valued functions from $S$ to $\mathbb{R}$) as a group generated by the orthogonal reflections across the walls of a simplicial cone $K$. For each $s \in S$, let $u_s$ be the unit vector which is normal to the wall corresponding to $s$ and which points into $K$. As explained in 7.3, to define the Coxeter polytope $C_W$, we need only choose a point $x$ in the interior of $K$. To this end, suppose $\mathbf{d} = (d_s)_{s\in S}$ is an $S$-tuple of positive real numbers and $x$ is the unique point in $\mathbb{R}^S$ defined by the linear system

$$\langle u_s, x\rangle = d_s \quad \text{for all } s \in S.$$

In other words, $x$ is the point in $K$ whose distance from the wall corresponding to $s$ is $d_s$. As in Definition 7.3.1, the *Coxeter polytope* $C_W(\mathbf{d})$ is the convex hull of $Wx$. Its vertex set is $Wx$. Its 1-skeleton is $\mathrm{Cay}(W, S)$. Thus, for each $s \in S$ and $w \in W$, there is an edge labeled $s$ between $wx$ and $wsx$. The length of an edge labeled $s$ is $2d_s$. For example, if $W = (\mathbf{C}_2)^S$, then $C_W(\mathbf{d})$ is the box, $\prod_{s\in S}[-d_s, d_s]$. (Compare Examples 7.3.2.)

### The Spherical Simplex $\sigma$

Consider the spherical polytope

$$\sigma := \mathrm{Lk}(x, C_W(\mathbf{d})).$$

By definition, it is the set of all inward-pointing unit tangent vectors at $x$. The unit vector, which is parallel to the edge at $x$ labeled $s$ and which points away from $x$, is $-u_s$. So, $\sigma$ is a spherical simplex in the unit sphere of $\mathbb{R}^S$ with $\mathrm{Vert}(\sigma) = \{-u_s\}_{s\in S}$. Since the antipodal map $v \to -v$ is an isometry of the unit sphere, we can identify $\mathrm{Lk}(x, C_W(\mathbf{d}))$ with the spherical simplex spanned by the $u_s$. (Note that $\sigma$ is independent of the choice of $\mathbf{d}$.)

**Lemma 12.1.1.** *Suppose $W$ is a finite Coxeter group. Then the spherical simplex $\sigma$ is the polar dual of the fundamental spherical simplex, $K \cap \mathbb{S}(\mathbb{R}^S)$, for the action of $W$ on the unit sphere in $\mathbb{R}^S$. Moreover, if $l_{st}$ denotes the length of the edge of $\sigma$ between $u_s$ and $u_t$, then the matrix $(c_{st})$ of cosines of edge lengths of $\sigma$, defined by $c_{st} := \cos(l_{st})$, is the cosine matrix from Definition 6.8.11 associated to $(W, S)$.*

*Proof.* Since the $u_s$ are the unit normals to the walls of the fundamental simplex, $\sigma$ is its polar dual. (For the definition of "polar dual" see I.5.2 in Appendix I.5.) As for the last sentence, $\cos(l_{st}) = \langle u_s, u_t\rangle = -\cos(\pi/m_{st})$. $\square$

### The Metric on $\Sigma$

We return to the situation where $(W, S)$ is an arbitrary Coxeter system and $\Sigma$ is the associated cell complex. Let $\mathbf{d} = (d_s)_{s\in S}$ be an element of $(0, \infty)^S$. For each spherical subset $T \in \mathcal{S}$, let $\mathbf{d}_T = (d_s)_{s\in T}$. Give $\Sigma$ a piecewise Euclidean cell structure by identifying the cell corresponding to $wW_T$ with the Coxeter polytope $C_{W_T}(\mathbf{d}_T)$. The resulting piecewise Euclidean cell complex is denoted $\Sigma(\mathbf{d})$. (The choice of $\mathbf{d}$ is usually irrelevant and we generally will omit it from our notation.)

### The Piecewise Spherical Structure on $L$

Let $L$ be the nerve of $(W, S)$. $W$ acts simply transitively on the vertex set of $\Sigma$ and as explained in Proposition 7.3.4, the link of each vertex is identified with $L$. As explained in Appendix I.3, the piecewise Euclidean structure on $\Sigma$ induces a piecewise spherical structure on $L$. The piecewise spherical structure on $L$ is defined as follows: the simplex $\sigma_T \subset L$ is identified with the $\mathrm{Lk}(x, C_T)$. As we saw in Lemma 12.1.1, the edge lengths of $\sigma_T$ are determined by the cosine matrix of $(W_T, T)$. (For the definition and general discussion of the notion of a piecewise spherical structure, see Appendix I.3.)

## 12.2. THE RIGHT-ANGLED CASE

As in Example 7.3.2 (iii), the Coxeter polytope associated to a product of cyclic groups of order 2 (and a constant $S$-tuple $\mathbf{d}$) is a cube. So, as in Example 7.3.6, if $(W, S)$ is right angled, $\Sigma$ is a cubical complex. Suppose this is the case and $\Sigma$ has the resulting piecewise Euclidean cubical structure. The link of each vertex in $\Sigma$ is isometric to $L$ with its "all right structure" (Definition I.5.7 of Appendix I.5). This means that each simplex of $L$ is identified with the regular spherical simplex of edge length $\pi/2$. By Lemma 7.1.8, $L$ is a flag complex. By Gromov's Lemma (Lemma I.6.1 of Appendix I.6), this implies that $L$ is CAT(1). This is precisely what is needed to show that $\Sigma$ is nonpositively curved. In fact, we have the following.

**THEOREM 12.2.1.** (Gromov [147].) *Suppose $(W, S)$ is right angled. Then*

*(i) The piecewise Euclidean cubical structure on $\Sigma$ is* CAT(0).

*(ii) The piecewise hyperbolic structure on $\Sigma$ in which each cube is a regular hyperbolic cube of edge length $\varepsilon$ is* CAT$(-1)$ *for some $\varepsilon > 0$ if and only if $L$ satisfies the no $\square$-condition of Appendix I.6.*

(The no $\square$-condition is that there are no "empty 4-circuits" in the 1-skeleton of $L$.)

*Proof.* (i) Since the link of each vertex is CAT(1), $\Sigma$ is nonpositively curved (Theorem I.3.5 in Appendix I.3). Since $\Sigma$ is simply connected (Lemma 7.3.5), it follows from Gromov's version of the Cartan-Hadamard Theorem (Theorem I.2.7 in Appendix I.2) that it is CAT(0).

(ii) The link of a vertex of a regular cube of edge length $\varepsilon$ in hyperbolic space is a regular spherical simplex with all edge lengths $< \pi/2$. As $\varepsilon$ increases from 0 this simplex is a deformation of the all right regular spherical simplex. So, the piecewise spherical structure on $L$ deforms from its usual all right structure. For small $\varepsilon$, will this deformation remain CAT(1)? As we explain at the end of Appendix I.6, it will if and only if in the all right structure on $L$, for each simplex $\sigma \subset L$ (including the empty simplex), the infimum of the lengths of all closed geodesics in $\mathrm{Lk}(\sigma, L)$ is strictly greater than $2\pi$ (in language from the end of Appendix I.3, $L$ must be "extra large"). Moreover, this condition holds for $L$ if and only if it satisfies the no $\square$-condition. Thus, the piecewise hyperbolic, cubical structure on $\Sigma$ has curvature $\leqslant -1$ for some $\varepsilon$ if and only if $L$ satisfies the no $\square$-condition. (See Proposition I.6.8.) $\square$

In Corollary 12.6.3, we will show that a right-angled $W$ is word hyperbolic (in the sense of [147]) if and only if $L$ satisfies the no $\square$-condition.

("Word hyperbolicity" will be discussed in 12.5.) In Appendix I.6 we use arguments of Vinberg from 6.11 to prove the following.

**PROPOSITION 12.2.2.** (Proposition I.6.6 of Appendix I.6.) *Suppose a flag complex $L$ is a* $\mathrm{GHS}^{n-1}$ *satisfying the no* $\square$*-condition. Then* $n \leqslant 4$.

**COROLLARY 12.2.3.** (Moussong [221].) *Suppose $W$ is right angled, word hyperbolic and type* $\mathrm{HM}^n$. *Then* $n \leqslant 4$.

("Type $\mathrm{HM}^n$" means that $\Sigma$ is a homology $n$-manifold; cf. Definition 10.6.2.)

In 8.5 we calculated the virtual cohomological dimension of any Coxeter group. In particular, if $\dim L = n-1$ and $\overline{H}^{n-1}(L) \neq 0$, then, by Corollary 8.5.5, $\mathrm{vcd}\, W = n$. So, Corollary 12.2.3 raises the question of whether there exist right angled Coxeter groups of arbitrarily high virtual cohomological dimension. (Similarly, Vinberg's Theorem 6.11.8 raises the same question for arbitrary Coxeter groups.) In contrast to Proposition 12.2.2, for pseudomanifolds, Januszkiewicz and Świątkowski proved the following result.

**THEOREM 12.2.4.** (Januszkiewicz and Świątkowski [166].) *For each integer* $n \geqslant 1$ *there is a finite* $(n-1)$*-dimensional flag complex $L$ such that*

- *$L$ is an orientable pseudomanifold.*
- *$L$ satisfies the no* $\square$*-condition.*

Since for such an $L$, $H^{n-1}(L) \neq 0$, we have the following.

**COROLLARY 12.2.5.** ([166].) *In each dimension $n$, there is a right-angled, word hyperbolic Coxeter group $W$ with* $\mathrm{vcd}\, W = n$.

(In Definition 13.3.1 of the next chapter, a Coxeter group with nerve a pseudomanifold of dimension $n-1$ will be called of "type $\mathrm{PM}^n$.")

## 12.3. THE GENERAL CASE

A simplicial complex $L$ with piecewise spherical structure has "simplices of size $\geqslant \pi/2$" if each of its edges has length $\geqslant \pi/2$ (Definition I.5.8 of Appendix I.5). In Definition I.7.1 such an $L$ is called a "metric flag complex" if it is "metrically determined by its 1-skeleton." This means that a finite set of vertices, which are pairwise connected by edges, spans a simplex of $L$ if and only if it is possible to find some spherical simplex (necessarily unique up to isometry) with the given edge lengths.

**LEMMA 12.3.1.** *Let L be the nerve of a Coxeter system* $(W, S)$ *with its natural piecewise spherical structure. Then L has simplices of size* $\geqslant \pi/2$. *Moreover, it is a metric flag complex.*

*Proof.* Let $u_s$ be the vertex of $L$ corresponding to $s$. There is an edge between distinct vertices $u_s$ and $u_t$ if and only if $m_{st} \neq \infty$ and its length is $\pi - \pi/m_{st}$ which is $\geqslant \pi/2$. So, $L$ has simplices of size $\geqslant \pi/2$. Suppose $T \subset S$ is a subset of vertices which are pairwise connected by edges. Let $(c_{st})$ be the $T \times T$ matrix defined by $c_{st} := \cos d(u_s, u_t) = -\cos(\pi/m_{st})$. Then $(c_{st})$ is the cosine matrix of $(W_T, T)$. Hence, it is positive definite if and only if $T$ is spherical (Theorem 6.12.9), i.e., if and only if $T$ corresponds to the vertex set of a simplex in $L$. But the condition of being a metric flag complex is precisely that any $T \subset S$ whose corresponding matrix of cosines of edge lengths is positive definite spans a simplex. (See the definitions of "nerve" in 7.1 and "metric flag complex" in Appendix I.5.) □

In his Ph.D. thesis [221], G. Moussong proved a beautiful generalization of Gromov's Lemma I.6.1: a piecewise spherical simplicial cell complex with simplices of size $\geqslant \pi/2$ is CAT(1) if and only if it is a metric flag complex. (The precise statement and a sketch of the proof are given in Appendix I.7.) So, a corollary to Lemma 12.3.1 and Moussong's Lemma I.7.4 is the following.

**COROLLARY 12.3.2.** *The nerve L of any Coxeter system, equipped with its natural piecewise spherical structure, is* CAT(1).

In the same way we deduced Theorem 12.2.1 in the right-angled case from this fact about the links, we get the following theorem of [221].

**THEOREM 12.3.3.** (Moussong's Theorem.) *For any Coxeter system, the associated cell complex* $\Sigma$, *equipped with its natural piecewise Euclidean structure, is* CAT(0).

Since CAT(0) spaces are contractible, this gives an alternative proof that $\Sigma$ is contractible (Theorem 8.2.13).

## Some Consequences for *W*

In the language of Appendix I.4, Moussong's Theorem means that any Coxeter group is a "CAT(0) group." In particular, any Coxeter group satisfies all the properties listed in Theorem I.4.1. We state some of these properties again in the following theorem.

**THEOREM 12.3.4.** *Suppose $(W, S)$ is a Coxeter system. Then*

*(i) Any finite subgroup $F \subset W$ is conjugate to a finite subgroup of a spherical special subgroup $W_T$.*

*(ii) $\Sigma$ is a finite model for $\underline{E}W$.*

*(iii) The Conjugacy Problem for $W$ is solvable.*

*(iv) Any virtually solvable subgroup of $W$ is finitely generated and virtually free abelian.*

*Proof.* (i) By construction, the isotropy subgroup at any point of $\Sigma$ is conjugate to some special spherical subgroup. Since $\Sigma$ is CAT(0), we can apply the Bruhat–Tits Fixed Point Theorem (Theorem I.2.11 in Appendix I.2), to conclude that the fixed point set of $F$ on $\Sigma$ is nonempty. Hence, $F$ is contained in the isotropy subgroup of any such fixed point and such an isotropy subgroup is a spherical parabolic subgroup.

(ii) $\Sigma$ has only finitely many orbits of cells since the fundamental domain $K$ is a finite simplicial complex. For the fact that $\Sigma$ is an $\underline{E}W$, see Theorem I.4.1 (i).

(iii) For a solution to the Conjugacy Problem for CAT(0) groups, see [37, pp. 445–446].

(iv) This property was stated already in Moussong's thesis [221]. As is explained in Appendix I.4.3, it is a consequence of the Solvable Subgroup Theorem I.4.3 (see [37, p. 249]). □

(See Definition 2.3.1 and Appendix I.2 for further discussion of $\underline{E}W$, the universal space for proper $W$-actions.)

Here is another consequence of Moussong's Theorem (together with the Flat Torus Theorem).

**THEOREM 12.3.5.** *Suppose $(W, S)$ is irreducible and $W$ is virtually abelian. Then $W$ is either finite or a cocompact Euclidean reflection group.*

*Proof.* By Theorem I.4.1 (iv), any abelian subgroup of $W$ is finitely generated; hence, $W$ is virtually free abelian. So $\mathbb{Z}^n \subset W$ is a subgroup of finite index for some integer $n$. If $n = 0$, then $W$ is finite. So, suppose $n > 0$. Then $W$ is a virtual $PD^n$-group. By Theorem 10.9.2, $W$ decomposes as $W = W_{S_0} \times W_{S_1}$, where $W_{S_1}$ is finite and $W_{S_0}$ is type HM$^n$. Since $W$ is irreducible and infinite, $S = S_0$. So $\Sigma$ is a homology $n$-manifold. By the Flat Torus Theorem in Appendix I.4, Min$(\mathbb{Z}^n)$ is isometric to $\mathbb{E}^n$. Therefore, $\Sigma = \mathbb{E}^n$ and $W$ acts as an isometric reflection group on it. □

## 12.4. THE VISUAL BOUNDARY OF $\Sigma$

Given a complete, locally compact, CAT(0)-space $X$, in Appendix I.8 we show how to to adjoin to $X$ a "visual boundary" $\partial X$, obtaining a compact space $\overline{X}$. Roughly, $\partial X$ consists of an "endpoint" for each geodesic ray emanating from any given base point. In the same appendix we define the notion of a "$Z$-set compactification" and we state, in Theorem I.8.3 (i), that $\partial X$ is a $Z$-set in $\overline{X}$. A consequence is that the (reduced) Cěch cohomology of $\partial X$ is isomorphic to the cohomology with compact supports of $X$ (with a dimension shift).

The visual boundaries of the CAT(0)-complexes $\Sigma$ are archetypes for boundaries of general CAT(0)-spaces. As one might suspect after reading 8.3 and 9.2, the topology of $\partial\Sigma$ is intimately connected to the topology of the finite simplicial complex $L$.

### The Case Where $L$ Is a PL Manifold

This case is particularly tractable. Suppose $X$ is a CAT(0) piecewise Euclidean or piecewise hyperbolic polyhedron and that the link of any vertex of $X$ is a PL manifold of dimension $n-1$. Let $B$ be a closed metric ball in $X$. We show in Theorem I.8.4 that $\partial B$ is the connected sum $\mathrm{Lk}(v_1)\sharp\cdots\sharp\,\mathrm{Lk}(v_m)$ of the links of those vertices $v_1,\dots v_m$ which are contained in the interior of $B$. It follows that $\partial X$ is the inverse limit of an inverse system of such connected sums.

This theorem applies to $\Sigma$ when $L$ is a PL manifold of dimension $(n-1)$. It follows that $\partial\Sigma$ is the inverse limit of an inverse sequence of connected sums of copies of $L$. The map (or "bond") from the $(m+1)$-fold connected sum to the $m$-fold sum is the obvious one which collapses the last factor to a disk. Properties of $\partial\Sigma$ throw into relief many of the properties of $\Sigma$ developed in Chapters 8, 9, and 10, for instance:

- For $i>0$, $\check{H}^i(\partial\Sigma)\cong H_c^{i+1}(\Sigma)$.
- If $L$ is a $\mathrm{GHS}^{n-1}$ (so that $W$ is a virtual $\mathrm{PD}^n$-group), then $\Sigma$ is a contractible homology manifold and the reduced (co)homology of $\partial\Sigma$ is concentrated in degree $n-1$.
- If $L$ is a PL homology sphere of dimension $n-1$, then $\partial\Sigma$ is the inverse limit of connected sums of copies of $L$ and $\partial\Sigma$ is a homology $(n-1)$-manifold with the same homology as $S^{n-1}$ (see Section 10.8). Moreover,

$$\pi_1(\partial\Sigma)\cong\pi_1^{\infty}(\Sigma)\cong\varprojlim(\pi_1(L)*\cdots*\pi_1(L)).$$

If $L$ is actually a sphere, then $\partial\Sigma$ is an $(n-1)$-sphere and $\overline{\Sigma}$ is PL homeomorphic to an $n$-disk (Theorem I.8.4 (iii)).

Various properties of $\partial\Sigma$ when $L$ is a PL manifold are developed by H. Fischer [128] (at least in the right-angled case). For example, he uses results of Jakobsche [165] to show that $\partial\Sigma$ is topologically homogeneous. In other words, for any two points $x, y \in \Sigma$, there is a self-homeomorphism of $\partial\Sigma$ taking $x$ to $y$.

**Example 12.4.1.** (*Continuation of Example 8.5.8.*) Suppose $L = \mathbb{R}P^2$. Then $\partial\Sigma$ is the inverse limit of $m$-fold connected sums of $\mathbb{R}P^2$, $m \in \mathbb{N}$. (Such spaces are called *Pontryagin surfaces.*) Similarly, if $L = M^{n-1}$ is any nonorientable manifold, as in 8.5.8, we get Coxeter groups $W$ with $\mathrm{vcd}_{\mathbb{Z}}(W) = n$ and $\mathrm{vcd}_{\mathbb{Q}}(W) = n - 1$ (since $H^n_c(\Sigma; \mathbb{Z}) = \mathbb{Z}/2$).

## 12.5. BACKGROUND ON WORD HYPERBOLIC GROUPS

Roughly, a quasi-isometry between two metric spaces is a function which is "bi-Lipschitz in the large." Here is the precise definition.

**Definition 12.5.1.** Suppose $(X_1, d_1)$ and $(X_2, d_2)$ are metric spaces. A (not necessarily continuous) function $f : X_1 \to X_2$ is a $(\lambda, \varepsilon)$-*quasi-isometric embedding* if there exists positive constants $\lambda$ and $\varepsilon$ such that

$$\frac{1}{\lambda} d_1(x, y) - \varepsilon \leqslant d_2(f(x), f(y)) \leqslant \lambda d_1(x, y) + \varepsilon$$

for all $x, y \in X_1$. $f$ is a $(\lambda, \varepsilon)$-*quasi-isometry* if, in addition, there is a constant $C$ so that every point of $X_2$ lies in a $C$-neighborhood of $\mathrm{Im} f$. If such an $f$ exists, then $X_1$ and $X_2$ are *quasi-isometric*.

Although the word metric on a finitely generated group $\Gamma$ (defined in 2.1) depends on the choice of generating set $S$, a basic fact is that any two such choices yield quasi-isometric metrics.

Geometric group theory begins with the observation (called the Švarc-Milnor Lemma in [37, p. 140]) that if $\Gamma$ is a discrete group of isometries acting properly and cocompactly on a length space $X$, then $\Gamma$, equipped with a word metric, is quasi-isometric to $X$. For example, $\Gamma$ is quasi-isometric to its Cayley graph. If $\Gamma = \mathbb{Z}^n$, then it is quasi-isometric to $\mathbb{E}^n$ and if $\Gamma$ is the fundamental group of a closed hyperbolic $n$-manifold, then it is quasi-isometric to $\mathbb{H}^n$. On the other hand, $\mathbb{E}^n$ and $\mathbb{E}^m$ (resp. $\mathbb{H}^n$ and $\mathbb{H}^m$) are not quasi-isometric if $n \neq m$ and $\mathbb{E}^n$ is not quasi-isometric to $\mathbb{H}^n$, $n \neq 1$. One is led to ask which properties of a group are quasi-isometry invariants? Also, to what extent do geometric notions such as curvature have group theoretic meanings? In a groundbreaking paper, Gromov [147] developed the theory of word hyperbolic groups. (Earlier this notion had been defined by I. Rips and D. Cooper, working independently of each other.) The idea is that "in the large" a word hyperbolic group should look like a negatively curved space.

The notion of a *geodesic triangle* in a metric space is defined in Appendix I.2. (It is a configuration of three points and three geodesic segments connecting them in pairs.) A triangle in a metric space is *$\delta$-slim* if each of its edges is contained in the $\delta$-neighborhood of the union of the other two. In other words, if $e_1, e_2, e_3$ are the three edges (in any order) and $x \in e_1$, then $d(x, e_2 \cup e_3) \leqslant \delta$. A geodesic space $X$ is $\delta$-hyperbolic if every triangle in $X$ is $\delta$-slim. A group $\Gamma$ is *word hyperbolic* if it has a finite set of generators $S$ so that $\mathrm{Cay}(\Gamma, S)$ is $\delta$-hyperbolic, for some $\delta \geqslant 0$.

**Example 12.5.2.** A tree, equipped with the natural metric in which each edge has length 1, is 0-hyperbolic. Conversely, a geodesic space is 0-hyperbolic if and only if it is an $\mathbb{R}$-tree (see [138, pp. 29–31]) for the definition of "$\mathbb{R}$-tree" and a proof of this fact). This is the basis for earlier statements by Gromov that in the large, the universal cover of any negatively curved space "looks like a tree." A consequence is that a finitely generated free group is word hyperbolic since the Cayley graph of a free group is a tree (see Figure 2.1 of Section 2.1).

A crucial property of the definition of $\delta$-hyperbolic is given in the following theorem. (For example, it implies that word hyperbolicity is independent of the choice of generating set.)

**THEOREM 12.5.3.** (Gromov [147].) *The property of being word hyperbolic is a quasi-isometry invariant for groups.*

A proof can be found in [37, pp. 400–405]. In outline it goes as follows. One shows that if two geodesic spaces are quasi-isometric and one is $\delta$-hyperbolic, then the other is $\delta'$-hyperbolic (for a different constant $\delta'$). To prove this, one needs to analyze "quasi-geodesics" (i.e., quasi-isometric embeddings of intervals) and "quasi-triangles" in a $\delta$-hyperbolic metric space. One can then apply the statement about geodesic spaces to the Cayley graphs of two groups to conclude Theorem 12.5.3.

**THEOREM 12.5.4.** *Suppose $\Gamma$ is a discrete group of isometries acting properly and cocompactly on a* $\mathrm{CAT}(-1)$*-space $X$. Then $\Gamma$ is word hyperbolic.*

*Sketch of Proof.* By Definition I.2.1, a geodesic space $X$ is $\mathrm{CAT}(-1)$ if and only if triangles in $X$ are "thinner" than comparison triangles in $\mathbb{H}^2$. So, to show $X$ is $\delta$-hyperbolic it suffices to show that any triangle in $\mathbb{H}^2$ is $\delta$-slim. This is a straightforward computation. Since $\Gamma$ is quasi-isometric to $X$ (by the Švarc-Milnor Lemma), it is word hyperbolic. □

In the next theorem we list some properties of word hyperbolic groups.

**THEOREM 12.5.5.** *Suppose* $\Gamma$ *is a word hyperbolic group. Then*

*(i)* $\mathbb{Z} \times \mathbb{Z} \not\subset \Gamma$.

*(ii) The Word Problem and the Conjugacy Problem for* $\Gamma$ *are solvable* ([37]).

*(iii) There is a finite model for* $\underline{E}\Gamma$ *namely, the "Rips complex" of* $(\Gamma, S)$ ([203]).

*(iv)* $\Gamma$ *satisfies a linear "isoperimetric inequality."*

Some comments about the meaning of some of the statements in this theorem are in order. First, consider (iii). Equip $\Gamma$ with the word metric with respect to a finite generating set $S$. Given a positive integer $r$, the *Rips complex*, $\mathrm{Rips}(\Gamma, r)$, is defined to be the simplicial complex with vertex set $\Gamma$ and with simplices the nonempty finite subsets of $\Gamma$ of diameter $\leqslant r$. It is proved in [203] that for $r$ sufficiently large (in fact, for $r > 4\delta + 6$), $\mathrm{Rips}(\Gamma, r)$ is a finite model for $\underline{E}\Gamma$. (Most of the consequences of this fact were known much earlier to Gromov and Rips.) Since $\mathrm{Rips}(\Gamma, r)$ has only a finite number of $\Gamma$-orbits of cells, $\Gamma$ has a finite presentation. Let $\langle S \mid \mathcal{R}\rangle$ be such a presentation.

Next, consider (iv). Let $X$ be a 2-complex associated to the presentation (cf. Section 2.2). A loop in the 1-skeleton of $X$ corresponds to a word $\mathbf{s}$ in $S \cup S^{-1}$ (that is, to an element of the free group on $S$). Suppose the loop is null-homotopic in the 2-skeleton of $X$. Then it can be shrunk to the basepoint by pushing it across 2-cells. This means that $\mathbf{s}$ can be written in the form

$$\mathbf{s} = \prod_{i=1}^{N} x_i r_i x_i^{-1}, \tag{12.1}$$

where $r_i \in \mathcal{R} \cup \mathcal{R}^{-1}$ and $x_i \in F_S$. The *area* of $\mathbf{s}$, denoted $\mathrm{Area}(\mathbf{s})$, is the smallest integer $N$ so that $\mathbf{s}$ can be expressed in the form (12.1). The presentation $\langle S \mid \mathcal{R}\rangle$ satisfies a *linear isoperimetric inequality* if there is a constant $C$ such that for any word $\mathbf{s}$ which represents the identity element of $\Gamma$, we have $\mathrm{Area}(\mathbf{s}) \leqslant Cl(\mathbf{s})$. The property of having a linear isoperimetric inequality is independent of the choice of presentation. (This is an easy exercise.) Moreover, this property immediately implies that the Word Problem for $\Gamma$ is solvable. (This is half of (ii).) As for the other half of (ii): the Conjugacy Problem asks if there is an algorithm for deciding if two elements $\gamma_1$ and $\gamma_2$ of $\Gamma$ are conjugate. A proof of the solvability of the Conjugacy Problem for word hyperbolic groups can be found in [37, pp. 451–454].

*Remark.* The converse to Theorem 12.5.5 (iv) is true: if a finitely generated group satisfies a linear isoperimetric inequality, then it is finitely presented and word hyperbolic (Gromov [147]). A proof can be found in [37, pp. 417–421].

## 12.6. WHEN IS Σ CAT(−1)?

As in 12.1, let $\mathbf{d} \in (0, \infty)^S$. For each $T \in \mathcal{S}$, we will now define a hyperbolic polytope $C_T(\mathbf{d})$, combinatorially isomorphic to the Coxeter polytope of type $W_T$. Let $Z_T \subset \mathbb{H}^n$, $n = \mathrm{Card}(T)$, be the fundamental simplicial cone for $W_T$ acting as a isometric reflection group on $\mathbb{H}^n$. Let $x_{\mathbf{d}} \in Z_T$ be the point which, for each $s \in T$, is distance $d_s$ from the $s$-wall of $Z_T$. Let $C_T(\mathbf{d})$ be the convex hull of the $W_T$-orbit of $x_{\mathbf{d}}$. $C_T(\mathbf{d})$ is the hyperbolic Coxeter polytope with edges of type $s$ having length $2d_s$. Let $\Sigma_{\mathbf{d}}$ be the piecewise hyperbolic structure on $\Sigma$ defined by identifying each Coxeter cell of type $W_T$ with $C_T(\mathbf{d})$.

**THEOREM 12.6.1.** (Moussong [221, Theorem 17.1].) *The piecewise hyperbolic structure on* $\Sigma_{\mathbf{d}}$ *is* CAT(−1) *if and only if there is no subset* $T \subset S$ *satisfying either of the following two conditions:*

*(a)* $W_T$ *is a cocompact Euclidean reflection group of dimension* $\geqslant 2$.

*(b)* $(W_T, T)$ *decomposes as* $(W_T, T) = (W_{T'} \times W_{T''}, T' \cup T'')$ *with both* $W_{T'}$ *and* $W_{T''}$ *infinite.*

The proof of Theorem 12.6.1 follows from the arguments of Appendices I.6 and I.7 in the same way as does the proof of Theorem 12.2.1 (ii).

Let us say that $(W, S)$ satisfies *Moussong's Condition* if neither (a) nor (b) holds for any $T \subset S$. The next result follows immediately from Lemma I.7.6.

**LEMMA 12.6.2.** *The natural piecewise spherical structure on* $L(W, S)$ *is extra large if and only if* $(W, S)$ *satisfies Moussong's Condition.*

*Proof.* Conditions (a) and (b) of Theorem 12.6.1 correspond to conditions (a) and (b) of Lemma I.7.6. □

*Proof of Theorem 12.6.1.* The link of a vertex in the piecewise hyperbolic structure on $\Sigma_{\mathbf{d}}$ is a $\delta$-change of the natural piecewise spherical structure on $L$, where $\delta \to 1$ as $\mathbf{d} \to 0$. (See Appendix I.6 for the definition of "$\delta$-change.") So, by Lemma I.6.7, $\Sigma_{\mathbf{d}}$ will be CAT(−1) for sufficiently small $\mathbf{d}$ if and only if $L$ is extra large. By Lemma 12.6.2, this is the case if and only if Moussong's Condition holds. □

**COROLLARY 12.6.3.** *The following conditions on* $(W, S)$ *are equivalent.*

*(i)* $W$ *is word hyperbolic.*

*(ii)* $\mathbb{Z} \times \mathbb{Z} \not\subset W$.

*(iii)* $(W, S)$ *satisfies Moussong's Condition.*

*(iv)* $\Sigma$ *admits a piecewise hyperbolic,* CAT(−1) *metric.*

*Proof.* The implication (i) $\Longrightarrow$ (ii) follows from Theorem 12.5.5 (i). If $W_T$ satisfies condition (a) of Theorem 12.6.1, then $W_T$ is a Euclidean reflection group and its translation subgroup of $W_T$ is free abelian of rank Card($T$) – 1; moreover, this rank is $\geqslant 2$. Similarly, if (b) holds for $W_T$ (so that it decomposes as the product of two infinite subgroups), then $\mathbb{Z} \times \mathbb{Z} \subset W_T$. Hence, (ii) $\Longrightarrow$ (iii). The implication (iii) $\Longrightarrow$ (iv) is Theorem 12.6.1; (iv) $\Longrightarrow$ (i) follows from Theorem 12.5.4. □

**Two-Dimensional Examples**

We expand on the discussion in Example 7.4.2. Let $J$ be a finite simplicial graph with vertex set $S$ and $m : \text{Edge}(J) \to \{2, 3, \dots\}$ a labeling of the edges. Use the notation $\{s, t\} \to m_{st}$ for this labeling. As in Example 7.1.6, there is an associated Coxeter group $W = W(J, m)$ with $J$ the 1-skeleton of $L(W, S)$. So, $J$ inherits a natural piecewise spherical structure in which the edge $\{s, t\}$ has length $\pi - \pi/m_{st}$. The proof of the next proposition is immediate.

**PROPOSITION 12.6.4.** *Suppose $(J, m)$ is a labeled simplicial graph as above.*

*(i) $J = L(W, S)$ if and only if the natural piecewise spherical structure on $J$ is that of a metric flag complex. This is the case if and only if for every 3-circuit of $L$, the sum over its edges satisfies*

$$\sum \frac{1}{m_{st}} \leqslant 1.$$

*(ii) $J$ is extra large if and only if for each 3-circuit, the above inequality is strict and furthermore, $J$ has no 4-circuits with each edge labeled 2.*

So, given $(J, m)$, the 2-skeleton of the resulting piecewise Euclidean complex $\Sigma(W, S)$ is CAT(0) if and only if condition (i) of the previous proposition holds and $\Sigma(W, S)$ admits a piecewise hyperbolic CAT($-1$)-structure if and only if condition (ii) holds. Next, we specialize to the case where $m$ is a constant function. In other words, suppose $m$ is an integer $\geqslant 2$ so that each Coxeter 2-cell is a $2m$-gon.

**COROLLARY 12.6.5.** (Gromov [147].) *Suppose $J$ is a simplicial graph and $m$ is an integer $\geqslant 2$.*

*(i) If $m = 2$, suppose $J$ has no 3-circuits. Then there exists a piecewise Euclidean,* CAT(0) *2-complex in which each 2-cell is a regular $2m$-gon and the link of each vertex is $J$.*

(ii) *If $m = 2$, suppose $J$ has no circuits of length $\leqslant 4$. If $m = 3$, suppose $J$ has no 3-circuits. Then there exists a piecewise hyperbolic, CAT(−1) 2-complex in which each 2-cell is a regular hyperbolic $2m$-gon and the link of each vertex is combinatorially equivalent to $J$.*

### Subgroups of $O_+(n, 1)$ Generated by Reflections

Suppose $(W, S)$ is a Coxeter system and that we have a representation $W \hookrightarrow O_+(n, 1)$ such that each $s \in S$ is sent to the isometric reflection across some hyperplane $H_s \subset \mathbb{H}^n$. (As in 6.2, $O_+(n, 1)$ is the group of isometries of $\mathbb{H}^n$.) Further suppose that for each $s \in S$ we can choose a half-space bounded by $H_s$ so that the intersection $C$ of all such half-spaces has nonempty interior and that for all $\{s, t\} \in \mathcal{S}^{(2)}$ the hyperplanes $H_s$ and $H_t$ make a dihedral angle of $\pi/m_{st}$. If $C$ is a compact polytope, then $W$ is a hyperbolic reflection group in the sense of Definition 6.4.4 and hence, is word hyperbolic (since it is quasi-isometric to $\mathbb{H}^n$). In other examples $C$ is not compact but still has finite volume. In this case the picture is that $C$ is a polytope with some "ideal" vertices which lie on the sphere at infinity. The isotropy subgroup at such an ideal vertex is a special subgroup $W_T$, which must be a Euclidean reflection group of rank $n$. If $n \geqslant 3$, then $W$ is not word hyperbolic since $\mathbb{Z}^{n-1} \subset W_T \subset W$. An element of infinite order in such a $W_T$ is a *parabolic isometry* of $\mathbb{H}^n$ (in the sense that it fixes a unique point in the sphere at infinity). Suppose $W_{\{s,t\}}$ is an infinite dihedral special subgroup (i.e., $m_{st} = \infty$). Let $u_s$ and $u_t$ be the unit inward-pointing vectors normal to $H_s$ and $H_t$. Then $\langle u_s, u_t \rangle \leqslant -1$, with equality if and only if $st$ is a parabolic element of $O_+(n, 1)$. So, $W_{\{s,t\}}$ contains parabolics if and only if $\langle u_s, u_t \rangle = -1$. On the other hand, if $W_T$ is a Euclidean reflection group of rank $\geqslant 3$, then it follows from the discussion in 6.12 that $W_T$ must fix a unique point on the sphere at infinity (and hence, contains parabolics). Here is another consequence of Theorem 12.6.1.

**PROPOSITION 12.6.6.** *Suppose, as above, that $W$ is represented as a group generated by hyperbolic reflections across the faces of a (not necessarily compact) convex set $C \subset \mathbb{H}^n$, given as the intersection of a finite number of half-spaces. Suppose further that $W$ contains no parabolic isometries of $\mathbb{H}^n$. Then $W$ is word hyperbolic.*

*Proof.* Condition (a) of Theorem 12.6.1 follows from the hypothesis of no parabolics. Condition (b), which does not allow special subgroups of the form $W_{T'} \times W_{T''}$ with both factors infinite, is also easily seen to hold. (For example, if we had such a decomposition, the Gram matrices associated to $T'$ and $T''$ must both be indefinite, contradicting the fact that the form on $\mathbb{R}^{n,1}$ is type $(n, 1)$.) □

## Word Hyperbolic Coxeter Groups of Type HM

Moussong observed that, just as Corollary 12.2.3 follows from Corollary 6.11.6, the arguments which prove Vinberg's Theorem 6.11.8 show the following.

**PROPOSITION 12.6.7.** (Vinberg, Moussong.) *Suppose W is word hyperbolic and type* $\mathrm{HM}^n$. *Then* $n \leqslant 29$.

If $L(W, S)$ is homeomorphic to $S^1$, then $W$ is a polygon group (cf. Section 6.5) and it is word hyperbolic if and only if it can be represented as a discrete subgroup of $\mathrm{Isom}(\mathbb{H}^2)$ generated by reflections across the sides of a polygon. A similar result holds in dimension 3 since, when $L$ is a triangulation of $S^2$, Moussong's Condition is identical with the conditions from Andreev's Theorem 6.10.2. This raises the question of whether any word hyperbolic Coxeter group of type $\mathrm{HM}^n$ can be realized as a geometric reflection group on $\mathbb{H}^n$. Moussong addressed this problem and gave some 4-dimensional counterexamples in [221, Section 18]. We discuss one of his examples below.

**Example 12.6.8.** (*A word hyperbolic Coxeter group of type* $\mathrm{HM}^4$.) Let $(W, S)$ be the Coxeter system corresponding to the diagram in Figure 12.1. We claim that $W$ is word hyperbolic, type $\mathrm{HM}^4$ and that it does not have a representation as a discrete group generated by reflections on $\mathbb{H}^n$. In order to better understand these claims, consider another Coxeter group $W'$ with diagram given by Figure 12.2. $W'$ is the product of two triangle groups; it acts cocompactly on $\mathbb{H}^2 \times \mathbb{H}^2$ and hence, is type $\mathrm{HM}^4$. Its nerve is $S^3$, triangulated as the join of two triangles. Identify the generating sets of $W$ and $W'$ via the obvious identifications of the nodes of their diagrams. Inspection shows $W$ and $W'$ have the same posets of spherical subsets; hence, they have the same nerve. This shows $W$ is also type $\mathrm{HM}^4$. $W'$ does not satisfy Moussong's Condition since it decomposes as the product of two infinite special subgroups (the triangle groups). However, $W$ does satisfy Moussong's Condition, the point being that the edge of the diagram connecting the two triangles groups destroys the product decomposition. How does one show that $W$ cannot be realized as a group generated by reflections across the faces of a hyperbolic polytope? Since any two elements of the generating set $S$ are connected by an edge (there being no $\infty$'s in the diagram), the Gram matrix of any such hyperbolic polytope must be equal to the cosine matrix of $(W, S)$. (See 6.8 for the definitions of the Gram matrix and the cosine matrix.) The final claim is that this cosine matrix is nonsingular (of rank 6) and type $(4, 2)$. On the other hand, if it could be realized in $\mathbb{H}^n$ it would be rank $k + 1$, $k = 4$ or 5 and type $(k, 1)$. This last claim is plausible because the cosine matrix of $W'$ is type $(4, 2)$. The cosine matrix of $W$ differs from this by changing only a single pair of entries from

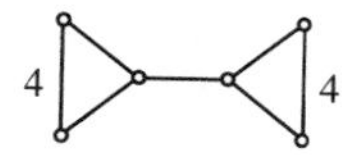


Figure 12.1. $W$ is word hyperbolic.

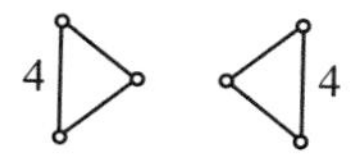


Figure 12.2. $W'$ is not word hyperbolic.

$0\ (= -\cos\pi/2)$ to $-1/2\ (= -\cos\pi/3)$. So, the claim follows by computing the determinant to be $> 0$.

## 12.7. FREE ABELIAN SUBGROUPS OF COXETER GROUPS

Call a subset $T$ of $S$ *affine* if $W_T$ is a Euclidean reflection group.

Let $T_1, \dots, T_n$ be the nonspherical, irreducible components of $S$ (i.e., each $W_{T_i}$ is an infinite, irreducible component of $W$). Choose subgroups $H_i \subset W_{T_i}$ as follows. If $T_i$ is affine, then $H_i$ is the translation subgroup. Otherwise, $H_i$ is any infinite cyclic subgroup. Any subgroup of $W$ which is conjugate to one of the form $\prod_i H_i$ is called a *standard free abelian subgroup*.

Moussong proved that (a) every solvable subgroup of $W$ is virtually free abelian (Theorem 12.3.4 (iv)) and (b) $W$ is word hyperbolic if and only if every standard free abelian subgroup is infinite cyclic (Theorem 12.6.1). In this section we discuss an important result of Krammer [181]: any free abelian subgroup of $W$ is virtually a subgroup of a standard one.

Following [181], for any subset $X \subset W$, define its *parabolic closure*, $\mathrm{Pc}(X)$, to be the smallest parabolic subgroup containing $X$. (See Section 4.5 for the definition of "parabolic subgroup.") An element $w \in W$ is *essential* if $\mathrm{Pc}(w) = W$. For $w \in W$, $Z(w)$ denotes its centralizer $(= \{v \in W \mid vw = wv\})$. The next lemma is one of the most important results in [181]. We will not give its proof since it would take us too far afield.

**LEMMA 12.7.1.** (Krammer [181, Cor. 6.3.10].) *Suppose $(W, S)$ is infinite, irreducible and not affine (i.e., $W$ is not a Euclidean reflection group) and that $w$ is an essential element of $W$. Then the infinite cyclic subgroup $\langle w\rangle$, generated by $w$, has finite index in $Z(w)$.*

**LEMMA 12.7.2.** (Krammer [181, Lemma 6.8.1].) *Suppose $W$ is irreducible and that $T \subset S$ is such that the parabolic closure of $N(W_T)$ (the normalizer of $W_T$) is all of $W$. Then either $T$ is spherical or $T = S$.*

Given $U \subset S$, define $U^{\perp} := \{s \in S \mid m_{st} = 2,\ \text{for all}\ t \in U\}$.

*Proof of Lemma 12.7.2.* The proof uses results and notation from 4.10. Suppose $T$ is not spherical. Let $U$ be a nonspherical irreducible component of $T$. By Proposition 4.10.2, $N(W_T) = W_T \rtimes G_T$. Let $w \in G_T$. Then $w^{-1}A_U = A_{U'}$ for some $U' \subset T$. ($A_U$ is the fundamental $U$-sector in $W$ as in Definition 4.5.1.) By Theorem 4.10.6, there is a directed path in $\Upsilon$:

$$U = T_0 \xrightarrow{s_0} T_1 \xrightarrow{s_1} \cdots \xrightarrow{s_k} T_{k+1} = U'$$

with $w = \nu(T_0, s_0) \cdots \nu(T_k, s_k)$. Since $\nu(T_0, s_0)$ is defined, $T_0 \cup \{s_0\}$ contains a spherical component. Since $U = T_0$ is irreducible and nonspherical, we must have $s_0 \in U^\perp$. So, $T_1 = U$ and $\nu(T_0, s_0) = s_0$. Continuing in this fashion we get $w = s_0 \cdots s_k \in W_{U^\perp}$. Therefore, $G_T \subset W_{U^\perp}$ and $N(W_T) \subset W_{T \cup U^\perp}$. So, $\mathrm{Pc}(N(W_T)) = W_{T \cup U^\perp}$. Since $\mathrm{Pc}(N(W_T)) = W$, this implies that $T \cup U^\perp = S$. Since $U$ is a component of $T \cup U^\perp$, we must have $U^\perp = \emptyset$ and $T = S$. $\square$

The following version of Theorem 12.3.5 was proved by Krammer.

**THEOREM 12.7.3.** (Krammer [181, Theorem 6.8.2].) *Suppose $W$ is infinite, irreducible, and not affine. Let $A \subset W$ be a free abelian subgroup with* $\mathrm{Pc}(A) =$ *$W$. Then $A$ is infinite cyclic.*

*Proof.* Let $a \in A - \{1\}$. After conjugating, we may suppose $\mathrm{Pc}(a)$ is a special subgroup, say, $W_T$. For any $b \in A$, $bW_Tb^{-1} = b\,\mathrm{Pc}(a)b^{-1} = \mathrm{Pc}(bab^{-1}) = \mathrm{Pc}(a) = W_T$; so, $A \subset N(W_T)$. Since $\mathrm{Pc}(A) = W$, it follows that $\mathrm{Pc}(N(W_T)) = W$. By Lemma 12.7.2, $T = S$. (Since $a$ has infinite order, $T$ is not spherical.) Hence, $\mathrm{Pc}(a) = W$. By Lemma 12.7.1, $[Z(a) : \langle a \rangle] < \infty$. Since $A \subset Z(a)$, it must have rank 1. $\square$

**THEOREM 12.7.4.** (Krammer [181, Theorem 6.8.3].) *Let $W$ be a Coxeter group. Then any free abelian subgroup has a subgroup of finite index which is a subgroup of a standard free abelian subgroup.*

*Proof.* Let $A \subset W$ be a free abelian subgroup. $A$ is finitely generated (Theorem I.4.1 (vii) in Appendix I.4). Let $N$ be a positive integer such that $w^N = 1$ for any torsion element $w \in W$. Put $B := \{a^N \mid a \in A\}$. Note that $[A : B] < \infty$. After conjugating, we may suppose that $\mathrm{Pc}(B)$ is a special subgroup, say, $W_T$. Let $T_1, \dots, T_k$ be the irreducible components of $T$ and let $B_i$ be the image of $B$ in $W_{T_i}$. Since $B_i$ is nontrivial, abelian, finitely generated and torsion-free (by our choice of $N$), it is free abelian. Moreover, $\mathrm{Pc}(B_i) = W_{T_i}$. So, each $T_i$ is nonspherical. If $T_i$ is not affine, then $B_i$ is rank 1, by Theorem 12.7.3. If $T_i$ is affine, then $B_i$ is contained in its translation subgroup by our choice of $N$. Hence, $B$ is contained in a standard free abelian subgroup. $\square$

**Question.** One way to get a flat subspace of $\Sigma$ is to take a product of geodesics in a subcomplex $\Sigma(W_{T_1}, T_1) \times \cdots \times \Sigma(W_{T_k}, T_k)$ where $W_{T_1} \times \cdots \times W_{T_k}$ is a special subgroup and each $W_{T_i}$ is infinite, irreducible and nonaffine. One can also take products with factors corresponding to spherical or affine subgroups. The translate of any such subspace is also a flat. Any such translate is a *standard flat*. Is every flat contained in a standard one?

## 12.8. RELATIVE HYPERBOLIZATION

Here we discuss a variant of the reflection group trick which preseves nonpositive curvature. This variant is a simultaneous generalization of the constructions in 11.1 and 12.2 (or 1.2). The input will consist of a pair $(X, B)$ of simplicial complexes so that $B$ has a piecewise Euclidean metric. As output we will produce a mirror structure on a simplicial complex $K(X, B)$ such that $B$ is a subcomplex of $K(X, B)$. (In fact, $K(X, B)$ will be homeomorphic to the first derived neighborhood of $B$ in $X$.) Let $W$ denote the right-angled Coxeter group associated to the mirror structure. The space $\mathcal{U}(W, K(X, B))$ will have a natural piecewise Euclidean cell structure in which each copy of $B$ is totally geodesic. Moreover, when $B$ is nonpositively curved, so is $\mathcal{U}(W, K(X, B))$. In the case where $X$ is the cone over a simplicial complex $L$ and $B$ is the cone point, this construction gives back the right-angled Coxeter group associated to $bL$ (the barycentric subdivision of $L$) and the complex $\Sigma$ with its usual cubical structure.

Here are the details of the construction. Suppose $B$ is a piecewise Euclidean cell complex. Subdividing if necessary, we may assume $B$ is a simplicial complex and that its simplices are isometrically identified with Euclidean simplices. Suppose further that $B$ is a subcomplex of another simplicial complex $X$. Possibly after another subdivision, we may assume $B$ is a full subcomplex. This means that if a simplex $\sigma$ of $X$ has nonempty intersection with $B$, then the intersection is a simplex of $B$ (and a face of $\sigma$). Note that while each simplex of $B$ is given a Euclidean metric, no metric is assumed on the simplices of $X$ which are not contained in $B$.

We will define a new cell complex $D(X, B)$ equipped with a piecewise Euclidean metric. We will also define a finite-sheeted covering space $\widetilde{D}(X, B)$ of $D(X, B)$. Each cell of $D(X, B)$ (or of $\widetilde{D}(X, B)$) will have the form $\alpha \times [-1, 1]^k$, for some integer $k \geqslant 0$, where $\alpha$ is a simplex of $B$ and where $\alpha \times [-1, 1]^k$ is equipped with the product metric. (Usually we will use $\alpha$ to stand for a simplex of $B$ and $\sigma$ for a simplex of $X$ which is not in $B$.) Define $\dim(X, B)$ to be the maximum dimension of any simplex of $X$ which intersects $B$ but which is not contained in $B$. In the next theorem we list some properties of the construction.

**THEOREM 12.8.1**

(i) *For $n = \dim(X, B)$, there are $2^n$ disjoint copies of $B$ in $D(X, B)$.*

(ii) *For each such copy and for each vertex $v$ in $B$, the link of $v$ in $D(X, B)$ is isomorphic to a subdivision of $Lk(v, X)$. In particular, if $X$ is a manifold, then so is $D(X, B)$.*

(iii) *If the metric on $B$ is nonpositively curved, then the metric on $D(X, B)$ is nonpositively curved and each copy of $B$ is a totally geodesic subspace of $D(X, B)$.*

(iv) *The group $(\mathbf{C}_2)^n$ acts as a reflection group on $D(X, B)$. ($\mathbf{C}_2$ denotes the cyclic group of order 2.) A fundamental chamber for this action is denoted by $K(X, B)$. It is homeomorphic to the first derived neighborhood of $B$ in $X$. Thus, $K(X, B)$ is a retract of $D(X, B)$ and $B$ is a deformation retract of $K(X, B)$.*

In fact, the entire construction depends only on a regular neighborhood of $B$ in $X$. More precisely, it depends only on the set of simplices of $X$ which intersect $B$. Let $\mathcal{P}$ denote the poset of simplices $\sigma$ in $X$ such that $\sigma \cap B \neq \emptyset$ and such that $\sigma$ is not a simplex of $B$. For each simplex $\alpha$ of $B$, let $\mathcal{P}_{>\alpha}$ be the subposet of $\mathcal{P}$ consisting of all $\sigma$ which have $\alpha$ as a face. Let $\mathcal{F} = \mathrm{Flag}(\mathcal{P})$ denote the poset of finite chains in $\mathcal{P}$. (A *chain* is a subset of $\mathcal{P}$ which happens to be totally ordered.) Given a chain $f = \{\sigma_0 < \cdots < \sigma_k\} \in \mathcal{F}$, let $\sigma_f$ denote its least element, i.e., $\sigma_f = \sigma_0$. Given a simplex $\alpha$ of $B$, let $\mathcal{F}_{>\{\alpha\}}$ denote the set of chains $f$ with $\sigma_f > \alpha$. The simplicial complex $\mathcal{F}$ can be identified with the first derived neighborhood of $B$ in $X$, i.e., the union of the stars of all simplices of $B$ in the barycentric subdivision of $X$. (See Example A.3.3 of Appendix A.3 for a more complete definition of $\mathrm{Flag}(\mathcal{P})$.)

Next we define the fundamental chamber $K$ $(= K(X, B))$. Each cell of $K$ will have the form $\alpha \times [0, 1]^f$, for some $f \in \mathcal{F}_{>\{\alpha\}}$. Here $[0, 1]^f$ means the set of functions from the set $f$ to $[0, 1]$. So, the number of interval factors of $\alpha \times [0, 1]^f$ is the number of elements of $f$. If $f \leqslant f'$, then we identify $[0, 1]^f$ with the face of $[0, 1]^{f'}$ defined by setting the coordinates $x_\sigma = 1$, for all $\sigma \in f' - f$. Define an order relation on the set of such cells as follows: $\alpha \times [0, 1]^f \leqslant \alpha' \times [0, 1]^{f'}$ if and only if $\alpha \leqslant \alpha'$ and $f \leqslant f'$. (Notice that if $\alpha \leqslant \alpha'$, then $\mathcal{F}_{>\{\alpha'\}} \subset \mathcal{F}_{>\{\alpha\}}$.) $K$ is defined to be the cell complex formed from the disjoint union $\coprod \alpha \times [0,1]^f$ by gluing together two such cells whenever they are incident. We leave it as an exercise for the reader to show that $K$ can be identified with a subcomplex of the second barycentric subdivision of $X$ and that it is homeomorphic to the first derived neighborhood of $B$ in $X$. The idea for proving this exercise is indicated in Figure 12.3 which illustrates the case where $X$ is a 2-simplex $\sigma$ and $B$ is an edge.

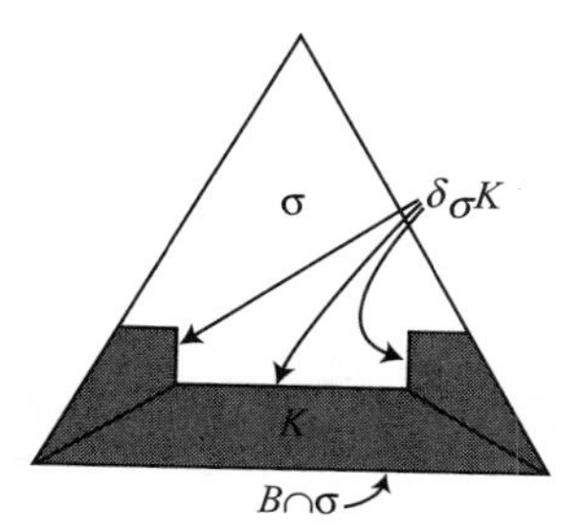


Figure 12.3. Intersection of $K$ with a simplex $\sigma$.

We define a mirror structure on $K$ indexed by the set $\mathcal{P}$. Instead of writing $K_\sigma$, we shall use the notation $\delta_\sigma K$ to denote the mirror corresponding to $\sigma$. For each $\sigma \in \mathcal{P}_{>\alpha}$ and each chain $f \in \mathcal{F}$ with $\sigma \in f$, define $\delta_\sigma(\alpha \times [0,1]^f)$ to be the face of $\alpha \times [0,1]^f$ defined by setting $x_\sigma = 0$, i.e.,

$$\delta_\sigma(\alpha \times [0,1]^f) = \alpha \times 0 \times [0,1]^{f-\{\sigma\}}.$$

The mirror $\delta_\sigma K$ is the subcomplex of $K$ consisting of all such cells. We note that $\delta_\sigma K \cap \delta_{\sigma'} K \neq \emptyset$ if and only if $\{\sigma, \sigma'\}$ is a chain. It follows that mirrors $\delta_{\sigma_1} K, \dots, \delta_{\sigma_k} K$ have nonempty intersection if and only if $\{\sigma_1, \dots, \sigma_k\}$ is a chain $f \in \mathcal{F}$ and if this is the case,

$$\delta_{\sigma_1} K \cap \dots \cap \delta_{\sigma_k} K = \bigcup_{f' \in \mathcal{F}_{\geqslant f}} (B \cap \sigma_f) \times 0 \times [0,1]^{f'-f}.$$

Hence, the nerve of the mirror structure is $\mathcal{F}$.

Next, apply the basic construction of Chapter 5. Define $\widetilde{D}$ $(= \widetilde{D}(X,B))$ by

$$\widetilde{D} := \mathcal{U}((\mathbf{C}_2)^{\mathcal{P}}, K).$$

*Remarks*

(i) Of course, there is a bigger Coxeter group in the background, namely, the right-angled Coxeter group $W$ associated to the 1-skeleton of the flag complex $\mathcal{F}$, as in Example 7.1.7. (There is one fundamental generator for each element of $\mathcal{P}$ and the nerve of $W$ is $\mathcal{F}$.) $\mathcal{U}(W,K)$ is a covering space of $\widetilde{D}$. The group $(\mathbf{C}_2)^{\mathcal{P}}$ is the abelianization of $W$.

(ii) Suppose $X$ is the cone on a simplicial complex $\partial X$ and $B$ is the cone point. Then $\widetilde{D}(X,B)$ coincides with the cubical complex $P_L$ defined in 1.2 (where $L$ is the barycentric subdivision of $\partial X$).

(iii) How does the above construction differ from the usual reflection group trick in the case when $X$ is a regular neighborhood of $B$ in some manifold? The point is that in the above construction the mirrors $\delta_\sigma K$ are not the top-dimensional dual cells to a triangulation of $\partial X$.

The definition of $D$ is similar to that of $\widetilde{D}$ only one uses the smaller group $(\mathbf{C}_2)^n$, $n = \dim(X, B)$, instead of $(\mathbf{C}_2)^{\mathcal{P}}$. More specifically, define a mirror structure $\{K_i\}_{1 \leq i \leq n}$ on $K$ by letting $K_i$ be the union of the $\delta_\sigma K$ with $\dim \sigma = i$. Define $D$ $(= D(X, B))$ by

$$D := \mathcal{U}((\mathbf{C}_2)^n, K). \tag{12.2}$$

In other words, if $\{r_1, \ldots, r_n\}$ is the standard set of generators for $(\mathbf{C}_2)^n$, then we identify the points $(gr_i, x)$ and $(g, x)$ of $(\mathbf{C}_2)^n \times K$ whenever $x$ belongs to a mirror $\delta_\sigma K$ with $i = \dim \sigma$.

At this stage it is clear that $D(X, B)$ satisfies properties (i), (ii) and (iv) of Theorem 12.8.1. Property (iii) (on nonpositive curvature) is more problematic. Checking this is easier if we use the alternate description of $D(X, B)$ given below.

### Background on Hyperbolization

The above variant of the reflection group trick has the additional benefit of providing a relative hyperbolization procedure. The term "hyperbolization" refers to an idea of Gromov [147, p.116] for starting with a simplicial complex $X$ and converting it into a new, nonpositively curved polyhedron $h(X)$. Such a procedure should satisfy at least some of the following properties.

- $h(X)$ should have nonpositively curved metric (often piecewise Euclidean).
- The construction should be functorial in the sense that if $X \hookrightarrow Y$ is a simplicial injection, then there should be associated a isometric embedding $h(X) \hookrightarrow h(Y)$.
- $h(X)$ should preserve local structure, e.g., if $X$ is a manifold, then $h(X)$ should also be a manifold.
- There should be a map $h(X) \to X$ and it should be "degree one," in the sense that it should induce a surjection on integral homology.

### Relationship with Relative Hyperbolization

Gromov also proposed the idea "relative hyperbolization"([147, pp.117–118]). Given $X$ and a subcomplex $B$, such a procedure should have as output a space $h(X, B)$ with $B \subset h(X, B)$ and satisfying appropriate properties. For example, if $B$ has nonpositive curvature, $h(X, B)$ should be nonpositively curved with $B$ as a totally geodesic subspace. Several such procedures have been proposed; however, none of them are completely satisfactory. One such procedure is described in [83, Section 1g]. Its major drawback is that it does not satisfy the previous property: $h(X, B)$ need not have nonpositive curvature even if $B$ does. (However, it is proved in [86] that if $B$ is aspherical, then so is $h(X, B)$.) In [162] and [59, pp. 135–138] a different relative hyperbolization procedure is introduced. It has the advantage of satisfying the previous property; however, its disadvantage is that property (d) fails. The variant of the reflection group trick described below gives a closely related relative hyperbolization procedure. The difference is that in [59, 162] the 1-skeleton of $X$ is not changed while in the procedure described below, it is. This procedure has the same advantages and disadvantages as the one in [59, 162].

### Another Definition of $D(X, B)$

Here we give a description of $D(X, B)$ similar to the definition of the "cross with interval" hyperbolization constructions in [59, 83]. We shall define a space $D^{(k)}(X, B)$ for any pair $(X, B)$ of simplicial complexes with $\dim(X, B) \leqslant k$. Moreover, these spaces come with commuting involutions $r_1, \dots, r_k$. The definition is by induction on $k$. First of all, $D^{(0)}(X, B)$ is defined to be $B$. Assume, by induction, that $D^{(n-1)}$ and the involutions $r_1, \dots, r_{n-1}$ have been defined. Suppose $\dim(X, B) = n$. Set

$$D^{(n)}(X^{n-1} \cup B, B) := D^{(n-1)}(X^{n-1} \cup B, B) \times \{-1, 1\}.$$

If $\sigma$ is an $n$-simplex in $\mathcal{P}$, define

$$D^{(n)}(\sigma, \sigma \cap B) := D^{(n-1)}(\partial\sigma, \partial\sigma \cap B) \times [-1, 1].$$

It is called a *hyperbolized n-simplex*. Note that the boundary of $D^{(n)}(\sigma, \sigma \cap B)$ is naturally a subcomplex of $D^{(n)}(X^{n-1} \cup B, B)$. Hence, we can glue in each hyperbolized simplex $D^{(n)}(\sigma, \sigma \cap B)$ to obtain $D^{(n)}(X, B) = D(X, B)$. For $1 \leqslant i < n$, $r_i$ is the given involution on $D^{(n-1)}(X^{n-1} \cup B, B)$ and the identity on the $[-1, 1]$ factor. The involution $r_n$ fixes $D^{(n-1)}(X^{n-1} \cup B, B)$ and flips the $[-1, 1]$ factor. We leave it for the reader to check that this definition agrees with the one given by (12.2).

*Proof of Theorem 12.8.1 (iii).* The proof is based on a Gluing Lemma of Reshetnyak [45, p. 316] or [147, p. 124]. This asserts that if we glue together two nonpositively curved spaces via an isometry of a common totally geodesic subspace, then the new metric space is nonpositively curved. The inductive hypothesis implies that the spaces $D^{(n-1)}(X^{n-1} \cup B, B)$ and $D^{(n-1)}(\partial\sigma, \partial\sigma \cap B)$ are nonpositively curved and that the subspaces $B$ and $\partial\sigma \cap B$ are totally geodesic. Using the Gluing Lemma, we get that $D^{(n)}(X, B)$ is also nonpositively curved. □

**COROLLARY 12.8.2.** *Suppose B is a finite, piecewise Euclidean polyhedron. Then there is a closed PL manifold M with a piecewise Euclidean metric such that M retracts onto B.*

*Proof.* Take $M = D(X, B)$ with $X$ any triangulation of a PL manifold containing $B$ as a full subcomplex. □

**Applications to the Borel and Novikov Conjectures**

We continue the line begun in 11.4. Farrell and Jones [121, 122, 123] have proved the Borel Conjecture in dimensions $\geqslant 5$ for the fundamental group of any nonpositively curved, closed Riemannian manifold. (In what follows we shall use the phrase the "Borel Conjecture" as a shorthand to stand for all three of the algebraic conjectures: the Reduced Projective Class group Conjecture, the Whitehead Group Conjecture and the Assembly Map Conjecture, discussed in Appendix H.) It seems plausible that the Farrell-Jones program can be adapted to prove the Borel Conjecture for the fundamental group of any closed *PL* manifold $M^n$ equipped with a nonpositively curved, piecewise Euclidean metric. We assume that the piecewise Euclidean structure on $M^n$ is compatible with its structure of a PL manifold. The reason for this assumption is that in order to apply the result of [120] we want the universal cover of $M^n$ to have a compactification which is homeomorphic to an $n$-disk so that the action of the fundamental group extends to the compactication and so that the key property from [120] (of having translates of compact sets shrink as they go to the boundary) holds. Assuming the polyhedral structure on $M$ is that of a PL manifold, the existence of such a compactification (formed by adding endpoints to geodesic rays) is explained in Theorem I.8.4 (iii) of Appendix I.8. Using Corollary 12.8.2, the proof of Theorem 11.4.1 from the previous chapter gives the next result.

**THEOREM 12.8.3.** *Suppose the Borel Conjecture is true for the fundamental group of any closed PL manifold with a nonpositively curved, piecewise Euclidean metric. Then it is also true for the fundamental group of any finite, nonpositively curved, piecewise Euclidean polyhedron.*

The Farrell-Jones Program for nonpositively curved, piecewise Euclidean manifolds has been partially carried out by B. Hu in [161, 162]. Hu first showed, in [161], that the Farrell-Jones arguments work to prove the vanishing the Whitehead group of the fundamental group of any closed *PL* manifold with a nonpositively curved, polyhedral metric. He then used a construction similar to the one in Theorem 12.8.1 to prove the Whitehead Group Conjecture for fundamental groups of nonpositively curved polyhedra. We state this as the following.

**THEOREM 12.8.4.** (Hu [161].) *For any finite polyhedron with nonpositively curved, piecewise Euclidean metric, the Whitehead group of its fundamental group vanishes.*

Hu [162, p. 146] also observed that the Farrell-Hsiang [120] proof of the Novikov Conjecture works for closed *PL* manifolds with nonpositively curved polyhedral metrics. (This uses the fact that the compactification of its universal cover is homeomorphic to a disk.) Hence, Corollary 12.8.2 yields the following.

**THEOREM 12.8.5.** (Hu [162].) *The Novikov Conjecture holds for the fundamental group $\pi$ of any finite polyhedron with a nonpositively curved piecewise Euclidean metric. Indeed, for such $\pi$, the assembly map, $A : H_n(B\pi;\mathbb{L}) \to L_n(\pi)$, is always an injection.*

## NOTES

**12.3.** Theorem 12.3.4 (i) is due to Tits. A sketch of his argument can be found in [29, Ex. 2d), p. 137] as well as in Corollary D.2.9 of Appendix D.2.

Given a Coxeter system $(W, S)$, Niblo and Reeves [228] show there is a finite dimensional, locally finite, CAT(0) cubical complex $\Lambda$ on which $W$ acts properly and isometrically. (However, the action need not be cocompact.) When $W$ is right-angled, $\Lambda = \Sigma$. The construction of [228] comes from Sageev [251]. The key idea is Sageev's notion of a "half-space system." Starting with $\Sigma$ (or with $\mathrm{Cay}(W, S)$ or with the interior $\mathcal{I}$ of the Tits cone), we have previously defined half-spaces and walls. All give the same half-space system. Let $\mathcal{H}$ be the set of half-spaces, $\mathcal{W}$ the set of walls and $\partial : \mathcal{H} \to \mathcal{W}$ the natural map. $\mathcal{H}$ is a poset (the partial order is inclusion). $\mathcal{H}$ admits an involution $H \to H^*$ which takes $H$ to its opposite half-space $H^*$. This poset with involution satisfies the additional axioms for it to be a "half-space system" in the sense of [251]. To such a system, Sageev associates a CAT(0) cubical complex $\Lambda$, as follows. A *vertex* of $\Lambda$ is a section of $\partial : \mathcal{H} \to \mathcal{W}$. In other words, a vertex specifies a "side" for each wall. Each element of $W$ ($:= \mathrm{Vert}(\mathrm{Cay}(W, S))$) defines a unique vertex in $\Lambda$; however, in the non-right-angled case, there are more vertices. For example, if $W = \mathbf{D}_3$, the dihedral group of order 6, there are 8 vertices. (In general, if $W$ is finite and has $m$ reflections, there are $2^m$ vertices and $2^m \geqslant \mathrm{Card}(W)$ with equality only if $W$ is right-angled.) Two vertices are *adjacent* if they agree except on a single wall. $\Lambda^1$ is constructed by connecting adjacent vertices by edges. One continues by filling in higher dimensional cubes in a fairly obvious fashion. The resulting cubical complex $\Lambda$ has dimension $\max\{m_T \mid T \in \mathcal{S}\}$, where $m_T$ is the number of reflections in $W_T$. ($\dim \Sigma = \max\{\mathrm{Card}(T) \mid T \in \mathcal{S}\}$.) It follows from the results of [50, 228, 302] that the $W$-action on $\Lambda$ is cocompact if and only if no special subgroup of $W$ is an irreducible Euclidean reflection group of rank $\geqslant 3$. Recently Haglund and Wise have proved that the cubical complex for any Coxeter system embeds virtually equivariantly as a subcomplex of $\Sigma(W, S)$ for some right-angled $(W, S)$.

**12.4.** In 1985 R. Ancel and L. Siebenmann [6] announced results concerning the visual boundary of $\Sigma$ in the case where $L$ is a PL homology sphere. In particular, they said that the space $\partial\Sigma$ was topologically homogeneous. Their results (at least in the right-angled case) were proved in the Ph.D. thesis of H. Fischer, published as [128].

**12.5.** A proof of the Švarc-Milnor Lemma can be found in [212], as well as Troyanov's article in [138, p. 60] or in [37, p. 140].

**12.6.** For further results on the two-dimensional examples, see [14, 18, 152, 275].

In Proposition 12.6.7 and in Vinberg's Theorem 6.11.8, it is not known whether the upper bound of 29 is the best possible. (In fact, everyone, who I have ever heard express an opinion, believes that it is not the best possible.)

**12.7.** The main result in Krammer's thesis [181] is a solution to the Conjugacy Problem for $W$. Although this follows from the fact that $W$ is a CAT(0) group (cf. Theorem 12.3.4 (iii)), Krammer gives an explicit algorithm which is faster than one would deduce from general CAT(0) properties. His arguments make extensive use of root systems and the Tits cone, described in Appendix D.

**12.8.** This section on relative hyperbolization is taken from [78, Section 17]. The construction of $D(X, B)$ was explained to me in the early 1990's by Lowell Jones. It is a variation of the "cross with interval" hyperbolization procedure which had been described previously by Januszkiewicz and me in [83, p. 377]. Relative versions of this were described by Hu [162, pp. 135–138] and by Charney and me in [59, pp. 333–335]. In these earlier versions of relative hyperbolization the 1-skeleton of $X$ was not changed. Jones realized that the construction is nicer if, as in this section, we also hyperbolize the 1-simplices.

# Chapter Thirteen

## RIGIDITY

The goal of this chapter is to prove Theorem 13.4.1 which asserts that any Coxeter group of type PM is strongly rigid. Roughly,"type PM" means the nerve of $(W, S)$ is an orientable pseudomanifold. "Strongly rigid" means any two fundamental sets of generators for $W$ are conjugate.

We begin with some basic definitions concerning rigidity in 13.1. Then we show how the "graph twisting" technique of [35] can be used to construct examples of nonrigid Coxeter groups. In 13.2 we explain the one-to-one correspondence between the set of spherical parabolic subgroups in $W$ and the set of their fixed subspaces in $\Sigma$. In 13.3 we define what it means for a Coxeter system to be "type PM$^n$." It is a generalization of the notion of type HM$^n$ from Definition 10.6.2. We prove, in Lemma 13.3.6, that if $(W, S)$ is type PM$^n$, then $H^n(W; \mathbb{Z}W) \cong \mathbb{Z}$, where $n = \text{vcd}(W)$. Similarly, for any spherical special subgroup $W_T$ and $k := \text{Card}(T)$, we have $\text{vcd}(N(W_T)) = n - k$ and that $H^{n-k}(N(W_T); \mathbb{Z}(W_T))$ has $\mathbb{Z}$ as a direct summand (Proposition 13.3.3). Using this, we show that the property of being type PM depends only on $W$ and not on the choice of a fundamental set of generators (Theorem 13.3.10). In 13.4 we use algebraic topology to prove rigidity for groups of type PM.

### 13.1. DEFINITIONS, EXAMPLES, COUNTEREXAMPLES

Recall Definition 3.3.2: a *fundamental set of generators* for a Coxeter group $W$ is a set $S$ of generators such that $(W, S)$ is a Coxeter system. Call $W$ *rigid* if any two fundamental sets of generators $S$ and $S'$ give isomorphic Coxeter diagrams. $W$ is *strongly rigid* if any two fundamental sets of generators for it are conjugate, i.e., if there is a $w \in W$ such that $S' = wSw^{-1}$.

**Example 13.1.1.** (*The dihedral group of order* 12, cf. [29, Ex. 8, p. 33].) The standard Coxeter diagram of $\mathbf{D}_6$, the dihedral group of order 12, is $\overset{6}{\circ\!\!-\!\!\!-\!\!\circ}$. $\mathbf{D}_6$ is isomorphic to $\mathbf{D}_3 \times \mathbf{C}_2$ which has diagram $\circ\!\!-\!\!\!-\!\!\circ \quad \circ$. So, $\mathbf{D}_6$ is not rigid.

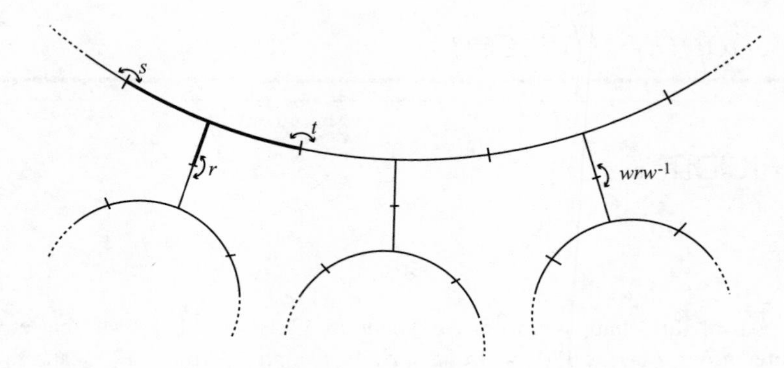


Figure 13.1. A trivalent tree.

**Example 13.1.2.** (*The free product of three cyclic groups of order* 2.) Suppose $W := \mathbf{C}_2 * \mathbf{C}_2 * \mathbf{C}_2$. Let $r$, $s$, $t$, be the generators of the three factors and $S := \{r, s, t\}$. $(W, S)$ is a Coxeter system. Let $w$ be any element in the special subgroup $W_{\{s,t\}}$ (an infinite dihedral group). Then $S' := \{wrw^{-1}, s, t\}$ is also a fundamental set of generators and $S'$ is not conjugate to $S$ unless $w = 1$. Thus, $W$ is not strongly rigid. We leave it as an exercise for the reader to show that $r$, $s$, and $t$ represent the three conjugacy classes of involutions in $W$ and that the only way to get a fundamental set of generators is to choose one from each conjugacy class. Hence, $W$ is rigid.

This example can be understood geometrically as follows. $\Sigma$ is the infinite trivalent tree shown in Figure 13.1; the fundamental chamber $K$ is the cone on three points indicated by the heavier lines. The infinite dihedral group $W_{\{s,t\}}$ stabilizes a subcomplex homeomorphic to a line (the axis of translation by $st$). A nontrivial element $w \in W_{\{s,t\}}$ takes the fixed point of $r$ to a different point, namely the fixed point of $wrw^{-1}$. $S' = \{wrw^{-1}, s, t\}$ is not conjugate to $S$, since the 3 points fixed by the elements of $S'$ are not mirrors of a common chamber of $\Sigma$.

## Automorphisms of Strongly Rigid Coxeter Groups

If $wSw^{-1} = S$ and the Coxeter diagram of $(W, S)$ has no spherical components, then it is proved by Qi [242] that $w = 1$. (The argument is similar to the proof of Theorem D.2.10 in Appendix D.2.) Recall Definition 9.1.6: a *diagram automorphism* of $(W, S)$ is an automorphism $\varphi$ of $W$ such that $\varphi(S) = S$. Since $S$ generates $W$, any automorphism is determined by its restriction to $S$. If $\varphi$ is a diagram automorphism, then $\varphi|_S$ induces an automorphism of the Coxeter diagram $\Gamma(W, S)$ (as a labeled graph). Hence, the diagram automorphisms of $(W, S)$ can be identified with the automorphisms of its Coxeter diagram.

**PROPOSITION 13.1.3.** *Suppose $W$ is strongly rigid. Then every automorphism of $W$ is the product of an inner automorphism and a diagram automorphism. Moreover, if every component of its Coxeter diagram is nonspherical, then the center of $W$ is trivial and there is a semidirect product decomposition:*

$$\mathrm{Aut}(W) = \mathrm{Inn}(W) \rtimes \mathrm{Diag}(W, S),$$

*where* $\mathrm{Aut}(W)$, $\mathrm{Inn}(W)$, *and* $\mathrm{Diag}(W, S)$ *denote, respectively, the groups of automorphisms, inner automorphisms, and diagram automorphisms of $W$.*

*Proof.* If $\alpha \in \mathrm{Aut}(W)$, then $\alpha(S)$ is another set of fundamental generators. By strong rigidity, $\alpha(S) = wSw^{-1}$. It follows that $\alpha$ can be factored as $\alpha = \gamma \circ \beta$, where $\gamma \in \mathrm{Diag}(W, S)$ and $\beta \in \mathrm{Inn}(W)$ is conjugation by $w$. If the diagram of $(W, S)$ has no irreducible spherical components, then $w = 1$. Consequently, the center is trivial, the factorization $\alpha = \gamma \circ \beta$ is unique and $\mathrm{Aut}(W)$ is the semidirect product. □

**Example 13.1.4.** (*The product of cyclic groups of order* 2.) $W = (\mathbf{C}_2)^n$ is a Coxeter group. Identify $\mathbf{C}_2$ with $\mathbb{F}_2$, the field with two elements. A fundamental set of generators for $W$ is the same thing as a basis for $(\mathbb{F}_2)^n$. So, $W$ is rigid but not strongly rigid (since $W$ is abelian, distinct bases are never conjugate). The group of diagram automorphisms of $(W, S)$ is the symmetric group $S_n$ on $n$ letters. For $n \geqslant 2$, the full automorphism group of $W$ is larger: $\mathrm{Aut}(W) = GL(n, \mathbb{F}_2)$.

## Reflections and Reflection Rigidity

Let $R_S$ be the set of reflections with respect to $S$, i.e., $R_S$ is the set of $r \in W$ such that $r$ is conjugate to some $s \in S$. $W$ has a *rigid reflection set* if $R_S = R_{S'}$ for any two fundamental sets of generators $S$ and $S'$. (This is called "reflection independence" in [13]).

**DEFINITION 13.1.5.** ([35].) $(W, S)$ is reflection rigid if any fundamental set of generators $S'$, with $S' \subset R_S$, has the same diagram as $S$. It is *strongly reflection rigid* if any such $S'$ is conjugate to $S$.

These notions separate the rigidity question into two: first, does $W$ have a rigid reflection set and, second, is a given Coxeter system reflection rigid?

**Example 13.1.6.** (*Reflection rigidity of dihedral groups.*) As in Example 3.1.2, let $\mathbf{D}_m$ be the dihedral group of order $2m$ generated by reflections $s$ and $t$ across lines $L_s$ and $L_t$ in $\mathbb{R}^2$ making an angle of $\pi/m$. If $m$ is odd, then $\mathbf{D}_m$ is reflection rigid (since every element of order two is a reflection). On the other hand, if $m = 2d$ is even, then, as in Example 13.1.1, $\mathbf{D}_{2d} \cong \mathbf{D}_d \times \mathbf{C}_2$; $\mathbf{D}_{2d}$ has $2d$ reflections while $\mathbf{D}_d \times \mathbf{C}_2$ has $d+1$; hence, if $d > 1$, $D_{2d}$ is *not* reflection rigid. Suppose $k$ is an integer prime to $m$ such that $k \not\equiv \pm 1 \pmod{m}$. Let $L$ be the line making an angle of $k\pi/m$ with $L_s$ and let $r_L$ be the corresponding reflection. Then $r_L \in \mathbf{D}_m$ and $S' = \{s, r_L\}$ is another fundamental set of generators not conjugate to $S = \{s, t\}$. Hence, $\mathbf{D}_m$ is not strongly reflection rigid whenever there is such an integer $k$.

*Remark.* It is observed in [35, Lemma 3.7] that if $S' \subset R_S$, then $R_{S'} = R_S$ and that it follows from a result of Dyer [107] that $\text{Card}(S') = \text{Card}(S)$ (see [35, Theorem 3.8]).

### Graph Twisting

For the remainder of this chapter it will be more convenient to use the labeled simplicial graph $\Upsilon$ associated to $(W, S)$ (cf. Example 7.1.6) instead of its Coxeter diagram. $\Upsilon$ is the 1-skeleton of the nerve of $(W, S)$. The edge $\{s, t\}$ is labeled by the integer $m_{st}$. The labeled graph $\Upsilon$ records contains exactly the same information as the Coxeter diagram $\Gamma$, the difference being that the edges in $\Upsilon$ with $m_{st} = 2$ are omitted from $\Gamma$, while those in $\Gamma$ with $m_{st} = \infty$ are omitted from $\Upsilon$.

Now suppose $\Upsilon$ is a labeled simplicial graph with vertex set $S$ and $(W, S)$ is the associated Coxeter system. Let $T$ and $U$ be disjoint subsets of $S$ such that

(a) $W_T$ is finite and

(b) each vertex of $S - (T \cup U)$ which is connected by an edge to a vertex in $U$ is also connected to all vertices in $T$ by edges labeled 2.

As in 4.6, $w_T$ denotes the element of longest length in $W_T$. Let $\iota_T : W \to W$ be conjugation by $w_T$. (Since $\iota_T(T) = T$, $\iota_T$ induces an automorphism of $\Gamma_T$.) Define a new labeled graph $\Upsilon'$ by changing each edge of $\Upsilon$ connecting a vertex $u \in U$ to $t \in T$ to an edge (with the same label) from $u$ to $\iota_T(t)$. Define a new set $S'$ of reflections in $W$ by replacing each element $u \in U$ by $\iota_T(u)$ and leaving the rest of $S$ unchanged, i.e.,

$$S' := \iota_T(U) \cup (S - U).$$

Clearly, $S'$ is also a set of generators for $W$. Following [35], call the operations $\Upsilon \to \Upsilon'$ and $S \to S'$ *twisting of $U$ around $T$.*

**PROPOSITION 13.1.7.** ([35, Theorem 4.5].) *If, as above, $S'$ and $\Upsilon'$ are obtained by twisting $U$ around $T$, then $S'$ is also a fundamental set of generators for $W$ and its associated labeled graph is $\Upsilon'$*

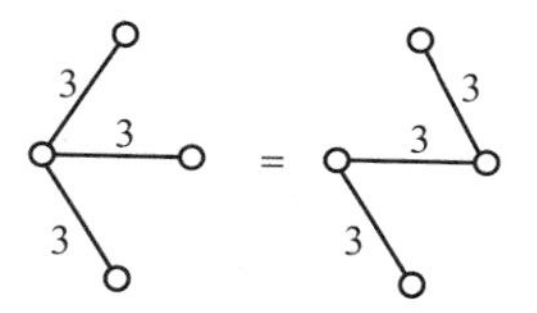


Figure 13.2. Twisted graphs for isomorphic Coxeter groups.

*Proof.* Let $f : S \to S'$ be the bijection defined by $f|_{T\cup U} := \iota_T|_{T\cup U} : T \cup U \to T \cup \iota_T(U)$ and $f|_{S-(T\cup U)} := \text{id}$. We claim

$$[f(s)f(t)]^{m_{st}} = 1 \tag{13.1}$$

for all $s, t \in S$ with $m_{st} \neq \infty$. We only need to consider the case where either $s$ or $t$ belongs to $T \cup U$. If they both do, we have $[f(s)f(t)]^m = \iota_T([st]^m) = 1$, where $m := m_{st}$. If $s \in U$ and $t \in S - (T \cup U)$ and $m := m_{st} \neq \infty$, then $\iota_T(t) = t$ by condition (b) above. So, $[f(s)f(t)]^m = [\iota_T(s)t]^m = [\iota_T(st)]^m = 1$. This proves (13.1). Hence, $f$ extends to a homomorphism $\tilde{f} : W \to W$. Since $S'$ is a set of generators, $\tilde{f}$ is onto. We remark that $W$ is Hopfian, i.e., any self-epimorphism is an automorphism. (This follows from the fact that $W$ has a faithful linear representation and hence, is residually finite, see Definition 14.1.9.) So, $\tilde{f}$ is an automorphism. Therefore, $S'$ is also a fundamental set of generators and the order of $[f(s)f(t)]$ is $m_{st}$. □

In Figure 13.2 we give two examples of how graph twisting can be used to produce nonisomorphic labeled graphs having isomorphic Coxeter groups. The first is from Mühlherr [222]. This was the first example of a infinite irreducible Coxeter group which was not rigid. The second figure is basically a random example of graph twisting taken from [35].

*Remark 13.1.8.* If the element $w_T \in W_T$ of longest length is central, then $\iota_T$ is the identity on $T$. Hence, twisting around $T$ leaves the graph $\Upsilon$ unchanged. Suppose $W_T$ is irreducible. Then $w_T$ is not in the center of $W_T$ only when its Coxeter diagram is either $\mathbf{A}_n$, $n \geqslant 2$, $\mathbf{D}_n$ with $n$ odd, $\mathbf{E}_6$ or $\mathbf{I}_2(m)$, $m$ odd. (Here, as in Table 6.1 of Section 6.9, $\mathbf{I}_2(m)$ stands for the dihedral group of order $2m$ and $\mathbf{D}_n$ for a subgroup of order 2 in the $n$-octahedral group.) Hence, we can only get a nontrivial twisting of $\Upsilon$ when an irreducible component of $W_T$ is one of these cases.

**Rigidity of 2-Spherical Groups**

Following [51], call a Coxeter system *2-spherical* if $m_{st}$ is never equal to $\infty$. In the next few sentences, we summarize, without proofs, results from [12, 35, 51, 130] on rigidity of 2-spherical Coxeter systems. In his thesis, Kaul [172] proved that if one fundamental set $S$ of generators is 2-spherical, then so is any other fundamental set $S'$; moreover, $\mathrm{Card}(S') = \mathrm{Card}(S)$. (He calls the 2-spherical condition "type $K_n$," $n = \mathrm{Card}(S)$, since if it holds, then $\Upsilon$ is the complete graph on $n$ vertices.) Kaul also proved that the set of edge labels, with multiplicity, is the same for $S$ and $S'$. So, for example, if all the edge labels are equal, then $W$ is rigid. It is observed in [35, Theorem 3.10] that irreducible spherical Coxeter groups are reflection rigid. As we saw in Example 13.1.6, most finite dihedral groups are not strongly reflection rigid. Franszen and Howlett [130, p. 330] proved that an irreducible spherical Coxeter group of rank $> 2$ is strongly reflection rigid if and only if no component of its diagram is type $\mathbf{H}_3$ or $\mathbf{H}_4$. (They go on to compute the entire automorphism group of any irreducible spherical Coxeter group.) In the nonspherical case we have the following.

**PROPOSITION 13.1.9.** (Caprace and Mühlherr [51].) *If $W$ is 2-spherical and no component of its Coxeter diagram is spherical, then it is strongly reflection rigid.*

## 13.2. SPHERICAL PARABOLIC SUBGROUPS AND THEIR FIXED SUBSPACES

Recall from 4.5 that a subgroup $G$ of $W$ is *parabolic* if it is conjugate to some special subgroup $W_T$. Of course, exactly which subgroups are parabolic might depend on the fundamental set of generators $S$. The *rank* of $G$ is defined by $\mathrm{rk}_S(G) := \mathrm{Card}(T)$. (This is well defined since, by Proposition 4.5.10, the special subgroups $W_T$ and $W_U$ are conjugate if and only if $T$ and $U$ are conjugate.) The parabolic subgroup $G$ is *spherical* if $W_T$ is spherical. The spherical parabolic subgroups are precisely the isotropy subgroups of $W$ on $\Sigma$. Let $\mathrm{Par}^{(k)}(W, S)$ denote the set of spherical parabolic subgroups of rank $k$ and let $\mathrm{Par}(W, S) := \bigcup \mathrm{Par}^{(k)}(W, S)$ be the set of all spherical parabolic subgroups.

A *subspace* of $\Sigma$ means the fixed point set of some spherical parabolic subgroup. (Note that if $G := wW_Tw^{-1}$ is such a subgroup, then $\mathrm{Fix}(G, \Sigma) = w\,\mathrm{Fix}(W_T, \Sigma)$.) By Lemma 7.5.4, any point of $\mathrm{Fix}(G, \Sigma)$ has a neighborhood of the form $F \times V$, where $F$ is open in $\mathrm{Fix}(G, \Sigma)$ and where $V$ is an open ball about the origin in the vector space $\mathbb{R}^T$ equipped with the action of $wW_Tw^{-1}$ as a linear reflection group. The *codimension* of the subspace fixed by $G$ is the rank of $G$ ($= \dim V$). A subspace of codimension one is the same thing as a wall of $\Sigma$ fixed by some reflection. If $E := \mathrm{Fix}(G, \Sigma)$ is a subspace, then, for any $x \in E$, $G$ is a subgroup of the isotropy group, $W_x$. Call $x$ a *generic point* of $E$ if $W_x = G$. The proof of the next lemma is immediate.

**LEMMA 13.2.1.** *Suppose $E$ is a subspace of $\Sigma$. A point $x \in E$ is generic if and only if for any wall $\Sigma^r$, with $r \in R_S$ and $E \not\subset \Sigma^r$, we have $x \notin \Sigma^r$. (Recall $\Sigma^r := \mathrm{Fix}(r, \Sigma)$.)*

Since each wall has empty interior in $\Sigma$ and since the walls form a locally finite family of closed subcomplexes of $\Sigma$, it follows that the set of generic points is open and dense in $E$. It also follows from this lemma that the notion of a generic point in $E$ can be defined without mentioning the parabolic subgroup. Hence, $G$ is determined by the subspace $E$: it is the isotropy subgroup at a generic point. So, we have proved the following.

**LEMMA 13.2.2.** *There is a canonical one-to-one correspondence between the set of subspaces of $\Sigma$ and $\mathrm{Par}(W, S)$. (The correspondence associates $\mathrm{Fix}(G, \Sigma)$ to the group $G \in \mathrm{Par}(W, S)$.)*

**LEMMA 13.2.3.** *Suppose $H$ is a finite subgroup of $W$. Then $\mathrm{Fix}(H, \Sigma)$ is a subspace of $\Sigma$ (i.e., $\mathrm{Fix}(H, \Sigma) = \mathrm{Fix}(G, \Sigma)$ for some $G \in \mathrm{Par}(W, S)$).*

*Proof.* As explained in Theorem 12.3.4 (ii), by the Bruhat-Tits Fixed Point Theorem (Theorem I.2.11), $\mathrm{Fix}(H, \Sigma)$ is nonempty. For any $x \in \mathrm{Fix}(H, \Sigma)$, $H$ is contained in the parabolic subgroup $W_x$. Let $G$ be a minimal spherical parabolic containing $H$. Then $\mathrm{Fix}(G, \Sigma) \subset \mathrm{Fix}(H, \Sigma)$. If $x \in \mathrm{Fix}(H, \Sigma) - \mathrm{Fix}(G, \Sigma)$, then $H \subset W_x \subsetneq G$, contradicting the minimality of $G$. So, $\mathrm{Fix}(H, \Sigma) = \mathrm{Fix}(G, \Sigma)$. □

Given a finite subgroup $H$ of $W$, put

$$R_S(H) := \{r \in R_S \mid \mathrm{Fix}(H, \Sigma) \subset \mathrm{Fix}(r, \Sigma)\}$$

and given a subspace $E$ of $\Sigma$, put

$$R_S(E) := \{r \in R_S \mid E \subset \mathrm{Fix}(r, \Sigma)\}.$$

Observe that the proper subspaces of $\Sigma$ are precisely the nonempty intersections of walls. Indeed, any subspace $E$ can be written as

$$E = \bigcap_{r \in R_S(E)} \mathrm{Fix}(r, \Sigma).$$

Conversely, suppose $F := \mathrm{Fix}(r_1, \Sigma) \cap \cdots \cap \mathrm{Fix}(r_k, \Sigma)$ is a nonempty intersection of walls. Then $F$ is the fixed set of the finite subgroup $\langle r_1, \ldots, r_k \rangle$ and by Lemma 13.2.3, any such fixed set is a subspace.

Given a finite subgroup $H \subset W$, define $\widehat{\mathrm{rk}}_S(H)$ to be the least integer $k$ such that there are reflections $r_1, \dots, r_k \in R_S(H)$ with $\mathrm{Fix}(H, \Sigma) = \mathrm{Fix}(r_1, \Sigma) \cap \cdots \cap \mathrm{Fix}(r_k, \Sigma)$.

**LEMMA 13.2.4.** *If $G \in \mathrm{Par}(W, S)$, then $\widehat{\mathrm{rk}}_S(G) = \mathrm{rk}_S(G)$.*

*Proof.* Without loss of generality, we can assume $G = W_T$ for some $T \in \mathcal{S}$. Set $k := \mathrm{Card}(T) = \mathrm{rk}_S(G)$. By Lemma 7.5.4, any point $x$ in $\mathrm{Fix}(W_T, \Sigma)$ has a neighborhood of the form $F \times V$ where $V$ is a neighborhood of the origin in $\mathbb{R}^k$. If $x$ is generic, then any wall containing the fixed set intersects the given neighborhood in a set of the form $F \times (\mathbb{R}^{k-1} \cap V)$, where $\mathbb{R}^{k-1}$ is some linear hyperplane in $\mathbb{R}^k$. Since the intersection of fewer than $k$ hyperplanes in $\mathbb{R}^k$ is a nonzero linear subspace, $\widehat{\mathrm{rk}}_S(W_T) \geqslant k$. Trivially, $\widehat{\mathrm{rk}}_S(W_T) \leqslant k$. So, $\widehat{\mathrm{rk}}_S(W_T) = k$. □

**LEMMA 13.2.5.** *Suppose $H$ and $G$ are finite subgroups of $W$ with $H \subset G$ and $H \in \mathrm{Par}(W, S)$. If $\widehat{\mathrm{rk}}_S(H) = \widehat{\mathrm{rk}}_S(G)$, then $H = G$.*

*Proof.* Since $H \subset G$, $\mathrm{Fix}(G, \Sigma) \subset \mathrm{Fix}(H, \Sigma)$. By Lemma 13.2.4, $\mathrm{Fix}(H, \Sigma)$ has codimension $k$, where $k := \widehat{\mathrm{rk}}_S(H)$. Since $k = \widehat{\mathrm{rk}}_S(G)$, the codimension of $\mathrm{Fix}(G, \Sigma)$ is $\leqslant k$. Hence, $\mathrm{Fix}(G, \Sigma) = \mathrm{Fix}(H, \Sigma)$. Since the parabolic subgroup $H$ is the maximal subgroup which fixes $\mathrm{Fix}(H, \Sigma)$, the lemma follows. □

**LEMMA 13.2.6.** *Suppose $S$ and $S'$ are two sets of fundamental generators for $W$. If $R_S = R_{S'}$, then $\mathrm{Par}^{(k)}(W, S) = \mathrm{Par}^{(k)}(W, S')$ for all $k \geqslant 1$.*

*Proof.* Set $\Sigma := \Sigma(W, S)$ and $\Sigma' := \Sigma(W, S')$. Suppose $H \in \mathrm{Par}^{(k)}(W, S)$. Then we can find $k$ reflections $r_1, \dots, r_k$ which generate $H$. Clearly, $r_i \in R_{S'}(H)$ and $\mathrm{Fix}(H, \Sigma') = \mathrm{Fix}(r_1, \Sigma') \cap \cdots \cap \mathrm{Fix}(r_k, \Sigma')$. Hence, $\widehat{\mathrm{rk}}_{S'}(H) \leqslant k = \mathrm{rk}_S(H)$. Let $G$ be the minimal $S'$-parabolic containing $H$ (so that $\mathrm{Fix}(G, \Sigma') = \mathrm{Fix}(H, \Sigma')$). By the same reasoning as above, $\widehat{\mathrm{rk}}_S(G) \leqslant \mathrm{rk}_{S'}(G)$. Therefore, $\mathrm{rk}_S(H) \geqslant \widehat{\mathrm{rk}}_{S'}(H) = \mathrm{rk}_{S'}(G) \geqslant \widehat{\mathrm{rk}}_S(G)$. Since $H \subset G$, $\mathrm{Fix}(G, \Sigma) \subset \mathrm{Fix}(H, \Sigma)$ and $\widehat{\mathrm{rk}}_S(G) \geqslant \mathrm{rk}_S(H)$; so, all of the previous inequalities are equalities. In particular, $\mathrm{rk}_S(H) = \widehat{\mathrm{rk}}_S(G)$. By Lemma 13.2.5, $G = H$, i.e., $H \in \mathrm{Par}^{(k)}(W, S')$. So, $\mathrm{Par}^{(k)}(W, S) \subset \mathrm{Par}^{(k)}(W, S')$. Since the argument is symmetric in $S$ and $S'$, $\mathrm{Par}^{(k)}(W, S') \subset \mathrm{Par}^{(k)}(W, S)$ and the lemma is proved. □

The proof of the next lemma is immediate and is omitted.

**LEMMA 13.2.7.** *Suppose $G_1, G_2$ are spherical parabolics and $G$ is the subgroup generated by $G_1$ and $G_2$. Then $\mathrm{Fix}(G_1, \Sigma) \cap \mathrm{Fix}(G_2, \Sigma) = \mathrm{Fix}(G, \Sigma)$. Hence, $\mathrm{Fix}(G_1, \Sigma) \cap \mathrm{Fix}(G_2, \Sigma)$ is empty if and only if $G$ is infinite.*

## 13.3. COXETER GROUPS OF TYPE PM

A locally finite convex cell complex $\Lambda$ is an *n-dimensional pseudomanifold* if (a) each maximal cell of $\Lambda$ is $n$ dimensional and (b) each $(n-1)$-cell is a face of precisely two $n$-cells. It follows that if $\sigma$ is a $k$-cell in an $n$-dimensional pseudomanifold $\Lambda$, then $\mathrm{Lk}(\sigma, \Lambda)$ is a pseudomanifold of dimension $n-k-1$. (The term "convex cell complex" is used as in Definition A.1.9 of Appendix A.1. The "link" of a cell is defined in Appendix A.5.)

A pseudomanifold $\Lambda$ is *orientable* if one can choose orientations for the top-dimensional cells so that their sum is a cycle (possibly an infinite cycle). If $\Lambda$ is orientable, then so is each link. Two top-dimensional cells of $\Lambda$ are *adjacent* if their intersection is a common face of codimension one. A *gallery* in $\Lambda$ is a sequence $(\sigma_1, \dots, \sigma_k)$ of top-dimensional cells such that any two successive ones are adjacent. $\Lambda$ is *gallery connected* if any two chambers can be connected by a gallery. (This is equivalent to the condition that the complement of the codimension two skeleton of $\Lambda$ be connected.)

**DEFINITION 13.3.1.** A Coxeter system $(W, S)$ is *type* $\mathrm{PM}^n$, $n > 1$, if its nerve $L$ is an orientable, gallery connected pseudomanifold of dimension $(n-1)$. (These conditions imply that $H_{n-1}(L) \cong \mathbb{Z}$.) When $n = 1$ we shall say $(W, S)$ is type $\mathrm{PM}^1$, if $L = S^0$ (so that $W$ is the infinite dihedral group).

**Example 13.3.2.** If $L$ is a triangulation of any orientable closed $(n-1)$-manifold, then $(W, S)$ is type $\mathrm{PM}^n$. In this case, the cellulation of $\Sigma$ by Coxeter polytopes, is an $n$-manifold except at vertices.

*We suppose for the remainder of this section that* $(W, S)$ *is type* $\mathrm{PM}^n$. Our first goal is to prove this condition is independent of the choice of the fundamental set of generators (Theorem 13.3.10). The proof makes use of our computations in 8.5 and 8.9 of the compactly supported cohomology of $\mathrm{Fix}(W_T, \Sigma)$ for any spherical subset $T$. By Corollary 7.5.3 (i), for each $T \in \mathcal{S}^{(k)}$ (the collection of spherical subsets with $k$ elements), $\mathrm{Fix}(W_T, \Sigma)$ is a pseudomanifold of dimension $n-k$.

**PROPOSITION 13.3.3.** *For each* $T \in \mathcal{S}^{(k)}$,

*(i)* $H_c^{n-k}(\mathrm{Fix}(W_T, \Sigma))$ *has* $\mathbb{Z}$ *as a direct summand.*

*(ii)* $\mathrm{Fix}(W_T, \Sigma)$ *is a contractible, orientable pseudomanifold of dimension* $n-k$.

*Proof.* $K_T$ is the cone on $\partial K_T$ ($\cong \mathrm{Lk}(\sigma_T, L)$), where $\sigma_T$ is the simplex of $L$ with vertex set $T$. Since $L$ is a orientable, so is $\mathrm{Lk}(\sigma_T, L)$. Hence, $H^{n-k}(K_T, \partial K_T)$

has $\mathbb{Z}$ as a direct summand (and is isomorphic to $\mathbb{Z}$ if $\mathrm{Lk}(\sigma_T, L)$ is gallery connected). The formula of Corollary 8.9.2 is

$$H_c^{n-k}(\mathrm{Fix}(W_T, \Sigma) \cong \bigoplus_{w \in \mathcal{F}(T)} H^{n-k}(K_{T_w}, \delta_w K_{T_w}).$$

Only $w = 1$ makes a nonzero contribution to the right hand side and since $\delta_1 K_T = \partial K_T$, its contribution is $H^{n-k}(K_T, \partial K_T)$. This proves (i).

We have already observed that $\mathrm{Fix}(W_T, \Sigma)$ is an $(n-k)$-dimensional pseudomanifold. It is contractible by Theorem 12.3.4 (ii). The argument in the previous paragraph shows it is orientable. □

**COROLLARY 13.3.4.** *If* $G \in \mathrm{Par}^{(n-1)}(W, S)$, *then* $\mathrm{Fix}(G, \Sigma)$ *is homeomorphic to a line.*

*Proof.* A one-dimensional pseudomanifold is a manifold; a contractible 1-manifold is a line. □

$N(W_T)$ denotes the normalizer of $W_T$ in $W$.

**COROLLARY 13.3.5.** *For any* $T \in \mathcal{S}^{(k)}$,

*(i)* $H^{n-k}(N(W_T); \mathbb{Z}N(W_T))$ *contains* $\mathbb{Z}$ *as a direct summand.*

*(ii)* $\mathrm{vcd}(N(W_T)) = n - k$.

In the special case $T = \emptyset$, we have the following slightly stronger statement.

**LEMMA 13.3.6.** $H_c^n(\Sigma) = H^n(W; \mathbb{Z}W) \cong \mathbb{Z}$.

*Proof.* The formula of Theorem 8.5.1 is $H_c^n(\Sigma) = \bigoplus H^n(K, K^{\mathrm{Out}(w)})$. The only nonzero term in this formula occurs when $w = 1$. The point is that in the definition of "type PM$^n$" we have the additional assumption that $\overline{H}^{n-1}(L) \cong \mathbb{Z}$. So, the nonzero term is $H^n(K, \partial K) = \overline{H}^{n-1}(L) \cong \mathbb{Z}$. □

**LEMMA 13.3.7.** *W does not have a nontrivial finite subgroup as a direct factor.*

*Proof.* Suppose $W = F \times W'$, with $F$ finite. Then $\mathrm{vcd}(N(F)) = \mathrm{vcd}(W) = n$. Since $\mathrm{Fix}(F, \Sigma)$ coincides with the fixed set of some parabolic subgroup of rank $k$, Corollary 13.3.5 (ii) implies $k = 0$. Therefore, $\mathrm{Fix}(F, \Sigma) = \Sigma$. So, $F$ is trivial. □

**LEMMA 13.3.8.** *Let* $S'$ *be another fundamental set of generators for* $W$. *Then for any* $s' \in S'$, $\mathrm{vcd}(N(s')) = n - 1$.

*Proof.* Set $\Sigma' := \Sigma(W, S')$, $K' := K(W, S')$, and $\delta_w K' = (K')^{\mathrm{Out}(w)}$. By Lemma 13.3.6, $H_c^n(\Sigma') = H_c^n(W; \mathbb{Z}W) \cong \mathbb{Z}$. So, applying Theorem 8.5.1 to $\Sigma'$, we get

$$\mathbb{Z} \cong \bigoplus_{w \in W} H^n(K', \delta_w K') \cong \bigoplus_{w \in W} \overline{H}^{n-1}(\delta_w K').$$

In Lemma 4.7.5 we proved that for any given spherical subset $T$ of $S'$, there is more than one $w \in W$ with $\mathrm{Out}(w) = S' - T$ except when $W_T$ is a direct factor of $W$ and $w = w_T$ (the element of longest length in $W_T$). Since $W$ does not have a nontrivial finite factor, the only possibility for a nonzero term is $w = 1$ and we have

$$\overline{H}^{n-1}(\delta_w K') = \begin{cases} \mathbb{Z} & \text{if } w = 1, \\ 0 & \text{if } w \neq 1. \end{cases} \tag{13.2}$$

Since $H^i_c(\Sigma') = 0$ for all $i > n$, $H^i(\delta_w K') = 0$ for all $i \geqslant n$ and all $w \in W$. Consider the special case $w = s'$. $\delta_{s'} K'$ is the closure of $\partial K' - K'_{s'}$. By excision, $H^*(\partial K', \delta_{s'} K') \cong H^*(K'_{s'}, \partial K'_{s'})$. Combining this with the exact sequence of the pair $(\partial K', \delta_{s'} K')$, we have the following commutative diagram:

$$\begin{array}{ccccc} H^{n-1}(\partial K', \delta_{s'} K') & \longrightarrow & \overline{H}^{n-1}(\partial K') & \longrightarrow & \overline{H}^{n-1}(\delta_{s'} K') \\ \cong \big\downarrow & & \big\downarrow \cong & & \cong \big\downarrow \\ H^{n-1}(K'_{s'}, \partial K'_{s'}) & \longrightarrow & \mathbb{Z} & \longrightarrow & 0 \end{array}$$

where the calculations $H^{n-1}(\partial K') \cong \mathbb{Z}$ and $H^{n-1}(\delta_{s'} K') = 0$ are by (13.2). Thus, $H^{n-1}(K'_{s'}, \partial K'_{s'})$ has rank at least 1. Similarly, $H^i(K'_{s'}, \partial K'_{s'}) = 0$ for $i \geqslant n$. Corollary 8.9.2 implies that $H^{n-1}_c(\mathrm{Fix}(s', \Sigma'))$ has rank at least 1 and that $H^i_c(\mathrm{Fix}(s', \Sigma')) = 0$ for $i \geqslant n$. Since $\mathrm{Fix}(s', \Sigma')$ is CAT(0), it is a model for $\underline{E}N(s')$; hence, $\mathrm{vcd}(N(s')) = n - 1$. □

**Proposition 13.3.9.** *Let $S'$ be another fundamental set of generators for $W$. Then, for each positive integer $k$,* $\mathrm{Par}^{(k)}(W, S') = \mathrm{Par}^{(k)}(W, S)$.

Set $\mathcal{S} := \mathcal{S}(W, S)$, $\mathcal{S}' := \mathcal{S}(W, S')$, $L := L(W, S)$, $L' := L(W, S')$, $K := K(W, S)$ and $K' := K(W, S')$.

*Proof.* A maximal spherical parabolic subgroup (with respect to either $S$ or $S'$) is just a maximal finite subgroup of $W$ (by Lemma 13.2.3). So the maximal elements of $\mathrm{Par}(W, S)$ and $\mathrm{Par}(W, S')$ coincide. Since $L$ is a pseudomanifold, any simplex is an intersection of maximal simplices, i.e., any $T \in \mathcal{S}$ is the intersection of elements in $\mathcal{S}^{(n)}$. It follows that any spherical $S$-parabolic subgroup is an intersection of maximal spherical parabolic subgroups. By Lemma 5.3.6, this implies that any $S$-parabolic is also an $S'$-parabolic, i.e., $\text{Par}(W,S) \subset \text{Par}(W,S')$. By Lemma 13.2.6, it suffices to show $\text{Par}^{(1)}(W,S) = \text{Par}^{(1)}(W,S')$. An $S$- (resp. $S'$-) parabolic subgroup lies in $\text{Par}^{(1)}(W,S)$ (resp. $\text{Par}^{(1)}(W,S')$) if and only if it is cyclic of order 2. Hence, $\text{Par}^{(1)}(W,S) \subset \text{Par}^{(1)}(W,S')$. By Lemma 13.3.8, $\text{vcd}(N(s')) = n-1$ for any $s' \in S'$. Hence, the fixed set of $s'$ on $\Sigma$ must have dimension $n-1$, i.e., $s' \in R_S$. So, $\text{Par}^{(1)}(W,S') \subset \text{Par}^{(1)}(W,S)$. Hence, they are equal. □

**THEOREM 13.3.10.** *Suppose $S'$ is another fundamental set of generators for $W$. Then $(W,S')$ is also type* $\text{PM}^n$.

*Proof.* We must show that $L'$ satisfies the conditions in Definition 13.3.1. The maximal elements of $\text{Par}(W,S)$ and $\text{Par}(W,S')$ are equal and $\text{Par}^{(n)}(W,S) = \text{Par}^{(n)}(W,S')$. So, $\text{Par}^{(n)}(W,S')$ is the set of maximal elements in $\text{Par}(W,S')$. In other words, every maximal simplex of $L'$ has dimension $n-1$. Let $T \in \mathcal{S}'^{(n-1)}$ (i.e., $\sigma_T$ is an $(n-2)$-simplex in $L'$). By Proposition 13.3.9, $W_T \in \text{Par}^{(n-1)}(W,S)$ and by Corollary 13.3.4, $\text{Fix}(W_T,\Sigma)$ is homeomorphic to $\mathbb{R}$. Thus,

$$H^i_c(\text{Fix}(W_T,\Sigma)) = \begin{cases} \mathbb{Z} & \text{if } i = 1, \\ 0 & \text{if } i \neq 1. \end{cases} \tag{13.3}$$

The same formula holds for $H^i(N(W_T);\mathbb{Z}N(W_T))$ (by Lemma F.2.2). Hence, (13.3) holds for $\text{Fix}(W_T,\Sigma')$ as well. For any $T$ with $\dim \sigma_T = n-2$, $K'_T$ is one dimensional and $\partial K'_T$ is a finite set of points. The formula of Corollary 8.9.2 is

$$H^1_c(\text{Fix}(W_T,\Sigma')) \cong \bigoplus_{w\in\mathcal{F}(T)} H^1(K'_{T_w}, \delta_w K'_{T_w}). \tag{13.4}$$

All the terms on the right-hand side of (13.4) occur an infinite number of times except the one corresponding to $w_T$, the element of longest length in $W_T$. Hence, $H^1(K'_T, \partial K'_T) \cong \mathbb{Z}$, that is, $\partial K'_T = S^0$. In other words, $\text{Lk}(\sigma_T, L') = S^0$. But this just means that $L'$ is an $(n-1)$-dimensional pseudomanifold.

By Lemmas 13.3.6, F.2.2, and Corollary 8.9.2,

$$\mathbb{Z} \cong H^n_c(\Sigma') \cong \bigoplus_{w\in W} H^n(K', \delta_w K').$$

A similar argument shows $H^n(K', \partial K') \cong \mathbb{Z}$; hence, $H_{n-1}(L') \cong \mathbb{Z}$. With $\mathbb{Z}/2$ coefficients, the same arguments give $H_{n-1}(L';\mathbb{Z}/2) \cong \mathbb{Z}/2$.

From these cohomology computations, we show, first, that $L'$ is gallery connected and second, that it is orientable. Call two $(n-1)$-simplices of $L'$, *equivalent* if they can be connected by a gallery in $L'$ and suppose $C_1, \dots, C_m$ are the equivalence classes of $(n-1)$-simplices of $L'$. Consider the chain with coefficients in $\mathbb{Z}/2$,

$$\zeta_i := \sum_{\sigma \in \mathbf{C}_i} \sigma.$$

It is a mod 2 cycle and $\{\zeta_1, \dots, \zeta_m\}$ is linearly independent. Since the group of top-dimensional cycles is equal to the top-dimensional homology group and since $H_{n-1}(L'; \mathbb{Z}/2) \cong H^{n-1}(L'; \mathbb{Z}/2) \cong \mathbb{Z}/2$, we must have $m = 1$, i.e., $L'$ is gallery connected. Secondly, since $H_{n-1}(L'; \mathbb{Z}) \cong \mathbb{Z}$, we can orient the $\sigma$ so that $\zeta := \sum \sigma$ is an integral cycle, i.e., $L'$ is orientable. So, $(W, S')$ is also type $\mathrm{PM}^n$. □

The argument in the last paragraph of the above proof can also be used to prove the following lemma which we shall need in the next section. Suppose a simplicial complex $L$ is a pseudomanifold. For $v \in \mathrm{Vert}(L)$, let $O(v, L)$ denote the open star of $v$ in $L$. Two top-dimensional simplices $\sigma, \sigma'$ are *$v$-connected* if they can be connected by a gallery in $L - O(v, L)$.

**LEMMA 13.3.11.** *For $n - 1 > 0$, suppose a finite simplicial complex $L$ is a gallery connected, orientable, $(n-1)$-dimensional pseudomanifold. (If $n - 1 = 0$, assume $L = S^0$.) If $v \in \mathrm{Vert}(L)$, then any two $(n-1)$-simplices in $L - O(v, L)$ are $v$-connected.*

We remark that the lemma is not true if we only assume that $L$ is connected (rather than gallery connected). To see this, consider the example where $L$ is the suspension of two disjoint circles and $v$ is one of the suspension points. The complement of $O(v, L)$ is then a wedge of two 2-disks. Two 2-simplices in this complement can be connected by a gallery within the complement if and only if they lie in the same 2-disk.

*Proof of Lemma 13.3.11.* As in the proof of Theorem 13.3.10, let $C_1, \dots C_m$ be the $v$-connected classes of $(n-1)$-simplices in $L - O(v, L)$. Identify $\mathrm{Lk}(v, L)$ with a subcomplex of $L$ in the usual fashion. Each $(n-2)$-simplex in $\mathrm{Lk}(v, L)$ is the face of exactly two $(n-1)$-simplices, one of which belongs to the star, $O(v, L)$, and the other to some $C_i$. Extend each $C_i$ to an equivalence class $D_i$ by adjoining the adjacent $(n-1)$-simplices of $O(v, L)$. Choose an orientation for $L$ and use it to orient each $(n-1)$-simplex. Define cycles $\zeta_i$ by

$$\zeta_i := \sum_{\sigma \in D_i} \sigma.$$

Then $\zeta_1, \dots, \zeta_m$ are $(n-1)$-cycles representing linearly independent classes in $\overline{H}_{n-1}(L)$. Since $\overline{H}_{n-1}(L) \cong \mathbb{Z}$, $m = 1$. □

## 13.4. STRONG RIGIDITY FOR GROUPS OF TYPE PM

Our goal in this section is a proof of the following.

**THEOREM 13.4.1.** *A Coxeter group $W$ of type* PM *is strongly rigid. More precisely, if $S$, $S'$ are two fundamental sets of generators for $W$, then there is a unique element $w \in W$ such that $S' = wSw^{-1}$.*

**COROLLARY 13.4.2.** *Suppose $W$ is a Coxeter group of type* PM. *Then*

$$\mathrm{Aut}(W) = \mathrm{Inn}(W) \rtimes \mathrm{Diag}(W, S).$$

As usual, $\mathcal{S} = \mathcal{S}(W, S)$, $\Sigma = \Sigma(W, S)$, $K = K(W, S)$, $L = L(W, S)$ and similarly, $\mathcal{S}' = \mathcal{S}(W, S')$, $\Sigma' = \Sigma(W, S')$, $K' = K(W, S')$, $L' = L(W, S')$. By Proposition 13.3.9, $R_S = R_{S'}$. So, each element of $S'$ acts as a reflection on $\Sigma$. In outline, the proof of Theorem 13.4.1 has two steps.

*Step 1*. For each $s' \in S'$, the corresponding wall separates $\Sigma$ into two half-spaces. We will show that we can choose one of these half-spaces, call it $H^{s'}$, so that $\bigcap_{s' \in S'} H^{s'}$ is a nonempty union of chambers of $\Sigma$. Denote this intersection by $D$.

*Step 2*. We will show that $D$ consists of a single chamber of $\Sigma$.

This second step will complete the proof since $D = wK$ implies $S' = wSw^{-1}$. (The uniqueness assertion of the theorem will be taken care of in Lemma 13.4.10 below.)

In the next two lemmas we begin the proof of Step 1 by showing how to choose the half-spaces $H^{s'}$.

**LEMMA 13.4.3.** *Suppose $s' \in S'$ and $T_1, T_2 \in \mathcal{S}'^{(n)}$ are distinct subsets such that $s' \notin T_1 \cup T_2$. Assume the corresponding $(n-1)$-simplices $\sigma_{T_1}$ and $\sigma_{T_2}$ are adjacent in $L'$. Then $(T_1 \cap T_2) \cup \{s'\}$ generates an infinite subgroup of $W$.*

*Proof.* This lemma is just a restatement of the hypothesis that $L'$ is a pseudomanifold. Given a $(n-2)$-simplex, exactly two $(n-1)$-simplices have it as a common face. Since $\sigma_{T_1 \cap T_2}$ is a codimension one face of both $\sigma_{T_1}$ and $\sigma_{T_2}$, $(T_1 \cap T_2) \cup \{s'\}$ cannot correspond to an $(n-1)$-simplex of $L'$, i.e., $W_{(T_1 \cap T_2) \cup \{s'\}}$ must be infinite. □

*Remark.* There is a dual picture to keep in mind. $W$ being type $\mathrm{PM}^n$ means that each codimension $k$ coface of $K'$ is the cone on a $(n-k)$-dimensional pseudomanifold (i.e., $K'_T \cong \mathrm{Cone}(\mathrm{Lk}(\sigma_T, L'))$. In particular, if $\mathrm{Card}(T) = n$, $K'_T$ is a point and if $\mathrm{Card}(T) = n-1$, $K'_T$ is an interval (the cone on $S^0$). Call these zero- and one-dimensional cofaces of $K'$ "vertices" and "edges," respectively. The maximal elements $T_1, T_2 \in \mathcal{S}'^{(n)}$ correspond to vertices $x'_1, x'_2$ of $K'$. The statement that $\sigma_{T_1}$ and $\sigma_{T_2}$ are adjacent means that $x_1'$ and $x_2'$ are connected by an edge of $K'$. The statement that $s' \notin T_i$, for $i = 1, 2$, means $x_i' \notin K_{s'}'$. The edge connecting $x_1$ and $x_2$ determines a one-dimensional subspace of $\Sigma'$, namely, $\mathrm{Fix}(W_{T_1 \cap T_2}, \Sigma')$. By Lemma 13.2.7, the meaning of the conclusion of Lemma 13.4.3 is that this one-dimensional subspace does not intersect the wall of $\Sigma'$ corresponding to $s'$. The next lemma shows that the corresponding assertion for $\Sigma$ is also true.

**LEMMA 13.4.4.** *Suppose $s' \in S'$ and $T_1, T_2$ are distinct maximal elements of $\mathcal{S}'$ such that $s' \notin T_1 \cup T_2$. For $i = 1, 2$, let $x_i$ be the point in $\Sigma$ corresponding to $W_{T_i}$, i.e., $x_i := \mathrm{Fix}(W_{T_i}, \Sigma)$. Then $x_1$ and $x_2$ lie on the same side of the wall* $\mathrm{Fix}(s', \Sigma)$.

*Proof.* By Lemma 13.3.11, any two top-dimensional simplices in $L' - O(s', L')$ are $s'$-connected. Hence, $\sigma_{T_1}$ and $\sigma_{T_2}$ are $s'$-connected. Without loss of generality we can assume they are adjacent. By Proposition 13.3.9, $W_{T_1 \cap T_2} \in \mathrm{Par}^{(n-1)}(W, S)$, so $E := \mathrm{Fix}(W_{T_1 \cap T_2}, \Sigma)$ is a one-dimensional subspace of $\Sigma$. By Lemmas 13.2.7 and 13.4.3, $E \cap \mathrm{Fix}(s', \Sigma) = \emptyset$. Hence, $E$ is contained in an open half-space bounded by the wall $\mathrm{Fix}(s', \Sigma)$. Since $x_1$ and $x_2$ are both in $E$, this completes the proof. $\square$

Let $\{T_1, \dots, T_l\} = \mathcal{S}'^{(n)}$ (i.e., $\sigma_{T_1}, \dots, \sigma_{T_l}$ are the $(n-1)$-simplices of $L'$). Denote the corresponding points of $\Sigma$ by $x_i$ ($:= \mathrm{Fix}(W_{T_i}, \Sigma)$). Let $s' \in S'$. If $s' \in T_i$, then $x_i \in \mathrm{Fix}(s', \Sigma)$. By Lemma 13.4.4, $\mathrm{Fix}(s', \Sigma)$ bounds a half-space $H^{s'}$ which contains in its interior all the $x_i$ with $s' \notin T_i$. So, $\{x_1, \dots, x_l\} \subset H^{s'}$. Set

$$D := \bigcap_{s' \in S'} H^{s'}.$$

The next lemma completes the proof of Step 1.

**LEMMA 13.4.5.** *$D$ is a nonempty union of chambers of $\Sigma$.*

*Proof.* $\{x_1, \dots, x_l\} \subset D$, so $D$ is nonempty. To prove an intersection of half-spaces (such as $D$) is a union of chambers it suffices to show that it is not contained in any proper subspace of $\Sigma$ (that is, it is not contained in the fixed point set of any nontrivial parabolic subgroup). Let $E$ be the smallest subspace of $\Sigma$ containing $\{x_1, \dots, x_l\}$. Since $\{x_1, \dots, x_l\} \subset D$, if $D$ is contained in a subspace, then this subspace must contain $E$. Clearly, $E = \mathrm{Fix}(W_T, \Sigma)$, where $T := T_1 \cap \dots \cap T_l$. If $T \neq \emptyset$, then every $(n-1)$-simplex of $L'$ contains $\sigma_T$ and this forces $L'$ to be the join $\sigma_T * \mathrm{Lk}(\sigma_T, L')$. But this is not the case, since the join is contractible, while the pseudomanifold $L'$ has $\overline{H}_{n-1}(L') \cong \mathbb{Z}$. Hence, $T = \emptyset$ and $E = \Sigma$. $\square$

Next, we define a mirror structure on $D$ indexed by $S'$. For each $s' \in S'$, put $D_{s'} := D \cap \mathrm{Fix}(s', \Sigma)$. The cofaces of $D$ are defined as follows. For any $T \subset S'$ put

$$D_T := \bigcap_{s' \in T} D_{s'}.$$

Also, put $D_\emptyset := D$ and

$$\partial D_T := \bigcup_{s' \in \mathrm{Vert}(\mathrm{Lk}(\sigma_T, L'))} D_{T \cup \{s'\}}.$$

An $n$-dimensional *pseudomanifold with boundary* is a pair $(A, B)$ of cell complexes such that

- $B$ is an $(n-1)$-dimensional pseudomanifold.
- If $\sigma$ is a $k$-cell in $A - B$, then $\mathrm{Lk}(\sigma, A)$ is an $(n-k-1)$-dimensional pseudomanifold (and is $S^0$ if $k = n-1$).
- Any point of $B$ has a neighborhood in $A$ homeomorphic to one of the form $U \times [0, \varepsilon)$, where $U$ is an open neighborhood of the point in $B$.

**LEMMA 13.4.6**

*(i) $D_T$ is nonempty if and only if $T \in \mathcal{S}'$.*

*(ii) For each $T \in \mathcal{S}'$, $(D_T, \partial D_T)$ is a contractible pseudomanifold with boundary. (Its dimension is $n - \mathrm{Card}(T)$.)*

*(iii) There are coface-preserving maps $\varphi : K' \to D$ and $\theta : D \to K'$ such that $\theta \circ \varphi$ and $\varphi \circ \theta$ are homotopic to the appropriate identity maps and the homotopies are through coface-preserving maps.*

*(iv) $D$ contains only a finite number of chambers of $\Sigma$.*

*Proof.* (i) If $T \in \mathcal{S}'$, then $D_T \neq \emptyset$ since it contains the point $x_i$ for any $i$ such that $T \subset T_i$. Conversely, since $D_T \subset \mathrm{Fix}(W_T, \Sigma)$, if $D_T \neq \emptyset$, then $W_T$ is finite.

(ii) By Corollary 7.5.3, $\mathrm{Fix}(W_T, \Sigma)$ is a pseudomanifold of dimension $n - \mathrm{Card}(T)$. Hence, the same is true for $D_T - \partial D_T$ since it is an open subset of $\mathrm{Fix}(W_T, \Sigma)$. Suppose $x \in \partial D_T$. Then $x \in D_U - \partial D_U$ for some $U \in \mathcal{S}'_{>T}$. By Lemma 7.5.4, $x$ has a neighborhood in $D_T$ of the form $F \times C$ where $F$ is an open set in $D_U - \partial D_U$ and $C$ is a simplicial cone in some Euclidean space. Since such a cone is homeomorphic to a half-space in the Euclidean space, $D_T$ is a pseudomanifold with boundary near $x$. Since half-spaces and subspaces of $\Sigma$ are geodesically convex (in the piecewise Euclidean metric of Chapter 12) and $D_T$ is the intersection of such convex subsets, it is also convex. Since $\Sigma$

is CAT(0) (Moussong's Theorem 12.3.3), so is $D_T$. Hence, it is contractible (Theorem I.2.6).

(iii) Since each coface $K'_T$, $T \in \mathcal{S}'$, is also nonempty and contractible, assertion (iii) is immediate from (ii).

(iv) By (iii), $\varphi : (K', \partial K') \to (D, \partial D)$ is a homotopy equivalence of pairs. So, $H_n(D, \partial D) \cong H_n(K', \partial K') \cong \mathbb{Z}$. Suppose $w_1 K, w_2 K, \dots$ are the chambers of $\Sigma$ contained in $D$. Let $\mu_i \in C_n(K_i)$ be the relative cycle representing the fundamental class in $H_n(w_i K, w_i \partial K)$ ($\cong \mathbb{Z}$). Then $\sum \mu_i$ is a relative $n$-cycle in $C_n(D, \partial D)$ representing the fundamental class in $H_n(D, \partial D)$. Since any such relative $n$-cycle is a finite sum of $n$-cells, there can be only finitely many $w_i K$ in $D$. □

We turn now to Step 2. We apply the basic construction of Chapter 5 to $(W, S')$ and the mirrored space $D$ (where the index set is $S'$) to obtain $\mathcal{U} := \mathcal{U}(W, D)$. By Lemma 13.4.6 (i), the mirror structure is $W$-finite; so, by Lemma 5.1.7, $W$ acts properly on $\mathcal{U}$. By Lemma 13.4.6 (ii), $\mathcal{U}$ is an orientable pseudomanifold. (Compare Proposition 10.7.5.)

**LEMMA 13.4.7.** *Suppose $\mathcal{U}$ ($:= \mathcal{U}(W, D)$) is defined as above. Then*

*(i) $\mathcal{U}$ is $W$-equivariantly homotopy equivalent to $\Sigma'$.*

*(ii) $\mathcal{U}$ is a model for $\underline{E}W$.*

*Proof.* By Lemma 5.2.5 (on the universal property of the basic construction), the maps $\varphi : K' \to D \subset \mathcal{U}(W, D)$ and $\theta : D \to K' \subset \Sigma'$ of Lemma 13.4.6 (iii) extend to maps $\tilde{\varphi} : \Sigma' = \mathcal{U}(W, K') \to \mathcal{U}(W, D)$ and $\tilde{\theta} : \mathcal{U}(W, D) \to \Sigma'$ and it follows that both compositions $\tilde{\varphi} \circ \tilde{\theta}$ and $\tilde{\theta} \circ \tilde{\varphi}$ are $W$-homotopic to the appropriate identity maps. This proves (i). Since $\Sigma'$ is a model for $\underline{E}W$ (by Theorem 12.3.4 (ii)), assertion (ii) follows. □

By Lemma 5.2.5, the inclusion $D \hookrightarrow \Sigma$ extends to a $W$-equivariant map, denoted $f : \mathcal{U} \to \Sigma$.

**LEMMA 13.4.8.** *$f : \mathcal{U} \to \Sigma$ is a $W$-equivariant homotopy equivalence.*

*Proof.* Indeed, any equivariant map between two models for $\underline{E}W$ is an equivariant homotopy equivalence. □

Let $\Sigma^{(2)}$ denote the union of the codimension two subspaces of $\Sigma$, i.e.,

$$\Sigma^{(2)} := \bigcup_{G \in \mathrm{Par}^{(2)}(W,S)} \mathrm{Fix}(G, \Sigma).$$

In the next lemma we show that $f : \mathcal{U} \to \Sigma$ is a branched covering.

**Lemma 13.4.9.** *The restriction of $f$ to $f^{-1}(\Sigma - \Sigma^{(2)})$ is a covering projection. Moreover the number of sheets of this covering is equal to the number $p$ of chambers of $\Sigma$ which are contained in $D$.*

*Proof.* Since $D$ is the union of a finite number of chambers of $\Sigma$, we have $D = w_1K \cup \cdots w_pK$, where $w_1, \dots, w_p$ are distinct elements of $W$. Let $x \in K - \Sigma^{(2)}$. For any $(w, y) \in W \times D$, $[w, y]$ denotes its image in $\mathcal{U}$. Note $[w, y] \in f^{-1}(x)$ if and only if $wy = x$. Since $y \in D$, we must have $y = w_ix$ for some $1 \leqslant i \leqslant p$. So, $w \in W_{S(x)}w_i^{-1}$. ($W_{S(x)}$ is the isotropy subgroup at $x$.) We must show that $x$ has a neighborhood $V$ which is evenly covered by $f$. The first case to consider is where $x \in K - \partial K$. In this case, take $V = K - \partial K$. For each $i$, the subset $w_i^{-1} \times w_iV$ of $W \times D$ projects homeomorphically onto a subset $V_i$ of $\mathcal{U}$ and $f$ takes $V_i$ homeomorphically onto $V$. This shows $f^{-1}(V)$ is the disjoint union of the $V_i$ and that $V$ is evenly covered by $f$.

The other case is where $x \in \partial K - \Sigma^{(2)}$. In other words, $x$ lies in a mirror, say $K_s$, $s \in S$, but not in any codimension two coface of $K$. Let $V_+$ be an open neighborhood of $x$ in $K$ such that $V^+ \cap K_t = \emptyset$ for all $t \neq s$. Then $V := V_+ \cup sV_+$ is an open neighborhood of $x$ in $\Sigma$. We next define, for each $1 \leqslant i \leqslant p$, an open subset $V_i$ of $f^{-1}(V)$. There are two subcases to consider.

*Subcase 1.* $w_isw_i^{-1} \notin S'$. In this case $w_iV \subset D$. Define $V_i$, as before, to be the image of $w_i^{-1} \times w_iV$ in $\mathcal{U}$.

*Subcase 2.* $w_isw_i^{-1} \in S'$, say, $w_isw_i^{-1} = s' \in S'$. We still have $w_iV_+ \subset D$ and $w_iV_+ \cap \partial D \subset D_{s'}$. So, the images of $w_i^{-1} \times w_iV_+$ and $w_i^{-1}s' \times w_iV_+$ in $\mathcal{U}$ fit together to give an open set $V_i$ in $\mathcal{U}$. Moreover, since $w_i^{-1} \times w_iV_+$ maps onto $V_+$ and $w_i^{-1}s' \times w_iV_+$ maps onto $w_i^{-1}s'w_iV_+ = sV_+$, we see that $f$ takes $V_i$ homeomorphically onto $V$.

So, in both subcases, $f^{-1}(V)$ is the disjoint union of the $V_i$ and hence, $V$ is evenly covered by $f$. Since any point in $\Sigma - \Sigma^{(2)}$ can be written in the form $wx$ for some $w \in W$ and $x \in K$, and since $f$ is equivariant, any point of $\Sigma - \Sigma^{(2)}$ has a neighborhood which is evenly covered by $f$. Moreover, the number of sheets is $p$. □

We can now complete the proof of Step 2. The argument goes roughly as follows. The spaces $\mathcal{U}$ and $\Sigma$ are both orientable, gallery connected, $n$-dimensional pseudomanifolds and $H_c^n(\mathcal{U}) \cong H^n(W; \mathbb{Z}W) \cong H_c^n(\Sigma) \cong \mathbb{Z}$. After picking orientations for $\mathcal{U}$ and $\Sigma$, it makes sense to speak of the *degree of $f$*, denoted $\deg(f)$; it is the integer $d$ such that $f^*$ takes the orientation class of $\Sigma$ to $d$ times the orientation class of $\mathcal{U}$. The fact (Lemmas 13.4.6 (iv) and 13.4.8) that $f$ is a proper homotopy equivalence means that $\deg(f) = 1$. On the other hand, a local computation using the fact that $f$ is a branched cover (Lemma 13.4.9) shows that $\deg(f) = p$ where $p$ is the number of chambers in $D$. Hence, $p = 1$, i.e., $D$ consists of a single chamber of $\Sigma$. The details follow.

*Proof of Step 2.* We begin by orienting $K, \Sigma, D, \mathcal{U}$ to get distinguished generators for the infinite cyclic groups $H^n(K, \partial K)$, $H^n_c(\Sigma)$, $H^n(D, \partial D)$ and $H^n_c(\mathcal{U})$. (Each of these spaces is a $n$-dimensional simplicial complex and we orient its $n$-simplices.) First orient $K$. (This can be done since, by Definition 13.3.1, $L$ is an orientable, $(n-1)$-dimensional pseudomanifold and $K$ is the cone on the barycentric subdivision of $L$.) Each $n$-simplex in $wK$ is then oriented by $(-1)^{l(w)}$ times the orientation of the corresponding simplex of $K$ (where $l(w)$ means word length with respect to $S$). This orients $\Sigma$ and $D$. Finally orient $\mathcal{U}$ by orienting each $n$-simplex in $wD$ as $(-1)^{l'(w)}$ times the orientation of the corresponding simplex of $D$ (where $l'(w)$ is word length with respect to $S'$).

As in the proof of Lemma 13.4.9, put $V = K - \partial K$ and $f^{-1}(V) = V_1 \cup \cdots \cup V_p$. Denote the orientation classes in $H^n(K, \partial K), H^n(\overline{V}_i, \partial\overline{V}_i), H^n_c(\Sigma)$ and $H^n_c(\mathcal{U})$ by $e, e_i, e_\Sigma$ and $e_\mathcal{U}$, respectively. By excision, $H^n(\Sigma, \Sigma - V) = H^n(K, \partial K) \cong \mathbb{Z}$. Since $H^n_c(\Sigma - V) = 0$, the exact sequence of the pair $(\Sigma, \Sigma - V)$ shows that the natural map $H^n(K, \partial K) = H^n(\Sigma, \Sigma - V) \to H^n_c(\Sigma)$ takes $e$ to $e_\Sigma$. Similarly,

$$H^n(\mathcal{U}, \mathcal{U} - (V_1 \cup \cdots \cup V_p)) \cong \bigoplus_{i=1}^{p} H^n(\overline{V}_i, \partial\overline{V}_i) \cong \mathbb{Z}^p$$

and the natural map $\oplus H^n(\overline{V}_i, \partial\overline{V}_i) \to H^n(\mathcal{U})$ sends $e_i$ to $e_\mathcal{U}$. Also, the map $(f|_{V_i})^* : H^n(K, \partial K) \to H^n(\overline{V}_i, \partial\overline{V}_i)$ sends $e$ to $e_i$. Consider the commutative diagram

$$\begin{array}{ccc} H^n_c(\Sigma) & \xrightarrow{f^*} & H^n_c(\mathcal{U}) \\ \uparrow & & \uparrow \\ H^n(K, \partial K) & \longrightarrow & \bigoplus H^n(\overline{V}_i, \partial\overline{V}_i). \end{array}$$

It follows that $f^*(e_\Sigma)$ is the sum of the images of the $e_i$ in $H^n_c(\mathcal{U})$, i.e., $f^*(e_\Sigma) = pe_\mathcal{U}$. So $\deg(f) = p$. As we explained earlier, since $f$ is a proper homotopy equivalence, $\deg(f) = 1$. So $p = 1$. □

So $D = wK$ and $S' = wSw^{-1}$. To complete the proof of Theorem 13.4.1 it remains to show that $w$ is unique. This is taken care of by the next lemma.

**LEMMA 13.4.10.** (cf. Theorem D.2.10.) *If $wSw^{-1} = S$, then $w = 1$.*

*Proof.* The argument is similar to the proof of Lemma 13.4.5. Let $\{T_1, \dots, T_l\} = \mathcal{S}^{(n)}$. For $1 \leqslant i \leqslant l$, let $y_i$ denote the unique fixed point of $W_{T_i}$. Then $y_i \in K$. Suppose $wSw^{-1} = S$. Since $wy_i$ is the unique fixed point of $W_{wT_iw^{-1}}$ and $wT_iw^{-1} \subset S$, we also have $wy_i \in K$. Hence, $w \in W_{T_i}$. Since this holds for all $i$, $w \in W_T$, where $T = T_1 \cap \cdots \cap T_l$. Since $L$ is a pseudomanifold, this implies, as in the proof of Lemma 13.4.5, that $T = \emptyset$. So, $w = 1$. □

## NOTES

**13.1.** The discussion in this section roughly follows that in [35]. Rather than the "associated labeled graph," the authors of [35] call $\Upsilon$ the "diagram" of $(W, S)$; however, this conflicts with the standard usage of the term "Coxeter diagram." This leads them to call the operation in this section "diagram twisting" rather than "graph twisting."

The proofs of the results in [51, 130] on 2-spherical Coxeter groups make extensive use of the "geometric representation" and "root systems" discussed in Appendix D.

Max Dehn introduced three algorithmic problems for finitely generated groups: the Word Problem, the Conjugacy Problem, and the Isomorphism Problem. We gave Tits' solution to the Word Problem for Coxeter groups in Section 3.4. As we pointed out in Theorem 12.3.4 (iii), since $\Sigma$ is CAT(0), the Conjugacy Problem for Coxeter groups is solvable. An algorithm is given in [181]. A solution to the Isomorphism Problem for Coxeter groups is not known, but these questions about rigidity are first steps in this direction. For overviews of work on the Isomorphism Problem for Coxeter groups see [12, 223].

**13.2.** The one-to-one correspondence between spherical parabolic subgroups and their fixed subspaces (Lemma 13.2.2) holds for all parabolics provided one replaces the complex $\Sigma$ by the Tits cone in the geometric representation. (The "Tits cone" is defined in Appendix D.2.) This observation is important in Krammer's thesis [181].

**13.3.** This section and the next are taken from [60]. The idea of using the virtual cohomological dimension of the normalizer of an element of $S'$ to prove that $W$ has a rigid reflection set (Proposition 13.3.9) was suggested by an argument of Rosas [249].

In 1977, in connection with his proof that a subgroup of infinite index in a PD$^n$-group $G$ has cohomological dimension $< n$, Strebel [272] wrote:

> ...one could require $G$ to be a finitely generated group of finite cohomological dimension $n$, $R$ to admit an $RG$-projective resolution which is finitely generated in the top dimension $n$ and $H^n(G; RG)$ to be a free cyclic $R$-module. ...*However, I know of no example which satisfies these assumptions without being a* PD-*group.* [emphasis added.]

Lemma 13.3.6 shows that any torsion-free subgroup of finite index in a Coxeter group which is type PM but not type HM gives such an example. For example, we could choose $W$ so that its nerve $L$ is any closed orientable $(n-1)$-manifold which is not a homology sphere.

**13.4.** A motivation for proving Theorem 13.4.1 was the work of Prassidis–Spieler [240] (also see [249]). They proved the following result. Suppose a Coxeter group $W$ of type HM acts as a proper, locally linear, cocompact reflection group on manifolds $M$ and $M'$ and that $f : M \to M'$ is a $W$-equivariant map. Further assume for both $M$ and $M'$ that

(a) The fixed point set of each spherical parabolic subgroup is contractible.

(b) If such a fixed point set is three dimensional, then it is homeomorphic to $\mathbb{R}^3$. (It follows from Perelman's Theorem [237, 239, 238] that this requirement is superfluous.)

Then $f$ is equivariantly homotopic to a $W$-equivariant homeomorphism. Let $C$ and $C'$ be fundamental chambers for $M$ and $M'$, respectively. (We note that by Theorem 8.2.7 each coface of $C$ or $C'$ is acyclic; condition (a) is equivalent to the requirement that each coface is contractible. Condition (b) is equivalent to the condition that each three-dimensional coface is homeomorphic to the 3-disk; hence, it is automatically satisfied provided that Perelman's proof, [237], of the three-dimensional Poincaré Conjecture holds up.) Prassidis-Spieler also prove an analogous result when $M$ and $M'$ are contractible manifolds with boundary and $f : (M, \partial M) \to (M', \partial M')$ is a $W$-equivariant homeomorphism on the boundary. In outline their argument goes as follows. Let $S$ and $S'$ be reflections across the walls of $C$ and $C'$, respectively. By Theorem 13.4.1, $S' = wSw^{-1}$; so, replacing $C'$ by $w^{-1}C$ we can assume that $S = S'$. The cofaces of $C$ and $C'$ are then indexed by the spherical subsets of $S$ and for each $T \in \mathcal{S}(S)$, $C_T$ and $C'_T$ are contractible manifolds with boundary (of the same dimension). After repeatedly applying the h-Cobordism Theorem, one concludes that there is a strata-preserving homeomorphism $g : C \to C'$. The induced map $\tilde{g} : \mathcal{U}(W, C) = M \to \mathcal{U}(W, C') = M'$ is the desired equivariant homeomorphism.

# Chapter Fourteen

# FREE QUOTIENTS AND SURFACE SUBGROUPS

In [147] Gromov asked if every word hyperbolic group is either virtually cyclic or "large" in the sense defined below. He also asked if every one-ended word hyperbolic group contains a surface group (see [21]). In this chapter these questions are answered for Coxeter groups. In 14.1 we prove a result of [196] (much of which was proved independently in [140]) that any Coxeter group is either virtually abelian or large. In 14.2 we prove the result of [141] that any Coxeter group is either virtually free or else it contains a surface group.

## 14.1. LARGENESS

References include [196, 139, 140, 65, 190, 210, 229].

**DEFINITION 14.1.1.** (Gromov [145].) A group $G$ is large if it virtually has a nonabelian free quotient (in other words, if there is a subgroup $\Gamma$ of finite index in $G$ such that $\Gamma$ maps onto a free group of rank $\geqslant 2$).

Our goal is to prove the following.

**THEOREM 14.1.2.** (Margulis and Vinberg [196], Gonciulea [140].) *Any Coxeter group is either virtually free abelian or large.*

By Theorem 12.3.5, if a Coxeter group is virtually abelian, then it is a product of a finite group and a cocompact Euclidean reflection group.

### Virtual Actions on Trees

For each reflection $r \in R$, let $\Sigma^r := \mathrm{Fix}(r, \Sigma)$ be the corresponding wall of $\Sigma$. Suppose $\Gamma$ is a subgroup of $W$ and $s \in S$. We say $(\Gamma, s)$ has the *trivial intersection property* if for all $\gamma \in \Gamma$, either $\gamma\Sigma^s = \Sigma^s$ or $\gamma\Sigma^s \cap \Sigma^s = \emptyset$. Note that $\gamma\Sigma^s = \Sigma^{\gamma s \gamma^{-1}}$. So, if $\gamma\Sigma^s = \Sigma^s$, then $\gamma s \gamma^{-1} = s$, i.e., $\gamma$ lies in $C_\Gamma(s)$ (the centralizer of $s$ in $\Gamma$). Conversely, when $\gamma \in C_\Gamma(s)$, it is obvious that $\gamma\Sigma^s = \Sigma^s$.

For the remainder of this section (except in Lemma 14.1.8) we suppose $(\Gamma, s)$ has the trivial intersection property. The union of walls $\Gamma\Sigma^s$ $(:= \bigcup_{\gamma\in\Gamma} \gamma\Sigma^s)$ divides $\Sigma$ into convex regions. (Such a region is an intersection of half-spaces bounded by walls of the form $\gamma\Sigma^s$; hence, is convex.) Define a graph $T$ with one vertex for each such region and with an edge connecting two vertices if the corresponding regions share a common wall.

**LEMMA 14.1.3.** *$T$ is a tree.*

*Proof.* Consider the cover of $\Sigma$ by the convex regions defined above. The nerve of the cover is $T$. Since $\Sigma$ is CAT(0) (Theorem 12.3.3), each convex region is contractible as is each wall. It follows that $T$ is homotopy equivalent to $\Sigma$, which is contractible. □

The set Edge$(T)$ of (unoriented) edges of $T$ is equal to $\{\gamma\Sigma^s\}_{\gamma\in\Gamma}$. Hence, $\Gamma$ is transitive on Edge$(T)$.

*Remark.* Of course, we get the same tree $T$ and the same action of $\Gamma$ on it if we replace $\Sigma$ by either: (a) the interior $\mathcal{I}$ of the Tits cone (Appendix D.2) or (b) the Cayley graph of $(W, S)$. In case (a) we know that $\Sigma$ and $\mathcal{I}$ are $W$-equivariantly homotopy equivalent. As for case (b), since Cay$(W, S)$ is the 1-skeleton of $\Sigma$ in its cellulation by Coxeter polytopes, walls in Cay$(W, S)$ have the same separation properties as those in $\Sigma$.

**DEFINITION 14.1.4.** An element $s \in S$ is two-sided with respect to $\Gamma$ if $C_\Gamma(s) \subset A_s$.

(Recall Definition 4.5.1: $A_s := \{w \in W \mid l(sw) > l(w)\}$, the fundamental half-space for $s$.)

Geometrically, two-sidedness means that the image of $\Sigma^s$ is a two-sided subspace in $\Sigma/\Gamma$. If $s$ is two-sided with respect to $\Gamma$, then a well-defined sign can be assigned to each half-space bounded by a wall $\gamma\Sigma^s$: the half-space is *positive* if it contains $\gamma K$ and *negative* otherwise. This definition is independent of the choice of $\gamma$ because (a) two such $\gamma$ lie in the same coset of $C_\Gamma(s)$ and (b) $C_\Gamma(s) \subset A_s$. So, when $s$ is two-sided, each edge of $T$ can be assigned an orientation which is preserved by the $\Gamma$-action: if $e$ is an edge corresponding to a wall $\gamma\Sigma^s$ and if $v$, $v'$ are the endpoints of $e$, then $e$ is oriented from $v$ to $v'$ if the region corresponding to $v$ is on the positive side of $\gamma\Sigma^s$. Since $\Gamma$ preserves these orientations, it acts on $T$ "without inversions," i.e., no edge of $T$ is flipped by an element of $\Gamma$ (Definition E.1.5). Thus, $s$ is two-sided with respect to $\Gamma$ if and only if the $\Gamma$-action on $T$ is without inversions. In this case there is a well-defined quotient graph $T/\Gamma$ with Edge$(T/\Gamma) = (\text{Edge}(T))/\Gamma$ and $\Gamma$ splits as a graph of groups over $T/\Gamma$. (See Appendix E.1.) Since $\Gamma$ is transitive on Edge$(T)$ there are only two possibilities for $T/\Gamma$; either it is a segment or a loop. (See Examples E.1.8 and E.1.9.)

Let $\theta_s : W \to \mathbb{Z}/2$ denote the number of times modulo 2 that a gallery from $K$ to $wK$ crosses a $\Gamma$-translate of $\Sigma^s$. In other words, $\theta_s(w)$ is the number, modulo 2, of reflections of the form $\gamma s \gamma^{-1}$, $\gamma \in \Gamma$, which separate 1 and $w$. Let $R_{(\Gamma,s)} := \{r \in R \mid r = \gamma s \gamma^{-1}, \gamma \in \Gamma\}$. As in 4.2, given $u, v \in W$, let $R(u, v)$ be the set of reflections separating them. (So, $\theta_s(w) \equiv \mathrm{Card}(R(1, w) \cap R_{(\Gamma,s)})$ (mod 2).) Although $\theta_s$ is not, in general, a homomorphism, we have the following.

**LEMMA 14.1.5.** ([139].) *For any $\gamma \in \Gamma$ and $w \in W$, $\theta_s(\gamma w) = \theta_s(\gamma) + \theta_s(w)$.*

*Proof.* By Lemma 4.2.1 (i), $R(1, w)$ and $R(\gamma, \gamma w)$ have the same cardinality. Conjugation by $\gamma$ maps the intersection of the first set with $R_{(\Gamma,s)}$ to its intersection with the second set. So, the number of $\Gamma$-translates of $\Sigma^s$ separating $K$ and $wK$ is the same as the number separating $\gamma K$ and $\gamma wK$. By Lemma 4.2.1 (ii), $\mathrm{Card}(R(1, \gamma w)) \equiv \mathrm{Card}(R(1, \gamma)) + \mathrm{Card}(R(\gamma, \gamma w))$ (mod 2) and the same is true after intersecting with $R_{(\Gamma,s)}$. So, $\theta_s(\gamma w) = \theta_s(\gamma) + \theta_s(w)$. □

A consequence is that $\theta_s|_\Gamma$ is a homomorphism. For any $\alpha \in C_\Gamma(s)$, $\Sigma^s$ is a wall of $\alpha K$ (since $s\alpha K = \alpha s K$, which is adjacent to $\alpha K$). So, for $\alpha \in C_\Gamma(s)$, the only wall of the form $\gamma \Sigma^s$ which can separate $K$ from $\alpha K$ is $\Sigma^s$ itself. It follows that $s$ is two sided with respect to $\Gamma'$, where $\Gamma' = \mathrm{Ker}(\theta_s|_\Gamma)$. So, by passing to a subgroup of $\Gamma$ of index one or two, we can assume $s$ is two sided. (More generally, for any group $G$ of automorphisms of a tree $T$ one can always find a subgroup $G_0$ of index one or two which acts without inversions [196, Lemma 3].)

**DEFINITION 14.1.6.** ([139].) An element $s \in S$ is $\Gamma$-*separating* if $\theta_s|_\Gamma = 0$.

Geometrically, $s$ being $\Gamma$-separating means that the image of $\Sigma^s$ separates $\Sigma/\Gamma$. Note that $s$ being $\Gamma$-separating implies that it is two-sided with respect to $\Gamma$. This discussion is summarized in the following.

**PROPOSITION 14.1.7.** *Let $v_0$ be the vertex of $T$ corresponding to the region containing $K$ and let $v_1$ be the vertex corresponding to the adjacent region across $\Sigma^s$. For $i = 0, 1$, let $\Gamma_i$ be the stabilizer of $v_i$ and $C = C_\Gamma(s)$ the stabilizer of the edge from $v_0$ to $v_1$.*

*(i) Suppose $s$ is $\Gamma$-separating. Then there are exactly two $\Gamma$-orbits of vertices in $T$ (one containing $v_0$ and the other $v_1$) and $\Gamma$ decomposes as an amalgamated product, $\Gamma = \Gamma_0 *_C \Gamma_1$.*

*(ii) Suppose $s$ is two-sided with respect to $\Gamma$ and not $\Gamma$-separating. Then $\Gamma$ acts transitively on* Vert($T$) *and $\Gamma$ splits as a HNN construction, $\Gamma = \Gamma_0 *_C$.*

("Amalgamated products" and "HNN constructions" are defined in E.1.1 and E.1.2, respectively. They correspond to the cases where $T/\Gamma$ is a segment or loop, respectively.) In terms of the orientations on the edges of $T$, the two possibilities in the proposition look as follows. In case (i) (where $s$ is $\Gamma$-separating), each vertex of $T$ is either a source or a sink. (All edges point out from $v_0$ and into $v_1$.) In case (ii), at any vertex, edges with both type of orientations occur.

We turn next to the existence of finite-index subgroups $\Gamma$ such that $\Gamma$ has the trivial intersection property.

**LEMMA 14.1.8.** ([210].) *If $\Gamma \subset W$ is normal and torsion-free, then $(\Gamma, s)$ has the trivial intersection property for any $s \in S$.*

The proof of this lemma appears in Millson's paper [210] (where it is attributed to H. Jaffe).

*Proof.* Given $\gamma \in \Gamma$, we must show that if $\gamma\Sigma^s \cap \Sigma^s \neq \emptyset$, then $\gamma\Sigma^s = \Sigma^s$. Suppose $x \in \gamma\Sigma^s \cap \Sigma^s$. Since the reflections $s$ and $\gamma s \gamma^{-1}$ both fix $x$, so does the commutator $[s, \gamma] := s\gamma s\gamma^{-1}$. Since the isotropy subgroup at any point of $\Sigma$ is finite, this implies $[s, \gamma]$ is torsion. Since $\Gamma$ is normal in $W$, $s\gamma s$ belongs to $\Gamma$ and hence, so does $[s, \gamma] = (s\gamma s)\gamma^{-1}$. Since $[s, \gamma]$ is torsion and $\Gamma$ is torsion-free, this means that $[s, \gamma] = 1$, i.e., $\gamma \in C_\Gamma(s)$. □

By Corollary D.1.4, any Coxeter group has a torsion-free subgroup $\Gamma$ of finite index. Replacing $\Gamma$ by the intersection of all its conjugates in $W$, we can assume $\Gamma$ is normal. Hence, one can always find a finite-index subgroup $\Gamma$ to which Lemma 14.1.8 applies.

### Residually Finite Actions

The following important generalization of a residually finite group is defined in [196].

**DEFINITION 14.1.9.** (Margulis and Vinberg [196, p. 172].) An action of a group $\Gamma$ on a set $X$ is residually finite if for any two distinct elements $x, x' \in X$, there is a $\Gamma$-action on a finite set $Y$ and a $\Gamma$-equivariant map $f : X \to Y$ such that $f(x) \neq f(x')$.

*Remark 14.1.10*

(i) A $\Gamma$-action on a set $X$ is residually finite if and only if its restriction to each orbit in $X$ is residually finite.

(ii) A $\Gamma$-action on $X$ is residually finite if and only if for any two distinct elements $x$ and $x'$ in $X$, there is a normal subgroup $N$ of finite index in $\Gamma$ such that $x, x'$ belong to different $N$-orbits. (Proof: Assume the $\Gamma$-action on $X$ is residually finite. Let $N$ be the kernel of the $\Gamma$-action on $Y$. Since $f(Nx) \neq f(Nx')$, $x, x'$ are in distinct $N$-orbits. For the converse, suppose $X$ is a single $\Gamma$-orbit and $Nx \neq Nx'$. Then take $Y := X/N$ and $f : X \to Y$ the natural projection.)

(iii) The standard definition of a *residually finite group* $\Gamma$ is equivalent to the condition that the action of $\Gamma$ on itself (by translation) is residually finite.

The proof of Selberg's Lemma shows any finitely generated linear group is residually finite. Margulis and Vinberg observe that a similar argument shows that if $G \hookrightarrow GL(V)$ is any real representation of a finitely generated group $G$, then the action of $G$ on the projective space $P(V)$ is residually finite. Apply this observation to the canonical representation (from 6.12) of $W$ on $\mathbb{R}^S$. Let $\mathbb{R}e_s$ denote the line spanned by the basis vector $e_s$. Restrict the representation to $\Gamma$. The isotropy subgroup at point represented by $\mathbb{R}e_s$ in $P(\mathbb{R}^S)$ is $C_\Gamma(s)$. Since $\Gamma/C_\Gamma(s) \cong \text{Edge}(T)$, we have the following.

**PROPOSITION 14.1.11.** (Margulis and Vinberg [196].) *The action of* $\Gamma$ *on* Edge($T$) *is residually finite.*

## Trees and Graphs

A tree is a *star* if all its edges have a common vertex; it is a *line* if it is homeomorphic to the real line. A graph is a *cycle* if it is homeomorphic to $S^1$. Suppose $\Lambda$ is a finite, connected graph. Its *Euler characteristic* $\chi(\Lambda)$ is Card(Vert($\Lambda$)) – Card(Edge($\Lambda$)). Since the reduced homology of $\Lambda$ is concentrated in dimension 1, we know that

- $\chi(\Lambda) \leqslant 1$,
- $\chi(\Lambda) = 1 \iff \Lambda$ is a tree (i.e., $\iff$ it is acyclic),
- $\chi(\Lambda) = 0 \iff$ it is homotopy equivalent to a circle,
- $\chi(\Lambda) < 0 \iff \pi_1(\Lambda)$ is a nonabelian free group (since any connected graph is homotopy equivalent to a bouquet of circles).

**LEMMA 14.1.12.** ([196, Lemma 4].) *Suppose* $\Lambda$ *is a finite, connected graph and that its automorphism group is transitive on* Edge($\Lambda$).

*(i) If* $\chi(\Lambda) = 1$, *then* $\Lambda$ *is a star.*

*(ii) If* $\chi(\Lambda) = 0$, *then* $\Lambda$ *is a cycle.*

*Proof.* Suppose $\Lambda$ has an extreme vertex. Then, by transitivity, every edge must contain an extreme vertex. By connectivity, the other vertices of the edges must all coincide. So, $\Lambda$ is a star.

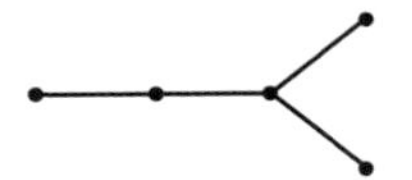

Figure 14.1. The tree $T_0$.

If $\chi(\Lambda) = 1$, then $\Lambda$ has an extreme vertex and so is a star. If $\chi(\Lambda) = 0$, $\Lambda$ contains exactly one cycle and since it also cannot have an extreme vertex, it must be that cycle. □

**LEMMA 14.1.13.** ([196, Proposition 2].) *Suppose $\Gamma$ acts without inversions on a tree $T$ and that its action on* Edge($T$) *is transitive and residually finite. If $T$ is not a star or a line, then $\Gamma$ is large (i.e., virtually has a nonabelian free quotient).*

*Proof.* Since $T$ is not a star or a line, it contains a subtree $T_0$ as in Figure 14.1. By the residual finiteness of the $\Gamma$-action on Edge($T$), there is a normal subgroup $N$ of finite index in $\Gamma$ such that the edges of $T_0$ lie in distinct $N$-orbits. This implies that the quotient graph $T/N$ is not a star or a cycle. Since $\Gamma/N$ is transitive on Edge($T/N$), Lemma 14.1.12 implies $\chi(T/N) < 0$. The natural map $\Gamma \to \pi_1(T/N)$ is the desired epimorphism onto a nonabelian free group. □

**Stars and Lines**

Suppose $\Gamma$ is a normal torsion-free subgroup of finite index in $W$ such that each $s \in S$ is two-sided with respect to $\Gamma$.

**LEMMA 14.1.14.** *For a given $s \in S$, suppose the corresponding tree $T$ is a star. Then $\Gamma = C_\Gamma(s)$.*

*Proof.* Suppose to the contrary that $\Gamma \neq C_\Gamma(s)$. Then there is an element $\gamma \in \Gamma$ such that $\gamma\Sigma^s \neq \Sigma^s$. Let $r$ be the reflection $\gamma s \gamma^{-1}$. Since $\Sigma^s$ and $\Sigma^r$ are disjoint, $r$ and $s$ generate an infinite dihedral group. The proof of Lemma 14.1.8 shows that the element $rs = \gamma s \gamma^{-1} s$ belongs to $\Gamma$. Put $\alpha := rs$. Since $\langle r, s\rangle$ is infinite dihedral, the walls $\alpha^m \Sigma^s$, $m \in \mathbb{Z}$, are disjoint and are arranged in a linear fashion. It follows that $T$ must contain a line. Since $T$ was assumed to be a star, this is a contradiction. □

**LEMMA 14.1.15.** *For a given $s \in S$, suppose $T$ is a line. For $i = 0, 1$, let $v_i$ and $\Gamma_i$ be as in Proposition 14.1.7 and let $C = C_\Gamma(s)$.*

*(i) If $s$ is $\Gamma$-separating, then $C$ is index 2 in both $\Gamma_0$ and $\Gamma_1$.*

*(ii) If $s$ is not $\Gamma$-separating, then $\Gamma_0 = C$.*

*Proof.* These are the only two possibilities when a group acts without inversions on a line and is transitive on the edge set. □

For a fixed $\Gamma$, $T(s)$ denotes the tree corresponding to $s$. Define three subsets of $S$:

$$\widetilde{S} := \{s \in S \mid T(s) \text{ is a line}\},$$

$$S' := \{s \in \widetilde{S} \mid s \text{ is } \Gamma\text{-separating}\},$$

$$S'' := \{s \in \widetilde{S} \mid s \text{is not } \Gamma\text{-separating}\}.$$

For each $s \in S'$, $T(s)/\Gamma$ is a segment and $\Gamma = \Gamma_0 *_C \Gamma_1$. By Lemma 14.1.15 (i), $\Gamma_0/C$ and $\Gamma_1/C$ are both equal to $\mathbf{C}_2$ (the cyclic group of order 2). So, we have a homomorphism $\varphi_s : \Gamma \to \Gamma_0/C * \Gamma_1/C = \mathbf{C}_2 * \mathbf{C}_2 = \mathbf{D}_\infty$, the infinite dihedral group. The kernel of $\varphi_s$ is $C$ and this is also the kernel of the action on $T(s)$. For each $s \in S''$, $T(s)/\Gamma$ is a loop and $\Gamma = \Gamma_0 *_C$. Let $\varphi_s : \Gamma \to \mathbb{Z}$ be the homomorphism $\Gamma \to \pi_1(T(s)/\Gamma)$. Once again, $\operatorname{Ker} \varphi_s = C$. The $\varphi_s$, $s \in \widetilde{S}$, fit together to define a homomorphism:

$$\Phi = (\varphi_s) : \Gamma \to (\mathbf{D}_\infty)^{S'} \times \mathbb{Z}^{S''}.$$

**PROPOSITION 14.1.16.** ([140]). *Suppose that for each $s \in S$, $T(s)$ is a star or a line. Then $\Gamma$ is virtually free abelian.*

*Proof.*

$$\operatorname{Ker} \Phi = \bigcap_{s \in \widetilde{S}} \operatorname{Ker} \varphi_s = \bigcap_{s \in \widetilde{S}} C_\Gamma(s).$$

On the other hand, by Lemma 14.1.14, for each $s \in S - \widetilde{S}$, $\Gamma = C_\Gamma(s)$. So, $\operatorname{Ker} \Phi \subset C_\Gamma(s)$ for all $s$. Since $s$ is two-sided with respect to $\Gamma$, $C_\Gamma(s) \subset A_s$. So,

$$\operatorname{Ker} \Phi \subset \bigcap_{s \in S} A_s = \{1\}.$$

Consequently, $\Phi$ is injective and $\Gamma$ is virtually free abelian. □

*Proof of Theorem 14.1.2.* If $T(s)$ is a star or line for all $s \in S$, then, by the previous proposition, $\Gamma$ is virtually free abelian. Otherwise, by Proposition 14.1.11 and Lemma 14.1.13, $\Gamma$ virtually has a nonabelian free quotient. □

## 14.2. SURFACE SUBGROUPS

A *surface group* means the fundamental group of a closed, orientable surface of genus $> 0$. Here we prove a theorem of [141] which completely answers the question of when a Coxeter group $W$ contains a surface subgroup.

Since the rational cohomological dimension of an (infinite) virtually free group is 1 while that of a surface group is 2, no virtually free group can contain a surface subgroup. In Proposition 8.8.5 we gave a condition on the nerve of $(W, S)$ for $W$ to be virtually free. The result of [141] is that this condition is necessary and sufficient for $W$ not to contain any surface subgroup.

**THEOREM 14.2.1.** (Gordon–Long–Reid, [141].) *The following two conditions on a Coxeter group $W$ are equivalent:*

*(i) $W$ does not contain a surface subgroup.*

*(ii) $W$ is virtually free.*

**Example 14.2.2.** (*Polygon groups.*) We repeat Examples 6.5.2 and 6.5.3. Suppose $W$ is a Coxeter group whose nerve $L$ is a $k$-cycle, $k > 3$. Denote the labeling, $\mathrm{Edge}(L) \to \{2, 3, \dots\}$, by $e \to m_e$ (if $s$ and $t$ are the endpoints of $e$, then $m_e := m_{st}$). Any such $W$ can be represented as a group of isometric reflections across the edges of a $k$-gon in either the hyperbolic plane $\mathbb{H}^2$ or the Euclidean plane $\mathbb{E}^2$. Moreover, $\mathbb{E}^2$ occurs only when $k = 4$ and all edge labels are 2. It follows that an orientation-preserving, torsion-free subgroup of finite index in $W$ is a surface group (and the surface has genus $\geqslant 2$ in the non-Euclidean case).

**Example 14.2.3.** (*Simplicial Coxeter groups.*) Here we repeat some of the discussion from 6.9 on Lannér groups. Suppose the nerve $L$ of $(W, S)$ is the boundary of the simplex on $S$, in other words, every proper subset of $S$ is spherical. Any such $W$ can be represented as a group generated by reflections across the faces of a simplex in either $\mathbb{H}^n$ or $\mathbb{E}^n$, where $n = \mathrm{Card}(S) - 1$. ($\mathbb{E}^n$ occurs only when the determinant of the cosine matrix is 0.) Thus, we can identify $K$ with a hyperbolic or Euclidean simplex and $\Sigma$ with $\mathbb{H}^n$ or $\mathbb{E}^n$. Suppose $n \geqslant 2$. In both cases $W$ contains a surface group. This can be seen as follows. Pick any $(n - 2)$-element subset $T$ of $S$. The fixed point set of $W_T$ on $\Sigma$ is a two-dimensional space isometric to $\mathbb{H}^2$ or $\mathbb{E}^2$. Let $C$ be the centralizer of $W_T$. Then $C$ acts isometrically and cocompactly on this fixed point set. So, any torsion-free subgroup of finite index in $C$ acts freely and cocompactly either on $\mathbb{H}^2$ or $\mathbb{E}^2$ and hence, is a surface group.

Recall that a Coxeter system is 2-spherical if the 1-skeleton of its nerve is a complete graph. The next lemma is a corollary of the previous example.

**LEMMA 14.2.4.** *Suppose $(W, S)$ is a 2-spherical Coxeter system. Then either $W$ is finite or it contains a surface group.*

*Proof.* If $W$ is not finite, we can find a subset $T \subset S$ such that $W_T$ is infinite and $W_U$ is finite for every proper subset $U \subsetneq T$. (Choose $T$ to be minimal with

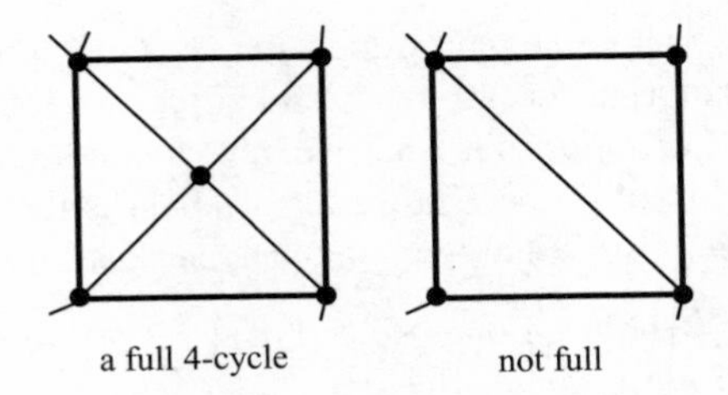


Figure 14.2. Full and non-full 4-cycles.

respect to the property that $W_T$ is infinite.) Then $W_T$ is one of the Lannér groups discussed in the previous example. Hence, it contains a surface group. □

For the remainder of this section *all graphs will be simplicial*, i.e., they will have no loops or multiple edges. A subgraph $\Omega'$ of a graph $\Omega$ is *full* if it is a full subcomplex. (In other words, if two vertices of $\Omega'$ are connected by an edge of $\Omega$, then the edge lies in $\Omega'$.) Given a subset $T \subset \mathrm{Vert}(\Omega)$, $\mathrm{Span}(T)$ denotes the full subgraph of $\Omega$ spanned by $T$. Let $\mathcal{C}$ be the smallest class of graphs such that:

(i) If $\Omega$ is in $\mathcal{C}$, then so is any graph isomorphic to $\Omega$.

(ii) $K_n$, the complete graph on $n$ vertices, is in $\mathcal{C}$ for any $n \geqslant 0$. (By convention, $K_0 := \emptyset$.)

(iii) If $\Omega = \Omega_1 \cup_{\Omega_0} \Omega_2$, where $\Omega_0 \cong K_n$ for some $n \geqslant 0$ and $\Omega_1, \Omega_2$ are in $\mathcal{C}$, then $\Omega \in \mathcal{C}$.

The proof of Theorem 14.2.1 is based on the following result of the graph theorist G. A. Dirac.

**THEOREM 14.2.5.** (Dirac [99].) *Let $\Omega$ be a finite graph. Then either $\Omega \in \mathcal{C}$ or $\Omega$ contains a full k-cycle for some $k \geqslant 4$.*

The proof of Dirac's Theorem is straightforward (it takes only half a page in [141]) and we omit it.

In 8.8 we gave a precise description of when $W$ is virtually free: it is if and only if it belongs to the smallest class $\mathcal{G}$ containing the spherical Coxeter groups and amalgamated products of groups in $\mathcal{G}$ along spherical special subgroups (Proposition 8.8.5).

*Proof of Theorem 14.2.1.* We need to show that (i) implies (ii). (The other implication was taken care of before the statement of the theorem.) $L^1$ denotes the 1-skeleton of the nerve of $(W, S)$. If $L^1$ contains a full $k$-cycle $C$, with $k \geqslant 4$, then, by Example 14.2.2, $W$ contains a surface group. (Since $C$ is full in $L^1$, the special subgroup $W_{\mathrm{Vert}(C)}$ has nerve $C$ and hence, contains a surface group.) So, by Dirac's Theorem, we can assume that $L^1 \in \mathcal{C}$. By Proposition 8.8.5, $W$ is virtually free if and only if $(W,S)$ is in the class $\mathcal{G}$. So, we must show that if $L^1 \in \mathcal{C}$, then either $W$ contains a surface group or it is in $\mathcal{G}$. We show this by induction on $n := \mathrm{Card}(S)$. If $L^1$ is the complete graph on $n$ vertices, the result follows from Lemma 14.2.4. If not, then by definition of $\mathcal{C}$, $L^1 = \Omega_1 \cup_{\Omega_0} \Omega_2$ where $\Omega_0$ is a complete graph and both $\Omega_1$ and $\Omega_2$ have fewer than $n$ vertices. This induces a splitting $W = W_1 *_{W_0} W_2$, where $W_i$ denotes the special subgroup corresponding to $\Omega_i$, $i = 0, 1, 2$. By Lemma 14.2.4 again, either $W_0$ is finite or it contains a surface subgroup. So, we can assume $W_0$ is finite and by induction, that both $W_1$ and $W_2$ are in $\mathcal{G}$. But this means $W \in \mathcal{G}$. $\square$

# Chapter Fifteen

# ANOTHER LOOK AT (CO)HOMOLOGY

Section 15.1 is separate from the rest of this chapter. It deals with the cohomology of $W$ with trivial coefficients in either $\mathbb{Q}$ (the rationals) or $\mathbb{F}_2$ (the field with two elements). We show that $W$ is rationally acyclic. In the right-angled case, we identify $H^*(W;\mathbb{F}_2)$ with the face ring of its nerve.

In the remaining sections we continue the line begun in Chapter 8. We answer a basic question left open there: what is the $W$-module structure on $H_*(\mathcal{U})$ and $H_c^*(\mathcal{U})$? The answer has three components:

- an identification of the homology (resp. compactly supported cohomology) of $\mathcal{U}$ with an equivariant homology group (resp. cohomology group) of $\mathcal{U}$ with coefficients in the group ring $\mathbb{Z}W$,
- a decreasing filtration of $W$-modules, $\mathbb{Z}W = F_0 \supset \cdots F_p \supset \cdots$, and
- a calculation of the associated graded modules in equivariant (co)homology.

In the end (Theorem 15.3.7), we determine the $W$-action on $H^*(W;\mathbb{Z}W)$ (or, at least, on its associated graded module). Along the way, in 15.2, we get new proofs for the formulas in Theorems 8.1.6 and 8.3.5. When $W$ is finite and the coefficients are in $\mathbb{Q}W$, there is no need to consider any associated graded object: the (co)homology decomposes as a direct sum of $W$-modules which we determine explicitly in Theorem 15.4.3.

## 15.1. COHOMOLOGY WITH CONSTANT COEFFICIENTS

By Theorem 8.2.13, $\Sigma$ is contractble. By Theorem 7.2.4, the orbit space $\Sigma/W$ can be identified with the fundamental chamber $K$, which is also contractible. Hence, Proposition F.1.3 from Appendix F yields the following.

**THEOREM 15.1.1.** *Any Coxeter group $W$ is rationally acyclic, i.e.,*

$$\overline{H}_*(W;\mathbb{Q}) = 0.$$

**The Face Ring of a Simplicial Complex**

Suppose $L$ is a simplicial complex with vertex set $S$. As in Examples A.2.3, $\mathcal{S}(L)$ is the poset of those subsets of $S$ which span simplices in $L$. Introduce indeterminates $(x_s)_{s \in S}$. Given a commutative ring $R$, let $R(S)$ denote the polynomial ring on $(x_s)_{s \in S}$. Let $I$ be the ideal in $R(S)$ generated by all square-free monomials of the form $x_{s_1} \cdots x_{s_p}$ where $\{s_1, \dots, s_p\} \notin \mathcal{S}(L)$. $R[L]$, the *face ring of L* (also called its *Stanley-Reisner ring*) is the quotient ring:

$$R[L] := R(S)/I. \tag{15.1}$$

Since the generators of $I$ are square-free monomials, $R[L]$ inherits the structure of a graded ring from $R(S)$.

As in Example 7.1.7, suppose $(W, S)$ is the right-angled Coxeter system associated to the 1-skeleton of $L$. As in the beginning of 10.8, let $K := |\mathcal{S}(L)|$ be the dual complex to $L$. $(K_s)_{s \in S}$ is the corresponding mirror structure on $K$ and $\mathcal{U}(W, K)$ the result of applying the basic construction to these data. (N.B. If $L$ is not a flag complex, then $\mathcal{U}(W, K) \neq \Sigma$.) Define

$$B_L := \mathcal{U}(W, K) \times_W EW. \tag{15.2}$$

(The twisted product $X \times_G Y$ was defined in 2.2: it is the quotient of $X \times Y$ by the diagonal $G$-action.) Let $\pi : B_L \to \mathcal{U}(W, K)/W = K$ be the map induced by the projection, $\mathcal{U}(W, K) \times EW \to \mathcal{U}(W, K)$.

Since $\mathcal{U}(W, K)$ is simply connected (Theorem 9.1.3), $\pi_1(B_L) = W$. Since $W$ is right-angled, its abelianization is $\mathbf{C}_2^S$ (where $\mathbf{C}_2$ is the cyclic group of order 2). Hence, there is a canonical map $p : B_L \to B\mathbf{C}_2^S = (B\mathbf{C}_2)^S$. Since $B\mathbf{C}_2 = \mathbb{R}P^\infty$, $H^*(B\mathbf{C}_2; \mathbb{F}_2) \cong \mathbb{F}_2[x]$, where $\mathbb{F}_2$ is the field with 2 elements and the indeterminate $x$ corresponds to the nonzero element in $H^1(B\mathbf{C}_2; \mathbb{F}_2)$. Hence, $H^*(B\mathbf{C}_2^S; \mathbb{F}_2) \cong \mathbb{F}_2(S)$, where the indeterminates $x_s$ lie in degree 1. So, we have an induced map on cohomology $p^* : \mathbb{F}_2(S) \to H^*(BL; \mathbb{F}_2)$.

**PROPOSITION 15.1.2.** ([82, Theorem 4.8, p. 436].) *The map $p^*$ induces an isomorphism, $\mathbb{F}_2[L] \cong H^*(B_L; \mathbb{F}_2)$.*

First we prove the proposition in the special cases where $L$ is a simplex or the boundary of a simplex.

**LEMMA 15.1.3.** *Suppose $\sigma$ is the simplex with vertex set $\{s_0, \dots, s_m\}$. Denote the corresponding set of indeterminates $\{x_0, \dots, x_m\}$. Then*

$$H^*(B_\sigma; \mathbb{F}_2) \cong \mathbb{F}_2[x_0, \dots, x_m] = \mathbb{F}_2[\sigma]$$

*and*

$$H^*(B_{\partial\sigma}; \mathbb{F}_2) \cong \mathbb{F}_2[x_0, \dots, x_m]/(x_0 \cdots x_m) = \mathbb{F}_2[\partial\sigma].$$

*Proof.* When $L = \sigma$, $W = \mathbf{C}_2^{m+1}$, $K = [0,1]^{m+1}$ and $\mathcal{U}(\mathbf{C}_2^{m+1}, [0,1]^{m+1}) = [-1,1]^{m+1}$. Hence, $B_\sigma \to B\mathbf{C}_2^{m+1}$ is a $[-1,1]^{m+1}$-bundle associated to a vector bundle which is a sum of line bundles. The first Stiefel-Whitney class of the $i^{\text{th}}$ line bundle is $x_i \in H^1(B\mathbf{C}_2^{m+1}; \mathbb{F}_2)$. Its top Stiefel-Whitney class $\chi$ (its "mod 2 Euler class") is $x_0 \cdots x_m$. Since $B_{\partial\sigma}$ is the sphere bundle associated to this vector bundle, we have the Gysin sequence:

$$\longrightarrow H^i(B_\sigma; \mathbb{F}_2) \xrightarrow{\cup\chi} H^{i+m+1}(B_\sigma; \mathbb{F}_2) \longrightarrow H^{i+m+1}(B_{\partial\sigma}; \mathbb{F}_2) \longrightarrow,$$

from which the lemma follows. □

*Proof of Proposition 15.1.2.* This follows from the Mayer-Vietoris sequence and the previous lemma by induction on the number of simplices in $L$. □

When $L$ is the nerve of a right-angled Coxeter system, $\mathcal{U}(W,K) = \Sigma$ and $B_L \sim BW$. So, Proposition 15.1.2 becomes the following.

**THEOREM 15.1.4.** ([82, Theorem 4.11, p. 437].) *Let $(W,S)$ be a right-angled Coxeter system with nerve $L$. Then $H^*(W;\mathbb{F}_2)$ is isomorphic to the face ring $\mathbb{F}_2[L]$.*

## 15.2. DECOMPOSITIONS OF COEFFICIENT SYSTEMS

Suppose a discrete group $G$ acts cellularly on a CW complex $Y$ and $M$ is a $G$-module. In Appendix F.2 we define the "equivariant (co)homology of $Y$ with coefficients in $M$" to be either the homology of $M \otimes_G C_*(Y)$ or the cohomology of $\mathrm{Hom}_G(C_*(Y), M)$. ($M$ is a right $G$-module in the case of homology and a left $G$-module in the case of cohomology.) If $G$ acts freely on $Y$, then $M$ is a "system of local coefficients" on $Y/G$ (see Appendix F.1 for the definition) and the equivariant (co)homology of $Y$ is just the (co)homology of $Y/G$ with local coefficients in $M$. Even when the action is not free, the same result holds, provided we use a general enough notion of coefficient system. In the case where $G = W$ and $Y = \mathcal{U}(W,X)$, a sufficiently general type of coefficient system on $Y/G$ is described below. (This notion of a "coefficient system" is applicable when $Y \to Y/G$ gives a "simple complex of groups" over $Y/G$ as defined in Appendix E.2. For an even more general notion of coefficient system, see [151].)

### Coefficient Systems

Given a CW complex $X$, let $X^{(i)}$ be its set of $i$-cells and $\mathcal{P}(X) := \bigcup X^{(i)}$. For $c \in X^{(i)}$, $d \in X^{(i-1)}$, $[d:c]$ denotes their incidence number. Write $d < c$ whenever $[d:c] \neq 0$. Extend this to a partial order on $\mathcal{P}(X)$.

**Definition 15.2.1.** A *system of coefficients* on a CW complex $X$ is a functor $\mathcal{F}$ from $\mathcal{P}(X)$ to the category of abelian groups. (Here the poset $\mathcal{P}(X)$ is regarded as a category with $\mathrm{Hom}_{\mathcal{P}(X)}(c, d)$ equal to a singleton whenever $c \leqslant d$ and empty otherwise.)

The functor $\mathcal{F}$ will be contravariant whenever we are dealing with chains or homology and covariant in the case of cochains or cohomology. Define chains and cochains with coefficients in $\mathcal{F}$ by

$$C_i(X;\mathcal{F}) := \bigoplus_{c\in X^{(i)}} \mathcal{F}(c) \quad \text{and} \quad C^i(X;\mathcal{F}) := \prod_{c\in X^{(i)}} \mathcal{F}(c). \tag{15.3}$$

We regard both $i$-chains and $i$-cochains as functions $f$ from $X^{(i)}$ to $\bigcup \mathcal{F}(c)$ such that $f(c) \in \mathcal{F}(c)$ for each $c \in X^{(i)}$. Boundary and coboundary maps are defined by the usual formulas:

$$\partial(f)(c) := \sum [c : d]\mathcal{F}_{cd}(f(d)),$$

$$\delta(f)(c) := \sum [d : c]\mathcal{F}_{dc}(f(d)),$$

where, given an $i$-cell $c$, the first sum is over all $(i + 1)$-cells $d$ which are incident to $c$ and the second sum is over all $(i - 1)$-cells $d$ which are incident to $c$ and where $\mathcal{F}_{cd} : \mathcal{F}(d) \to \mathcal{F}(c)$ is the homomorphism corresponding to $c < d$ (in the first case) or $d < c$ (in the second). From this we get the corresponding (co)homology groups, $H_*(X;\mathcal{F})$ and $H^*(X;\mathcal{F})$.

### Invariants and Coinvariants

Given a left $W$-module $M$ and a subset $T \subset S$, define the $W_T$*-invariants* of $M$ by

$$M^T := M^{W_T} := \{x \in M \mid wx = x \text{ for all } w \in W_T\}. \tag{15.4}$$

$M^T$ is a $\mathbb{Z}$-submodule of $M$. If $M$ is a right $W$-module, define its $W_T$-*coinvariants* by

$$M_T := M_{W_T} := M \otimes_{W_T} \mathbb{Z} \cong M/MI_T, \tag{15.5}$$

where $I_T$ denotes the augmentation ideal in $\mathbb{Z}W_T$ and where $\mathbb{Z}$ has the trivial $W_T$-action.

Now suppose $X$ is a mirrored CW complex over $S$. For each cell $c$ of $X$, let $S(c) := \{s \in S \mid c \subset X_s\}$. Given a left $W$-module $M$, define $\mathcal{I}(M)$ to be the system of coefficients on $X$ which assigns to each cell $c$ the abelian group $M^{S(c)}$. If $d < c$, then $S(c) \subset S(d)$ and we have an inclusion $M^{S(d)} \hookrightarrow M^{S(c)}$. So, $\mathcal{I}(M)$ is a functor on the poset of cells in $X$. Similarly, if $M$ is a right $W$-module, let $\mathcal{C}(M)$ be the system of coefficients which assigns $M_{S(c)}$ to $c$. If $d < c$, we have the projection $M_{S(c)} \to M_{S(d)}$; so, $\mathcal{C}(M)$ is a cofunctor on the poset of cells. Let

$X^{(p)}$ be the set of $p$-cells in $X$. Define

$$C^p(X;\mathcal{I}(M)) := \{f : X^{(p)} \to M \mid f(c) \in M^{S(c)}\} \subset \mathrm{Hom}(C_p(X), M),$$

$$C_p(X;\mathcal{C}(M)) := \bigoplus_{c\in X^{(p)}} M_{S(c)}.$$

For any $\mathbb{Z}$-submodule $N$ of a left $W$-module $M$ and any subset $T$ of $S$, put $N^T := N \cap M^T$. This gives a subcoefficient system $\mathcal{I}(N)$ of $\mathcal{I}(M)$ defined by $\mathcal{I}(N)(c) := N^{S(c)}$. Suppose we have a direct sum decomposition (of $\mathbb{Z}$-modules), $M = N \oplus P$, satisfying the following condition:

$$M^T = N^T \oplus P^T \quad \text{for all } T \in \mathcal{P}(S) \text{ (or } \mathcal{S}). \tag{15.6}$$

This gives a direct sum decomposition of coefficient systems: $\mathcal{I}(M) = \mathcal{I}(N) \oplus \mathcal{I}(P)$. (Here $\mathcal{P}(S)$ is the power set of $S$.)

Similarly, for any $\mathbb{Z}$-submodule $N$ of a right $W$-module $M$ and any $T \subset S$, let $N_T$ denote the image of $N$ in $M_T$. This gives us a subcoefficient system $\mathcal{C}(N)$ of $\mathcal{C}(M)$. As before, a direct sum decomposition $M = N \oplus P$ satisfying

$$M_T = N_T \oplus P_T \quad \text{for all } T \in \mathcal{P}(S) \text{ (or } \mathcal{S}) \tag{15.7}$$

gives a direct sum decomposition of coefficient systems: $\mathcal{C}(M) = \mathcal{C}(N) \oplus \mathcal{C}(P)$.

The key observation in this section is that such direct sum decompositions of coefficient systems give direct sum decompositions of (co)chain complexes and of their (co)homology groups.

Let $\mathbb{Z}(W/W_T)$ denote the (left) permutation module defined by the $W$-action on $W/W_T$.

**LEMMA 15.2.2.** *For any $T \subset S$ and any $W$-module $M$,*

*(i)* $\mathrm{Hom}_W(\mathbb{Z}(W/W_T), M) \cong M^T$,

*(ii)* $M \otimes_W \mathbb{Z}(W/W_T) \cong M_T$.

*($M$ is a left $W$-module in (i) and a right $W$-module in (ii).)*

*Proof*

(i) $\mathrm{Hom}_W(\mathbb{Z}(W/W_T), M)$ can be identified with the set of $W$-equivariant functions $f : W/W_T \to M$. The isotropy subgroup at the coset of the identity element is $W_T$. So, for any equivariant $f$, $f(W_T) \in M^T$. Conversely, given any $x_0 \in M^T$, the formula $f(wW_T) = wx_0$, gives a well-defined $f : W/W_T \to M$.

(ii) $M \otimes_W \mathbb{Z}(W/W_T) = M \otimes_W \mathbb{Z}W \otimes_{W_T} \mathbb{Z} = M \otimes_{W_T} \mathbb{Z} = M_T$. □

*Remark.* Suppose $M$ is a $W$-bimodule. Then the right $W$-action on $M$ gives $\mathrm{Hom}_W(\mathbb{Z}(W/W_T), M)$ and $M^T$ the structure of right $W$-modules and (i) is an isomorphism of right $W$-modules. Similarly, (ii) is an isomorphism of left $W$-modules.

The basic construction gives a $W$-complex $\mathcal{U}(W,X)$ $(=\mathcal{U})$. Let $H^*_W(\mathcal{U};M)$ and $H^W_*(\mathcal{U};M)$ be the equivariant cohomology and homology groups, respectively, as defined in Appendix F.2. A corollary to Lemma 15.2.2 is the following.

**COROLLARY 15.2.3.** *There are natural identifications:*

$$C^W_*(\mathcal{U};M) = C_*(X;\mathcal{C}(M)) \quad \textit{and} \quad C^*_W(\mathcal{U};M) = C^*(X;\mathcal{I}(M)).$$

## Symmetrization and Alternation Again

Let $\{e_w\}_{w\in W}$ be the standard basis for $\mathbb{Z}W$. For each spherical subset $T$ of $S$, define elements $\tilde{a}_T$ and $\tilde{h}_T$ in $\mathbb{Z}W$ by formula (8.1) of Section 8.1:

$$\tilde{a}_T := \sum_{w\in W_T} e_w \quad \text{and} \quad \tilde{h}_T := \sum_{w\in W_T} (-1)^{l(w)} e_w.$$

For any subset $T$ of $S$, let $(\mathbb{Z}W)^T$ denote the $W_T$-invariants in $\mathbb{Z}W$, defined in (15.4). Notice that $(\mathbb{Z}W)^T$ is 0 if $T \notin \mathcal{S}$ and is equal to the right ideal $\tilde{a}_T(\mathbb{Z}W)$ if $T \in \mathcal{S}$. Similarly, for $T \subset S$, define

$$H^T := \{x \in \mathbb{Z}W \mid xe_u = (-1)^{l(u)}x, \text{ for all } u \in W_T\}.$$

One checks that $H^T$ is the left ideal $(\mathbb{Z}W)\tilde{h}_T$ when $T$ is spherical and that $H^T = 0$ when it is not.

Let $(\mathbb{Z}W)_T$ be the $W_T$-coinvariants, defined in (15.5), and $I_T$ the augmentation ideal of $\mathbb{Z}(W_T)$. For any $s \in S$, note that $(\mathbb{Z}W)I_{\{s\}} = H^{\{s\}}$. Hence, $(\mathbb{Z}W)_{\{s\}} = (\mathbb{Z}W)/H^{\{s\}}$. More generally, for any $T \subset S$,

$$(\mathbb{Z}W)I_T = \sum_{s\in T} H^{\{s\}} \quad \text{so,} \quad (\mathbb{Z}W)_T = (\mathbb{Z}W)\Big/\sum_{s\in T} H^{\{s\}}. \tag{15.8}$$

## Two Bases for $\mathbb{Z}W$

As in 4.7, for each $w \in W$, $\mathrm{In}(w)$ denotes the set of letters with which $w$ can end (Definition 4.7.1). It is a spherical subset of $S$ (Lemma 4.7.2). Put

$$\mathrm{In}'(w) := \mathrm{In}(w^{-1}) = \{s \in S \mid l(sw) < l(w)\}. \tag{15.9}$$

Define elements $b'_w, b_w \in \mathbb{Z}W$ by

$$b'_w := \tilde{a}_{\mathrm{In}'(w)} e_w \quad \text{and} \quad b_w := e_w \tilde{h}_{\mathrm{In}(w)}. \tag{15.10}$$

**LEMMA 15.2.4.** *$\{b'_w\}_{w\in W}$ is a basis for $\mathbb{Z}W$ (as a $\mathbb{Z}$-module). More generally, for any $T \in \mathcal{S}$, $\{b'_w \mid T \subset \mathrm{In}'(w)\}$ is a basis for $(\mathbb{Z}W)^T$.*

Recall from 4.5 that for any $T \subset S$, $A_T$ (resp. $B_T$) is the set of elements of $W$ which are $(T, \emptyset)$-reduced (resp. $(\emptyset, T)$-reduced) (see Definition 4.3.2). For any spherical subset $T$, $w_T \in W_T$ is the element of longest length.

*Proof of Lemma 15.2.4.* The point is that the matrix which expresses the $b'_w$ in terms of the $e_w$ has 1's on the diagonal and is upper triangular when the elements of $W$ are ordered compatibly with word length. In detail: suppose $\sum \beta_w b'_w = 0$ is a nontrivial linear relation. Let $v \in W$ be an element with $\beta_v \neq 0$ and $l(v)$ maximum. Note that $b'_v$ is the sum of $e_v$ with various $e_w$ having $l(w) < l(v)$. Since the coefficient of $e_v$ in the linear relation must be 0, we have $\beta_v = 0$; so, $\{b'_w\}_{w\in W}$ is linearly independent. Similarly, one shows, by induction on word length, that each $e_v$ is a linear combination of $b'_w$ with $l(w) \leqslant l(v)$. Hence, $\{b'_w\}$ spans $\mathbb{Z}W$.

To prove the second sentence, we first show that $b'_w \in (\mathbb{Z}W)^T$ whenever $T \subset \mathrm{In}'(w)$. For any $U \in \mathcal{S}$ with $T \subset U$, computation gives $\tilde{a}_U = \tilde{a}_T c_{U,T}$, where

$$c_{U,T} := \sum_{u\in A_T \cap W_U} e_u.$$

So, when $T \subset \mathrm{In}'(w)$, $b'_w = \tilde{a}_{\mathrm{In}'(w)} e_w = \tilde{a}_T c_{\mathrm{In}'(w),T} e_w \in (\mathbb{Z}W)^T$. Note $T \subset \mathrm{In}'(w)$ if and only if $w \in w_T A_T$. By the previous paragraph, $\{b'_w \mid w \in w_T A_T\}$ is linearly independent. It remains to show it spans $(\mathbb{Z}W)^T$. Since $w_T A_T$ is a set of coset representatives for $W_T\backslash W$, a basis for $(\mathbb{Z}W)^T$ is $\{\tilde{a}_T e_w \mid w \in w_T A_T\}$. Let $\overline{e}_w := c_{\mathrm{In}'(w),T} e_w$. For $w \in w_T A_T$, the matrix which expresses $\{\overline{e}_w \mid w \in w_T A_T\}$ in terms of $\{e_w \mid w \in w_T A_T\}$ has 1's on the diagonal is upper triangular with respect to word length. So,

$$\{\tilde{a}_T \overline{e}_w \mid w \in w_T A_T\} = \{b'_w \mid T \subset \mathrm{In}'(w)\}$$

is also a basis for $(\mathbb{Z}W)^T$. □

**LEMMA 15.2.5.** *$\{b_w\}_{w\in W}$ is a basis for $\mathbb{Z}W$. More generally, for any subset $U$ of $S$, the projection $\mathbb{Z}W \to (\mathbb{Z}W)_{S-U}$ maps $\{b_w \mid U \supset \mathrm{In}(w)\}$ injectively to a basis for $(\mathbb{Z}W)_{S-U}$.*

*Proof.* The proof of the first sentence is omitted since it is similar to that of the first sentence of the previous lemma.

Fix a subset $U \subset S$ and let $p : \mathbb{Z}W \to (\mathbb{Z}W)_{S-U}$ denote the projection. Since $(\mathbb{Z}W)_{S-U} = \mathbb{Z}(W/W_{S-U})$, $\{p(e_w) \mid w \in B_{S-U}\}$ is an obvious basis for $(\mathbb{Z}W)_{S-U}$ (as a $\mathbb{Z}$-module). Any element $y \in \mathbb{Z}W$ can be written in the form

$$y = \sum_{w\in B_{S-U}} \sum_{u\in W_{S-U}} \alpha_{wu} e_{wu}. \tag{15.11}$$

An element $y$ belongs to $(\mathbb{Z}W)I_{S-U} = \mathrm{Ker}(p)$ if and only if $\sum_{u \in W_{S-U}} \alpha_{wu} = 0$ for each $w \in B_{S-U}$. Let $y$ be an element in the $\mathbb{Z}$-submodule spanned by $\{b_w \mid U \supset \mathrm{In}(w)\}$ $(= \{b_w \mid w \in B_{S-U}\})$, i.e., let

$$y = \sum_{w \in B_{S-U}} y_w b_w.$$

Suppose $p(y) = 0$. Let $v \in B_{S-U}$ be such that $y_v \neq 0$ and $l(v)$ is maximum with respect to this property. Since $b_v$ is the sum of $e_v$ and $\pm 1$ times various $e_w$ with $l(w) < l(v)$, the coefficients $\alpha_{vu}$ in (15.11) are 0 for all $u \neq 1$ in $W_{S-U}$. Then $\sum \alpha_{vu} = 0$ forces $\alpha_v = 0$; so, $\{p(b_w) \mid w \in B_{S-U}\}$ is linearly independent in $(\mathbb{Z}W)_{S-U}$. The usual argument, using induction on word length, shows that $\{p(b_w) \mid w \in B_{S-U}\}$ spans $(\mathbb{Z}W)_{S-U}$. □

For each $T \in \mathcal{S}$, define $\mathbb{Z}$-submodules of $\mathbb{Z}W$:

$$\widehat{A}^T := \mathrm{Span}\{b'_w \mid \mathrm{In}'(w) = T\}, \tag{15.12}$$

$$\widehat{H}^T := \mathrm{Span}\{b_w \mid \mathrm{In}(w) = T\}. \tag{15.13}$$

A corollary to Lemma 15.2.4 is the following.

**COROLLARY 15.2.6.** *We have direct sum decompositions of $\mathbb{Z}$-modules:*

$$\mathbb{Z}W = \bigoplus_{T \in \mathcal{S}} \widehat{A}^T \qquad \text{and for any } U \in \mathcal{S} \qquad (\mathbb{Z}W)^U = \bigoplus_{T \in \mathcal{S}_{\geqslant U}} \widehat{A}^T.$$

Consequently, given $T \in \mathcal{S}$, for any $U \subset S$ we have

$$(\widehat{A}^T)^U = \begin{cases} \widehat{A}^T & \text{if } U \subset T, \\ 0 & \text{if } U \cap (S - T) \neq \emptyset. \end{cases} \tag{15.14}$$

So, the decomposition satisfies condition (15.6). Hence, the direct sum decomposition in Corollary 15.2.6 gives a decomposition of coefficient systems:

$$\mathcal{I}(\mathbb{Z}W) = \bigoplus_{T \in \mathcal{S}} \mathcal{I}(\widehat{A}^T). \tag{15.15}$$

For the $\widehat{H}^T$, we have

$$(\widehat{H}^T)_U \cong \begin{cases} \widehat{H}^T & \text{if } U \subset S - T, \\ 0 & \text{if } U \cap T \neq \emptyset. \end{cases} \tag{15.16}$$

In the above formula, by writing $(\widehat{H}^T)_U \cong \widehat{H}^T$, we mean that the projection $\mathbb{Z}W \to (\mathbb{Z}W)_U$ is injective on $\widehat{H}^T$. To see that $(\widehat{H}^T)_U = 0$ when $U \cap T \neq \emptyset$, note that if $s \in U \cap T$, then $\widehat{H}^T \subset H^s \subset (\mathbb{Z}W)I_U$. The $\widehat{H}^T$ version of Corollary 15.2.6 is the following.

**COROLLARY 15.2.7.** *We have direct sum decompositions of $\mathbb{Z}$-modules:*

$$\mathbb{Z}W = \bigoplus_{T \in \mathcal{S}} \widehat{H}^T \qquad \text{and for any } U \in \mathcal{S} \qquad (\mathbb{Z}W)_{S-U} = \bigoplus_{T \in \mathcal{S}_{\leqslant U}} (\widehat{H}^T)_{S-U}.$$

So the decomposition in Corollary 15.2.7 satisfies (15.7) and gives a decomposition of coefficient systems:

$$\mathcal{C}(\mathbb{Z}W) = \bigoplus_{T \in \mathcal{S}} \mathcal{C}(\widehat{H}^T). \tag{15.17}$$

## Chapter 8 Revisited

We can use the previous two corollaries to give different proofs of the two main results in Chapter 8, the formulas in Theorems 8.1.6 and 8.3.5. For the remainder of this chapter, in all formulas involving cochains or cohomology we shall assume $X$ is compact and that its mirror structure is $W$-finite (Definition 5.1.6), i.e., $X_T = \emptyset$ whenever $W_T$ is infinite. These assumptions imply that $W$ acts properly and cocompactly on $\mathcal{U}$ (Lemma 5.1.7). In the case of homology no such assumptions are needed.

By Lemma F.2.1,

$$C_*(\mathcal{U}) = C_*^W(\mathcal{U}; \mathbb{Z}W), \tag{15.18}$$

$$C_c^*(\mathcal{U}) = C_W^*(\mathcal{U}; \mathbb{Z}W). \tag{15.19}$$

Recall Definition 4.7.4: $W^T := \{w \in W \mid \operatorname{In}(w) = T\}$. Since $w \to w^{-1}$ is a bijection from $W^T$ to $\{w \in W \mid \operatorname{In}'(w) = T\}$, we see from (15.12) and (15.13) that both $\widehat{A}^T$ and $\widehat{H}^T$ have one basis element for each element of $W^T$. So, the next theorem is the same calculation as in Chapter 8.

**THEOREM 15.2.8.** (= Theorems 8.1.6 and 8.3.5.)

$$H_c^i(\mathcal{U}) \cong \bigoplus_{T \in \mathcal{S}} H^i(X, X^{S-T}) \otimes \widehat{A}^T,$$

$$H_i(\mathcal{U}) \cong \bigoplus_{T \in \mathcal{S}} H_i(X, X^T) \otimes \widehat{H}^T.$$

*Proof.* To prove the first formula, note that by (15.19) and Corollaries 15.2.3 and 15.2.7,

$$C_c^i(\mathcal{U}) = C_*^W(\mathcal{U}; \mathbb{Z}W) = C^i(X; \mathcal{I}(\mathbb{Z}W)) = \bigoplus_{T \in \mathcal{S}} C^i(X; \mathcal{I}(\widehat{A}^T)).$$

Given a cell $c \in X^{(i)}$, by (15.14),

$$(\widehat{A}^T)^{S(c)} = \begin{cases} 0 & \text{if } c \subset X^{S-T}, \\ \widehat{A}^T & \text{otherwise.} \end{cases}$$

Hence,

$$\begin{aligned} C^i(X; \mathcal{I}(\widehat{A}^T)) &= \{f : X^{(i)} \to \widehat{A}^T \mid f(c) = 0 \text{ if } c \subset X^{S-T}\} \\ &= C^i(X, X^{S-T}) \otimes \widehat{A}^T. \end{aligned}$$

Taking cohomology, we get the first formula.

To prove the second formula, note that by (15.18) and Corollaries 15.2.3 and 15.2.6,

$$C_i(\mathcal{U}) = C_*^W(\mathcal{U}; \mathbb{Z}W) = C_i(X; \mathcal{C}(\mathbb{Z}W)) = \bigoplus_{T \in \mathcal{S}} C_i(X; \mathcal{C}(\widehat{H}^T)).$$

By (15.16), given a cell $c \in X^{(i)}$,

$$(\widehat{H}^T)_{S(c)} \cong \begin{cases} 0 & \text{if } c \subset X^T, \\ \widehat{H}^T & \text{otherwise.} \end{cases}$$

Hence,

$$C_i(X; \mathcal{C}(\widehat{H}^T)) = \bigoplus_{\substack{c \in X^{(i)} \\ c \not\subset X^T}} \widehat{H}^T \cong C_i(X, X^T) \otimes \widehat{H}^T.$$

Taking homology, we get the second formula. □

## 15.3. THE $W$-MODULE STRUCTURE ON (CO)HOMOLOGY

The $W$-action on $\mathcal{U}$ makes $H_*(\mathcal{U})$ into a left $W$-module and $H_c^*(\mathcal{U})$ into a right $W$-module. Since $\mathbb{Z}W$ is a bimodule, $\mathcal{I}(\mathbb{Z}W)$ is a system of right $W$-modules and hence, $H^*(X; \mathcal{I}(\mathbb{Z}W))$ is a right $W$-module. Similarly, $\mathcal{C}(\mathbb{Z}W)$ is a system of left $W$-modules and $H_*(X; \mathcal{C}(\mathbb{Z}W))$ is a left $W$-module. Moreover, the isomorphisms,

$$H_*(\mathcal{U}) \cong H_*^W(\mathcal{U}; \mathbb{Z}W) = H_*(X; \mathcal{C}(\mathbb{Z}W))$$

and

$$H_c^*(\mathcal{U}) \cong H_W^*(\mathcal{U}; \mathbb{Z}W) = H^*(X; \mathcal{I}(\mathbb{Z}W)),$$

of (15.18), (15.19), and Corollary 15.2.3 are compatible with the $W$-actions.

Our plan is to define a decreasing filtration, $\mathbb{Z}W = F_0 \supset F_1 \supset \cdots$ of right $W$-modules, leading to a filtration of $H^*(X;\mathcal{I}(\mathbb{Z}W))$ by $W$-submodules. We will then compute the associated graded terms. Similarly, we will define a decreasing filtration $(F'_p)_{p\geqslant 0}$ of left $W$-modules and use it to compute an associated graded module of $H_*(X;\mathcal{C}(\mathbb{Z}W))$.

For each nonnegative integer $p$, define

$$F_p := \sum_{\mathrm{Card}(T)\geqslant p} (\mathbb{Z}W)^T, \qquad E_p := \bigoplus_{\mathrm{Card}(T)<p} \widehat{A}^T, \tag{15.20}$$

$$F'_p := \sum_{\mathrm{Card}(T)\geqslant p} H^T, \qquad E'_p := \bigoplus_{\mathrm{Card}(T)<p} \widehat{H}^T, \tag{15.21}$$

where $H^T$ is defined by alternation over $W_T$ and $\widehat{A}^T$, $\widehat{H}^T$ are defined in (15.12) and (15.13), respectively. $(F_p)_{p\geqslant 0}$ and $(F'_p)_{p\geqslant 0}$ are clearly decreasing filtrations. $F_p$ is a right $W$-module and $\mathcal{I}(F_p)$ is a subsystem of $\mathcal{I}(\mathbb{Z}W)$ of right $W$-modules. Similarly, $\mathcal{C}(F'_p)$ is a subsystem of left $W$-modules. (However, since $E_p$ and $E'_p$ only have the structure of $\mathbb{Z}$-submodules of $\mathbb{Z}W$, $\mathcal{I}(E_p)$ and $\mathcal{C}(E'_p)$ are *not* systems of $W$-modules.) As we shall see below, $E_p$ (resp. $E'_p$) is a complementary $\mathbb{Z}$-submodule to $F_p$ (resp. $F'_p$).

**LEMMA 15.3.1.** *We have decompositions (as $\mathbb{Z}$-modules):*

*(i)* $\mathbb{Z}W = F_p \oplus E_p$ *and this induces a decomposition of coefficient systems,* $\mathcal{I}(\mathbb{Z}W) = \mathcal{I}(F_p) \oplus \mathcal{I}(E_p)$.

*(ii)* $\mathbb{Z}W = F'_p \oplus E'_p$ *and this induces a decomposition of coefficient systems,* $\mathcal{C}(A) = \mathcal{C}(F'_p) \oplus \mathcal{C}(E'_p)$.

*Proof.* (i) By the second formula of Corollary 15.2.6, $F_p = \bigoplus_{\mathrm{Card}(T)\geqslant p} \widehat{A}^T$; hence, by the first formula of the same corollary, $\mathbb{Z}W = F_p \oplus E_p$. To get a decomposition of coefficient systems, we must show $(\mathbb{Z}W)^U = (F_p)^U \oplus (E_p)^U$ for all $U \subset S$. Since $(\mathbb{Z}W)^U = \bigoplus_{T\supset U} \widehat{A}^T$,

$$(\mathbb{Z}W)^U = \bigoplus_{\substack{T\supset U\\ \mathrm{Card}(T)\geqslant p}} \widehat{A}^T \oplus \bigoplus_{\substack{T\supset U\\ \mathrm{Card}(T)<p}} \widehat{A}^T. \tag{15.22}$$

Denote the first summation in (15.22) by $B$ and the second one by $C$.

*Claim.* $B = (F_p)^U$.

*Proof of Claim.* Obviously, $B \subset (F_p)^U$. Let $x \in (F_p)^U$. Since $x \in F_p$,

$$x = \sum_{\mathrm{Card}(T) \geqslant p} \alpha^T,$$

where $\alpha^T \in \widehat{A}^T$. Since $x \in (\mathbb{Z}W)^U$,

$$x = \sum_{T \supset U} \beta^T,$$

where $\beta^T \in \widehat{A}^T$. But $A = \bigoplus_{T \subset S} \widehat{A}^T$, so the two decompositions of $x$ coincide. Therefore, $\alpha^T = 0$ unless $T \supset U$. This gives

$$x = \sum_{\substack{T \supset U \\ \mathrm{Card}(T) \geqslant p}} \alpha^T \in B,$$

which proves that $(F_p)^U \subset B$. □

We continue with the proof of the lemma by noting that a similar argument shows $(E_p)^U = C$. Hence, $(\mathbb{Z}W)^U = (F_p)^U \oplus (E_p)^U$ and (i) is proved.

(ii) As before, by Corollary 15.2.7, $\mathbb{Z}W = F'_p \oplus E'_p$. To get the decomposition of coefficient systems, we must show that $(\mathbb{Z}W)_{S-U} = (F'_p)_{S-U} \oplus (E'_p)_{S-U}$ for all $U \subset S$. Since $(\mathbb{Z}W)_{S-U} = \bigoplus_{T \subset U} (\widehat{H}^T)_{S-U}$,

$$(\mathbb{Z}W)_{S-U} = \bigoplus_{\substack{T \subset U \\ \mathrm{Card}(T) \geqslant p}} (\widehat{H}^T)_{S-U} \oplus \bigoplus_{\substack{T \subset U \\ \mathrm{Card}(T) < p}} (\widehat{H}^T)_{S-U}. \tag{15.23}$$

Denote the first summation in (15.23) by $B'$ and the second by $C'$. We claim that $(F'_p)_{S-U} = B'$. Obviously, $B' \subset (F'_p)_{S-U}$. Any $x \in F_p$ can be written in the form

$$x = \sum_{\mathrm{Card}(T) \geqslant p} \gamma^T,$$

where $\gamma^T \in \widehat{H}^T$. Since $\gamma^T \in I_{S-U}$ whenever $T \cap (S - U) \neq \emptyset$, for those $T$ with $T \not\subset U$, we can set $\gamma^T = 0$ without changing the congruence class of $x$ modulo $I_{S-U}$. So, putting

$$y = \sum_{\substack{T \subset U \\ \mathrm{Card}(T) \geqslant p}} \gamma^T,$$

we have $y \equiv x \mod I_{S-U}$ and $y \in B'$. So, $(F'_p)_{S-U} \subset B'$. A similar argument shows $(E'_p)_{S-U} = C'$. Hence, $(\mathbb{Z}W)_{S-U} = (F'_p)_{S-U} \oplus (E'_p)_{S-U}$ and (ii) is proved. □

**COROLLARY 15.3.2**

*(i)* $F_p \hookrightarrow \mathbb{Z}W$ *induces an embedding* $H^i(X;\mathcal{I}(F_p)) \hookrightarrow H^i(X;\mathcal{I}(\mathbb{Z}W))$ *which is W-equivariant. Its image is a* $\mathbb{Z}$*-module direct summand.*

*(ii)* $F'_p \hookrightarrow \mathbb{Z}W$ *induces an embedding* $H_i(X;\mathcal{C}(F'_p)) \hookrightarrow H_i(X;\mathcal{C}(\mathbb{Z}W))$ *which is W-equivariant. Its image is a* $\mathbb{Z}$*-module direct summand.*

It follows that $F_{p+1} \hookrightarrow F_p$ induces $H^*(X;\mathcal{I}(F_{p+1})) \hookrightarrow H^*(X;\mathcal{I}(F_p))$, an embedding of right $W$-modules. This gives an associated graded group of right $W$-modules,

$$H^*(X;\mathcal{I}(F_p))/H^*(X;\mathcal{I}(F_{p+1})).$$

Similarly, there is an embedding, $H_*(X;\mathcal{C}(F'_{p+1})) \hookrightarrow H_*(X;\mathcal{C}(F'_p))$ and an associated graded group of left $W$-modules. The goal of this section is to prove Theorem 15.3.4 below. It gives a complete computation of these graded $W$-modules.

For each $T \in \mathcal{S}$, put

$$(\mathbb{Z}W)^{>T} := \sum_{U \supsetneq T} (\mathbb{Z}W)^U \quad \text{and} \quad H^{>T} := \sum_{U \supsetneq T} H^U.$$

$(\mathbb{Z}W)^T/(\mathbb{Z}W)^{>T}$ is a right $W$-module and $H^T/H^{>T}$ is a left $W$-module.

**Example 15.3.3.** (*The sign representation.*) $(\mathbb{Z}W)^{\emptyset}/(\mathbb{Z}W)^{>\emptyset}$ is isomorphic to $\mathbb{Z}$ as an abelian group. We can take the image $\overline{b}'_1$ of the basis element $b'_1$ $(= e_1)$ as the generator. Since $a_s b'_1 \in (\mathbb{Z}W)^{>\emptyset}$, $\overline{b}'_1 \cdot a_s = 0$ for all $s \in \mathcal{S}$. Hence, $\overline{b}'_1 \cdot s = -\overline{b}'_1$. It follows that $W$ acts on $(\mathbb{Z}W)^{\emptyset}/(\mathbb{Z}W)^{>\emptyset}$ via the sign representation:

$$\overline{b}'_1 \cdot w = (-1)^{l(w)} \overline{b}'_1.$$

**THEOREM 15.3.4.** *Let p be a nonnegative integer.*

*(i) There is an isomorphism of right W-modules,*

$$H^*(X;\mathcal{I}(F_p))/H^*(X;\mathcal{I}(F_{p+1})) \cong \bigoplus_{\mathrm{Card}(T)=p} H^*(X, X^{S-T}) \otimes ((\mathbb{Z}W)^T/(\mathbb{Z}W)^{>T}).$$

*(ii) There is an isomorphism of left $W$-modules,*

$$H_*(X;\mathcal{C}(F'_p))/H^*(X;\mathcal{C}(F'_{p+1})) \cong \bigoplus_{\mathrm{Card}(T)=p} H_*(X,X^T)\otimes(H^T/H^{>T}).$$

**LEMMA 15.3.5.** *There are isomorphisms of $W$-modules*

$$\psi : F_p/F_{p+1} \xrightarrow{\cong} \bigoplus_{\mathrm{Card}(T)=p} (\mathbb{Z}W)^T/(\mathbb{Z}W)^{>T}$$

*and*

$$\psi' : F'_p/F'_{p+1} \xrightarrow{\cong} \bigoplus_{\mathrm{Card}(T)=p} H^T/H^{>T}.$$

*Proof.* Suppose $\mathrm{Card}(T)=p$. The inclusion $(\mathbb{Z}W)^T \hookrightarrow F_p$ induces $(\mathbb{Z}W)^T \to F_p/F_{p+1}$ and $(\mathbb{Z}W)^{>T}$ is in the kernel. Let $(\mathbb{Z}W)^T/(\mathbb{Z}W)^{>T} \to F_p/F_{p+1}$be the induced map. This gives a map of right $W$-modules:

$$\varphi : \bigoplus_{\mathrm{Card}(T)=p} (\mathbb{Z}W)^T/(\mathbb{Z}W)^{>T} \to F_p/F_{p+1}.$$

By Corollary 15.2.6, the inclusion $\widehat{A}^T \hookrightarrow (\mathbb{Z}W)^T$ induces an isomorphism (of $\mathbb{Z}$-modules), $\widehat{A}^T \to (\mathbb{Z}W)^T/(\mathbb{Z}W)^{>T}$. Also, $F_p = \bigoplus_{\mathrm{Card}(T)=p}\widehat{A}^T \oplus F_{p+1}$. So, we have a commutative diagram (of maps of $\mathbb{Z}$-modules):

$$\begin{array}{ccc} \bigoplus_{\mathrm{Card}(T)=p}(\mathbb{Z}W)^T/(\mathbb{Z}W)^{>T} & \xrightarrow{\varphi} & F_p/F_{p+1} \\ \nwarrow & & \nearrow \\ & \bigoplus_{\mathrm{Card}(T)=p}\widehat{A}^T & \end{array}$$

Since the two slanted arrows are bijections, so is $\varphi$. Therefore, $\varphi$ is an isomorphism of right $W$-modules. Put $\psi := \varphi^{-1}$.

The definition of the second isomorphism $\psi'$ is similar. □

**NOTATION.** $D_T := (\mathbb{Z}W)^T/(\mathbb{Z}W)^{>T}$ *and* $G_T = H^T/H^{>T}$.

*Remark.* If $T$ is nonspherical, then both $D_T$ and $G_T$ are 0 (since the same is true for $(\mathbb{Z}W)^T$ and $H^T$).

Since the right $W$-module $D_T$ is neither a left $W$-module or even a $\mathbb{Z}$-submodule of a left $W$-module, the definition of its (left) $W_U$-invariants from (15.4) cannot be applied directly. Similarly, the definition of (right) coinvariants from (15.5) does not apply directly to $G_T$. Nevertheless, for each $U \subset S$, define

$$(D_T)^U := ((\mathbb{Z}W)^T \cap (\mathbb{Z}W)^U)/((\mathbb{Z}W)^{>T} \cap (\mathbb{Z}W)^U),$$
$$(G_T)_U := (H^T)_U/(H^{>T})_U.$$

These give a systems of $W$-modules on $X$ defined by

$$\mathcal{I}(D_T)(c) := (D_T)^{S(c)},$$
$$\mathcal{C}(G_T)(c) := (G_T)_{S(c)},$$

respectively. As in (15.14) and (15.16),

$$(D_T)^U = \begin{cases} (\mathbb{Z}W)^T/(\mathbb{Z}W)^{>T} & \text{if } U \subset T, \\ 0 & \text{if } U \cap S_T \neq \emptyset; \end{cases} \tag{15.24}$$

$$(G_T)_U = \begin{cases} H^T/H^{>T} & \text{if } U \subset S - T, \\ 0 & \text{if } U \cap T \neq \emptyset. \end{cases} \tag{15.25}$$

**Lemma 15.3.6.** *Suppose $U \subset S$.*

*(i) The following sequence of right $W$-modules is exact,*

$$0 \longrightarrow (F_{p+1})^U \longrightarrow (F_p)^U \xrightarrow{\tilde{\psi}} \bigoplus_{\mathrm{Card}(T)=p} (D_T)^U \longrightarrow 0,$$

*where $\tilde{\psi}$ is the map induced by $\psi$ and*

*(ii) The following sequence of left $W$-modules is exact,*

$$0 \longrightarrow (F'_{p+1})_{S-U} \longrightarrow (F'_p)_{S-U} \xrightarrow{\tilde{\psi}'} \bigoplus_{\mathrm{Card}(T)=p} (G_T)_{S-U} \longrightarrow 0,$$

*where $\tilde{\psi}'$ is the map induced by $\psi'$.*

*Proof.* In the proof of Lemma 15.3.1, in (15.22), we showed

$$(F_p)^U = \bigoplus_{\substack{\mathrm{Card}(T) \geqslant p \\ T \supset U}} \widehat{A}^T.$$

Put

$$B := \bigoplus_{\substack{\mathrm{Card}(T) = p \\ T \supset U}} \widehat{A}^T.$$

$B$ is a $\mathbb{Z}$-submodule of $(F_p)^U$ and it maps isomorphically onto $(F_p)^U/(F_{p+1})^U$.

The image of $B$ under $\tilde{\psi}$ is

$$\bigoplus_{\substack{\mathrm{Card}(T)=p \\ T\supset U}} (\mathbb{Z}W)^T/(\mathbb{Z}W)^{>T} = \bigoplus_{\substack{\mathrm{Card}(T)=p \\ T\supset U}} D_T.$$

This proves (i).

The proof that the sequence in (ii) is short exact is similar. □

*Proof of Theorem 15.3.4.* (i) By Lemma 15.3.6 (i), we have a short exact sequence of coefficient systems on $X$:

$$0 \longrightarrow \mathcal{I}(F_{p+1}) \longrightarrow \mathcal{I}(F_p) \longrightarrow \bigoplus_{\mathrm{Card}(T)=p} \mathcal{I}(D_T) \longrightarrow 0$$

inducing a short exact sequence of cochain complexes:

$$0 \longrightarrow C^*(X;\mathcal{I}(F_{p+1})) \longrightarrow C^*(X;\mathcal{I}(F_p)) \longrightarrow \bigoplus_{\mathrm{Card}(T)=p} C^*(X;\mathcal{I}(D_T)) \longrightarrow 0.$$

By the argument for Corollary 15.3.2, $H^*(X;\mathcal{I}(F_{p+1})) \to H^*(X;\mathcal{I}(F_p))$ is an injection onto a $\mathbb{Z}$-module direct summand. Hence, the long exact sequence in cohomology decomposes into short exact sequences and we get

$$\begin{aligned} H^i(X;\mathcal{I}(F_p))/H^i(X;\mathcal{I}(F_{p+1})) &\cong \bigoplus_{\mathrm{Card}(T)=p} H^i(X;\mathcal{I}(D_T)) \\ &\cong \bigoplus_{\mathrm{Card}(T)=p} H^i(X, X^{S-T}) \otimes D_T, \end{aligned}$$

where the second isomorphism comes from using (15.24), to get

$$\begin{aligned} C^i(X;\mathcal{I}(D_T)) &= \{ f : X^{(i)} \to D_T \mid f(c) \in (D_T)^{S(c)} \text{ for all } c \in X^{(i)} \} \\ &= \{ f : X^{(i)} \to D_T \mid f(c) = 0 \text{ if } c \subset X^{S-T} \} \\ &= C^i(X, X^{S-T}) \otimes D_T. \end{aligned}$$

(ii) The proof is similar to that of (i). □

*Remark.* The decreasing filtration $\cdots \supset F_p \supset F_{p+1} \cdots$ of (15.20) gives a filtration of cochain complexes

$$\cdots \supset C^*(X;\mathcal{I}(F_p)) \supset C^*(X;\mathcal{I}(F_{p+1})) \cdots.$$

The quotient cochain complexes have the form $C^*(X;\mathcal{I}(F_p)/\mathcal{I}(F_{p+1}))$. As explained in the beginning of Appendix E.3, by taking homology, we get a

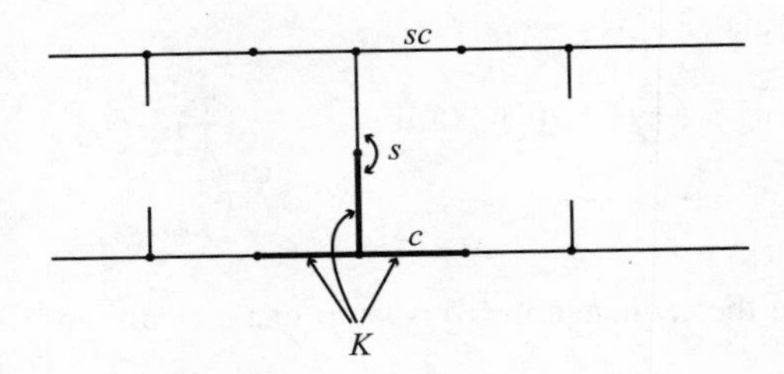


Figure 15.1. Cocycles $x$ and $sx$.

spectral sequence with $E_1$-term:

$$E_1^{pq} := H^{p+q}(X; \mathcal{I}(F_p)/\mathcal{I}(F_{p+1})).$$

It converges to

$$E_\infty^{pq} := \frac{H^{p+q}(X; \mathcal{I}(F_p))}{\mathrm{Im}(H^{p+q}(X; \mathcal{I}(F_{p+1})))}. \tag{15.26}$$

So the meaning of Theorem 15.3.4 is that $E_1^{pq} = E_\infty^{pq}$.

### The $W$-Module Structure on $H^*(W; \mathbb{Z}W)$

As above, we have a filtration of $H_c^*(\Sigma) = H^*(K; \mathcal{I}(\mathbb{Z}W))$ by $W$-submodules. As in (15.26), $E_\infty^{pq}$ is the associated right $W$-module in filtration degree $p$ associated to $H^{p+q}(K; \mathcal{I}(\mathbb{Z}W))$. So Theorem 15.3.4 has the following corollary.

**THEOREM 15.3.7.** *As a right $W$-module, $H^{p+q}(W; \mathbb{Z}W)$ $(= H_c^{p+q}(\Sigma))$ has as its associated graded module,*

$$E_\infty^{pq} = \bigoplus_{\mathrm{Card}(T)=p} H^{p+q}(K, K^{S-T}) \otimes D_T.$$

By Theorem 15.2.8, we have a direct sum decomposition of $\mathbb{Z}$-modules:

$$H_c^*(\Sigma) \cong \bigoplus_{T \in \mathcal{S}} H^*(K, K^{S-T}) \otimes D_T.$$

In view of Theorem 15.4.1, one might naively conjecture that $H_c^*(\Sigma)$ decomposes into a direct sum of right $W$-modules as above. In fact there is, in general, no such decomposition as can be seen from the following example.

**Example 15.3.8.** Suppose $W$ is the free product of three copies of $\mathbb{Z}/2$. Then $K$ is the cone on three points. It has three edges. By Theorem 15.3.4, $H^1(K, K^S) \otimes D_\emptyset$ is a quotient of $H_c^1(\Sigma)$. Let $x \in C^1(K)$ be a cochain (= cocycle) which evaluates to 1 on one of the edges, call it $c$, and to 0 on the other two edges. Choose an element $s$ in $S$ which is not a vertex of $c$. Let $y$ denote the image of $x \otimes 1$ in $H^1(K, K^S) \otimes D_\emptyset$. By Example 15.3.3, $D_\emptyset$ has rank 1 as an abelian group and the $W$-action on it is given by the sign representation. Hence, $sy = -y$ in $H^1(K, K^S) \otimes D_\emptyset$. Suppose we had a $W$-equivariant splitting $\varphi : H^1(K, K^S) \otimes D_\emptyset \to H^1_c(\Sigma)$. When regarded as an element of $C^1_c(\Sigma)$, $x + sx$ represents $\varphi(y + sy)$ in $H^1_c(\Sigma)$, i.e., it represents 0. But $x$ and $-sx$ are not cohomologous cocycles in $C^1_c(\Sigma)$. (One can see this by noting that there is a line (= infinite 1-cycle) on which $x$ evaluates to 1 and $sx$ to 0; see Figure 15.1.) Hence, there can be no such splitting $\varphi$.

## 15.4. THE CASE WHERE $W$ IS FINITE

In this section, *$W$ is finite*. We first deal with the analog of Theorem 15.3.4 when the coefficients are in $\mathbb{R}$ instead of $\mathbb{Z}$. In this case there is no need to consider quotient modules or associated graded modules in (co)homology. The reason is that every short exact sequence of finite dimensional $W$-representations splits. (Indeed, suppose $V$ is a real representation of $W$. Since $W$ is finite, we can choose of $W$-invariant inner product. If $V' \subset V$ is a subrepresentation, then $V = V' \oplus (V')^\perp$, where $(V')^\perp$ denotes the orthogonal complement.) Before continuing we need to set up some notation.

Let $\langle\ ,\ \rangle$ denote the standard inner product on $\mathbb{R}W$ in which $\{e_w\}_{w \in W}$ is an orthonormal basis. Let $x \to x^*$ be the anti-involution of $\mathbb{R}W$ which sends $e_w$ to $e_{w^{-1}}$. It is the adjoint with respect to the inner product, i.e., $\langle xy, z\rangle = \langle y, x^*z\rangle = \langle x, zy^*\rangle$. For each $T \subset S$, we have normalized versions of symmetrization and alternation:

$$a_T := \frac{\tilde{a}_T}{\mathrm{Card}(W_T)} = \frac{1}{\mathrm{Card}(W_T)} \sum_{w \in W_T} e_w, \tag{15.27}$$

$$h_T := \frac{\tilde{h}_T}{\mathrm{Card}(W_T)} = \frac{1}{\mathrm{Card}(W_T)} \sum_{w \in W_T} (-1)^{l(w)} e_w. \tag{15.28}$$

It is readily checked that $a_T$ and $h_T$ are self-adjoint idempotents, i.e., $a_T^* = a_T$, $h_T^* = h_T$, $(a_T)^2 = a_T$ and $(h_T)^2 = h_T$. That is to say, left or right multiplication by $a_T$ or $h_T$ on $\mathbb{R}W$ is an orthogonal projection. (We will provide more details about these formulas in Chapter 20.)

Define $(\mathbb{R}W)^T$ and $(\mathbb{R}W)^{>T}$ as before, i.e., $(\mathbb{R}W)^T := a_T(\mathbb{R}W)$. Also, $H^T_{\mathbb{R}}$ and $H^{>T}_{\mathbb{R}}$ are the analogs of $H^T$ and $H^{>T}$, i.e., $H^T_{\mathbb{R}} := (\mathbb{R}W)h_T$. The appropriate analogs of $D_T$ and $G_T$ are defined as follows: $D^{\mathbb{R}}_T$ is the orthogonal complement of $(\mathbb{R}W)^{>T}$ in $(\mathbb{R}W)^T$ and $G^{\mathbb{R}}_T$ is the orthogonal complement of $H^{>T}_{\mathbb{R}}$ in $H^T_{\mathbb{R}}$.

By Corollaries 15.2.6 and 15.2.7, we have direct sum decompositions:

$$(\mathbb{R}W)^T = \bigoplus_{U \supset T} D_U^{\mathbb{R}} \quad \text{and} \quad H_{\mathbb{R}}^T = \bigoplus_{U \supset T} G_U^{\mathbb{R}}. \tag{15.29}$$

Since a basis for $\widehat{A}^T$ (resp. $\widehat{H}^T$) projects to a basis for $D_T^{\mathbb{R}}$ (resp. $G_T^{\mathbb{R}}$), we have $\dim D_T^{\mathbb{R}} = \dim G_T^{\mathbb{R}} = \text{Card}(W^T)$. The proof of Theorem 15.3.4 immediately yields the following.

**THEOREM 15.4.1.** *There is an isomorphism of (left) W-representations:*

$$H_*(\mathcal{U}; \mathbb{R}) \cong \bigoplus_{T \subset S} H_*(X, X^T) \otimes_{\mathbb{R}} G_T^{\mathbb{R}}.$$

*Similarly, there is an isomorphism of (right) W-representations:*

$$H^*(\mathcal{U}; \mathbb{R}) \cong \bigoplus_{T \subset S} H^*(X, X^{S-T}) \otimes_{\mathbb{R}} D_T^{\mathbb{R}}.$$

### Rational Coefficients

In $\mathbb{Q}W$ we cannot take orthogonal complements complements with impunity; however, we can use the following result of L. Solomon [261] instead.

**THEOREM 15.4.2.** (Solomon [261, Theorem 2].) *For any subset $T \subset S$, there are direct sum decompositions*

$$(\mathbb{Q}W)a_T = \bigoplus_{U \supset T} (\mathbb{Q}W)h_{S-U}a_U,$$

$$(\mathbb{Q}W)h_T = \bigoplus_{U \supset T} (\mathbb{Q}W)a_{S-U}h_U.$$

*There are also dual, right-hand versions of both decompositions.*

(In Chapter 20 we will prove a Hilbert space version of this, valid for infinite Coxeter groups.)

Define

$$D_T^{\mathbb{Q}} := a_T h_{S-T}(\mathbb{Q}W),$$

$$G_T^{\mathbb{Q}} := (\mathbb{Q}W)a_{S-T}h_T.$$

Taking $T = \emptyset$ in Solomon's Theorem, we get decompositions of $\mathbb{Q}W$:

$$\mathbb{Q}W = \bigoplus_{U \subset S} D_U^{\mathbb{Q}} \quad \text{and} \quad \mathbb{Q}W = \bigoplus_{U \subset S} G_U^{\mathbb{Q}}. \tag{15.30}$$

(The first is a decomposition of the left regular representation, the second of the right.) By Solomon's Theorem, we have $(\mathbb{Q}W)^{>T} = \bigoplus_{U \supsetneq T} D_U^{\mathbb{Q}}$. Therefore, $D_T^{\mathbb{Q}}$ is a complementary subspace for $(\mathbb{Q}W)^{>T}$ in $(\mathbb{Q}W)^T$. Similarly, $G_T^{\mathbb{Q}}$ is a complementary subspace for $H_{\mathbb{Q}}^{>T}$ in $H_{\mathbb{Q}}^T$. (In general, they are not orthogonal complements.) It follows that we have $W$-module isomorphisms:

$$D_T^{\mathbb{R}} \cong D_T^{\mathbb{Q}} \otimes_{\mathbb{Q}} \mathbb{R} \quad \text{and} \quad G_T^{\mathbb{R}} \cong G_T^{\mathbb{Q}} \otimes_{\mathbb{Q}} \mathbb{R}.$$

The direct sum decompositions in (15.30) induce direct sum decompositions of coefficient systems (as $W$-modules) and direct sum decompositions of (co)chain complexes of $W$-modules:

$$C_*(\mathcal{U};\mathbb{Q}) = \bigoplus_{T \subset S} C_*(K, K^T) \otimes_{\mathbb{Q}} G_T^{\mathbb{Q}},$$

$$C^*(\mathcal{U};\mathbb{Q}) = \bigoplus_{T \subset S} C_*(K, K^T) \otimes_{\mathbb{Q}} D_T^{\mathbb{Q}}$$

Taking (co)homology we get the following.

**THEOREM 15.4.3.** ([73, Theorem B].) *There is an isomorphism of rational $W$-representations*

$$H_*(\mathcal{U};\mathbb{Q}) \cong \bigoplus_{T \subset S} H_*(X, X^T) \otimes_{\mathbb{Q}} G_T^{\mathbb{Q}}.$$

*Similarly, there is an isomorphism of $W$-representations*

$$H^*(\mathcal{U};\mathbb{Q}) \cong \bigoplus_{T \subset S} H^*(X, X^{S-T}) \otimes_{\mathbb{Q}} D_T^{\mathbb{Q}}.$$

*Remark.* Since $H_*(\mathcal{U};\mathbb{Q})$ and $H^*(\mathcal{U};\mathbb{Q})$ are dual $W$-representations, the two formulas in Theorem 15.4.3 would lead one to suspect that $G_T^{\mathbb{Q}}$ and $D_{S-T}^{\mathbb{Q}}$ are dual representations (and hence are isomorphic) This is indeed the case, since it follows from arguments in [261] that $(\mathbb{Q}W)(a_{S-T}h_Ta_{S-T}) = (\mathbb{Q}W)h_Ta_{S-T}$ and that left multiplication by $a_{S-T}$ maps $G_T^{\mathbb{Q}} = (\mathbb{Q}W)a_{S-T}h_T$ isomorphically onto $(\mathbb{Q}W)h_Ta_{S-T}$. This last expression is just the dual representation to $D_{S-T}^{\mathbb{Q}} = a_{S-T}h_T(\mathbb{Q}W)$. We shall return to this topic in Chapter 20 (see Theorems 20.6.19 and 20.6.20).

## NOTES

**15.1.** The identification of the face ring of $L$ with $H^*(W;\mathbb{F}_2)$ in the right-angled case is taken from my paper with Januszkiewicz [82]. More information about the face ring of a simplicial complex can be found in [268].

**15.2, 15.3, 15.4.** The material in these sections is taken from [80]. Theorem 15.4.3 is essentially proved in [73] by an argument which avoids explicit mention of the type of coefficient systems discussed here.

# Chapter Sixteen

# THE EULER CHARACTERISTIC

Algebraic topology began with Euler's result that for any cellulation of the 2-sphere, the alternating sum of the number of cells is equal to 2. This leads to the notion of the "Euler characteristic," defined in 16.1. Given a group of type *VF*, we define, in the same section, a rational number called its "Euler characteristic" and give two explicit formulas for it in the case of a Coxeter group. In 16.2 we discuss a conjecture about the Euler characteristic of any even dimensional, closed, aspherical manifold $M^{2k}$. It asserts that $(-1)^k\chi(M^{2k}) \geqslant 0$. The special case for a torsion-free subgroup of finite index in a right-angled Coxeter group of type $\mathrm{HM}^{2k}$ is called the "Charney–Davis Conjecture." It can be reformulated as a conjecture about the sign of a certain number associated to any triangulation of $S^{2k-1}$ as a flag complex. This is explained in 16.3.

## 16.1. BACKGROUND ON EULER CHARACTERISTICS

Suppose $X$ is a space with finitely generated homology. Its $i^{\text{th}}$-*Betti number* $b_i(X)$ is defined by

$$b_i(X) := \mathrm{rk}_{\mathbb{Z}}(H_i(X)),$$

where the *rank* of a finitely generated abelian group is given by $\mathrm{rk}_{\mathbb{Z}}(A) := \dim_{\mathbb{Q}}(A \otimes \mathbb{Q})$. The *Euler characteristic* of $X$ is the integer $\chi(X)$ defined by

$$\chi(X) := \sum (-1)^i b_i(X). \tag{16.1}$$

If $X$ is a finite CW complex, we can calculate $\chi(X)$ in terms of the number of its cells. Let $c_i(X)$ denote the number of $i$-cells in $X$, i.e., $c_i(X) = \mathrm{rk}_{\mathbb{Z}}(C_i(X))$, where $C_i(X)$ is the group of cellular chains on $X$. A standard argument (e.g., see [113]) shows that the alternating sum of the ranks of the $C_i$ in any chain complex is equal to the alternating sum of the ranks after taking homology. Hence,

$$\chi(X) = \sum (-1)^i c_i(X) = \sum_{\text{cells } \sigma} (-1)^{\dim \sigma}.$$

A group $G$ is of *finite homological type* if $\operatorname{vcd} G < \infty$ and if for every $G$-module $M$ which is finitely generated as an abelian group and for each nonnegative integer $i$, $H_i(G; M)$ is finitely generated. The most obvious examples of groups of finite homological type are those of type *VFP* (where *VFP* stands for "virtually type *FP*"). (The general reference for homology of groups is [42]; for a quick summary, see Appendix F. In particular, the notion of "vcd" or "virtual cohomological dimension" is defined in Appendix F.3 while "type *FP*" or "*VFP*" is defined in Appendix F.4.) If $\Gamma$ is torsion-free and of finite homological type, then its *Euler characteristic* is defined by

$$\chi(\Gamma) := \chi(B\Gamma) := \sum (-1)^i \operatorname{rk}_{\mathbb{Z}}(H_i(\Gamma)). \tag{16.2}$$

When $G$ is only required to be virtually torsion-free, this definition is not the correct one. (For one reason, it is not multiplicative upon passing to subgroups of finite index.) Instead, $\chi(G)$ should be defined as the rational number,

$$\chi(G) := \frac{\chi(\Gamma)}{[G : \Gamma]}, \tag{16.3}$$

where $\Gamma \subset G$ is a torsion-free subgroup of finite index, $[G : \Gamma]$.

## Orbihedral Euler Characteristics

Suppose a discrete group $G$ acts properly, cellularly, and cocompactly on a cell complex $Y$. As we explain in Appendix E.2, the quotient space $Y/G$ has the structure of the geometric realization of a small category without loop (a "scwol"). Over this scwol we get a "complex of groups," denoted by $Y/\!/G$. When all the local groups are finite (as they are if the action is proper), this complex of groups is an *orbihedron.*

The *orbihedral Euler characteristic* of $Y/\!/G$ is the rational number

$$\chi^{orb}(Y/\!/G) := \sum_{\text{orbits of cells}} \frac{(-1)^{\dim \sigma}}{\operatorname{Card}(G_\sigma)}, \tag{16.4}$$

where the summation is over a set $\{\sigma\}$ of representatives for the $G$-orbits of cells and where $G_\sigma$ denotes the stabilizer of $\sigma$. If the action is free (so that each cell stabilizer is the trivial subgroup), then the orbihedral Euler characteristic is the usual Euler characteristic of the quotient space $Y/G$.

An important feature of the Euler characteristics is that they are multiplicative with respect to finite sheeted covers. Thus, if $X' \to X$ is an $m$-sheeted covering space, then $\chi(X') = m\,\chi(X)$. This property persists for the orbihedral Euler characteristic: if $G' \subset G$ is a subgroup of index $m$, then

$$\chi^{orb}(Y/\!/G') = m\,\chi^{orb}(Y/\!/G). \tag{16.5}$$

In particular, if $Y$ is acyclic and $G$ is type *VFL*, then $\chi(G) = \chi^{orb}(Y/\!/G)$.

### Another Definition of the Rational Euler Characteristic of a Group

Suppose $\mathbb{Q}G$ denotes the rational group algebra of a group $G$. Denote the standard basis for $\mathbb{Q}G$ as a rational vector space by $(e_g)_{g\in G}$. The *standard inner product* on $\mathbb{Q}G$ is defined by setting $\langle e_g, e_h\rangle$ equal to the Kronecker delta $\delta(g,h)$. For any $\lambda \in \mathbb{Q}G$, the coefficient of the identity element in $\lambda$ is $\langle \lambda, e_1\rangle$. Let $\varphi$ be an endomorphism of a free $\mathbb{Q}G$-module $F$ of rank $n$. After choosing a basis, $\varphi$ is represented by an $(n \times n)$ matrix $(\varphi_{ij})$ with entries in $\mathbb{Q}G$. Define $\mathrm{tr}_G(\varphi)$, the *Kaplansky trace* of $\varphi$, by

$$\mathrm{tr}_G(\varphi) := \sum_{i=1}^{n} \langle \varphi_{ii}, e_1\rangle.$$

A standard argument shows this is independent of the choice of basis.

Next, suppose $P$ is a finitely generated projective $\mathbb{Q}G$-module. So, $P$ is a direct summand of some finitely generated free module $F$. Let $p : F \to F$ be a projection onto $P$. Define $\rho(P)$, the *rank* of $P$, by $\rho(P) := \mathrm{tr}_G(p)$. Standard arguments show this is a well-defined, nonnegative, rational number. Moreover, $\rho(P) = 0$ if and only if $P = 0$.

A group $G$ is has *type FP over* $\mathbb{Q}$ if there is a finite length resolution of $\mathbb{Q}$ by finitely generated projective $\mathbb{Q}G$-modules. (Groups of type $FP$ over $\mathbb{Z}$ are discussed in Appendix F.4.) Suppose $0 \to P_n \to \cdots P_0 \to \mathbb{Q} \to 0$ is such a projective resolution for $G$. Define $\chi(G)$, the *Euler characteristic of* $G$, by

$$\chi(G) := \sum (-1)^i \rho(P_i). \tag{16.6}$$

If $G$ acts properly, cellularly, and cocompactly on a rationally acyclic CW complex $Y$, then it is of type $FP$ over $\mathbb{Q}$. The cellular chain complex $C_*(Y;\mathbb{Q})$ provides the desired projective resolution. (Proof: since the action is proper and cellular, each $C_i(Y;\mathbb{Q})$ is a sum of $G$-modules of the form $\mathbb{Q}(G/G_\sigma)$, where $G_\sigma$ is finite. Since $G_\sigma$ is finite, such a module is a direct summand of $\mathbb{Q}G$ and hence, is projective. Since there are only finitely many orbits of cells, $C_i(Y;\mathbb{Q})$ is a finitely generated projective $\mathbb{Q}G$-module.)

**PROPOSITION 16.1.1.** *Suppose $G$ acts properly, cellularly, and cocompactly on a CW complex $Y$ which is acyclic over $\mathbb{Q}$. Then*

$$\chi(G) = \chi^{orb}(Y/\!/G).$$

*Proof.* Since $\overline{H}_*(Y;\mathbb{Q}) = 0$, the cellular chain complex of $Y$ gives a finitely generated projective resolution of $\mathbb{Q}$:

$$\to C_i(Y;\mathbb{Q}) \to C_{i-1}(Y,\mathbb{Q}) \to \cdots \to C_0(Y;\mathbb{Q}) \to \mathbb{Q} \to 0.$$

For any finite subgroup $H$ of $G$, the $\mathbb{Q}G$-module $\mathbb{Q}(G/H)$ is induced from the trivial representation of $H$. It follows that

$$\rho(\mathbb{Q}(G/H)) = \frac{1}{\text{Card}(H)}.$$

Hence,

$$\rho(C_i(Y;\mathbb{Q})) = \sum \frac{1}{\text{Card}(G_\sigma)},$$

where the sum is over a set of representatives for the orbits of $i$-cells. Comparing this with the definiton of the orbihedral Euler characteristic in (16.5), the result follows. □

**COROLLARY 16.1.2.** *Suppose we are given a model for $\underline{E}G$ with a finite number of orbits of cells. Then $\chi(G) = \chi^{orb}(\underline{E}G/\!/G)$.*

*Remark.* As explained in Appendix F.4, a group $G$ is "type *VFL*" if it has a torsion-free subgroup $\Gamma$ of finite index such that $\mathbb{Z}$ admits a finite length resolution by finitely generated free $\mathbb{Z}G$-modules. If $G$ is type *VFL*, then it can be shown that the two definitions of $\chi(G)$ given by (16.3) and (16.6) are equal ([42, Exercise 1, p. 252]). However, if $G$ is only assumed to be type *VFP*, the equality of the two definitions is an open question (related to a weak version of the Bass Conjecture). Since Coxeter groups are type *VFL*, it will not be necessary for us to introduce notation distinguishing (16.3) from (16.6).

## Two Formulas for the Euler Characteristic of a Coxeter Group

Given a Coxeter system $(W, S)$, we have the simplicial complex $\Sigma$ of Chapter 7. It is a model for $\underline{E}W$ (Theorem 12.3.4 (ii)). Each $W$-orbit of simplices intersects the fundamental chamber $K$ in a single simplex. The stabilizer of such a simplex $\sigma$ contains the spherical subgroup $W_T$ if and only if $\sigma \subset K_T$ and the inclusion of $W_T$ in the stabilizer is strict if and only if $\sigma \subset \partial K_T$. Hence, the contribution of the open simplices with stabilizer $W_T$ to the orbihedral Euler characteristic of $\Sigma/\!/W$ is the product of $[\chi(K_T) - \chi(\partial K_T)]$ with $1/\text{Card}(W_T)$. Since $K_T$ is a cone, its Euler characteristic is 1. $\partial K_T$ is the barycentric subdivision of $\text{Lk}(\sigma_T, L)$, where $\sigma_T$ is the simplex in the nerve $L$ corresponding to $T \in \mathcal{S}$. (As usual, if $T = \emptyset$, we interpret this link to be all of $L$.) So, $\chi(\partial K_T) = \chi(\text{Lk}(\sigma_T, L)$.) Hence,

$$\chi(W) = \chi^{orb}(\Sigma/\!/W) = \sum_{T\in\mathcal{S}} \frac{1 - \chi(\text{Lk}(\sigma_T, L))}{\text{Card}(W_T)}. \tag{16.7}$$

The second formula for $\chi(W)$ comes from the cellulation of $\Sigma$ by Coxeter cells. There is one $W$-orbit of such cells for each $T \in \mathcal{S}$. The dimension of the cell corresponding to $T$ is Card($T$) and its stabilizer is $W_T$. Hence, $\chi^{orb}(\Sigma /\!/ W)$ can also be computed as

$$\chi(W) = \sum_{T \in \mathcal{S}} \frac{\varepsilon(T)}{\text{Card}(W_T)}, \tag{16.8}$$

where $\varepsilon(T) := (-1)^{\text{Card}(T)}$.

**The $f$-Polynomial of a Simplicial Complex**

Given a finite simplicial complex $L$, denote by $f_i(L)$ (or simply by $f_i$) the number of $i$-simplices of $L$. Put $f_{-1}(L) = 1$ (corresponding to the one empty simplex). $(f_{-1}, f_0, \dots, f_{\dim L})$ is the $f$-*vector* of $L$. The $f$-*polynomial* of $L$ is defined by

$$f(t) := \sum_{i=-1}^{\dim L} f_i t^{i+1}. \tag{16.9}$$

Now suppose $(W, S)$ is a right angled. Then, for each $T \in \mathcal{S}$, $W_T$ is the $T$-fold product, $(\mathbf{C}_2)^T$. So $\text{Card}(W_T) = 2^{\text{Card}(T)} = 2^{\dim \sigma_T + 1}$ and (16.8) gives the following formula for the Euler characteristic:

$$\chi(W) = \sum_{i=0}^{\dim L+1} \frac{(-1)^i}{2^i} f_{i-1} = f(-\tfrac{1}{2}). \tag{16.10}$$

## 16.2. THE EULER CHARACTERISTIC CONJECTURE

By Poincaré duality, the Euler characteristic of any odd-dimensional, closed manifold is 0. A closed surface $M^2$ is aspherical if and only if it is not $S^2$ or $\mathbb{R}P^2$, i.e., if and only if $\chi(M^2) \leqslant 0$. Since $\chi(X \times Y) = \chi(X)\chi(Y)$, it follows that if a closed $2k$-manifold is the product of $k$ aspherical surfaces, then the sign of its Euler characteristic is $(-1)^k$. Such considerations led to the following conjecture about even dimensional, aspherical manifolds.

**CONJECTURE 16.2.1.** (The Euler Characteristic Conjecture.) *Suppose $M^{2k}$ is a closed, aspherical manifold. Then $(-1)^k \chi(M^{2k}) \geqslant 0$.*

We might as well formulate the following two mild generalizations of the Euler Characteristic Conjecture.

**CONJECTURE 16.2.2.** *Suppose $Q^{2k}$ is a closed, aspherical $2k$-dimensional orbifold (i.e., suppose $Q^{2k}$ is the quotient orbifold of a proper cocompact action of a group $G$ on a contractible manifold). Then $(-1)^k \chi^{orb}(Q^{2k}) \geqslant 0$.*

**CONJECTURE 16.2.3.** *Suppose $G$ is a virtual Poincaré duality group of formal dimension $2k$. Then $(-1)^k\chi(G) \geqslant 0$.*

(The notion of a "virtual Poincaré duality group" is defined in Appendix F.5.)

### The Chern-Gauss-Bonnet Theorem

Suppose $M^{2k}$ is a closed Riemannian manifold. Then

$$\chi(M^{2k}) = \int_{M^{2k}} \omega,$$

where $\omega$ is a certain $2k$-form called the "Euler form." (It is a constant multiple of the Pfaffian of the curvature.) In dimension 2, $\omega$ is just the Gaussian curvature multiplied by $1/2\pi$.

### The Combinatorial Version of the Gauss-Bonnet Theorem

In Definition I.2.1 of Appendix I.2, we define what it means for a metric space $X$ to be CAT($\kappa$), $\kappa \in \mathbb{R}$. $X$ is *nonpositively curved* if it is locally CAT(0). If, in addition, it is complete, then it is aspherical (Corollary I.2.9). Polyhedral metrics of piecewise constant curvature are discussed in Appendix I.3. In particular, there we define what is meant by a "piecewise Euclidean metric" on a cell complex. The link of a vertex in a piecewise Euclidean cell complex is naturally a piecewise spherical cell complex. One of the main results in Appendix I.3, Theorem I.3.5, states that a piecewise Euclidean cell complex is nonpositively curved if and only if the link of each of its vertices is CAT(1).

Suppose $\sigma^{n-1}$ is a convex polytope in $\mathbb{S}^{n-1}$ and $u_1, \dots, u_l$ are the inward-pointing unit normal vectors to the codimension-one faces of $\sigma^{n-1}$ (see 6.3 and Appendix I.5). The intersection of the cone in $\mathbb{R}^n$ spanned by the $u_i$ with $\mathbb{S}^{n-1}$ is a convex polytope in $\mathbb{S}^{n-1}$, denoted by $\sigma^*$ and called the *dual polytope* of $\sigma$. (This is explained in more detail in Definition I.5.2.) The *angle* determined by $\sigma$ is its $(n-1)$-dimensional volume, normalized so that the volume of $\mathbb{S}^{n-1}$ is 1, i.e., $a(\sigma) := \mathrm{vol}(\sigma)/\mathrm{vol}(\mathbb{S}^{n-1})$. Its *exterior angle* $a^*(\sigma)$ is defined by $a^*(\sigma) := a(\sigma^*)$.

**Example 16.2.4.** If $\sigma \subset \mathbb{S}^1$ is a circular arc of length $\theta$, then $a(\sigma) = \theta/2\pi$. Moreover, $\sigma^*$ is an arc of length $\pi - \theta$; so $a^*(\sigma) = \frac{1}{2} - a(\sigma)$.

**Example 16.2.5.** (*The case of an all right simplex.*) For the notion of an "all right" spherical simplex see Definition I.5.7 of Appendix I.5. ($\sigma^{n-1}$ is *all right* if it is isometric to the intersection of the unit sphere $\mathbb{S}^{n-1}$ with the positive "quadrant," $[0, \infty)^n \subset \mathbb{R}^n$.) Suppose $\sigma^{n-1}$ is all right. Since $2^n$ copies of $\sigma^{n-1}$ tessellate $\mathbb{S}^{n-1}$, $a(\sigma^{n-1}) = 2^{-n}$. Since $\sigma^{n-1}$ is isometric to its dual, we also have $a^*(\sigma^{n-1}) = 2^{-n}$.

**Example 16.2.6.** (*A fundamental simplex.*) Suppose $W$ is a spherical reflection group on $\mathbb{S}^{n-1}$ and that $\sigma_W$ is a fundamental simplex. Then $a(\sigma_W) = 1/|W|$, where $|W| = \mathrm{Card}(W)$.

Now suppose $L$ is a piecewise spherical cell complex. Define

$$\omega(L) := \sum_{\sigma \in \mathcal{F}(L)} (-1)^{\dim \sigma + 1} a^*(\sigma), \tag{16.11}$$

Where the sum is over all cells $\sigma$ in $L$, including the empty cell, which contributes $+1$ to the sum. (It is proved in [63] that $\omega(L)$ depends only on the piecewise spherical metric on $L$ and not on the particular cell structure.) If $\Lambda$ is a piecewise Euclidean cell complex, then for each $v \in \mathrm{Vert}(\Lambda)$, put

$$\omega_v := \omega(\mathrm{Lk}(v, \Lambda)), \tag{16.12}$$

where the piecewise spherical structure on the "link of a vertex, " $\mathrm{Lk}(v, \Lambda)$, is explained in Appendix I.3.

**THEOREM 16.2.7.** (The Combinatorial Gauss-Bonnet Theorem of [63].) *If $\Lambda$ is a finite piecewise Euclidean cell complex, then*

$$\chi(\Lambda) = \sum_{v \in \mathrm{Vert}(\Lambda)} \omega_v.$$

Thus, for piecewise Euclidean cell complexes, $\omega_v$ plays the role of the Gauss-Bonnet integrand.

**Exercise 16.2.8.** Prove this theorem by first showing that if $P$ is any convex Euclidean polytope, then $\sum a^*(\mathrm{Lk}(v, P) = 1$, where the summation is over all vertices of $P$.

**Example 16.2.9.** (*The case of an all right simplicial complex.*) Suppose $L$ is an all right simplicial complex (Definition I.5.7) and $f(t)$ its $f$-polynomial defined in (16.9). By Example 16.2.5, for each $(i-1)$-simplex $\sigma$ of $L$, $a^*(\sigma) = 2^{-i}$. So, the contribution of the $(i-1)$-simplices to $\omega(L)$ is $f_{i-1}2^{-i}$ and hence, $\omega(L) = f(-\frac{1}{2})$. (Compare this with formula (16.10).)

It is conjectured in [55] that the Euler Characteristic Conjecture for nonpositively curved, piecewise Euclidean manifolds should hold for local reasons. More precisely, there is the following.

**CONJECTURE 16.2.10.** *Suppose $L^{2k-1}$ is a piecewise spherical cell complex homeomorphic to the $(2k-1)$-sphere (or even to a generalized homology sphere) and that $L$ is* CAT(1). *Then $(-1)^k \omega(L^{2k-1}) \geqslant 0$.*

The Combinatorial Gauss-Bonnet Theorem shows that this conjecture implies the Euler Characteristic Conjecture for any nonpositively curved, piecewise Euclidean manifold $M^{2k}$. The conjecture is obviously true in dimension 1. For then, $L$ is a circle, $\omega(L) = 1 - (2\pi)^{-1}l(L)$ (where $l(L)$ means the length of $L$) and $L$ is CAT(1) if and only if $l(L) \geqslant 2\pi$ (Exercise I.2.2 in Appendix I.2).

## 16.3. THE FLAG COMPLEX CONJECTURE

Let $L$ be a simplicial complex endowed with its all right piecewise spherical structure (in which each simplex is all right). By Gromov's Lemma (Lemma I.6.1), $L$ is CAT(1) if and only if it is a flag complex. So, in this case, by using Example 16.2.9, Conjecture 16.2.10 becomes the following purely combinatorial conjecture.

**CONJECTURE 16.3.1.** (The Flag Complex Conjecture of [55].) *Suppose $L$ is a triangulation of a* $(2k-1)$*-dimensional generalized homology sphere as a flag complex. Let $f(t)$ be its $f$-polynomial from* (16.9). *Then* $(-1)^k f(-\frac{1}{2}) \geqslant 0$.

Recall Theorem 10.6.1: $\Sigma$ is a *PL* $n$-manifold if and only if $L$ is a *PL* triangulation of $S^{n-1}$; moreover, $\Sigma$ is a homology $n$-manifold if and only if $L$ is a $\mathrm{GHS}^{n-1}$. By (16.10) and Example 16.2.9, the Euler characteristic of the associated right-angled Coxeter group is $\omega(L) = f(-\frac{1}{2})$. So, the Flag Complex Conjecture (also known as the Charney–Davis Conjecture) is just the Euler Characteristic Conjecture (particularly, 16.2.3) in the special case of a right-angled Coxeter group of type *HM*.

Suppose a manifold $M^{2k}$ is cellulated by cubes. By declaring each cube to be a regular Euclidean cube, we get an induced piecewise Euclidean metric on $M^{2k}$ (Examples I.3.2 (ii)). Since the link of a vertex in a cube is an all right simplex, the link of a vertex in $M^{2k}$ is cellulated by all right simplices; so, by Gromov's Lemma, $M^{2k}$ is nonpositively curved if and only if the link of each vertex is a flag complex.

**PROPOSITION 16.3.2.** ([55].) *The Euler Characteristic Conjecture for (homology) manifolds with nonpositively curved, piecewise Euclidean metrics coming from a cubical structures is equivalent to the Flag Complex Conjecture.*

*Proof.* As explained above, given a flag triangulation $L$ of a $\mathrm{GHS}^{2k-1}$, the Euler Characteristic Conjecture for the associated right-angled Coxeter group $W_L$ implies $(-1)^k\omega(L) = (-1)^k\chi(W_L) \geqslant 0$. Conversely, if the Flag Complex Conjecture holds for the link of each vertex in a cubical, nonpositively curved, homology manifold $M^{2k}$, then by the Combinatorial Gauss-Bonnet Theorem we have $(-1)^k\chi(M^{2k}) \geqslant 0$. □

**THEOREM 16.3.3.**

*(i)* ([55], [269].) *The Flag Complex Conjecture is true whenever $L$ is the barycentric subdivision of the boundary of any convex polytope. More generally, it holds whenever $L = \mathrm{Flag}(\Lambda)$ (Definition A.3.2) where $\Lambda$ is any cellulation of $S^{2k-1}$ as a regular CW complex which is "shellable."*

*(ii)* ([91].) *The Flag Complex Conjecture is true whenever $L$ is a triangulation of a rational homology 3-sphere as a flag complex.*

(It is explained in [55], how statement (i) follows from a result of Stanley [269]. In Corollary 20.5.3 it is explained how statement (ii) follows from results on $L^2$-cohomology.)

## NOTES

**16.1.** The material on "rational Euler characteristics" is from [42, Ex. 1, p.252].

**16.2.** In the case of Riemannian manifolds of nonpositive sectional curvature, the Euler Characteristic Conjecture is called the Chern–Hopf Conjecture. It first appeared in print in Chern's paper [64] on the general Gauss–Bonnet Theorem. This question (and a similar question for positive sectional curvature) were asked by H. Hopf. Underlying Hopf's question is the question of whether the conjecture follows directly from the Gauss-Bonnet Theorem, that is, does the nonpositivity of sectional curvature imply that the sign of the Gauss-Bonnet integrand is $(-1)^k$. Of course, this is exactly what happens in dimension 2. The same is true in dimension 4. In the four-dimensional case Chern gives a proof in [64] and attributes the argument to Milnor. (Presumably, Hopf asked the question after his work relating the Euler characteristic of a hyperbolic manifold to its volume, but before Chern had formulated the general Gauss–Bonnet Theorem.) In 1976 R. Geroch [137] showed there is no such result in dimensions $\geqslant 6$ (the sign of the sectional curvature does not determine the sign of the Gauss-Bonnet integrand). On the other hand, if the "curvature operator" is negative semidefinite, then the sign of Gauss–Bonnet integrand is $(-1)^k$.

In the mid 1970s Thurston asked if the Euler Characteristic Conjecture were true for all aspherical manifolds. In the special case of dimension 4, this conjecture was included in Kirby's well-known list of problems in low-imensional topology. In the case of manifolds with piecewise Euclidean metrics of nonpositive sectional curvature, the Euler Characteristic Conjecture was explicitly formulated in [55]. The Euler Characteristic Conjecture is known to hold for closed manifolds which

- are locally symmetric manifolds (e.g., hyperbolic manifolds) or more generally,
- have negative semidefinite curvature operator, or
- are Kähler manifolds with negative sectional curvature ([148]).

# Chapter Seventeen

## GROWTH SERIES

Suppose $S$ a finite set of generators for a group $G$. As in 2.1, $l : G \to \mathbb{N}$ is word length. Define a power series $f(t)$ (the *growth series* of $G$) by $f(t) := \sum_{g \in G} t^{l(g)}$. Thus, $f(t) = \sum_{n=0}^{\infty} a_n t^n$, where $a_n$ is the number of vertices in a sphere of radius $n$ in $\mathrm{Cay}(G, S)$. If $G$ is finite, $f(t)$ is a polynomial. Under favorable circumstances (for example, when $G$ is an "automatic group"), $f(t)$ is a rational function. One of the first results in this line was the proof of the rationality of growth series of Coxeter groups. We give the argument in Corollary 17.1.6. It turns out that in the case of a Coxeter system $(W, S)$, the definition of the growth series can be extended to a power series $W(\mathbf{t})$ in a certain vector $\mathbf{t}$ of indeterminates. Again, it is a rational function of $\mathbf{t}$. (This improvement will be important in Chapters 18 and 20.) In 17.1 we give several explicit formulas for $W(\mathbf{t})$.

### 17.1. RATIONALITY OF THE GROWTH SERIES

As usual, $(W, S)$ is a Coxeter system. Suppose we are given an index set $I$ and a function $i : S \to I$ so that $i(s) = i(s')$ whenever $s$ and $s'$ are conjugate in $W$. (The largest possible choice for the image of $i$ is the set of conjugacy classes of elements in $S$ and the smallest possible choice is a singleton.) Let $\mathbf{t} := (t_i)_{i \in I}$ stand for an $I$-tuple of indeterminates and let $\mathbf{t}^{-1} := (t_i^{-1})_{i \in I}$. Write $t_s$ instead of $t_{i(s)}$. If $s_1 \cdots s_l$ is a reduced expression for $w$, define $t_w$ to be the monomial

$$t_w := t_{s_1} \cdots t_{s_l}. \tag{17.1}$$

It follows from Tits' solution to the word problem (Theorem 3.4.2 (ii)), that $t_w$ is independent of the choice of reduced expression for $w$. Indeed, two reduced expressions for $w$ differ by a sequence of elementary $M$-operations of type (II) (Definition 3.4.1). Such an operation replaces an alternating subword $ss' \cdots$ of length $m_{ss'}$ by the alternating word $s's \cdots$ of the same length but in the other order. If $m_{ss'}$ is even, $s$ and $s'$ occur the same number of times in these subwords, so the monomial $t_w$ stays the same. If $m_{ss'}$ is odd, such an operation does change the number of occurrences of $s$ and $s'$ in the reduced expression. However, when $m_{ss'}$ is odd, $s$ and $s'$ are conjugate in the dihedral subgroup which they generate (Lemma 3.1.8) and so, *a fortiori*, are conjugate in $W$. Thus, $i(s) = i(s')$ and the monomial again remains unchanged.

Similarly, define a monomial in the $(t_i)^{-1}$ by

$$t_w^{-1} := (t_{s_1})^{-1} \cdots (t_{s_l})^{-1}. \tag{17.2}$$

The *growth series* of $W$ is the power series in $\mathbf{t}$ defined by

$$W(\mathbf{t}) := \sum_{w \in W} t_w. \tag{17.3}$$

For any subset $X$ of $W$, put

$$X(\mathbf{t}) := \sum_{w \in X} t_w. \tag{17.4}$$

For any subset $T$ of $S$, we have the the special subgroup $W_T$ and its growth series $W_T(\mathbf{t})$. Note that if $T$ is spherical, then $W_T(\mathbf{t})$ is a polynomial in $\mathbf{t}$.

**Lemma 17.1.1.** *Suppose $W$ is finite and $w_S$ is the element of longest length (Section 4.6). Put $t_S := t_{w_S}$. Then $W(\mathbf{t}) = t_S W(\mathbf{t}^{-1})$.*

*Proof.* By Lemma 4.6.1, for any $w \in W$, $l(w_S w) = l(w_S) - l(w)$. So concantenation of a reduced expression for $w_S w$ with one for $w^{-1}$ gives a reduced expression for $w_S$. Hence, $t_S = t_{w_S w} t_{w^{-1}}$. If $s_1 \cdots s_l$ is a reduced expression for $w$, then $s_l \cdots s_1$ is a reduced expression for $w^{-1}$; so $t_{w^{-1}} = t_w$ and therefore, $t_{w_S w} = t_S t_w^{-1}$. This gives

$$W(\mathbf{t}) = \sum_{w_S w \in W} t_{w_S w} = \sum_{w \in W} t_S t_w^{-1} = t_S W(\mathbf{t}^{-1}).$$

□

*Remark.* This lemma means that when $W$ is finite, the coefficients of the polynomial $W(\mathbf{t})$ have a certain symmetry. If we write $W(\mathbf{t}) = \sum a_J t_J$, where the $t_J$ are monomials in the $t_i$ and the $a_J$ are positive integers, then the lemma implies that each $t_J$ is a factor of $t_S$ and that if $J'$ denotes the index of the monomial $t_S/t_J$, then $a_J = a_{J'}$. In the case of a single indeterminate $\mathbf{t} = t$, this condition means that polynomial $W(t)$ is palindromic, i.e., if

$$W(t) = 1 + a_1 t + \cdots + a_{m-1} t^{m-1} + t^m,$$

then $a_j = a_{m-j}$.

As in 4.5, for each $T \subset S$, $B_T$ denotes the $(\emptyset, T)$-reduced elements in $W$.

**Lemma 17.1.2.** ([29, Ex. 26 c), p. 43].) *For each $T \subset S$, $W(\mathbf{t}) = B_T(\mathbf{t}) W_T(\mathbf{t})$.*

*Proof.* By Lemma 4.3.3, each element $w \in W$ can be written uniquely as $w = uv$, where $u \in B_T$, $v \in W_T$ and $l(w) = l(u) + l(v)$. So, $t_w = t_u t_v$. The lemma follows. □

Given a finite set $T$, put $\varepsilon(T) := (-1)^{\mathrm{Card}(T)}$. In the next lemma we recall a basic formula from combinatorics. (A generalization is sometimes called the Möbius Inversion Formula.) The proof is left as an exercise.

**LEMMA 17.1.3.** ([29, Ex. 25, p. 42].) *Suppose $f, g$ are two functions from the power set of a finite set $S$ to an abelian group such that for any $T \subset S$,*

$$f(T) = \sum_{U \subset T} g(U).$$

*Then for any $T \subset S$,*

$$g(T) = \sum_{U \subset T} \varepsilon(T - U) f(U).$$

Recall Definition 4.7.4: if $T \subset S$, $W^T$ is the set of $w \in W$ with $\mathrm{In}(w) = T$.

**LEMMA 17.1.4.** ([29, Ex. 26, pp. 42–43].) *For any $T \subset S$,*

$$W^T(\mathbf{t}) = W(\mathbf{t}) \sum_{U \subset T} \frac{\varepsilon(T - U)}{W_{S-U}(\mathbf{t})}.$$

*Proof.* By Lemma 4.3.3 (ii), an element $w \in W$ lies in $B_{S-T}$ if and only if $\mathrm{In}(w) \subset T$. Hence, $B_{S-T}$ can be decomposed as a disjoint union:

$$B_{S-T} = \bigcup_{U \subset T} W^U.$$

So

$$B_{S-T}(\mathbf{t}) = \sum_{U \subset T} W^U(\mathbf{t}). \tag{17.5}$$

Apply Lemma 17.1.3 with $f(T) = B_{S-T}(\mathbf{t})$ and $g(T) = W^T(\mathbf{t})$, to get

$$W^T(\mathbf{t}) = \sum_{U \subset T} \varepsilon(T - U) B_{S-U}(\mathbf{t}). \tag{17.6}$$

By Lemma 17.1.2, $B_{S-U}(\mathbf{t}) = W(\mathbf{t})/W_{S-U}(\mathbf{t})$. Substituting this into (17.6), we get the formula of the lemma. □

By Lemma 4.7.2, $W^T$ is nonempty if and only if $T$ is spherical; so the left-hand side of the formula in Lemma 17.1.4 is 0 whenever $T$ is not spherical. The special case $T = S$ is the following.

**COROLLARY 17.1.5**

*(i) Suppose $W$ is finite and $t_S$ is the monomial corresponding to its element of longest length. Then*

$$t_S = W(\mathbf{t}) \sum_{T \subset S} \frac{\varepsilon(T)}{W_T(\mathbf{t})}.$$

*(ii) If $W$ is infinite, then*

$$0 = \sum_{T \subset S} \frac{\varepsilon(T)}{W_T(\mathbf{t})}.$$

*Proof.* Apply Lemma 17.1.4 in the case $T = S$, to get

$$W^S(\mathbf{t}) = W(\mathbf{t}) \sum_{U \subset S} \frac{\varepsilon(S - U)}{W_{S-U}(\mathbf{t})}.$$

Suppose $W$ is finite. Then, by Lemma 4.6.1, $W^S = \{w_S\}$; so, $W^S(\mathbf{t}) = t_S$. Reindex the sum by setting $T = S - U$, to get (i).

If $W$ is infinite, then $W^S = \emptyset$ and $W^S(\mathbf{t}) = 0$. Reindex the sum by setting $T = S - U$ and then divide by $W(\mathbf{t})$ to get (ii). □

**COROLLARY 17.1.6.** *$W(\mathbf{t}) = f(\mathbf{t})/g(\mathbf{t})$, where $f, g \in \mathbb{Z}[\mathbf{t}]$ are polynomials with integral coefficients.*

*Proof.* When $W$ is finite, $W(\mathbf{t})$ is a polynomial with integral coefficients. When $W$ is infinite, we can use Corollary 17.1.5 (ii) to express $1/W(\mathbf{t})$ as an integral linear combination of the $1/W_T(\mathbf{t})$ with $T \subsetneq S$. So, the result follows by induction on Card$(S)$. □

For any $T \in \mathcal{S}$, set $B'_T := B_T w_T$.

**LEMMA 17.1.7.** *An element $w \in W$ lies in $B'_T$ if and only if it is the longest element in $wW_T$. (In particular, $B'_T$ is a set of coset representatives for $W/W_T$.) Moreover, $B'_T = \{w \in W \mid T \subset \text{In}(w)\}$.*

*Proof.* By Lemma 4.3.3 (i), $w$ can be written uniquely in as $w = uv$, where $u \in B_T$, $v \in W_T$ and where $l(w) = l(u) + l(v)$ (by Lemma 4.3.1). So, $w$ is the longest element in its coset if and only if $v$ is the longest element of $W_T$. This proves the first sentence of the lemma.

To prove the final sentence, write $w$ as $uw_{\text{In}(w)}$, where $u$ is $(\emptyset, \text{In}(w))$-reduced. If $w \in B'_T$, then $w$ ends in the letters of $T$; so, $T \subset \text{In}(w)$. Conversely, if $T \subset \text{In}(w)$, then, by Lemma 4.6.1, we can write $w_{\text{In}(w)} = aw_T$ with $a$ being $(\emptyset, T)$-reduced and $l(w_{\text{In}(w)}) = l(a) + l(w_T)$. So, $l(w) = l(ua) + l(w_T)$, $ua \in B_T$ and therefore, $w \in B'_T$. □

**LEMMA 17.1.8.** *For any $T \in \mathcal{S}$,*

$$W^T(\mathbf{t}) = W(\mathbf{t}) \sum_{U \in \mathcal{S}_{\geqslant T}} \frac{\varepsilon(U - T)}{W_U(\mathbf{t}^{-1})}.$$

*Proof.* By Lemma 17.1.7, $B'_T(\mathbf{t}) = t_T B_T(\mathbf{t})$. Since $B'_T = \{w \in W \mid T \subset \text{In}(w)\}$ (also by Lemma 17.1.7), $B'_T$ can be decomposed as a disjoint union

$$B'_T = \bigcup_{U \in \mathcal{S}_{\geqslant T}} W^U. \tag{17.7}$$

Hence,

$$B'_T(\mathbf{t}) = \sum_{U \in \mathcal{S}_{\geqslant T}} W^U(\mathbf{t}). \tag{17.8}$$

If a subset $T$ of $S$ is not spherical, put $B'_T := \emptyset$ and $B'_T(\mathbf{t}) = 0$. We can apply Lemma 17.1.3 with $f(T) = B'_T(\mathbf{t})$ and $g(T) = W^T(\mathbf{t})$, to get

$$W^T(\mathbf{t}) = \sum_{U \in \mathcal{S}_{\geqslant T}} \varepsilon(U - T) B'_U(\mathbf{t}). \tag{17.9}$$

By Lemma 17.1.1,

$$\frac{t_T}{W_T(\mathbf{t})} = \frac{1}{W_T(\mathbf{t}^{-1})}.$$

So, by Corollary 17.1.5 (i) and Lemma 17.1.2,

$$B'_T(\mathbf{t}) = t_T B_T(\mathbf{t}) = \frac{t_T W(\mathbf{t})}{W_T(\mathbf{t})} = \frac{W(\mathbf{t})}{W_T(\mathbf{t}^{-1})}.$$

Substituting this into (17.9) yields the formula in the lemma. □

**THEOREM 17.1.9.** ([270].)

$$\frac{1}{W(\mathbf{t}^{-1})} = \sum_{T \in \mathcal{S}} \frac{\varepsilon(T)}{W_T(\mathbf{t})}.$$

*Proof.* It is immediate from the definitions that $W^{\emptyset} = \{1\}$ and $W^{\emptyset}(\mathbf{t}) = 1$. When $T = \emptyset$ the formula of Lemma 17.1.8 is

$$1 = W(\mathbf{t}) \sum_{U \in \mathcal{S}} \frac{\varepsilon(U)}{W_U(\mathbf{t}^{-1})}.$$

Replacing $\mathbf{t}$ by $\mathbf{t}^{-1}$ and $U$ by $T$, we get the result. (A different proof can be found in [53].) □

As usual, $L$ denotes the nerve of $(W, S)$. For each $T \in \mathcal{S}$, let $L_T := \mathrm{Lk}(\sigma_T, L)$, where $\sigma_T$ is the simplex of $L$ corresponding to $T$ and where the "link" of $\sigma_T$ is as defined in Appendix A.6. The face poset of $L_T$ is isomorphic with $\mathcal{S}_{>T}$. (The face poset is defined in Example A.2.3.) By convention, $L_\emptyset := L$. By definition, the Euler characteristic of $L_T$ is given by the formula

$$\chi(L_T) := \sum_{U \in \mathcal{S}_{>T}} (-1)^{\dim \sigma_{U-T}} = - \sum_{U \in \mathcal{S}_{>T}} \varepsilon(U - T). \tag{17.10}$$

**THEOREM 17.1.10.** ([53].)

$$\frac{1}{W(\mathbf{t})} = \sum_{T \in \mathcal{S}} \frac{1 - \chi(L_T)}{W_T(\mathbf{t})}.$$

The proof of this depends on the following standard fact.

**LEMMA 17.1.11.** *Suppose $U$ is a subset of a finite set $S$. Then*

$$\sum_{U \subset T \subset S} \varepsilon(T) = 0.$$

*Proof of Theorem 17.1.10.* When $W$ is finite, $L_T$ is a simplex for each $T \neq S$ and $L_S = \emptyset$. Hence,

$$1 - \chi(L_T) = \begin{cases} 0 & \text{if } T \neq S, \\ 1 & \text{if } T = S. \end{cases}$$

So when $W$ is finite the theorem is the tautology $1/W(\mathbf{t}) = 1/W(\mathbf{t})$.

Suppose $W$ is infinite. We can rewrite Corollary 17.1.5 (ii) as

$$\frac{1}{W(\mathbf{t})} = -\varepsilon(S) \sum_{T \subsetneq S} \frac{\varepsilon(T)}{W_T(\mathbf{t})}. \tag{17.11}$$

The proof is by induction on Card$(T)$. For any $T \subset S$, let $\mathcal{S}(T)$ be the set of spherical subsets of $T$ and for any $U \in \mathcal{S}(T)$, let $L_U(T)$ be the simplicial complex corresponding to $\mathcal{S}(T)_{>U}$. Using (17.11) and the inductive hypothesis, we get

$$\frac{1}{W(\mathbf{t})} = -\varepsilon(S) \sum_{T \subsetneq S} \varepsilon(T) \sum_{U \in \mathcal{S}(T)} \frac{1 - \chi(L_U(T))}{W_U(\mathbf{t})}. \tag{17.12}$$

The coefficient of $1/W_U(\mathbf{t})$ on the right hand side of (17.12) is

$$-\varepsilon(S) \sum_{U \subset T \subsetneq S} \varepsilon(T)[1 - \chi(L_U(T))].$$

We want to prove this coefficient is equal to $1 - \chi(L_U)$, i.e.,

$$\sum_{U \subset T \subset S} \varepsilon(T)[1 - \chi(L_U(T))] = 0.$$

By Lemma 17.1.11, the previous equation is equivalent to

$$\sum_{U \subset T \subset S} \varepsilon(T)\chi(L_U(T)) = 0. \tag{17.13}$$

Let $T' \in \mathcal{S}_{>U}$ and let $\sigma_{T'}$ be the corresponding simplex in $L_U$. The contribution of $T'$ to the left-hand side of (17.13) is

$$(-1)^{\dim \sigma_{T'}} \sum_{T' \subset T \subset S} \varepsilon(T),$$

which, by Lemma 17.1.11, is 0. Thus, (17.13) holds and (17.12) can be rewritten as the formula of the theorem. □

### The Region of Convergence $\mathcal{R}$

Let $\mathbf{z} := (z_i)_{i \in I}$ be a point in $\mathbb{C}^I$. Put $\mathbf{z}^{-1} := (z_i^{-1})_{i \in I}$ and for each $T \subset S$, put

$$\mathcal{R}_T := \{\mathbf{z} \in \mathbb{C}^I \mid W_T(\mathbf{z}) \text{ converges}\}, \tag{17.14}$$

$$\mathcal{R}_T^{-1} := \{\mathbf{z} \in (\mathbb{C}^*)^I \mid W_T(\mathbf{z}^{-1}) \text{ converges}\}. \tag{17.15}$$

Write $\mathcal{R}$ and $\mathcal{R}^{-1}$ instead of $\mathcal{R}_S$ and $\mathcal{R}_S^{-1}$. Note that when $T$ is spherical, $W_T(\mathbf{t})$ is a polynomial; hence, $\mathcal{R} = \mathbb{C}^I$.

**Example 17.1.12.** *(The infinite dihedral group.)* Suppose $S = \{s_1, s_2\}$, $m_{s_1 s_2} = \infty$, so that $W$ is the infinite dihedral group $\mathbf{D}_\infty$. Its nerve is the 0-sphere. Suppose $I = \{1, 2\}$ and $S \to I$ sends $s_j$ to $j$. Using Theorem 17.1.9 and Lemma 17.1.1, we compute

$$\frac{1}{W(\mathbf{t})} = \frac{1 - t_1 t_2}{(1 + t_1)(1 + t_2)}.$$

So, $\mathcal{R} = \{(z_1, z_2) \mid |z_1||z_2| < 1\}$. In particular, $(0, 1)^2 \subset \mathcal{R}$.

The proof of the next lemma is an easy exercise.

**LEMMA 17.1.13.** *Suppose $(W, S)$ decomposes as a product $(W_1 \times W_2, S_1 \cup S_2)$. For $i = 1, 2$, let $I_i$ be the index set for $S_i$ and $\mathbf{t}_i$ an $I_i$-tuple of indeterminates. Then $W(\mathbf{t}_1, \mathbf{t}_2) = W_1(\mathbf{t}_1)W_2(\mathbf{t}_2)$. Moreover, $\mathcal{R} = \mathcal{R}_1 \times \mathcal{R}_2$, where $\mathcal{R}$, $\mathcal{R}_1$, and $\mathcal{R}_2$ are the regions of convergence for $W(\mathbf{t}_1, \mathbf{t}_2)$, $W_1(\mathbf{t}_1)$, and $W_2(\mathbf{t}_2)$, respectively.*

**Example 17.1.14.** Suppose $W = (\mathbf{D}_\infty)^n$, the $n$-fold product of infinite dihedral groups. Its nerve $L$ is then the $n$-fold join of copies of $S^0$, i.e., it is the boundary complex of an $n$-octahedron. By Example 17.1.12 and Lemma 17.1.13, $(0,1)^I \subset \mathcal{R}$.

### The Radius of Convergence $\rho$

When $I$ is a singleton, $\mathbf{t}$ is a single indeterminate $t$, $W_T(t)$ is a power series in one variable and the interior of the region of convergence is a disk about the origin in $\mathbb{C}$. Its radius (the *radius of convergence*) is denoted $\rho_T$. Write $\rho$ instead of $\rho_S$. As in Corollary 17.1.6, express $W(t)$ as a rational function $f(t)/g(t)$ where $f$ and $g$ are polynomials with no common factors. Then $\rho$ is the smallest modulus of a root of $g(t)$. Since the coefficients of the power series $W(t)$ are positive, $\rho$ must be a positive real root of this modulus, i.e., $\rho$ is the smallest root of $g(t)$.

**Example 17.1.15.** (*Right-angled polygon groups.*) Suppose $W$ is right angled with nerve a $k$-gon, $k \geqslant 4$, and that $\mathbf{t}$ is a single indeterminate $t$. Using Theorem 17.1.9 and Lemma 17.1.1 as before, we get

$$\frac{1}{W(t)} = 1 - \frac{kt}{1+t} + \frac{kt^2}{(1+t)^2} = \frac{t^2 + (2-k)t + 1}{(1+t)^2}.$$

The roots of the numerator are $\rho$ and $\rho^{-1}$; so

$$\rho^{\pm 1} = \frac{(k-2) \mp \sqrt{k^2 - 4k}}{2},$$

e.g., $\rho = \frac{3-\sqrt{5}}{2}$ when $k = 5$.

## 17.2. EXPONENTIAL VERSUS POLYNOMIAL GROWTH

We recall what it means for a finitely generated group $G$ to be "amenable." Let $\Omega$ denote its Cayley graph with respect to some finite set of generators. For any subset $F \subset G$, let $\partial F$ denote the set of $g \in F$ such that there is an edge of $\Omega$ joining $g$ to an element not in $F$. $G$ satisfies the *Følner Condition* (and is an *amenable group*) if for any positive number $\varepsilon$ there is a finite subset $F \subset G$ such that

$$\frac{\operatorname{Card}(\partial F)}{\operatorname{Card}(F)} < \varepsilon.$$

In the next proposition we list six other conditions equivalent to the condition that the radius of convergence of $W(t)$ is 1.

**PROPOSITION 17.2.1.** *The following conditions on an infinite Coxeter system $(W, S)$ are equivalent.*

*(i)* $W$ *is amenable.*

*(ii)* $W$ *does not contain a free group on two generators.*

*(iii)* $W$ *is not large (i.e.,* $W$ *does not contain a finite index subgroup* $\Gamma$ *which maps onto* $F_2$*; cf. Definition 14.1.1).*

*(iv)* $W$ *is virtually abelian.*

*(v)* $(W, S)$ *decomposes as* $(W_0 \times W_1, S_0 \cup S_1)$ *where* $W_1$ *is finite and* $W_0$ *is a cocompact Euclidean reflection group.*

*(vi)* $\rho = 1$.

*(vii)* $W$ *has subexponential growth.*

*Proof.* The implication (i) $\Longrightarrow$ (ii) is a standard fact.

(ii) $\Longrightarrow$ (iii). Suppose for some subgroup $\Gamma$ of $W$ we have a surjection $f : \Gamma \to F_2$ where $F_2$ is the free group on $\{x_1, x_2\}$. Choose $\gamma_1 \in f^{-1}(x_1)$, $\gamma_2 \in f^{-1}(x_2)$. Then $\langle \gamma_1, \gamma_2 \rangle$ is a free subgroup of $W$.

(iii) $\Longrightarrow$ (iv). By Theorem 14.1.2, if $W$ is not virtually abelian, then it is large.

(iv) $\Longrightarrow$ (v). This implication was proved as Theorem 12.3.5.

(v) $\Longrightarrow$ (vi). Since a Euclidean reflection group is virtually free abelian, it has polynomial growth and therefore, the radius of convergence of its growth series is 1. (In fact, the poles of its growth series are all roots of unity; see Remark 17.2.2 below.)

(vi) $\Longrightarrow$ (vii) is obvious.

(vii) $\Longrightarrow$ (i) by the Følner Condition for amenability. □

*Remark 17.2.2.* Suppose $W$ is a (cocompact) Euclidean reflection group. Consider the case where $(W, S)$ is irreducible. Let $W'$ be the finite linear reflection group obtained by dividing out the translation subgroup of $W$ and let $m_1, \dots, m_n$ be the exponents of $W'$. (The *exponents* are the degrees of generators for the invariant polynomials in the ring of invariant polynomials on the canonical representation of $W'$; see [29].) According to [29, Ex. 10, p. 245], the growth series of $W$ is given by the following formula of Bott [28]:

$$W(t) = \prod_{i=1}^{n} \frac{1 + t + \cdots + t^{m_i}}{1 - t^{m_i}}.$$

In particular, all poles of $W(t)$ are roots of unity. We can reach the same conclusion without the assumption of irreducibility, since the growth series of $(W, S)$ is the product of the growth series of its irreducible factors.

## 17.3. RECIPROCITY

**DEFINITION 17.3.1.** Let $\delta = \pm 1$. The rational function $W(\mathbf{t})$ is $\delta$-*reciprocal* if $W(\mathbf{t}^{-1}) = \delta W(\mathbf{t})$.

**DEFINITION 17.3.2.** A locally finite cell complex $\Lambda$ is an *Euler complex of type* $\delta$ if for each (nonempty) cell $\sigma$ in $\Lambda$, $\mathrm{Lk}(\sigma, \Lambda)$ has the same Euler characteristic as $S^{n-\dim\sigma-1}$, with $\delta = (-1)^n$. In other words, we require

$$\chi(\mathrm{Lk}(\sigma, \Lambda)) = 1 - \delta(-1)^{\dim\sigma}.$$

If, in addition, $\Lambda$ is a finite complex and the above also holds for $\sigma = \emptyset$, then it is an *Euler sphere of type* $\delta$ (By convention, $\dim\emptyset = -1$ and $\mathrm{Lk}(\emptyset, \Lambda) = \Lambda$.)

Thus, $\Lambda$ is a Euler complex of type $\delta$ if and only if the link of each vertex is an Euler sphere of type $-\delta$. If $\Lambda$ is a homology $n$-manifold, then it is an Euler complex of type $(-1)^n$; if it is a generalized homology $n$-sphere, then it is an Euler sphere of type $(-1)^n$. The intuition behind the next lemma is that any closed odd-dimensional manifold has Euler characteristic 0 and hence, is an Euler sphere of type $-1$.

**LEMMA 17.3.3.** *If $\Lambda$ is finite Euler complex of type $-1$, then $\chi(\Lambda) = 0$ (and hence, $\Lambda$ is automatically an Euler sphere of type $-1$).*

*Proof.* As in 7.2, partition $\Lambda$ into "dual cones." To do this, first recall from Appendix A.3 that the barycentric subdivision of $\Lambda$ is the geometric realization $|\mathcal{F}(\Lambda)|$ of its face poset. For each cell $\sigma \in \Lambda$, set

$$D_\sigma := |\mathcal{F}(\Lambda)_{\geqslant\sigma}|, \quad \partial D_\sigma := |\mathcal{F}(\Lambda)_{>\sigma}| \quad \dot{D}_\sigma := D_\sigma - \partial D_\sigma.$$

($D_\sigma$ is the *dual cone* to $\sigma$.) Since $D_\sigma$ is a cone (on $\partial D_\sigma$), $\chi(D_\sigma) = 1$ and since $\partial D_\sigma$ is (the barycentric subdivision of) $\mathrm{Lk}(\sigma, \Lambda)$, $\chi(\partial D_\sigma) = \chi(\mathrm{Lk}(\sigma, \Lambda))$. Since (the barycentric subdivision of) $\Lambda$ can be partitioned as the disjoint union of the open dual cones $(\dot{D}_\sigma)_{\sigma\in\mathcal{F}(\Lambda)}$, we have

$$\chi(\Lambda) = \sum 1 - \chi(\mathrm{Lk}(\sigma, \Lambda) = \sum 1 - (1 + (-1)^{\dim\sigma})$$
$$= -\sum(-1)^{\dim\sigma} = -\chi(\Lambda).$$

So $\chi(\Lambda) = 0$. □

**THEOREM 17.3.4.** ([53].) *Let $(W, S)$ be a Coxeter system and $L$ its nerve. If $L$ is an Euler sphere of type $-\delta$, with $\delta \in \{\pm 1\}$, then $W(\mathbf{t})$ is $\delta$-reciprocal.*

*Proof.* By hypothesis, for each $T \in \mathcal{S}$, $1 - \chi(\mathrm{Lk}(T, L)) = 1 - (1 - \delta(-1)^{\dim T}) = \delta(-1)^{\mathrm{Card}(T)} = \delta\varepsilon(T)$. So, combining the formulas in Theorems 17.1.9 and 17.1.10, we get $1/W(\mathbf{t}) = \delta/W(\mathbf{t}^{-1})$. □

**COROLLARY 17.3.5.** *If $W$ is type* $\mathrm{HM}^n$, *then $W(\mathbf{t})$ is $(-1)^n$-reciprocal.*

*Remarks 17.3.6*

(i) When $\mathbf{t} = t$, a single indeterminate, if we substitute $t = 0$ into the formula of Theorem 17.1.9, we get

$$\frac{1}{W(\infty)} = \sum_{T \in \mathcal{S}} \varepsilon(T) = 1 - \sum_{\sigma \in L} (-1)^{\dim(\sigma)} = 1 - \chi(L).$$

(ii) If $W(t)$ is $\delta$-reciprocal, then

$$1 = 1/W(0) = \delta/W(\infty) = \delta(1 - \chi(L)),$$

Hence, $\chi(L) = 1 - \delta$.

(iii) If the nerve $L$ is an Euler sphere of type $-\delta$, then $\Sigma$ is an Euler complex of type $\delta$.

## 17.4. RELATIONSHIP WITH THE $h$-POLYNOMIAL

Suppose $L$ is a simplicial complex with vertex set $S$. As in Example A.2.3 and in Sections 7.2 and 10.8, $\mathcal{S}(L)$ denotes the face poset of $L$ (including the empty simplex). As usual, we identify a simplex in $\mathcal{S}(L)$ with its vertex set. If $T \in \mathcal{S}(L)$, $\sigma_T$ denotes the corresponding geometric simplex in $L$ spanned by $T$. $\mathcal{S}^{(m)}(L) := \{T \in \mathcal{S}(L) \mid \mathrm{Card}(T) = m\}$ is the set of $(m-1)$-simplices. Suppose $\dim L = n - 1$.

In formula (16.9) of 16.1, we defined the "$f$-polynomial," $f_L(t)$, of $L$ by

$$f_L(t) := \sum_{T \in \mathcal{S}(L)} t^{\mathrm{Card}(T)} = \sum_{i=0}^{n} f_{i-1} t^i,$$

where $f_m$ is the number of $m$-simplices of $L$ and $f_{-1} = 1$. Its *$h$-polynomial*, $h_L(t)$, is defined by

$$h_L(t) := (1 - t)^n f_L\left(\frac{t}{1 - t}\right). \tag{17.16}$$

Using the $h$-polynomial, the Flag Complex Conjecture 16.3.1 can be reformulated as follows.

**CONJECTURE 17.4.1.** (Reformulation of the Flag Complex Conjecture.) *Suppose $L$ is a triangulation of a* $\mathrm{GHS}^{2k-1}$ *as a flag complex. Then* $(-1)^k h_L(-1) \geqslant 0$.

To see that this is equivalent to Conjecture 16.3.1, substitute $t = -1$ into (17.16) to get $h_L(-1) = 2^n f_L(-\frac{1}{2})$. So, $h_L(-1)$ and $f_L(-\frac{1}{2})$ have the same sign.

Now suppose that $(W, S)$ is right-angled. Then for each spherical subset $T$, $W_T \cong (\mathbb{Z}/2)^{\mathrm{Card}(T)}$. So, $W_T(t) = (1+t)^{\mathrm{Card}(T)}$ and hence,

$$\frac{1}{W_T(t)} = \left(\frac{1}{1+t}\right)^{\mathrm{Card}(T)} \quad \text{and} \quad \frac{1}{W_T(t^{-1})} = \left(\frac{t}{1+t}\right)^{\mathrm{Card}(T)}. \tag{17.17}$$

**PROPOSITION 17.4.2.** *Suppose $(W, S)$ is a right-angled Coxeter system and its nerve $L$ has dimension $n-1$. Then*

$$\frac{1}{W(t)} = \frac{h_L(-t)}{(1+t)^n}.$$

*Proof.* By Theorem 17.1.9 and (17.17),

$$\frac{1}{W(t)} = \sum_{T \in \mathcal{S}} \frac{\varepsilon(T)}{W_T(t^{-1})} = \sum_{T \in \mathcal{S}} \left(\frac{-t}{1+t}\right)^{\mathrm{Card}(T)}$$

$$= f_L\left(\frac{-t}{1+t}\right) = \frac{h_L(-t)}{(1+t)^n}.$$

□

**Example 17.4.3.** (*Right-angled three-dimensional polytope groups.*) Suppose $(W, S)$ is right angled with nerve $L$ a flag triangulation of $S^2$. (It is a classical result that the dual of $L$ is isomorphic to the boundary complex of a simple, convex polytope.) By Proposition 17.4.2, $1/W(t) = h_L(-t)/(1+t)^3$. By the definition (17.16), $h_L(-t) = (1+t)^3 f_L(\frac{-t}{1+t})$, which simplifies to

$$h_L(-t) = 1 - (f_0 - 3)t + (f_1 - 2f_0 + 3)t^2 - (f_0 - f_1 + f_2 - 1)t^3$$

$$= 1 - (f_0 - 3)t + (f_0 - 3)t^2 - t^3$$

$$= -(t-1)(t^2 - (f_0 - 4)t + 1).$$

(To get the second equality, use $3f_2 = 2f_1$ and Euler's formula $f_0 - f_1 + f_2 = 2$.) If $f_0 \geqslant 6$, there are three positive real roots, $\rho$, 1, and $\rho^{-1}$, where

$$\rho^{\pm 1} = \frac{(f_0 - 4) \mp \sqrt{(f_0 - 4)^2 - 4}}{2}.$$

***Exercise***: Show that if $L$ is a flag triangulation of $S^2$, then $f_0 \geqslant 6$.

## NOTES

**17.1.** In the case where **t** is a single indeterminate, most of the results of this section come from [29, Ex. 26, pp. 42–43]. The idea of extending the results from this exercise to an $I$-tuple of indeterminates comes from [255]. Lemma 17.1.8 is due to Steinberg [270]. For more about growth series of Coxeter groups, see [48, 129, 234, 235] or [26, §7.1].

In Chapter 20 we will explain a connection between this material on growth series and "weighted $L^2$-Betti numbers." The "weight" is a certain $I$-tuple $\mathbf{q}$ of positive real numbers. The weighted $L^2$-Euler characteristic of $\Sigma$ turns out to be $1/W(\mathbf{q})$ (Proposition 20.2.4). This number also turns up in Section 18.4 of the next chapter as the "Euler Poincaré measure" of a building (Theorem 18.4.3). In Theorem 20.4.2 we give a different proof of reciprocity for growth series of Coxeter groups of type HM using Poincaré duality for weighted $L^2$-cohomology (Corollary 17.3.5). The regions $\mathcal{R}$ and $\mathcal{R}^{-1}$ of 17.1 also play a key role in the results of Chapter 20.

**17.2.** The usual definition of an "amenable" group $G$ is that it admits an invariant "mean" $M : L^\infty(G) \to \mathbb{R}$ satisfying certain conditions. For finitely generated groups this definition is equivalent to the Følner Condition.

# Chapter Eighteen

# BUILDINGS

Buildings were introduced by Tits as an abstraction of certain incidence geometries which had been studied earlier in the context of algebraic groups (often over finite fields). As a combinatorial object, a building is a generalization of a Coxeter system. It is first of all a "chamber system" as defined below. The additional data for a chamber system to be a building consists of a Coxeter system $(W, S)$ and a "$W$-valued distance function" on the set of chambers. (This is different from, but equivalent to, the classical definition of a building in [43] as a certain type simplicial complex together with a system of subcomplexes called "apartments," subject to certain axioms.)

One can define the "geometric realization of a building" analogously to the simplicial complex $\Sigma(W, S)$ in Chapter 7. In fact, the geometric realization of a building of type $(W, S)$ will contain many copies of $\Sigma(W, S)$, each of which is called an "apartment." The basic picture to keep in mind is the case where $W$ is the infinite dihedral group $\mathbf{D}_\infty$. $\Sigma$ is then a copy of the real line cellulated by edges of unit length. The geometric realization of a building of type $\overset{\infty}{\circ\!\!-\!\!\!-\!\!\circ}$ is a tree (and any tree without extreme points is such a building). In $\Sigma$ exactly two chambers (= edges) meet along a mirror (= vertex), while in a building (= tree) there can be more than two chambers along a mirror. In 18.2, we discuss other possible realizations of a building as a topological space using a construction similar to the one in Chapter 5. In 18.3 we use Moussong's Theorem to prove that the geometric realization of any building is CAT(0).

## 18.1. THE COMBINATORIAL THEORY OF BUILDINGS

### Chamber Systems

A *chamber system over a set* $S$ is a set $\Phi$ together with a family of equivalence relations on $\Phi$ indexed by $S$. The elements of $\Phi$ are *chambers*. Two chambers are *s-equivalent* if they are equivalent via the equivalence relation corresponding to $s$; they are *s-adjacent* if they are $s$-equivalent and not equal.

**Example 18.1.1.** (*The chamber system of a barycentric subdivision.*) As in Appendix B.1, suppose that $P$ is a $(n + 1)$-dimensional convex polytope, that

$b(\partial P)$ is the barycentric subdivision of its boundary and that Cham$(b(\partial P))$ is the set of top-dimensional simplices of $b(\partial P)$. Each vertex of $P$ has a *type* in $\{0, 1, \dots, n\}$—it is the dimension of the face of which the vertex is a barycenter. Each codimension-one face of a chamber has a *cotype*—it is the type of an opposite vertex. Call two chambers *i-adjacent* if they are adjacent simplices across a face of cotype $i$. This gives Cham$(b(\partial P))$ the structure of a chamber system over $\{0, 1, \dots, n\}$.

**Example 18.1.2.** (*The chamber system associated to a family of subgroups.*) Let $G$ be a group, $B$ a subgroup, and $(G_s)_{s\in S}$ a family of subgroups such that each $G_s$ contains $B$. Define a chamber system $\Phi = \Phi(G, B, (G_s)_{s\in S})$ as follows: $\Phi := G/B$ and chambers $gB$ and $g'B$ are *s-equivalent* if they have the same image in $G/G_s$.

Let $\Phi$ be a chamber system over $S$. A *gallery* in $\Phi$ is a finite sequence of chambers $(\varphi_0, \dots, \varphi_k)$ such that $\varphi_{j-1}$ is adjacent to $\varphi_j$, $1 \leqslant j \leqslant k$. The *type* of the gallery is the word $\mathbf{s} = (s_1, \dots, s_k)$ where $\varphi_{j-1}$ is $s_j$-adjacent to $\varphi_j$. If each $s_j$ belongs to a given subset $T$ of $S$, then it is a *T-gallery*. A chamber system is *connected* (or *T-connected*) if any two chambers can be joined by a gallery (or a $T$-gallery). The $T$-connected components of a chamber system $\Phi$ are its *T-residues*. For example, an $\{s\}$-residue is the same thing as an $s$-equivalence class; so, in Example 18.1.2, an $\{s\}$-residue is a coset $gG_s$.

The *rank* of a chamber system over $S$ is the cardinality of $S$. A *morphism* $f : \Phi \to \Phi'$ of chamber systems over the same set $S$ is a function which preserves $s$-equivalence classes, for all $s \in S$. A group $G$ of automorphisms of $\Phi$ is *chamber transitive* if it is transitive on $\Phi$. Suppose $G$ is chamber transitive. Fix $\varphi \in \Phi$ and for each $s \in S$ choose an $\{s\}$-residue containing $\varphi$. Let $B$ denote the stabilizer of $\varphi$ and $G_s$ the stabilizer of the $\{s\}$-residue. Clearly, $\Phi$ is isomorphic to the chamber system $\Phi(G, B, (G_s)_{s\in S})$ of Example 18.1.2.

Suppose $\Phi_1, \dots, \Phi_k$ are chamber systems over $S_1, \dots, S_k$. Their *direct product* $\Phi_1 \times \cdots \times \Phi_k$ is a chamber system over the disjoint union $S_1 \cup \cdots \cup S_k$. Its chambers are $k$-tuples $(\varphi_1, \dots, \varphi_k)$ with $\varphi_i \in \Phi_i$. For $s \in S_i$, the chamber $(\varphi_1, \dots, \varphi_k)$ is *s-adjacent* to $(\varphi'_1, \dots, \varphi'_k)$ if $\varphi_j = \varphi'_j$ for $j \neq i$ and $\varphi_i$ and $\varphi'_i$ are $s$-adjacent.

### Buildings

Suppose $(W, S)$ is a Coxeter system and $M = (m_{st})$ is its Coxeter matrix (Definition 3.3.1). A *building of type M* (or of *type* $(W, S)$) is a chamber system $\Phi$ over $S$ such that

(i) for all $s \in S$, each $s$-equivalence class contains at least two chambers, and

(ii) there exists a *W-valued distance function* $\delta : \Phi \times \Phi \to W$, meaning that given $\mathbf{s}$ an $M$-reduced word (Section 3.4), chambers $\varphi$ and $\varphi'$ can be joined by a gallery of type $\mathbf{s}$ if and only if $\delta(\varphi, \varphi') = w(\mathbf{s})$ (where if $\mathbf{s} = (s_1, \dots, s_k)$ is a word, then $w(\mathbf{s}) := s_1 \cdots s_k$ is its *value*).

A building of type $(W, S)$ is *spherical* if $W$ is finite; it is *Euclidean* (or *affine type*) if $W$ is a Euclidean reflection group. A residue of type $T$ in a building is *spherical* if $T$ is spherical subset. A building is *thick* if for all $s \in S$, each $s$-equivalence class has at least three elements.

**Example 18.1.3.** (*Thin buildings.*) Let $\mathbf{W} = \Phi(W, \{1\}, (W_{\{i\}})_{s \in S})$, where the notation is as in Example 18.1.2. In other words, the set of chambers is $W$ and two chambers $w$ and $w'$ are $s$-adjacent if and only if $w' = ws$. There is a $W$-valued distance function $\delta : W \times W \to W$ defined by $\delta(w, w') = w^{-1}w'$. Thus, $\mathbf{W}$ is a building, called the *abstract Coxeter complex* of $W$.

**Example 18.1.4.** (*Rank-1 buildings.*) Suppose that $\Phi$ is a building of rank 1. $W$ is the cyclic group of order 2. $\Phi$ can be an arbitrary set with more than two elements. There is only one possibility for $\delta : \Phi \times \Phi \to W$; it must map the diagonal of $\Phi \times \Phi$ to the identity element and the complement of the diagonal to the nontrivial element. Thus, any two chambers are adjacent.

**Example 18.1.5.** (*Direct products.*) Suppose that $M_1, \dots, M_k$ are Coxeter matrices over $S_1, \dots, S_k$, respectively. Let $S$ denote the disjoint union $S_1 \cup \cdots \cup S_k$ and define a Coxeter matrix $M$ over $S$ by setting $m_{st}$ equal to the corresponding entry of $M_p$ whenever $s, t$ belong to the same component $S_p$ of $S$ and $m_{st} = 2$ when they belong to different components. Then $W = W_1 \times \cdots \times W_k$ where $W = W(M)$ and $W_p = W(M_p)$. Suppose $\Phi_1, \cdots, \Phi_k$ are buildings over $S_1, \dots, S_k$. Their direct product $\Phi = \Phi_1 \times \cdots \times \Phi_k$ is a chamber system over $\Phi$. Moreover, the direct product of the $W_p$-valued distance functions gives a $W$-valued distance function on $\Phi$. Hence, $\Phi$ is a building of type $M$.

**Example 18.1.6.** (*Trees.*) Suppose $W$ is the infinite dihedral group. Any tree is bipartite, i.e., its vertices can be labeled by the two elements of $S$ so that the vertices of any edge have distinct labels. Suppose $T$ is a tree with such a labeling and suppose no vertex of $T$ is extreme. Let $\Phi$ be the set of edges of $T$. Given $s \in S$, call two edges *s-equivalent* if they meet at a vertex of type $s$. An $\{s\}$-residue is the set of edges in the star of a vertex of type $s$. A gallery in $\Phi$ is a sequence of adjacent edges in $T$. The type of a gallery is a word in $S$. The word is reduced if and only if the gallery is minimal. Given two edges $\varphi, \varphi'$ of $T$, there is a (unique) minimal gallery connecting them. The corresponding word represents an element of $w \in W$ and $\delta(\varphi, \varphi') := w$. Thus, every such tree $T$ defines a building of type $\overset{\infty}{\circ\!\!-\!\!\!-\!\!\circ}$. Not surprisingly, in the next section we will define the "geometric realization of a building" so that for the building $\Phi$ corresponding to $T$ it is $T$.

**Example 18.1.7.** (*Generalized polygons.*) A *generalized m-gon* is a bipartite graph of diameter $m$ and girth $2m$. (If each edge has length 1, the *diameter* of a graph is the maximum distance between two points; its *girth* is the length of the shortest circuit.) Generalized 3-gons are also called *projective planes*. In the same way as trees are the geometric realizations of buildings of type $\circ\overset{\infty}{\text{—}}\circ$, generalized $m$-gons are the geometric realizations of buildings of type $\circ\overset{m}{\text{—}}\circ$. To be more specific, given a generalized $m$-gon $P$, put $\Phi =$ Edge($P$) and color the 2 different types of vertices by the fundamental set of generators, $\{s, t\}$, of the dihedral group $\mathbf{D}_m$. As before, a gallery is a sequence of adjacent edges and it is minimal if and only if its type $\mathbf{s}$ is an $M$-reduced word.

For a proof of the next result we refer the reader to [248, p. 29].

**PROPOSITION 18.1.8.** *In a thick generalized m-gon, vertices of the same type have the same valence and if m is odd, all vertices have the same valence.*

We denote the valencies at vertices of type $s$ and $t$ by $q_s + 1$ and $q_t + 1$, respectively.

**THEOREM 18.1.9.** (Feit-Higman [126].) *Finite, thick, generalized m-gons exist only for m* $= 2, 3, 4, 6,$ *or* $8$. *Moreover, there are the following restrictions on the parameters* $q_s$ *and* $q_t$*:*

$$\begin{cases} \dfrac{q_s q_t(q_s q_t + 1)}{q_s + q_t} \text{ is an integer} & \text{for } m = 4, \\ st \text{ is a perfect square} & \text{for } m = 6, \\ 2st \text{ is a perfect square} & \text{for } m = 8. \end{cases}$$

**Example 18.1.10.** (*Graph products and right-angled buildings.*) Let $\Upsilon$ be a simplicial graph with vertex set $S$. As in Example 7.1.7, we have an associated right-angled Coxeter system $(W, S)$ and its Coxeter matrix $M$. Suppose we are given a family of groups $(G_s)_{s \in S}$. For each $T \in \mathcal{S}$, let $G_T$ denote the direct product

$$G_T := \prod_{s \in T} G_s$$

($G_\emptyset = \{1\}$). For $T \leqslant T' \in \mathcal{S}$, let $f^T_{T'}\colon G_T \to G_{T'}$ be the natural inclusion. The *graph product* $G$ of the family $(G_s)_{s\in S}$ (with respect to $\Upsilon$) is defined to be the direct limit:

$$G := \varinjlim_{T\in\mathcal{S}} G_T.$$

Equivalently, we could have defined it as the direct limit of the smaller family $\{G_T \mid T \in \mathcal{S}_{\leqslant 2}\}$, where $\mathcal{S}_{\leqslant k} := \{T \in \mathcal{S} \mid \mathrm{Card}(T) \leqslant k\}$. (See the beginning of Appendix G.1 for the definition of the direct limit of a poset of groups.) Alternatively, $G$ could have been defined as the quotient of the free product of the $(G_s)_{s\in S}$ by the normal subgroup generated by all commutators of the form $[g_s, g_t]$, with $g_s \in G_s$, $g_t \in G_t$ and $m_{st} = 2$.

It is shown in [76] that the chamber system $\Phi(G, \{1\}, (G_s)_{s\in S})$ is a building of type $(W, S)$. The $W$-valued distance function $\delta : G \times G \to W$ can be defined as follows. First, note that it is clear that any $g$ in $G$ can be written in the form $g = g_{s_1} \cdots g_{s_k}$, with $g_{s_i} \in G_{s_i} - \{1\}$ and with $\mathbf{s} = (s_1, \ldots, s_k)$ $M$-reduced. Moreover, if $g' = g_{s'_1} \cdots g_{s'_k}$ is another such representation, then $g = g'$ if and only if we can get from one representation to the other by a sequence of replacements of the form, $g_s g_t \to g_t g_s$ with $m_{st} = 2$. In particular, the words $\mathbf{s} = (s_1, \ldots, s_k)$ and $\mathbf{s}' = (s'_1, \ldots, s'_k)$ must have the same image in $W$. If $g = g_{s_1} \cdots g_{s_k}$, then define $\delta(1, g) = w(\mathbf{s})$ and then extend this to $G \times G$ by $\delta(g, g') = \delta(1, g^{-1}g')$. Then $\delta$ has the desired properties.

**Example 18.1.11.** (*Spherical buildings of type* $\mathbf{A}_n$.) Here we describe one of the most well-known and important examples of a building. Let $\mathbb{F}$ be a field and $V = \mathbb{F}^{n+1}$. Define $\Phi$ to be the set of all complete flags of subspaces of $V$:

$$V_1 \subset V_2 \subset \cdots \subset V_n \subset V,$$

where $\dim_{\mathbb{F}}(V_k) = k$. We can give this the structure of a chamber system over $I := \{1, \ldots, n\}$ by calling two complete flags $i$-adjacent if they differ only in their $i^{\text{th}}$ term (i.e., the subspace of dimension $i$). Fix the base chamber to be the standard flag: $V_k = \mathbb{F}^k$. The group $G = GL(n+1, \mathbb{F})$ acts transitively on $\Phi$. So, as in Example 18.1.2, $\Phi = \Phi(G, B, (G_i)_{i\in I})$, where $G = GL(n+1, \mathbb{F})$, $B$ is the stabilizer of the standard flag (i.e., $B$ is the subgroup of upper triangular matrices) and $G_i$ is the stabilizer of the standard flag after the $i^{\text{th}}$ subspace has been deleted. It turns out that this is a building associated to the Coxeter system of type $\mathbf{A}_n$. The Coxeter group $W$ is the symmetric group $S_{n+1}$ (see Example 6.7.1). Its $W$-valued distance function is defined as follows. Suppose $\varphi = \{V_1, \ldots, V_n\}$, $\varphi' = \{V'_1, \ldots, V'_n\}$ are two chambers. Put $V_0 = V'_0 = \{0\}$ and $V_{n+1} = V'_{n+1} = V$. Define $\sigma(i) := \min\{j \mid V'_i \subset V'_{i-1} + V_j\}$. One checks that $\sigma$ is a permutation in $S_{n+1}$ and that $\delta(\varphi, \varphi') := \sigma$ is a $W$-valued distance function and hence, that $\Phi$ is a building. When $n = 2$, such buildings are generalized 3-gons.

### Apartments and Retractions

Suppose $\Phi$ is a building of type $(W, S)$. Let $\mathbf{W}$ be the abstract Coxeter complex of Example 18.1.3. A $W$-isometry of $\mathbf{W}$ into $\Phi$ is a map $\alpha : W \to \Phi$ which preserves $W$-distances, i.e.,

$$\delta_\Phi(\alpha(w), \alpha(w')) = \delta_{\mathbf{W}}(w, w')$$

for all $w, w' \in W$. ($\delta_\Phi$ is the $W$-valued distance function on $\Phi$ and $\delta_{\mathbf{W}}(w, w') = w^{-1}w'$.) An *apartment* in $\Phi$ is an isometric image $\alpha(W)$ of $\mathbf{W}$ in $\Phi$.

The set of $W$-isometries of $\mathbf{W}$ with itself is bijective with $W$. Indeed, given $w \in W$, there is a unique isometry $\alpha_w : W \to W$ sending 1 to $w$, defined by $\alpha_w(w') = ww'$. It follows that an isometry $\alpha : W \to \Phi$ is uniquely determined by its image $A = \alpha(W)$ together with a chamber $\varphi = \alpha(1)$.

Fix an apartment $A$ in $\Phi$ and a base chamber $\varphi$ in $A$. Define a map $\rho_{\varphi,A} : \Phi \to A$, called the *retraction of* $\Phi$ *onto* $A$ *with center* $\varphi$, as follows. Suppose $A = \alpha(W)$ with $\alpha(1) = \varphi$. Set $\rho_{\varphi,A}(\varphi') = \alpha(\delta(\varphi, \varphi'))$. The next lemma is an easy exercise for the reader.

**Lemma 18.1.12.** *The map $\rho_{\varphi,A} : \Phi \to A$ is a morphism of chamber systems.*

In other words, if $\delta(\varphi', \varphi'') = s$, then either $\rho(\varphi') = \rho(\varphi'')$ or $\delta\big(\rho(\varphi'), \rho(\varphi'')\big) = s$, where $\rho = \rho_{\varphi,A}$.

**Corollary 18.1.13.** *Let $\rho = \rho_{\varphi,A}$. If $\delta(\varphi', \varphi'') \in W_T$, then $\delta(\rho(\varphi'), \rho(\varphi'')) \in W_T$.*

**Lemma 18.1.14.** *Let $\rho = \rho_{\varphi,A} : \Phi \to A$. If $A'$ is another apartment containing $\varphi$, then $\rho|_{A'} : A' \to A$ is a $W$-isometry (i.e., $\rho|_{A'}$ is an isomorphism).*

*Proof.* Let $\alpha : W \to A$ and $\beta : W \to A'$ be $W$-isometries such that $\alpha(1) = \varphi = \beta(1)$. If $\varphi' \in A'$, then $\rho(\varphi') = \alpha(\delta(\varphi, \varphi'))$ and $\beta(\delta(\varphi, \varphi')) = \varphi'$. Therefore, $\rho(\varphi') = \alpha \circ \beta^{-1}(\varphi')$, i.e., $\rho|_{A'} = \alpha \circ \beta^{-1}$. □

### The Thickness Vector

A building $\Phi$ of type $(W, S)$ has *finite thickness* if, for all $s \in S$, each $s$-equivalence class is finite. If $\Phi$ has finite thickness, then it follows from the existence of a $W$-distance function that each of its spherical residues is finite.

Let us say that $\Phi$ is *regular* if, for each $s \in S$, each $s$-equivalence class has the same number of elements. When finite, we denote this number by $q_s + 1$. For other examples, note that Proposition 18.1.8 asserts that every finite, thick, generalized $m$-gon is regular. The proof Proposition 18.1.8 also gives the following lemma valid when the generalized $m$-gon is regular but not necessarily thick.

**LEMMA 18.1.15.** *Suppose $P$ is a finite, regular, generalized $m$-gon with $m$ odd. Then any two vertices of $P$ have the same valence.*

*Proof.* Let $x$ and $y$ be opposite vertices in $P$ (i.e., $d(x,y) = m$). Let $e$ be an edge at $y$ and $y'$ its other vertex. Since $d(x,y) > d(x,y')$ and the girth is $m$, this implies that there is a unique edge path of length $m-1$ from $y'$ to $x$. This defines a bijection from the set of edges at $y$ to those at $x$. Since $m$ is odd, $x$ and $y$ have different types, so $q_s = q_t$. □

**COROLLARY 18.1.16.** *Suppose $\Phi$ is a regular building (of finite thickness) of type $(W,S)$. If $s$ and $s'$ are conjugate elements of $S$, then $q_s = q_{s'}$.*

*Proof.* Suppose $s$ and $s'$ are conjugate. By Lemma 3.3.3 there is a sequence, $s_0, s_1, \dots, s_n$, with $s_0 = s$, $s_n = s'$ and with each $m(s_i, s_{i+1})$ odd. By the previous lemma, $q_s = q_{s_1} \cdots = q_{s_{n-1}} = q_{s'}$. □

Let $I$ be the set of conjugacy classes of elements in $S$. For any regular building $\Phi$, the integers $q_s$ define an $I$-tuple $\mathbf{q}$ called the *thickness vector* of $\Phi$. Obviously, if $G$ is a chamber transitive automorphism group of $\Phi$, then $\Phi$ is regular. For example, the abstract Coxeter complex $\mathbf{W}$ (Example 18.1.3) is regular and its thickness vector is $\mathbf{1}$, the $I$-tuple with all entries equal to 1. As another example, if $\Phi$ is a finite building of rank one, then its thickness vector is the number $q := \mathrm{Card}(\Phi) - 1$. A third example is where $\mathbb{F}$ is a finite field with $q$ elements and $\Phi$ is the corresponding building of type $\mathbf{A}_n$: the thickness vector is the number $q$. The reason is that there are $q+1$ points in the projective line over $\mathbb{F}$. (The vector $\mathbf{q}$ is a single number in the $\mathbf{A}_n$ case, since the elements of $S$ are all conjugate.) Finally, in the case of a regular right-angled building, Example 18.1.10 shows that $\mathbf{q}$ can be an arbitrary element of $\mathbb{N}^S$.

As in formula (17.1), given an $I$-tuple $\mathbf{q}$ and an element $w \in W$, put $q_w = q_{s_1} \cdots q_{s_l}$, where $s_1 \cdots s_l$ is any reduced expression for $w$.

Suppose $\Phi$ is a regular building of thickness $\mathbf{q}$. Choose a base chamber $\varphi_0 \in \Phi$ and define $\rho : \Phi \to W$ by $\rho(\varphi) := \delta(\varphi_0, \varphi)$. (Essentially, $\rho$ is a retraction onto an apartment centered at $\varphi_0$.)

**LEMMA 18.1.17.** *For any $w \in W$, $\mathrm{Card}(\rho^{-1}(w)) = q_w$.*

*Proof.* This is proved by induction on $l = l(w)$. It is trivially true for $l = 0$. If $l > 0$, choose a reduced expression $\mathbf{s} = (s_1, \dots, s_l)$ for $w$. Put $w' = s_1 \cdots s_{l-1}$ and $s = s_l$. It follows from the definition of a building, that for any $\varphi \in \rho^{-1}(w)$, there is a unique gallery of type $\mathbf{s}$ from $\varphi_0$ to $\varphi$. The penultimate chamber in this gallery lies in $\rho^{-1}(w')$. Thus, each chamber in $\rho^{-1}(w)$ is $s$-adjacent to a unique one in $\rho^{-1}(w')$. By induction, $\mathrm{Card}(\rho^{-1}(w')) = q_{w'}$. So

$$\mathrm{Card}(\rho^{-1}(w)) = q_s\,\mathrm{Card}(\rho^{-1}(w')) = q_s q_{w'} = q_w.$$

□

**COROLLARY 18.1.18.** *Let $\Phi$ be a finite, spherical building of type $(W, S)$. Suppose it is regular with thickness vector $\mathbf{q}$. Then $\mathrm{Card}(\Phi) = W(\mathbf{q})$, where $W(\mathbf{q})$ is the growth polynomial from Chapter 17 evaluated at $\mathbf{t} = \mathbf{q}$.*

*Proof.* Let $\rho : \Phi \to W$ be as in the previous lemma. Then

$$\mathrm{Card}(\Phi) = \sum_{w \in W} \mathrm{Card}(\rho^{-1}(w)) = \sum_{w \in W} q_w = W(\mathbf{q}).$$

□

### Complement on Tits Systems and $BN$-Pairs

A *Tits system* (or a *BN-pair*) is a quadruple $(G, B, N, S)$, where $G$ is a group, $B$ and $N$ are subgroups of $G$, $W := N/N \cap B$, $S$ is a subset of $W$ and the following axioms hold:

(T1) The set $B \cup N$ generates $G$ and $B \cap N$ is a normal subgroup of $N$.

(T2) The set $S$ generates the group $W := N/(B \cap N)$ and consists of elements of order 2.

(T3) $sBw \subset BwB \cup BswB$ for $s \in S$ and $w \in W$.

(T4) For all $s \in S$, $sBs \not\subset B$.

Given $w \in W$, put $C(w) := BwB$. The axioms imply that

- For each $s \in S$, $G_s := B \cup C(s)$ is a subgroup of $G$.
- $(W, S)$ is a Coxeter system.
- The chamber system $\Phi(G, B, (G_s)_{s \in S})$ is a building of type $(W, S)$.
- Suppose $r : G/B \to W$ is defined by $gB \to \delta(B, gB)$ where $\delta$ is $W$-distance in the building. Then the induced map $\overline{r} : B\backslash G/B \to W$ is a bijection.

For example, suppose, as in 18.1.11, $G = GL(n+1, \mathbb{F})$ and $B$ is the set of upper triangular matrices. Let $T \subset B$ be the subgroup of all diagonal matrices and $N \subset G$ the monomial matrices (i.e., those with only one nonzero entry in each column). Then $W := N/N \cap B = N \cap T$ is isomorphic to $S_{n+1}$. Let $S$ be the usual fundamental set of generators for $S_{n+1}$. Then $(G, B, N, S)$ is a Tits system.

## 18.2. THE GEOMETRIC REALIZATION OF A BUILDING

Given a building $\Phi$ of type $(W, S)$, we will define a simplicial complex $\text{Geom}(\Phi)$, called its "geometric realization," so that for each apartment $\alpha(W)$, the corresponding subcomplex will be isomorphic to the simplicial complex $\Sigma(W, S)$ of Chapter 7. The definition is similar to the one in 7.2. As in Definition A.3.4, given a poset $\mathcal{P}$, its *geometric realization* $|\mathcal{P}|$ is defined to be the simplicial complex $\text{Flag}(\mathcal{P})$. Let $\mathcal{C}$ $(= \mathcal{C}(\Phi))$ denote the poset of all spherical residues in $\Phi$. Define the *geometric realization* of $\Phi$ to be

$$\text{Geom}(\Phi) := |\mathcal{C}|.$$

For each $\varphi \in \Phi$ let $|\varphi|$ denote the maximal coface $|\mathcal{C}_{\geqslant \{\varphi\}}|$. The map Type: $\mathcal{C} \to \mathcal{S}$ which associates to each residue its type induces a map of geometric realizations $\text{Geom}(\Phi) \to K$. Moreover, the restriction of this map to each chamber $|\varphi|$ is a homeomorphism.

There is a more illuminating approach using the construction of Chapter 5. Let $(X_s)_{s \in S}$ be a mirror structure on a space $X$. Define an equivalence relation on $\Phi \times X$ by $(\varphi, x) \sim (\varphi', x')$ if and only if $x = x'$ and $\delta(\varphi, \varphi') \in W_{S(x)}$. The *X-realization* of $\Phi$, denoted $\mathcal{U}(\Phi, X)$, is defined by

$$\mathcal{U}(\Phi, X) := (\Phi \times X)/\sim . \tag{18.1}$$

(Here $\Phi$ has the discrete topology.)

Given $\varphi \in \Phi$, $T \in \mathcal{S}$, let $\varphi \cdot T \in \mathcal{C}$ denote the $T$-residue containing $\varphi$. The next result is a straightforward generalization of Theorem 7.2.4. Its proof is left to the reader.

**PROPOSITION 18.2.1.** *Suppose $\Phi$ is a building of type $(W, S)$ and, as in 7.2, that $K$ is the geometric realization of $\mathcal{S}$. Then the map $\Phi \times \mathcal{S} \to \mathcal{C}$ defined by $(\varphi, T) \to \varphi \cdot T$ induces a homeomorphism $\mathcal{U}(\Phi, K) \to \text{Geom}(\Phi)$.*

If $\varphi$ is a chamber in $\Phi$, then let $X(\varphi)$ denote the image of $\varphi \times X$ in $\mathcal{U}(\Phi, X)$. For any subset $\Omega$ of $\Phi$, put

$$\mathcal{U}(\Omega, X) := \bigcup_{\varphi \in \Omega} X(\varphi). \tag{18.2}$$

Note that, if $A \subset \Phi$ is an apartment, then $\mathcal{U}(A, X) \cong \mathcal{U}(W, X)$. In particular, $\mathcal{U}(A, K) \cong \Sigma$.

**Example 18.2.2.** (*The usual definition.*) As in Example 5.2.7, suppose $\Delta$ is a simplex of dimension $\text{Card}(S) - 1$ and that its codimension one faces are indexed by $S$. The "usual definition" of the geometric realization of the building $\Phi$ is $\mathcal{U}(\Phi, \Delta)$ (see [43, 248]). If $W$ is irreducible and affine type, then

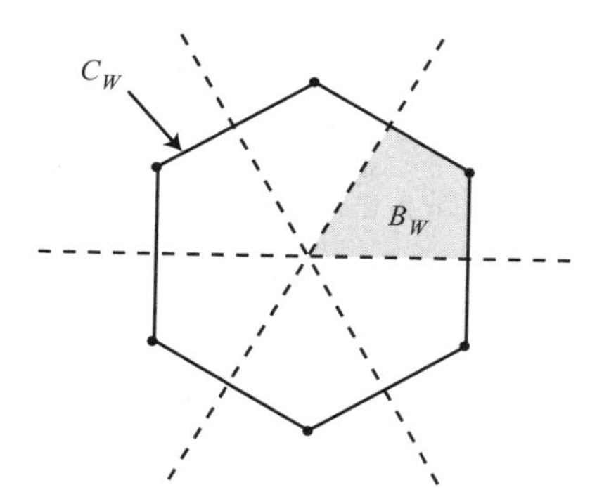


Figure 18.1. Coxeter cell and Coxeter block.

$K = \Delta$ and this agrees with the previous definition. However, in general, they are different.

### The Piecewise Euclidean Structure on Geom($\Phi$)

First, suppose $W$ is spherical. Recall that its associated Coxeter polytope $C_W$ is the convex hull of a generic $W$-orbit in the canonical representation on $\mathbb{R}^S$ (Definition 7.3.1). The intersection of $C_W$ with the fundamental simplicial cone bounded by the walls corresponding to $S$ is denoted $B_W$ and called a *Coxeter block of type* $(W, S)$. $B_W$ is combinatorially equivalent to a cube (of dimension Card($S$)).

We return to the case where $(W, S)$ is arbitrary. Given $T \in \mathcal{S}$, regard $W_T$ as the coset of the identity element in $W/W_T \subset W\mathcal{S}$. As in 7.3, the cell of $\Sigma$ corresponding to $W_T$ is a Coxeter polytope $C_{W_T}$ of type $(W_T, T)$. Its intersection with a fundamental chamber $K$ is a Coxeter block $B_{W_T}$.

We put a piecewise Euclidean structure on Geom($\Phi$) similarly to the way we put one on $\Sigma$ in 12.1. Let $\mathbf{d} \in (0, \infty)^S$. For each $T \in \mathcal{S}$, $\mathbf{d}_T \in (0, \infty)^T$ denotes the restriction of $\mathbf{d}$ to $T$ and $C_{W_T}(\mathbf{d}_T)$ is the Coxeter polytope of type $(W_T, T)$ corresponding to $\mathbf{d}_T$ (i.e., each edge of type $s$ has length $2d_s$). Let $B_{W_T}(\mathbf{d}_T)$ be the corresponding Coxeter block. Identify the intersection of $C_{W_T}(\mathbf{d}_T)$ and $K$ with $B_{W_T}(\mathbf{d}_T)$. This defines a piecewise Euclidean structure on $K$. Since Geom($\Phi$) ($= \mathcal{U}(\Phi, K)$) is tessellated by copies of $K$, there is an induced a piecewise Euclidean structure on it. On each apartment it has the desired property of being isometric to the natural piecewise Euclidean structure on $\Sigma$.

### Spherical Buildings

Suppose $W$ is finite. Then $\Sigma$ is the Coxeter polytope $C_W$ and $K$ is the Coxeter block $B_W$. $K$ has a unique minimal coface $K_S$; it is the point $v_S := \text{Fix}(W, C_W)$.

The point $v_S$ is a vertex of $B_W$. $\mathrm{Lk}(v_S, B_W)$ is a spherical simplex $\Delta$ in the unit sphere $\mathbb{S}^{n-1}$ of $\mathbb{R}^S$. $\Delta$ can be identified with the fundamental simplex for $W$ on $\mathbb{S}^{n-1}$; its Gram matrix is the cosine matrix of $(W, S)$ (Definition 6.8.11). Thus, $\mathcal{U}(W, \Delta) = \mathbb{S}^{n-1}$. If $\Phi$ is a spherical building of type $(W, S)$, then there is a unique residue of type $S$; it is the unique minimal element of $\mathcal{C}$ and hence, specifies a cone point $v \in \mathrm{Geom}(\Phi)$. It follows that

$$\mathrm{Lk}(v, \mathrm{Geom}(\Phi)) = \mathcal{U}(\Phi, \Delta). \tag{18.3}$$

$\mathcal{U}(\Phi, \Delta)$ is the *spherical realization* of the spherical building $\Phi$. The spherical realization of each apartment of $\Phi$ is isometric to $\mathbb{S}^{n-1}$.

## 18.3. BUILDINGS ARE CAT(0)

Our purpose in this section is to prove the following generalization of Moussong's Theorem (12.3.3).

**THEOREM 18.3.1.** *The geometric realization of any building is a complete* CAT(0) *space.*

**COROLLARY 18.3.2.** *For any spherical building $\Phi$, its spherical realization $\mathcal{U}(\Phi, \Delta)$ is* CAT(1).

*Proof of the corollary.* By (18.3), $\mathrm{Lk}(v, \mathrm{Geom}(\Phi)) = \mathcal{U}(\Phi, \Delta)$. Since $\mathrm{Geom}(\Phi)$ is CAT(0), every link in $\mathrm{Geom}(\Phi)$ is CAT(1) (Theorem I.3.5 in Appendix I.3). In particular, $\mathrm{Lk}(v, \mathrm{Geom}(\Phi))$ is CAT(1). □

When the building $\Phi$ is of irreducible affine type, Theorem 18.3.1 is well known. A proof can be found in [43, Ch. VI §3]. Our proof combines the argument given there with Moussong's Theorem.

Suppose $A$ is an apartment in a building $\Phi$, $\varphi \in A$ and $\rho = \rho_{\varphi,A} : \Phi \to A$ is the retraction onto $A$ with center $\varphi$. Since $\rho$ is morphism of chamber systems (Lemma 18.1.12), it induces a map of spaces, $\overline{\rho} : \mathrm{Geom}(\Phi) \to \mathrm{Geom}(A)$, which takes each chamber of $\mathrm{Geom}(\Phi)$ isometrically onto a chamber of $\mathrm{Geom}(A)$. Hence, $\overline{\rho}$ maps a geodesic segment in $\mathrm{Geom}(\Phi)$ to a piecewise geodesic segment in $\mathrm{Geom}(A)$ of the same length. From this observation we conclude the following.

**LEMMA 18.3.3.** *The retraction $\overline{\rho} : \mathrm{Geom}(\Phi) \to \mathrm{Geom}(A)$ is distance decreasing, i.e., for all $x, y \in \mathrm{Geom}(\Phi)$,*

$$d_A(\overline{\rho}(x), \overline{\rho}(y)) \leqslant d_\Phi(x, y),$$

*where $d_A$ and $d_\Phi$ denote distance in* Geom($A$) *and* Geom($\Phi$), *respectively. In particular, if $x, y \in \mathrm{Geom}(A)$, then $d_A(x, y) = d_\Phi(x, y)$.*

**LEMMA 18.3.4.** *There is a unique geodesic between any two points in* $\mathrm{Geom}(\Phi)$.

*Proof.* A basic fact about buildings is that any two chambers are contained in a common apartment ([248, Theorem 3.11, p. 34]). This implies that, given $x, y \in \mathrm{Geom}(\Phi)$, there is an apartment $A$ such that $x, y \in \mathrm{Geom}(A)$. (Choose chambers $\varphi, \varphi' \in \Phi$ so that $x \in K(\varphi), y \in K(\varphi')$ and an apartment $A$ containing $\varphi$ and $\varphi'$; then $x, y \in \mathrm{Geom}(A)$.) A basic fact about CAT(0)-spaces is that any two points are connected by a unique geodesic segment (Theorem I.2.5). Since $\mathrm{Geom}(A)$ is CAT(0) (Moussong's Theorem), there is a unique geodesic segment in $\mathrm{Geom}(A)$ from $x$ to $y$. Let $\gamma : [0, d] \to \mathrm{Geom}(A)$ be a parameterization of this segment by arc length, where $d = d_A(x, y)$. By the last sentence of the previous lemma, $\gamma$ is also a geodesic in $\mathrm{Geom}(\Phi)$. Let $\gamma' : [0, d] \to \mathrm{Geom}(\Phi)$ be another geodesic from $x$ to $y$. If $\overline{\rho} : \mathrm{Geom}(\Phi) \to \mathrm{Geom}(A)$ is the geometric realization of any retraction onto $A$, then $\overline{\rho} \circ \gamma'$ is a geodesic in $\mathrm{Geom}(A)$ from $x$ to $y$ (since it is a piecewise geodesic of length $d$). Hence, $\overline{\rho} \circ \gamma' = \gamma$. Let $t_0 = \sup\{t \mid \gamma|_{[0,t]} = \gamma'|_{[0,t]}\}$. Suppose $t_0 < d$. Then, for sufficiently small positive values of $\varepsilon$, $\gamma(t_0 + \varepsilon)$ lies in the relative interior of some coface $K(\varphi)_T$, with $\varphi \in A$, while $\gamma'(t_0 + \varepsilon)$ lies in the relative interior of a different coface, say, $K(\varphi')_T$ , where $K(\varphi')_T \not\subset \mathrm{Geom}(A)$. Set $\rho = \rho_{\varphi,A}$. Since $K(\varphi)_T \neq K(\varphi')_T, \delta(\varphi, \varphi') \notin W_T$. Hence, $\overline{\rho}(\gamma'(t_0 + \varepsilon)) \neq \gamma(t_0 + \varepsilon)$, a contradiction. Therefore, $t_0 = d$ and $\gamma = \gamma'$. □

**LEMMA 18.3.5.** *Suppose* $\rho = \rho_{\varphi,A}$. *If* $x \in K(\varphi)$, *then* $d(x, \overline{\rho}(y)) = d(x, y)$ *for all* $y \in \mathrm{Geom}(\Phi)$.

*Proof.* Choose an apartment $A'$ so that $\mathrm{Geom}(A')$ contains both $x$ and $y$. By the previous lemma, the image of the geodesic $\gamma$ from $x$ to $y$ is contained in $\mathrm{Geom}(A')$. By Lemma 18.1.14, $\rho|_{A'} : A' \to A$ is a $W$-isometry. It follows that $\overline{\rho}|_{\mathrm{Geom}(A')} : \mathrm{Geom}(A') \to \mathrm{Geom}(A)$ is an isometry. Hence, $\overline{\rho} \circ \gamma$ is actually a geodesic (of the same length as $\gamma$). □

*Proof of Theorem 18.3.1.* Suppose $\Phi$ is a building and $x, y, z \in \mathrm{Geom}(\Phi)$. For $t \in [0, 1]$, let $p_t$ be the point on the geodesic segment from $y$ to $x$ such that $d(y, p_t) = td(x, y)$. By Lemma I.2.15 in Appendix I.2, to prove that $\mathrm{Geom}(\Phi)$ is CAT(0) we must show

$$d^2(z, p_t) \leqslant (1 - t)d^2(z, x) + td^2(z, y) - t(1 - t)d^2(x, y).$$

Choose an apartment $A$ so that $x, y \in \mathrm{Geom}(A)$. Since the geodesic segment from $x$ to $y$ lies in $\mathrm{Geom}(A)$, $p_t \in \mathrm{Geom}(A)$. Hence, we can choose a chamber $\varphi$ in $A$ so that $p_t \in K(\varphi)$. Let $\rho = \rho_{\varphi,A}$. By Lemma 18.3.5, $d(z, p_t) = d(\overline{\rho}(z), p_t)$.

Hence,

$$\begin{aligned} d^2(z,p_t) &= d^2(\overline{\rho}(z),p_t) \\ &\leqslant (1-t)d^2(\overline{\rho}(z),x) + td^2(\overline{\rho}(z),y) - t(1-t)d^2(x,y) \\ &\leqslant (1-t)d^2(z,x) + td^2(z,y) - t(1-t)d^2(x,y). \end{aligned}$$

The first inequality holds since $x$, $y$ and $\overline{\rho}(z)$ all lie in Geom($A$), which is is CAT(0). The second inequality is from Lemma 18.3.3. Therefore, Geom($\Phi$) is CAT(0). □

Since CAT(0)-spaces are contractible (Theorem I.2.6), we have the following.

**COROLLARY 18.3.6.** *For any building* $\Phi$, Geom($\Phi$) *is contractible.*

The Bruhat-Tits Fixed Point Theorem (I.2.11) states that if a group of isometries of a complete CAT(0) space has a bounded orbit, then it has a fixed point. Applying this result to the case of a building with a chamber transitive automorphism group, we get the following.

**COROLLARY 18.3.7.** *Suppose* $\Phi = \Phi(G, B, (G_s)_{s\in S})$ *is a building as in Example 18.1.2. Let* $H$ *be a subgroup of* $G$ *which has a bounded orbit in* Geom($\Phi$). *Then* $H$ *is conjugate to a subgroup of* $G_T$, *for some* $T \in \mathcal{S}$.

The next result also follows from Theorem 18.3.1 (by Corollary I.2.13).

**COROLLARY 18.3.8.** (Meier [202]). *Suppose* $G$ *is a graph product of finite groups* $(G_s)_{s\in S}$ *and, as in 18.1, that* $\Phi = \Phi(G, \{1\}, (G_s)_{s\in S}$ *is the associated building. Then* Geom($\Phi$) *is a model for* $\underline{E}G$.

## CAT(−1)-Structures on Buildings

A finite Coxeter group $W_T$ acts as a group of isometries on hyperbolic space of dimension Card($T$) as a group generated by isometric reflections across the faces of a simplicial cone. Hence, given $\mathbf{d} \in (0,\infty)^S$, there is a *hyperbolic Coxeter polytope* $\mathbb{H}C_{W_T}(\mathbf{d}_T)$ such that each edge of type $s$ has length $2d_s$. Its intersection with the fundamental simplicial cone is a *hyperbolic Coxeter block* $\mathbb{H}B_{W_T}(\mathbf{d}_T)$. So, for each $\mathbf{d} \in (0,\infty)^S$ we can put a piecewise hyperbolic structure on Geom($\Phi$) by identifying each Coxeter block of type $T$ with $\mathbb{H}B_{W_T}(\mathbf{d}_T)$. By Theorem 12.6.1, for sufficiently small $\mathbf{d}$, this procedure results in a CAT(−1) metric on $\Sigma$ if and only if $(W,S)$ satisfies Moussong's Condition. When the metric on $\Sigma$ is CAT(−1), then the arguments proving Theorem 18.3.1 show that Geom($\Phi$) is also CAT(−1). So, we have the following.

**THEOREM 18.3.9.** *Suppose $\Phi$ is a building of type $(W,S)$. Then* $\mathrm{Geom}(\Phi)$ *can be given a piecewise hyperbolic structure which is* $\mathrm{CAT}(-1)$ *if and only if $(W,S)$ satisfies Moussong's Condition (from Theorem 12.6.1).*

We also have the following analog of Corollary 12.6.3.

**COROLLARY 18.3.10.** *Suppose that $G$ is a graph product of finite groups with associated right-angled Coxeter system $(W,S)$ and associated building $\Phi$. Then the following conditions are equivalent.*

(i) *$G$ is word hyperbolic.*

(ii) *$\mathbb{Z}+\mathbb{Z} \not\subset G$.*

(iii) *$(W,S)$ satisfies Moussong's Condition.*

(iv) *$L(W,S)$ satisfies the no $\square$-condition (from Appendix I.6).*

(v) $\mathrm{Geom}(\Phi)$ *admits a piecewise hyperbolic* $\mathrm{CAT}(-1)$ *metric.*

*Proof.* In the case of a right-angled Coxeter system, Moussong's Condition is equivalent to the no $\square$-condition. So (iii) $\Longleftrightarrow$ (iv). The remainder of the argument is exactly the same as the proof of Corollary 12.6.3. $\square$

## 18.4. EULER-POINCARÉ MEASURE

Suppose $G$ is a locally compact, unimodular group and $\mu$ is a (not necessarily positive) $G$-invariant measure on $G$. (If the measure is positive, it is a *Haar measure*.) If $\Gamma$ is a discrete subgroup of $G$, we get a measure on $G/\Gamma$, which we also denote by $\mu$.

**DEFINITION 18.4.1.** ([255]). A measure $\mu$ on $G$ is an *Euler-Poincaré measure* if every discrete, cocompact, torsion-free subgroup $\Gamma \subset G$ has the following two properties:

(a) $\Gamma$ is type *FL*. (The definition is given in Appendix F.4.)

(b) $\chi(\Gamma) = \mu(G/\Gamma)$. ($\chi(\Gamma)$ was defined in (16.2) of Section 16.1.)

Now suppose $\Phi$ is a building with chamber-transitive automorphism group $G$ and thickness vector $\mathbf{q}$. Let $\mu$ be any nonzero $G$-invariant measure on $G$. Define

$$\chi(\mu) := \sum_{\sigma\subset K} \frac{(-1)^{\dim\sigma}}{\mu(G_{S(\sigma)})} = \sum_{T\in\mathcal{S}} \frac{1-\chi(L_T)}{\mu(G_T)}, \tag{18.4}$$

where the first sum ranges over all simplices $\sigma$ in the fundamental chamber $K$, where $G_{S(\sigma)}$ is the stabilizer of $\sigma$ and where, as in Theorem 17.1.10, $L_T := \mathrm{Lk}(\sigma_T, L)$. As in [255, p. 139], define the *canonical measure* $\mu_G$ by $\mu_G := \chi(\mu)\mu$. It is independent of the choice of $\mu$ (provided $G$ is unimodular).

**PROPOSITION 18.4.2.** ([255, Prop. 24].) *$\mu_G$ is an Euler-Poincaré measure.*

*Proof.* As explained in [255], this follows from the following facts:

(a) Geom($\Phi$) is contractible (Corollary 18.3.6),

(b) Geom($\Phi$) is locally finite, and

(c) the $G$-action is cocompact.

Indeed, by (a) and (b),

$$\chi(\Gamma) = \sum_{\sigma \subset K} (-1)^{\dim \sigma} \operatorname{Card}(\Gamma \backslash G / G_{S(\sigma)}).$$

The number of $\Gamma$-orbits of simplices of type $\sigma$ is $\operatorname{Card}(\Gamma \backslash G / G_{S(\sigma)})$. Since $\Gamma$ is torsion-free and $G_{S(\sigma)}$ is compact, $G_{S(\sigma)}$ acts freely on $\Gamma \backslash G$. So, if $\mu$ is a Haar measure, we have $\mu(\Gamma \backslash G) = \mu(G_{S(\sigma)}) \operatorname{Card}(\Gamma \backslash G / G_{S(\sigma)})$. Hence,

$$\begin{aligned} \chi(\Gamma) &= \mu(\Gamma \backslash G) \sum_{\sigma \subset K} \frac{(-1)^{\dim \sigma}}{\mu(G_{S(\sigma)})} \\ &= \mu(\Gamma \backslash G) \chi(\mu) = \mu_G(G/\Gamma). \end{aligned}$$

□

**THEOREM 18.4.3.** ([255, Théorème 6].) $\mu_G(B) = 1/W(\mathbf{q})$.

*Proof.* For $T \in \mathcal{S}$, let $\Phi_T$ denote the $T$-residue containing the chamber corresponding to $B$. It is a finite spherical building. We have $[G_T : B] = \operatorname{Card}(\Phi_T) = W_T(\mathbf{q})$ (Corollary 18.1.18). Also, $\mu(G_T) = [G_T : B]\mu(B)$. Substituting these formulas into (18.4), we get

$$\chi(\mu) = \sum_{T \in \mathcal{S}} \frac{1 - \chi(L_T)}{\mu(G_T)} = \frac{1}{\mu(B)} \sum_{T \in \mathcal{S}} \frac{1 - \chi(L_T)}{W_T(\mathbf{q})} = \left(\frac{1}{\mu(B)}\right)\left(\frac{1}{W(\mathbf{q})}\right),$$

where the last equation is Theorem 17.1.10. So, $\mu_G(B) = \chi(\mu)\mu(B) = 1/W(\mathbf{q})$.

□

## NOTES

The fundamental ideas in this chapter go back to Tits [284, 285, 287]. Much of our exposition is taken from [76]. Two basic references are [43, 248].

Some people in representation theory believe only two types of buildings are of any importance, those of spherical or affine type. The examples of right-angled buildings described in Example 18.1.10 should convince the reader that there are many other interesting examples of buildings. More examples come from the theory of Kac-Moody groups e.g., see [52]. As we mentioned previously, spherical buildings are often associated with algebraic groups over finite fields. Euclidean buildings are associated with algebraic groups over fields with discrete valuations, such as the $p$-adic rationals. (See [43, V.8] for the standard example of such a Euclidean building.) Regular buildings of type $(W, S)$, when $W$ is a hyperbolic polygon group, are called *Fuchsian*. Fuchsian buildings, as well as, other buildings associated to hyperbolic reflection groups are discussed in [30, 31, 32, 111, 133, 245].

**18.1.** Chamber systems were introduced by Tits [286]. As Weiss [298] points out, the information in a chamber system is exactly the same as that in an edge colored graph (without loops): the vertices are the chambers, two such are $s$-adjacent if and only if they are connected by an edge colored $s$. (An example is $\mathrm{Cay}(W, S)$.) Expositions of the theory of buildings based on the notion of chamber system are given in the books [248, 298]. In earlier versions of the theory Tits used a certain type of simplicial complexes called "chamber complexes," in place of chamber systems. (The chambers are the top-dimensional simplices.) The relationship between these two notions is described in Example 18.2.2.

The condition that $m = 2$, 3, 4, 6 or 8 which occurs in the Feit-Higman Theorem (18.1.9) is called the "crystallographic condition" (this condition should also include $m = \infty$). The crystallographic condition also occurs in the theory of Kac-Moody groups and Lie algebras.

The definition of a "Tits system" is taken from [29, p. 15].

**18.2.** In [248, p. 184], Ronan comments to the effect that in the general case, the geometric realization of a building should coincide with our definition, rather than the "usual one" from Example 18.2.2.

**18.3.** Although Theorem 18.3.1 was known to Moussong, it is not stated in [221]. The theorem is stated in [76] and it is shown there how it follows from Moussong's Theorem 12.3.3.

**18.4.** The material in this section is essentially due to Serre [255, §7]. In [255] Theorem 18.4.3 is stated only for $(W, S)$ a simplicial Coxeter system. The reason for this restriction was that the contractible complex $\Sigma$ had not been defined when [255] was written.

# Chapter Nineteen

## HECKE–VON NEUMANN ALGEBRAS

In 19.1 we define the Hecke algebra associated to a Coxeter system $(W, S)$ and a certain tuple of numbers $\mathbf{q}$. It is a deformation of the group algebra $\mathbb{R}W$ and is isomorphic to $\mathbb{R}W$. Hecke algebras arise in the study of buildings of type $(W, S)$ with thickness vector $\mathbf{q}$. Given $\mathbf{q}$, an $I$-tuple of positive real numbers, in 19.2 we define a nonstandard inner product on $\mathbb{R}W$ and then complete it to a Hilbert space $L^2_{\mathbf{q}}(W)$. The Hecke algebra $\mathbb{R}_{\mathbf{q}}W$ is a $*$-algebra of operators on $L^2_{\mathbf{q}}(W)$ and it can be completed to a von Neumann algebra $\mathcal{N}_{\mathbf{q}}$. We also define certain $\mathbb{R}_{\mathbf{q}}W$-stable subspaces $A^T$ and $H^T$ of $L^2_{\mathbf{q}}(W)$. Their von Neumann dimensions (with respect to $\mathcal{N}_{\mathbf{q}}$) are calculated by using the growth series of $W_T$.

### 19.1. HECKE ALGEBRAS

Let $A$ be a commutative ring with unit and $A^{(W)}$ the free $A$-module on $W$ with basis $\{e_w\}_{w \in W}$. $AW$ denotes this $A$-module equipped with its structure as the group ring of $W$. As in Chapter 17, $i : S \to I$ is a function such that $i(s) = i(s')$ whenever $s$ and $s'$ are conjugate in $W$ and given an $I$-tuple $\mathbf{q} = (q_i)_{i \in I} \in A^I$, write $q_s$ for $q_{i(s)}$.

**PROPOSITION 19.1.1.** (Compare [29, Exercise 23, p. 57].) *Given $I$-tuples $\mathbf{q}$ and $\mathbf{r}$ in $A^I$and a function $i : S \to I$ as above, there is a unique ring structure on $A^{(W)}$ such that*

$$e_s e_w = \begin{cases} e_{sw} & \text{if } l(sw) > l(w), \\ q_s e_{sw} + r_s e_w & \text{if } l(sw) < l(w), \end{cases}$$

*for all $s \in S$ and $w \in W$.*

We will use the notation $A_{\mathbf{q},\mathbf{r}}W$ to denote $A^{(W)}$ with this ring structure. Note that $A_{\mathbf{1},\mathbf{0}}W = AW$, where $\mathbf{1}$ and $\mathbf{0}$ denote the constant $I$-tuples $\mathbf{1} := (1, \dots, 1)$ and $\mathbf{0} := (0, \dots, 0)$ respectively. So $A_{\mathbf{q},\mathbf{r}}W$ is a deformation of the group ring.

The function $e_w \to e_{w^{-1}}$ induces a linear involution $*$ of $A^{(W)}$, i.e.,

$$\left(\sum x_w e_w\right)^* := \sum x_{w^{-1}} e_w. \tag{19.1}$$

*Proof of Proposition 19.1.1.* For each $s \in S$, define an $A$-linear endomorphism $L_s$ of $A^{(W)}$ by the formula in the proposition. Let $M_s$ be the endomorphism defined by $M_s(x) := (L_s(x^*))^*$.

*Claim. For all $s, t \in S$, $L_s \circ M_t = M_t \circ L_s$.* The proof of this claim is a computation. One shows $L_s(M_t(e_w)) = M_t(L_s(e_w))$ for all $w \in W$. This is proved by computing the cases $l(swt) = l(w) - 2$, $l(swt) = l(w) + 2$ and $l(swt) = l(w)$, separately (as well as some subcases). When $l(swt) = l(w) + 2$, one needs to recall Remark 3.2.15 that in the subcase $l(sw) = l(wt) = l(w) + 1$, the Folding Condition (F) implies that $sw = wt$ and since $s$ and $t$ are then conjugate that $q_s = q_t$ and $r_s = r_t$.

Let $\mathcal{A}$ denote the subalgebra of End $(A^{(W)})$ generated by the $L_s$. By the claim, for any $g \in \mathcal{A}$ and $t \in S$, we have $g \circ M_t = M_t \circ g$. This implies that if $g(e_1) = 0$, then $g = 0$. (Indeed, if $g(e_1) = 0$, then $g(e_t) = g(M_t(e_1)) = M_t(g(e_1)) = 0$ for all $t \in S$.)

For each $w \in W$, choose a reduced expression $w = s_1 \cdots s_l$ and set

$$L_w := L_{s_1} \circ \cdots \circ L_{s_l}.$$

Then $L_w \in \mathcal{A}$ and $L_w(e_1) = e_w$. If $l(sw) > l(w)$, then $L_s \circ L_w(e_1) = L_{sw}(e_1)$, while if $l(sw) < l(w)$, then $L_s \circ L_w(e_1) = q_s L_{sw}(e_1) + r_s L_w(e_1)$. By the previous paragraph, if two elements of $\mathcal{A}$ agree on $e_1$, then they are equal. Hence,

$$L_s \circ L_w = \begin{cases} L_{sw} & \text{if } l(sw) > l(w), \\ q_s L_{sw} + r_s L_w & \text{if } l(sw) < l(w). \end{cases}$$

Given any $x = \sum x_w e_w$ in $A^{(W)}$, there is a unique element $f \in \mathcal{A}$ such that $f(e_1) = x$, namely, $f = \sum x_w L_w$. Hence, $f \to f(e_1)$ is an $A$-linear isomorphism $\mathcal{A} \to A^{(W)}$ which transports the algebra structure on $\mathcal{A}$ to the desired structure on $A^{(W)}$. □

**LEMMA 19.1.2.** *The following statements about $A_{\mathbf{q},\mathbf{r}}W$ are true.*

*(i) For all $u, v \in W$ with $l(uv) = l(u) + l(v)$, $e_u e_v = e_{uv}$.*

*(ii) For all $s \in S$, $e_s^2 = r_s e_s + q_s$.*

*(iii)* (Artin relations.) *For any two distinct elements $s, t \in S$ with $m_{st} \neq \infty$,*

$$\underbrace{e_s e_t \ldots}_{m_{st}} = \underbrace{e_t e_s \ldots}_{m_{st}} .$$

*Proof.* The formulas in (i) and (ii) follow immediately from the formula in Proposition 19.1.1. To prove (iii), suppose $u$ is the element of longest length in the dihedral group $W_{\{s,t\}}$ (i.e., $u$ is an alternating word of length $m_{st}$ in $s$ and $t$). Then both sides of the formula in (iii) are equal to $e_u$. □

From this lemma we get the following right-hand version of the formula in Proposition 19.1.1.

**COROLLARY 19.1.3.** *For all $s \in S$ and $w \in W$,*

$$e_w e_s = \begin{cases} e_{ws} & \text{if } l(ws) > l(w), \\ q_s e_{ws} + r_s e_w & \text{if } l(ws) < l(w). \end{cases}$$

*Proof.* If $l(ws) > l(w)$, then, by Lemma 19.1.2 (i), $e_w e_s = e_{ws}$. If $l(ws) < l(w)$, then $w = w's$ with $l(w') = l(w) - 1$ and we have

$$\begin{aligned} e_w e_s &= e_{w's} e_s = e_{w'}(e_s)^2 = e_{w'}(r_s e_s + q_s) \\ &= r_s e_{w's} + q_s e_{w'} = r_s e_w + q_s e_{ws}, \end{aligned}$$

where the last equation on the first line is from Lemma 19.1.2 (iii). □

**LEMMA 19.1.4.** *Formula (19.1) defines an anti-involution of the ring $A_{\mathbf{q},\mathbf{r}}W$. In other words, for all $x, y \in A_{\mathbf{q},\mathbf{r}}W$, $(xy)^* = y^* x^*$.*

*Proof.* For each $w \in W$, let $L_w$ (resp. $R_w$) denote left (resp. right) translation by $e_w$ defined by $L_w(x) = e_w x$ (resp. $R_w(x) = x e_w$). A calculation using Corollary 19.1.3 gives: $R_s = *L_s* = M_s$, for all $s \in S$ (where $M_s$ is as in the proof of Proposition 19.1.1). Indeed,

$$\begin{aligned} (*L_s*)(e_w) &= (L_s(e_{w^{-1}}))^* = (e_s e_{w^{-1}})^* \\ &= \begin{cases} (e_{sw^{-1}})^* & \text{if } l(sw^{-1}) > l(w^{-1}), \\ (q_s e_{sw^{-1}} + r_s e_{w^{-1}})^* & \text{if } l(sw^{-1}) < l(w^{-1}) \end{cases} \\ &= \begin{cases} e_{ws} & \text{if } l(ws) > l(w), \\ q_s e_{ws} + r_s e_w & \text{if } l(ws) < l(w) \end{cases} \\ &= e_w e_s = R_s(e_w), \end{aligned}$$

where the next to last equality is Corollary 19.1.3. If $s_1 \cdots s_l$ is a reduced expression for $w$, then $R_w = R_{s_l} \cdots R_{s_1} = *L_{s_l} \cdots L_{s_1}* = *L_{w^{-1}}*$. Therefore, $xe_w = R_w(x) = (*L_{w^{-1}}*)(x) = (e_{w^{-1}} x^*)^*$. Hence, $R_w = *L_{w^{-1}}*$, for all $w \in W$. So, $(xe_w)^* = (e_{w^{-1}} x^*)^{**} = e_{w^{-1}} x^* = e_w^* x^*$. The lemma follows. □

### The Special Case $\mathbf{r} = \mathbf{q} - 1$

For the remainder of this book we shall only be interested in the case where $r_s = q_s - 1$, for all $s \in S$. We will write $A_{\mathbf{q}}W$ instead of $A_{\mathbf{q},\mathbf{q}-\mathbf{1}}W$ and call it the *Hecke algebra* associated to the *multiparameter* $\mathbf{q}$. ($A_{\mathbf{q}}W$ is also called the "Hecke-Iwahori algebra.")

If $s_1 \cdots s_l$ is a reduced expression for $w \in W$, then, as in formula (17.1) for the monomial $t_w$, define $q_w \in A$ by

$$q_w := q_{s_1} \cdots q_{s_l}. \tag{19.2}$$

As before, it is independent of the choice of reduced expression for $w$. Set $\varepsilon_w := (-1)^{l(w)}$. The maps $e_w \to q_w$ and $e_w \to \varepsilon_w$ extend linearly to ring homomorphisms $A_{\mathbf{q}}W \to A$.

**NOTATION.** *As in the previous chapter, use* $\mathbf{q}^{-1}$ *to denote the I-tuple* $(1/q_i)_{i \in I}$. *When we want to use* $\mathbf{q}^{-1}$ *as a subscript or superscript we will write* $\mathbf{1/q}$ *instead.*

### The j-isomorphism

Following Kazhdan and Lusztig [173], define an isomorphism of algebras, $j_{\mathbf{q}} : A_{\mathbf{q}}W \to A_{\mathbf{1/q}}W$, by the formula

$$j_{\mathbf{q}}(e_w) := \varepsilon_w q_w e_w. \tag{19.3}$$

It is called the *j-isomorphism* and denoted by $j$ when there is no ambiguity.

### Hecke Algebras and Functions on $B \backslash G/B$

Suppose $G$ is a topological group and $B$ is a compact open subgroup. Let $C(G)$ denote the vector space of continuous real-valued functions on $G$. Let $\alpha : G \to G/B$ and $\beta : G \to B\backslash G/B$ be the natural projections. Define subspaces $H \subset L \subset C(G)$ by

$$L := \alpha^*(\mathbb{R}^{(G/B)}) \quad \text{and} \quad H := \beta^*(\mathbb{R}^{(B\backslash G/B)}),$$

where, as before, for any set $X$, $\mathbb{R}^{(X)}$ is the vector space of finitely supported functions on $X$.

For each $gB \in G/B$, define $a_{gB} \in L$ by $a_{gB}(x) = 1$ for $x \in gB$ and $a_{gB}(x) = 0$ for $x \notin gB$ (i.e., $a_{gB}$ is the characteristic function of $gB$). Since $(a_{gB})$ is a basis for $L$, there is a unique linear form on $L$ such that $a_{gB} \to 1$ for all $gB \in G/B$. We denote this form by $\varphi \to \int \varphi$ (since it coincides with the Haar integral normalized by the condition that $\int a_B = 1$).

If $\varphi \in L$ and $\psi \in H$, then for each $x \in G$, the function $\theta_x : G \to \mathbb{R}$, defined by $\theta_x(y) = \varphi(y)\psi(y^{-1}x)$, belongs to $L$. The function $\varphi * \psi : x \to \int \varphi(y)\psi(y^{-1}x)dy$ also belongs to $L$. Moreover, if $\varphi \in H$, then $\varphi * \psi \in H$. The function $\varphi * \psi$ is the *convolution* of $\varphi$ and $\psi$. The map $(\varphi, \psi) \to \varphi * \psi$ makes $H$ into an algebra and $L$ into a right $H$-module. $H$ is called the *Hecke algebra* of $G$ with respect to $B$.

Next, suppose that $G$ is a chamber transitive automorphism group on a building and that $r : G/B \to W$ is defined by taking the $W$-distance from the chamber corresponding to $B$. Let $\gamma := r \circ \alpha : G \to W$ and $J := \gamma^*(\mathbb{R}^{(W)}) \subset H$.

*Remark.* Suppose $(G, B, N, S)$ is a Tits system (as in the complement at the end of 18.1). Then $\overline{r} : B\backslash G/B \to W$ is a bijection and hence, $J = H$.

**LEMMA 19.1.5.** *Suppose, as above, that a given building admits a chamber transitive automorphism group $G$ (so $G/B$ is the set of chambers). Let $\mathbf{q}$ be the thickness vector. Then*

(i) *$J$ is a subalgebra of $H$ and*

(ii) *$J \cong \mathbb{R}_{\mathbf{q}}W$, the Hecke algebra defined previously.*

*Proof.* Since $G$ is chamber transitive, $\gamma^* : \mathbb{R}^{(W)} \to J$ is an isomorphism of vector spaces. So, we only need to check that $\gamma^*$ is an algebra homomorphism from $\mathbb{R}_q W$ to $H$. Let $f_w = \gamma^*(e_w)$. Then $f_w$ is the characteristic function of $\{g \in G \mid r(gB) = w\}$. In particular, for each $s \in S, f_s$ is the characteristic function of $G_s - B$. We want to show that

$$f_w * f_s = \begin{cases} f_{ws} & \text{if } l(ws) > l(w), \\ q_s f_{ws} + (q_s - 1)f_w & \text{if } l(ws) < l(w). \end{cases}$$

By definition of convolution,

$$\begin{aligned}(f_w * f_s)(g) &= \int_G f_w(x)f_s(x^{-1}g)dx = \int_G f_w(gu)f_s(u^{-1})du \\ &= \int_{G_s - B} f_w(gu)du,\end{aligned}$$

which is equal to the Haar measure of the set

$$U_g := \{u \in G_s - B \mid r(guB) = w\}.$$

Let $C_0 := g_0 B$ be the chamber $s$-adjacent to $gB$ which is closest to $B$. There are $q_s$ other chambers adjacent to $gB$. We list them as: $C_1 = g_1 B, \dots, C_{q_s} = g_{q_s} B$. So, for $i > 0$, $r(C_i) = r(C_0)s$. Notice that if $u \in G_s - B$, then $guB$ is $s$-adjacent to $gB$ and therefore, $guB$ is equal to some $C_i$. So, if $r(guB) = w$, then $r(gB) = w$ or $ws$. In other words, if $r(gB) \notin \{w, ws\}$, then $(f_w * f_s)(g) = 0$. We now consider two cases. Each case further divides into two subcases depending on whether $r(gB) = w$ or $ws$.

*Case 1.* $l(w) < l(ws)$. In this case $r(C_0) = w$ and $r(C_i) = ws$ for $i > 0$.

(a) Suppose $r(gB) = w$. Then $gB = C_0$ and $guB = C_i$ for $i > 0$, so that $r(guB) = ws$. Thus, $U_g = \emptyset$ and $(f_w * f_s)(g) = 0$.

(b) Suppose $r(gB) = ws$. Then $gB = C_k$, for some $k > 0$, and

$$U_g = \{u \in G_s - B \mid guB = C_0\} = (G_s - B) \cap g^{-1}g_0B.$$

Since $gB$ and $g_0B$ are $s$-adjacent and not equal, $g^{-1}g_0B \subset G_s - B$, so that $U_g = g^{-1}g_0B$ has measure 1. Therefore, $(f_w * f_s)(g) = 1$. So, in Case 1, $f_w * f_s = f_{ws}$.

*Case 2.* $l(w) > l(ws)$. In this case $r(C_0) = ws$ and $r(C_i) = w$ for $i > 0$.

(a) Suppose $r(gB) = w$. Then $gB = C_k$ for some $k > 0$. So the set

$$U_g = \bigcup_{0<i}\{u \in G_s - B \mid guB = C_i\} = \bigcup_{0<i\neq k} g^{-1}g_iB$$

has measure $q_s - 1$.

(b) Suppose $r(gB) = ws$. Then $gB = C_0$, and the set

$$U_g = \bigcup_{0<i}\{u \in G_s - B \mid guB = C_i\} = \bigcup_{0<i} g^{-1}g_iB$$

has measure $q_s$. So, in Case 2, $f_w * f_s = q_s f_{ws} + (q_s - 1)f_w$. □

## 19.2. HECKE–VON NEUMANN ALGEBRAS

For the remainder of this chapter, $A$ will be the field of real numbers $\mathbb{R}$ and the components of the multiparameter $\mathbf{q} = (q_i)_{i \in I}$ will be positive real numbers. As explained in Appendix J.2, associated to the group algebra $\mathbb{R}W$ we have the Hilbert space $L^2(W)$ and the von Neumann algebra $\mathcal{N}(W)$ of $W$-equivariant bounded linear operators on it. ($\mathcal{N}(W)$ is the weak closure of $\mathbb{R}W$.) Here we give an analogous construction associated to the Hecke algebra $\mathbb{R}_{\mathbf{q}}W$.

Define an inner product $\langle\ ,\ \rangle_{\mathbf{q}}$ on $\mathbb{R}^{(W)}$ $(= \mathbb{R}_{\mathbf{q}}W)$ by

$$\left\langle \sum a_w e_w, \sum b_w e_w \right\rangle_{\mathbf{q}} := \sum a_w b_w q_w, \tag{19.4}$$

where $q_w$ is defined by (19.2). As in [194], sometimes it is convenient to normalize $(e_w)_{w \in W}$ to an orthonormal basis for $\mathbb{R}_{\mathbf{q}}W$ by setting

$$\tilde{e}_w := q_w^{-1/2} e_w. \tag{19.5}$$

The formula in Proposition 19.1.1 can then be rewritten as

$$(\tilde{e}_s + q_s^{-1/2})(\tilde{e}_s - q_s^{-1/2}) = 0. \tag{19.6}$$

The (Hilbert space) completion of $\mathbb{R}^{(W)}$ with respect to the inner product $\langle\ ,\ \rangle_{\mathbf{q}}$ is denoted $L^2_{\mathbf{q}}(W)$, or simply $L^2_{\mathbf{q}}$, when there is no ambiguity.

**PROPOSITION 19.2.1.** (Dymara [109, Proposition 2.1].) *The inner product defined by (19.4), multiplication defined by the formula in Proposition 19.1.1, and the anti-involution $*$ defined by (19.1), give $\mathbb{R}_{\mathbf{q}}W$ the structure of a Hilbert algebra structure in the sense of* [100, A.54]. *This means that*

*(i)* $(xy)^* = y^*x^*$,

*(ii)* $\langle x, y\rangle_{\mathbf{q}} = \langle y^*, x^*\rangle_{\mathbf{q}}$,

*(iii)* $\langle xy, z\rangle_{\mathbf{q}} = \langle y, x^*z\rangle_{\mathbf{q}}$,

*(iv) for any $x \in \mathbb{R}_{\mathbf{q}}W$, left translation by $x$, $L_x : \mathbb{R}_{\mathbf{q}}W \to \mathbb{R}_{\mathbf{q}}W$, defined by $L_x(y) = xy$, is continuous,*

*(v) the products $xy$ over all $x, y \in \mathbb{R}_{\mathbf{q}}W$ are dense in $\mathbb{R}_{\mathbf{q}}W$.*

Since the action of $\mathbb{R}_{\mathbf{q}}W$ on itself by multiplication is continuous, $L^2_{\mathbf{q}}$ is a $\mathbb{R}_{\mathbf{q}}W$-bimodule. An element $x \in L^2_{\mathbf{q}}$ is *bounded* if right multiplication by $x$ is bounded on $\mathbb{R}_{\mathbf{q}}W$ (or equivalently, if left multiplication by $x$ is bounded) . Let $\mathbb{R}^b_{\mathbf{q}}W$ be the set of all bounded elements in $L^2_{\mathbf{q}}$. As in [100] there are two von Neumann algebras associated with this situation. Initially we denote them $\mathcal{N}^L_{\mathbf{q}}$ and $\mathcal{N}^R_{\mathbf{q}}$. $\mathcal{N}^R_{\mathbf{q}}(W)$ acts from the right on $L^2_{\mathbf{q}}$ and $\mathcal{N}^L_{\mathbf{q}}$ from the left. Here are two equivalent definitions of $\mathcal{N}^R_{\mathbf{q}}$:

(a) $\mathcal{N}^R_{\mathbf{q}}$ is the algebra of all bounded linear endomorphisms of $L^2_{\mathbf{q}}$ that commute with the left $\mathbb{R}_{\mathbf{q}}W$-action.

(b) $\mathcal{N}^R_{\mathbf{q}}$ is the weak closure of $\mathbb{R}^b_{\mathbf{q}}W$ acting from the right on $L^2_{\mathbf{q}}$.

If we interchange the roles of left and right in the above, we get the two equivalent definitions of $\mathcal{N}^L_{\mathbf{q}}$. At cost of introducing some confusion, but with the benefit of simplifying notation we will denote both von Neumann algebras simply by $\mathcal{N}_{\mathbf{q}}$.

**LEMMA 19.2.2.** *If $T \subset S$, then the inclusion $\mathbb{R}_{\mathbf{q}}[W_T] \hookrightarrow \mathbb{R}_{\mathbf{q}}W$ induces inclusions $\mathbb{R}^b_{\mathbf{q}}[W_T] \hookrightarrow \mathbb{R}^b_{\mathbf{q}}W$ and $\mathcal{N}_{\mathbf{q}}(W_T) \hookrightarrow \mathcal{N}_{\mathbf{q}}$.*

*Proof.* Let $L^2_{\mathbf{q}}(wW_T) \subset L^2_{\mathbf{q}}(W)$ denote the subspace of functions supported on the coset $wW_T$. Then $L^2_{\mathbf{q}}(W)$ decomposes as an orthogonal direct sum of spaces of the form $L^2_{\mathbf{q}}(wW_T)$. Suppose $\lambda \in \mathcal{N}_{\mathbf{q}}(W_T)$. Right multiplication by $\lambda$ preserves the summands and acts on each summand in the same way. The norm in the space $L^2_{\mathbf{q}}(wW_T)$ is the norm in $L^2_{\mathbf{q}}(W_T)$ rescaled by a factor of $(q_w)^{-1/2}$, so the operator norms of right multiplication by $\lambda$ on each of these subspaces is bounded. Hence, $\lambda \in \mathcal{N}_{\mathbf{q}}$. □

### The $j$-isomorphism

It follows from the definitions that the isomorphism $j_{\mathbf{q}} : \mathbb{R}_{\mathbf{q}}W \to \mathbb{R}_{1/\mathbf{q}}W$ of (19.3) takes the orthonormal basis $(\tilde{e}^{\mathbf{q}}_w)$ for $\mathbb{R}_{\mathbf{q}}W$, defined by (19.5), to the orthonormal basis $(\varepsilon_w \tilde{e}^{1/\mathbf{q}}_w)$ for $\mathbb{R}_{1/\mathbf{q}}W$ (where we have put superscripts on the $\tilde{e}_w$ to indicate the dependence on $\mathbf{q}$). So, $j_{\mathbf{q}}$ is an isometry. Hence, it extends to an isometry of Hilbert spaces $j : L^2_{\mathbf{q}} \to L^2_{1/\mathbf{q}}$. From this, it is obvious that $j$ takes a bounded element of $L^2_{\mathbf{q}}$ to a bounded element of $L^2_{1/\mathbf{q}}$. Hence, it extends to an isomorphism of von Neumann algebras $j : \mathcal{N}_{\mathbf{q}} \to \mathcal{N}_{1/\mathbf{q}}$.

### The Von Neumann Trace

Define the *trace* of an element $\varphi \in \mathcal{N}_{\mathbf{q}}$ by

$$\operatorname{tr}_{\mathcal{N}_{\mathbf{q}}}(\varphi) := \langle e_1\varphi, e_1\rangle_{\mathbf{q}},$$

where $e_1$ denotes the basis element of $L^2_{\mathbf{q}}$ corresponding to the identity element of $W$. If $\Phi : \bigoplus_{i=1}^n L^2_{\mathbf{q}} \to \bigoplus_{i=1}^n L^2_{\mathbf{q}}$ is a bounded linear map of left $\mathbb{R}_{\mathbf{q}}W$-modules, then we can represent it as right multiplication by an $n \times n$ matrix $(\varphi_{ij})$ with entries in $\mathcal{N}_{\mathbf{q}}$. Define

$$\operatorname{tr}_{\mathcal{N}_{\mathbf{q}}}(\Phi) := \sum_{i=1}^{n} \operatorname{tr}_{\mathcal{N}_{\mathbf{q}}}(\varphi_{ii}).$$

### Hilbert $\mathcal{N}_{\mathbf{q}}$-Modules and von Neumann Dimension

**DEFINITION 19.2.3.** A subspace $V$ of an orthogonal direct sum of a finite number of copies of $L^2_{\mathbf{q}}$ is a *Hilbert $\mathcal{N}_{\mathbf{q}}$-module* if it is a closed subspace stable under the diagonal left action of $\mathbb{R}_{\mathbf{q}}W$.

By a *map* of Hilbert $\mathcal{N}_{\mathbf{q}}$-modules we will mean a bounded linear map of left $\mathbb{R}_{\mathbf{q}}W$-modules. A map is *weakly surjective* if it has dense image; it is a *weak isomorphism* if it is injective and weakly surjective.

Let $V \subset \bigoplus_{i=1}^n L^2_{\mathbf{q}}$ be a Hilbert $\mathcal{N}_{\mathbf{q}}$-module and let $p_V : \bigoplus_{i=1}^n L^2_{\mathbf{q}} \to \bigoplus_{i=1}^n L^2_{\mathbf{q}}$ be the orthogonal projection onto $V$. The *von Neumann dimension* of $V$ is the nonnegative real number defined by

$$\dim_{\mathcal{N}_{\mathbf{q}}} V = \operatorname{tr}_{\mathcal{N}_{\mathbf{q}}}(p_V). \tag{19.7}$$

As usual, $\dim_{\mathcal{N}_{\mathbf{q}}} V$ does not depend on the choice of embedding of $V$ into a finite direct sum of copies of $L^2_{\mathbf{q}}$. If a subspace $V \subset \bigoplus L^2_{\mathbf{q}}$ is $\mathbb{R}_{\mathbf{q}}W$-stable but not necessarily closed, one defines $\dim_{\mathcal{N}_{\mathbf{q}}} V := \dim_{\mathcal{N}_{\mathbf{q}}} \overline{V}$. As in Proposition J.2.8 of Appendix J.2, this dimension function satisfies the usual list of properties:

(i) $\dim_{\mathcal{N}_{\mathbf{q}}} V = 0$ if and only if $V = 0$.

(ii) For any two Hilbert $\mathcal{N}_{\mathbf{q}}$-modules $V$ and $V'$,

$$\dim_{\mathcal{N}_{\mathbf{q}}}(V \oplus V') = \dim_{\mathcal{N}_{\mathbf{q}}} V + \dim_{\mathcal{N}_{\mathbf{q}}} V'.$$

(iii) $\dim_{\mathcal{N}_{\mathbf{q}}} L^2_{\mathbf{q}} = 1$.

(iv) Suppose $f : V \to V'$ is a weak isomorphism of Hilbert $\mathcal{N}_{\mathbf{q}}$-modules. Then $\dim_{\mathcal{N}_{\mathbf{q}}} V = \dim_{\mathcal{N}_{\mathbf{q}}} V'$.

(v) Suppose $(W', S')$ and $(W'', S'')$ are Coxeter systems, that $S' \to I'$ and $S'' \to I''$ are indexing functions, that $\mathbf{q}'$ and $\mathbf{q}''$ are $I'$ and $I''$-tuples, that $S = S' \cup S''$ and $I = I' \cup I''$ are disjoint unions, that $(W, S) = (W' \times W'', S' \cup S'')$ and that $\mathbf{q}$ is the multiparameter for $(W, S)$ formed by combining $\mathbf{q}'$ and $\mathbf{q}''$. Let $V'$ (resp. $V''$) be a Hilbert $\mathcal{N}_{\mathbf{q}'}(W')$-module (resp. $\mathcal{N}_{\mathbf{q}''}(W'')$-module). Then the completed tensor product $V := V' \otimes V''$ is naturally a Hilbert $\mathcal{N}_{\mathbf{q}}$-module and

$$\dim_{\mathcal{N}_{\mathbf{q}}(W)}(V' \otimes V'') = (\dim_{\mathcal{N}_{\mathbf{q}}(W')} V')(\dim_{\mathcal{N}_{\mathbf{q}}(W'')} V'').$$

(vi) Suppose $T \subset S$ and $V_T$ is a Hilbert $\mathcal{N}_{\mathbf{q}}(W_T)$-module. The *induced* Hilbert $\mathcal{N}_{\mathbf{q}}$-module $V$ is defined to be the completion of the tensor product

$$V := L^2_{\mathbf{q}}(W) \otimes_{\mathbb{R}_{\mathbf{q}}(W_T)} V_T.$$

Its dimension is given by $\dim_{\mathcal{N}_{\mathbf{q}}} V = \dim_{\mathcal{N}_{\mathbf{q}}(W_T)} V_T$.

**LEMMA 19.2.4.** *Suppose $V \subset \bigoplus L^2_{\mathbf{q}}$ is a Hilbert $\mathcal{N}_{\mathbf{q}}$-module. Extend the j-isomorphism to the linear isometry $j : \bigoplus L^2_{\mathbf{q}} \to \bigoplus L^2_{\mathbf{1/q}}$ which is $j$ on each summand. Then $j(V) \subset \bigoplus L^2_{\mathbf{1/q}}$ is a Hilbert $\mathcal{N}_{\mathbf{1/q}}$-module and*

$$\dim_{\mathcal{N}_{\mathbf{q}}} V = \dim_{\mathcal{N}_{\mathbf{1/q}}} j(V).$$

*Proof.* If $x \in \mathbb{R}_{\mathbf{q}}W$ and $v \in V$, then $j(x)j(v) = j(xv) \in j(V)$; so, $j(V)$ is a Hilbert $\mathcal{N}_{\mathbf{1/q}}$-module. If $p_V : \bigoplus L^2_{\mathbf{q}} \to \bigoplus L^2_{\mathbf{q}}$ is orthogonal projection onto $V$, then $j \circ p_V \circ j^{-1} : \bigoplus L^2_{\mathbf{1/q}} \to \bigoplus L^2_{\mathbf{1/q}}$ is orthogonal projection onto $j(V)$. We calculate its trace:

$$\begin{aligned}\operatorname{tr}_{\mathcal{N}_{1/\mathbf{q}}}(j \circ p_V \circ j^{-1}) &= \langle (j \circ p_V \circ j^{-1})(e_1), e_1 \rangle_{1/\mathbf{q}} \\ &= \langle (p_V \circ j^{-1})(e_1), j^{-1} e_1 \rangle_{\mathbf{q}} = \langle p_V(e_1), e_1 \rangle_{\mathbf{q}} \\ &= \operatorname{tr}_{\mathcal{N}_{\mathbf{q}}}(p_V).\end{aligned}$$

Hence, $\dim_{\mathcal{N}_{\mathbf{q}}} V = \dim_{\mathcal{N}_{1/\mathbf{q}}} j(V)$. □

## Idempotents in $\mathcal{N}_{\mathbf{q}}$ and Growth Series

Given a subset $T$ of $S$, recall $\mathcal{R}_T$ is the region of convergence for $W_T(\mathbf{t})$ and $\mathcal{R}_T^{-1} := \{\mathbf{q} \mid \mathbf{q}^{-1} \in \mathcal{R}_T\}$.

**LEMMA 19.2.5.** *Given a subset $T$ of $S$ and $\mathbf{q} \in \mathcal{R}_T$, there is an idempotent $a_T$ in $\mathcal{N}_{\mathbf{q}}$ defined by*

$$a_T := \frac{1}{W_T(\mathbf{q})} \sum_{w \in W_T} e_w,$$

*Proof.* Define

$$\tilde{a}_T = \sum_{w \in W_T} e_w. \tag{19.8}$$

Then $\langle \tilde{a}_T, \tilde{a}_T \rangle_{\mathbf{q}} = \sum q_w = W_T(\mathbf{q})$; hence, $\tilde{a}_T \in L^2_{\mathbf{q}}(W_T)$ if and only if $\mathbf{q} \in \mathcal{R}_T$. So, assume $\mathbf{q} \in \mathcal{R}_T$. As in 4.5, for each $s \in S$, $B_s$ denotes the set of $(\emptyset, \{s\})$-reduced elements in $W$. Using the formula in Corollary 19.1.3, we calculate, for each $s \in T$, that

$$\begin{aligned}\tilde{a}_T e_s &= \sum_{w \in B_s \cap W_T} e_w e_s + e_{ws} e_s \\ &= \sum e_{ws} + q_s e_w + (q_s - 1) e_{ws} \\ &= q_s \tilde{a}_T.\end{aligned}$$

Hence, for $w \in W_T$,

$$\tilde{a}_T e_w = q_w \tilde{a}_T \quad \text{and} \quad \tilde{a}_T \tilde{e}_w = q_w^{1/2} \tilde{a}_T. \tag{19.9}$$

We claim that $\tilde{a}_T$ is a bounded element of $L^2_{\mathbf{q}}(W_T)$ (and hence, by Lemma 19.2.2, it lies in $\mathcal{N}_{\mathbf{q}}$). To see this, note that if $x = \sum x_w \tilde{e}_w \in \mathbb{R}_{\mathbf{q}}[W_T]$, then (19.9) can be rewritten as

$$\tilde{a}_T \sum x_w \tilde{e}_w = \left( \sum x_w q_w^{1/2} \right) \tilde{a}_T$$

and hence $\|\tilde{a}_T x\|_{\mathbf{q}} \le \|\tilde{a}_T\|_{\mathbf{q}} \|x\|_{\mathbf{q}}$. It also follows from (19.9) that $(\tilde{a}_T)^2 = W_T(\mathbf{q})\tilde{a}_T$. So

$$a_T = \frac{1}{W_T(\mathbf{q})}\tilde{a}_T$$

is an idempotent. □

**LEMMA 19.2.6.** *Given a subset $T$ of $S$ and an $I$-tuple $\mathbf{q} \in \mathcal{R}_T^{-1}$, there is an idempotent $h_T \in \mathcal{N}_{\mathbf{q}}$ defined by*

$$h_T := \frac{1}{W_T(\mathbf{q}^{-1})} \sum_{w \in W_T} \varepsilon_w q_w^{-1} e_w,$$

*where $\varepsilon_w := (-1)^{l(w)}$.*

*Proof.* The proof is similar to the previous one. Define

$$\tilde{h}_T := \sum_{w \in W_T} \varepsilon_w q_w^{-1} e_w. \tag{19.10}$$

Then $\langle \tilde{h}_T, \tilde{h}_T \rangle_{\mathbf{q}} = \sum q_w^{-1} = W_T(\mathbf{q}^{-1})$, so $\tilde{h}_T \in L^2_{\mathbf{q}}(W_T)$ if and only if $\mathbf{q}^{-1} \in \mathcal{R}_T$. Assume this. For $s \in T$, we calculate

$$\begin{aligned}
\tilde{h}_T e_s &= \sum_{w \in B_s \cap W_T} \varepsilon_w q_w^{-1} e_w e_s + \varepsilon_{ws} q_{ws}^{-1} e_{ws} e_s \\
&= \sum \varepsilon_w q_w^{-1} e_{ws} + \varepsilon_{ws} q_w^{-1} q_s^{-1} (q_s e_w + (q_s - 1) e_{ws}) \\
&= -\sum \varepsilon_w q_w^{-1} e_w + \varepsilon_{ws} q_w^{-1} q_s^{-1} e_{ws} \\
&= -\tilde{h}_T.
\end{aligned}$$

Hence, for $w \in W_T$,

$$\tilde{h}_T e_w = \varepsilon_w \tilde{h}_T \tag{19.11}$$

and

$$(\tilde{h}_T)^2 = \sum_{w \in W_T} \varepsilon_w q_w^{-1} \tilde{h}_T e_w = W_T(\mathbf{q}^{-1}) \tilde{h}_T. \tag{19.12}$$

As before, $\tilde{h}_T \in \mathbb{R}^b_{\mathbf{q}}[W_T]$; hence, $\tilde{h}_T \in \mathcal{N}_{\mathbf{q}}$. By (19.12), we get an idempotent

$$h_T := \frac{1}{W_T(\mathbf{q}^{-1})} \tilde{h}_T.$$

□

Of course, when $\mathbf{q} = \mathbf{1}$ and $W_T$ is finite, $a_T$ and $h_T$ reduce to the familiar symmetrization and alternation elements in $\mathbb{R}W_T$ defined by formulas (15.27) and (15.28) in Section 15.4.

Using the formula in Proposition 19.1.1 (instead of Corollary 19.1.3), we get the following right hand versions of (19.9) and (19.11) for $T \subset S$ and $w \in W_T$:

$$e_w a_T = q_w a_T \tag{19.13}$$

and

$$e_w h_T = \varepsilon_w h_T. \tag{19.14}$$

What is the effect of the $j$-isomorphism on these idempotents? It follows from definitions (19.3), (19.8), and (19.10) that

$$j(\tilde{a}_T^{\mathbf{q}}) = \tilde{h}_T^{\mathbf{1/q}} \quad \text{and} \quad j(\tilde{h}_T^{\mathbf{q}}) = \tilde{a}_T^{\mathbf{1/q}},$$

where we have used the superscripts $\mathbf{q}$ and $\mathbf{1/q}$ to keep track of the dependence on $\mathbf{q}$. By the definitions in Lemmas 19.2.5 and 19.2.6,

$$j(a_T^{\mathbf{q}}) = h_T^{\mathbf{1/q}} \quad \text{and} \quad j(h_T^{\mathbf{q}}) = a_T^{\mathbf{1/q}}.$$

Using (19.9), (19.11), (19.13), and (19.14), we calculate that for $U \subset T \subset S$:

$$a_U a_T = a_T = a_T a_U \quad \text{whenever } \mathbf{q} \in \mathcal{R}_T,$$

$$h_U h_T = h_T = h_T h_U \quad \text{whenever } \mathbf{q} \in \mathcal{R}_T^{-1}.$$

If $s_1 \cdots s_l$ is a reduced expression for $w$, then $s_l \cdots s_1$ is a reduced expression for $w^{-1}$. Hence,

$$q_{w^{-1}} = q_w \quad \text{and} \quad \varepsilon_{w^{-1}} = \varepsilon_w. \tag{19.15}$$

So

$$a_T^* = a_T \quad \text{and} \quad h_T^* = h_T, \tag{19.16}$$

whenever $a_T$ and $h_T$ are defined. In other words, the maps $x \to x a_T$ and $x \to x h_T$ are orthogonal projections from $L^2_{\mathbf{q}}$ onto sub-$\mathcal{N}_{\mathbf{q}}$-modules.

*Remarks on terminology.* The "$a$" in $a_T$ is for "average," while the "$h$" in $h_T$ is for "harmonic."

**LEMMA 19.2.7.**

*(i) For $\mathbf{q} \in \mathcal{R}$, $L^2_{\mathbf{q}} a_S$ is the line in $L^2_{\mathbf{q}}$ consisting of all multiples of $a_S$.*

*(ii) For $\mathbf{q} \in \mathcal{R}^{-1}$, $L^2_{\mathbf{q}} h_S$ is the line in $L^2_{\mathbf{q}}$ consisting of all multiples of $h_S$.*

*Proof.* Suppose $x = \sum x_w e_w \in L^2_{\mathbf{q}} a_S$. Then

$$\begin{aligned} x_w &= q_w^{-1} \langle x, e_w \rangle_{\mathbf{q}} = q_w^{-1} \langle x a_S, e_w \rangle_{\mathbf{q}} = q_w^{-1} \langle x, a_S e_w \rangle_{\mathbf{q}} \\ &= q_w^{-1} \langle x, q_w a_S \rangle_{\mathbf{q}} = \langle x a_S, e_1 \rangle_{\mathbf{q}} = \langle x, e_1 \rangle_{\mathbf{q}} = x_1. \end{aligned}$$

So the coefficients are constant and $x = x_1 \tilde{a}_S$. The proof of (ii) is similar. (It also follows from an application of the $j$-isomorphism.) □

**DEFINITION 19.2.8.** For each $T \subset S$, let $\alpha_T : \mathbb{R}[W_T] \to \mathbb{R}$ and $\beta_T : \mathbb{R}_{\mathbf{q}}[W_T] \to \mathbb{R}$ be the algebra homomorphisms defined by $e_w \to q_w$ and $e_w \to \varepsilon_w$, respectively. $\alpha_T$ is the *symmetric character* and $\beta_T$ is the *alternating character*.

**COROLLARY 19.2.9.** *For $\mathbf{q} \in \mathcal{R}$, $L^2_{\mathbf{q}} a_S$ is stable under right Hecke multiplication. (Hence, it is a Hecke bimodule.) The right action of $\mathbb{R}_{\mathbf{q}} W$ on $L^2_{\mathbf{q}} a_S$ is via the symmetric character $\alpha_S$. Similarly, for $\mathbf{q} \in \mathcal{R}^{-1}$, $L^2_{\mathbf{q}} h_S$ is stable under right Hecke multiplication and the action is via the alternating character $\beta_S$.*

For any $T \subset S$ and $\mathbf{q} \in \mathcal{R}_T$, it is obvious that $L^2_{\mathbf{q}} a_T$ is the induced representation from $L^2_{\mathbf{q}}(W_T) a_T$. Similarly, for $\mathbf{q} \in \mathcal{R}_T^{-1}$, $L^2_{\mathbf{q}} h_T$ is induced from $L^2_{\mathbf{q}}(W_T) h_T$. Hence, we also have the following corollary.

**COROLLARY 19.2.10.** *For any $T \subset S$ and $\mathbf{q} \in \mathcal{R}_T$, $L^2_{\mathbf{q}} a_T$ is stable under right Hecke multiplication by $\mathbb{R}_{\mathbf{q}}[W_T]$ and the action is via the symmetric character $\alpha_T$. Similarly, for $\mathbf{q} \in \mathcal{R}_T^{-1}$, $L^2_{\mathbf{q}} h_T$ is stable under right multiplication by $\mathbb{R}_{\mathbf{q}}[W_T]$ and the action is via $\beta_T$.*

## Some $\mathcal{N}_q$-Hilbert -Submodules of $L^2_{\mathbf{q}}$

To simplify notation, for each $s \in S$, write $a_s$ and $h_s$ for the idempotents $a_{\{s\}}$ and $h_{\{s\}}$. Let $A^s = L^2_{\mathbf{q}} a_s$ and $H^s = L^2_{\mathbf{q}} h_s$ be the corresponding Hilbert $\mathcal{N}_{\mathbf{q}}$-submodules of $L^2_{\mathbf{q}}$.

**LEMMA 19.2.11.** *For each $s \in S$, the subspaces $A^s$ and $H^s$ are the orthogonal complements of each other in $L^2_{\mathbf{q}}$.*

*Proof*

$$\begin{aligned} a_s + h_s &= \frac{1}{1 + q_s}(1 + e_s) + \frac{1}{1 + q_s^{-1}}(1 - q_s^{-1} e_s) \\ &= 1. \end{aligned}$$

So $a_s$ and $h_s$ are orthogonal projections onto complementary subspaces. □

For each $T \subset S$, set

$$A^T := \bigcap_{s \in T} A^s \quad \text{and} \quad H^T := \bigcap_{s \in T} H^s. \tag{19.17}$$

For any subspace $E \subset L^2_{\mathbf{q}}$, let $E^{\perp}$ denote its orthogonal complement. Since $\perp$ takes sums to intersections and intersections to closures of sums,

$$\left(\sum_{s \in T} A^s\right)^{\perp} = H^T, \qquad \left(\sum_{s \in T} H^s\right)^{\perp} = A^T, \tag{19.18}$$

$$\overline{\sum_{s \in T} A^s} = (H^T)^{\perp}, \qquad \overline{\sum_{s \in T} H^s} = (A^T)^{\perp}. \tag{19.19}$$

**Lemma 19.2.12.** *Suppose $T \subset S$ and $A^T$ is defined by (19.17). Then*

*(i) $A^T$ is stable under right Hecke multiplication by $\mathbb{R}_{\mathbf{q}}[W_T]$: for any $x \in A^T$ and $u \in W_T$, $x \cdot e_u = q_u x$.*

*(ii) If $\mathbf{q} \notin \mathcal{R}_T$, then $A^T = 0$, while if $\mathbf{q} \in \mathcal{R}_T$, $A^T = L^2_{\mathbf{q}} a_T$. In the second case,*

$$\dim_{\mathcal{N}_{\mathbf{q}}} A^T = \frac{1}{W_T(\mathbf{q})}.$$

There is also the following version for $H^T$.

**Lemma 19.2.13.** *Suppose $T \subset S$ and $H^T$ is defined by (19.17). Then*

*(i) $H^T$ is stable under right Hecke multiplication by $\mathbb{R}_{\mathbf{q}}[W_T]$: for any $x \in H^T$ and $u \in W_T$, $x \cdot e_u = \varepsilon_u x$.*

*(ii) If $\mathbf{q} \notin \mathcal{R}_T^{-1}$, then $H^T = 0$, while if $\mathbf{q} \in \mathcal{R}_T^{-1}$, $H^T = L^2_{\mathbf{q}} h_T$. In the second case,*

$$\dim_{\mathcal{N}_{\mathbf{q}}} H^T = \frac{1}{W_T(\mathbf{q}^{-1})}.$$

We prove only the first version, the second version being entirely similar.

*Proof of Lemma 19.2.12.* (i) The $\alpha_{\{s\}}$-eigenspace for the right action of $\mathbb{R}_{\mathbf{q}}[W_{\{s\}}]$ on $L^2_{\mathbf{q}}$ is $\operatorname{Ker}(q_s - e_s) = \operatorname{Ker} h_s = L^2_{\mathbf{q}} a_s = A^s$. So $A^T$, being the intersection of the $A^s$, is in the $\alpha_{\{s\}}$-eigenspace for all $s \in T$. In particular, for any $x \in A^T$ and $s \in T$, $x \cdot e_s = q_s x$. So, if $s_1 \cdots s_k$ is a reduced expression for $u \in W_T$,

$$x \cdot e_u = x \cdot e_{s_1} \cdots e_{s_k} = q_{s_1} \cdots q_{s_k} x = q_u x,$$

which proves (i).

(ii) Let $x = \sum x_w e_w \in A^T$. Write $w = vu$ where $v$ is $(\emptyset, T)$-reduced and $u \in W_T$. We have

$$\begin{aligned} q_{vu} x_{vu} &= \langle e_{vu}, x \rangle_{\mathbf{q}} = \langle e_v e_u, x \rangle_{\mathbf{q}} = \langle e_v, x e_u^* \rangle_{\mathbf{q}} \\ &= \langle e_v, x e_{u^{-1}} \rangle_{\mathbf{q}} = \langle e_v, q_u x \rangle_{\mathbf{q}} = q_v q_u x_v, \end{aligned}$$

where the penultimate inequality is from part (i) and (19.15). So $x_{vu} = x_v$. Thus,

$$\langle x, x \rangle_{\mathbf{q}} = \sum_{v \in B_T} \sum_{u \in W_T} q_v q_u x_{vu}^2 = W_T(\mathbf{q}) \sum_{v \in B_T} q_v x_v^2,$$

where $B_T$ denotes the set of $(\emptyset, T)$-reduced elements. So $x \in L^2_{\mathbf{q}}$ if and only if $x = 0$ or $\mathbf{q} \in \mathcal{R}_T$. Since $x_{vu} = x_v$ for all $u \in W_T$, we also have

$$x = \sum_{v \in B_T} x_v e_v \sum_{u \in W_T} e_u = \Big( \sum_{v \in B_T} x_v e_v \Big) \tilde{a}_T \in L^2_{\mathbf{q}} a_T,$$

which proves (ii). □

## NOTES

**19.1.** Most of the material in this section is taken from [29, Exercise 23, p. 57]. The terminology "*j*-isomorphism" comes from [173]. The final paragraphs on Hecke algebras and buildings are taken from [29, Ex. 22, pp. 56–57].

**19.2.** This section comes from [109] and [79]. I first learned about completing Hecke algebras to von Neumann algebras from Jan Dymara. The idea may have been known earlier, at least in the case of Euclidean reflection groups.

If $W$ has no spherical or affine components, it can be shown that each nontrivial conjugacy class in $W$ is infinite. When this is the case, it follows that the von Neumann algebra $\mathcal{N}(W)$ $(= \mathcal{N}_{\mathbf{1}})$ associated to the group algebra is a factor (i.e., it has trivial one-dimensional center).

As $\mathbf{q}$ varies, the Hecke algebras $\mathbb{R}_{\mathbf{q}} W$ are all isomorphic to the group algebra $\mathbb{R}_{\mathbf{1}} W$. One way to see this is to note that $\mathbb{R}_{\mathbf{q}} W$ is generated by the idempotents $a_s^{\mathbf{q}}$ (where the superscript denotes the dependence on $\mathbf{q}$) and the correspondence $a_s^{\mathbf{1}} \to a_s^{\mathbf{q}}$ induces an isomorphism $\mathbb{R}_{\mathbf{1}} W \cong \mathbb{R}_{\mathbf{q}} W$. However, the same is not true for the Hecke – von Neumann algebras $\mathcal{N}_{\mathbf{q}}$. For example, if $\mathbf{q} \in \mathcal{R}$, the center of $\mathcal{N}_{\mathbf{q}}$ is at least 2-dimensional as it contains 1 and $a_S$. The reason is that, by (19.9) and (19.13), $a_s$ commutes with every element of $\mathbb{R}_{\mathbf{q}} W$ and hence, with every element of $\mathcal{N}_{\mathbf{q}}$. Similarly, for $\mathbf{q} \in \mathcal{R}^{-1}$ the center of $\mathcal{N}_{\mathbf{q}}$ contains 1 and $h_S$. Thus, $\mathcal{N}_{\mathbf{q}}$ is never a factor in this range of $\mathbf{q}$. On the other hand, if $W$ has no spherical or affine components, then, by the previous paragraph, $\mathcal{N}_{\mathbf{1}}$ is a factor. So *the isomorphism type of $\mathcal{N}_{\mathbf{q}}$ depends on* $\mathbf{q}$. A classification of these von Neumann algebras up to isomorphism would be very interesting.

# Chapter Twenty

## WEIGHTED $L^2$-(CO)HOMOLOGY

Suppose $\Gamma$ is a discrete group acting properly and cellularly on a CW complex $Y$. $C_*(Y;\mathbb{R})$ and $C^*(Y;\mathbb{R})$ denote the vector spaces of real-valued cellular chains and cochains, respectively. $C^i(Y;\mathbb{R})$ is the set of all functions on the set of $i$-cells of $Y$ while $C_i(Y;\mathbb{R})$ consists of the finitely supported ones. Let $L^2C^i(Y) \subset C^i(Y;\mathbb{R})$ be the subspace of square summable cochains. ($L^2C^i(Y)$ $(= L^2C_i(Y))$ is the Hilbert space completion of $C_i(Y;\mathbb{R})$ with respect to a standard inner product.) Taking (co)homology, we get the $L^2$-(co)homology spaces of $Y$. If we are careful to use only closed subspaces in these Hilbert spaces (by taking closures of the spaces of boundaries and coboundaries), we get the "reduced" $L^2$-(co)homology spaces of $Y$. When nonzero, these Hilbert spaces tend to be infinite dimensional. More information is available. The group $\Gamma$ acts on these (co)homology spaces. Using the von Neumann algebra associated to the group algebra $\mathbb{R}\Gamma$, it is possible to define the "$\Gamma$-dimension" of such a Hilbert space; it is a nonnegative real number. Hence, we have the notion of an "$L^2$-Betti number," namely, the $\Gamma$-dimension of a reduced $L^2$-(co)homology space. All this is explained in Appendix J. Jan Dymara [109] discovered a refinement of this theory in the case where $\Gamma$ is a Coxeter group $W$ and the CW complex has the form $\mathcal{U}(W,X)$ $(= \mathcal{U})$. This theory was further developed in [79]. It is the subject of this chapter.

Suppose we are given an $I$-tuple $\mathbf{q}$ of positive real numbers as in the previous two chapters. This gives a deformation of the group algebra to a Hecke algebra $\mathbb{R}_{\mathbf{q}}W$ and a corresponding von Neumann algebra $\mathcal{N}_{\mathbf{q}}$. The multiparameter $\mathbf{q}$ gives a deformation of the standard inner product on $C_*(\mathcal{U};\mathbb{R})$ so that the resulting square summable cochains $L^2_{\mathbf{q}}C^*(\mathcal{U}) \subset C^*(\mathcal{U};\mathbb{R})$ become $\mathcal{N}_{\mathbf{q}}$-modules and so that the coboundary maps are maps of $\mathcal{N}_{\mathbf{q}}$-modules. As in the case of the standard inner product on the group algebra, this yields (reduced) "weighted $L^2$-cohomology" spaces $L^2_{\mathbf{q}}\mathcal{H}^*(\mathcal{U})$, whose "von Neumann dimensions" are the "weighted $L^2$-Betti numbers" $b^i_{\mathbf{q}}(\mathcal{U})$ of $\mathcal{U}$. (N.B. Although the usual coboundary maps are $\mathcal{N}_{\mathbf{q}}$-equivariant, the usual boundary maps are *not*. To remedy this one defines the "weighted $L^2$-homology spaces" $L^2_{\mathbf{q}}\mathcal{H}_*(\mathcal{U})$

by using for boundary maps the adjoints of the coboundary maps rather than those defined by the usual formula.) When $\mathbf{q}$ is the constant $I$-tuple $\mathbf{1}$, the Hecke algebra is the group algebra and the weighted $L^2$-cohomology spaces are the standard ones referred to in the first paragraph and in Appendix J.

The study of the weighted $L^2$-cohomology of $\mathcal{U}$ is closely related to the growth series $W(\mathbf{t})$ of Chapter 17. For example, for any nonzero constant, the constant 0-cocycle on $\Sigma$ is square summable if and only if $\mathbf{q}$ lies in the region of convergence $\mathcal{R}$ for $W(\mathbf{t})$. Moreover, provided $\mathcal{U}$ is connected, in this same range of $\mathbf{q}$, $b^0_{\mathbf{q}}(\mathcal{U}) = 1/W(\mathbf{q})$ (Proposition 20.3.1). Particularly easy to compute is the "$L^2_{\mathbf{q}}$-Euler characteristic" $\chi_{\mathbf{q}}(\mathcal{U})$, defined as the alternating sum of the $b^i_{\mathbf{q}}(\mathcal{U})$. An explicit formula is given in 20.2. In the case of principal interest, where $\mathcal{U} = \Sigma$, we have $\chi_{\mathbf{q}}(\Sigma) = 1/W(\mathbf{q})$ (Proposition 20.2.4).

When $W$ is type HM$^n$ so that $\Sigma$ is a homology $n$-manifold, we have Poincaré duality: $b^i_{\mathbf{q}}(\Sigma) = b^{n-i}_{\mathbf{1/q}}(\Sigma)$. For general values of $\mathbf{q}$ we cannot say much more about the $b^i_{\mathbf{q}}$, other than the function $\mathbf{q} \to b^i_{\mathbf{q}}(\mathcal{U})$ is continuous (Theorem 20.2.1). However, for $\mathbf{q} \in \mathcal{R}$ or for $\mathbf{q}^{-1} \in \mathcal{R}$, we give a complete calculation of these weighted Betti numbers in Theorem 20.7.1. The method for doing this follows the line laid down in 15.2. $L^2_{\mathbf{q}}\mathcal{H}^*(\mathcal{U})$ can be thought of as a type of equivariant cohomology of $\mathcal{U}$. A direct sum decomposition of coefficient systems on $X$ yields a decomposition of (co)chains and (co)homology. We find such a decompositon of coefficient systems so that each term in the sum essentially becomes a relative (co)homology group of $X$ with constant coefficients. The desired decompositions of coefficient systems (i.e., of $L^2_{\mathbf{q}}$) are proved in 20.6. (When $W$ is finite and $\mathbf{q} = \mathbf{1}$, this decompositon is Solomon's Theorem 15.4.2.) Our proof of the decomposition theorems ultimately comes down to showing the vanishing, except in the bottom dimension, of the relative $L^2_{\mathbf{q}}$-homology of certain subcomplexes of $\Sigma$ for $\mathbf{q} \in \mathcal{R}$. (These subcomplexes are called "ruins" in 20.6.)

The calculation shows that as $\mathbf{q}$ goes from 0 to $\infty$, $L^2_{\mathbf{q}}\mathcal{H}^*(\mathcal{U})$ interpolates between ordinary cohomology groups and cohomology with compact supports. (In Chapters 8 and 15 we calculated both the ordinary cohomology and cohomology with compact supports of $\mathcal{U}$. The answers are wildly different.) Roughly speaking, for $\mathbf{q} \in \mathcal{R}$, $L^2_{\mathbf{q}}\mathcal{H}^*(\mathcal{U})$ looks like ordinary cohomology while for $\mathbf{q} \in \mathcal{R}^{-1}$, it looks like cohomology with compact supports. The precise statement is given in Theorem 20.7.6.

When $\mathbf{q} = \mathbf{1}$, $L^2_{\mathbf{q}}\mathcal{H}^*(\Sigma)$ is the ordinary $L^2$-cohomology of $\Sigma$. In this case, when $(W, S)$ is type HM$^n$, there is a pre-eminent question, the Singer Conjecture (see Appendix J.7). It implies that $L^2\mathcal{H}^k(\Sigma)$ should vanish except for $k = \frac{n}{2}$. (In particular, it should always vanish when $n$ is odd.) This conjecture is proved for $W$ right-angled and $n \leqslant 4$ in [91]. A generalization for weighted $L^2$-cohomology is explained in 20.5.

## 20.1. WEIGHTED $L^2$-(CO)HOMOLOGY

As usual, $X$ is a finite, mirrored CW complex over $S$ and $\mathcal{U} = \mathcal{U}(W, X)$. For any cell $c$ of $X$, $S(c) := \{s \in S \mid c \subset X_s\}$. $X^{(i)}$ and $\mathcal{U}^{(i)}$ denote the set of $i$-cells in $X$ and $\mathcal{U}$, respectively. As in 19.2, $\mathbf{q}$ is a multiparameter of positive real numbers and $\mathcal{N}_{\mathbf{q}}$ is the von Neumann algebra associated to the Hecke algebra $\mathbb{R}_{\mathbf{q}}W$.

We begin by defining a chain complex of Hilbert $\mathcal{N}_{\mathbf{q}}$-modules for the CW complex $\mathcal{U}$ with its cell structure induced from that of $X$. In this case, each orbit of cells contributes an $\mathcal{N}_{\mathbf{q}}$-module of the form $A^T$, for $T = S(c)$. Next, we consider the chain complexes arising from the cellulation of $\Sigma$ by Coxeter polytopes. These chain complexes are also $\mathcal{N}_{\mathbf{q}}$-modules. In this case, each orbit of cells contributes a $\mathcal{N}_{\mathbf{q}}$-module isomorphic to $H^T$, for some $T = S(c) \in \mathcal{S}$. When $\mathcal{U} = \Sigma$, this chain complex is chain homotopy equivalent to the first one.

### Weighted (Co)chain Complexes for $\mathcal{U}(W, X)$

Orient the cells of $X$ arbitrarily and then orient the remaining cells of $\mathcal{U}$ so that for each positively oriented cell $c$ of $X$ and each $w \in W$, $wc$ is positively oriented.

Define a measure $\mu_{\mathbf{q}}$ on the orbit of a cell $c \in X^{(i)}$ by

$$\mu_{\mathbf{q}}(wc) := q_u, \tag{20.1}$$

where $u$ is the shortest element in $wW_{S(c)}$ (i.e., $u$ is the $(\emptyset, S(c))$-reduced element in this coset). Formula (20.1) extends to a measure, also denoted $\mu_{\mathbf{q}}$, on $\mathcal{U}^{(i)}$. As in [109], define the $\mathbf{q}$-*weighted $L^2$-(co)chains* on $\mathcal{U}$ in dimension $i$ to be the Hilbert space

$$L^2_{\mathbf{q}}C_i(\mathcal{U}) = L^2_{\mathbf{q}}C^i(\mathcal{U}) := L^2(\mathcal{U}^{(i)}, \mu_{\mathbf{q}}). \tag{20.2}$$

We have coboundary and boundary maps, $\delta^i : L^2_{\mathbf{q}}C^i(\mathcal{U}) \to L^2_{\mathbf{q}}C^{i+1}(\mathcal{U})$ and $\partial_i : L^2_{\mathbf{q}}C_i(\mathcal{U}) \to L^2_{\mathbf{q}}C_{i-1}(\mathcal{U})$ defined by the usual formulas:

$$\delta^i(f)(\gamma) := \sum [\beta : \gamma] f(\beta), \tag{20.3}$$

$$\partial_i(f)(\alpha) := \sum [\alpha : \beta] f(\beta), \tag{20.4}$$

where the first sum is over all $i$-cells $\beta$ incident to the $(i+1)$-cell $\gamma$ and the second is over all $\beta$ whose boundary contains the $(i-1)$-cell $\alpha$. In contrast to the usual situation, $\delta^i$ and $\partial_{i+1}$ are not adjoint to one another. In fact, if one defines $\partial^{\mathbf{q}}_i : L^2_{\mathbf{q}}C_i(\mathcal{U}) \to L^2_{\mathbf{q}}C_{i-1}(\mathcal{U})$ by

$$\partial^{\mathbf{q}}_i(f)(\alpha) := \sum [\alpha : \beta] \mu_{\mathbf{q}}(\beta) \mu_{\mathbf{q}}(\alpha)^{-1} f(\beta), \tag{20.5}$$

then a simple calculation shows $\partial^{\mathbf{q}}$ is the adjoint of $\delta$ (with respect to the $\mu_{\mathbf{q}}$ inner product), We state this as the next lemma.

**LEMMA 20.1.1.** ([109, §1].) $\delta^* = \partial^{\mathbf{q}}$.

Since $\delta^2 = 0$, by taking adjoints, we have $(\partial^{\mathbf{q}})^2 = 0$. Hence, $(L^2_{\mathbf{q}}C_*(\mathcal{U}), \partial^{\mathbf{q}})$ is a chain complex. Define the **q**-*weighted* $L^2$-*(co)homology* of $\mathcal{U}$ by

$$L^2_{\mathbf{q}}H^i(\mathcal{U}) := H^i((L^2_{\mathbf{q}}C^*(\mathcal{U}), \delta)) = \operatorname{Ker}\delta^i / \operatorname{Im}\delta^{i-1},$$

$$L^2_{\mathbf{q}}H_i(\mathcal{U}) := H_i((L^2_{\mathbf{q}}C_*(\mathcal{U}), \partial^{\mathbf{q}})) = \operatorname{Ker}\partial^{\mathbf{q}}_i / \operatorname{Im}\partial^{\mathbf{q}}_{i+1}.$$

There is a standard problem with these (co)homology groups: the above quotients need not be Hilbert spaces. To remedy this, define the *reduced* **q**-weighted $L^2$-(co)homology by

$$L^2_{\mathbf{q}}\mathcal{H}^i(\mathcal{U}) := \operatorname{Ker}\delta^i / \overline{\operatorname{Im}\delta^{i-1}},$$

$$L^2_{\mathbf{q}}\mathcal{H}_i(\mathcal{U}) := \operatorname{Ker}\partial^{\mathbf{q}}_i / \overline{\operatorname{Im}\partial^{\mathbf{q}}_i}.$$

### The Isomorphism $\theta$

The next lemma describes an important trick that we will need to use several times.

**LEMMA 20.1.2.** *The chain complexes* $(L^2_{\mathbf{q}}C_*(\mathcal{U}), \partial^{\mathbf{q}})$ *and* $(L^2_{\mathbf{1/q}}C_*(\mathcal{U}), \partial)$ *are isomorphic.*

*Proof.* Given a chain $f$ on $\mathcal{U}$, define another chain $\theta(f)$ by $\theta(f)(\beta) := \mu_{\mathbf{q}}(\beta)f(\beta)$. Note that $\theta(f)$ is $\mathbf{q}^{-1}$-square summable if and only if $f$ is **q**-square summable. Hence, we get a linear isomorphism $\theta : L^2_{\mathbf{1/q}}C_*(\mathcal{U}) \to L^2_{\mathbf{q}}C_*(\mathcal{U})$. Using (20.4) and (20.5), a computation shows $\theta \circ \partial = \partial^{\mathbf{q}} \circ \theta$. So $\theta$ is a chain isomorphism. □

*Remark 20.1.3.* We have canonical inclusions of chain complexes:

$$C_*(\mathcal{U}; \mathbb{R}) \hookrightarrow (L^2_{\mathbf{q}}C_*(\mathcal{U}), \partial) \hookrightarrow C^{lf}_*(\mathcal{U}; \mathbb{R}). \tag{20.6}$$

Using the isomorphism in Lemma 20.1.2 we get inclusions:

$$C_*(\mathcal{U}; \mathbb{R}) \hookrightarrow (L^2_{\mathbf{1/q}}C_*(\mathcal{U}), \partial^{\mathbf{1/q}}) \hookrightarrow C^{lf}_*(\mathcal{U}; \mathbb{R}). \tag{20.7}$$

Similarly, we have inclusions of cochain complexes:

$$C^*_c(\mathcal{U}; \mathbb{R}) \hookrightarrow L^2_{\mathbf{q}}C^*(\mathcal{U}) \hookrightarrow C^*(\mathcal{U}; \mathbb{R}). \tag{20.8}$$

(Here $C^{lf}_*(\,)$ and $C^*_c(\,)$ stand for, respectively, infinite cellular chains and finitely supported cellular cochains.) The second map in (20.6) (or the second map in

(20.7)) is obtained by dualizing the first map in (20.8). Similarly, the second map in (20.8) is obtained by dualizing the first map in (20.6). Replacing $\mathbf{1/q}$ by $\mathbf{q}$, the first map in (20.7) induces

$$\text{can} : H_i(\mathcal{U};\mathbb{R}) \to L^2_{\mathbf{q}}\mathcal{H}_i(\mathcal{U}). \tag{20.9}$$

Similarly, the first map in (20.8) induces

$$\text{can} : H^i_c(\mathcal{U};\mathbb{R}) \to L^2_{\mathbf{q}}\mathcal{H}^i(\mathcal{U}). \tag{20.10}$$

We will show in Theorem 20.7.6 that, for $\mathbf{q} \in \mathcal{R}$, the canonical map (20.9) is a monomorphism with dense image, while for $\mathbf{q} \in \mathcal{R}^{-1}$, the map (20.10) is a monomorphism with dense image. This is the sense in which $L^2_{\mathbf{q}}\mathcal{H}^*(\mathcal{U})$ interpolates between ordinary cohomology and cohomology with compact supports. This is reminiscent of a well-known result of Cheeger-Gromov [62] that if a discrete amenable group $A$ acts properly on a CW complex $X$, then the canonical map $L^2H^*(X) \to H^*(X;\mathbb{R})$ is injective (Theorem J.8.2). So, for $\mathbf{q} \in \mathcal{R}$, weighted $L^2$-cohomology behaves as if $W$ were amenable.

### Hodge Decomposition

Since $\delta^* = \partial^{\mathbf{q}}$ and $(\partial^{\mathbf{q}})^* = \delta$, we get a Hodge decomposition

$$L^2_{\mathbf{q}}C^i(\mathcal{U}) = (\operatorname{Ker}\delta^i \cap \operatorname{Ker}\partial^{\mathbf{q}}_i) \oplus \overline{\operatorname{Im}\delta^{i-1}} \oplus \overline{\operatorname{Im}\partial^{\mathbf{q}}_{i+1}},$$

as in Appendix J.3. It follows that both $L^2_{\mathbf{q}}\mathcal{H}^i(\mathcal{U})$ and $L^2_{\mathbf{q}}\mathcal{H}_i(\mathcal{U})$ can be identified with the space $\operatorname{Ker}\delta^i \cap \operatorname{Ker}\partial^{\mathbf{q}}_i$ of *harmonic cochains*. In particular, $L^2_{\mathbf{q}}\mathcal{H}^i(\mathcal{U}) \cong L^2_{\mathbf{q}}\mathcal{H}_i(\mathcal{U})$.

### The Hilbert $\mathcal{N}_{\mathbf{q}}$-Module Structure on $L^2_{\mathbf{q}}C^*(\mathcal{U})$

Terminology is taken from 19.2. $L^2_{\mathbf{q}} = L^2(W, \nu_{\mathbf{q}})$, where $\nu_{\mathbf{q}}$ is the measure on $W$ defined by $\nu_{\mathbf{q}}(w) = q_w$. For each subset $T$ of $S$, the Hilbert $\mathcal{N}_{\mathbf{q}}$-submodule $A^T \subset L^2_{\mathbf{q}}$, defined by (19.17), can be identified with $L^2(W, \nu_{\mathbf{q}})^{W_T}$, the subspace of functions which are constant on each right coset $wW_T$. Since each cell of $\mathcal{U}$ has the form $wc$ for some cell $c$ of $X$,

$$L^2_{\mathbf{q}}C^i(\mathcal{U}) = \bigoplus_{c \in X^{(i)}} L^2(Wc, \mu_{\mathbf{q}}),$$

where the sum ranges over all $i$-cells $c$ of $X$. $L^2(Wc, \mu_{\mathbf{q}})$ can be identified with $A^{S(c)}$ via the isometry $\varphi_c : L^2(Wc, \mu_{\mathbf{q}}) \to A^{S(c)}$ defined by

$$\varphi_c(f) = \sqrt{W_{S(c)}(\mathbf{q})}\left(\sum_{u \in B_{S(c)}} f(uc)e_u a_{S(c)}\right),$$

where the summation is over all $(\emptyset, S(c))$-reduced elements $u$ and where $a_{S(c)}$ is the idempotent in $\mathcal{N}_{\mathbf{q}}$ defined in Lemma 19.2.5. So we get an isometry

$$\oplus\varphi_c : L^2_{\mathbf{q}}C^i(\mathcal{U}) = \bigoplus_{c\in X^{(i)}} L^2(Wc, \mu_{\mathbf{q}}) \xrightarrow{\cong} \bigoplus_{c\in X^{(i)}} A^{S(c)}. \tag{20.11}$$

Since each $A^{S(c)}$ is a left $\mathbb{R}_{\mathbf{q}}W$-submodule of $L^2_{\mathbf{q}}$, this gives $L^2_{\mathbf{q}}C^i(\mathcal{U})$ the structure of a Hilbert $\mathcal{N}_{\mathbf{q}}$-module as in Definition 19.2.3 (provided we assume, as we shall, that $X$ is a finite complex). It also gives an isometric embedding

$$\Phi : L^2_{\mathbf{q}}C^i(\mathcal{U}) \hookrightarrow \bigoplus_{c\in X^{(i)}} L^2_{\mathbf{q}} = C^i(X) \otimes L^2_{\mathbf{q}}. \tag{20.12}$$

The next result is a straightforward computation.

**Lemma 20.1.4.** ([109, Lemma 3.2].) *$\delta$ and $\partial^{\mathbf{q}}$ are maps of Hilbert $\mathcal{N}_{\mathbf{q}}$-modules.*

(It is *not* true that $\partial$ is a map of Hilbert $\mathcal{N}_{\mathbf{q}}$-modules.)

It follows that Ker $\delta$, Ker $\partial^{\mathbf{q}}$, $\overline{\text{Im}\,\delta}$ and $\overline{\text{Im}\,\partial^{\mathbf{q}}}$ are Hilbert $\mathcal{N}_{\mathbf{q}}$-modules. Hence, $L^2_{\mathbf{q}}\mathcal{H}^i(\mathcal{U})$ and $L^2_{\mathbf{q}}\mathcal{H}_i(\mathcal{U})$) are also Hilbert $\mathcal{N}_{\mathbf{q}}$-modules.

### Weighted (Co)chain Complexes for the Cellulation by Coxeter Polytopes

The cellulation of $\Sigma$ by Coxeter polytopes gives a different (co)chain complex, which also has the structure of a Hilbert $\mathcal{N}_{\mathbf{q}}$-module. Let $C_T$ denote the Coxeter polytope in $\Sigma$ corresponding to $W_T \in W\mathcal{S}$ (the $W_T$ coset of the identity element). $WC_T$ denotes the $W$-orbit of $C_T$, i.e., the set of all Coxeter polytopes in $\Sigma$ of type $W_T$ (see 7.3). Let $\Sigma_{\text{cc}}$ stand for $\Sigma$ with this cell structure. Define a measure $\mu_{\mathbf{q}}$ on $\Sigma^{(i)}_{\text{cc}}$ by $\mu_{\mathbf{q}}(wC_T) = q_u$, where $u$ $(= p_T(w))$ is the shortest element in $wW_T$. Define the *$\mathbf{q}$-weighted $L^2$-(co)chains* on $\Sigma$ in dimension $i$ to be the Hilbert space,

$$L^2_{\mathbf{q}}C_i(\Sigma_{\text{cc}}) = L^2_{\mathbf{q}}C^i(\Sigma_{\text{cc}}) := L^2(\Sigma^{(i)}_{\text{cc}}, \mu_{\mathbf{q}}).$$

We have

$$L^2_{\mathbf{q}}C_i(\Sigma_{\text{cc}}) = \bigoplus_{T\in\mathcal{S}^{(i)}} L^2(WC_T, \mu_{\mathbf{q}}).$$

For each $T \in \mathcal{S}$, arbitrarily choose an orientation for $C_T$. Orient the remaining cells in $WC_T$ as follows: if $u \in B_T$, orient $uC_T$ so that left translation by $u$ is an orientation-preserving map $C_T \to uC_T$. (As in 4.5 and 17.1, $B_T$ is the set of $(\emptyset, T)$-reduced elements in $W$.)

As in (20.3), $\delta : L^2_{\mathbf{q}}C^i(\Sigma_{\text{cc}}) \to L^2_{\mathbf{q}}C^{i+1}(\Sigma_{\text{cc}})$ is the usual coboundary map. Its adjoint $\partial^{\mathbf{q}} : L^2_{\mathbf{q}}C_{i+1}(\Sigma_{\text{cc}}) \to L^2_{\mathbf{q}}C_i(\Sigma_{\text{cc}})$ is defined in (20.5). Next, we determine the formula for the restriction of $\partial^{\mathbf{q}}$ to the summand $L^2(WC_U, \mu_{\mathbf{q}})$, where $U \in \mathcal{S}^{(i+1)}$. Suppose $T \in \mathcal{S}^{(i)}$ is obtained by deleting one element of $U$ and that $w \in B_T$. Any $w \in W$ can be uniquely decomposed as $w = uv$ with $u \in B_U$ and $v \in W_U$. If $w \in B_T$, then $v \in W_U \cap B_T$. For any $f \in L^2(WC_U, \mu_{\mathbf{q}})$, we have the following formula for $\partial^{\mathbf{q}}$:

$$\partial^{\mathbf{q}} f(wC_T) = \varepsilon_v q_v^{-1} f(uC_U). \tag{20.13}$$

$W_T$ acts nontrivially on the cell $C_T$ and $v \in W_T$ is $\varepsilon_v$-orientation-preserving. Hence, the right $\mathbb{R}_{\mathbf{q}}[W_T]$-action on $L^2(WC_T, \mu_{\mathbf{q}})$ is via the alternating character $\beta_T$ of Definition 19.2.8, i.e., $L^2(WC_T, \mu_{\mathbf{q}})$ is identified with $H^T$. A specific isometry $\psi : L^2(WC_T, \mu_{\mathbf{q}}) \to H^T$ is defined by

$$\psi_T(f) = \sqrt{W_T(\mathbf{q}^{-1})}\left(\sum_{u \in B_T} f(uC_T)e_u\right)h_T, \tag{20.14}$$

where $h_T$ is the idempotent of $\mathcal{N}_{\mathbf{q}}$ defined in Lemma 19.2.6. This gives an isometry

$$L^2_{\mathbf{q}} C_i(\Sigma_{\mathrm{cc}}) = \bigoplus_{T \in \mathcal{S}^{(i)}} L^2(WC_T, \mu_{\mathbf{q}}) \xrightarrow{\cong} \bigoplus_{T \in \mathcal{S}^{(i)}} H^T. \tag{20.15}$$

Since each $H^T$ is a left $\mathbb{R}_{\mathbf{q}}W$-submodule of $L^2_{\mathbf{q}}$, this gives $L^2_{\mathbf{q}}C^i(\Sigma_{\mathrm{cc}})$ the structure of a Hilbert $\mathcal{N}_{\mathbf{q}}$-module. It also gives an isometric embedding,

$$\Psi : L^2_{\mathbf{q}} C_i(\Sigma_{\mathrm{cc}}) \hookrightarrow \bigoplus_{c \in \mathcal{S}^{(i)}} L^2_{\mathbf{q}} = C_i(K) \otimes L^2_{\mathbf{q}}.$$

Use the isomorphism in (20.15) to transport the Hilbert $\mathcal{N}_{\mathbf{q}}$-module structure from the right hand side of (20.15) to $L^2_{\mathbf{q}}C_i(\Sigma_{\mathrm{cc}})$. It is proved in [109, Lemma 4.3] that $\delta$ and $\partial^{\mathbf{q}}$ are maps of Hilbert $\mathcal{N}_{\mathbf{q}}$-modules. We shall give the argument in Lemma 20.6.21, below. Hence, the reduced $L^2$-(co)homology groups,

$$L^2_{\mathbf{q}}\mathcal{H}^i(\Sigma_{\mathrm{cc}}) = \operatorname{Ker}\delta^i/\overline{\operatorname{Im}\delta^{i-1}} \quad \text{and} \quad L^2_{\mathbf{q}}\mathcal{H}_i(\Sigma_{\mathrm{cc}}) = \operatorname{Ker}\partial^{\mathbf{q}}_i/\overline{\operatorname{Im}\partial^{\mathbf{q}}_i}$$

are also Hilbert $\mathcal{N}_{\mathbf{q}}$-modules. It is proved in [109, §5] that the (co)homology groups of $L^2_{\mathbf{q}}C_*(\Sigma_{\mathrm{cc}})$ and of $L^2_{\mathbf{q}}C_*(\Sigma)$ are the same, i.e., $L^2_{\mathbf{q}}H_*(\Sigma_{\mathrm{cc}}) \cong L^2_{\mathbf{q}}H_*(\Sigma)$, $L^2_{\mathbf{q}}H^*(\Sigma_{\mathrm{cc}}) \cong L^2_{\mathbf{q}}H^*(\Sigma)$ and $L^2_{\mathbf{q}}\mathcal{H}^*(\Sigma_{\mathrm{cc}}) \cong L^2_{\mathbf{q}}\mathcal{H}^*(\Sigma)$. (The point is that the simplicial structure on $\Sigma$ is a subdivision of $\Sigma_{\mathrm{cc}}$.)

The chain complex $(L^2_{\mathbf{q}}C_*(\Sigma_{\mathrm{cc}}), \partial^{\mathbf{q}})$ looks like this:

$$L^2_{\mathbf{q}} \longleftarrow \bigoplus_{s \in S} H^s \longleftarrow \bigoplus_{T \in \mathcal{S}^{(2)}} H^T \longleftarrow \cdots.$$

We shall explicitly describe the boundary maps in Lemma 20.6.21.

## 20.2. WEIGHTED $L^2$-BETTI NUMBERS AND EULER CHARACTERISTICS

In Appendix J.5, in the case of ordinary $L^2$-(co)homology ($\mathbf{q} = \mathbf{1}$), we define $L^2$-Betti numbers and $L^2$-Euler characteristics and we prove Atiyah's Formula which identifies the $L^2$-Euler characteristic with the orbihedral Euler characteristic of the quotient. (The orbihedral Euler characteristic is defined by formula (16.4) in Section 16.1).) Here we give the analogous definitions for weighted $L^2$-(co)homology.

In the previous section we showed that the weighted $L^2$-(co)chains are Hilbert $\mathcal{N}_{\mathbf{q}}$-modules. As in 19.2, they have a von Neumann dimension, $\dim_{\mathcal{N}_{\mathbf{q}}}(\ )$, defined by (19.7). Set

$$c^i_{\mathbf{q}}(\mathcal{U}) := \dim_{\mathcal{N}_{\mathbf{q}}} L^2_{\mathbf{q}}C^i(\mathcal{U}),$$

The stabilizer of an $i$-cell $c \subset X$ is the special subgroup $W_{S(c)}$ and the summand of $L^2_{\mathbf{q}}C^i(\mathcal{U})$ corresponding to the orbit of $c$ is isomorphic to $A^{S(c)}$. Its dimension is $\operatorname{tr}_{\mathcal{N}_{\mathbf{q}}}(a_{S(c)}) = 1/W_{S(c)}(\mathbf{q})$. Hence,

$$c^i_{\mathbf{q}}(\mathcal{U}) = \sum_{c \in X^{(i)}} \frac{1}{W_{S(c)}(\mathbf{q})}. \tag{20.16}$$

By the Hodge decomposition, the reduced weighted $L^2$-(co)homology spaces are closed, $\mathbb{R}_{\mathbf{q}}W$-stable subspaces of the weighted $L^2$-(co)chains; so, they are also Hilbert $\mathcal{N}_{\mathbf{q}}$-modules. The $i^{th}$ *weighted $L^2_{\mathbf{q}}$-Betti number* of $\mathcal{U}$ is defined by

$$b^i_{\mathbf{q}}(\mathcal{U}) := \dim_{\mathcal{N}_{\mathbf{q}}} L^2_{\mathbf{q}}\mathcal{H}^i(\mathcal{U}). \tag{20.17}$$

**THEOREM 20.2.1.** ([79, Theorem 7.7].) *For each i, the function $\mathbf{q} \to b^i_{\mathbf{q}}(\mathcal{U})$ is continuous.*

The proof can be found in [79, Section 7]. It is fairly long and we will not reproduce it here.

*Remark 20.2.2.* In Theorem 20.7.1 we will give explicit formulas for the $b^i_{\mathbf{q}}$, for $\mathbf{q}$ in $\mathcal{R} \cup \mathcal{R}^{-1}$ (in fact, for $\mathbf{q} \in \overline{\mathcal{R}} \cup \overline{\mathcal{R}^{-1}}$). It follows from these formulas that $\lim b^i_{\mathbf{q}}$ exists as $\mathbf{q} \to 0$ or $\mathbf{q} \to \infty$ and that

$$\lim_{\mathbf{q}\to 0} b^i_{\mathbf{q}}(\mathcal{U}) = b^i(X) \quad \text{and} \quad \lim_{\mathbf{q}\to \infty} b^i_{\mathbf{q}}(\mathcal{U}) = b^i(X, \partial X), \tag{20.18}$$

where for any pair of spaces, $(Y, Z)$, $b^i(Y, Z) := \dim_{\mathbb{R}}(H^i(Y, Z; \mathbb{R}))$ is the ordinary $i^{th}$-Betti number. (See Remark 20.7.3.)

By the additive property of $\dim_{\mathcal{N}_{\mathbf{q}}}$, the alternating sum of the Hecke–von Neumann dimensions of a chain complex of Hilbert $\mathcal{N}_{\mathbf{q}}$-modules is equal to the alternating sum of the dimensions of its reduced homology groups. This gives the following version of of Atiyah's Formula (Theorem J.5.3) for $L^2_{\mathbf{q}}$-cohomology:

$$\sum (-1)^i b^i_{\mathbf{q}}(\mathcal{U}) = \sum (-1)^i c^i_{\mathbf{q}}(\mathcal{U}). \tag{20.19}$$

We denote either side of this equation by $\chi_{\mathbf{q}}(\mathcal{U})$ and call it the *$L^2_{\mathbf{q}}$-Euler characteristic* of $\mathcal{U}$. Not only is $\mathbf{q} \to \chi_{\mathbf{q}}(\mathcal{U})$ continuous, it is a rational function of $\mathbf{q}$. As we show below, this is a easy corollary of (20.19).

**PROPOSITION 20.2.3.** (Rationality of Euler characteristics.) *$\chi_{\mathbf{q}}(\mathcal{U}) = f(\mathbf{q})/g(\mathbf{q})$ where $f$ and $g$ are polynomials in $\mathbf{q}$ with integral coefficients.*

*Proof.* For each $T \in \mathcal{S}$, $X_T$ (resp. $\partial X_T$) is defined as the subcomplex consisting of those cells $c$ such that $T \subset S(c)$ (resp. $T \subsetneq S(c)$). By (20.16) and (20.19),

$$\chi_{\mathbf{q}}(\mathcal{U}) = \sum_{T \in \mathcal{S}} \frac{\chi(X_T) - \chi(\partial X_T)}{W_T(\mathbf{q})}, \tag{20.20}$$

where each term in the sum has numerator an integer and denominator a polynomial in $\mathbf{q}$. (Compare formula (16.7) in Section 16.1.) □

**PROPOSITION 20.2.4.** (Dymara [109, Cor. 3.4] or [79].)

$$\chi_{\mathbf{q}}(\Sigma) = \frac{1}{W(\mathbf{q})}.$$

We give two proofs.

*First proof.* By (20.20),

$$\chi_{\mathbf{q}}(\Sigma) = \sum_{T \in \mathcal{S}} \frac{\chi(K_T) - \chi(\partial K_T)}{W_T(\mathbf{q})},$$

and since $\chi(K_T) = 1$ (because $K_T$ is a cone), Theorem 17.1.10 shows that the right-hand side is equal to $1/W(\mathbf{q})$. □

*Second proof.* This time use the cellulation of $\Sigma$ by Coxeter cells. We have

$$\dim_{\mathcal{N}_{\mathbf{q}}} L^2(WC_T, \mu_{\mathbf{q}}) = \dim_{\mathcal{N}_{\mathbf{q}}} H^T = \frac{1}{W_T(\mathbf{q}^{-1})}.$$

Hence,

$$c^i_{\mathbf{q}}(\Sigma_{cc}) = \sum_{T \in \mathcal{S}^{(i)}} \frac{1}{W_T(\mathbf{q}^{-1})},$$

so

$$\chi_{\mathbf{q}}(\Sigma) = \sum_{T \in \mathcal{S}} \frac{\varepsilon(T)}{W_T(\mathbf{q}^{-1})} = \frac{1}{W(\mathbf{q})},$$

where the last equality is Theorem 17.1.9. □

## 20.3. CONCENTRATION OF (CO)HOMOLOGY IN DIMENSION 0

$\mathrm{Vert}(\Sigma)$ can be identified with $W$. So, $L^2_{\mathbf{q}}C^0(\Sigma_{cc}) \cong L^2_{\mathbf{q}}$. A 0-cochain is a cocycle if and only if it is the constant function on $W$. (See the discussion preceding formula (J.9) of Appendix J.4.) If $c$ denotes the constant, then the norm, with respect to the inner product $\langle\, ,\rangle_{\mathbf{q}}$, of the cocycle is

$$|c| \sum_{w \in W} q_w$$

and this is $< \infty$ if and only if $c = 0$ or $\mathbf{q} \in \mathcal{R}$ (and, if this is the case, its norm is $|c|\, W(\mathbf{q})$). The subspace of constants in $L^2_{\mathbf{q}}$ is $A^S$. When $\mathbf{q} \in \mathcal{R}$, the orthogonal projection $L^2_{\mathbf{q}} \to A^S$ is the idempotent $a_S$. So, $b^0_{\mathbf{q}}(\Sigma) = \dim_{\mathcal{N}_{\mathbf{q}}}(A^S) = \mathrm{tr}_{\mathcal{N}_{\mathbf{q}}}\, a_S = 1/W(\mathbf{q})$. (As a real vector space, $A^S$ has dimension 1 or 0; however, in our setup, its dimension varies continuously from 1 ($= \chi(K)$) to 0.) This proves the following.

**PROPOSITION 20.3.1.** (Dymara [109].) *$L^2_{\mathbf{q}}H^0(\Sigma)$ is nonzero if and only if $\mathbf{q} \in \mathcal{R}$. Moreover, when $\mathbf{q} \in \mathcal{R}$, $b^0_{\mathbf{q}}(\Sigma) = 1/W(\mathbf{q})$.*

*Remark 20.3.2.* $\mathcal{U}$ is connected if and only if $X$ is connected and $X_s \neq \emptyset$ for each $s \in S$ (Theorem 8.2.7). Suppose this. An argument similar to the one above shows $L^2_{\mathbf{q}}H^0(\mathcal{U})$ is nonzero if and only if $\mathbf{q} \in \mathcal{R}$ and when this is the case, $b^0_{\mathbf{q}}(\mathcal{U}) = 1/W(\mathbf{q})$.

### A Chain Contraction

The following result of Dymara is fundamental to the remaining computations in this chapter.

**THEOREM 20.3.3.** (Dymara [109].) *Suppose $\mathbf{q} \in \mathcal{R}$. Then $L^2_{\mathbf{q}}H_*(\Sigma)$ is concentrated in dimension* 0.

The idea for the proof is straightforward, but the details involve some delicate estimates. When $\mathbf{q} \in \mathcal{R}$, we will define a chain contraction $H : (L^2_{\mathbf{q}}C_*(\Sigma), \partial^{\mathbf{q}}) \to (L^2_{\mathbf{q}}C_{*+1}(\Sigma), \partial^{\mathbf{q}})$. By Lemma 20.1.2, this is equivalent to defining a chain contraction $H : (L^2_{\mathbf{1/q}}C_*(\Sigma), \partial) \to (L^2_{\mathbf{1/q}}C_{*+1}(\Sigma), \partial)$ (with respect to the usual boundary map). For ordinary chains this follows from the contractibility of $\Sigma$. However, now we need to show $H$ is a bounded operator. In order to get the required estimates, one needs to use the fact that natural piecewise Euclidean metric on $\Sigma$ is CAT(0) (Moussong's Theorem 12.3.3). Dymara first uses a simplicial version of geodesic contraction to define a chain contraction, $H : (C_*(\Sigma), \partial) \to (C_{*+1}(\Sigma), \partial)$, on the ordinary real-valued simplicial $k$-chains. Next he shows that for $\mathbf{q} \in \mathcal{R}^{-1}$, $H$ extends to a bounded linear operator, $H : L^2_{\mathbf{q}}C_*(\Sigma) \to L^2_{\mathbf{q}}C_{*+1}(\Sigma)$, which is a chain homotopy with respect to the usual boundary map, $\partial$. (N.B., $H$ is bounded for large values of $\mathbf{q}$ not small ones.) Finally, one uses the isomorphism $\theta$ from Lemma 20.1.2 to convert this to the desired chain contraction (with respect to $\partial^{\mathbf{1/q}}$), $H : L^2_{\mathbf{1/q}}C_*(\Sigma) \to L^2_{\mathbf{1/q}}C_{*+1}(\Sigma)$, valid for all $\mathbf{1/q} \in \mathcal{R}$.

Let $x_0 \in K$ be the central point in the fundamental chamber (i.e., $x_0$ is the vertex corresponding to the element $W_\emptyset \in W\mathcal{S}$). The lemmas Dymara needs are stated below. In the first one, he defines the chain contraction on $(C_*(\Sigma), \partial)$ and shows it has the required properties.

**LEMMA 20.3.4.** (Dymara [109, Theorem 9.1].) *There exists a chain contraction $H : C_*(\Sigma) \to C_{*+1}(\Sigma)$ and constants $C$ and $R$ with the following properties:*

*(a) If $v \in \Sigma^{(0)}$, then $\partial H(v) = v - x_0$.*

*(b) If $\alpha$ is a simplex of positive dimension, then $\partial(H(\alpha)) = \alpha - H(\partial\alpha)$.*

*(c) For every simplex $\alpha$, $\|H(\alpha)\|_{L^\infty} < C$.*

*(d) If $v$ is a vertex of a simplex $\alpha$ and $\gamma : [0, d] \to \Sigma$ is the geodesic segment from $v$ to $x_0$ (where $d := d(v, x_0)$ is the distance from $v$ to $x_0$), then* $\operatorname{supp}(H(\alpha)) \subset N_R(\operatorname{Im}\gamma)$. *(Here, for a simplicial chain $f$,* $\operatorname{supp}(f)$ *means the union of simplices which have nonzero coefficient in $f$. $N_R(\operatorname{Im}\gamma)$ means the $R$-neighborhood of the geodesic segment $\gamma$).*

$H$ is defined on vertices (and hence on 0-chains) by approximating geodesic contraction to $x_0$. Using induction on dimension, the definition is extended to $k$-chains in a straightforward fashion. Using the fact that the metric is CAT(0), Dymara shows $H$ has the desired properties. In the second lemma, he shows that properties (a)–(d) above are enough to guarantee that $H$ extends to a bounded linear operator.

**LEMMA 20.3.5.** (Dymara [109, Theorem 10.1].) *Suppose $\mathbf{q} \in \mathcal{R}^{-1}$. Then the map $H$ of Lemma 20.3.4 extends to a bounded operator $H : L^2_{\mathbf{q}}C_k(\Sigma) \to L^2_{\mathbf{q}}C_{k+1}(\Sigma)$.*

We will not give the proofs here.

## 20.4. WEIGHTED POINCARÉ DUALITY

When $W$ is type PM$^n$ (Definition 13.3.1), $\Sigma$ is an orientable pseudomanifold and so has a fundamental cycle (an infinite cycle in the top dimension $n$). This cycle is not square summable and so is never an ordinary $L^2$-chain (see the discussion near the end of Appendix J.4). However, the analogous weighted $L^2$-chain is square summable with respect to $\langle , \rangle_{\mathbf{q}}$ provided $\mathbf{q}$ is sufficiently large. To see this, first consider what it means for an $n$-chain $f$ to be a cycle with respect to $\partial^{\mathbf{q}}$. The coefficient in $f$ of each $n$-simplex of $K$ must be the same, say, 1. Similarly, all $n$-simplices in $wK$ must have the same coefficient. Let us determine it. First consider an $n$-simplex $\alpha$ of $sK$ which has a codimension one face contained in $K_s$. Since $\partial^{\mathbf{q}} f = 0$, the coefficient of $\alpha$ must be $q_s^{-1}$ (see (20.5)). So, the coefficent of each $n$-simplex in $sK$ is also $q_s^{-1}$. Continuing in this fashion we see that the coefficient of each $n$-simplex of $wK$ is $q_w^{-1}$. So, the norm of $f$ is $N \sum q_w^{-1}$, where $N$ is the number of $n$-simplices in $K$ and this norm is $< \infty$ if and only if $\mathbf{q} \in \mathcal{R}^{-1}$. It follows that there are nontrivial $n$-cycles in $\Sigma$ if and only if $\mathbf{q} \in \mathcal{R}^{-1}$ and when this is the case, the space of such $n$-cycles $(= L^2_{\mathbf{q}} H_n(\Sigma))$ is a 1-dimensional real vector space isomorphic to $H^S$ as an $\mathcal{N}_{\mathbf{q}}$-module. Hence, we have proved the following result, analogous to Proposition 20.3.1.

**PROPOSITION 20.4.1.** (Compare [108].) *Suppose $W$ is type* PM$^n$. *Then $L^2_{\mathbf{q}} H_n(\Sigma)$ is nonzero if and only if $\mathbf{q} \in \mathcal{R}^{-1}$. Moreover, when $\mathbf{q} \in \mathcal{R}^{-1}$,*

$$b^n_{\mathbf{q}}(\Sigma) = \frac{1}{W(\mathbf{q}^{-1})}.$$

For the remainder of this section, $W$ will be type HM$^n$ (Definition 10.6.2). Recall this means $\Sigma$ is a homology $n$-manifold, i.e., its nerve is a GHS$^{n-1}$ (a "generalized homology sphere," cf. Definition 10.4.5). Equivalently, it means that $K$ is a "generalized polytope," i.e., for each $T \in \mathcal{S}$, $(K_T, \partial K_T)$ is a GHD ("generalized homology disk") of codimension Card($T$) (see Lemma 10.8.1). For the purpose of computing (co)homology, the decomposition of $\Sigma$ into translates of the $K_T$ behaves exactly as if it were a decomposition into cells.

**THEOREM 20.4.2.** (Dymara [109, Theorem 6.1].) *Suppose $W$ is type* HM$^n$. *Then there is an isometry $\mathcal{D} : L^2_{\mathbf{q}} \mathcal{H}_i(\Sigma) \to L^2_{\mathbf{1/q}} \mathcal{H}^{n-i}(\Sigma)$, which is a $j$-equivariant isomorphism from an $\mathcal{N}_{\mathbf{q}}$-module to an $\mathcal{N}_{\mathbf{1/q}}$-module. Hence,*

$$b^i_{\mathbf{q}}(\Sigma) = b^{n-i}_{\mathbf{1/q}}(\Sigma).$$

(The $j$-isomorphism was defined in Chapter 19.)

The proof is along the same lines as the classical proof of Poincaré duality for orientable PL manifolds, which we now recall. Let $M^n$ be a PL manifold. For each $i$-cell $c$ in $M^n$, there is a dual $(n-i)$-cell $D_c \subset M^n$. (See Definition A.5.4 for the meaning of "dual cone; " in the case of a PL manifold, a dual cone is a cell.) One defines the duality isomorphism $\mathcal{D} : C_i(M^n) \to C^{n-i}(M^n)$ by sending $c$ to $\mathcal{D}(c)$ (the characteristic function of) the dual cell $D_c$. The reason this works is that $\mathcal{D}$ is a map of complexes. In fact, up to sign, it intertwines the boundary and coboundary maps: $\mathcal{D}(\partial c) = \pm\delta\mathcal{D}(c)$.

In the case at hand, the cells are the Coxeter polytopes of the form $wC_T$. The dual (generalized) cell to $wC_T$ is $wK_T$. The problem is that the standard duality: $wC_T \leftrightarrow wK_T$ interwines $\delta$ with the ordinary boundary map $\partial$ not with $\partial^{\mathbf{q}}$. What saves us is the $j$-isomorphism (defined by formula (19.3)) and the trick in Lemma 20.1.2.

*Proof of Theorem 20.4.2.* We have two cell structures on $\Sigma$: its cellulation by the Coxeter polytopes $wC_T$ and the dual "cellulation" by the generalized homology polytopes, $wK_T$. To distinguish them we use the notation $\Sigma_{\mathrm{cc}}$ in the first case and $\Sigma_{\mathrm{dual}}$ in the second. Next, we need to fix some orientation conventions. Orient $\Sigma$. Orient the $K_T$ arbitrarily subject only to the conditions that the orientation of $K$ agrees with that of $\Sigma$. (If $T \in \mathcal{S}^{(n)}$, then $K_T$ is a 0-cell so there is no choice.) Extend this equivariantly to orientations on the orbits of the $K_T$. The effect is to give an equality of incidence numbers: $[wK_U : wK_T] = [K_U : K_T]$, for all $w \in W$. We give the Coxeter cells the dual orientations: if $u$ is $(\emptyset, T)$-reduced, then $uC_T$ is oriented dually to $uK_T$. This means that the orientation on $uC_T$ followed by the orientation on $uK_T$ gives the orientation on $\Sigma$. (If $w = uv$, with $v \in W_T$, then the orientation on $wC_T$ is $\varepsilon_v$ times the orientation on $uC_T$.)

Define $\mathcal{D} : L^2_{\mathbf{q}}C_i(\Sigma_{\mathrm{cc}}) \to L^2_{\mathbf{1/q}}C^{n-i}(\Sigma_{\mathrm{dual}})$ by

$$\mathcal{D}f(wK_T) := \varepsilon_u q_u f(wC_T), \tag{20.21}$$

where $\mathrm{Card}(T) = i$ and $u$ is the $(\emptyset, T)$-reduced element of $wW_T$. Use $c$ and $d$ to denote generic Coxeter cells of dimension $i$ and $(i-1)$, respectively, and let $D_c$, $D_d$ be their dual cells. If $c = wC_T$, then, as in 20.1, use the notation $\mu_{\mathbf{q}}(c) = q_u$ and $\varepsilon_{\mathbf{q}}(c) := \varepsilon_u = (-1)^{l(u)}$. Formula (20.21) becomes $\mathcal{D}f(D_c) = \varepsilon_{\mathbf{q}}(c)\mu_{\mathbf{q}}(c)f(c)$. We have the usual formula for $\delta$ and formula (20.13) for $\partial^{\mathbf{q}}$:

$$\delta(g)(D_d) = \sum_{D_c \subset D_d} [D_c : D_d] g(D_c)$$

$$\partial^{\mathbf{q}}(f)(d) = \sum_{c \supset d} [d : c]\varepsilon_{\mathbf{q}}(c)\varepsilon_{\mathbf{q}}(d)\mu_{\mathbf{q}}(c)\mu_{\mathbf{q}}(d)^{-1}f(c).$$

Let us check that $\mathcal{D}$ intertwines $\delta$ and $\partial^{\mathbf{q}}$. The key point is that the incidence numbers satisfy, $[D_c : D_d] = \pm[d : c]$, where the sign depends only on $i$ and $n$.

Therefore,

$$\delta(\mathcal{D}f)(D_d) = \sum_{c \supset d} [D_c : D_d](\mathcal{D}f)(c) = \pm \sum_{c \supset d} [d : c]\varepsilon_{\mathbf{q}}(c)\mu_{\mathbf{q}}(c)f(c),$$

while

$$\begin{aligned}\mathcal{D}(\partial^{\mathbf{q}}f)(D_d) &= \varepsilon_{\mathbf{q}}(d)\mu_{\mathbf{q}}(d)(\partial^{\mathbf{q}}f)(d)\\ &= \varepsilon_{\mathbf{q}}(d)\mu_{\mathbf{q}}(d)\sum_{c \supset d}[d : c]\varepsilon_{\mathbf{q}}(c)\varepsilon_{\mathbf{q}}(d)\mu_{\mathbf{q}}(c)\mu_{\mathbf{q}}(d)^{-1}f(c)\\ &= \pm\delta(\mathcal{D}f)(D_d),\end{aligned}$$

as claimed.

Next we check that $\mathcal{D}$ takes the $\mathcal{N}_{\mathbf{q}}$-module $L^2_{\mathbf{q}}C_i(\Sigma_{\mathrm{cc}})$ isometrically onto the $\mathcal{N}_{\mathbf{1/q}}$-module $L^2_{\mathbf{1/q}}C^{n-i}(\Sigma_{\mathrm{dual}})$. Since $\mathcal{D}$ clearly respects the decompositions into Coxeter cells and their duals, this amounts to checking that the following diagram commutes:

$$\begin{array}{ccc} L^2(WC_T, \mu_{\mathbf{q}}) & \xrightarrow{\psi_T} & H^{\mathbf{q}}_T \\ \downarrow{\scriptstyle \mathcal{D}} & & \downarrow{\scriptstyle j} \\ L^2(WK_T, \mu_{\mathbf{1/q}}) & \xrightarrow{\varphi_T} & A^{\mathbf{1/q}}_T. \end{array}$$

Here $\psi_T$ and $\varphi_T$ are as in 20.1:

$$\psi_T(f) := \sqrt{W_T(\mathbf{q}^{-1})}\sum_{u \in B_T} f(uC_T)e_u h_T,$$

$$\varphi_T(g) := \sqrt{W_T(\mathbf{q}^{-1})}\sum_{u \in B_T} g(uK_T)e_u a_T.$$

The sums are over the set $B_T$ of $(\emptyset, T)$-reduced elements. Compute:

$$\begin{aligned}\varphi_T(\mathcal{D}f) &= \sqrt{W_T(\mathbf{q}^{-1})}\sum \mathcal{D}f(uK_T)e_u a_T^{\mathbf{q}^{-1}}\\ &= \sqrt{W_T(\mathbf{q}^{-1})}\sum \varepsilon_u q_u f(uC_T)e_u a_T^{\mathbf{q}^{-1}}\\ &= \sqrt{W_T(\mathbf{q}^{-1})}\sum f(uC_T)j(e_u)j(h^{\mathbf{q}}_T)\\ &= j(\psi_T(f)),\end{aligned}$$

where again the sums range over $u \in B_T$. So, the diagram commutes. □

We can use Theorem 20.4.2 to calculate the $L^2_{\mathbf{q}}$-Euler characteristic of $\Sigma$ in two different ways. This gives the following alternative proof of

Corollary 17.3.5 (which states that when $W$ is type HM$^n$ its growth series is $(-1)^n$-reciprocal).

*Proof.* Writing $b^i_{\mathbf{q}}$ for $b^i_{\mathbf{q}}(\Sigma)$, we have:

$$\frac{1}{W(\mathbf{q})} = \sum(-1)^i b^i_{\mathbf{q}} = \sum(-1)^i b^{n-i}_{\mathbf{1/q}} = (-1)^n \sum(-1)^i b^i_{\mathbf{1/q}}$$
$$= \frac{(-1)^n}{W(\mathbf{q}^{-1})} .$$

□

Combining Theorems 20.3.3 and 20.4.2, we get the following.

**COROLLARY 20.4.3.** *Suppose $W$ is type* HM$^n$ *and $\mathbf{q} \in \mathcal{R}^{-1}$. Then $L^2_{\mathbf{q}}H_*(\Sigma)$ is concentrated in dimension $n$.*

It follows from the continuity of the weighted Betti numbers that $L^2_{\mathbf{q}}\mathcal{H}^*(\Sigma)$ vanishes identically for $\mathbf{q} \in \overline{\mathcal{R}} - \mathcal{R}$ or $\mathbf{q} \in \overline{\mathcal{R}^{-1}} - \mathcal{R}^{-1}$.

Write $\mathbf{q} \geqslant \mathbf{1}$ to mean each $q_s$ is $\geqslant 1$.

**COROLLARY 20.4.4.** (Affine groups). *Suppose $W$ is a Euclidean reflection group (i.e., an affine Coxeter group) and $\mathbf{q} \geqslant \mathbf{1}$. Then $L^2_{\mathbf{q}}\mathcal{H}^*(\Sigma)$ is concentrated in the top dimension.*

*Proof.* Let $W(t)$ be the growth series in a single indeterminate $t$. Its radius of convergence is 1 (Remark 17.2). It follows that $\mathcal{R} \supset \{\mathbf{q} \mid \mathbf{q} < \mathbf{1}\}$ and that $\overline{\mathcal{R}^{-1}} \supset \{\mathbf{q} \mid \mathbf{q} \geqslant \mathbf{1}\}$. □

**COROLLARY 20.4.5.** (Polygon groups.) *Suppose $W$ is a nonspherical polygon group (see 6.5). Then $L^2_{\mathbf{q}}\mathcal{H}^*_{\mathbf{q}}(\Sigma)$ is concentrated in dimension*

$$\begin{cases} 0 & \textit{if } \mathbf{q} \in \mathcal{R}, \\ 1 & \textit{if } \mathbf{q} \notin \mathcal{R} \cup \mathcal{R}^{-1}, \\ 2 & \textit{if } \mathbf{q} \in \mathcal{R}^{-1}. \end{cases}$$

*So, whenever $b^i_{\mathbf{q}}(\Sigma)$ is nonzero it is equal to $|1/W(\mathbf{q})|$, the absolute value of the $L^2_{\mathbf{q}}$-Euler characteristic.*

*Proof.* If $\mathbf{q} \notin \mathcal{R} \cup \mathcal{R}^{-1}$, then, by Propositions 20.3.1 and 20.4.1, $L^2_{\mathbf{q}}\mathcal{H}^i(\Sigma) = 0$ for $i = 0$ or 2; hence, it is concentrated in dimension 1. □

**Example 20.4.6.** (*Right-angled polygon groups.*) To get a better feel for what is going on in Corollary 20.4.5, let us return to Example 17.1.15: $W$ is right angled with nerve a $k$-gon, $k \geqslant 4$, and $\mathbf{q}$ is a singleton $q$. Then $\chi_q = 1/W(q) = (q^2 - (k-2)q + 1)/(1+q)^2$, which has roots

$$\rho^{\pm 1} = \frac{(k-2) \mp \sqrt{k^2 - 4k}}{2}.$$

$1/W(q)$ is positive on the intervals $[0, \rho)$ and $(\rho^{-1}, \infty)$ and negative on $(\rho, \rho^{-1})$. (This example also should be compared with Example J.5.4 in Appendix J.5.)

## 20.5. A WEIGHTED VERSION OF THE SINGER CONJECTURE

The Singer Conjecture ([102]) asserts that the ordinary reduced $L^2$-cohomology of the universal cover of any even-dimensional, closed, aspherical manifold vanishes except in the middle dimension and in the odd-dimensional case, that it vanishes in all dimensions. It is explained in Proposition J.7.2 of Appendix J.7 how this conjecture implies the Euler Characteristic Conjecture of 16.2. The Singer Conjecture for Coxeter groups is the following.

**CONJECTURE 20.5.1.** ([91].) *Suppose $W$ is a Coxeter group of type* $\mathrm{HM}^n$. *Then*

$$b^i_{\mathbf{1}}(\Sigma) = 0, \quad \textit{for all } i \neq \frac{n}{2},$$

*where $b^i_{\mathbf{1}}(\Sigma)$ denotes the ordinary $L^2$-Betti number of $\Sigma$ with respect to $W$.*

For $n$ odd, this means $b^i_{\mathbf{1}}(\Sigma)$ should equal 0 for all $i$.

In [91] B. Okun and I described a program (not yet successfully completed) for proving Conjecture 20.5.1 by induction on dimension in the case where $W$ is right-angled. We did succeed in proving the following.

**THEOREM 20.5.2.** ([91, Theorem 11.1.1].) *Conjecture 20.5.1 is true for $n \leqslant 4$ when $W$ is right-angled.*

Actually, this was proved under the weaker assumption that $W$ is "rationally type $\mathrm{HM}^n$," meaning that the nerve $L$ is a $\mathrm{GHS}^{n-1}$ with rational coefficients. (This guarantees Poincaré duality holds for $L^2_{\mathbf{q}}\mathcal{H}^*(\Sigma)$.) In outline, the proof goes as follows. Since $W$ is type $\mathrm{HM}^n$, $L$ is a $\mathrm{GHS}^{n-1}$. For $n \leqslant 3$ this means $L$ is an actual $(n-1)$-sphere. For $n = 1$, the only possibility is $\Sigma = \mathbb{E}^1$ and since the reduced $L^2$-cohomology of Euclidean space vanishes identically, the conjecture holds. For $n = 2$, we proved it as Corollary 20.4.5. Dodziuk [102] showed that the reduced $L^2$-cohomology of hyperbolic space $\mathbb{H}^n$ vanishes for $i \neq \frac{n}{2}$ (Theorem J.8.7). So, if $W$ is a cocompact, geometric reflection group on $\mathbb{H}^3$, the reduced $L^2$-cohomology of $\Sigma$ vanishes. Hence, if $L$ is a flag triangulation of $S^2$ satisfing the conditions of Andreev's Theorem in 6.10, the conjecture holds. (In the right-angled case, Andreev's Conditions amount to the requirement that $L$ has no empty 4-circuits and that it is not the suspension of a 4- or 5-gon.) In the general three-dimensional case, one argues that $L$ can be decomposed as a connected sum, at vertices of valence 4, into pieces each satisfying Andreev's Conditions. One can then use the Mayer-Vietoris sequence (from Appendix J.4) to conclude the conjecture holds for $n = 3$. (Alternatively, the conjecture in dimension 3 follows from the result of Lott–Lück [189].) The inductive arguments of [91] show the case $n = 3$ implies the case $n = 4$.

A corollary of Theorem 20.5.2 is the Flag Complex Conjecture in dimension 3.

**COROLLARY 20.5.3.** ([91, Theorem 11.2.1].) *The Flag Complex Conjecture of 16.3 holds for flag triangulations of (rational) homology 3-spheres.*

*Proof.* If $L$ is a flag triangulation of a rational homology 3-sphere and $W$ the assoiated right–angled Coxeter group, then $\chi(W) = \chi^{orb}(\Sigma /\!/ W)$ is the number $f(-\frac{1}{2})$ in the Flag Complex Conjecture (by Example 16.2.9). As observed in the proof of Proposition J.7.2, it follows from Atiyah's Formula (20.19) (or Theorem J.5.3) that $\chi^{orb}(\Sigma /\!/ W) = \chi_{\mathbf{1}}(\Sigma)$. By Theorem 20.5.2, this number is $b^2_{\mathbf{1}}(\Sigma)$ which is $\geqslant 0$. □

By Proposition 16.3.2, this implies the following.

**COROLLARY 20.5.4.** *The Euler Characteristic Conjecture holds for any (rational) homology 4-manifold $M^4$ with a nonpositively curved cubical cell structure. (In other words, for any such $M^4$, $\chi(M^4) \geqslant 0$.)*

For general $\mathbf{q}$, the appropriate generalization of Conjecture 20.5.1 is the following ([79, Conj. 14.7]).

**CONJECTURE 20.5.5.** (The Weighted Singer Conjecture.) *Suppose $W$ is type* HM$^n$. *If $\mathbf{q} \leqslant \mathbf{1}$ and $k > \frac{n}{2}$, then $b^k_{\mathbf{q}}(\Sigma) = 0$.*

By weighted Poincaré duality (Theorem 20.4.2), this is equivalent to the conjecture that if $\mathbf{q} \geqslant \mathbf{1}$ and $k < \frac{n}{2}$, then $b^k_{\mathbf{q}}(\Sigma) = 0$. In [79] we improved the proof in [91] of Theorem 20.5.2 to get the corresponding result for arbitrary $\mathbf{q}$, which we state below.

**THEOREM 20.5.6.** ([79, Theorem 16.13].) *In the right-angled case, the Weighted Singer Conjecture (20.5.5) is true for $n \leqslant 4$.*

Assume, for simplicity, $\mathbf{q} = q$, a single indeterminate. By Theorem 17.3.4, the roots of $\chi_q$ $(= 1/W(q))$ are symmetric about 1, i.e., if $q$ is a root, then so is $q^{-1}$.

**Example 20.5.7.** (*Right-angled 3-dimensional polytope groups.*) We contine Example 17.4.3: $L$ is a flag triangulation of $S^2$ and $(W, S)$ is the associated right-angled Coxeter system. In 17.4.3 we computed

$$\frac{1}{W(q)} = \frac{-(q-1)(q^2 - (f_0 - 4)q + 1)}{(1+q)^3},$$

which has three positive real roots, $\rho$, 1, and $\rho^{-1}$ ($\rho^{\pm 1}$ also is computed in 17.4.3). So $\chi_q$ is positive on the intervals $[0, \rho)$, $(1, \rho^{-1})$ and negative on $(\rho, 1)$, $(\rho^{-1}, \infty)$. By Theorem 20.3.3, on $(0, \rho)$, $L^2_{\mathbf{q}}\mathcal{H}^*(\Sigma)$ is concentrated in dimension 0 and on $(\rho^{-1}, \infty)$ it is concentrated in dimension 3. By Proposition 20.3.1 and weighted Poincaré duality, it vanishes in the intermediate range, $q \in (\rho, \rho^{-1})$, in dimensions 0 and 3. Using Theorem 20.5.6 we see that on $(\rho, 1)$ it is concentrated in dimension 1 and on $(1, \rho^{-1})$ it is concentrated in dimension 2.

At one point, the following scenario seemed plausible:

(a) $\chi_q$ has exactly $n$ positive real roots (counted with multiplicity) and

(b) $L^2_q\mathcal{H}_*(\Sigma)$ is always concentrated in a single dimension. The dimension jumps each time $q$ passes a root and the size of the jump is the multiplicity of the root.

In fact, both (a) and (b) are false. Gal [134] gave counterexamples to (a) in dimensions $\geqslant 6$. Counterexamples to (b) are given in [79, Section 17] in dimensions $\geqslant 4$.

## 20.6. DECOMPOSITION THEOREMS

In 15.4, in the case where $W$ is finite, we gave two direct sum decompositions of $\mathbb{R}W$ (in formulas (15.29) and equivalent versions in Solomon's Theorem 15.4.2). The goal of this section is to establish analogous decompositions for $L^2_{\mathbf{q}}$.

For each $T \subset S$, define

$$A^{>T} := \overline{\sum_{U \supsetneq T} A^U} \quad \text{and} \quad H^{>T} := \overline{\sum_{U \supsetneq T} H^U}.$$

Put $D_T := A^T \cap (A^{>T})^{\perp}$ and $G_T := H^T \cap (H^{>T})^{\perp}$. Call a subset $U$ of $S$ *cospherical* if $S - U$ is spherical. The decompositon theorems we are aiming for are the following.

**THEOREM 20.6.1.** (First Decomposition Theorem.) *If* $\mathbf{q} \in \overline{\mathcal{R}} \cup \overline{\mathcal{R}^{-1}}$, *then*

$$\sum_{T \subset S} D_T$$

*is a direct sum and a dense subspace of* $L^2_{\mathbf{q}}$. *Moreover, if* $\mathbf{q} \in \overline{\mathcal{R}}$, *the only nonzero terms in the sum are those with* $T$ *cospherical (i.e., with* $S - T \in \mathcal{S}$*) and if* $\mathbf{q} \in \overline{\mathcal{R}^{-1}}$, *the only nonzero terms in the sum are those with* $T$ *spherical.*

**THEOREM 20.6.2.** (Second Decomposition Theorem.) *If* $\mathbf{q} \in \overline{\mathcal{R}} \cup \overline{\mathcal{R}^{-1}}$, *then*

$$\sum_{T \subset S} G_T$$

*is a direct sum and a dense subspace of* $L^2_{\mathbf{q}}$. *Moreover, if* $\mathbf{q} \in \overline{\mathcal{R}}$, *the only nonzero terms in the sum are those with* $T$ *spherical, and if* $\mathbf{q} \in \overline{\mathcal{R}^{-1}}$, *the only nonzero terms in the sum are those with* $T$ *cospherical.*

(Note that the roles of spherical and cospherical are reversed in the two theorems.) When $W$ is finite and $\mathbf{q} = \mathbf{1}$, the theorems reduce to the formulas (15.29) in Section 15.4.

Before beginning work in earnest, we prove the following lemma which states that in both cases the sums are dense in $L^2_{\mathbf{q}}$.

**LEMMA 20.6.3**

$$A^T = \overline{\sum_{U \supset T} D_U} \quad \textit{and} \quad H^T = \overline{\sum_{U \supset T} G_U}.$$

*Proof.* By definition of $D_T$,

$$A^T := D_T + A^{>T} = D_T + \overline{\sum_{U \supsetneq T} A^U},$$

and the first formula follows by induction on the size of $S - T$. Similarly, for the second formula. □

The issue is to show the sums in Theorems 20.6.1 and 20.6.2 are direct. What made the argument tractable for a finite $W$ was the fact that we had a partition of a basis for $\mathbb{R}W$ such that each subset of the partition projected to a basis for a corresponding subspace in the direct sum decomposition. In essence, the proof was a dimension count. Although this argument is not available when $W$ is infinite, we can give an analogous dimension count using the von Neumann dimension. In rough outline the proof has three steps:

(1) For $\mathbf{q} \in \overline{\mathcal{R}}$, we compute $\dim_{\mathcal{N}_{\mathbf{q}}}(G_T)$ and show $\sum \dim_{\mathcal{N}_{\mathbf{q}}}(G_T) = 1$ $(= \dim_{\mathcal{N}_{\mathbf{q}}} L^2_{\mathbf{q}})$. (This implies the Second Decomposition Theorem for $\mathbf{q} \in \overline{\mathcal{R}}$.)

(2) For $\mathbf{q} \in \overline{\mathcal{R}}$, we show that $D_{S-T}$ is the closure of $G_T a_{S-T}$. From this we get the First Decomposition Theorem for $\mathbf{q} \in \overline{\mathcal{R}}$.
(3) The $j$-isomorphism (formula (19.3) in 19.1) switches $G_T$ and $D_T$. Applying it to (1) and (2) we get that both theorems also hold for $\mathbf{q} \in \overline{\mathcal{R}^{-1}}$.

The difficult step is (1). Most of the work comes down to showing that a certain chain complex of Hilbert $\mathcal{N}_{\mathbf{q}}$-modules, defined below, is acyclic. Fix a spherical subset $T \in \mathcal{S}^{(k)}$ (i.e., $\mathrm{Card}(T) = k$). Put

$$C_i(H^T) := \bigoplus_{U \in (\mathcal{S}_{\geqslant T})^{(i+k)}} H^U \tag{20.22}$$

and abbreviate it $C_i$. Whenever $U \subset V$, we have an inclusion $\iota^U_V : H^V \hookrightarrow H^U$. Fix some ordering of $\{s \in S - T \mid T \cup \{s\} \in \mathcal{S}\}$. The boundary map, $\partial : C_{i+1} \to C_i$ corresponds to a matrix $(\partial_{UV})$, where $\partial_{UV} = 0$ unless $U \subset V$ and is equal to $(-1)^j \iota^U_V$ if $U$ is obtained by deleting the $j^{\text{th}}$ element of $V$. This gives a chain complex of Hilbert $\mathcal{N}_{\mathbf{q}}$-modules which starts out like this:

$$0 \longleftarrow H^T \longleftarrow \bigoplus_{(T\cup\{s\}) \in (\mathcal{S}_{\geqslant T})^{(k+1)}} H_{T\cup\{s\}} \longleftarrow \cdots . \tag{20.23}$$

**THEOREM 20.6.4.** *For any $T \in \mathcal{S}$ and $\mathbf{q} \in \mathcal{R}$, $H_*(C_*(H^T))$ is concentrated in dimension 0. Therefore (by continuity of weighted $L^2$-Betti numbers), for $\mathbf{q} \in \partial\mathcal{R}$ $(:= \overline{\mathcal{R}} - \mathcal{R})$, the reduced homology, $\mathcal{H}_*(C_*(H^T))$, is also concentrated in dimension 0.*

This means that, for $\mathbf{q} \in \overline{\mathcal{R}}$, the family of subspaces $(H^T)_{T\in\mathcal{S}}$ is "in general position" in $L^2_{\mathbf{q}}$.

The proof is algebraic topological: we identify $C_*(H^T)$ with the relative $L^2_{\mathbf{q}}$-chains on a certain pair of subcomplexes of $\Sigma$ and then use an inductive argument starting from Lemma 20.3.4 to show its only nonvanishing weighted $L^2$-Betti number is in the lowest possible dimension. The details of the proof will be postponed until the end of this section.

Assuming Theorem 20.6.4, let us continue with the proof of Step (1). If $\mathbf{q} \in \mathcal{R}$, then $H^T = 0$ for all nonspherical $T$ (because $T$ is spherical whenever $\mathcal{R}_T \cap \mathcal{R}_T^{-1} \neq \emptyset$). So, for $T \notin \mathcal{S}$, $G_T = 0$ (since it is a submodule of $H^T$). For $T \in \mathcal{S}$, $G_T$ is the orthogonal complement of the image of $\partial : C_1(H^T) \to C_0(H^T) = H^T$; hence, $\mathcal{H}_0(H^T) = G_T$.

Denote by $R(\mathcal{N}_{\mathbf{q}})$ the Grothendieck group of Hilbert $\mathcal{N}_{\mathbf{q}}$-modules. If $F$ is such a Hilbert module, $[F]$ denotes its class in $R(\mathcal{N}_{\mathbf{q}})$. By additivity of dimension, the function $F \to \dim_{\mathcal{N}_{\mathbf{q}}} F$ induces a homomorphism $\dim_{\mathcal{N}_{\mathbf{q}}} : R(\mathcal{N}_{\mathbf{q}}) \to \mathbb{R}$. We derive some corollaries of Theorem 20.6.4.

**COROLLARY 20.6.5.** *For* $\mathbf{q} \in \overline{\mathcal{R}}$ *and* $T \in \mathcal{S}$*, the following formulas hold in* $R(\mathcal{N}_{\mathbf{q}})$*:*

$$[G_T] = \sum_{U \in \mathcal{S}_{\geqslant T}} \varepsilon(U - T)[H^U],$$

$$[H^T] = \sum_{U \in \mathcal{S}_{\geqslant T}} [G_U].$$

*Proof.* The boundary maps in $C_*(H^T)$ are maps of Hilbert $\mathcal{N}_{\mathbf{q}}$-modules. So, the first formula follows from Theorem 20.6.4 by taking the Euler characteristics. The second formula follows from this and Möbius Inversion (see Lemma 17.1.3). □

**COROLLARY 20.6.6.** *Suppose* $\mathbf{q} \in \overline{\mathcal{R}}$ *and* $T \in \mathcal{S}$*. Then* $\dim_{\mathcal{N}_{\mathbf{q}}} G_T = W^T(\mathbf{q})/W(\mathbf{q})$.

*Proof.* By Lemma 19.2.13 (ii), $\dim_{\mathcal{N}_{\mathbf{q}}} H^U = 1/W_U(\mathbf{q}^{-1})$. So

$$\dim_{\mathcal{N}_{\mathbf{q}}} G_T = \sum_{U \in \mathcal{S}_{\geqslant T}} \frac{\varepsilon(U - T)}{W_U(\mathbf{q}^{-1})} = \frac{W^T(\mathbf{q})}{W(\mathbf{q})},$$

where the first equality comes from the previous corollary and the second from Lemma 17.1.8. □

**COROLLARY 20.6.7.** *The Second Decomposition Theorem holds for* $\mathbf{q} \in \overline{\mathcal{R}}$*:*

$$\sum_{T \subset S} G_T$$

*is a direct sum and a dense subspace of* $L^2_{\mathbf{q}}$*. Moreover,* $G_T = 0$ *whenever* $T$ *is not spherical.*

*Proof.* Since $G_T \subset H^T$, the last sentence follows from Lemma 19.2.13. Consider the natural epimorphism

$$f : \bigoplus_{T \in \mathcal{S}} G_T \to \sum_{T \in \mathcal{S}} G_T,$$

where $\bigoplus$ means the external direct sum and where, as usual, $\sum$ means the internal sum. The statement that the internal sum is direct means that $f$ is injective. Since $(W^T)_{T \in \mathcal{S}}$ is a partition of $W$, the dimension of the domain of $f$ is

$$\sum_{T \in \mathcal{S}} \dim_{\mathcal{N}_{\mathbf{q}}} G_T = \sum_{T \in \mathcal{S}} \frac{W^T(\mathbf{q})}{W(\mathbf{q})} = 1.$$

So $\dim_{\mathcal{N}_{\mathbf{q}}}(\mathrm{Ker} f)$ is $1 - 1 = 0$; hence, $f$ is injective. □

**NOTATION.** *Use the symbol* $\uplus$ *to denote an internal sum of submodules of* $L^2_{\mathbf{q}}$, *which happens to be a direct sum.*

The fact that the sum of $G_T$ in Corollary 20.6.7 is direct has the following two consequences.

**COROLLARY 20.6.8.** *Let* $\mathcal{A}$ *and* $\mathcal{B}$ *be collections of subsets of* $S$. *If* $\mathbf{q} \in \overline{\mathcal{R}}$, *then*

$$\overline{\biguplus_{T \in \mathcal{A}} G_T} \cap \overline{\biguplus_{T \in \mathcal{B}} G_T} = \overline{\biguplus_{T \in \mathcal{A} \cap \mathcal{B}} G_T}.$$

**COROLLARY 20.6.9.** *If* $\mathbf{q} \in \overline{\mathcal{R}}$ *and* $T \subset S$, *then*

$$H^T = \overline{\biguplus_{U \supset T} G_U}.$$

We turn to the proof of Step (2).

**LEMMA 20.6.10.** (Compare [261, Lemma 1].) *Suppose* $T, U$ *are subsets of* $S$ *and* $\mathbf{q} \in \mathcal{R}_U \cap \mathcal{R}_T^{-1}$ *(so that* $h_T$ *and* $a_U$ *are defined). If* $T \cap U \neq \emptyset$, *then* $h_T a_U = 0$.

*Proof.* Let $s \in T \cap U$. Then $h_T a_U = h_T h_s a_s a_U = 0$ (since $h_s a_s = 0$ by Lemma 19.2.11). □

**LEMMA 20.6.11.** *Suppose* $\mathbf{q} \in \mathcal{R}$ *and* $T \not\subset U$. *Then*

$$G_T a_{S-U} = 0.$$

*Proof.* Since $G_T \subset H^T$, the assertion follows from the previous lemma. □

**LEMMA 20.6.12.** *If* $\mathbf{q} \in \overline{\mathcal{R}}$ *and* $U \subset S$, *then*

$$\sum_{\substack{T \in \mathcal{S} \\ T \subset U}} G_T a_{S-U}$$

*is a dense subspace of* $A^{S-U}$ *and a direct sum. Moreover, if* $T \in \mathcal{S}$, *then right multiplication by* $a_{S-T}$ *induces a weak isomorphism* $G_T \to G_T a_{S-T}$.

*Proof.* As usual, $B_{S-U}$ is the set of $(\emptyset, S-U)$-reduced elements in $W$. By formula (17.5) in 17.1,

$$B_{S-U}(\mathbf{q}) = \sum_{T \subset U} W^T(\mathbf{q}).$$

Dividing by $W(\mathbf{q})$ and using Lemma 17.1.2, we get

$$\frac{1}{W_{S-U}(\mathbf{q})} = \sum_{T \subset U} \frac{W^T(\mathbf{q})}{W(\mathbf{q})}.$$

Applying Lemma 20.6.3 to $A_\emptyset = L^2_\mathbf{q}$ yields

$$L^2_\mathbf{q} = \overline{\sum_{T \in \mathcal{S}} G_T}.$$

Multiplying on the right by $a_{S-U}$ and using Lemma 20.6.11, we obtain

$$A^{S-U} = \overline{\sum_{T \in \mathcal{S}} G_T a_{S-U}} = \overline{\sum_{\substack{T \in \mathcal{S} \\ T \subset U}} G_T a_{S-U}}.$$

So

$$\begin{aligned} \dim_{\mathcal{N}_\mathbf{q}} A^{S-U} &\leqslant \sum_{\substack{T \in \mathcal{S} \\ T \subset U}} \dim_{\mathcal{N}_\mathbf{q}} G_T a_{S-U} \leqslant \sum_{\substack{T \in \mathcal{S} \\ T \subset U}} \dim_{\mathcal{N}_\mathbf{q}} G_T = \sum_{T \subset U} \frac{W^T(\mathbf{q})}{W(\mathbf{q})} \\ &= \frac{1}{W_{S-U}(\mathbf{q})} = \dim_{\mathcal{N}_\mathbf{q}} A^{S-U}, \end{aligned}$$

where the last equality is from Lemma 19.2.12 (ii). Hence, both inequalities are equalities and

$$\dim_{\mathcal{N}_\mathbf{q}} \overline{G_T a_{S-T}} = \dim_{\mathcal{N}_\mathbf{q}} G_T.$$

It follows that right multiplication by $a_{S-T}$ is a weak isomorphism from $G_T$ to $G_T a_{S-T}$ and that the sum is direct. □

**LEMMA 20.6.13.** *Suppose $\mathbf{q} \in \overline{\mathcal{R}}$ and $T \subset S$. Then $D_{S-T} = \overline{G_T a_{S-T}}$. In particular, $D_{S-T} = 0$ if $T \notin \mathcal{S}$.*

*Proof.* Since, by definition, $D_{S-T} \subset A^{S-T}$, we have $D_{S-T} = D_{S-T} a_{S-T}$ and since $D_{S-T} \subset (A^{>(S-T)})^\perp$, we have

$$D_{S-T} \subset (A^{>(S-T)})^\perp a_{S-T}.$$

Using equations (19.19), we compute

$$(A^{>(S-T)})^\perp = \bigcap_{U \subsetneq T} (A^{S-U})^\perp = \bigcap_{U \subsetneq T} \overline{\sum_{s \in S-U} H^s}.$$

By Corollary 20.6.9, $H^s = \overline{\biguplus_{V \ni s} G_V}$. Therefore,

$$(A^{>(S-T)})^\perp = \bigcap_{U \subsetneq T} \overline{\sum_{s \in (S-U)} \biguplus_{V \ni s} G_V} = \bigcap_{U \subsetneq T} \overline{\biguplus_{V \subset U} G_V}.$$

Using Corollary 20.6.8,

$$(A^{>(S-T)})^{\perp} = \overline{\biguplus_{V \not\subset U\ \forall U \subsetneq T} G_V} = \overline{G_T \uplus \biguplus_{V \not\subset T} G_V}.$$

Thus,

$$D_{S-T} \subset \overline{\Big(G_T \uplus \biguplus_{V \not\subset T} G_V\Big) a_{S-T}} = \overline{G_T a_{S-T} + \sum_{V \not\subset T} G_V a_{S-T}}.$$

By Lemma 20.6.11, the only nonzero term on the far right is $G_T a_{S-T}$. Therefore, $D_{S-T} \subset \overline{G_T a_{S-T}}$.

For the opposite inclusion, note that, for $U \subsetneq T$, $G_T a_{S-T} a_{S-U} = G_T a_{S-U} = 0$ (by Lemma 20.6.11). Since $\operatorname{Ker} a_{S-U} = (A^{S-U})^{\perp}$, we have, $G_T a_{S-T} \subset (A^{S-U})^{\perp}$ for all $U \subsetneq T$. Since $G_T a_{S-T} \subset A^{S-T}$, it follows from the definition of $D_{S-T}$ that $G_T a_{S-T} \subset D_{S-T}$. □

We now complete Step (2).

**COROLLARY 20.6.14.** *The First Decomposition Theorem holds for* $\mathbf{q} \in \overline{\mathcal{R}}$*:*

$$\sum_{T \subset S} D_{S-T}$$

*is a direct sum and a dense subspace of* $L^2_{\mathbf{q}}$*. Moreover,* $D_{S-T} = 0$ *whenever* $T$ *is not spherical (and* $\mathbf{q} \in \overline{\mathcal{R}}$*).*

*Proof.* The last sentence is from Lemma 20.6.13. By Lemmas 20.6.13 and 20.6.12 and the proof of Corollary 20.6.7, the dimensions of the nontrivial terms sum to 1. □

The final step is now a triviality.

*Proofs of Theorems 20.6.1 and 20.6.2.* The $j$-isomorphism switches $\mathbf{q}$ with $\mathbf{q}^{-1}$, $H^T$ with $A^T$, and $G_T$ with $D_T$. So the First Decomposition Theorem for $\mathbf{q} \in \overline{\mathcal{R}^{-1}}$ follows from the Second for $\mathbf{q} \in \overline{\mathcal{R}}$ and vice versa. □

We state some further corollaries of the First Decomposition Theorem.

**COROLLARY 20.6.15.** *Let* $\mathcal{A}$ *be a collection of subsets of* $S$ *and* $U \subset S$ *a given subset. If* $\mathbf{q} \in \overline{\mathcal{R}} \cup \overline{\mathcal{R}^{-1}}$*, then*

$$D_U \cap \overline{\biguplus_{U \in \mathcal{A}} D_V} = \begin{cases} 0 & \text{if } U \notin \mathcal{A}, \\ D_U & \text{if } U \in \mathcal{A}. \end{cases}$$

**COROLLARY 20.6.16.** *If* $\mathbf{q} \in \overline{\mathcal{R}} \cup \overline{\mathcal{R}^{-1}}$ *and* $U \subset S$, *then*

$$A^U = \overline{\biguplus_{V \subset U} D_V}.$$

**COROLLARY 20.6.17.** (Compare Lemmas 17.1.4 and 17.1.8 and Corollary 20.6.6.) *Suppose* $T \in \mathcal{S}$.

*(i) For* $\mathbf{q} \in \mathcal{R}$, $\dim_{\mathcal{N}_\mathbf{q}} D_{S-T} = W^T(\mathbf{q})/W(\mathbf{q})$.

*(ii) For* $\mathbf{q} \in \mathcal{R}^{-1}$, $\dim_{\mathcal{N}_\mathbf{q}} D_T = W^T(\mathbf{q}^{-1})/W(\mathbf{q}^{-1})$.

*Proof*

(i) By Lemma 20.6.13, $a_{S-T}$ maps $G_T$ monomorphically onto a dense subspace of $D_{S-T}$. So, $\dim_{\mathcal{N}_\mathbf{q}} D_{S-T} = \dim_{\mathcal{N}_\mathbf{q}} G_T = W^T(\mathbf{q})/W(\mathbf{q})$, where the second equality is Corollary 20.6.6.

(ii) For $\mathbf{q} \in \mathcal{R}^{-1}$, the following formulas hold in $R(\mathcal{N}_\mathbf{q})$,

$$[A^T] = \sum_{U \in \mathcal{S}_{\geqslant T}} [D_U],$$

$$[D_T] = \sum_{U \in \mathcal{S}_{\geqslant T}} \varepsilon(U-T)[A^U],$$

where the first formula is from Corollary 20.6.16 and the second follows from the first by the Möbius Inversion Formula. So, as in Corollary 20.6.6,

$$\dim_{\mathcal{N}_\mathbf{q}} D_T = \sum_{U \in \mathcal{S}_{\geqslant T}} \frac{\varepsilon(U-T)}{W_U(\mathbf{q})} = \frac{W^T(\mathbf{q}^{-1})}{W(\mathbf{q}^{-1})},$$

where the second equality is Lemma 17.1.8. □

Of course, similar corollaries hold for the $G_T$ for all $\mathbf{q} \in \overline{\mathcal{R}} \cup \overline{\mathcal{R}^{-1}}$. We will also need the following version of Lemmas 20.6.12 and 20.6.13. Its proof is essentially the same as the proofs of these lemmas, except that we use Theorem 20.6.1 and its corollaries instead of the corresponding statements involving the $G_U$.

**LEMMA 20.6.18.** (Compare Lemmas 20.6.12 and 20.6.13.) *Suppose* $\mathbf{q} \in \overline{\mathcal{R}}$ *and* $U \subset S$. *Then*

$$\sum_{\substack{T \in \mathcal{S} \\ T \subset U}} D_{S-T} h_U$$

*is a dense subspace of $H^U$ and a direct sum decomposition. If $T \in \mathcal{S}$, then the right multiplication by $h_T$ induces a weak isomorphism $D_{S-T} \to D_{S-T}h_T$. Moreover, $G_U = \overline{D_{S-U}h_U}$.*

### A Generalization of Solomon's Theorem

When $W$ is finite, Theorem 15.4.2 gives a decomposition of the rational (or real) group algebra (formula (15.29) in 15.4). Here we give an analogous decomposition of $L^2_{\mathbf{q}}$. First, we need the following.

**THEOREM 20.6.19.** *Suppose $T \in \mathcal{S}$.*

*(i) If $\mathbf{q} \in \mathcal{R}$, then $\overline{L^2_{\mathbf{q}}a_{S-T}h_T} = G_T$ and $\overline{L^2_{\mathbf{q}}h_Ta_{S-T}} = D_{S-T}$.*

*(ii) If $\mathbf{q} \in \mathcal{R}^{-1}$, then $\overline{L^2_{\mathbf{q}}a_Th_{S-T}} = G_{S-T}$ and $\overline{L^2_{\mathbf{q}}h_{S-T}a_T} = D_T$.*

*Proof*

(i) Suppose $\mathbf{q} \in \mathcal{R}$. By Lemma 20.6.11, right multiplication by $a_{S-T}$ annihilates $G_U$ if $U \not\subset T$ and by Lemma 20.6.12, it is a weak isomorphism from $G_T$ to $G_Ta_{S-T}$. So, by Lemma 20.6.13, $\overline{L^2_{\mathbf{q}}h_Ta_{S-T}} = \overline{G_Ta_{S-T}} = D_T$. Similarly, by Lemma 20.6.18, $\overline{L^2_{\mathbf{q}}a_{S-T}h_T} = G_T$.

(ii) Applying the $j$-isomorphism to the two equations in (i), we get the two equations in (ii). □

*Remark.* It seems probable that $\overline{L^2_{\mathbf{q}}a_{S-U}h_U} = G_U$ and $\overline{L^2_{\mathbf{q}}h_Ua_{S-U}} = D_U$ whenever $\mathbf{q} \in \mathcal{R}_{S-U} \cap \mathcal{R}_U^{-1}$ (so that $h_U$ and $a_{S-U}$ are both defined).

An immediate consequence of the First and Second Decomposition Theorems and the previous theorem is the following generalization of Solomon's result.

**THEOREM 20.6.20**

*(i) If $\mathbf{q} \in \mathcal{R}$, then*

$$\sum_{T\in\mathcal{S}} \overline{L^2_{\mathbf{q}}h_Ta_{S-T}} \quad \text{and} \quad \sum_{T\in\mathcal{S}} \overline{L^2_{\mathbf{q}}a_{S-T}h_T}$$

*are direct sum decompositions and dense subspaces of $L^2_{\mathbf{q}}$.*

*(ii) If $\mathbf{q} \in \mathcal{R}^{-1}$, then*

$$\sum_{T\in\mathcal{S}} \overline{L^2_{\mathbf{q}}h_{S-T}a_T} \quad \text{and} \quad \sum_{T\in\mathcal{S}} \overline{L^2_{\mathbf{q}}a_Th_{S-T}}$$

*are direct sum decompositions and dense subspaces of $L^2_{\mathbf{q}}$.*

### Ruins and the Proof of Theorem 20.6.4

It remains to prove Theorem 20.6.4 (which states that the homology of $C_*(H^T)$ is concentrated in dimension 0). Given $U \subset S$, put $\mathcal{S}(U) := \{T \in \mathcal{S} \mid T \subset U\}$ and define $\Sigma(U)$ to be the subcomplex of $\Sigma_{cc}$ consisting of all (closed) Coxeter cells of type $T$ with $T \in \mathcal{S}(U)$. Put $K(U) := \Sigma(U) \cap K$. Clearly, $K(U) = K(W_U, U)$, the standard fundamental chamber for $\Sigma(W_U, U)$. Thus,

$$\Sigma(U) = \mathcal{U}(W, K(U))) = W \times_{W_U} \Sigma(W_U, U). \tag{20.24}$$

Our goal is to identify the chain complex $C_*(H^T)$ of (20.22) with the relative $L^2_{\mathbf{q}}$-chains on a pair of subcomplexes of $\Sigma$. To this end, given $T \in \mathcal{S}(U)$, define three subcomplexes of $\Sigma(U)$:

$\Omega_{UT}$ : the union of closed cells of type $T'$, with $T' \in \mathcal{S}(U)_{\geqslant T}$,

$\widehat{\Omega}_{UT}$ : the union of closed cells of type $T''$, $T'' \in \mathcal{S}(U)$, $T'' \notin \mathcal{S}(U)_{\geqslant T}$,

$\partial\Omega_{UT}$ : the cells of $\Omega_{UT}$ of type $T''$, with $T'' \notin \mathcal{S}(U)_{\geqslant T}$.

$\Omega_{UT}$ is the union of all cells of type $T''$, where $T'' \leqslant T'$ for some $T' \in \mathcal{S}(U)_{\geqslant T}$. So,

$$\partial\Omega_{UT} = \Omega_{UT} \cap \widehat{\Omega}_{UT}$$

and

$$\Sigma(U) = \Omega_{UT} \cup \widehat{\Omega}_{UT}.$$

The pair $(\Omega_{UT}, \partial\Omega_{UT})$ is called the $(U,T)$-*ruin*. For example, for $T = \emptyset$, we have $\Omega_{U\emptyset} = \Sigma(U)$ and $\partial\Omega_{U\emptyset} = \emptyset$.

By (20.24), the $\mathcal{N}_{\mathbf{q}}$-modules $L^2_{\mathbf{q}}C_*(\Sigma(U))$ and $L^2_{\mathbf{q}}\mathcal{H}_*(\Sigma(U))$ are induced from the $\mathcal{N}_{\mathbf{q}}[W_U]$-modules $L^2_{\mathbf{q}}C_*(\Sigma(W_U, U))$ and $L^2_{\mathbf{q}}\mathcal{H}_*(\Sigma(W_U, U)$; hence, we can calculate von Neumann dimensions over $\mathcal{N}_{\mathbf{q}}[W]$ by calculating them with respect to $\mathcal{N}_{\mathbf{q}}[W_U]$.

**LEMMA 20.6.21**

*(i) There is a isomorphism of chain complexes of $\mathcal{N}_{\mathbf{q}}$-modules*

$$\Psi' : L^2_{\mathbf{q}}C_*(\Sigma_{cc}) \to C_*(H^{\emptyset}),$$

*where $C_*(H^{\emptyset})$ is the chain complex from* (20.22).

*(ii) Suppose $T \in \mathcal{S}^{(k)}$. Then $\Psi'$ induces an isomorphism of chain complexes of $\mathcal{N}_{\mathbf{q}}$-modules*

$$L^2_{\mathbf{q}}C_*(H^T) \xrightarrow{\cong} L^2_{\mathbf{q}}C_{*+k}(\Omega_{ST}, \partial\Omega_{ST}).$$

*In particular, $L^2_{\mathbf{q}}C_m(\Omega_{ST}, \partial\Omega_{ST}) = 0$ for $m < k$.*

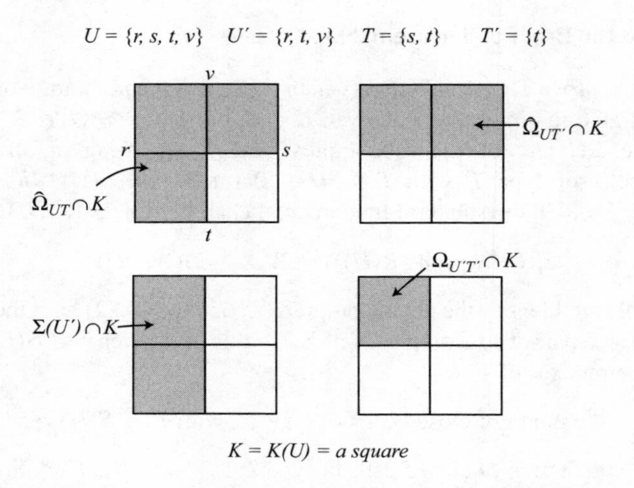


Figure 20.1. Complements of ruins intersected with $K$.

*Proof.* (i) For each $T \in \mathcal{S}$ modify the isometry $\psi_T$ of (20.14) to another Hilbert $\mathcal{N}_{\mathbf{q}}$-module isomorphism, $\psi'_T : L^2(W\langle T\rangle, \mu_{\mathbf{q}}) \to H^T$ as follows:

$$\psi'_T(f) := \sqrt{W_T(\mathbf{q}^{-1})}\psi_T(f) = W_T(\mathbf{q}^{-1}) \left( \sum_{u \in B_T} f(uC_T)e_u \right) h_T. \tag{20.25}$$

$\Psi'$ is defined to be the direct sum of the $\psi'_T$. Suppose $U \in \mathcal{S}_{>\emptyset}$ and $T \subset U$ is obtained by deleting one element from $U$. Look at Figure 20.1 and define

$$K(U,T) := \Omega_{UT} \cap K,$$
$$\partial K(U,T) := \partial\Omega_{UT} \cap K,$$
$$\widehat{K}(U,T) := \widehat{\Omega}_{UT} \cap K.$$

So

$$\Omega_{UT} = \mathcal{U}(W, K(U,T)),$$
$$\partial\Omega_{UT} = \mathcal{U}(W, \partial K(U,T)),$$
$$\widehat{\Omega}_{UT} = \mathcal{U}(W, \widehat{K}(U,T)).$$

Statement (i) follows immediately from the next claim.

*Claim.* The following diagram commutes:

$$\begin{array}{ccc} L^2(WC_U, \mu_{\mathbf{q}}) & \xrightarrow{\psi'_U} & H^U \\ {\scriptstyle \partial^{\mathbf{q}}_T}\downarrow & & \downarrow{\scriptstyle i} \\ L^2(WC_T, \mu_{\mathbf{q}}) & \xrightarrow{\psi'_T} & H^T \end{array}$$

where $\partial^{\mathbf{q}}_T$ denotes the $L^2(W\langle C_T\rangle, \mu_{\mathbf{q}})$-component of $\partial^{\mathbf{q}}$ and $i$ is the natural inclusion.

*Proof of Claim.* Using (20.25) and (20.13), we get

$$\begin{aligned} \psi'_T(\partial^{\mathbf{q}}_T f) &= W_T(\mathbf{q}^{-1}) \left( \sum_{w \in B_T} (\partial^{\mathbf{q}}_T f)(wC_T) e_w \right) h_T \\ &= W_T(\mathbf{q}^{-1}) \left( \sum_{u \in B_U} \sum_{v \in W_U \cap B_T} \varepsilon_v q_v^{-1} f(uC_U) e_u e_v \right) h_T \\ &= W_T(\mathbf{q}^{-1}) \left( \sum_{u \in B_U} f(uC_U) e_u \right) \left( \sum_{v \in W_U \cap B_T} \varepsilon_v q_v^{-1} e_v \right) h_T \\ &= W_U(\mathbf{q}^{-1}) \left( \sum_{u \in B_U} f(uC_U) e_u \right) h_U \\ &= i(\psi'_U(f)), \end{aligned}$$

where the next to last equality is from the following formula for $h_U$, valid whenever $T \subset U$ and $\mathbf{q} \in \mathcal{R}_U^{-1}$:

$$h_U = \left( \sum_{v \in W_U \cap B_T} \varepsilon_v q_v^{-1} e_v \right) h_T.$$

(This formula holds since $W_U \cap B_T$ is a set of coset representatives for $W_U/W_T$ and since for any $v \in W_U \cap B_T$ and $w \in W_T$, we have $e_v e_w = e_{vw}$ and $q_v q_w = q_{vw}$.) This completes the proof of the claim. □

(ii) The second part of the lemma follows from part (i). The point is that if we write $\Omega$ for $\Omega(S, T)$, then the cells of form $WC_{T'}$, $T' \in (\mathcal{S}_{\geqslant T})^{(i+1)}$, are a basis for $C_i(\Omega, \partial\Omega)$. Hence,

$$L^2_{\mathbf{q}} C_i(\Omega, \partial\Omega) = \bigoplus_{T' \in (\mathcal{S}_{\geq T})^{(i+1)}} L^2(WC_{T'}, \mu_{\mathbf{q}}) \cong \bigoplus H^{T'}$$

and (i) shows that the $\partial^{\mathbf{q}}$ maps are induced by the inclusions $H^{T''} \hookrightarrow H^{T'}$, with $T' \subset T''$. □

**THEOREM 20.6.22.** *Suppose $T \in \mathcal{S}^{(k)}$. If $\mathbf{q} \in \mathcal{R}$, then $L^2_{\mathbf{q}}H_*(\Omega_{ST}, \partial\Omega_{ST})$ is concentrated in dimension k. If $\mathbf{q} \in \partial\mathcal{R}$, the same holds for $L^2_{\mathbf{q}}\mathcal{H}_*(\Omega_{ST}, \partial\Omega_{ST})$.*

The final sentence of the theorem follows from the second and the continuity of the $b^i_{\mathbf{q}}$ (Theorem 20.2.1). In the special case $T = \emptyset$, we have $\Omega_{ST} = \Sigma$ and the theorem asserts that for $\mathbf{q} \in \mathcal{R}$, $L^2_{\mathbf{q}}H_*(\Sigma)$ is concentrated in dimension 0. This is Theorem 20.3.3. It is the first step in an inductive proof.

Before beginning the proof, note that there is an excision

$$L^2_{\mathbf{q}}C_*(\Omega_{UT}, \partial\Omega_{UT}) \cong L^2_{\mathbf{q}}C_*(\Sigma(U), \widehat{\Omega}_{UT}). \tag{20.26}$$

Also, for any $s \in T$, if we set $T' := T - s$ and $U' = U - s$, there is an excision

$$L^2_{\mathbf{q}}C_*(\Sigma(U'), \widehat{\Omega}_{U'T'}) \cong L^2_{\mathbf{q}}C_*(\widehat{\Omega}_{UT}, \widehat{\Omega}_{UT'}). \tag{20.27}$$

(See Figure 20.1.)

*Proof of Theorem 20.6.22.* Suppose $U \subset S$ and $T \in \mathcal{S}(U)^{(k)}$. We shall prove, by induction on $k$, that $L^2_{\mathbf{q}}H_*(\Omega_{UT}, \partial\Omega_{UT})$ is concentrated in dimension $k$. When $k = 0$ (i.e., when $T = \emptyset$) this is Theorem 20.3.3 (and the fact that $L^2_{\mathbf{q}}C_*(\Sigma(U))$ is induced from $L^2_{\mathbf{q}}C_*(\Sigma(W_U, U))$). Assume by induction, that our assertion holds for $k - 1$, with $k - 1 \geqslant 0$. By (20.26), the assertion is equivalent to showing that $L^2_{\mathbf{q}}H_*(\Sigma(U), \widehat{\Omega}_{UT})$ is concentrated in dimension $k$. Choose an element $s \in T$ and set $T' := T - s$, $\widehat{\Omega} := \widehat{\Omega}_{UT}$, $\widehat{\Omega}' := \widehat{\Omega}_{UT'}$. Consider the long exact sequence of the triple $(\Sigma(U), \widehat{\Omega}, \widehat{\Omega}')$:

$$L^2_{\mathbf{q}}H_*(\Sigma(U), \widehat{\Omega}') \to L^2_{\mathbf{q}}H_*(\Sigma(U), \widehat{\Omega}) \to L^2_{\mathbf{q}}H_{*-1}(\widehat{\Omega}, \widehat{\Omega}')$$

By (20.27), the right-hand term excises to the homology of the $(U', T')$-ruin, while the middle term is that of the $(U, T)$-ruin and the left-hand term is that of the $(U, T')$-ruin. By induction, the left-hand and right-hand terms are concentrated in dimension $k - 1$. So the middle term can only be nonzero in dimensions $k - 1$ and $k$. On the other hand, by Lemma 20.6.21(ii), the middle term vanishes in dimensions $< k$. □

*Proof of Theorem 20.6.4.* By Lemma 20.6.21,

$$H_*(C_*(H^T)) = L^2_{\mathbf{q}}H_{*+k}(\Omega_{ST}, \partial\Omega_{ST}).$$

By Theorem 20.6.22, the right-hand side is concentrated in dimension $k$; so, the left-hand side is concentrated in dimension 0. □

**Exercise 20.6.23.** Use Corollary 8.1.5 to compute the ordinary homology of the $(S,T)$-ruin. Show that $H_*(\Omega_{ST}, \partial\Omega_{ST})$ is concentrated in dimension $k$.

## 20.7. DECOUPLING COHOMOLOGY

In this section we use the First Decomposition Theorem (20.6.1) to get a decomposition of weighted $L^2$-cohomology and a computation of $L^2_{\mathbf{q}}$-Betti numbers. We retain notation of 20.1: $X$ is a CW complex with mirror structure $(X_s)_{s\in S}$, $X^V = \bigcup_{s\in V} X_s$, $\mathcal{U} = \mathcal{U}(W,X)$ and for any cell $c$ of $X$, $S(c) = \{s \in S \mid c \subset X_s\}$. Given a pair of spaces $(Y,Z)$, $b^i(Y,Z)$ ($:= \dim_{\mathbb{R}}(H^i(Y,Z)\otimes\mathbb{R})$) is the ordinary $i^{\text{th}}$ Betti number. We will prove the following.

**THEOREM 20.7.1.** ([79, Theorem 10.3].)

*(i) For $\mathbf{q} \in \overline{\mathcal{R}}$, there is an isomorphism of $\mathcal{N}_{\mathbf{q}}$-modules*

$$L^2_{\mathbf{q}}\mathcal{H}^*(\mathcal{U}) \cong \bigoplus_{T\in\mathcal{S}} H^*(X, X^T) \otimes D_{S-T}.$$

*So*

$$b^i_{\mathbf{q}}(\mathcal{U}) = \sum_{T\in\mathcal{S}} b^i(X, X^T)\frac{W^T(\mathbf{q})}{W(\mathbf{q})}.$$

*(ii) For $\mathbf{q} \in \overline{\mathcal{R}^{-1}}$, there is an isomorphism of $\mathcal{N}_{\mathbf{q}}$-modules*

$$L^2_{\mathbf{q}}\mathcal{H}^*(\mathcal{U}) \cong \bigoplus_{T\in\mathcal{S}} H^*(X, X^{S-T}) \otimes D_T.$$

*So*

$$b^i_{\mathbf{q}}(\mathcal{U}) = \sum_{T\in\mathcal{S}} b^i(X, X^{S-T})\frac{W^T(\mathbf{q}^{-1})}{W(\mathbf{q}^{-1})}.$$

The special case $\mathcal{U} = \Sigma$ is the following corolllary (the first part of which is equivalent to Theorem 20.3.3).

**COROLLARY 20.7.2.** ([79, Theorem 10.4].)

*(i) If $\mathbf{q} \in \overline{\mathcal{R}}$, then $L^2_{\mathbf{q}}\mathcal{H}^*(\Sigma)$ is concentrated in dimension 0 and is $\cong A^S$. So, $b^0_{\mathbf{q}}(\Sigma) = 1/W(\mathbf{q})$.*

*(ii) If* $\mathbf{q} \in \overline{\mathcal{R}^{-1}}$, *then*

$$L^2_{\mathbf{q}}\mathcal{H}^*(\Sigma) \cong \bigoplus_{T\in\mathcal{S}} H^*(K, K^{S-T}) \otimes D_T.$$

*So,*

$$b^i_{\mathbf{q}}(\Sigma) = \sum_{T\in\mathcal{S}} b^i(K, K^{S-T}) \frac{W^T(\mathbf{q}^{-1})}{W(\mathbf{q}^{-1})}.$$

*Proof.* If $T$ is a nonempty spherical subset, then $K^T$ is contractible and therefore, $b^i(K, K^T) = 0$ for all $i$. So, the only nonzero term in Theorem 20.7.1 (i) is $H^0(K) \otimes D_S$ $(= A^S)$, which gives (i). Statement (ii) is the same as Theorem 20.7.1 (ii) with $X$ replaced by $K$. □

*Remark 20.7.3.* For $T \neq \emptyset$, as $\mathbf{q} \to 0$, $W^T(\mathbf{q})/W(\mathbf{q}) \to 0/1 = 0$; hence, as $\mathbf{q} \to \infty$, $W^T(\mathbf{q}^{-1})/W(\mathbf{q}^{-1}) \to 0$. On the other hand, for $T = \emptyset$,

$$\lim_{\mathbf{q}\to 0} 1/W(\mathbf{q}) = \lim_{\mathbf{q}\to\infty} 1/W(\mathbf{q}^{-1}) = 1.$$

Hence, only the limits with $T = \emptyset$ are nonzero. Combining this with the formulas in Theorem 20.7.1, we get formula (20.18) of Remark 20.2.2,

$$\lim_{\mathbf{q}\to 0} b^i_{\mathbf{q}}(\mathcal{U}) = b^i(X) \quad \text{and} \quad \lim_{\mathbf{q}\to\infty} b^i_{\mathbf{q}}(\mathcal{U}) = b^i(X, \partial X).$$

We will prove Theorem 20.7.1 by applying techniques from Chapter 15. As in 15.2, by taking invariants, $L^2_{\mathbf{q}}$ (or any other Hilbert $\mathcal{N}_{\mathbf{q}}$-module) defines a simple system of coefficients $\mathcal{I}(L^2_{\mathbf{q}})$ on $X$. It assigns to each $i$-cell $c$, the submodule $A^{S(c)}$ $(= (L^2_{\mathbf{q}})^{\mathbb{R}_{\mathbf{q}}[W_{S(c)}]})$. Just as in Corollary 15.2.3, the map $\Phi$ from (20.12) gives an identification

$$L^2_{\mathbf{q}}C^i(\mathcal{U}) = C^i(X; \mathcal{I}(L^2_{\mathbf{q}})). \tag{20.28}$$

By the First Decomposition Theorem (20.6.1), for $\mathbf{q} \in \overline{\mathcal{R}} \cup \overline{\mathcal{R}^{-1}}$,

$$L^2_{\mathbf{q}} = \overline{\biguplus_{V\subset S} D_V}.$$

For the same range of $\mathbf{q}$, given subsets $U$, $V$ of $S$, by Corollaries 20.6.15 and 20.6.16,

$$D_V \cap A^U = \begin{cases} D_V & \text{if } U \subset V, \\ 0 & \text{if } U \not\subset V. \end{cases} \tag{20.29}$$

It follows that, for $\mathbf{q} \in \overline{\mathcal{R}} \cup \overline{\mathcal{R}^{-1}}$, there is a direct sum decomposition of coefficient systems,

$$\mathcal{I}(L^2_{\mathbf{q}}) = \overline{\biguplus_{V \subset S} \mathcal{I}(D_V)}, \tag{20.30}$$

where $\mathcal{I}(D_V)$ is the coefficient system which associates $D_V \cap A^{S(c)}$ to a cell $c$. This gives a decomposition of cochain complexes,

$$C^*(X; \mathcal{I}(L^2_{\mathbf{q}})) = \overline{\biguplus_{V \subset S} C^*(X; \mathcal{I}(D_V))}. \tag{20.31}$$

**LEMMA 20.7.4.** *For $V \subset S$, $C^*(X; \mathcal{I}(D_V)) \cong C^*(X, X^{S-V}) \otimes D_V$.*

*Proof.* As in the proof of Theorem 15.3.4,

$$\begin{aligned} C^i(X; \mathcal{I}(D_V)) &= \{f : X^{(i)} \to D_V \mid f(c) \in D_V \cap A^{S(c)},\ \forall c \in X^{(i)}\} \\ &= \{f : X^{(i)} \to D_V \mid f(c) = 0,\ \forall c \subset X^{S-V}\} \\ &= C^i(X, X^{S-V}) \otimes D_V. \end{aligned}$$

□

*Proof of Theorem 20.7.1.* By (20.31), $\bigoplus C^*(X; \mathcal{I}(D_V)) \hookrightarrow C^*(X; L^2_{\mathbf{q}})$ is a weak isomorphism of cochain complexes. By the proof of Lemma J.3.2 in Appendix J.3, this induces a weak isomorphism in reduced cohomology,

$$\bigoplus_{V \subset S} \mathcal{H}^*(X; \mathcal{I}(D_V)) \to \mathcal{H}^*(X; \mathcal{I}(L^2_{\mathbf{q}})) \tag{20.32}$$

and by the proof of Lemma J.2.6, if two Hilbert $\mathcal{N}_{\mathbf{q}}$-modules are weakly isomorphic, they are isomorphic. By Lemma 20.7.4, $\mathcal{H}^*(X; \mathcal{I}(D_V)) = H^*(X, X^{S-V}) \otimes D_V$. By (20.28), the right hand side of (20.32) is $L^2_{\mathbf{q}}\mathcal{H}^*(\mathcal{U})$. By Theorem 20.6.1, when $\mathbf{q} \in \overline{\mathcal{R}}$, only cospherical $V$ contribute nonzero terms to the left hand side of (20.32), while if $\mathbf{q} \in \overline{\mathcal{R}^{-1}}$ only spherical $V$ appear. This proves the theorem except for the formulas for $L^2_{\mathbf{q}}$- Betti numbers. These follow from the calculations of $\dim_{\mathcal{N}_{\mathbf{q}}} D_V$ in Corollary 20.6.17. □

There is a version of Theorem 20.7.1 in $L^2_{\mathbf{q}}$-homology. Although $L^2_{\mathbf{q}}$-homology is isomorphic to $L^2_{\mathbf{q}}$-cohomology, the natural decomposition of it comes from the Second Decomposition Theorem 20.6.2 (as in Theorem 15.3.4) and involves the $G_V$ rather than the $D_V$. We state the homology version below.

**THEOREM 20.7.5**

(i) *If $\mathbf{q} \in \overline{\mathcal{R}}$, then*

$$L^2_{\mathbf{q}}\mathcal{H}_*(\mathcal{U}) \cong \bigoplus_{T \in \mathcal{S}} H_*(X, X^T) \otimes G_T.$$

*(ii) If* $\mathbf{q} \in \overline{\mathcal{R}^{-1}}$, *then*

$$L^2_{\mathbf{q}}\mathcal{H}_*(\mathcal{U}) \cong \bigoplus_{T \in \mathcal{S}} H_*(X, X^{S-T}) \otimes G_{S-T}.$$

## The Relationship with Ordinary Homology and Cohomology with Compact Supports

**THEOREM 20.7.6**

*(i) For* $\mathbf{q} \in \mathcal{R}$, *the canonical map* $\mathrm{can} : H_*(\mathcal{U}; \mathbb{R}) \to L^2_{\mathbf{q}}\mathcal{H}_*(\mathcal{U})$ *is an injection with dense image.*

*(ii) For* $\mathbf{q} \in \mathcal{R}^{-1}$, *the canonical map* $\mathrm{can} : H^*_c(\mathcal{U}; \mathbb{R}) \to L^2_{\mathbf{q}}\mathcal{H}^*(\mathcal{U})$ *is an injection with dense image.*

In Chapter 15 we carefully distinguished between left and right modules: left modules should be used as coefficient modules for cohomology and right modules for homology. However, in this chapter, in effort to keep notation from becoming hopelessly complicated, we have dropped the distinction and used right $\mathcal{N}_{\mathbf{q}}$-modules as coefficients for both homology and cohomology. We need to slightly modify our notation from Section 15.2 to be consistent with the notation here. First, (15.10) should be changed to $b'_w = e_w \tilde{a}_{\mathrm{In}(w)}$ (the $\tilde{a}$ term is on the right instead of the left). $\{b'_w\}_{w \in W}$ is a basis $\mathbb{R}W$. Let $A^T_{\mathbb{R}}$ be the subspace of $\mathbb{R}W$ spanned by $\{b'_w \mid T \subset \mathrm{In}(w)\}$ and as in (15.12), $\widehat{A}^T_{\mathbb{R}}$ is the span of $\{b'_w \mid T = \mathrm{In}(w)\}$. As usual, $A^T = L^2_{\mathbf{q}} a_T$. Let $i : A^T_{\mathbb{R}} \to A^T$ be the inclusion induced by $b'_w \to b'_w$. (The element $\tilde{a}_T$ is defined by the same formula either as an element of $\mathbb{R}W$ in (8.1) or as an element of $\mathcal{N}_{\mathbf{q}}$ in (19.8). So, using Hecke multiplication, we can also regard $b'_w$ as an element of $\mathcal{N}_{\mathbf{q}}$.) Let $p_T : A^T \to D_T$ be orthogonal projection.

**LEMMA 20.7.7.** *Suppose* $T \in \mathcal{S}$. *For* $\mathbf{q} \in \mathcal{R}^{-1}$, *the map* $p_T \circ i : \widehat{A}^T_{\mathbb{R}} \to D_T$ *is injective with dense image. Similarly, for* $\mathbf{q} \in \mathcal{R}$, $\widehat{H}^T_{\mathbb{R}} \to G_T$ *is injective with dense image.*

*Proof.* First, $i : A^T_{\mathbb{R}} \to A^T$ is injective with dense image. By Corollary 20.6.16, the kernel of $p_T$ is $\sum_{U \in \mathcal{S}_{>T}} A^U$. So, $i$ takes $\bigoplus_{U \in \mathcal{S}_{>T}} \widehat{A}^U_{\mathbb{R}}$ onto a dense subspace of $\mathrm{Ker}\, p_T$. Since, by Corollary 15.2.6, $A^T_{\mathbb{R}} = \bigoplus_{U \in \mathcal{S}_{\geqslant T}} \widehat{A}^U_{\mathbb{R}}$, it follows that $i$ takes $\widehat{A}^T_{\mathbb{R}}$ injectively to a subspace whose closure is complementary subspace for $\mathrm{Ker}\, p_T$. This proves the statement concerning $\mathbf{q} \in \mathcal{R}^{-1}$. The proof of the last sentence of the lemma is similar. □

*Proof of Theorem 20.7.6.* We first prove (ii). Suppose $\mathbf{q} \in \mathcal{R}^{-1}$. We have a commutative diagram:

$$\begin{array}{ccc} \bigoplus H^*(X, X^{S-T}) \otimes \widehat{A}^T_{\mathbb{R}} & \longrightarrow & H^*_c(\mathcal{U};\mathbb{R}) = H^*(X;\mathcal{I}(\mathbb{R}W)) \\ & & \downarrow \text{can} \\ \downarrow d & & L^2_{\mathbf{q}}\mathcal{H}^*(\mathcal{U}) = \mathcal{H}^*(X;\mathcal{I}(L^2_{\mathbf{q}})) \\ & & \downarrow \oplus\pi_T \\ \bigoplus H^*(X, X^{S-T}) \otimes D_T & & \bigoplus H^*(X;\mathcal{I}(D_T)). \end{array}$$

Here the sums range over all $T \in \mathcal{S}$ and the horizontal maps are the natural identifications from Theorem 15.2.8 and Lemma 20.7.4. Also, $\pi_T : \mathcal{H}^*(X;\mathcal{I}(L^2_{\mathbf{q}})) \to H^*(X;\mathcal{I}(D_T))$ is the coefficient homomorphism induced by orthogonal projection $L^2_{\mathbf{q}} \to D_T$ and $d := \oplus(p_T \circ i)$ is the coefficient homomorphism induced from the maps $p_T \circ i : \widehat{A}^T_{\mathbb{R}} \to D_T$ of Lemma 20.7.7. By Theorem 20.7.1, $\oplus\pi_T$ is a weak isomorphism. In other words, up to a weak isomorphism of Hilbert $\mathcal{N}_{\mathbf{q}}$-modules, the canonical map $H^*_c(\mathcal{U};\mathbb{R}) \to L^2_{\mathbf{q}}\mathcal{H}^*(\mathcal{U})$ is identified with $d$. By Lemma 20.7.7, each $p_T \circ i$ is injective with dense image. This proves (ii).

The canonical map in (i) is induced by the composition of chain maps:

$$(C_*(\mathcal{U};\mathbb{R}), \partial) \hookrightarrow (L^2_{\mathbf{1/q}}C_*(\mathcal{U}), \partial) \xrightarrow{\cong} L^2_{\mathbf{q}}(C_*(\mathcal{U}), \partial^{\mathbf{q}}),$$

where the second map is the isomorphism of Lemma 20.1.2. Suppose $\mathbf{q} \in \mathcal{R}$. Again we have a commutative diagram:

$$\begin{array}{ccc} \bigoplus H_*(X, X^{S-T}) \otimes \widehat{H}^T_{\mathbb{R}} & \longrightarrow & H_*(\mathcal{U};\mathbb{R}) = H_*(X;\mathcal{C}(\mathbb{R}W)) \\ & & \downarrow \text{can} \\ \downarrow g & & L^2_{\mathbf{q}}\mathcal{H}_*(\mathcal{U}) = \mathcal{H}_*(X;\mathcal{C}(L^2_{\mathbf{q}})) \\ & & \downarrow \oplus\pi'_T \\ \bigoplus H_*(X, X^{S-T}) \otimes G_{S-T} & \longleftarrow & \bigoplus H_*(X;\mathcal{C}(G_{S-T})). \end{array}$$

where the horizontal maps are the natural identifications, $\pi'_T$ is coefficient homomorphism induced by the orthogonal projection $L^2_{\mathbf{q}} \to G_{S-T}$ and $g$ is the sum of coefficient homomorphisms from Lemma 20.7.7. So, as before, up to a weak isomorphism of Hilbert $\mathcal{N}_{\mathbf{q}}$-modules, the canonical map $H_*(\mathcal{U};\mathbb{R}) \to L^2_{\mathbf{q}}\mathcal{H}_*(\mathcal{U})$ is identified with $g$. By Lemma 20.7.7, $g$ is a sum of maps, each of which is injective with dense image, proving (i). □

## 20.8. $L^2$-COHOMOLOGY OF BUILDINGS

As in Chapter 18, let $\Phi$ be a building of type $(W, S)$ with a chamber transitive automorphism group $G$ and thickness vector $\mathbf{q}$. Its geometric realization, $\mathrm{Geom}(\Phi)$ $(= \mathcal{U}(\Phi, K))$, is defined in 18.2. $G$ is a group of homeomorphisms of $\mathrm{Geom}(\Phi)$. Give it the compact open topology. The stabilizer $B$ of a given chamber $\varphi$ is a compact open subgroup. As in 18.4, let $\mu$ be Haar measure on $G$, normalized by the condition that $\mu(B) = 1$.

### The von Neumann Algebra of $G$

$G$ acts on $L^2(G)$ via the left regular representation. The *von Neumann algebra* $\mathcal{N}(G)$ consists of all $G$-equivariant bounded linear endomorphisms of $L^2(G)$. Any $\alpha \in \mathcal{N}(G)$ is represented by convolution with some distribution $f_\alpha$. This distribution need not be a function. For example, if $\alpha$ is the identity map on $L^2(G)$, then $f_\alpha = \delta_1$ (the Dirac delta). One would like to define the "trace" of $\alpha$ to be $f_\alpha(1)$ whenever $f_\alpha$ is a function. However, since $f_\alpha$ is well-defined only up to sets of measure 0, we must proceed differently. Suppose $\alpha \in \mathcal{N}(G)$ is nonnegative and self-adjoint. Let $\beta$ be its square root. If $f_\beta$ is a $L^2$ function, put

$$\mathrm{tr}_{\mathcal{N}(G)}\,\alpha := \|f_\beta\| := \left(\int_G f_\beta(x)^2 d\mu\right)^{1/2}.$$

This extends in the usual fashion to give a "trace" on $(n \times n)$-matrices with coefficients in $\mathcal{N}(G)$. If $V$ is a closed, $G$-stable subspace of $\bigoplus L^2(G)$ and $\pi_V : \bigoplus L^2(G) \to \bigoplus L^2(G)$ is orthogonal projection, then the *von Neumann dimension* of $V$ is defined by $\dim_{\mathcal{N}(G)} V := \mathrm{tr}_{\mathcal{N}(G)}\,\pi_V$.

We identify $L^2(\Phi) = L^2(G/B)$ with the subspace of $L^2(G)$ consisting of the functions which are constant on each left coset $gB$, $g \in G$. Orthogonal projection from $L^2(G)$ onto $L^2(G/B)$ is given by convolution with the characteristic function of $B$. Since $\mu(B) = 1$, $\dim_{\mathcal{N}(G)} L^2(G/B) = 1$. The retraction $r : G/B \to W$, defined by taking the $W$-distance to a base chamber, induces a bounded linear map $L^2_{\mathbf{q}}(W) \to L^2(G/B)$ which we also denote by $r$. Since this map takes bounded elements of $L^2_{\mathbf{q}}(W)$ to bounded elements of $L^2(G/B)$, we get the following version of Lemma 19.1.5.

**LEMMA 20.8.1.** *The map $r : L^2_{\mathbf{q}}(W) \to L^2(G/B)$ induces a monomorphism of von Neumann algebras $r : \mathcal{N}_{\mathbf{q}} \to \mathcal{N}(G)$. (In particular, $r$ commutes with the $*$ anti-involutions on $\mathcal{N}_{\mathbf{q}}$ and $\mathcal{N}(G)$.)*

$L^2C^*(\Phi)$ denotes the Hilbert space of square summable simplicial cochains on Geom$(\Phi)$ and $\mathcal{H}^*(\Phi)$ denotes the subspace of harmonic cocycles. (Of course, $\mathcal{H}^*(\Phi)$ is isomorphic to reduced cohomology of the cochain complex $L^2C^*(\Phi)$.) We have

$$L^2C^i(\Phi) = \bigoplus_{c\in K^{(i)}} L^2(G/G_{S(c)}) \subset \bigoplus_{c\in K^{(i)}} L^2(G),$$

where $G_{S(c)}$ is the stabilizer of the $i$-simplex $c$. Define the *$L^2$-Betti numbers of $\Phi$ with respect to $G$* by

$$b^i(\Phi; G) = \dim_{\mathcal{N}(G)} \mathcal{H}^i(\Phi),$$

The retraction $r : \text{Geom}(\Phi) \to \Sigma$ induces a map on cochains which we again denote by $r : L^2_{\mathbf{q}}C^*(\Sigma) \to L^2C^*(\Phi)$. We also have "transfer maps" on chains and cochains. On the level of chains, the transfer map sends a cell $c$ of $\Sigma$ to $r^{-1}(c)/\operatorname{Card}(r^{-1}(c))$. On the level of cochains, the transfer map $t : L^2C^*(\Phi) \to L^2_{\mathbf{q}}C^*(\Sigma)$ is defined by

$$t(f)(c) := \frac{1}{\operatorname{Card}(r^{-1}(c))} \sum f(c'),$$

where the sum is over all $c' \in r^{-1}(c)$. (The orientations on the $c'$ are induced from the orientation of $c$.) Note that $\operatorname{Card}(r^{-1}(c)) = \mu_{\mathbf{q}}(c)$, where $\mu_{\mathbf{q}}$ is the measure on $Wc$ defined in (20.1).

*Remark.* Suppose Geom$(\Phi)$ is the geometric realization of a building associated to a Tits system $(G, B, N, S)$. Then $L^2_{\mathbf{q}}C^*(\Sigma)$ can be identified with the $B$-invariant cochains $L^2C^*(\Phi)^B$ and the map $r : L^2_{\mathbf{q}}C^*(\Sigma) \to L^2C^*(\Phi)$ with the inclusion of the $B$-invariant cochains. The map $t : L^2C^*(\Phi) \to L^2_{\mathbf{q}}C^*(\Sigma)$ is then averaging over $B$. In other words, if $\Sigma$ is identified with a subspace of Geom$(\Phi)$ via some section of $r : \text{Geom}(\Phi) \to \Sigma$, then

$$t(f)(c) = \int_{x\in B} f(xc)d\mu.$$

The proofs of the next two lemmas can be found in [79] and we omit them.

**LEMMA 20.8.2.** ([79, Lemma 13.6].)

*(i)* $t \circ r = id : L^2_{\mathbf{q}}C^i(\Sigma) \to L^2_{\mathbf{q}}C^i(\Sigma)$.

*(ii) The maps r and t are adjoint to each other.*

*(iii) These maps take harmonic cocycles to harmonic cocycles.*

Consider the diagram

$$\begin{array}{ccc} L^2_{\mathbf{q}}C^*(\Sigma) & \xrightarrow{\ r\ } & L^2C^*(\Phi) \\ {\scriptstyle p}\big\downarrow & & \big\downarrow{\scriptstyle P} \\ L^2_{\mathbf{q}}\mathcal{H}^*(\Sigma) & \xrightarrow{\ r\ } & \mathcal{H}^*(\Phi) \end{array}$$

where $p$ and $P$ denote the projections onto harmonic cocycles.

**LEMMA 20.8.3.** ([79, Lemma 13.7].) $P \circ r = r \circ p$.

**THEOREM 20.8.4.** ([79, Theorem 13.8].) *Suppose $\Phi$ is a building with a chamber transitive automorphism group $G$ and with thickness vector $\mathbf{q}$. Then the $L^2$-Betti numbers of* $\mathrm{Geom}(\Phi)$ *$(= \mathcal{U}(\Phi, K))$ equal the $L^2_{\mathbf{q}}$-Betti numbers of $\Sigma$, i.e.,*

$$b^i(\Phi; G) = b^i_{\mathbf{q}}(\Sigma).$$

*Remark.* This theorem is proved in [109, Fact 3.5] in the case where the building comes from a Tits system. Here we use Lemma 20.8.3 to weaken the hypothesis to the case of an arbitrary chamber transitive group $G$. The technique in [109] of integrating over $B$ is replaced by the use of the transfer map $t$.

*Proof of Theorem 20.8.4.* For each simplex $c$ in the fundamental chamber $K$, consider the commutative diagram

$$\begin{array}{ccc} L^2_{\mathbf{q}}(W) & \xrightarrow{\ r\ } & L^2(G/B) \\ \big\downarrow & & \big\downarrow \\ L^2_{\mathbf{q}}(W/W_{S(c)}) & \xrightarrow{\ r\ } & L^2(G/G_{S(c)}) \end{array}$$

where $W_{S(c)}$ and $G_{S(c)}$ are the isotropy subgroups of $c$ in $W$ and $G$, respectively, the vertical maps are orthogonal projections and $r$ $(= r^*)$ is the map induced by $r : G/B \to W$. Let $e_B \in L^2(G/B)$ denote the characteristic function of $B$ and let $e_c$ be its image under orthogonal projection to $L^2(G/G_{S(c)})$. ($e_c$ is the characteristic function of $G_{S(c)}$ renormalized to have norm 1.) Note that $e_B$ is the image of the basis vector $e_1 \in L^2(W)$ under $r$ and $e_c$ is the image of $a_{S(c)}$.

We have the commutative diagram:

$$\begin{array}{ccc}
\bigoplus L^2_{\mathbf{q}}(W) & \xrightarrow{\quad r \quad} & \bigoplus L^2(G) \\
\downarrow & & \downarrow \\
\bigoplus L^2_{\mathbf{q}}(W/W_{S(c)}) = L^2_{\mathbf{q}}C^i(\Sigma) & \xrightarrow{r} & L^2C^i(\Phi) = \bigoplus L^2(G/G_{S(c)}) \\
{\scriptstyle p}\downarrow & & \downarrow{\scriptstyle P} \\
L^2_{\mathbf{q}}\mathcal{H}^i(\Sigma) & \xrightarrow{r} & \mathcal{H}^i(\Phi),
\end{array}$$

where the sums are over all $c \in K^{(i)}$. Let $\mathbf{e} \in \bigoplus L^2(G/G_{S(c)})$ be the vector $(e_c)_{c \in K^{(i)}}$ and let $\mathbf{a} \in \bigoplus L^2_{\mathbf{q}}(W/W_{S(c)})$ be the vector $(a_{S(c)})_{c \in K^{(i)}}$. (So, $r(\mathbf{a}) = \mathbf{e}$.) Using Lemma 20.8.2, we get

$$\begin{aligned}
b^i(\Phi; G) &:= \dim_{\mathcal{N}(G)} \mathcal{H}^i(\Phi) \\
&= \langle P(\mathbf{e}), \mathbf{e} \rangle = \langle Pr(\mathbf{a}), r(\mathbf{a}) \rangle = \langle rp(\mathbf{a}), r(\mathbf{a}) \rangle \\
&= \langle p(\mathbf{a}), tr(\mathbf{a}) \rangle_{\mathbf{q}} = \langle p(\mathbf{a}), \mathbf{a} \rangle_{\mathbf{q}} = \dim_{\mathcal{N}_{\mathbf{q}}} L^2_{\mathbf{q}}\mathcal{H}^i(\Sigma) \\
&:= b^i_{\mathbf{q}}(\Sigma).
\end{aligned}$$

□

## The Decomposition Theorem for $L^2(G/B)$

For each $T \in \mathcal{S}$, let

$$\widetilde{A}^T := L^2(G/G_T) = L^2(G)^{G_T}$$

be the subspace of $L^2(G/B)$ consisting of the square summable functions on $G$ which are constant on each coset $gG_T$. Set

$$\widetilde{D}_T := \widetilde{A}^T \cap \left( \sum_{U \in \mathcal{S}_{>T}} \widetilde{A}^U \right)^{\perp}.$$

$\widetilde{D}_T$ is a closed $G$-stable subspace in the regular representation. (It corresponds to the $\mathcal{N}_{\mathbf{q}}$-module $D_T$ defined in 20.6.)

**THEOREM 20.8.5.** (The Decomposition Theorem for $L^2(G/B)$.) *Suppose $G$ is a chamber transitive automorphism group of a building $\Phi$ and $B$ is the* *stabilizer of a given chamber. If the thickness vector* $\mathbf{q}$ *lies in* $\mathcal{R}^{-1}$, *then*

$$\sum_{T \in \mathcal{S}} \widetilde{D}_T$$

*is a dense subspace of* $L^2(G/B)$ *and a direct sum decomposition.*

Given a module $M$ and a collection of submodules $(M_\alpha)_{\alpha \in \mathcal{A}}$, the statement that $(M_\alpha)_{\alpha \in \mathcal{A}}$ gives a direct sum decomposition of $M$ can be interpreted as a statement about chain complexes as follows. Set

$$C_1 := \bigoplus_{\alpha \in \mathcal{A}} M_\alpha \quad \text{and} \quad C_0 := M,$$

where $\bigoplus$ means external direct sum. Let $\partial : C_1 \to C_0$ be the natural map. This gives a chain complex, $C_* := \{C_0, C_1\}$, with nonzero terms only in degrees 0 and 1. The statement that the internal sum $\sum M_\alpha$ is direct is equivalent to the statement that $\partial$ is injective, i.e., that $H_*(C_*)$ vanishes in dimension 1. The statement that the $M_\alpha$ span $M$ is equivalent to the statement that $\partial$ is onto, i.e., that $H_*(C_*)$ vanishes in dimension 0. Similarly, if $M$ and the $M_\alpha$ are Hilbert spaces, then the statement that $M_\alpha$ is dense in $M$ is equivalent to the statement that the reduced homology $\mathcal{H}_*(C_*)$ vanishes in dimension 0.

*Proof of Theorem 20.8.5.* The map $r$ from Lemma 20.8.1 takes $A^T$ to $\widetilde{A}^T$ and $D_T$ to $\widetilde{D}_T$. Define chain complexes $\widehat{C}_* = \{\widehat{C}_0, \widehat{C}_1\}$ and $C_* = \{C_0, C_1\}$ by

$$\widehat{C}_1 := \bigoplus_{T \in \mathcal{S}} \widetilde{D}_T \qquad \text{and} \qquad \widehat{C}_0 := L^2(G/B),$$

$$C_1 := \bigoplus_{T \in \mathcal{S}} D_T \qquad \text{and} \qquad C_0 := L^2_{\mathbf{q}}(W),$$

where the boundary maps $\widehat{C}_1 \to \widehat{C}_0$ and $C_1 \to C_0$ are the natural maps. By the Decomposition Theorem for $L^2_{\mathbf{q}}$ (Theorem 20.6.1), $\mathcal{H}_*(C_*)$ vanishes identically. So, by the proof of Theorem 20.8.4, $\mathcal{H}_*(\widehat{C}_*)$ has dimension 0 with respect to $\mathcal{N}(G)$ and hence, also vanishes identically. The theorem then follows from the previous paragraph. □

*Remark.* The representations $\widetilde{D}_T$ are defined by Dymara and Januszkiewicz in [110]. Their notation is different; they use something like $L^2(G)^\sigma$ instead of $\widetilde{D}_T$. They prove Theorem 20.8.5 (as well as, Theorem 20.8.6 below) but only under the assumption that the thickness $\mathbf{q}$ is extremely large. They deal only with the case where $\mathbf{q}$ is a singleton $q$ and they need to assume $q > \frac{1}{25}(1764^{\dim \Sigma})$.

### Decoupling Cohomology

The proof of Theorem 20.7.1 goes through to give the following.

**THEOREM 20.8.6.** ([79, Cor. 13.11], [110, Cor. 8.2 and Prop. 8.5].) *Suppose $\Phi$ is a building with a chamber transitive automorphism group G and that its thickness vector $\mathbf{q}$ lies in $\mathcal{R}^{-1}$. Then there is an isomorphism of orthogonal G-representations:*

$$\mathcal{H}^*(\Phi) \cong \bigoplus_{T\in\mathcal{S}} H^*(K, K^{S-T}) \otimes \widetilde{D}_T.$$

### Euclidean Buildings

From Corollary 20.4.4 we get the following (known) result.

**COROLLARY 20.8.7.** *Suppose that $\Phi$ is a Euclidean building with a chamber transitive automorphism group. Then its reduced $L^2$-cohomology is concentrated in the top dimension.*

### Application to Fuchsian Buildings

A building $\Phi$ is *Fuchsian* if its associated Coxeter group is a hyperbolic polygon group (see Example 6.5.3) and its automorphism group is chamber transitive. When the polygon is a right-angled $m$-gon, $m \geqslant 5$, we construct can construct Fuchsian buildings as in Example 18.1.10. The thickness vector $\mathbf{q}$ can be an arbitrary $k$-tuple of positive integers. From Corollary 20.4.5 and Theorem 20.8.6, we get the following.

**COROLLARY 20.8.8.** *Suppose $\Phi$ is a Fuchsian building, with a chamber transitive automorphism group G and with thickness vector $\mathbf{q}$. Then $\mathcal{H}^*(\Phi)$ is concentrated in dimension*

$$\begin{cases} 1 & \text{if } \mathbf{q} \notin \mathcal{R}^{-1}, \\ 2 & \text{if } \mathbf{q} \in \mathcal{R}^{-1}. \end{cases}$$

*Moreover, when $b^i(\Phi; G))$ is nonzero it is equal to $|1/W(\mathbf{q})|$.*

**Example 20.8.9.** (*Examples 17.1.15 and 20.4.6 continued.*) Suppose $K$ is a right-angled $k$-gon, $k \geqslant 4$, and the thickness of $\Phi$ is a constant integer $q$. As in Example 20.4.6,

$$\rho^{-1} = \frac{(k-2) + \sqrt{k^2 - 4k}}{2}$$

and $\mathcal{H}^*(\Phi)$ is concentrated in dimension

$$\begin{cases} 1 & \text{if } q < \rho^{-1}, \\ 2 & \text{if } q > \rho^{-1}. \end{cases}$$

## NOTES

This chapter comes from [79, 109].

**20.2.** As explained in Theorem 18.4.3, the relationship between Euler characteristics of groups acting on buildings and growth series of Coxeter groups was first pointed out by Serre [255].

**20.5.** In [91] it is suggested that there is a close analogy between Coxeter groups of type $\mathrm{HM}^n$, with $n$ odd, and 3-manifold theory (at least in the right-angled case). A well-known conjecture of Thurston asserts that any aspherical 3-manifold $M^3$ should "virtually fiber" over $S^1$. This means that there should be a finite-sheeted coveri $\widetilde{M}^3 \to M^3$ so that $\widetilde{M}^3$ fibers over $S^1$. Similarly, it is conjectured in [91, §14] that if $W$ is right-angled and $L$ is a PL triangulation of $S^{n-1}$, with $n$ odd, then there should be a finite index, torsion-free subgroup $\Gamma \subset W$ so that $\Sigma/\Gamma$ fibers over $S^1$. Lück [191] has proved the vanishing of the $L^2$-Betti numbers for the universal cover of any mapping torus (Theorem J.8.1 in Appendix J.8). Since $L^2$-Betti numbers are multiplicative with respect to finite index subgroups (Proposition J.5.1 (ii)), the above conjecture implies Conjecture 20.5.1 (at least for $n$ odd). In the right-angled case, the arguments of [91] would also give us the even-dimensional case.

# Appendix A

# CELL COMPLEXES

## A.1. CELLS AND CELL COMPLEXES

### Affine Spaces and Affine Maps

An *affine space* is a set $\mathbb{A}$ together with a simply transitive action of a real vector space $V$ on it. (In fancier language, $\mathbb{A}$ is a "torsor" for $V$.) $V$ is the *tangent space* of $\mathbb{A}$. An *affine subspace* of $\mathbb{A}$ is the orbit of a point under a linear subspace of $V$. For example, if $E$ is a vector space and $V$ is a linear subspace, then any coset of the form $\mathbb{A} = e + V$, with $e \in E$, is an affine subspace of $E$; its tangent space is $V$. The standard example of an affine space is Euclidean $n$-space $\mathbb{E}^n$; its tangent space is $\mathbb{R}^n$. (See 6.2.)

Given a point $p$ in $\mathbb{A}$ and a vector $v$ in $V$, the translate of $p$ by $v$ is denoted by $p + v$. If $p$ and $q$ are two points in $\mathbb{A}$, then, since the $V$-action is simply transitive, there is a unique $v \in V$ such that $q = p + v$. We write $v = q - p$. Hence, once a base point $p$ in $\mathbb{A}$ has been chosen we get a bijection $\alpha_p : V \to \mathbb{A}$ defined by $v \to p + v$.

Suppose $\mathbb{A}$ and $\mathbb{A}'$ are affine spaces with tangent spaces $V$ and $V'$. A function $f : \mathbb{A} \to \mathbb{A}'$ is an *affine map* if there is a linear map $F : V \to V'$ such that $f(p + v) - f(p) = F(v)$ for all $v \in V$. So, a map $f : V \to V'$ between two vector spaces is affine if and only if $f(v) = u_0 + F(v)$ for some fixed $u_0 \in V'$ and some linear map $F$. $F$ is the *linear part* of $f$ and $u_0$ is its *translational part*. Thus, an affine map is a linear map plus a translation. It follows that the group of affine automorphisms of $\mathbb{A}$ is the semidirect product $V \rtimes GL(V)$.

If $p_0, p_1, \dots, p_n$ are points in an affine space $\mathbb{A}$ and $t_0, t_1, \dots, t_n$ are real numbers such that $\sum t_i = 1$, then the *affine combination* $t_0p_0 + t_1p_1 + \cdots + t_np_n$ is defined to be the point $p_0 + t_1(p_1 - p_0) + \cdots + t_n(p_n - p_0)$. (This is independent of the choice of which $p_i$ is used as base point.) The set of all affine combinations of a set is its *affine span*. The points $p_0, p_1, \dots, p_n$ are *affinely independent* if the vectors $p_1 - p_0, \dots, p_n - p_0$ are linearly independent. If $p_0, p_1, \dots, p_n$ are affinely independent, then the coordinates $(t_0, t_1, \dots, t_n)$ of any point $p = \sum t_ip_i$ in the affine span of $p_i$ are uniquely determined by $p$. An affine combination $t_0p_0 + \cdots + t_np_n$ is a *convex combination* if each $t_i \geqslant 0$. Two points $p_0$ and $p_1$ determine a *line segment* $[p_0, p_1]$, defined as the set of all convex combinations of $p_0$ and $p_1$, i.e., $[p_0, p_1] = \{(1 - t)p_0 + tp_1 \mid t \in [0, 1]\}$.

Similarly, $(p_0, p_1) = \{(1-t)p_0 + tp_1 \mid t \in (0,1)\}$ is the *open line segment* determined by $p_0$ and $p_1$.

A subset $X$ of $\mathbb{A}$ is *convex* if given any two points in $X$, the line segment between them is contained in $X$. Equivalently, $X$ is convex if and only if any convex combination of points in $X$ is contained in $X$. The *convex hull* of a subset $S$ of $\mathbb{A}$ is the smallest convex set which contains it. In other words, it is the set of all convex combinations of points in $S$.

A *hyperplane* in $\mathbb{A}$ is an affine subspace of codimension one. Thus, a hyperplane is defined by an equation of the form $\varphi(x) = 0$, where $\varphi : \mathbb{A} \to \mathbb{R}$ is a nonconstant affine map. A *half-space* in $\mathbb{A}$ is defined by an inequality of the form $\varphi(x) \geqslant 0$. If the inequality is strict, we get an *open half-space*. Any hyperplane separates $\mathbb{A}$ into two half-spaces.

Suppose $X$ is a closed convex subset of $\mathbb{A}$. An affine hyperplane $H$ is a *supporting hyperplane* of $X$ if $X \cap H \neq \emptyset$ and if $X$ is contained in one of the half-spaces bounded by $H$. (That is to say, *$X$ lies on one side of $H$*.) Obviously, any closed convex subset $X$ is the intersection of the half-spaces which contain it and are bounded by supporting hyperplanes.

### Convex Polytopes

**DEFINITION A.1.1.** A *convex polytope* in $\mathbb{A}$ is the convex hull of a finite subset. Its *dimension* is the dimension of the affine subspace it spans. (We will use the term *convex cell* interchangeably with convex polytope.)

Equivalently, a convex polytope is a compact intersection of a finite number of half-spaces.

If a convex polytope is 0-dimensional, it is a point; if it is 1-dimensional, it is an interval; if it is 2-dimensional, it is a polygon.

Suppose $P$ is a convex polytope and $H$ a supporting hyperplane. Then $P \cap H$ is also a convex polytope. (If $P$ is the convex hull of $S$, then $P \cap H$ is the convex hull of $S \cap H$.) $P \cap H$ is called a *face* of $P$. A 0-dimensional face of $P$ is a *vertex*; a one-dimensional face is an *edge*. The set of vertices is denoted $\mathrm{Vert}(P)$ and called the *vertex set*.

**DEFINITION A.1.2.** A *simplex* $\sigma$ in $\mathbb{A}$ is the convex hull of a set $\{p_0, \dots, p_n\}$ of affinely independent points. Equivalently, $\sigma$ is the set of all convex combinations of the $p_i$. The points $p_0, \dots p_n$ are the *vertices* of $\sigma$.

It follows that a simplex is a convex polytope of dimension $n$ (where $n$ is the number of vertices minus 1). A 2-simplex is a triangle; a 3-simplex is a tetrahedron. The following basic lemma explains why simplices are easier to work with than general convex polytopes.

**LEMMA A.1.3.** *Suppose that $\sigma$ (resp. $\sigma'$) is a simplex in some affine space, that $I$ (resp. $I'$) is its vertex set and that $\mathbb{A}$ (resp. $\mathbb{A}'$) is the affine subspace spanned by $I$ (resp. $I'$). Let $\varphi : I \to I'$ be any function. Then there is a unique affine map $f : \mathbb{A} \to \mathbb{A}'$ such that $f(p) = \varphi(p)$ for each vertex $p \in I$ (in particular, $f(\sigma) \subset \sigma'$).*

In other words, the set of affine maps from a simplex $\sigma$ to another one $\sigma'$ is naturally identifed with the set of functions $\mathrm{Vert}(\sigma) \to \mathrm{Vert}(\sigma')$.

*Proof.* Suppose $I = \{p_0, \dots, p_n\}$. Then $\{p_1 - p_0, \dots, p_n - p_0\}$ is a basis for the tangent space of $\mathbb{A}$. Hence, there is a unique linear map $F$ between tangent spaces defined by $p_i - p_0 \to \varphi(p_i) - \varphi(p_0)$. The required affine map is then defined by $f(x) = F(x - p_0) + \varphi(p_0)$. □

**Example A.1.4.** The *standard n-simplex* $\Delta^n$ is the convex hull of the standard basis, $e_1, \dots, e_{n+1}$, in $\mathbb{R}^{n+1}$. More generally, given a set $S$, let $\mathbb{R}^S$ denote the real vector space of all finitely supported functions $S \to \mathbb{R}$. ($\mathbb{R}^S$ is topologized as the direct limit of its finite-dimensional subspaces, i.e., a subset of $\mathbb{R}^S$ is *closed* if and only if its intersection with each finite-dimensional subspace is closed.) For each $s \in S$, let $e_s$ denote the characteristic function of $\{s\}$. Then $\{e_s\}_{s \in S}$ is the *standard basis* of $\mathbb{R}^S$. The *standard simplex* on $S$, denoted by $\Delta^S$, is the convex hull of the standard basis in $\mathbb{R}^S$. In other words, it is the set of all convex combinations of the $e_s$.

**Example A.1.5.** The *standard n-dimensional cube* $\square^n$ is the convex polytope $[-1, 1]^n \subset \mathbb{R}^n$. Its vertex set is $\{\pm 1\}^n$. Let $I_n = \{1, \dots, n\}$ and let $\{e_i\}_{i \in I_n}$ be the standard basis for $\mathbb{R}^n$. For each subset $J$ of $I_n$, $\mathbb{R}^J$ denotes the linear subspace spanned by $\{e_i\}_{i \in J}$ and $\square_J$ denotes the standard cube in $\mathbb{R}^J$. (If $J = \emptyset$, then $\mathbb{R}^\emptyset = \square_\emptyset = \{0\}$.) Each face of $[-1, 1]^n$ is a translate of $\square_J$ for some $J \subset I_n$. We say such a face is *type J*.

**Example A.1.6.** (*Cartesian products.*) If $P$ and $P'$ are convex polytopes in affine spaces $\mathbb{A}$ and $\mathbb{A}'$, then $P \times P'$ is a convex polytope in $\mathbb{A} \times \mathbb{A}'$. More generally, if, for $1 \leqslant i \leqslant n$, $P_i$ is a convex polytope in $\mathbb{A}_i$, then $P_1 \times \cdots \times P_n$ is a convex polytope in $\mathbb{A}_1 \times \cdots \times \mathbb{A}_n$. In the special case where each $P_i$ is an interval and each $\mathbb{A}_i$ is one dimensional, we get a convex polytope affinely isomorphic to an $n$-cube (and we also call it an "$n$-cube").

**Example A.1.7.** (*The n-dimensional octahedron.*) The *n-octahedron* $O^n$ is the convex hull of the set of points in $\mathbb{R}^n$ of the form $\varepsilon_i e_i$ where $\{e_i\}_{i \in I_n}$ is the standard basis for $\mathbb{R}^n$ and each $\varepsilon_i \in \{\pm 1\}$. The group $G$ generated by all sign changes and permutations of the coordinates acts orthogonally on $\mathbb{R}^n$. The set of all $\varepsilon_i e_i$ is a single $G$-orbit. It follows that $G$ is a group of symmetries of

$O^n$ and that each of the points $\varepsilon_i e_i$ is a vertex of $O^n$. (If one point is a vertex, then so is any other point in its orbit.) So, $O^n$ has $2n$ vertices. The group $G$ is called the *n-octahedral group*; it is the semidirect product of the symmetric group $S_n$ on $n$ letters with the group $(\mathbf{C}_2)^n$ of all sign changes. (In fact, $G$ is the Coxeter group of type $\mathbf{B}_n$ discussed in Example 6.7.2.) It is easy to see that $\{\varepsilon_{i_1} e_{i_1}, \dots, \varepsilon_{i_k} e_{i_k}\}$ is the vertex set of a face of $O^n$ if and only if the elements of $\{e_{i_1}, \dots, e_{i_k}\}$ are distinct. Thus, each proper face of $O^n$ is a simplex. In the literature the $n$-octahedron is often called the $n$-dimensional "cross polytope."

**DEFINITION A.1.8.** A (linear) *convex polyhedral cone* $C$ in a finite-dimensional real vector space $V$ is the intersection of a finite number of linear half-spaces in $V$ (a half-space is *linear* if its bounding hyperplane is a linear subspace). $C$ is *essential* if contains no line (or equivalently, if the intersection of $C$ with a sphere about the origin does not contain any pair of antipodal points). If $C$ is essential, then the origin is its *vertex* (or *cone point*). An essential polyhedral cone $C$ is a *simplicial cone* if any $m$ of its codimension one faces intersect in a codimension $m$ face (in other words, if its intersection with a sphere about the origin is a spherical simplex).

### Cell Complexes

We begin with the classical definition of a cell complex in an affine space $\mathbb{A}$.

**DEFINITION A.1.9.** A *convex cell complex* is a collection $\Lambda$ of convex polytopes in $\mathbb{A}$ such that

(i) if $P \in \Lambda$ and $F$ is a face of $P$, then $F \in \Lambda$ and

(ii) for any two polytopes $P$ and $Q$ in $\Lambda$ either $P \cap Q = \emptyset$ or $P \cap Q$ is a common face of both polytopes.

Traditionally, the elements of $\Lambda$ are called *cells* (instead of "polytopes"). A subset $\Lambda'$ of $\Lambda$ is a *subcomplex* if it is closed under the operation of taking faces (i.e., if $\Lambda'$ satisfies condition (i) above). If each cell of $\Lambda$ is a simplex, then $\Lambda$ is a *simplicial complex*. If each cell of $\Lambda$ is a cube, then $\Lambda$ is a *cubical complex*. $\Lambda$ is *locally finite* if each cell in $\Lambda$ is a face of only finitely many other cells in $\Lambda$.

Associated to $\Lambda$, there is a topological space, called its *underlying space*. For the moment we denote it $X(\Lambda)$. As a set, $X(\Lambda)$ is

$$X(\Lambda) := \bigcup_{P \in \Lambda} P.$$

If $\Lambda$ is locally finite, then $X(\Lambda)$ is given the induced topology as a subspace of $\mathbb{A}$. If $\Lambda$ is not locally finite, this topology is not the correct one. Instead

$X(\Lambda)$ is given the "CW topology." This means that $X(\Lambda)$ is topologized as the direct limit of the underlying spaces of its finite subcomplexes. (See the last paragraph of this section for the definition of "CW complex.")

Perhaps unfortunately, it is traditional in topology to blur the distinction between a cell complex and its underlying space and use the same symbol, say, $\Lambda$ to stand for both. Sometimes we will follow this tradition and sometimes we will not. However, henceforth, will not use the notation $X(\Lambda)$. When we want to emphasize the set of cells in $\Lambda$ we will write something like $\mathcal{F}(\Lambda)$ and call it the "face poset" of $\Lambda$. (See Examples A.2.3 in the next section.)

**Definition A.1.10.** A space $X$ is a *polyhedron* if it is homeomorphic to (the underlying space of) a convex cell complex.

**Definition A.1.11.** A *cellulation* of a space $X$ is a homeomorphism $f$ from a cell complex $\Lambda$ onto $X$; $f$ is a *triangulation* if $\Lambda$ is a simplicial complex. By an abuse of language we will sometimes say that $\Lambda$ is a cellulation (or triangulation) of $X$ without specifying the homeomorphism.

**Example A.1.12.** Given a convex polytope $P$, the set of all its nonempty faces is a convex cell complex. We also denote this convex cell complex by $P$. A slightly less trivial example is the subcomplex $\partial P$ of $P$ consisting of all proper faces. It is the *boundary complex* of $P$. If $P$ is $n$-dimensional, then the underlying space of $P$ is an $n$-disk and the underlying space of $\partial P$ is an $(n-1)$-sphere.

**Definition A.1.13.** The *k-skeleton* of a convex cell complex $\Lambda$ is the subcomplex $\Lambda^k$ consisting of all cells of dimension $\leqslant k$.

## CW Complexes

Given a space $Y$ and a map $\varphi$ from $\partial D^n$ $(= S^{n-1})$ to $Y$, let $Y \cup_\varphi D^n$ denote the quotient space of the disjoint union $Y \coprod D^n$ by the equivalence relation which identifies a point $x \in \partial D^n$ with its image $\varphi(x)$. $Y \cup_\varphi D^n$ is called the space formed by *attaching an n-cell* to $Y$ via $\varphi$.

A *CW complex* is a space $X$ which can be constructed by successively attaching cells to the union of the lower-dimensional cells. A more intrinsic definition goes as follows. A *CW complex* is a Hausdorff space $X$ together with an increasing filtration by closed subspaces: $X^{-1} \subset X^0 \subset X^1 \subset \cdots$, where $X^{-1} := \emptyset$ and $X^n - X^{n-1}$ is (homeomorphic to) a disjoint union of open $n$-cells $\{e^n_\alpha\}$, $\alpha$ ranging over some index set. Moreover, for each such cell there must be a *characteristic map* $\Phi_\alpha : (D^n, \partial D^n) \to (X^n, X^{n-1})$ which maps the interior of $D^n$ homeomorphically onto $e^n_\alpha$. (The restriction of $\Phi_\alpha$ to $\partial D^n$ is the *attaching map* for $e^n_\alpha$.) If there are an infinite number of cells, then it is further required that $X$ is the "closure finite" (i.e., the image of each attaching map is contained in a finite number of lower dimensional cells) and has the "weak topology" (i.e., a subset $C \subset X$ is closed if and only if its intersection with each cell is closed). $X^n$ is called the *n-skeleton* of $X$. $X$ is a *regular* CW complex if each attaching map is an embedding. For a more detailed discussion of CW complexes, see [153, Appendix].

## A.2. POSETS AND ABSTRACT SIMPLICIAL COMPLEXES

A *poset* is a partially ordered set. Given a poset $\mathcal{P}$ and an element $p \in \mathcal{P}$, put

$$\mathcal{P}_{\leqslant p} := \{x \in \mathcal{P} \mid x \leqslant p\}. \tag{A.1}$$

Define $\mathcal{P}_{\geqslant p}$, $\mathcal{P}_{<p}$ and $\mathcal{P}_{>p}$ similarly. Given elements $p, q \in \mathcal{P}$, define $\mathcal{P}_{[p,q]}$, the *interval* from $p$ to $q$, by

$$\mathcal{P}_{[p,q]} := \{x \in \mathcal{P} \mid q \leqslant x \leqslant p\}. \tag{A.2}$$

The *opposite* or *dual* poset to $\mathcal{P}$ is the poset $\mathcal{P}^{op}$ with the same underlying set but with the order relations reversed.

**Example A.2.1.** (*Power sets.*) Given a set $I$, its *power set* $\mathcal{P}(I)$ is the set of all subsets of $I$. It is partially ordered by inclusion.

**Example A.2.2.** (*The poset of intervals.*) Given a poset $\mathcal{P}$, let $\mathcal{I}(\mathcal{P})$ denote its set of intervals, partially ordered by inclusion.

**Example A.2.3**

- (*The poset of faces of a polytope.*) Given a convex polytope $P$, let $\widetilde{\mathcal{F}}(P)$ denote its set of faces (including the empty face), partially ordered by inclusion. Let $\mathcal{F}(P) = \widetilde{\mathcal{F}}(P)_{>\emptyset}$ denote the poset of nonempty faces.
- (*The poset of cells of a cell complex.*) More generally, if, as in Definition A.1.9, $\Lambda$ is a cell complex, then let $\mathcal{F}(\Lambda)$ denote the poset of cells of $\Lambda$. In order to be consistent with the notation of the previous paragraph, we let $\widetilde{\mathcal{F}}(\Lambda) = \mathcal{F}(\Lambda) \cup \{\emptyset\}$ be the poset obtained by adjoining the "empty cell" to $\mathcal{F}(\Lambda)$.
- (*The poset of simplices of a simplicial complex.*) Suppose $\Lambda = L$, a simplicial complex. For consistentcy with the notation in 7.1, write

  $$\mathcal{S}(L) := \widetilde{\mathcal{F}}(L), \tag{A.3}$$

  for the poset of simplices of $L$ (including the "empty simplex").

**Example A.2.4.** (*The face poset of an n-simplex.*) $\mathcal{S}(\Delta^n) \cong \mathcal{P}(I_{n+1})$, where $I_{n+1} = \{1, \dots, n+1\}$.

**Example A.2.5.** (*The face poset of an n-cube.*) Consider the $n$-cube $\square^n := [-1,1]^n$. Its vertex set is $\{\pm 1\}^n$. Explicitly, for each $J \in \mathcal{P}(I_n)$, define a function $\varepsilon_J : I_n \to \{\pm 1\}$ by

$$\varepsilon_J(i) := \begin{cases} -1 & \text{if } i \in J, \\ +1 & \text{if } i \notin J. \end{cases}$$

and set

$$e_J := \sum_{i=1}^{n} \varepsilon_J(i) e_i = \sum_{i \notin J} e_i - \sum_{i \in J} e_i \tag{A.4}$$

$\{e_J\}_{J \in \mathcal{P}(I_n)}$ is clearly the vertex set of $\square^n$. In other words, the vertex set of $\square^n$ can be identified with the power set $\mathcal{P}(I_n)$. It is also clear that for each interval $[J_1, J_2] \in \mathcal{I}(\mathcal{P}(I_n))$, the set of vertices $\{e_J\}_{J \in [J_1, J_2]}$ is the vertex set of the face of $\square^n$ defined by the equations

$$x_i = -1 \qquad \text{for } i \in J_1,$$
$$x_i = +1 \qquad \text{for } i \in I_n - J_2.$$

Denote this face by $\square_{[J_1, J_2]}$. It is a face of type $J_2 - J_1$. Thus, $\widetilde{\mathcal{F}}(\square^n) \cong \mathcal{I}(\mathcal{P}(I_n))$ via the correspondence $\square_{[J_1, J_2]} \leftrightarrow [J_1, J_2]$.

Two polytopes $P$ and $P'$ are *combinatorially isomorphic* if their face posets are isomorphic. $P$ and $P'$ are *combinatorially dual* (or simply *dual*) if $\widetilde{\mathcal{F}}(P) \cong \widetilde{\mathcal{F}}(P')^{op}$. Given an $n$-dimensional convex polytope $P$ in Euclidean space $\mathbb{E}^n$, there is a simple construction (cf. [39, pp. 37–43]) of another polytope $P' \subset \mathbb{E}^n$ which is combinatorially dual to it. As examples, an $n$-cube is dual to an $n$-dimensional octahedron; an $n$-simplex is dual to itself; any polygon is also self-dual.

### Abstract Simplicial Complexes

**Definition A.2.6.** An *abstract simplicial complex* consists of a set $S$ (the *vertex set*) and a collection $\mathcal{S}$ of finite subsets of $S$ such that

(i) for each $s \in S$, $\{s\} \in \mathcal{S}$ and

(ii) if $T \in \mathcal{S}$ and if $T' \subset T$, then $T' \in \mathcal{S}$.

An abstract simplicial complex $\mathcal{S}$ is a poset: the partial order is inclusion. Condition (ii) means that if $T \in \mathcal{S}$, then $\mathcal{S}_{\leqslant T}$ is the power set of $T$. An element of $\mathcal{S}$ is a *simplex* of $\mathcal{S}$. If $T$ is a simplex of $\mathcal{S}$ and $T' \leqslant T$, then $T'$ is a *face* of $T$.

The *dimension* of a simplex $T$ is defined by

$$\dim T := \mathrm{Card}(T) - 1. \tag{A.5}$$

A simplex of dimension $k$ is a *$k$-simplex*. A 0-simplex is also called a *vertex*. (We will often blur the distinction between an element $s \in S$ and the singleton $\{s\}$ and write $\mathrm{Vert}(\mathcal{S}) := S$.) A 1-simplex is an *edge*.

Sometimes we will say a poset $\mathcal{P}$ is an abstract simplicial complex if there is an (order-preserving) isomorphism from $\mathcal{P}$ to an abstract simplicial complex $\mathcal{S}$.

**DEFINITION A.2.7.** A subset $\mathcal{S}'$ of an abstract simplicial complex $\mathcal{S}$ is a *subcomplex* of $\mathcal{S}$ if it is also an abstract simplicial complex (in other words, if $T \in \mathcal{S}'$ and $T' < T$, then $T' \in \mathcal{S}'$). A subcomplex $\mathcal{S}'$ of $\mathcal{S}$ is a *full subcomplex* if $T \in \mathcal{S}$ and $T \subset \mathrm{Vert}(\mathcal{S}')$ implies that $T \in \mathcal{S}'$.

The next definition is justified by Lemma A.1.3.

**DEFINITION A.2.8.** Suppose $\mathcal{S}$ and $\mathcal{S}'$ are abstract simplicial complexes with vertex sets $S$ and $S'$, respectively. A *simplicial map* from $\mathcal{S}$ to $\mathcal{S}'$ is a function $\varphi : S \to S'$ such that whenever $T \in \mathcal{S}$, $\varphi(T) \in \mathcal{S}'$.

**DEFINITION A.2.9.** The *$k$-skeleton* of an abstract simplicial complex $\mathcal{S}$ is the subcomplex $\mathcal{S}^k$ consisting of all simplices of dimension $\leqslant k$.

**Example A.2.10.** Suppose $\Lambda$ is a simplicial complex as in Definition A.1.9 (i.e., $\Lambda$ is a convex cell complex in which each cell is a simplex). We can associate to $\Lambda$ an abstract simplicial complex $\mathcal{S}(\Lambda)$ as follows. The vertex set $S$ is the set of 0-dimensional cells in $\Lambda$. $\mathcal{S}(\Lambda)$ is defined by declaring that a subset $T \subset S$ belongs to $\mathcal{S}(\Lambda)$ if and only if $T$ spans a simplex $\sigma_T$ in $\Lambda$. We will show below that all abstract simplicial complexes arise from a convex cell complex in this fashion. That is to say, any abstract simplicial complex $\mathcal{S}$ is associated to a geometric simplicial complex $\Lambda$ so that $\mathcal{S} = \mathcal{S}(\Lambda)$. Such a $\Lambda$ is called a *geometric realization* of $\mathcal{S}$.

## The Geometric Realization of an Abstract Simplicial Complex

Suppose $\mathcal{S}$ is an abstract simplicial complex with vertex set $S$. As in Example A.1.4, let $\Delta^S$ denote the standard simplex on $S$. For each nonempty finite subset $T$ of $S$, $\sigma_T$ denotes the face of $\Delta^S$ spanned by $T$. Define a subcomplex $\mathrm{Geom}(\mathcal{S})$ of $\Delta^S$ by

$$\sigma_T \in \mathrm{Geom}(\mathcal{S}) \quad \text{if and only if} \quad T \in \mathcal{S}_{>\emptyset}. \tag{A.6}$$

The convex cell complex $\mathrm{Geom}(\mathcal{S})$ is the *standard* geometric realization of $\mathcal{S}$.

**DEFINITION A.2.11.** A convex polytope $P$ is a *simple* polytope if the boundary complex of its dual polytope is a simplicial complex, in other words, if $\mathcal{F}(\partial P)^{op}$ is an abstract simplicial complex. For $n = \dim P$, this is equivalent to the condition that exactly $n$ codimension one faces meet at each vertex.

For example, a cube is simple, an octahedron is not.

**DEFINITION A.2.12.** A poset $\mathcal{P}$ is an *abstract convex cell complex* if satisfies the following two conditions:

(a) For each $p \in \mathcal{P}$, $\mathcal{P}_{\leqslant p}$ is isomorphic to the poset of faces of some convex polytope.

(b) If $p, p' \in \mathcal{P}$ are such that $\mathcal{P}_{\leqslant p} \cap \mathcal{P}_{\leqslant p'} \neq \emptyset$, then $\mathcal{P}_{\leqslant p} \cap \mathcal{P}_{\leqslant p'}$ contains a greatest element (which we shall denote by $p \cap p'$).

If each of the convex polytopes in condition (a) is a cube, then $\mathcal{P}$ is an *abstract cubical complex.*

Condition (b) in the above definition is the analog of condition (ii) in Definition A.1.9, which requires that if two cells have nonempty intersection, then the intersection is a common face of both.

It is clear that the poset of cells in a convex cell complex is an abstract convex cell complex. Conversely, it is not hard to prove that for any abstract cell complex $\mathcal{P}$, there is a cell complex which realizes it (see Example A.5.1).

## A.3. FLAG COMPLEXES AND BARYCENTRIC SUBDIVISIONS

**DEFINITION A.3.1.** An *incidence relation* on a set $S$ is a symmetric and reflexive relation.

**DEFINITION A.3.2.** Suppose $S$ is a set equipped with an incidence relation. A *flag* in $S$ is a subset of pairwise incident elements. Let Flag($S$) denote the set of all finite flags in $S$, partially ordered by inclusion. Obviously, it is an abstract simplicial complex with vertex set $S$.

**Example A.3.3.** (*Chains in a poset.*) Suppose $\mathcal{P}$ is a poset. We can symmetrize the partial order to get an incidence relation on $\mathcal{P}$: two elements $p$ and $q$ of $\mathcal{P}$ are *incident* if and only if $p \leqslant q$ or $q \leqslant p$. In a poset any finite set of pairwise incident elements is totally ordered. So, a flag in $\mathcal{P}$ is the same thing as a finite *chain*, i.e., a finite, totally ordered subset. When $\mathcal{P}$ is a poset, Flag($\mathcal{P}$) denotes the abstract simplicial complex of all finite chains in $\mathcal{P}$. We call it the *flag complex* of $\mathcal{P}$. In the literature this simplicial complex is also called the "derived complex" or the "order complex" of $\mathcal{P}$.

**DEFINITION A.3.4.** The *geometric realization of a poset* $\mathcal{P}$ is the geometric realization of the simplicial complex Flag($\mathcal{P}$). We use the notation

$$|\mathcal{P}| := \text{Geom}(\text{Flag}(\mathcal{P}))$$

especially when we are interested in the underlying space of Flag($\mathcal{P}$).

**DEFINITION A.3.5.** An abstract simplicial complex $\mathcal{S}$ with vertex set $S$ is a *flag complex* if given a finite nonempty subset $T \subset S$, any two elements of which are connected by an edge in $\mathcal{S}$, then $T$ is a simplex in $\mathcal{S}$. We will also call the geometric realization $L$ of such an $\mathcal{S}$ a *flag complex*.

Flag complexes play an important role throughout this book, particularly, in the discussion of right-angled Coxeter groups (e.g., Example 7.1.7) and in the theory of nonpositively curved cubical cell complexes (e.g., Gromov's Lemma in Appendix I.6). Flag complexes are also needed in 12.2 and 16.3.

**Example A.3.6.** (*Polygons.*) Suppose $L_m$ is the boundary complex of an $m$-gon, i.e., suppose $L_m$ is a triangulation of a circle into $m$ edges. Then $L_m$ is a flag complex if and only if $m \neq 3$.

In combinatorics literature flag complexes are called "clique complexes." In [147] Gromov called them simplicial complexes which satisfy the "no $\Delta$ condition." In [75] I called them complexes which "were determined by their 1-skeleton." (See Example A.3.8 below.)

It is worth recording the following lemma which asserts that the complexes of Definition A.3.2 are all flag complexes and that every flag complex arises from this construction. (The proof is a triviality.)

**LEMMA A.3.7.** (Compare [43, p. 29].) *Suppose $L$ is an abstract simplicial complex with vertex set $S$. Then $L$ is a flag complex if and only if there is an incidence relation on $S$ so that $L =$* Flag($S$).

*Proof.* Suppose $S$ is equipped with an incidence relation. Distinct vertices $v_0$ and $v_1$ span an edge in Flag($S$) if and only if they are incident. Hence, if $T$ is a nonempty subset of vertices which are pairwise connected by edges, then the elements of $T$ are pairwise incident. So, such a $T$ is a flag; hence, Flag($S$) is a flag complex. Conversely, suppose $L$ is a flag complex. Call two distinct vertices *incident* if and only if they are connected by an edge. Then, clearly, $L = \text{Flag}(S)$. □

**Example A.3.8.** (*Simplicial graph = incidence relation.*) A simplicial graph with vertex set $S$ carries exactly the same information as an incidence relation on $S$. Indeed, given such a simplicial graph $J$, define vertices $v$ and $v'$ to be *incident* if $v = v'$ or $\{v, v'\}$ is an edge of $J$. The associated flag complex

$L = \text{Flag}(S)$ can be described as follows: $T \subset S$ spans a simplex $\sigma_T$ of $L$ if and only if $T$ spans a complete subgraph of $J$ (i.e., if any two distinct vertices of $T$ are connected by an edge in $J$.) The 1-skeleton of $L$ is $J$ and $L$ is the smallest full subcomplex of the full simplex on $S$ which contains $J$. Thus, a flag complex is a simplicial complex which is "determined by its 1-skeleton."

### Barycentric Subdivision

Let $P$ be a convex polytope. For each face $F$ of $P$ choose a point $v_F$ in the relative interior of $F$. This is enough data to define a certain subdivision of $P$ into simplices. Given a flag $\alpha = (F_1 < \cdots < F_k) \in \text{Flag}(\mathcal{F}(P))$, the associated points $v_{F_1}, \ldots, v_{F_k}$ are affinely independent. Hence, we can define $\langle\alpha\rangle$ to be the $k$-simplex in $P$ spanned by $v_{F_1}, \ldots, v_{F_k}$.

**PROPOSITION A.3.9.** *Let $P$ be a convex polytope. Each point of $P$ lies in the relative interior of some simplex $\langle\alpha\rangle$ for some unique $\alpha \in \text{Flag}(\mathcal{F}(P))$.*

The proposition means that $\{\langle\alpha\rangle\}_{\alpha\in\text{Flag}(\mathcal{F}(P))}$ defines a *subdivision* of $P$ in the sense that the relative interiors of the simplices $\langle\alpha\rangle$ give a partition of $P$ into disjoint subsets. The convex cell complex $\{\langle\alpha\rangle\}_{\alpha\in\text{Flag}(\mathcal{F}(P))}$ is a simplicial complex. It will be denoted by $bP$ and called the *barycentric subdivision* of $P$. Another way of saying this is that $bP$ is a geometric realization of the abstract simplicial complex $\text{Flag}(\mathcal{F}(P))$.

More generally, if $\Lambda$ is a convex cell complex, one can define its *barycentric subdivision* to be the simplicial complex $b\Lambda$ obtained by barycentrically subdividing all the cells of $\Lambda$. The associated abstract simplicial complex is $\text{Flag}(\mathcal{F}(\Lambda))$. In other words, the barycentric subdivision of $\Lambda$ is the geometric realization of the poset of cells of $\Lambda$, that is, $b\Lambda = |\mathcal{F}(\Lambda)|$.

*Proof of Proposition A.3.9.* The proof is by induction on $\dim P$. For $\dim P = 0$ there is nothing to prove. So, suppose that $\dim P = n$ and the proposition is true in dimensions $< n$. Roughly, the proof comes down to the fact that $P$ is the cone on $\partial P$. More precisely, suppose $x \in P$. If $x = v_P$, then $\langle\alpha\rangle$ is the 0-simplex $v_P$. If $x \neq v_P$, then the ray from $v_P$ through $x$ intersects $\partial P$ in a unique point $x'$. Since $x'$ lies in some proper face $F$ of $P$, it follows from the inductive hypothesis that $x'$ lies in the relative interior of some simplex $\langle\alpha'\rangle$ for some unique $\alpha' \in \text{Flag}(\mathcal{F}(F))$. If $x = x'$, we are done. Otherwise, $x$ is in the relative interior of the simplex $\alpha$ spanned by $v_P$ and the vertices of $\alpha'$. $\square$

*Remark A.3.10.* Since the barycentric subdivision of any convex cell complex is a flag complex, the condition that a simplicial complex $L$ be a flag complex imposes no restrictions on its topology: it can be any polyhedron.

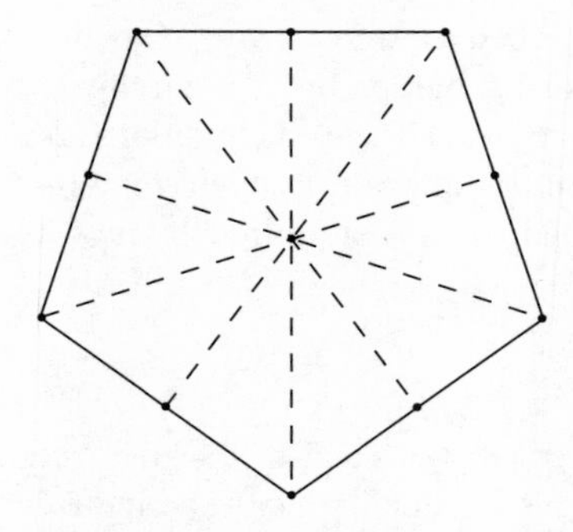

Figure A.1. Barycentric subdivision of a polygon.

**DEFINITION A.3.11.** A *combinatorial equivalence* between two cell complexes $\Lambda$ and $\Lambda'$ is an isomorphism of face posets $\varphi : \mathcal{F}(\Lambda) \to \mathcal{F}(\Lambda')$. Let $\mathrm{Aut}(\Lambda)$ denote the group of combinatorial self-equivalences of $\Lambda$.

A combinatorial equivalence $\varphi : \mathcal{F}(\Lambda) \to \mathcal{F}(\Lambda')$ induces a simplicial isomorphism $\mathrm{Flag}(\mathcal{F}(\Lambda)) \to \mathrm{Flag}(\mathcal{F}(\Lambda'))$. The geometric realization of this simplicial isomorphism is a homeomorphism $b\varphi : b\Lambda \to b\Lambda'$ of barycentric subdivisions. This shows that any combinatorial equivalence induces a cell-preserving homeomorphism of the underlying cell complexes.

## A.4. JOINS

Suppose $E$ and $E'$ are disjoint subspaces of an affine space $\mathbb{A}$. If $S$ (resp. $S'$) is a maximal affinely independent set of points in $E$ (resp. $E'$), then $S \cup S'$ might or might not be affinely independent in $\mathbb{A}$. If it is, then $E$ and $E'$ are *in general position.* It is easy to see that this condition does not depend on the choice of $S$ and $S'$. Here is an equivalent definition in the case where $E$ and $E'$ are finite dimensional. Let $E''$ denote the affine subspace spanned by $E$ and $E'$. Since we can find a maximal affinely independent subset of $E''$ having each of its points either in $E$ or $E'$, it is clear that $\dim E'' \leqslant \dim E + \dim E' + 1$ and that we have equality if and only if $E$ and $E'$ are in general position. If they are in general position, then $E''$ is called the *join* of $E$ and $E'$ and we will write $E'' = E * E'$.

**DEFINITION A.4.1.** Suppose $P$ and $P'$ are convex polytopes in general position in some affine space (more precisely, the affine subpaces which they span are in general position). The *join* of $P$ and $P'$ is the convex polytope $P * P'$ defined as the convex hull of $P \cup P'$.

Here are two other equivalent definitions of the join:

- $P * P'$ is the convex hull of $\mathrm{Vert}(P) \cup \mathrm{Vert}(P')$.
- $P * P'$ is the union of all line segments connecting a point in $P$ to one in $P'$.

By the first paragraph of this section,

$$\dim(P * P') = \dim P + \dim P' + 1. \tag{A.7}$$

**Example A.4.2.** The join of two simplices is a simplex.

**DEFINITION A.4.3.** Suppose $\Lambda$ (resp. $\Lambda'$) is a convex cell complex in an affine subspace $E$ (resp. $E'$) of some affine space. Suppose further that $E$ and $E'$ are in general position. Their *join* $\Lambda * \Lambda'$ is the cell complex consisting of all cells obtained by taking the join of a cell in $\Lambda$ with one in $\Lambda'$. (Here it is important to include the cells which lie entirely in one cell complex or the other. This is done by remembering to take into account the empty cell. For example, a cell $P \in \Lambda$ can be regarded as the join of $P$ and $\emptyset$.)

The *cone* on a cell complex $\Lambda$, denoted $\mathrm{Cone}(\Lambda)$, is defined to be the join of $\Lambda$ with a disjoint point.

**Example A.4.4.** (Compare Example A.1.7.) Let $S^0$ denote the cell complex consiting of two points ($S^0$ is the 0-sphere). The $n$-octahedron is the $n$-fold join of $S^0$ with itself: $O^n = S^0 * \cdots * S^0$.

If $\Lambda$ and $\Lambda'$ are locally finite, then the underlying space of $\Lambda * \Lambda'$ is the union of all line segments connecting a point in $\Lambda$ to one in $\Lambda'$. This motivates the following definition.

**DEFINITION A.4.5.** Suppose $X$ and $Y$ are nonempty topological spaces. Their *join* $X * Y$ is defined to be the quotient of $X \times Y \times [0, 1]$ by the equivalence relation $\sim$ generated by

$$(x, y, 0) \sim (x, y', 0) \quad \text{and} \quad (x, y, 1) \sim (x', y, 1)$$

where $x, x' \in X$ and $y, y' \in Y$. In particular, $\mathrm{Cone}(X)$, the *cone* on $X$ is defined to be the join of $X$ and a point.

What is the poset of faces of the join of two convex polytopes $P$ and $P'$? The answer is obvious:

$$\widetilde{\mathcal{F}}(P * P') \cong \widetilde{\mathcal{F}}(P) \times \widetilde{\mathcal{F}}(P'), \tag{A.8}$$

where the isomorphism is given by $F * F' \leftrightarrow (F, F')$. Here $F$ and $F'$ are faces of $P$ and $P'$, respectively, and either can be $\emptyset$. The isomorphism in (A.8) is order preserving if we put the natural partial order on the product $\widetilde{\mathcal{F}}(P) \times \widetilde{\mathcal{F}}(P')$: $(F_1, F_1') \leqslant (F_2, F_2')$ if and only if $F_1 \leqslant F_2$ and $F_1' \leqslant F_2'$. Similarly, if $\Lambda$ and $\Lambda'$ are convex cell complexes, then

$$\widetilde{\mathcal{F}}(\Lambda * \Lambda') \cong \widetilde{\mathcal{F}}(\Lambda) \times \widetilde{\mathcal{F}}(\Lambda'). \tag{A.9}$$

Formula (A.9) makes it clear how to define the join of two abstract simplicial complexes $L$ and $L'$, we should have

$$\mathcal{S}(L * L') \cong \mathcal{S}(L) \times \mathcal{S}(L'), \tag{A.10}$$

where, as in formula (A.3) of Example A.2.3, $\mathcal{S}(L)$ is the poset of simplices in $L$. More explicitly, suppose $S$ and $S'$ are the vertex sets of $L$ and $L'$ and that $S \cap S' = \emptyset$. The simplicial complex $L * L'$ can be defined as follows: its vertex set is the disjoint union $S \cup S'$ and a nonempty subset $T \cup T'$ of $S \cup S'$ is the vertex set of a simplex in $L * L'$ if and only if $T \in \mathcal{S}(L)$ and $T' \in \mathcal{S}(L')$.

**Example A.4.6.** (*The cone on a barycentric subdivision.*) If $L$ is the geometric realization of an abstract simplicial complex $\mathcal{S}$ $(= \mathcal{S}(L))$, then $|\mathcal{S}|$ $(= \mathrm{Flag}(\mathcal{S}))$, the geometric realization of the poset $\mathcal{S}$, is the cone on $bL$, the barycentric subdivision of $L$. (By Proposition A.3.9, $|\mathcal{S}_{>\emptyset}|$ is $bL$. So, $|\mathcal{S}(L)|$ is the join of $bL$ and a point, where $\emptyset$ corresponds to the cone point.)

The following lemma, which is used in 12.2 and Appendix I.6, is left as an exercise for the reader.

**LEMMA A.4.7.** *The join of two flag complexes is a flag complex.*

**Example A.4.8.** (*The standard subdivision of an n-cube.*) We take notation from Examples A.2.1, A.2.3, A.2.4, and particularly A.2.5. Let $\square^n = [-1, 1]^n$ be the standard $n$-cube. We are going to define a subdivision of $\square^n$ into simplices, without introducing any new vertices, so that the corresponding abstract simplicial complex is $\mathrm{Flag}(\mathcal{P}(I_n))$. As in Example A.2.5, $\mathrm{Vert}(\square^n) \cong \mathcal{P}(I_n)$ via the correspondence $e_J \leftrightarrow J$. Let $\alpha = (J_1 < \cdots < J_k) \in \mathrm{Flag}(\mathcal{P}(I_n))$. The corresponding geometric simplex $\langle\alpha\rangle$ of the subdivision is the simplex spanned by $\{e_{J_1}, \dots, e_{J_k}\}$. The simplex $\langle\alpha\rangle$ is contained in the face $\square_{[J_1,J_k]}$ of $\square^n$. It is easy to see that this actually gives a subdivision of the cube. See Figure A.2.

Since by Example A.2.4, $\mathcal{S}(\Delta^{n-1}) \cong \mathcal{P}(I_n)$, and since by Example A.4.6, $|\mathcal{S}(\Delta^{n-1})|$ is the cone on the barycentric subdivision of $\Delta^{n-1}$, we have established the following.

**LEMMA A.4.9.** *The standard subdivision of $\square^n$ is combinatorially isomorphic to the cone on the barycentric subdivision of $\Delta^{n-1}$.*

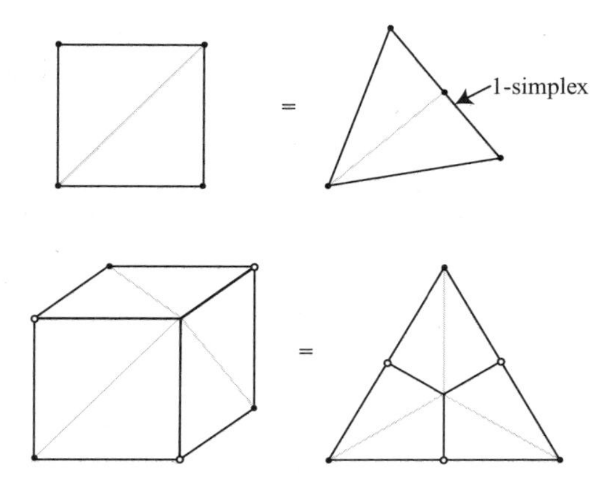


Figure A.2. Standard subdivisions of cubes.

## A.5. FACES AND COFACES

Let $\mathcal{P}$ be a poset and $|\mathcal{P}|$ its geometric realization. There are two decompositions of $|\mathcal{P}|$ into closed subspaces. Both decompositions are indexed by $\mathcal{P}$. For the first, take the geometric realizations of the subposets $\mathcal{P}_{\leqslant p}$, with $p \in \mathcal{P}$. $|\mathcal{P}_{\leqslant p}|$ is a *face* of $|\mathcal{P}|$. For the second decomposition, take the geometric realizations of the subposets $\mathcal{P}_{\geqslant p}$. $|\mathcal{P}_{\geqslant p}|$ is a *coface* of $|\mathcal{P}|$.

Since a simplex in Flag($\mathcal{P}$) is a chain in $\mathcal{P}$, the vertex set of any simplex is totally ordered. In particular, each simplex in Flag($\mathcal{P}$) has a minimum vertex as well as a maximum one. The face $|\mathcal{P}_{\leqslant p}|$ is the union of all simplices with maximum vertex $p$. Similarly, the coface $|\mathcal{P}_{\geqslant p}|$ is the union of all simplices with minimum vertex $p$.

**Example A.5.1.** (*The geometric realization of an abstract convex cell complex.*) If $\mathcal{P} = \mathcal{F}(\Lambda)$ for some convex cell complex $\Lambda$ and $P$ a cell of $\Lambda$, then $\mathcal{F}(\Lambda)_{\leqslant P} = \mathcal{F}(P)$ and the geometric realization of the poset $\mathcal{F}(\Lambda)_{\leqslant P}$ is $bP$, the barycentric subdivision of $P$.

On the other hand, suppose $\mathcal{P}$ is an abstract convex cell complex in the sense of Definition A.2.12. In view of condition (b) of that definition, for each $p \in \mathcal{P}$, the face $|\mathcal{P}_{\leqslant p}|$ is (the barycentric subdivision of) a convex cell. So, the faces of $|\mathcal{P}|$ are convex cells and they are glued together by (geometric realizations of) combinatorial equivalences of their faces (these are homeomorphisms). So any abstract convex cell complex $\mathcal{P}$ has a geometric realization by a cell complex, namely $|\mathcal{P}|$.

*Remark.* If $\Lambda$ is a PL manifold and $P$ is a cell in $\Lambda$, then $|\mathcal{F}(\Lambda)_{\geqslant P}|$ is called the *dual cell* to $P$.

For the remainder of this section, $L$ is a simplicial complex with vertex set $S$. As in formula (A.3) of Example A.2.3, $\mathcal{S}(L)$ (or simply $\mathcal{S}$) denotes the poset of abstract simplices in $L$ (including the empty simplex). In other words, $\mathcal{S}$ is the poset of those subsets $T$ of $S$ such that $T$ spans a simplex of $L$. Write $\sigma_T$ for the corresponding (closed) geometric simplex in $L$. The abstract simplicial complex $\mathrm{Flag}(\mathcal{S}_{>\emptyset})$ is the poset of (nonempty) simplices in the barycentric subdivision $bL$ of $L$. Suppose $\alpha = \{T_0, \dots, T_k\} \in \mathrm{Flag}(\mathcal{S}_{>\emptyset})$, where $T_i \in \mathcal{S}_{>\emptyset}$ and $T_0 < \cdots < T_k$. Denote the corresponding closed simplex in $bL$ by $\langle \alpha \rangle$.

Here is another example in which each face is (combinatorially isomorphic to) a convex polytope.

**Example A.5.2.** (*The cubical structure on* $|\mathcal{S}(L)|$.) By Example A.4.6, $|\mathcal{S}|$ $(= |\mathcal{S}(L)|)$ is the cone on the barycentric subdivision $bL$. By Lemma A.4.9, for each $T \in \mathcal{S}$, $|\mathcal{S}_{\leqslant T}|$ is isomorphic to (the standard subdivision of) a cube. The dimension of the cube is $\mathrm{Card}(T)$ $(= \dim \sigma_T + 1)$. So, each face of $|\mathcal{S}|$ is a cube. It follows that the poset $\mathcal{S}$ is an abstract cubical cell complex in the sense of Definition A.2.12.

**Example A.5.3.** (*Cofaces of* $|\mathcal{S}(L)|$.) Set $K = |\mathcal{S}|$ and for each $s \in S$ $(= \mathrm{Vert}(L))$, let $K_s$ be the coface $|\mathcal{S}_{\geqslant \{s\}}|$. $K_s$ is the union of all closed simplices with minimum vertex $\{s\}$. In other words, $K_s$ is the closed star of the vertex $\{s\}$ in $bL$. If a simplex $\alpha \in \mathrm{Flag}(\mathcal{S})$ has minimum vertex $T$ for some $T \in \mathcal{S}$, then $\alpha$ is a face of a simplex with minimum vertex $\{s\}$ for each $s \in T$. It follows that

$$|\mathcal{S}_{\geqslant T}| = \bigcap_{s \in T} K_s.$$

(This should be compared to formula (5.2) in Section 5.1.)

**DEFINITION A.5.4.** For each $T \in \mathcal{S}$, let $K_T$ be the coface $|\mathcal{S}_{\geqslant T}|$ and $\partial K_T := |\mathcal{S}_{>T}|$. In particular, $\partial K := |\mathcal{S}_{>\emptyset}|$ is $bL$. We shall sometimes say the coface $K_T$ is the *dual cone* to the simplex $\sigma_T$ and that $K$, together with its partition into cofaces, is the *dual complex* to $L$.

### The Relationship Between $\partial K$ and $L$

As in 5.1 and 7.2, for each $T \subset S$, put

$$K^T := \bigcup_{s \in T} K_s. \tag{A.11}$$

The next lemma is used several times in Chapters 8, 9, and 10.

**LEMMA A.5.5.** *Suppose $L$ is a simplicial complex. Then, for each $T \in \mathcal{S}(L)$, $L - \sigma_T$ is homotopy equivalent to $K^{S-T}$.*

$K^{S-T}$ and $L - \sigma_T$ can both be regarded as subspaces of $bL$. We will prove the lemma by showing that $K^{S-T}$ is a deformation retract of $L - \sigma_T$. We need some more notation. Suppose $\alpha = \{T_0, \dots, T_k\} \in \text{Flag}(\mathcal{S}_{>\emptyset})$. Denote the corresponding closed simplex in $bL$ by $\langle\alpha\rangle$. A point $x \in \langle\alpha\rangle$, can be written as a unique convex linear combination

$$x = \sum_{T\in\alpha} x_T v_T,$$

where $v_T$ is the vertex of $bL$ corresponding to $T \in \mathcal{S}_{>\emptyset}$. (The $x_T$ are nonnegative real numbers satisfying: $\sum x_T = 1$). If $U \notin \alpha$, put $x_U := 0$. The $(x_T)_{T\in\mathcal{S}_{>\emptyset}}$ are the *barycentric coordinates* of $x$. The *support* of $x$, denoted $\text{supp}(x)$, is the set of $T \in \mathcal{S}_{>\emptyset}$ such that $x_T \neq 0$.

*Proof of Lemma A.5.5.* As a subset of $bL$,

$$\begin{aligned}\sigma_T &= \{x \in bL \mid (U \in \text{supp}(x)) \implies (U \subset T)\}\\ &= \{x \in bL \mid \text{supp}(x) \subset \text{Flag}(\mathcal{S}_{\leqslant T})\}.\end{aligned}$$

So

$$L - \sigma_T = \{x \in bL \mid \exists\, U \in \text{supp}(x) \text{ with } U \not\subset T\}$$

and

$$K^{S-T} = \{x \in bL \mid \text{supp}(x) \cap \text{Flag}(\mathcal{S}_{\leqslant T}) = \emptyset\}.$$

Hence, $K^{S-T} \subset L - \sigma_T$ and $(L - \sigma_T) - K^{S-T} = \{x \mid \text{supp}(x) \cap \text{Flag}(\mathcal{S}_{\leqslant T}) \neq \emptyset,\ \text{supp}(x) \not\subset \text{Flag}(\mathcal{S}_{\leqslant T})\}$. Put $\partial K^{S-T} := K^{S-T} \cap K^T$, i.e.,

$$\partial K^{S-T} = \{x \mid (U \in \text{supp}(x)) \implies (U \cap T \neq \emptyset,\ U \cap S - T \neq \emptyset)\}.$$

Put $\beta(x) := \text{supp}(x) - (\text{supp}(x) \cap \text{Flag}(\mathcal{S}_{\leqslant T}))$. Define $r : (L - \sigma_T) - K^{S-T} \to \partial K^{S-T}$ by setting all barycentric coordinates $x_U$ with $U \notin \beta(x)$ equal to 0 and then renormalizing so the sum of the coordinates is 1, i.e.,

$$r(x) := \sum_{U\in\beta(x)} x_U v_U \Bigg/ \left|\sum_{U\in\beta(x)} x_U\right| .$$

Extend $r$ to a retraction $(L - \sigma_T) \to K^{S-T}$ by setting it equal to the identity map on $K^{S-T}$. We claim $r$ is a deformation retraction, i.e., if $i : K^{S-T} \to L - \sigma_T$ is the inclusion, then $i \circ r$ is homotopic to the identity map of $K^{S-T}$. For $x \in (L - \sigma_T) - K^{S-T}$, the restriction of $r$ to the complement of the face $\langle\text{supp}(x) \cap \text{Flag}(\mathcal{S}_{\leqslant T})\rangle$ in the simplex $\langle\text{supp}(x)\rangle$ is just the standard projection onto the opposite face $\langle\beta(x)\rangle$ and we have the usual straight line homotopy defined by $(x, t) \to tx + (1 - t)r(x)$. These homotopies fit together to give the homotopy between $i \circ r$ and $id$. $\square$

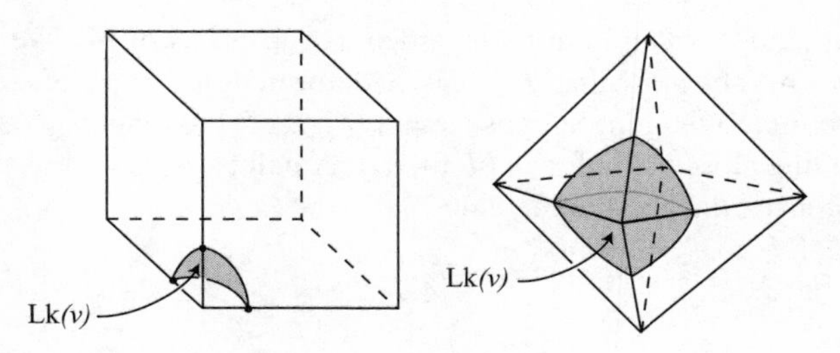


Figure A.3. Two links.

## A.6. LINKS

Suppose $P$ is a convex polytope in a finite-dimensional affine space $\mathbb{A}$ with tangent space $V$. Let $TP$ denote the tangent space of $P$, i.e., $TP$ is the linear subspace of $V$ consisting of all vectors of the form $t(x - y)$, where $x, y \in P$ and $t \in \mathbb{R}$. Given $x \in P$, define $C_xP$, the *inward-pointing tangent cone* at $x$, to be the set of all $v \in TP$ such that $x + tv \in P$ for all $t \in [0, \varepsilon)$ for some $\varepsilon > 0$. $C_xP$ is a linear polyhedral cone in $TP$. Suppose $F$ is a proper face of $P$. If $x \in \text{int}(F)$ and $y \in F$, then $C_yP \subset C_xP$, with equality if and only if $y \in \text{int}(F)$. (Here $\text{int}(F)$ is the relative interior of $F$.) Also, if $x, y$ are both in $\text{int}(F)$, then $C_xP$ and $C_yP$ have the same image in $TP/TF$. This common image is an essential polyhedral cone, denoted by $\text{Cone}(F, P)$.

One way to define the "link" of $F$ in $P$ is as the image of $\text{Cone}(F, P) - 0$ in the unit sphere of $TP/TF$. Equivalently, the *link* of $F$ in $P$, denoted $\text{Lk}(F, P)$ is the intersection of $\text{Cone}(F, P)$ with an affine hyperplane in $TP/TF$ which intersects every nonzero face of $\text{Cone}(F, P)$. $\text{Lk}(F, P)$ is a convex polytope in this affine hyperplane, well defined up to an affine isomorphism. Its dimension is $\dim P - \dim F - 1$. For example, the link of a vertex in a three-dimensional cube is a triangle, while the link of a vertex in an octahedron is a square. (See Figure A.3.)

If $F_1 < F_2 < P$, then the inclusions $TF_2 \subset TP$, $TF_2/TF_1 \subset TP/TF_1$ and $C_yF_2 \subset C_yP$ induce a natural identification of $\text{Lk}(F_1, F_2)$ with a face of $\text{Lk}(F_1, P)$. So, the poset of faces of $\text{Lk}(F, P)$ is naturally isomorphic to $\mathcal{F}(P)_{>F}$, where $\mathcal{F}(P)$ is the poset of faces of $P$, discussed in Example A.2.3.

**DEFINITION A.6.1.** Suppose $\Lambda$ is a convex cell complex and $F$ is a cell of $\Lambda$, then $\text{Lk}(F, \Lambda)$, the *link* of $F$ in $\Lambda$ is the convex cell complex consisting of all cells of the form $\text{Lk}(F, P)$, where $F < P \in \Lambda$ and where, whenever $F_1 < F_2 < P \in \Lambda$, we identify $\text{Lk}(F_1, F_2)$ with the corresponding face of $\text{Lk}(F_1, P) \in \text{Lk}(F_1, \Lambda)$.

As before, the poset of cells in $\text{Lk}(F, \Lambda)$ is identified with $\mathcal{F}(\Lambda)_{>F}$.

**PROPOSITION A.6.2.** *Suppose a point $x$ in (the geometric realization of) a locally finite convex cell complex $\Lambda$ lies in the relative interior of some cell $F$. Then there is an open neighborhood of $x$ in $\Lambda$ of the form* $\text{int}(F) \times N_\varepsilon$, *where $N_\varepsilon$ denotes an $\varepsilon$ neighborhood of the cone point in* $\text{Cone}(\text{Lk}(F, \Lambda))$. *Equivalently, a neighborhood of $x$ is homeomorphic to the cone on the $k$-fold suspension of* $\text{Lk}(F, \Lambda)$, *where $k = \dim F$.*

*Proof.* The *open star* of $F$ in $\Lambda$, denoted $\text{Star}(F, \Lambda)$, is defined as the union of all open cells $\text{int}(F')$, where $F' \in \mathcal{F}(\Lambda)$ and $F$ is a face of $F'$. Given such an $F'$, let $\sigma_{F'}$ denote the corresponding cell in $\text{Lk}(F, \Lambda)$. ($\dim \sigma_{F'} = \dim F' - \dim F - 1$.) Since $\text{int}(F') \cong \text{int}(F * \sigma_{F'})$, it follows that $\text{Star}(F, \Lambda) \cong \text{int}(F) \times N_\varepsilon$. To prove the last sentence of the proposition, let $D^k$ be a disk neighborhood of $x \in \text{int}(F)$. Then $D^k \times \text{Cone}(\text{Lk}(F, \Lambda)) \cong S^{k-1} * \text{Lk}(F, \Lambda)$. (Compare equation (A.10).) Finally, taking the join with $S^{k-1}$ is the same as taking the $k$-fold suspension. □

**LEMMA A.6.3.** *Suppose $v$ is a vertex of a cell $F$ in some convex cell complex $\Lambda$. Then* $\text{Lk}(\sigma_F, \text{Lk}(v, \Lambda)) = \text{Lk}(F, \Lambda)$, *where $\sigma_F$ denotes the cell of* $\text{Lk}(v, \Lambda)$ *corresponding to $F$.*

*Proof.* This is simply a matter of unwinding the definitions. On the level of posets of cells, we have

$$\begin{aligned}\mathcal{F}(\text{Lk}(\sigma_F, \text{Lk}(v, \Lambda)) &= \mathcal{F}(\text{Lk}(v, \Lambda))_{>\sigma_F} = (\mathcal{F}(\Lambda)_{>\{v\}})_{>F} \\ &= \mathcal{F}(\Lambda)_{>F} = \mathcal{F}(\text{Lk}(F, \Lambda)),\end{aligned}$$

from which the lemma follows. □

*Remark.* When, as in Chapter 6, $F$ is a face of a convex polytope $P$ in a space of constant curvature, $\text{Cone}(F, P)$ is naturally a polyhedral cone in the orthogonal complement of $T_xF$ in $T_xP$ and the intersection of this cone with the unit sphere in $T_xP$ is naturally a spherical polytope. So, in this geometric context, we will always regard $\text{Lk}(F, P)$ as a spherical polytope.

# Appendix B

## REGULAR POLYTOPES

### B.1. CHAMBERS IN THE BARYCENTRIC SUBDIVISION OF A POLYTOPE

Suppose $P$ is an $(n+1)$-dimensional convex polytope and $\partial P$ is its boundary complex (Example A.1.12). As in Appendix A.3, $b(\partial P)$ is the barycentric subdivision of $\partial P$.

**DEFINITION B.1.1.** A top-dimensional simplex of $b(\partial P)$ is a *chamber*. The set of all chambers in $b(\partial P)$ is denoted $\mathrm{Cham}(b(\partial P))$.

A simplex of $b(\partial P)$ corresponds to a chain (or flag) of cells in $\partial P$. We will identify a simplex with the corresponding chain of cells. In particular, a chamber $\sigma^n$ of $b(\partial P)$ is identified with a maximal chain $\{F_0, F_1, \dots, F_n\}$, where $F_i < F_{i+1}$. Since the chain is maximal, we must have $\dim F_i = i$. In fact, this will be our convention, when dealing with such chains: the cells will be indexed by dimension, i.e., $\dim F_i = i$.

**DEFINITION B.1.2.** Suppose $\alpha = (F_{i(1)} < \cdots < F_{i(k)})$ is a (not necessarily maximal) chain of cells in $\partial P$, with $\dim F_{i(j)} = i(j)$. Then $\alpha$ has *type* $(i(1), \dots, i(k))$ The corresponding geometric simplex of $b(\partial P)$ spanned by the barycenters of the $F_{i(j)}$ is also said to have *type* $(i(1), \dots, i(k))$.

As examples, a chain consisting of a single $i$-dimensional face is a chain of type $(i)$; a maximal chain is of type $(0, 1, \dots, n)$.

As in Definition A.3.11, $\mathrm{Aut}(P)$ denotes the group of combinatorial automorphisms of $P$. Note that $\mathrm{Aut}(P)$ acts on the poset $\mathrm{Flag}(\mathcal{F}(\partial P))$ of Appendix A.3. ($\mathrm{Flag}(\mathcal{F}(\partial P))$ is the set of chains of cells in $\partial P$.) Moreover, the action preserves type. Hence, $\mathrm{Aut}(P)$ acts as a group of type-preserving simplicial automorphisms of $b(\partial P)$. In particular, $\mathrm{Aut}(P)$ acts on $\mathrm{Cham}(b(\partial P))$.

Let $\Delta^n$ denote the standard $n$-simplex on the vertex set $\{0, 1, \dots, n\}$. There is a natural simplicial projection $d : b(\partial P) \to \Delta^n$ defined on the vertices by sending the barycenter $v_F$ of a cell $F$ to the integer, $\dim F \in \{0, 1, \dots, n\}$. Note that the restriction of $d$ to any chamber is an isomorphism. As explained in

Example 18.1.1, $d$ gives Cham($b\partial P$) the structure of a chamber system over $S$ in the sense of 18.1. (Here $S$ is the set of codimension one faces of $\Delta^n$.)

**LEMMA B.1.3.** Aut($P$) *acts freely on* Cham($b(\partial P)$).

*Proof.* The point is that $b(\partial P)$ is a (gallery connected) pseudomanifold (in fact, an $n$-sphere). (Although the 0-sphere is not connected, the case $n = 0$ is easily dealt with by a separate argument.) Suppose $\varphi \in$ Aut($P$) and $\varphi$ stabilizes a chamber $\sigma \in$ Cham($b(\partial P)$). Since the action of Aut($P$) preserves dimension, $\varphi$ must fix each vertex of $\sigma$. Hence, it must fix $\sigma$ pointwise. Since $b(\partial P)$ is a pseudomanifold, there is exactly one other chamber across any given codimension-one face of $\sigma$. This chamber must also be fixed by $\varphi$. Since any two chambers can be connected by a sequence of adjacent chambers (a "gallery"), the geometric realization of $\varphi$ must actually be the identity map. □

Suppose $P^{n+1}$ is a convex polytope in $\mathbb{E}^{n+1}$ and that Isom($P$) is its group of symmetries. If an element $g \in$ Isom($P$) stabilizes a face of $P$, then it must fix the center (= barycenter) of that face. It follows that the action of Isom($P$) on $b(\partial P)$ is through simplicial automorphisms and that the natural map Isom($P$) $\to$ Aut($P$) is injective.

Fix a chamber $\sigma = \{F_0, \dots, F_n\} \in$ Cham($b(\partial P^{n+1})$). For $1 \leqslant i \leqslant n$, define one-dimensional cell complex $L_i(\sigma)$ as follows:

$$L_i(\sigma) = \begin{cases} \partial F_2 & \text{if } i = 1, \\ \mathrm{Lk}(F_{i-2}, \partial F_{i+1}) & \text{if } 2 \leqslant i \leqslant n-1, \\ \mathrm{Lk}(F_{n-2}, \partial P) & \text{if } i = n \text{ and } n \geqslant 2. \end{cases} \tag{B.1}$$

Since $\mathrm{Lk}(F_{i-2}, \partial F_{i+1})$ is the boundary complex of the polygon $\mathrm{Lk}(F_{i-2}, F_{i+1})$, each $L_i(\sigma)$ is the boundary of a polygon. In other words, it is a triangulation of $S^1$. (By an abuse of language, we will often blur the distinction between a polygon and its boundary complex and say that $L_i(\sigma)$ "is a polygon.") Let $m_i(\sigma)$ be the number of edges of $L_i(\sigma)$. The $n$-tuple $\mathbf{m}(\sigma) := (m_1(\sigma), \dots, m_n(\sigma))$ is the *Schläfli symbol* of $P$ at $\sigma$. Since a convex polygon has at least three sides, $m_i(\sigma) \geqslant 3$.

For $0 \leqslant i \leqslant n$, let $\sigma_i$ be the codimension one face of $\sigma$ opposite to the barycenter of $F_i$. Let us say that $\sigma_i$ is a simplex of *cotype i*. We can identify $\sigma_i$ with the (nonmaximal) chain of cells $\{F_0, \dots, F_n\} - \{F_i\}$. Let $\sigma(i)$ denote the adjacent chamber to $\sigma$ across its face of cotype $i$. We say that $\sigma(i)$ is *i-adjacent* to $\sigma$. Thus, the $i$-adjacent chamber to $\sigma$ has the form

$$\sigma(i) = (\{F_0, \dots, F_n\} - \{F_i\}) \cup F_i', \tag{B.2}$$

for some (uniquely determined) $i$-dimensional face $F_i'$.

Next we want to consider links of codimension two simplices in the barycentric subdivision of $\partial P$. The following lemma clarifies the situation.

**Lemma B.1.4.** *Fix a chamber $\sigma$. As above, let $F'_i$ and $F'_j$ be the cells of $\partial P$ determined by the chambers which are i-adjacent and j-adjacent to $\sigma$. Suppose $|i-j| \geqslant 2$. Then*

$$\sigma(i,j) := (\{F_0, \dots, F_n\} - \{F_i, F_j\}) \cup \{F'_i, F'_j\}$$

*is a chamber of $b(\partial P)$.*

*Proof.* We must show that $\sigma(i,j)$ is a chain of cells. We know that

$$F_{i-1} < F'_i < F_{i+1} \quad \text{and} \quad F_{j-1} < F'_j < F_{j+1}.$$

Without loss of generality we can assume $j > i$. Since $|i-j| \geqslant 2$, we have $j-1 \geqslant i+1$; so

$$F_0 < \cdots F_{i-1} < F'_i < F_{i+1} \cdots F_{j-1} < F'_j < F_{j+1} \cdots < F_n,$$

which proves the lemma. □

**Lemma B.1.5.** *Suppose $0 \leqslant i < j \leqslant n$ and $\sigma_{ij} = \sigma_i \cap \sigma_j$ is a codimension-two face of the chamber $\sigma$. Let $\mathrm{Lk}(\sigma_{ij})$ denote the link of $\sigma_{ij}$ in $b(\partial P)$. ($\mathrm{Lk}(\sigma_{ij})$ is a triangulation of $S^1$.) Then*

*(a) If $j = i+1$, then $\mathrm{Lk}(\sigma_{ij})$ is the barycentric subdivision of the polygon $L_j(\sigma)$ defined in (B.1). Hence, $\mathrm{Lk}(\sigma_{ij})$ is a $2m_j(\sigma)$-gon.*

*(b) If $j \geqslant i+2$, then $\mathrm{Lk}(\sigma_{ij})$ is a 4-gon.*

*Proof.* We have $\sigma_{ij} = \{F_0, \dots, F_n\} - \{F_i, F_j\}$.

*Case* (a). $j = i+1$. An edge of $\mathrm{Lk}(\sigma_{ij})$ is a pair of cells $\{G_{j-1}, G_j\}$, where

$$F_{j-2} < G_{j-1} < G_j < F_{j+1},$$

but this means that $\mathrm{Lk}(\sigma_{ij}) = bL_j(\sigma)$.

*Case* (b). $j \geqslant i+2$. An edge of $\mathrm{Lk}(\sigma_{ij})$ is a pair of cells $\{G_i, G_j\}$ where

$$F_{i-1} < G_i < F_{i+1} \quad \text{and} \quad F_{j-1} < G_j < F_{j+1}.$$

By the sentence containing equation (B.2), there are precisely two such $G_i$, namely, $F_i$ and $F'_i$ and precisely two such $G_j$, namely, $F_j$ and $F'_j$. Using Lemma B.1.4, there are exactly four edges in $\mathrm{Lk}(\sigma_{ij})$: $\{F_i, F_j\}$, $\{F'_i, F_j\}$, $\{F'_i, F'_j\}$ and $\{F_i, F'_j\}$. So $\mathrm{Lk}(\sigma_{ij})$ is a 4-gon. □

The *Coxeter matrix of $P$ at $\sigma$* is the symmetric, $(n+1) \times (n+1)$ matrix $(m_{ij}(\sigma))$ with all diagonal entries $= 1$ and with the entries above the diagonal defined by

$$m_{ij}(\sigma) = \begin{cases} m_j(\sigma) & \text{if } j = i+1, \\ 2 & \text{if } j \geqslant i+2. \end{cases} \tag{B.3}$$

The Coxeter diagram associated to such a Coxeter matrix is a "straight line" in the following sense. It is a graph with $n+1$ vertices, say, $0, 1, \dots, n$. Two vertices are connected by an edge if and only if they differ by 1 and the edge between $i-1$ and $i$ is labeled by an integer $m_i \geqslant 3$ (where, as usual, we omit the labels from the edges with $m_i = 3$). We shall say that such a diagram is a *straight line Coxeter diagram*.

## B.2. CLASSIFICATION OF REGULAR POLYTOPES

**DEFINITION B.2.1.** A convex polytope $P$ is *combinatorially regular* if $\text{Aut}(P)$ acts transitively on $\text{Cham}(b(\partial P))$. A polytope in Euclidean space is *regular* if $\text{Isom}(P)$ acts transitively on $\text{Cham}(b(\partial P))$.

*Remarks B.2.2*

(i) Regularity implies combinatorial regularity.

(ii) Suppose $P$ is regular and $G = \text{Isom}(P)$ or $\text{Aut}(P)$. Since any chain of faces can be extended to a maximal chain, it follows that $G$ acts transitively on the set of chains of a given type (as defined in Definition B.1.2). In particular, $G$ acts transitively on the set of $i$-dimensional faces for $0 \leqslant i \leqslant n$ and each face of $P$ is also a regular polytope.

(iii) Definition B.2.1 is what one arrives at by trying to make precise the notion that $P$ is as "symmetric as possible." To see this, suppose $G = \text{Aut}(P)$. If we want $P$ to be highly symmetric, the first condition we might impose is that $G$ acts transitively on $\text{Vert}(P)$. Next we might want the condition that for each $v \in \text{Vert}(P)$, $\text{Lk}(v, P)$ should be (combinatorially) regular. By induction on dimension, "regular" should mean that the isotropy subgroup $G_v$ acts transitively on the chambers in $b(\partial \text{Lk}(v, P))$. An $i$-cell of $\text{Lk}(v, P)$ corresponds to an $(i+1)$-cell of $P$ which contains $v$, so a chamber of $b(\partial \text{Lk}(v, P))$ gives a chamber $\{F_0, \dots, F_n\}$ of $b(\partial P)$, with $F_0 = v$. Thus, $G$ should act transitively on the set of maximal chains of cells of $P$.

If $P$ is combinatorially regular, then the Schläfli symbol $(m_1(\sigma), \dots, m_n(\sigma))$ and the Coxeter matrix $(m_{ij}(\sigma))$ are obviously independent of the choice of chamber $\sigma$. We denote them by $\mathbf{m} := (m_1, \dots, m_n)$ and $(m_{ij})$ and call them, respectively, the *Schläfli symbol* and *Coxeter matrix* of $P$.

In dimension 2 we have the regular polygons. The Schläfli symbol of an $m$-gon is $(m)$; its symmetry group is the dihedral group $\mathbf{D}_m$ (Example 3.1.2). In each dimension $n > 2$ there are three "obvious" regular polytopes: the $n$-simplex $\Delta^n$ (Example A.1.4), the $n$-cube $\square^n$ (Example A.1.5) and the $n$-octahedron $O^n$ (see Example A.1.7). The symmetry group of $\Delta^n$ is the symmetric group $S_{n+1}$ on its vertex set (Example 6.7.1). Its Schläfli symbol is $(3, 3, \dots, 3)$. The $n$-cube and the $n$-octahedron are dual polytopes and hence, have the same group of symmetries, namely, the octahedral group of Example 6.7.2. The Schläfli symbol of $\square^n$ is $(4, 3, \dots, 3)$, while the Schläfli symbol of $O^n$ is $(3, \dots, 3, 4)$. (The Schläfli symbols of a regular polytope and its dual are the reverses of each other: if $(m_1, \dots, m_n)$ is the Schläfli symbol of $P$, then the Schläfli symbol of its dual is $(m_n, \dots, m_1)$. Thus, $P$ is self-dual if and only if its Schläfli symbol is palindromic.) All of the facts in this paragraph are easy to prove and are left as exercises for the reader.

If $P$ is a regular polytope with Schläfli symbol $(m_1, \dots, m_n)$, then it each of its codimension one faces is regular with Schläfli symbol $(m_1, \dots, m_{n-1})$ and the link of each of its vertices is (combinatorially) regular with Schläfli symbol $(m_2, \dots, m_n)$.

Of course, in dimension 3 there are two other regular polytopes known to the ancient Greeks: the dodecahedron and its dual, the icosahedron. These have Schläfli symbols $(5, 3)$ and $(3, 5)$, respectively.

One of the remarkable discoveries of nineteenth century (due to Schläfli in 1850) is that the above examples almost constitute the complete list of regular polytopes. As we shall see below, each regular polytope is determined (up to an isometry and a homothety) by its Schläfli symbol. Furthermore, it turns out that there are exactly three other "nonobvious" regular polytopes and all three occur in dimension 4. Their Schläfli symbols are $(3, 4, 3)$, $(5, 3, 3)$, and $(3, 3, 5)$. They are called by Coxeter in [69] "the 24-cell," "the 120-cell," and "the 600-cell," respectively. The codimension one faces of a 24-cell are octahedra; there are 24 such faces; the link of each vertex is a cube. It is self-dual. The 120-cell and the 600-cell are dual to one another. The 120-cell has dodecahedra as its three-dimensional faces; there are 120 such faces; the link of each vertex is a tetrahedron (a 3-simplex). The 600-cell has tetrahedra as its three-dimensional faces; there are 600 such faces; the link of each vertex is an icosahedron.

In the next few pages we will establish a one-to one correspondence between each of the following:

- the set of dual pairs of combinatorially regular $(n + 1)$-dimensional polytopes (up to combinatorial equivalence),
- the set of dual pairs of regular $(n + 1)$-dimensional polytopes in $\mathbb{E}^{n+1}$ (up to isometry and homothety), and

- the set of Coxeter systems $(W, S)$ of rank $(n + 1)$ with $W$ finite and with a straight line Coxeter diagram (up to isomorphism of Coxeter systems).

Thus, the classification of regular polytopes reduces to the classification of finite Coxeter groups (which will be accomplished in Appendix C). Here is the theorem.

**THEOREM B.2.3.** *If $P^{n+1} \subset \mathbb{E}^{n+1}$ is a regular polytope, then its symmetry group* $\mathrm{Isom}(P^{n+1})$ *is a geometric reflection group on $\mathbb{S}^n$. Conversely, if $(W, S)$ is a Coxeter system of rank $(n + 1)$ with $W$ finite and with diagram a straight line, then, up to isometry and homothety, there are at most two regular polytopes in $\mathbb{E}^{n+1}$ (a polytope and its dual) which have $(W, S)$ as their associated Coxeter system. Moreover, these dual polytopes are isomorphic if and only if the Coxeter diagram admits a nontrivial involution (i.e., the Schläfli symbol is palindromic).*

In the course of proving this the following facts will come out in the wash:

- Any combinatorially regular polytope is combinatorially equivalent to a regular polytope in Euclidean space (a result of [201]).
- A sequence $(m_1, \dots, m_n)$ of integers $\geqslant 3$ is the Schläfli symbol of a regular polytope if and only if its associated Coxeter matrix (defined in (B.3)) determines a finite Coxeter group. Furthermore, if this is the case, then the Schläfli symbol determines the regular polytope (up to isomorphism).

References for this material include [69, 74, 171, 201] and [257, Chapter 5 §3].

## B.3. REGULAR TESSELLATIONS OF SPHERES

Given a straight line spherical Coxeter diagram, we show here how to assemble the chambers into spherical polytopes to get a regular tessellation of the sphere and then by taking the convex hull of its vertices a regular polytope in $\mathbb{E}^{n+1}$.

**THEOREM B.3.1.** (Compare [74, Theorem 3.9].) *Let $(W, S)$ be a Coxeter system with $S = \{s_0, s_1, \dots, s_n\}$ and with Coxeter matrix $(m_{ij})_{0 \leqslant i,j \leqslant n}$. Suppose $W$ is finite and that the Coxeter diagram of $(W, S)$ is a straight line in the precise sense that $m_{ij} = 2$ when $|i - j| \geqslant 2$ and $2 < m_{ij} < \infty$ when $|i - j| = 1$. For $1 \leqslant i \leqslant n$, put $m_i = m_{i-1,i}$. Then there is a regular tessellation of $\mathbb{S}^n$ with Schläfli symbol $(m_1, \dots, m_n)$. Moreover, this tessellation is unique up to isometries of $\mathbb{S}^n$.*

*Proof.* We know from Section 6.6 that $W$ can be represented as a geometric reflection group on $\mathbb{S}^n$ as follows. There is a spherical $n$-simplex $\sigma$ with codimension one faces $(\sigma_i)_{0\leqslant i\leqslant n}$ such that the dihedral angle between $\sigma_i$ and $\sigma_j$ is $\pi/m_{ij}$ and such that $s_i$ is reflection across $\sigma_i$. Let $v_i$ be the vertex opposite to $\sigma_i$. We need to understand how to construct the top-dimensional cells of the tessellation. One such cell $F_n$ is the union of translates of $\sigma$ by the elements of the isotropy subgroup at $v_n$. The other $n$-cells are then the translates of $F_n$ by the elements of $W$. The claim is that $F_n$ is a spherical polytope and that its subdivision into translates of $\sigma$ is the barycentric subdivision. In proving this, it is no harder to reconstruct an entire maximal chain $F_0 < \cdots < F_n$ of the tessellation.

The isotropy subgroup of $W$ at $v_i$ is the special subgroup $W_{(i)}$ generated by $S - s_i$. The diagram of $W_{(i)}$ is either a line segment (if $i = 0$ or $n$) or two line segments (if $1 \leqslant i \leqslant n-1$). Thus, $W_{(i)} = G_i \times H_i$ where $G_i$ is the special subgroup generated by $\{s_0, \dots, s_{i-1}\}$ and $H_i$ is the special subgroup generated by $\{s_{i+1}, \dots, s_n\}$. ($G_i$ is trivial if $i = 0$; $H_i$ is trivial if $i = n$.) Let $\sigma(i)$ be the face of $\sigma$ spanned by $v_0, \dots, v_i$ and let

$$F_i = \bigcup_{w\in G_i} w\sigma(i).$$

We claim that, for $0 \leqslant i \leqslant n$,

(a) $F_i$ is a convex spherical polytope in the $i$-sphere $\mathbb{S}^i$ fixed by $H_i$.

(b) The triangulation of $F_i$ by the $i$-simplices $\{w\sigma(i)\}_{w\in W_{(i)}}$ is the barycentric subdivision.

Note that $G_{i+1}$ is a reflection group on $\mathbb{S}^i$. Claims (a) and (b) imply that the $G_{i+1}$-translates of the $F_i$ are the cells of the regular tessellation. To prove the claims, first note that $\sigma(i) \subset \mathbb{S}^i$. Since the $G_i$ and $H_i$ actions commute, $F_i \subset \mathbb{S}^i$. Since the sum of two numbers of the form $\pi/m_{ij}$ is $\leqslant \pi$, two codimension one faces of $F_i$ intersect at an angle $\leqslant \pi$. Hence, $F_i$ is convex. This proves (a). Suppose, by induction, that for $i < k$ the triangulation of $F_i$ by the translates of $\sigma(i)$ is its barycentric subdivision. (The case $i = 0$ is trivial.) So, this holds for $F_{k-1}$ and all its translates under $G_k$. Since the union of these translates is $\partial F_k$, they give the barycentric subdivision of $\partial F_k$. Since $v_k$ is the center of $F_k$, the translates of $\sigma(k)$ give the cone on the barycentric subdivision of $\partial F_k$, i.e., the barycentric subdivision of $F_k$. This proves (b). Since the simplex $\sigma$ is determined, up to isometry, by its dihedral angles, the tessellation is also unique up to isometry. □

*Proof of Theorem B.2.3.* Suppose $P^{n+1}$ is regular. Declare the center of each face to be its barycenter and then take the barycentric subdivision of $P$. Since the centers are preserved, this is stable under $\mathrm{Isom}(P^{n+1})$. Radially project from the center of $P^{n+1}$ onto a sphere $\mathbb{S}^n$ centered at the same point. Choose $\sigma \in \mathrm{Cham}(b(\partial P))$. The element of $\mathrm{Isom}(P)$ that takes $\sigma$ to an adjacent chamber must act on $\mathbb{S}^n$ as reflection across their common face (since it fixes the face pointwise). It follows from the analysis in the previous section that the dihedral angle between $\sigma_i$ and $\sigma_j$ must be $\pi/m_{ij}$ where the $m_{ij}$ are defined by (B.3). So, $\mathrm{Isom}(P)$ acts on $\mathbb{S}^n$ as the geometric reflection group generated by the reflections across the faces of $\sigma$ and by, Theorem 6.4.3, it is a Coxeter group $W$ with a straight line diagram.

Conversely, suppose $(W, S)$ is a spherical Coxeter system with a straight line diagram. As in Theorem B.3.1, we get a cellulation of $\mathbb{S}^n$. Take the convex hull of the set of 0-cells in this cellulation to get the regular polytope $P^{n+1} \subset \mathbb{R}^{n+1}$. □

## B.4. REGULAR TESSELLATIONS

The results of the previous three sections can be generalized to "regular tessellations" of pseudomanifolds, following the lines laid down by Lannér [184]. We sketch the theory here; more details can be found in [74]. Throughout this section, $\Lambda$ is a $n$-dimensional convex cell complex satisfying the following conditions.

**Conditions B.4.1**

- Each cell of $\Lambda$ is a face of some $n$-dimensional cell (i.e., $\Lambda$ is a "pure" cell complex).
- $\Lambda$ is connected.
- The link of each codimension one cell is $S^0$ (i.e., $\Lambda$ is a pseudomanifold) and the link of each cell of codimension $\geqslant 2$ is connected. (This implies that $\mathrm{Cham}(b\Lambda)$ is "gallery connected" in the sense of 18.1.)

Sometimes we will also want to impose the following additional conditions.

**Conditions B.4.2**

- $\Lambda$ is simply connected.
- The link of any codimension two cell is a triangulation of $S^1$.
- The link of each cell of codimension $\geqslant 3$ is simply connected.

We proceed as in B.1. A top-dimensional simplex in $b\Lambda$ is a *chamber*. Let $\mathrm{Cham}(b\Lambda)$ denote the set of all chambers in $b\Lambda$. A chamber $\sigma$ is the same thing as a maximal flag of cells, $(F_0 < \cdots < F_n)$, where $\dim F_i = i$. As before, we have a simplicial projection $d : b\Lambda \to \Delta^n$, defined on vertices by sending the barycenter of a cell $F$ to the number $\dim F$. Aut($\Lambda$) acts on $b\Lambda$ and the action commutes with the projection $d$. As in Lemma B.1.3, Aut($\Lambda$) acts freely on Cham($b\Lambda$).

Fix a chamber $\sigma = (F_0 < \cdots < F_n) \in \text{Cham}(b\Lambda)$. As in formula (B.1), for $1 \leqslant i \leqslant n$, define a triangulation $L_i(\sigma)$ of a connected 1-manifold by

$$L_i(\sigma) := \begin{cases} \partial F_2 & \text{if } i = 1, \\ \text{Lk}(F_{i-2}, \partial F_{i+1}) & \text{if } 2 \leqslant i \leqslant n-1; \\ \text{Lk}(F_{n-2}, \Lambda) & \text{if } i = n \text{ and } n \geqslant 2. \end{cases}$$

For $i < n$, $L_i(\sigma)$ is a triangulation of $S^1$; $m_i(\sigma)$ denotes its number of edges. $L_n(\sigma)$ is either a circle or a line. When it is a circle, define $m_n(\sigma)$ to be the number of its edges. When it is the line, $m_n(\sigma) := \infty$. The $n$-tuple $\mathbf{m}(\sigma) := (m_1, \ldots, m_n)$ is the *Schläfli symbol* at $\sigma$. $\Lambda$ is *regular* if $\mathbf{m}(\sigma)$ is a constant function of $\sigma$; it is *symmetrically regular* if Aut($\Lambda$) is transitive on Cham($b\Lambda$).

*Remarks on terminology.* Our terminology here is somewhat inconsistent with that in B.2. There we used "combinatorially regular" to mean the same as "symmetrically regular" here. Also, in B.2 we did not have a term meaning that the Schläfli symbol was constant on chambers (here called "regular").

Given an $n$-tuple $\mathbf{m} = (m_1, \ldots, m_n)$, define a Coxeter matrix $(m_{ij})$ on $I := \{0, 1, \ldots, n\}$ as in formula (B.3). All diagonal entries are equal to 1 and the entries above the diagonal are given by

$$m_{ij}(\sigma) = \begin{cases} m_j(\sigma) & \text{if } j = i+1, \\ 2 & \text{if } j \geqslant i+2. \end{cases}$$

Let $W_{\mathbf{m}}$ be the resulting Coxeter group. Its diagram is the straight line

$$\overset{m_1}{\circ\!\!-\!\!-\!\!-\!\!\circ} \cdots\cdots \overset{m_n}{\circ\!\!-\!\!-\!\!-\!\!\circ}\ .$$

Let $W_{(i)}$ be the special subgroup obtained by omitting the $i^{\text{th}}$ vertex of this diagram. The *initial part* of $\mathbf{m}$ is the $(n-1)$-tuple $(m_1, \ldots, m_{n-1})$; its *final part* is the $(n-1)$-tuple $(m_2, \ldots, m_n)$. If $\mathbf{m}$ is the Schläfli symbol of a regular cell complex, then its initial part is the symbol of a regular $n$-dimensional polytope, i.e., the corresponding Coxeter group $W_{(n)}$ is spherical (as in the previous two sections).

We assume for the rest of this appendix that $\Lambda$ is regular. There is a natural action (from the right) of $W_{\mathbf{m}}$ on Cham($b\Lambda$), the $i^{\text{th}}$ reflection $s_i$ takes a chamber $\sigma$ to the adjacent chamber across its codimension one face $\sigma_i$ of cotype $i$. Since $b\Lambda$ is gallery connected, $W_{\mathbf{m}}$ is transitive on Cham($b\Lambda$). Let $\pi(b\Lambda, \sigma)$ denote the isotropy subgroup at $\sigma$. (Here we are thinking of $b\Lambda \to \Delta^n$ as an "orbihedral covering," $W_{\mathbf{m}}$ as the "orbihedral fundamental group" of $\Delta^n$, the action of $W_{\mathbf{m}}$ on $\text{Cham}(b\Lambda)$ as the analog of the action of the fundamental group on the inverse image of a base point in a covering space, so that $\pi(b\Lambda, \sigma)$ is the analog of the fundamental group of $b\Lambda$.)

Suppose $\Lambda$ and $\Lambda'$ are regular cell complexes with the same Schläfli symbol. In keeping with the above analogy, call a simplicial map $b\Lambda \to b\Lambda'$ an "orbihedral covering" if $d = p \circ d'$, where $d : b\Lambda \to \Delta^n$ and $d' : b\Lambda' \to \Delta^n$ are the canonical projections. The proofs of the next three results are omitted, since they are very close to the proofs of the corresponding results in covering space theory.

**LEMMA B.4.3.** ([74, Prop. 2.9].) *Suppose $\Lambda$ and $\Lambda'$ are regular cell complexes with the same Schläfli symbol and $\sigma \in \text{Cham}(b\Lambda)$, $\sigma' \in \text{Cham}(b\Lambda')$. Then there is an orbihedral covering $\Lambda \to \Lambda'$ taking $\sigma$ to $\sigma'$ if and only if $\pi(b\Lambda, \sigma)$ is a subgroup of $\pi(b\Lambda', \sigma')$.*

**COROLLARY B.4.4.** ([74, Prop. 3.5].) *Regular cell complexes $\Lambda$ and $\Lambda'$ are isomorphic if and only if they have the same Schläfli symbol $\mathbf{m}$ and $\pi(b\Lambda, \sigma)$ and $\pi(b\Lambda', \sigma')$ are conjugate subgroups of $W_{\mathbf{m}}$.*

**COROLLARY B.4.5.** ([74, Prop. 3.6].) *A regular cell complex $\Lambda$ is symmetrically regular if and only if $\pi(b\Lambda, \sigma)$ is a normal subgroup of $W_{\mathbf{m}}$ for some (in fact, for any) $\sigma \in \text{Cham}(b\Lambda)$. Moreover, $\text{Aut}(\Lambda) = W_{\mathbf{m}}/\pi$ where $\pi := \pi(b\Lambda, \sigma)$.*

The role of the universal cover of $\Lambda$ is played by the Coxeter complex

$$\mathcal{U}_{\mathbf{m}} := \mathcal{U}(W_{\mathbf{m}}, \Delta^n)$$

of Example 5.2.7 in Section 5.2. A vertex $v$ in $\mathcal{U}_{\mathbf{m}}$ has *type i* if it projects to the vertex $i \in \{0, \dots, n\} = \text{Vert}(\Delta^n)$. Suppose $v_n$ is type $n$. As in Theorem B.3.1, by assembling together all top-dimensional simplices which meet at $v_n$, we obtain a regular polytope $F_n$ of dimension $n$. In this way, we see that $\mathcal{U}_{\mathbf{m}}$ is the barycentric subdivision of a regular cell complex which we denote by the same symbol $\mathcal{U}_{\mathbf{m}}$. Since $\mathcal{U}_{\mathbf{m}}$ exists for any $n$-tuple $\mathbf{m}$, an $n$-tuple $\mathbf{m}$ occurs as the Schläfli symbol of a regular cell complex if and only if its initial part is spherical. Since $W_{\mathbf{m}}$ acts freely on $\text{Cham}(\mathcal{U}_{\mathbf{m}})$, Lemma B.4.3 shows that $\mathcal{U}_{\mathbf{m}}$ is the universal orbihedral cover. This gives the following.

**THEOREM B.4.6.** ([74].) *Suppose $\Lambda$ is a regular cell complex with Schläfli symbol $\mathbf{m}$. Then there is an orbihedral cover $p : \mathcal{U}_{\mathbf{m}} \to b\Lambda$ and $b\Lambda \cong \mathcal{U}_{\mathbf{m}}/\pi$ for some subgroup $\pi \subset W_{\mathbf{m}}$.*

We turn now to Conditions B.4.2.

**LEMMA B.4.7.** ([74, Lemma 2.25].) *Let $\Lambda$ be a regular cell complex of dimension $n \geqslant 2$ and* $\mathbf{m}$ *its Schläfli symbol. Suppose $\Lambda$ is a finite complex. Then the following statements are equivalent:*

(a) *$\Lambda$ satisfies Conditions B.4.2.*

(b) *$\Lambda$ is PL homeomorphic to $S^n$.*

(c) *$W_{\mathbf{m}}$ is spherical and $p : \mathcal{U}_{\mathbf{m}} \to b\Lambda$ is an isomorphism.*

Obviously, (c) $\Longrightarrow$ (b) $\Longrightarrow$ (a). The proof that (a) $\Longrightarrow$ (c) is by induction on dimension using the fact that the link of any cell in a regular cell complex is also regular. A fairly immediate corollary is the following.

**THEOREM B.4.8.** ([74, Theorem 2.26].) *Let $\Lambda$ be a regular cell complex of dimension $n \geqslant 2$ and* $\mathbf{m}$ *its Schläfli symbol. Suppose $\Lambda$ is simply connected. Then the following statements are equivalent.*

(a) *$\Lambda$ satisfies Conditions B.4.2.*

(b) *$\Lambda$ is a PL manifold.*

(c) *The final part $(m_2, \dots, m_n)$ of* $\mathbf{m}$ *is the symbol of a spherical Coxeter group ($= W_{(0)}$) and the projection $p : \mathcal{U}_{\mathbf{m}} \to \Lambda$ is an isomorphism.*

If we drop the simple connectivity requirement, then we see that $\Lambda$ is a PL manifold if and only if $\mathcal{U}_{\mathbf{m}}$ is a PL manifold and the subgroup $\pi$ $(= \pi(b\Lambda, \sigma))$ acts freely on $\mathcal{U}_{\mathbf{m}}$. Moreover, this is the case if and only if the intersection of $\pi$ with any conjugate of the subgroup $W_{(0)}$ is trivial. (Its intersection with conjugates of $W_{(n)}$, the subgroup corresponding to the initial part of $\mathbf{m}$, is automatically trivial.)

If both the initial and final parts of $\mathbf{m}$ are spherical, then $W_{\mathbf{m}}$ is a simplicial Coxeter group (or a "Lannér group") as in Section 6.9. It follows that $\mathcal{U}(W_{\mathbf{m}}, \Delta^n)$ can be identified with $\mathbb{S}^n$, $\mathbb{E}^n$ or $\mathbb{H}^n$ as the determinant of the cosine matrix is positive, zero or negative, respectively. (Following Section 6.2, we denote these constant curvature spaces by $\mathbb{X}^n_\kappa$ for $\kappa = +1$, $0$ or $-1$, respectively.) Since $\mathbb{E}^n$ and $\mathbb{H}^n$ are contractible, it follows that when $\kappa = 0$ or $-1$, $\pi$ acts freely if and only if it is torsion-free.

The following theorem summarizes the above discussion.

**THEOREM B.4.9.** ([74].)

(i) *For $n \geqslant 2$, an $n$-tuple* $\mathbf{m} = (m_1, \dots, m_n)$*, of positive integers each $\geqslant 3$, is a Schläfli symbol of a regular tessellation of a PL manifold if and only if its initial and final parts are both spherical.*

(ii) *Suppose* $\Lambda$ *is a PL manifold and a regular cell complex with Schläfli symbol* $\mathbf{m}$. *Then* $\Lambda$ *is equivalent to a classical geometric tessellation of a space of constant curvature. That is to say,* $W_{\mathbf{m}}$ *is a simplicial Coxeter group acting isometrically on* $\mathcal{U}_{\mathbf{m}} = \mathbb{X}^n_\kappa$ *for some* $\kappa \in \{+1, 0, -1\}$ *and* $\Lambda \cong \mathbb{X}^n_\kappa/\pi$, *where* $\pi \subset W_m$ *is some subgroup which acts freely on* $\mathbb{X}^n_\kappa$.

## Euclidean Tessellations

According to Table 6.1 in Section 6.9, there is only one infinite family of Euclidean reflection groups with straight line diagrams, namely the family of diagrams $\widetilde{\mathbf{C}}_n$. $\widetilde{\mathbf{C}}_n$ has Schläfli symbol $(4, 3, \dots, 3, 4)$. It corresponds to the standard tessellation of $\mathbb{E}^n$ by $n$-cubes; the link of each vertex is the boundary complex of an $n$-octahedron. The tessellation is self-dual. (In dimensions 1 and 2 this tessellation corresponds to the diagrams $\widetilde{\mathbf{A}}_1$ and $\widetilde{\mathbf{B}}_2$, respectively.)

There are only two other Euclidean reflection groups with straight line diagrams, one in dimension 2, the other in dimension 4. In dimension 2, we have $\widetilde{\mathbf{G}}_2$ with corresponding Schläfli symbols $(6, 3)$ or $(3, 6)$. The first is the usual tiling of the plane by regular hexagons, the second is the dual tiling by equilateral triangles. In dimension 4 we have the diagram $\widetilde{\mathbf{F}}_4$ corresponding to $(3, 4, 3, 3)$ and $(3, 3, 4, 3)$. These give the only two interesting regular tessellations of Euclidean space. The first is a tessellation of $\mathbb{E}^4$ by 24-cells; the link of each vertex is the boundary of a 4-cube. The second is the dual tessellation by four-dimensional octahedra; the link of each vertex being the boundary of a 24-cell.

## Hyperbolic Tessellations

Examining Table 6.2 in Section 6.9, we see that simplicial hyperbolic reflection groups occur only in dimensions 2, 3, and 4. In dimension 2 there are an infinite number with straight line diagrams. There are three in dimension 3 and three more in dimension 4. In dimension 2 we have the $(p, q, 2)$ triangle groups with Schläfli symbols $(p, q)$, with $\frac{1}{p} + \frac{1}{q} < \frac{1}{2}$, giving a tessellation of $\mathbb{H}^2$ by $p$-gons, $q$ meeting at each vertex. (See [115].) In dimension 3 we have the symbols $(3, 5, 3)$, $(5, 3, 4)$, $(4, 3, 5)$, and $(5, 3, 5)$. In each case the cells and vertex links can be read off from the symbol. For example, $(4, 3, 5)$ corresponds to a tessellation of $\mathbb{H}^3$ by cubes; the link of each vertex being the boundary of an icosahedron. In dimension 4 we have $(5, 3, 3, 3)$, $(3, 3, 3, 5)$, $(5, 3, 3, 4)$, $(4, 3, 3, 5)$, and $(5, 3, 3, 5)$. Again in each case we can read off the structure from the symbol. For example, $(5, 3, 3, 5)$ is the self-dual tessellation of $\mathbb{H}^4$ by 120-cells in which the link of each vertex is the boundary of a 600-cell. These regular tessellations of $\mathbb{H}^n$, $n \geqslant 3$, were explained by Coxeter [68].

# Appendix C

## THE CLASSIFICATION OF SPHERICAL AND EUCLIDEAN COXETER GROUPS

Here we prove the classical classification results for simplicial Coxeter groups. That is to say, we show that Tables 6.1 and 6.2 of Section 6.9 are correct.

### C.1. STATEMENTS OF THE CLASSIFICATION THEOREMS

In view of Theorem 6.8.12 the classification of spherical reflection groups reduces to the classification of those Coxeter matrices $M$ such that the corresponding cosine matrix $(c_{ij})$ (defined by equation (6.21) of Section 6.8) is positive definite. By Theorem 6.12.9, classifying spherical reflection groups is the same as classifying finite Coxeter groups.

Recall that $M$ is *reducible* if its Coxeter diagram is not connected. Equivalently, $M$ is reducible if the index set $I$ has a nontrivial partition as $I = I' \cup I''$, with $m_{ij} = 2$ for all $i \in I'$ and $j \in I''$. A third way to say this is that its cosine matrix is decomposable (Definition 6.3.6). A decomposable matrix is positive definite if and only if each of its indecomposable principal submatrices is positive definite. So, it suffices to consider the case where $M$ is irreducible.

**DEFINITION C.1.1.** Let $\Gamma$ be a Coxeter diagram. We say that $\Gamma$ is *positive semidefinite* (or *positive definite*) as its cosine matrix is positive semidefinite (or positive definite). Similarly, $\Gamma$ is of *type* $(n, 1)$ if its cosine matrix is of this type.

Here is the famous classification of finite Coxeter groups (see [67]).

**THEOREM C.1.2.** (Classification of finite Coxeter groups.) *The connected, positive definite Coxeter diagrams are those listed in the left-hand column of Table 6.1. (In other words, this is the complete list of diagrams of the irreducible Coxeter systems* $(W, S)$ *with* $W$ *finite.)*

In the course of proving this we will classify the other simplicial Coxeter groups as well. In the Euclidean case, by Theorem 6.8.12, this amounts to classifying the Coxeter diagrams which are connected and positive semidefinite degenerate (and therefore of corank 1).

**THEOREM C.1.3.** (Classification of Euclidean Coxeter groups). *The connected, positive semidefinite Coxeter diagrams, which are not positive definite, are those listed in the right hand column of Table 6.1. (In other words, this is the complete list of diagrams of the irreducible Euclidean reflection groups.)*

By Theorem 6.8.12, the classification of hyperbolic simplicial Coxeter groups amounts to the classification of Coxeter diagrams of type $(n, 1)$ with the property that every proper subdiagram is positive definite. The result is the following.

**THEOREM C.1.4.** (Classification of hyperbolic simplicial Coxeter groups.) *The connected, Coxeter diagrams of type* $(n, 1)$ *with the property that every proper subdiagram is positive definite exist only for* $n \leqslant 4$ *and they are listed in Table 6.2. (In other words, this is the complete list of diagrams of hyperbolic simplicial reflection groups.)*

Our proof of these theorems is essentially the one given by Humphreys [163] (which follows the treatment in [303]). This proof is not as elegant as the one given in [29] and it gives no hint as to how one might discover the list in Table 6.1. However, the basic argument of [163] reduces to straightforward calculations of determinants, which the reader can easily carry out himself.

## C.2. CALCULATING SOME DETERMINANTS

Suppose $A$ is an $n \times n$ symmetric matrix. If $A$ is positive definite, then so is its principal submatrix $A_{(n)}$ (where, by definition, $A_{(n)}$ is obtained by deleting the last row and column of $A$). Conversely, if $A_{(n)}$ is positive definite, then there are only three possibilities for $A$: it is either positive definite, positive semidefinite of corank 1 or nondegenerate of type $(n, 1)$. We can distinguish these possibilities by calculating $\det(A)$: it is $> 0$, $= 0$ or $< 0$, respectively. Thus, $A$ is positive definite if and only if $A_{(n)}$ is positive definite and $\det(A) > 0$. So, ultimately, we can decide the positivity of $A$ by calculating determinants.

Define a *principal minor* of $A$ to be the determinant of a matrix obtained by deleting the last $k$ rows and columns of $A$, $0 \leqslant k \leqslant n$. The well-known positivity test alluded to above is the following: $A$ is positive definite if and only if each principal minor is positive.

### Positivity of Certain Cosine Matrices

In the next few paragraphs we show that the diagrams in the left hand column of Table 6.1 are all positive definite. First consider the rank two case, i.e., the cosine matrices of the $\mathbf{I}_2(m)$. We have

$$\det A = \begin{pmatrix} 1 & -\cos(\pi/m) \\ -\cos(\pi/m) & 1 \end{pmatrix} = \sin^2(\pi/m) > 0.$$

So $A$ is positive definite.

Suppose the rank of $A$ is $\geqslant 3$. Looking at the diagrams on the left hand side of Table 6.1, we see that whenever there are more than three vertices, the greatest possible label on any edge is 5. So, the relevant values of the cosine function are given by

$$\cos(\pi/m) = \begin{cases} 1/2 & \text{if } m = 3, \\ \sqrt{2}/2 & \text{if } m = 4, \\ (1+\sqrt{5})/4 & \text{if } m = 5. \end{cases}$$

Since the denominator 2 often occurs in the cosine matrices of these diagrams, it makes more sense to calculate the determinant of the matrix $2A$.

Another thing we can see from Table 6.1 is that it is always possible to number the vertices of the diagram so that the $n^{\text{th}}$ vertex is connected to only one other vertex, the $(n-1)^{\text{st}}$, and with this edge labeled by $m = 3$ or 4. Let $d_i$ be the determinant of the upper left $i \times i$ matrix ($d_i$ is a principal minor). Expanding $2A$ along its last row we find that

$$\det 2A = 2d_{n-1} - d_{n-2} \qquad \text{if } m = 3, \tag{C.1}$$

$$\det 2A = 2d_{n-1} - 2d_{n-2} \qquad \text{if } m = 4. \tag{C.2}$$

Using these two formulas it is easy to inductively verify that the values of $\det 2A$, where $A$ is the cosine matrix of a diagram of rank $> 2$ in the left-hand column of Table 6.1, are given by Table C.1. So, we have proved the following lemma.

**LEMMA C.2.1.** *Each of the diagrams in the left-hand column of Table 6.1 is positive definite.*

### Some Nonpositive Determinants

We turn our attention to the diagrams on the right hand side of Table 6.1. As was pointed out in 6.9, these diagrams have the property that the cosine matrix of every proper subdiagram is positive definite. So, if we can show that in each case the determinant is 0, we will have proved the following.

**LEMMA C.2.2.** *Each of the diagrams in the right-hand column of Table 6.1 is positive semidefinite of corank 1.*

| $\mathbf{A}_n$ | $\mathbf{B}_n$ | $\mathbf{D}_n$ | $\mathbf{E}_6$ | $\mathbf{E}_7$ | $\mathbf{E}_8$ | $\mathbf{F}_4$ | $\mathbf{H}_3$ | $\mathbf{H}_4$ |
|---|---|---|---|---|---|---|---|---|
| $n+1$ | 2 | 4 | 3 | 2 | 1 | 1 | $3-\sqrt{5}$ | $(7-3\sqrt{5})/2$ |

**Table C.1.** Determinant of $2A$.

In the case of $\widetilde{\mathbf{A}}_n$, the sum of all the rows of $A$ is 0; hence, $\det A = 0$. In all other cases, we can choose one of the terminal vertices of the diagram (which is a tree) and then use (C.1) or (C.2) together with Table C.1 to immediately calculate that the determinant is 0. As a random example, for $\widetilde{\mathbf{E}}_7$, the relevant subdiagrams are $\mathbf{E}_7$ and $\mathbf{D}_6$ and (C.1) reads: $\det 2A = 4 - 4 = 0$.

**LEMMA C.2.3.** *The cosine matrices of the diagrams*

$\mathbf{Z}_4$ (5)

$\mathbf{Z}_5$ (5)

*have negative determinants. (These diagrams occur in Table 6.2.)*

*Proof.* If $A$ is the cosine matrix of $\mathbf{Z}_4$ or $\mathbf{Z}_5$, then we can compute $\det 2A$ by using (C.1) and the values of the determinant for $\mathbf{H}_3$ and $\mathbf{H}_4$ in Table C.1. In the case of $\mathbf{Z}_4$, we get $\det 2A = 3 - 2\sqrt{5}$. For $\mathbf{Z}_5$, we get $\det 2A = 4 - 2\sqrt{5}$. □

## C.3. PROOFS OF THE CLASSIFICATION THEOREMS

Suppose $\Gamma$ and $\Gamma'$ are Coxeter diagrams and that $\Gamma$ is connected. Say that $\Gamma$ *dominates* $\Gamma'$ if the underlying graph of $\Gamma'$ is a subgraph of $\Gamma$ and if the label on each edge of $\Gamma'$ is $\leqslant$ the label on the corresponding edge of $\Gamma$. If, in addition, $\Gamma \neq \Gamma'$, then say $\Gamma$ *strictly dominates* $\Gamma'$ and write $\Gamma \succ \Gamma'$. The next lemma is a corollary of Lemma 6.3.7.

**LEMMA C.3.1.** ([163, p. 36].) *Suppose an irreducible Coxeter diagram $\Gamma$ is positive semidefinite. If $\Gamma \succ \Gamma'$, then $\Gamma'$ is positive definite.*

*Proof.* Let $A = (a_{ij})$ be the $n \times n$ cosine matrix for $\Gamma$ and $A' = (a'_{ij})$ the cosine matrix for $\Gamma'$. Since $\Gamma'$ is a subgraph of $\Gamma$, after reordering the vertices of $\Gamma$ we can assume that $A'$ is the $k \times k$ matrix in the upper left corner of $A$ for some $k \leqslant n$. Since $\Gamma$ dominates $\Gamma'$, $a_{ij} \leqslant a'_{ij} \leqslant 0$ for all $1 \leqslant i, j \leqslant k$, $i \neq j$. Suppose $A'$ is not positive definite, i.e., suppose there is a nonzero vector $x \in \mathbb{R}^k$ with $x^t A' x \leqslant 0$. We claim that

$$0 \leqslant \sum a_{ij}|x_i||x_j| \leqslant \sum a'_{ij}|x_i||x_j| \leqslant \sum a'_{ij}x_i x_j \leqslant 0, \qquad \text{(C.3)}$$

where the sums are over $1 \leqslant i,j \leqslant k$. The first inequality in (C.3) is because $A$ is positive definite, the second because $a_{ij} \leqslant a'_{ij}$, for $1 \leqslant i,j \leqslant k$, the third because $a'_{ij} \leqslant 0$ for $i \neq j$ and the last is by assumption. So the inequalities in (C.3) are all equalities. If we write

$$y = (|x_1|, \dots, |x_k|) \in \mathbb{R}^k,$$

$$z = (|x_1|, \dots, |x_k|, 0, \dots, 0) \in \mathbb{R}^n,$$

then (C.3) reads $0 \leqslant z^tAz \leqslant y^tA'y \leqslant x^tA'x \leqslant 0$. Since $z^tAz = 0$, $z \in \operatorname{Ker} A$. By Lemma 6.3.7, its coordinates are all nonzero. Hence, $k = n$ and the coordinates of $y$ are all nonzero. Since $\Gamma \succ \Gamma'$, $a_{ij} < a'_{ij}$, for at least one pair $\{i,j\}$. Hence, $z^tAz < y^tA'y$, a contradiction. $\square$

*Proofs of Theorems C.1.2 and C.1.3.* ([163, p. 37].) We want to show Table 6.1 is the complete list of the connected Coxeter diagrams which have positive semidefinite cosine matrices. Suppose, to the contrary, that $\Gamma$ is a connected, positive semidefinite Coxeter diagram not on this list. Let $n$ be the rank of $\Gamma$ and $m$ the maximum label on any edge. By using Lemma C.3.1 repeatedly, one now establishes the following statements, showing that no such $\Gamma$ exists:

- $n \geqslant 3$ (since all Coxeter diagrams of rank $\leqslant 2$ are positive semidefinite).
- $m \neq \infty$ (since $\Gamma \not\succeq \widetilde{\mathbf{A}}_1$).
- $\Gamma$ is a tree ($\Gamma$ contains no circuits since $\Gamma \not\succeq \widetilde{\mathbf{A}}_n$).

Suppose $m = 3$. Then

- $\Gamma$ must have a branch vertex (since $\Gamma \neq \mathbf{A}_n$).
- The branch vertex is unique (since $\Gamma \not\succeq \widetilde{\mathbf{D}}_n$, $n > 4$).
- Each branch vertex is valence 3 (since $\Gamma \not\succeq \widetilde{\mathbf{D}}_4$).

So, suppose $\Gamma$ has three branches with $a \leqslant b \leqslant c$ edges along each of the three branches.

- $a = 1$ (since $\Gamma \not\succeq \widetilde{\mathbf{E}}_6$).
- $b \leqslant 2$ (since $\Gamma \not\succeq \widetilde{\mathbf{E}}_7$).
- $b = 2$ ($b \neq 1$ since $\Gamma \neq \mathbf{D}_n$).
- $c \leqslant 4$ (since $\Gamma \not\succeq \widetilde{\mathbf{E}}_8$).

Therefore,

- $m \geqslant 4$ (the case $m = 3$ is impossible, since $\Gamma \neq \mathbf{E}_6, \mathbf{E}_7, \mathbf{E}_8$).
- Only one edge of $\Gamma$ has a label $> 3$ (since $\Gamma \not\geq \widetilde{\mathbf{C}}_n$).
- $\Gamma$ has no branch vertices (since $\Gamma \not\geq \widetilde{\mathbf{B}}_n$).

Suppose $m = 4$. Then

- The two extreme edges of $\Gamma$ are labeled 3.
- $n = 4$ (since $\Gamma \not\geq \widetilde{\mathbf{F}}_4$).

Therefore,

- $m = 5$ ($m = 4$ is impossible since $\Gamma \neq \mathbf{F}_4$ and $m = 6$ is impossible since $\Gamma \not\geq \widetilde{\mathbf{G}}_2$).
- The edge labeled 5 must be an extreme edge (since $\Gamma \not\geq \mathbf{Z}_4$, where $\mathbf{Z}_4$ is the diagram $\circ\!-\!\circ\overset{5}{-}\circ\!-\!\circ$ of Table 6.2).
- Contradiction (since $\Gamma \neq \mathbf{H}_3, \mathbf{H}_4$, which are the only remaining possibilities). □

*Proof of Theorem C.1.4.* $\Gamma$ is the diagram of a hyperbolic simplicial Coxeter group if and only if it satisfies the following two conditions:

(a) Every proper subdiagram is positive definite.

(b) $\Gamma$ does not appear in Table 6.1.

It is simple to check that the diagrams in Table 6.2 satisfy both conditions.

It is also easy to check that there are no further possibilities. To do this, a couple of observations are useful. First, in any connected spherical diagram with $n \geqslant 3$, the largest label $m$ which can occur is 5. Second, $\Gamma$ can contain no proper cycle (since every connected spherical diagram is a tree). So $\Gamma$ is either a cycle or a tree. Now consider the ways of adding a node to a diagram in the left hand column of Table 6.1 so that (a) holds. Inspection shows that in every case we either get another diagram in Table 6.1 or a diagram in Table 6.2. □

# Appendix D

## THE GEOMETRIC REPRESENTATION

In Chapter 7 we associated to a Coxeter system $(W, S)$ a geometric object: the cell complex $\Sigma$. Here we associate another geometric object to it: the interior $\mathcal{I}$ of the "Tits cone." $\Sigma$ and $\mathcal{I}$ have several important properties in common. $W$ acts properly on both. Both are models for $\underline{E}W$. (In particular, both are contractible.) The main advantage of $\Sigma$ is that its $W$-action is cocompact. The advantage of $\mathcal{I}$ is that it is a $W$-stable open set in a faithful linear representation. For example, the existence of this representation allows one to conclude that $W$ is virtually torsion-free (Corollary D.1.4 below).

### D.1. INJECTIVITY OF THE GEOMETRIC REPRESENTATION

In 6.12 we started with a Coxeter matrix $M = (m_{ij})$ over a index set $I$, then defined an associated Coxeter system $(W, S)$ (where $S := \{s_i\}_{i \in I}$) and a bilinear form $B_M$ on $\mathbb{R}^I$ given by $B_M(e_i, e_j) := -\cos(\pi/m_{ij})$. (For the notion of a Coxeter matrix and its associated Coxeter system, see Definitions 3.3.1 and 3.3.2, respectively.) We go on to define the "canonical representation" $\rho : W \to GL(\mathbb{R}^I)$ (Definition 6.12.5). It is induced from the map which sends the generator $s_i$ to the linear reflection $\rho_i$ defined by equation (6.33) of 6.12.

To simplify notation, put $E := \mathbb{R}^I$. We are interested in the dual of the canonical representation $\rho^* : W \to GL(E^*)$. We call it the *geometric representation* of $(W, S)$. In many ways it is more important than the canonical representation. We denote vectors of $E$ by Roman letters such as $e_i$ or $x$ and vectors in $E^*$ by Greek letters such as $\xi$. The natural pairing of $E^*$ and $E$ will be denoted either by $(\xi, x) \to \xi(x)$ or $(\xi, x) \to x(\xi)$. Let $\xi_i \in E^*$ be the linear form $x \to B_M(x, e_i)$. Then $\rho^*(s_i)$ is the linear reflection $\rho_i^*$ on $E^*$ defined by

$$\rho_i^*(\varphi) = \varphi - 2e_i(\varphi)\xi_i. \tag{D.1}$$

To check that this actually is the formula for $\rho^*(s_i)$, compute

$$\begin{aligned}(\rho^*(s_i)(\varphi))(x) &= \varphi(\rho_i(x)) = \varphi(x - 2B_M(x, e_i)e_i) \\ &= \varphi(x) - 2B_M(x, e_i)\varphi(e_i) = (\varphi - 2e_i(\varphi)\xi_i)(x) \\ &= \rho_i^*(\varphi)(x).\end{aligned}$$

(Since $e_i$ is a linear form on $E^*$ and $\xi_i \in E^*$ is a vector with $e_i(\xi_i) = 1$, it follows from formula (6.1) of 6.1 that (D.1) actually defines a linear reflection.)

Let $C$ be the simplicial cone in $E^*$ defined by the inequalities

$$e_i(\varphi) \geqslant 0, \qquad i \in I,$$

and let $\overset{\circ}{C}$ be its interior (defined by the strict inequalities, $e_i(\varphi) > 0$).

**THEOREM D.1.1.** (Tits, see [29, p. 97].) *Let $w \in W$. If $w\overset{\circ}{C} \cap \overset{\circ}{C} \neq \emptyset$, then $w = 1$.*

A corollary is Theorem 6.12.10 which states that $\rho^* : W \to GL(E^*)$ is faithful. If $\rho^*$ is faithful, then so is $\rho$. This gives the following (stated earlier as Corollary 6.12.11).

**COROLLARY D.1.2.** *The canonical representation $\rho$ is faithful.*

Another corollary is the following.

**COROLLARY D.1.3.** ([29, Cor. 3, p. 97].) *$\rho^*(W)$ is a discrete subgroup of $GL(E^*)$. Similarly, $\rho(W)$ is a discrete subgroup of $GL(E)$.*

*Proof of Corollary D.1.3.* Pick $\varphi \in \overset{\circ}{C}$. Let $V := \{g \in GL(E^*) \mid g(\varphi) \in \overset{\circ}{C}\}$. $V$ is an open neighborhood of 1 in $GL(E^*)$. By Theorem D.1.1, $V \cap \rho^*(W) = \{1\}$. So $\rho^*(W)$ is a discrete subgroup. The second sentence follows from the first. □

Selberg's Lemma [254] states that any finitely generated subgroup of $GL(n, \mathbb{C})$ is virtually torsion-free. (Recall that a group *virtually* has some property if it has a subgroup of finite index which is torsion-free.) So a consequence of Corollary D.1.2 is the following.

**COROLLARY D.1.4.** *Any (finitely generated) Coxeter group is virtually torsion-free.*

When $s = s_i$, write $e_s$ instead of $e_i$. Let $H_s$ (or $H_s(C)$) be the open half-space in $E^*$ defined by $e_s(\xi) > 0$. Set

$$A_s := \{w \in W \mid w\overset{\circ}{C} \subset H_s(C)\}.$$

The proof of Theorem D.1.1 is based on Tits' Lemma from 4.8. To apply this lemma we must verify a certain Property (P) holds for each pair of distinct elements in $S$. We do so below.

**LEMMA D.1.5.** ([29, Lemma 1, p. 98].) *Given distinct elements $s, t \in S$ and an element $v \in W_{\{s,t\}}$, $v(A_s \cap A_t)$ is contained either in $A_s$ or in $sA_s$ and in the second case, $l(sv) = l(v) - 1$.*

*Proof.* Let $E_{\{s,t\}} := \mathbb{R}e_s \oplus \mathbb{R}e_t \subset E$ and let $E^*_{\{s,t\}}$ be the dual 2-plane given by $E^*_{\{s,t\}} := E^*/\mathrm{Ann}(E_{\{s,t\}})$, where $\mathrm{Ann}(E_{\{s,t\}})$ is the annihilator of $E_{\{s,t\}}$ (i.e., $\mathrm{Ann}(E_{\{s,t\}}) := \{\xi \in E^* \mid e_s(\xi) = 0 = e_t(\xi)\}$). $W_{\{s,t\}}$ acts naturally on $E_{\{s,t\}}$ and is naturally identified with the geometric representation of the Coxeter system $(W_{\{s,t\}}, \{s,t\})$. The dual of the inclusion $E_{\{s,t\}} \hookrightarrow E$ is the $W_{\{s,t\}}$-equivariant surjection $p : E^* \to E^*_{\{s,t\}}$. The subsets $H_s(C)$, $H_t(C)$ and $H_s(C) \cap H_t(C)$ are the inverse images under $p$ of the corresponding subsets of $E_{\{s,t\}}$. So, we are reduced to proving the lemma in the case where $W$ is the dihedral group $W_{\{s,t\}}$ of order $2m$, $m := m(s,t)$. We shall show this reduces to checking the lemma in the two standard pictures described in Examples 3.1.2 and 3.1.3. We distinguish two cases.

*Case 1.* $m = \infty$. Let $(\xi, \varphi)$ be the dual basis to $(e_s, e_t)$. Then

$$\begin{aligned} s \cdot \xi &= -\xi + 2\varphi, & t \cdot \xi &= \xi, \\ s \cdot \varphi &= \varphi, & t \cdot \varphi &= 2\xi - \varphi. \end{aligned}$$

Let $L$ be the affine line in $E^*$ spanned by $\xi$ and $\varphi$. The above formulas show $L$ is stable under $W$. Let $f : \mathbb{R} \to L$ be the affine isomorphism defined by $t \to t\xi + (1-t)\varphi$. Transporting the action to $\mathbb{R}$ via this map we obtain the standard action of the infinite dihedral group on $\mathbb{R}$, where $s$ and $t$ act by reflections across 0 and 1, respectively. Let $I_n \subset L$ be the image under $f$ of $[n, n+1]$ and let $C_n \subset E^*$ be the union of all positive real multiples of $I_n$. Then $C = C_0$. Since the $I_n$ are permuted simply transitively by $W$ so are the $C_n$. If $v \in W$, then $vC$ is equal to one of the $C_n$ and hence, is on the positive side of 0 (if $n \geqslant 0$) or on the negative (if $n < 0$). It follows that $v(A_s \cap A_t)$ is contained either in $A_s$ (if $n \geqslant 0$) or in $sA_s$ (if $n < 0$). In the second case $I_0$ and $I_n$ are on opposite sides of $\xi$ and hence, $l(sv) = l(v) - 1$.

*Case 2.* $m < \infty$. The bilinear form $B_M$ has matrix

$$\begin{pmatrix} 1 & -\cos(\pi/m) \\ -\cos(\pi/m) & 1 \end{pmatrix}$$

Since its determinant is $1 - \cos^2(\pi/m) = \sin^2(\pi/m) > 0$, $B_M$ is positive definite. Hence, we can use the form $B_M$ to identify both $E$ and $E^*$ with the Euclidean plane $\mathbb{R}^2$. After this identification, $e_s$ and $e_t$ become unit vectors making an angle of $\pi - \pi/m$ and $C$ becomes a sector bounded by two lines making an angle of $\pi/m$ ($e_s$ and $e_t$ are the inward-pointing unit normal vectors). Hence, $C$ is a fundamental chamber for the dihedral group. The lemma follows easily from this (for example, from the results of 6.6). $\square$

*Proof of Theorem D.1.1.* $(W, S, \{A_s\}_{s\in S})$ satisfies Property (P) by the previous lemma together with Tits' Lemma 4.8.3. By Remark 4.8.1 (i), $\bigcap A_s$ is prefundamental for $W$, i.e., $w\overset{\circ}{C} \cap \overset{\circ}{C} \neq \emptyset \implies w = 1$. □

## D.2. THE TITS CONE

Given $s \in S$, let $C_s$ denote the codimension one face of $C$ defined by $e_s(x) = 0$. (We are altering our notation and using the Roman letter $x$ for a point in $C$ rather than a Greek letter.) This defines a tautological mirror structure on $C$ indexed by $S$. (See 5.1 for the definition of this and other constructions used below.) As in (5.1), $S(x) := \{s \in S \mid x \in C_s\}$ and as in (5.2), for each $T \subset S$, $C_T$ is the intersection of the $C_s$, $s \in T$. Let $\overset{\circ}{C}_T := \{x \in C \mid S(x) = T\}$. $\overset{\circ}{C}_T$ is the relative interior of $C_T$. Put

$$C^f := \bigcup_{T\in\mathcal{S}} \overset{\circ}{C}_T$$

(where $\mathcal{S}$ is the set of spherical subsets of $S$) and put

$$U := \bigcup_{w\in W} wC \quad \text{and} \quad \mathcal{I} := \bigcup_{w\in W} wC^f.$$

$U$ is the *Tits cone*. (Since $C$ is a cone, for $x \in C$ and $\lambda \in [0, \infty)$, $\lambda x \in C$; hence, if $wx \in U$, then so is $\lambda wx = w\lambda x$, i.e., $U$ is a cone.)

**Examples D.2.1**

(i) (*The infinite dihedral group* $\mathbf{D}_\infty$.) Since $S = \{s_1, s_2\}$, $\dim E^* = 2$. Introduce coordinates $(x_1, x_2)$ by $x_i = e_i(x)$. $C$ is the positive quadrant bounded by the coordinate lines. As in Case 1 in the proof of Lemma D.1.5, the affine line $x_1 + x_2 = 1$ is stable under $\mathbf{D}_\infty$ and the action on it can be identified with the standard $\mathbf{D}_\infty$-action on $\mathbb{E}^1$. $C^f = C - \{0\}$, $U$ is the half-plane $x_1 + x_2 \geqslant 0$ and $\mathcal{I}$ is the corresponding open half-plane. Similarly, if $W$ is any irreducible Euclidean reflection group, then $C$ is a simplicial cone, $C^f = C - \{0\}$, $U$ is a half-space and $\mathcal{I}$ is an open half-space. (See Proposition 6.8.8.)

(ii) (*Hyperbolic triangle groups.*) Suppose $W$ is a hyperbolic triangle group generated by the reflections $s_1, s_2, s_3$ across the faces of a triangle $\Delta \subset \mathbb{H}^2$. Since bilinear form $B_M$ is associated to the cosine matrix, it is nondegenerate of type (2, 1). So both $E$ and $E^*$ can be identified with Minkowski space $\mathbb{R}^{2,1}$. $C$ is the cone on $\Delta$, $C^f = C - \{0\}$, $U$ is the positive light cone and $\mathcal{I}$ is its interior.

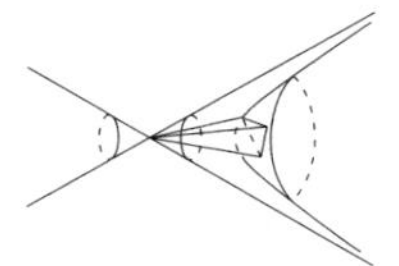

Figure D.1. The cone on a hyperbolic triangle group.

(See Figure D.1.) Similarly, if $W$ is one of the simplicial hyperbolic Coxeter groups with diagram listed in Table 6.2, then $\mathcal{I}$ is the interior of the positive light cone in $\mathbb{R}^{n,1}$ $(n \leqslant 4)$.

On the other hand, if, as in Example 6.5.3, $W$ is a hyperbolic $m$-gon group, with $m > 3$, we again get a representation $W \hookrightarrow O(2,1)$ with fundamental chamber the cone on the $m$-gon. However, this is *not* the geometric representation. (The cone on an $m$-gon, $m > 3$, is not a simplicial cone; moreover, $\dim E^* = m \neq \dim \mathbb{R}^{2,1}$.)

Once we choose a point $x \in \overset{\circ}{C}$ we get an embedding of the Cayley graph of $(W, S)$ in $U$: the vertex set of $\mathrm{Cay}(W, S)$ is identified with $Wx$ and vertices $wx$ and $wsx$ are connected by the line segment between them. Let $E^*_s \subset E^*$ denote the hyperplane fixed by $\rho^*(s)$. By Lemma 4.2.2, the vertices $x$ and $wx$ are on opposite sides of $E^*_s$ if and only if $l(w) > l(sw)$. Hence, we have the following.

**LEMMA D.2.2.** *Given $s \in S$ and $w \in W$, the chambers $C$ and $wC$ lie on opposite sides of the hyperplane $E^*_s$ if and only if $l(w) > l(sw)$.*

Let $-C \subset E^*$ denote the image of $C$ under the antipodal map $x \to -x$.

**LEMMA D.2.3.** *Suppose $y \in -\overset{\circ}{C}$. Then $y \in U$ if and only if $W$ is finite.*

*Proof.* Suppose $y = wx$ for some $x \in \overset{\circ}{C}$. For each $s \in S$, $l(ws) < l(w)$, since $y$ and $x$ lie on opposite sides of $E^*_s$. Hence, by Lemma 4.6.2, $W$ is finite and $w$ is the element of longest length. Conversely, if $W$ is finite and $w_0$ is its element of longest length, then $w_0 y \in \overset{\circ}{C}$ and $y \in U$. □

**COROLLARY D.2.4.** *The point $0 \in E^*$ lies in the interior of $U$ if and only if $W$ is finite.*

*Proof.* If $W$ is finite, then $E^* = U$. If $W$ is infinite, then, by the previous lemma, there is a point $y \in -\overset{\circ}{C}$ such that $ty \notin U$ for all $t \in (0, \infty)$. As $t \to 0$, $\lim ty = 0$; so $0$ is not in the interior. □

We omit the proof of the next lemma since it essentially identical to the proof of Lemma 6.6.8.

**LEMMA D.2.5.** ([29, Prop. 5, p. 101].) *Suppose $x, y \in C$ and $w \in W$ are such that $wx = y$. Then $x = y$ and $w \in W_{S(x)}$.*

Since we have mirror structures on $C$ and $C^f$, we can apply the basic construction of Chapter 5 to get $W$-spaces $\mathcal{U}(W, C)$ and $\mathcal{U}(W, C^f)$. It is a tautology that the mirror structure on $C^f$ is $W$-finite (Definition 5.1.6). It follows from the universal property of the basic construction (Lemma 5.2.5) that the inclusion $\iota : C \to U$ induces an equivariant map $\tilde{\iota} : \mathcal{U}(W, C) \to U$. Similarly, the restriction of $\iota$ to $C^f$ induces $\mathcal{U}(W, C^f) \to \mathcal{I}$.

**THEOREM D.2.6**

*(i) The natural map $\tilde{\iota} : \mathcal{U}(W, C) \to U$ is a bijection.*

*(ii) Its restriction $\mathcal{U}(W, C^f) \to \mathcal{I}$ is a $W$-equivariant homeomorphism.*

*(iii) $\mathcal{I}$ is the interior of $U$.*

*Proof.* (i) By definition $U = WC$, so, $\tilde{\iota}$ is surjective. By Lemma D.2.5, $wx = w'x'$ if and only if $x = x'$and $w^{-1}w' \in W_{S(x)}$, i.e., if and only if $[w, x] = [w', x']$ in $\mathcal{U}(W, C)$. So $\tilde{\iota}$ is injective.

Given $x \in C$, let $V_x$ be a neighborhood of $x$ in $C$ which intersects only those $C_s$ with $s \in S(x)$. (We could take $V_x$ to be the complement in $C$ of those $C_s$ which do not contain $x$.) Then $\mathcal{U}(W_{S(x)}, V_x)$ is a neighborhood of $[1, x]$ in $\mathcal{U}(W, C)$ and $W_{S(x)} V_x$ is an open neighborhood of $x$ in $U$.

(ii) Suppose $W_{S(x)}$ is finite. By Theorem 6.6.3 (iv), $\tilde{\iota}$ takes $\mathcal{U}(W_{S(x)}, V_x)$ homeomorphically onto $W_{S(x)} V_x$. Similarly, it takes a neighborhood of $[w, x]$ homeomorphically onto its image. This shows that the bijection $\tilde{\iota} : \mathcal{U}(W, C^f) \to \mathcal{I}$ is a homeomorphism.

(iii) The argument for (ii) shows that $\mathcal{I} \subset \text{int}(U)$. Conversely, suppose $x \in C - C^f$ so that $W_{S(x)}$ is infinite. Then Corollary D.2.4 (applied to the geometric representation of $W_{S(x)}$) shows that $x \notin \text{int}(U)$. So $\mathcal{I} = \text{int}(U)$. □

*Remark.* In regard to part (i) of the theorem, when $W$ is infinite, $\mathcal{U}(W, C)$ and $U$ are definitely *not* homeomorphic. The topology on $\mathcal{U}(W, C)$ is that of a cell complex (the CW topology) while $U$ has the induced topology as a subset of $E^*$. These are not the same. For example, if one picks a point in the interior of each chamber intersected with a sphere about the origin in $E^*$, the result is not discrete in $U$; however, its inverse image in $\mathcal{U}(W, C)$ is discrete.

Let $\mathcal{F}$ denote the set of all subsets of $U$ of the form $w\mathring{C}_T$ for some $w \in W$ and $T \subset S$.

**THEOREM D.2.7.** ([29, Prop. 6, p. 102].)

*(i) $U$ is convex.*

*(ii) Any closed line segment in $U$ meets only finitely many elements of $\mathcal{F}$.*

*Proof.* Suppose $x, y \in U$. We will prove that the segment $[x, y]$ is covered by finitely many elements of $\mathcal{F}$. This implies both (i) and (ii). After translating both $x$ and $y$ by the same element of $W$, we can assume $x \in C$. Let $w \in W$ be such that $y \in wC$. The proof is by induction on $l(w)$. For each $s \in S$, let $H_s$ be the half-space of $E^*$ bounded by $E^*_s$ and containing $C$. The relation $wC \not\subset H_s$ is equivalent to $\overset{\circ}{C} \not\subset \mathrm{int}(H_s)$ and hence, by Lemma D.2.2, to $l(sw) < l(w)$. The intersection $[x, y] \cap C$ is a closed segment $[x, z]$ for some $z \in C$. First suppose $z = y$, i.e., $y \in C$. Then there are subsets $X, Y$ of $S$ such that $x \in \overset{\circ}{C}_X$, $y \in \overset{\circ}{C}_Y$. The open segment $(x, y)$ is contained in $C_{X \cap Y}$; so, $[x, y] \subset \overset{\circ}{C}_X \cup \overset{\circ}{C}_Y \cup \overset{\circ}{C}_{X \cap Y}$, which proves the assertion in this case. Next, suppose $z \neq y$. Then $z \in E^*_s$ for some $s \in S$. So, $\overset{\circ}{C}$ and $w\overset{\circ}{C}$ are on opposite sides of $E^*_s$ and $l(sw) < l(w)$. By the inductive hypothesis $[z, y]$ $(= s[z, sy])$ is covered by finitely many elements of $\mathcal{F}$ and hence, so is $[x, y] = [x, z] \cup [z, y]$ since $[x, z] \subset C$. □

## Some Consequences

Since $U$ is convex, so is its interior $\mathcal{I}$. Since the mirror structure on $C^f$ is $W$-finite, it follows from Lemma 5.1.7 that $W$ acts properly on $\mathcal{U}(W, C^f)$ and hence, also on $\mathcal{I}$. From this we get the following.

**THEOREM D.2.8.** (Compare Theorem 12.3.4.) *$\mathcal{I}$ is a model for $\underline{E}W$.*

(The definition of $\underline{E}W$ is given in 2.3.1. Further discussion can be found in Corollary I.2.12 of Appendix I.2.)

*Proof.* For any finite subgroup $F \subset W$ and $x \in \mathcal{I}$, define the $\mathrm{Av}_F(x)$, the *$F$-average of $x$*, to be the convex combination

$$\mathrm{Av}_F(x) := \frac{1}{\mathrm{Card}(F)} \sum_{f \in F} fx.$$

(The notion of a convex combination of points in an affine space is explained in the beginning of Appendix A.1.) Since $\mathcal{I}$ is convex, $\mathrm{Av}_F(x) \in \mathcal{I}$. Since $\mathrm{Av}_F(x)$ is fixed by $F$, this shows that $\mathrm{Fix}(F, \mathcal{I}) \neq \emptyset$. Since $W$ acts on $\mathcal{I}$ via affine transformations, $\mathrm{Fix}(F, \mathcal{I})$ is convex; hence, contractible. □

Recall that a conjugate of a spherical special subgroup is called a "spherical parabolic subgroup" in 13.2. Since each isotropy group on $\mathcal{I}$ is a conjugate of some spherical special subgroup, we get the following result of Tits (also proved as Theorem 12.3.4 (i)).

**COROLLARY D.2.9.** (Tits, [29, Ex. 2d), p. 137].) *Each finite subgroup of $W$ is contained in a spherical parabolic subgroup.*

**THEOREM D.2.10.** *The center of an infinite, irreducible Coxeter group is trivial.*

*Proof.* Suppose $w$ is a nontrivial element of the center of $W$. By Lemma 6.12.2 (i), $\rho^*(w)$ acts as a homothety. Since it cannot map the fundamental cone $C$ to itself (since $w \neq 1$), it must map it to $-C$. If $C$ and $-C$ are both contained in the Tits cone, then we must have $\mathcal{I} = E^*$. But this can only happen if $W$ is finite. So $w$ must be trivial. □

### $\Sigma$ Is the Cocompact Core of $\mathcal{I}$

Let $\mathcal{P}$ be the power set of $S$ (Example A.2.1). Its geometric realization $|\mathcal{P}|$ has a mirror structure over $S$ defined by $|\mathcal{P}|_s := |\mathcal{P}_{\geqslant \{s\}}|$. For each $T \subset S$ choose a "barycenter" $v_T$ in the relative interior of the face $C_T$. The vertices corresponding to any chain in $\mathcal{P}$ span a simplex in $C$. In this way we get any embedding of the barycentric subdivision of the cone on a simplex of dimension Card($S$) – 1 as a neighborhood of the origin in $C$. (Call this the "barycentric subdivision" of $C$.) $\mathcal{U}(W, |\mathcal{P}|)$ is the cone on the Coxeter complex and $\Sigma = \mathcal{U}(W, |\mathcal{S}|)$ is a subcomplex. As in Theorem D.2.6, the inclusion $|\mathcal{P}| \hookrightarrow C$ induces an injection $\mathcal{U}(W, |\mathcal{P}|) \hookrightarrow U$ and its restriction to $\Sigma$ is an embedding $\Sigma \hookrightarrow \mathcal{I}$. There is a coface-preserving deformation retraction $C^f \to |\mathcal{S}|$. This induces a $W$-equivariant retraction $\mathcal{I} \to \Sigma$.

## D.3. COMPLEMENT ON ROOT SYSTEMS

A *root basis* is a triple $(E, \langle\, ,\rangle, \Pi)$, where, $\langle\, ,\rangle$ is a symmetric bilinear form on a real vector space $E$ and $\Pi$ is a finite subset of $E$ such that

(a) For all $p \in \Pi$, $\langle p, p\rangle = 1$.

(b) For any two distinct elements $p, q \in \Pi$,

$$\langle p, q\rangle \in \{-\cos(\pi/m) \mid m \text{ an integer } \geqslant 2\} \cup (-\infty, -1].$$

(c) There is an element $\xi \in E^*$ with $\xi(p) > 0$, for all $p \in \Pi$.

For each $p \in \Pi$, let $r_p \in GL(E)$ be the reflection defined by $r_p(x) = x - 2\langle x, p\rangle p$. Let $W \subset GL(E)$ be the subgroup generated by $S := \{r_p\}_{p \in \Pi}$. It follows from the arguments in the previous two sections of this appendix that $W$ acts as a discrete reflection group on the dual space $E^*$ with fundamental chamber $C$, the simplicial cone defined by the inequalities $\xi(p) \geqslant 0$, $p \in \Pi$ and that $(W, S)$ is a Coxeter system.

A *root* is an element $x \in E$ of the form $x = wp$ for some $w \in W$ and $p \in \Pi$. The set $\Phi$ of all roots in $E$ is called the *root system* associated to the root basis. The elements of $\Pi$ are *simple roots*. A root $x \in \Phi$ is *positive* if $\xi(x) \geqslant 0$ for all $\xi \in C$. When $E = \mathbb{R}^I$, $\langle\, ,\rangle = B_M(\, ,)$ and $\Pi = \{e_s\}_{s \in S}$ as in D.1, we get the *classical* root basis and root system.

For any $w \in W$, the condition that $\xi(we_s) < 0$ for all $\xi \in \overset{\circ}{C}$ means that $wC$ and $C$ are separated by the wall in $E^*$ defined by $\xi(e_s) = 0$, i.e., that $l(sw) > l(w)$. In other words, if $\widehat{A}_s := \{w \in W \mid \xi(we_s) < 0\}$, then $\widehat{A}_s = A_s$, the half-space defined in 4.5. Thus, a root carries essentially the same information as a half-space. (In fact, in [248] a "root" is synonymous with a half-space in $W$.)

## NOTES

**D.1, D.2.** The material in these sections is due to Tits in [281]. A published version appeared in Bourbaki [29] and our presentation follows the treatment there. In [29] the canonical representation $\rho$ is called the "geometric representation" and what we call the geometric representation (that is, $\rho^*$) is called the "contragredient representation." When the bilinear form $B_M(\ ,\ )$ is nonsingular, $\rho$ and $\rho^*$ are isomorphic.

Vinberg wrote an important paper [290] expanding on Tits' results. Among other things, he considered linear representations similar to the geometric one but with fundamental chambers allowed to be a polyhedral cones rather than being restricted simplicial cones. For example, if we start with a hyperbolic polygon group (Example 6.5.3), then we get a linear representation into $O(2, 1) \subset GL(3, \mathbb{R})$ with fundamental chamber the cone on an $m$-gon. This is a simplicial cone only for $m = 3$. (See Examples D.2.1 (ii).)

A different argument for Theorem D.2.10 is given in [242].

**D.3.** Root systems are underemphasized in this book. They play an imporant role in most of the rest of the literature on Coxeter groups allowing the tools of linear algebra to be brought into play. Our discussion in this section is taken from [181].

# Appendix E

## COMPLEXES OF GROUPS

The theory of complexes of groups was developed in the early 1990s as a natural outgrowth of Serre's theory of graphs of groups and groups acting on trees (usually called "Bass-Serre theory"). A graph of groups is a special case of a complex of groups. The ideas are somewhat easier to understand in this case and we discuss it first in E.1. One way in which a complex of groups can arise is from the action of a discrete group $G$ on a cell complex $Y$. The underlying cell complex of the complex groups is the orbit space $K := Y/G$. The group associated to a cell $\sigma$ of $K$ is the stabilizer of a preimage of $\sigma$ in $Y$. When $Y$ is simply connected, $G$ is the "fundamental group" and $Y$ the "universal cover" of the associated complex of groups. A basic question is whether every complex of groups arises in this fashion. (Is it "developable?") This turns out always to be the case for graphs of groups (the universal cover is a tree), but is not always the case for general complexes of groups. An extra hypothesis which insures this in the general case is one of nonpositive curvature. A prototypical example of a complex of groups is given by the system of spherical special subgroups in a Coxeter system $(W, S)$ (see Example E.2.2). Its underlying cell complex is the geometric realization of $\mathcal{S}$. This complex of groups is developable, its fundamental group is $W$ and its universal cover is $\Sigma$. By Moussong's Theorem, it is nonpositively curved.

Roughly speaking, a "complex of spaces" is a space $X$ together with a projection map to a cell complex. A complex of spaces gives a complex of groups: the group associated to a cell is the fundamental group of the inverse image of the cell in $X$. The notion of the "fundamental group of a complex of groups" $\mathcal{G}$ is best understood in terms of complexes of spaces: when $\mathcal{G}$ is associated to a complex of spaces $X$, $\pi_1(\mathcal{G})$ is simply $\pi_1(X)$. One of the motivations for developing the theory is the following problem. Suppose a collection of aspherical cell complexes are glued together in the same combinatorial pattern as some cell complex (in other words, it gives a complex of spaces over the cell complex). Is the resulting space aspherical? When the cell complex is a graph the answer is affirmative (Whitehead's Theorem E.1.15). In the general case the issue is whether the universal cover of the associated complex of groups is contractible.

In E.3 we discuss a different topic, the "Mayer-Vietoris spectral sequence" associated to the cover of a CW complex $X$ by a family of subcomplexes. The

$E^2$-term involves the homology of the nerve of the cover with coefficients in a system which associates to each simplex of the nerve, the homology of the associated subspace. (This type of coefficient system also occurs in a spectral sequence calculating the homology of a complex of spaces with coefficients in any "module" over an associated complex of groups.) The main result of E.3 is the Acyclic Covering Lemma which was needed in 8.2.

## E.1. BACKGROUND ON GRAPHS OF GROUPS

Here we summarize some of the main the points of the Bass-Serre theory of graphs of groups from [256, 252].

**DEFINITION E.1.1.** Suppose $G_1$, $G_2$ and $A$ are groups and that $i_1 : A \hookrightarrow G_1$, $i_2 : A \hookrightarrow G_2$ are monomorphisms. The *amalgamated product* of $G_1$ and $G_2$ along $A$ is the quotient of the free product of $G_1$ and $G_2$ by the normal subgroup generated by $\{i_1(a)i_2(a)^{-1} \mid a \in A\}$. It is denoted $G_1 *_A G_2$.

**DEFINITION E.1.2.** Suppose $A$ and $G$ are groups and $i_1 : A \hookrightarrow G$, $i_2 : A \hookrightarrow G$ are monomorphisms. Form $G * \mathbf{C}_\infty$, the free product of $G$ and the infinite cyclic group. Let $t$ be a generator of $\mathbf{C}_\infty$. The *HNN construction* on these data is the quotient of $G * \mathbf{C}_\infty$ by the normal subgroup generated by $\{ti_1(a)t^{-1}i_2(a)^{-1} \mid a \in A\}$. It is denoted $G*_A$. (In other words, we impose the condition that $t$ conjugates $i_1(a)$ to $i_2(a)$ for all $a \in A$.)

Throughout this section all graphs will be connected. Given a graph $\Gamma$, $\mathrm{edge}(\Gamma)$ is the set of its directed edges. Given $e \in \mathrm{edge}(\Gamma)$, $t(e)$ denotes its terminal vertex; $\overline{e}$ is the same edge in the reverse direction.

**DEFINITION E.1.3.** ([256, p. 38].) A *graph of groups* $\mathcal{G}$ over $\Gamma$, is an assignment of a group $\mathcal{G}(v)$ to each $v \in \mathrm{Vert}(\Gamma)$ and a group $\mathcal{G}(e)$ to each $e \in \mathrm{edge}(\Gamma)$, together with a monomorphism $\psi_e : \mathcal{G}(e) \hookrightarrow \mathcal{G}(t(e))$. In addition, it is required that $\mathcal{G}(e) = \mathcal{G}(\overline{e})$.

**DEFINITION E.1.4.** ([256, pp. 42–43].) Suppose $\mathcal{G}$ is a graph of groups over $\Gamma$ and $T$ is a maximal tree in $\Gamma$. Introduce a symbol $\lambda_e$ for each $e \in \mathrm{edge}(\Gamma)$. The *fundamental group* of the graph of groups $\mathcal{G}$, denoted $\pi_1(\mathcal{G}, T)$, is the quotient of the free product of the $\mathcal{G}(v)$, $v \in \mathrm{Vert}(\Gamma)$, and the free group on the $\lambda_e$, $e \in \mathrm{edge}(\Gamma)$ by the normal subgroup generated by all relations of the form:

$$\lambda_e \psi_e(a) \lambda_e^{-1} = \psi_{\overline{e}}(a) \quad \lambda_{\overline{e}} = \lambda_e^{-1} \quad \text{if } e \in \mathrm{edge}(\Gamma),\ a \in \mathcal{G}(e)$$

$$\lambda_e = 1, \quad \text{if } e \in \mathrm{edge}(T)$$

(Usually, we will drop the $T$ from our notation and write simply $\pi_1(\mathcal{G})$.)

Another way to define $\pi_1(\mathcal{G})$ is via the notion of a "graph of spaces" explained in Definition E.1.13, below. If $X$ is a graph of spaces with associated graph of fundamental groups $\mathcal{G}$, then $\pi_1(\mathcal{G}) := \pi_1(X)$.

In the case where no edge of $\Gamma$ connects a vertex to itself, it is unnecessary to consider directed edges and the definition of a graph of groups can be simplified. Let $\mathcal{F}(\Gamma)$ be the poset of cells in $\Gamma$ and $\mathcal{F}(\Gamma)^{op}$ the dual (or opposite) poset, thought of as a category. When $\Gamma$ has no loops of length one, a graph of groups over $\Gamma$ is simply a functor $\mathcal{G}$ from $\mathcal{F}(\Gamma)^{op}$ to the category of groups and monomorphisms. So, associated to each vertex $v$ of $\Gamma$ there is a group $\mathcal{G}(v)$ and to each (undirected) edge $e$, a group $\mathcal{G}(e)$. Moreover, when $v$ is a vertex of $e$, there is a monomorphism $\mathcal{G}(e) \hookrightarrow \mathcal{G}(v)$.

When $\Gamma$ is a tree, $\mathcal{G}$ is called a *tree of groups* and $\pi_1(\mathcal{G})$ is the direct limit of the family of groups consisting of the edge groups and vertex groups ("direct limit" is defined in Appendix G.1).

**DEFINITION E.1.5.** The action of a group $G$ on a simplicial graph $\Omega$ is *without inversions* if no edge of $\Omega$ is flipped by an element of $G$. (In other words, given an edge of $\Omega$ there is no element of $G$ which switches its vertices.)

When $G$ acts on $\Omega$ without inversions, there is a well-defined quotient graph $\Omega/G$ (not necessarily a simplicial graph).

**Example E.1.6.** (*Actions on simplicial graphs.*) Suppose $G$ acts on a connected simplicial graph $\Omega$ without inversions. This gives the data for a graph of groups $\mathcal{G}$ on the quotient graph $\Gamma := \Omega/G$. For each vertex $v$ (resp. oriented edge $e$) of $\Gamma$, choose a vertex $\tilde{v}$ (resp. oriented edge $\tilde{e}$) of $\Omega$ projecting onto it. Set $\mathcal{G}(v) := G_{\tilde{v}}$ and $\mathcal{G}(e) := G_{\tilde{e}}$, the isotropy subgroups at $\tilde{v}$ and $\tilde{e}$, respectively. If $v$ is the terminal vertex of the oriented edge $e$, then we can find $g \in G$ so that $\tilde{v}$ is the terminal vertex of $g\tilde{e}$. Define $\mathcal{G}(e) \hookrightarrow \mathcal{G}(v)$ to be the composition of inner automorphism by $g$ with the natural inclusion, $G_{g\tilde{e}} \hookrightarrow G_{\tilde{v}}$.

The main result of Bass-Serre theory is the following.

**THEOREM E.1.7.** ([256].) *Suppose that $\mathcal{G}$ is a graph of groups over $\Gamma$ and $G = \pi_1(\mathcal{G})$. Then there exists a tree $T$ with $G$-action with the following properties.*

*(a)* $T/G = \Gamma$.

*(b) Suppose $e$ is an edge of $\Gamma$, $v$ a vertex of $e$, $\tilde{e}$ a lift of $e$ to $T$ and $\tilde{v}$ a vertex of $\tilde{e}$ lying above $v$. Then there are isomorphisms $G_{\tilde{v}} \cong \mathcal{G}(v)$ and $G_{\tilde{e}} \cong \mathcal{G}(e)$ taking the inclusion $G_{\tilde{e}} \hookrightarrow G_{\tilde{v}}$ to the monomorphism $\mathcal{G}(e) \hookrightarrow \mathcal{G}(v)$.*

*Moreover, $T$ is unique up to isomorphism.*

$T$ is the *universal cover of $\mathcal{G}$*.

**Example E.1.8.** (*A segment of groups*). Suppose $\Gamma$ is the graph with one edge and two vertices (so that $\Gamma$ is a segment). A graph of groups over $\Gamma$ is a *segment of groups*. It gives the data for an amalgamated product: the vertex groups are $G_1$ and $G_2$, the edge group is $A$. The fundamental group of the segment of groups is $G := G_1 *_A G_2$. Let $T$ be the graph with vertex set the disjoint union of $G/G_1$ and $G/G_2$. Two vertices $g_1G_1$ and $g_2G_2$ are connected by an edge if and only if $g_1^{-1}g_2 \in A$. The proof that $T$ is a tree follows from the standard normal form in which an element of the amalgamated product can be written. (See [256].)

**Example E.1.9.** (*A loop of groups*). Suppose $\Gamma$ is the graph with one vertex and one edge. ($\Gamma$ is homeomorphic to a circle.) A graph of groups over $\Gamma$ is a *loop of groups*. It gives the data for a HNN construction: the vertex group is $G$ and the edge group is $A$. The fundamental group of the loop of groups is $G*_A$.

A consequence of Theorem E.1.7 (b) is that the natural map $\mathcal{G}(v) \to \pi_1(\mathcal{G})$ is injective (since it is isomorphic to the inclusion $G_{\tilde{v}} \hookrightarrow G$). In the language of complexes of groups in [37, 151], this means $\mathcal{G}$ is *developable*. In contrast with the general theory of complexes of groups, an important feature of a graph of groups is that its universal cover $T$ is not only simply connected but contractible. One reason for this is that any tree is nonpositively curved (in sense which will be explained in Appendix I).

A graph of groups $\mathcal{G}$ is *uninteresting* if for some vertex $v$ the natural map $\mathcal{G}(v) \to \pi_1(\mathcal{G})$ is an isomorphism. (In other words, if $\pi_1(\mathcal{G})$ has a fixed point on $T$.) Otherwise, $\mathcal{G}$ is *interesting*.

**DEFINITION E.1.10.** A *splitting* of a group $G$ is an interesting graph of groups $\mathcal{G}$ together with an isomorphism $G \cong \pi_1(\mathcal{G})$.

**DEFINITION E.1.11.** ([256, p. 58].) A group $G$ has *property* FA if any $G$-action on any tree has a fixed point.

If $G$ has property FA, then every graph of groups for it is uninteresting; hence, such a group cannot split.

**Example E.1.12.** ([256, Exercise 3, p. 66].) Every 2-spherical Coxeter group has property FA. (Recall that "2-spherical" means that no entry of the Coxeter matrix is $\infty$.)

Let $\Gamma$ be a graph and $b\Gamma$ its barycentric subdivision. For each cell $\sigma$ in $\Gamma$, there is a vertex $v_\sigma$ in $b\Gamma$. There are two types of vertices in $b\Gamma$: if $w \in \mathrm{Vert}(\Gamma)$, we have the corresponding vertex $v_w$ of $b\Gamma$ and if $e \in \mathrm{Edge}(\Gamma)$ we have its midpoint $v_e$. For each vertex $v$ in $b\Gamma$, let $D(v)$ denote its closed star in $b\Gamma$. (There is one exception to this if $v$ is the midpoint of an edge $e$ and if $e$ is a loop, then we split the single vertex of the loop into two and define $D(v)$ to be a copy of the interval projecting onto $e$.)

**DEFINITION E.1.13.** A *graph of spaces* over a graph $\Gamma$ is a space $X$ together with a map $p : X \to b\Gamma$ such that

(a) For each vertex $v_\sigma$ of $b\Gamma$, $p^{-1}(v_\sigma)$ is path connected.

(b) For each vertex $v_\sigma$, there is a projection map $p^{-1}(D(v_\sigma)) \to p^{-1}(v_\sigma)$, which is a homotopy equivalence.

(c) For $e$ an edge and $w$ an endpoint of $e$, let $\phi_{e,w} : \pi_1(p^{-1}(v_e)) \to \pi_1(p^{-1}(v_w))$ be the map on fundamental groups induced by the composition of the inclusion $p^{-1}(v_e) \hookrightarrow p^{-1}(D(v_w))$ with the projection $p^{-1}(D(v_w)) \to p^{-1}(v_w)$. Then it is required that each $\phi_{e,w}$ be a monomorphism.

If, in addition, $X$ is a CW complex, the map $p$ is cellular and each fiber $p^{-1}(v_\sigma)$ is a subcomplex of $X$, then $X$ is a *cellular* graph of spaces.

To a graph of spaces $p : X \to b\Gamma$ and a section $s : b\Gamma \to X$ of $p$, one can associate a graph of groups $\mathcal{G}$ over $\Gamma$ as follows. If $\sigma \in \mathcal{F}(\Gamma)$, then $\mathcal{G}(\sigma) = \pi_1(p^{-1}(v_\sigma, s(v_\sigma)))$. Furthermore, if $e$ is an edge of $\Gamma$ and $w$ is an endpoint of $e$, then the monomorphism $\mathcal{G}(v_e) \to \mathcal{G}(v_w)$ is defined to be $\phi_{e,w}$. If $p : X \to b\Gamma$ is a graph of spaces with associated graph of groups $\mathcal{G}$, then it follows from van Kampen's Theorem that $\pi_1(X) \cong \pi_1(\mathcal{G})$.

**DEFINITION E.1.14.** An *aspherical realization* of $\mathcal{G}$ is a graph of spaces $p : X \to b\Gamma$ with $\mathcal{G}$ as its associated graph of groups such that for each vertex $v_\sigma$ of $b\Gamma$, $p^{-1}(v_\sigma)$ is aspherical.

Given a graph of groups $\mathcal{G}$, there are two methods for constructing an aspherical realization of it. The first is the most direct: glue together the appropriate aspherical spaces. Start with the disjoint union of $B\mathcal{G}(v)$, $v \in \mathrm{Vert}(\Gamma)$. For each edge $e$ of $\Gamma$ take a copy of $B\mathcal{G}(e) \times [0, 1]$. If $v_0$ and $v_1$ are the endpoints of $e$, then the data for a complex of groups give us monomorphisms $\mathcal{G}(e) \to \mathcal{G}(v_0)$ and $\mathcal{G}(e) \to \mathcal{G}(v_1)$. By the universal property of classifying spaces, we can realize these monomorphisms by maps $B\mathcal{G}(e) \times 0 \to B\mathcal{G}(v_0)$ and $B\mathcal{G}(e) \times 1 \to B\mathcal{G}(v_1)$ and then use these maps to glue $B\mathcal{G}(e) \times [0, 1]$ onto $B\mathcal{G}(v_0)$ and $B\mathcal{G}(v_1)$ for each $e \in \mathrm{Edge}(\Gamma)$. If $X$ is the resulting space, then the obvious projection map $p : X \to b\Gamma$ is an aspherical realization of $\mathcal{G}$.

The second construction of an aspherical realization goes as follows. Let $G = \pi_1(\mathcal{G})$ and let $EG$ denote the universal cover of $BG$. Let $T$ be the tree associated to $\mathcal{G}$ and let $EG \times_G T$ denote the quotient space of the diagonal $G$-action. Projection on the second factor $EG \times T \to T$ induces $p : EG \times_G T \to b\Gamma$, which is clearly an aspherical realization of $\mathcal{G}$.

A cellular aspherical realization $X$ of $\mathcal{G}$ has the following universal property. Given another cellular graph of spaces $Y$ over $\Gamma$ with associated graph of groups $\mathcal{G}'$ and a homorphism $\varphi : \mathcal{G}' \to \mathcal{G}$, there is a map $f : Y \to X$, compatible with the projections to $b\Gamma$, and which realizes $\varphi$. (A *homomorphism* between graphs of groups means a natural transformation of functors.) Moreover, $f$ is unique up to a homotopy through such maps. It follows that a cellular aspherical realization of $\mathcal{G}$ is unique up to homotopy equivalence. We denote it by $B\mathcal{G}$ and call it the *classifying space* of $\mathcal{G}$.

There is an important application of Theorem E.1.7 to the problem of deciding when a space constructed by gluing together various aspherical spaces is aspherical. The following classical theorem of J.H.C. Whitehead provides the answer.

**THEOREM E.1.15.** (Whitehead). *Let $\mathcal{G}$ be a graph of groups, $B\mathcal{G}$ an aspherical realization and $G = \pi_1(\mathcal{G})$. Then $B\mathcal{G}$ is aspherical (i.e., $B\mathcal{G}$ is homotopy equivalent to $BG$).*

*Proof.* Consider the aspherical realization $EG \times_G T$ of $\mathcal{G}$. Its universal cover is $EG \times T$, which is contractible. □

## E.2. COMPLEXES OF GROUPS

In this section we summarize the theory of complexes of groups. This theory is mainly due to Haefliger [150, 151]. (Also, see [66].) A fairly complete treatment can be found in the book of Bridson and Haefliger [37].

### Simple Complexes of Groups

We begin by discussing a generalization of graphs of groups over graphs without a loop consisting of a single edge.

**DEFINITION E.2.1.** A *simple complex of groups* $\mathcal{G}$ over a poset $\mathcal{P}$ is a functor from $\mathcal{P}$ to the category of groups and monomorphisms. In other words, $\mathcal{G}$ is a family $\{G_p\}_{p\in\mathcal{P}}$ of groups together with monomorphisms $\psi_{pq} : G_p \hookrightarrow G_q$, defined whenever $p < q$, such that $\psi_{pr} = \psi_{qr} \circ \psi_{pq}$ for all $p < q < r$. If $\mathcal{P}$ is the dual poset to the face poset of a simplex, then a simple complex of groups over it is a *simplex of groups*.

(In the case of a graph of groups over a graph $\Gamma$ without loops consisting of one edge, the poset is $\mathcal{F}(\Gamma)^{op}$.)

**Example E.2.2.** (*Simple complex of groups associated to a Coxeter system.*) We previously have been dealing with the prototypical example of a simple complex of groups: the one associated to a Coxeter system $(W, S)$. Let $\mathcal{S}$ be the poset of spherical subsets of $S$. Define a simple complex of groups $\mathcal{W}$ over $\mathcal{S}$ by setting $\mathcal{W}_T := W_T$, for each $T \in \mathcal{S}$. (For $T' \subset T$, the required monomorphism $W_{T'} \hookrightarrow W_T$ is inclusion.)

Associated to a simple complex of groups we have the direct limit:

$$\widehat{\mathcal{G}} := \varinjlim_{p \in \mathcal{P}} G_p.$$

(The direct limit of a family of groups is a generalization of the amalgamated product; see [256, p. 1], as well as, the beginning of Appendix G.1.)

In Definition A.3.4 we explained $|\mathcal{P}|$, the "geometric realization" of the poset $\mathcal{P}$. In Appendix A.5 we showed how to decompose $|\mathcal{P}|$ into its "cofaces," $\{|\mathcal{P}_{\geqslant p}|\}_{p \in \mathcal{P}}$.

By using a slight variation of the basic construction of Chapter 5, we can associate a cell complex $\mathcal{U}(\mathcal{G})$ to a simple complex of groups $\mathcal{G}$. For each $x \in |\mathcal{P}|$, let $p(x)$ be the smallest $p \in \mathcal{P}$ such that $x \in |\mathcal{P}_{\geqslant p}|$. Put $G := \widehat{\mathcal{G}}$ and

$$\mathcal{U}(\mathcal{G}) := (G \times |\mathcal{P}|)/\sim, \tag{E.1}$$

where $\sim$ is the equivalence relation defined by $(g, x) \sim (h, y) \iff x = y$ and $gh^{-1} \in G_{p(x)}$. If $|\mathcal{P}|$ is simply connected (e.g., if $\mathcal{P}$ has an initial element), then $\widehat{\mathcal{G}}$ coincides with what we call below the *fundamental group of* $\mathcal{G}$ and $\mathcal{U}(\mathcal{G})$ is the *universal cover* of $\mathcal{G}$. (See [37, Examples 3.11 (1), p. 551].)

## Scwols

A category is *small* if its objects form a set. A *small* **c***ategory* **w***ithout* **l***oop* (abbreviated *scwol*) is a small category in which the only morphism from a given object to itself is the identity morphism.

A poset is naturally a scwol. The objects are the elements of the poset. There is exactly one morphism from $p$ to $q$ if $p \leqslant q$ and $\mathrm{Hom}(p, q) = \emptyset$ if $p$ and $q$ are incomparable.

Suppose $\mathcal{X}$ is a scwol. Its *vertex set*, $\mathrm{Vert}(\mathcal{X})$, is the set of objects in $\mathcal{X}$. Define $\mathrm{Edge}(\mathcal{X})$ to be the set of morphisms which are not the identity morphism of an object. If $a \in \mathrm{Edge}(\mathcal{X})$ is a morphism from $v$ to $y$, then its *initial vertex* $i(a)$ (resp. its *terminal vertex* $t(a)$) is $v$ (resp. $y$). For example, if $\Gamma$ is a graph without loops consisting of one edge, then $\mathcal{F}(\Gamma)^{op}$, the dual poset of its face poset, is a scwol. Its vertex set is the set of vertices in the barycentric subdivision $b\Gamma$; its edge set is the set of edges in $b\Gamma$. The edges of $b\Gamma$ are oriented so that the terminal vertex of any edge is one of the original vertices (and its initial vertex is the midpoint of one of the original edges). On the other hand, if $\Gamma$ has one-edged loops, then it gives a scwol which is not a poset. As before, one introduces a midpoint in each edge, but now we can have two morphisms connecting vertices.

The *geometric realization* of a scwol $\mathcal{X}$ can be defined in a similar fashion to that of a poset (Definition A.3.4). It is a cell complex denoted by $|\mathcal{X}|$. Each cell a simplex. There is an $n$-simplex for each $n$-tuple of composable edges. (However, unlike the case of a poset, $|\mathcal{X}|$ might not be a simplicial complex: a nonempty intersection of two simplices is a union of common faces, but it need not be a single simplex. For example, $|\mathcal{X}|$ could be a digon.)

A reader, who is unfamiliar with this material, probably expects the definition of a "complex of groups" over a scwol $\mathcal{Y}$ to be a functor from $\mathcal{Y}$ to the category of groups and monomorphisms. In fact, the correct definiton is a slightly following weaker notion (called a "lax functor" by homotopy theorists).

**Definition E.2.3.** (See [37, p. 535].) Suppose $\mathcal{Y}$ is a scwol. A *complex of groups* $\mathcal{G}$ over $\mathcal{Y}$ is a family $(G_u, \psi_a, g_{a,b})$ where $u$ ranges over Vert($\mathcal{Y}$), $a$ ranges over Edge($\mathcal{Y}$) and $(a, b)$ ranges over pairs of composable edges in Edge($\mathcal{Y}$) such that

- $\psi_a : G_{i(a)} \to G_{t(a)}$ is an injective homomorphism,
- $g_{a,b} \in G_{t(a)}$, and
- $\mathrm{Ad}(g_{a,b})\psi_{ab} = \psi_a \circ \psi_b$ (where $ab$ denotes composition in $\mathcal{Y}$ and $\mathrm{Ad}(g_{a,b})$ means conjugation by $g_{a,b}$).

Moreover, the elements $g_{a,b}$ must satisfy the "cocycle condition:"

$$\psi_a(g_{b,c})g_{a,bc} = g_{a,b}g_{ab,c}.$$

$G_u$ is called the *local group* at $u$.

**Example E.2.4.** (*The trivial complex of groups over a scwol.*) There is a trivial complex of groups over any scwol $\mathcal{Y}$: all the local groups are defined to be the trivial group.

### Actions on Scwols

Suppose a group $G$ acts on a scwol $\mathcal{X}$, that $\mathcal{Y}$ is the quotient scwol and that $p : \mathcal{X} \to \mathcal{Y}$ is the projection. One can associate to such an action, a complex of groups $\mathcal{G} = \mathcal{X}/\!/G$ over $\mathcal{Y}$, as follows. For each $v \in \mathrm{Vert}(\mathcal{Y})$, choose $\tilde{v} \in \mathrm{Vert}(\mathcal{X})$ with $p(\tilde{v}) = v$. Given $a \in \mathrm{Edge}(\mathcal{Y})$ with $i(a) = v$, it follows from Definition E.2.3 that there is a unique edge $\tilde{a} \in \mathrm{Edge}(\mathcal{X})$ with $i(\tilde{a}) = \tilde{v}$. If $y = t(a)$, then $\tilde{y}$ might not be to equal $t(\tilde{a})$. So, choose $h_a \in G$ with $h_a \cdot t(\tilde{a}) = \tilde{y}$. For each $v \in \mathrm{Vert}(\mathcal{Y})$, the local group $G_v$ is defined to be the isotropy subgroup at $\tilde{v}$. For $a \in \text{Edge}(\mathcal{Y})$, $\psi_a : G_{i(a)} \to G_{t(a)}$ is defined by $\psi_a(g) := h_a g h_a^{-1}$. For each pair $(a, b)$ of composable edges in $\mathcal{Y}$, put $g_{a,b} := h_a h_b h_{ab}^{-1}$. One checks that $(G_v, \psi_a, g_{a,b})$ is a complex of groups, well-defined up to isomorphism. (See[37].)

**DEFINITION E.2.5.** A complex of groups $\mathcal{G}$ over a scwol $\mathcal{Y}$ is developable if it is isomorphic to the complex of groups associated to the action of a group $G$ on some scwol $\mathcal{X}$ with $\mathcal{Y} \cong \mathcal{X}/G$.

Unlike the case of a graph of groups, not every complex of groups is developable.

### The Fundamental Group of a Complex of Groups

Given a complex of groups $\mathcal{G}$ over $\mathcal{Y}$, one can define its *fundamental group*, $\pi_1(\mathcal{G})$, as in Definition E.1.4 (see [37, pp. 546–553]). Using generators and relations, it can be defined in a fairly straightforward way from the local groups $G_v$, $v \in \text{Vert}(\mathcal{Y})$, by adjoining generators for the edges in $\mathcal{Y}$ and then adding relations for the 2-simplices in $|\mathcal{Y}|$ and the 1-simplices of a maximal tree $T$ in the 1-skeleton of $|\mathcal{Y}|$. (We will suppress this dependence on the choice of $T$.) It follows from the definition that there are natural maps $G_v \to \pi_1(\mathcal{G})$.

As we mentioned previously, if $\mathcal{G}$ is a simple complex of groups over a poset $\mathcal{P}$ whose geometric realization is simply connected, then $\pi_1(\mathcal{G})$ is just the direct limit group, $\widehat{\mathcal{G}}$. For another example, if we have a trivial complex of groups over a scwol $\mathcal{Y}$ (Example E.2.4), its fundamental group is $\pi_1(|\mathcal{Y}|)$.

**PROPOSITION E.2.6.** ([37, Prop. 3.9, p. 550].) *$\mathcal{G}$ is developable if and only if for each local group, the natural map $G_v \to \pi_1(\mathcal{G})$ is injective.*

**PROPOSITION E.2.7.** (Haefliger [150, 151].) *Let $\mathcal{G} = (G_v, \psi_a, g_{a,b})$ be a complex of groups over a scwol $\mathcal{Y}$. Then there is an action of $G = \pi_1(\mathcal{G})$ on a scwol $\widetilde{\mathcal{Y}}$ with quotient $\mathcal{Y}$ and associated complex of groups $(G'_v, \psi'_a, g'_{a,b})$ such that*

- *(a) $|\widetilde{\mathcal{Y}}|$ is simply connected.*
- *(b) $G'_v = \text{Im}(G_v \to G)$; $\psi'_a$ is induced by $\psi_a$ and $g'_{a,b}$ is the image of $g_{a,b}$.*
- *(c) In particular, if $\mathcal{G}$ is developable, then $(G'_v, \psi'_a, g'_{a,b})$ is identified with $\mathcal{G}$.*

*Moreover, $\widetilde{\mathcal{Y}}$ is unique up to G-equivariant isomorphism.*

The scwol $\widetilde{\mathcal{Y}}$ is the *universal cover* of $\mathcal{G}$. One of the main accomplishments in [37] is a proof that in the presence of nonpositive curvature any complex of groups is developable. Moreover, its universal cover is contractible. Here, as in Appendix I, nonpositive curvature means "locally CAT(0)." To be slightly more precise, to say that a complex of groups $\mathcal{G}$ over $\mathcal{Y}$ is nonpositively curved means that each $v \in \text{Vert}(\mathcal{Y})$ has a neighborhood in $|\mathcal{Y}|$ which is the quotient of an isometric action of $G_v$ on a CAT(0) "$\mathbb{X}_\kappa$-polyhedral complex," with $\kappa \leqslant 0$. (See [37], as well as, Appendix I.3 for more precise definitions.) Here is a statement of the theorem.

**THEOREM E.2.8.** (Haefliger [37, Thm. 4.17, p. 562].) *Suppose $\mathcal{G}$ is a nonpositively curved complex of groups over a scwol $\mathcal{Y}$. Then $\mathcal{G}$ is developable. Moreover, if $\widetilde{\mathcal{Y}}$ denotes its universal cover, then $|\widetilde{\mathcal{Y}}|$ is* CAT(0) *(and hence, by Theorem I.2.6, is contractible).*

## Complexes of Spaces

The following material can be found in [150, 151] (but not in [37]). Roughly speaking, a complex of spaces over a scwol $\mathcal{X}$ consists of a space $Y$ and a projection map $\pi : Y \to |\mathcal{X}|$ such that the inverse image of each cell of $\mathcal{X}$ is homotopy equivalent to the inverse image of its barycenter. The actual definition, in which we must keep track of base points, is somewhat more complicated.

Given $v \in \text{Vert}(\mathcal{X})$, Haefliger defines its "dual cone," $D_v$. In the case where the scwol is a poset $\mathcal{P}$, $D_v := |\mathcal{P}_{\geqslant v}|$, the subcomplex of $|\mathcal{P}|$ which we earlier called a "coface." In the general case, $D_v$ is not a subcomplex of $|\mathcal{X}|$; however, morally, it is. In particular, there is a canonical map $j_v : D_v \to |\mathcal{X}|$ and one can "restrict" any complex of groups over $\mathcal{X}$ to $D_v$. Given a space $Y$, a scwol $\mathcal{X}$ and a projection map $\pi : Y \to \mathcal{X}$, for each $v \in \text{Vert}(\mathcal{X})$, put $Y(v) := \pi^{-1}(v)$ and let $Y(D_v)$ denote the subspace of $D_v \times Y$ consisting of all $(x, y)$ such that $j_v(x) = \pi(y)$.

A *complex of spaces* over a scwol $\mathcal{X}$ is a space $Y$, a projection map $\pi : Y \to |\mathcal{X}|$ and a section $s : |\mathcal{X}|^1 \to Y$ of $\pi$ defined over the 1-skeleton of $|\mathcal{X}|$ satisfying certain conditions (see [151, p. 290]). These conditions insure that if we put $G_v := \pi_1(Y(v), s(v))$, then it is possible to define for each $a \in \text{Edge}(\mathcal{X})$, $\psi_a : G_{i(a)} \to G_{t(a)}$ and for each pair $(a, b)$ of composable edges, $g_{a,b} \in G_{t(a)}$ so that $\mathcal{G} := (G_v, \psi_a, g_{a,b})$ is a complex of groups over $\mathcal{X}$. $\mathcal{G}$ is the complex of groups *associated* to the complex of spaces $Y$. Once this definition is accomplished it is straightforward consequence of van Kampen's Theorem that $\pi_1(\mathcal{G}) = \pi_1(Y)$.

**Example E.2.9.** Suppose $(X_s)_{s\in S}$ is a mirror structure on a CW complex $X$ and $N(X)$ denotes its nerve. ($N(X)$ is the abstract simplicial complex from

Definition 5.1.1.) Put $X_\emptyset := X$. This gives the data for a complex of spaces over the poset $\mathcal{S}(N(X))^{op}$ (where $\mathcal{S}(N(X))$, the poset of simplices of $N(X)$, is defined in Example A.2.3).

We have the following generalization of Definition E.1.14.

**DEFINITION E.2.10.** A (cellular) *aspherical realization* of a complex of groups $\mathcal{G}$ over a scwol $\mathcal{X}$ is complex of CW complexes $Y \to |\mathcal{X}|$ such that each $Y(v)$ is aspherical and $\mathcal{G}$ is the associated complex of groups.

It is proved in [151, Theorems 3.4.1 and 3.52] that aspherical realizations exist and are unique up to homotopy equivalence.

## E.3. THE MAYER-VIETORIS SPECTRAL SEQUENCE

### Background

Spectral sequences were invented by Leray around 1946. They are a tool for computing (co)homology. To fix ideas, we stick to homology. Roughly speaking, a *spectral sequence* is family $\{E^r_{pq}\}$ of abelian groups (or $R$-modules) where $p, q$ and $r$ are integers and $r \geqslant 0$. The number $p + q$ is the *total degree*. Each of the $E^r_{**}$ is equipped with a *differential* $d_r$ of total degree $-1$ and "bidegree" $(-r, r-1)$, i.e., $d_r$ maps $E^r_{pq}$ to $E^r_{p-r,q+r-1}$. It is required that $E^{r+1}_{**}$ is obtained by taking the homology of $E^r_{**}$, i.e.,

$$E^{r+1}_{pq} = \frac{\operatorname{Ker}(d_r : E^r_{pq} \to E^r_{p-r,q+r-1})}{\operatorname{Im}(d_r : E^r_{p+r,q-r+1} \to E^r_{pq})}.$$

A spectral sequence is *first quadrant* if $E^r_{pq} = 0$ whenever $p < 0$ or $q < 0$. If this is the case (and henceforth, we will always assume that it is), then the spectral sequence *converges to* $E^\infty$, i.e., for a fixed $p$ and $q$, there is an $r$ so that for all $s \geqslant r$, all the incoming and outgoing differentials $d_s$ vanish and $E^r_{pq} = E^{r+1}_{pq} = \cdots =: E^\infty_{pq}$.

The most important examples of spectral sequences arise in association with filtrations of chain complexes. Given an abelian group (or $R$-module or chain complex) $A$, recall that an (increasing) *filtration* is a sequence $F_pA$ of subgroups (or submodules or subcomplexes) such that $F_pA \subset F_{p+1}A$. Given a filtration $\{F_pA\}$, define the *associated graded group* $\operatorname{Gr} A = \{\operatorname{Gr}_p A\}$ by $\operatorname{Gr}_p A := F_pA/F_{p-1}A$.

Suppose $C$ is a chain complex with an increasing filtration $\{F_pC\}$. There is an induced filtration of homology: $F_pH_*(C) := \operatorname{Im}(H_*(F_pC) \to H_*(C))$. Define subgroups of $C_{p+q}$:

$$Z^r_{pq} := F_pC_{p+q} \cap \partial^{-1}(F_{p-r}C_{p+q-1}),$$

$$Z^\infty_{pq} := F_pC_{p+q} \cap \operatorname{Ker} \partial,$$

$$B^r_{pq} := F_pC_{p+q} \cap \partial F_{p+r-1}C_{p+q+1},$$

$$B^\infty_{pq} := F_pC_{p+q} \cap \partial C_{p+q+1}.$$

The associated spectral sequence is defined by

$$E^r_{pq} := Z^r_{pq}/(B^r_{pq} + Z^{r-1}_{p-1,q+1})$$

and

$$E^\infty_{pq} := Z^\infty_{pq}/(B^\infty_{pq} + Z^\infty_{p-1,q+1}) = \operatorname{Gr}_p H_{p+q}(C).$$

(The differentials are induced by $\partial$.)

In particular, $E^0_{pq} = F_pC_{p+q}/F_{p-1}C_{p+q} := \operatorname{Gr}_p C_{p+q}$. So, the $E^0$-terms are the associated graded groups of the filtration of the chain complex, while the $E^\infty$-terms are the graded groups associated to the filtration of homology. In practice, this means that after $E^\infty_{**}$ has been calculated, we can read off the homology of $C$ in dimension $n$ (or at least its associated graded group) by looking at the terms on the "isodiagonal," $p + q = n$.

A *double complex* is a family of abelian groups, $\{C_{pq}\}$, $p, q \in \mathbb{Z}$, equipped with a *horizontal differential* $\partial' : C_{pq} \to C_{p-1,q}$ of bidegree $(-1, 0)$ and a *vertical differential* $\partial'' : C_{pq} \to C_{p,q-1}$ of bidegree $(0, -1)$, so that $\{C_{*,q}, \partial'\}$ and $\{C_{p,*}, \partial''\}$ are both chain complexes (i.e., $\partial' \circ \partial' = 0$, $\partial'' \circ \partial'' = 0$). It is also required that $\partial'$ and $\partial''$ commute in the graded sense: $\partial'\partial'' = -\partial''\partial'$. The *total complex* $TC$ is defined by

$$TC_n := \bigoplus_{p+q=n} C_{pq}\,.$$

It has a differential, $\partial := \oplus(\partial' + \partial'')$, which makes it into a chain complex (i.e., $(\partial' + \partial'')^2 = 0$).

There are two filtrations of $TC$, leading to two different spectral sequences. The first filtration is defined by $F_p(TC_n) := \bigoplus_{i \leqslant p} C_{i,n-i}$. The associated spectral sequence $\{E^r\}$ converges to $H_*(TC)$. It has $E^0_{pq} = C_{pq}$ with $d_0 = \pm\partial'$; so, $E^1_{pq} = H_q(C_{p,*})$ and $d_1$ is induced by $\partial'$. As Ken Brown writes in [42, p. 165], "$E^2$ can be described as the horizontal homology of the vertical homology of $C$." Similarly, the second filtration is defined by $F_p(TC_n) := \bigoplus_{j \leqslant p} C_{n-j,j}$. Its spectral sequence again converges to $H_*(TC)$. It has $E^0_{pq} = C_{qp}$ and $E^1_{pq} = H_q(C_{*,p})$.

### Coefficient Systems

In 15.2 we considered coefficient systems $\mathcal{F}$ on a simplicial complex $L$ of the following type: $\mathcal{F}$ is a functor (either covariant or contravariant) from the poset $\mathcal{S}(L)_{>\emptyset}$ to the category of abelian groups. For example, a contravariant coefficient system is a family $(\mathcal{F}(T))$ of abelian groups indexed by $\mathcal{S}(L)_{>\emptyset}$ and homomorphisms $\mathcal{F}_{UT} : \mathcal{F}(T) \to \mathcal{F}(U)$ whenever $U < T$ such that $\mathcal{F}_{VT} = \mathcal{F}_{VU} \circ \mathcal{F}_{UT}$ whenever $V < U < T$. As in 15.2, define a chain complex

$$C_p(L; \mathcal{F}) := \bigoplus_{T \in \mathcal{S}^{(p+1)}(L)} \mathcal{F}(T),$$

where, as usual, we are identifying the $p$-simplices in $L$ with their vertex sets (which have cardinality $p + 1$). The corresponding homology groups are denoted $H_*(L; \mathcal{F})$. (See [87, 151, 243, 294] for more details.)

**Example E.3.1.** (*The nerve of a cover.*) Suppose $X$ is a mirrored CW complex with mirror structure $(X_s)_{s \in S}$. Further suppose $X = \bigcup_{s \in S} X_s$. So $X$ is covered by its mirrors. As in Example E.2.9, let $L = N(X)$ be the nerve of cover. For each nonnegative integer $q$ define a contravariant coefficient system $\mathfrak{h}_q$ on $L$ by

$$\mathfrak{h}_q(T) := H_q(X_T).$$

Whenever $U < T, X_T \subset X_U$ and the inclusion induces a map on homology; so, $\mathfrak{h}_q$ is a contravariant functor.

### The Spectral Sequence

We continue with the notation of the above example. For each nonnegative integer $p$, define $C_p$ to be the following direct sum of chain complexes:

$$C_p := \bigoplus_{\sigma_T \in L^{(p)}} C(X_T),$$

where $L^{(p)}$ means the set of $p$-simplices in $L$ and $\sigma_T$ means the simplex with vertex set $T$. Totally order the vertex set $S$ of $L$. If the vertex set of $\sigma_T$ is $\{t_0, \dots, t_p\}$, with $t_0 < \cdots < t_p$, define $\partial_i T$ to be $\{t_0, \dots, \widehat{t_i}, \dots, t_p\}$ and $\partial_i \sigma_T := \sigma_{\partial_i T}$ to be the corresponding face. The inclusion $X_T \subset X_{\partial_i T}$ induces a chain map $C(X_T) \hookrightarrow C(X_{\partial_i T})$ and we denote the sum of these maps over $p$-simplices by $\partial_i : C_p \to C_{p-1}$. Define the boundary map $\partial : C_p \to C_{p-1}$ by

$$\partial := \sum_{i=0}^{p} (-1)^i \partial_i.$$

The inclusions $X_s \hookrightarrow X$ induce an *augmentation map*, $\varepsilon : C_0 \to C(X)$. This defines a chain complex in the category of chain complexes

$$\longrightarrow C_p \xrightarrow{\partial} C_{p-1} \cdots \longrightarrow C_0 \xrightarrow{\varepsilon} C(X) \longrightarrow 0 \tag{E.2}$$

and hence a double complex

$$C_{pq} := \bigoplus_{\sigma_T \in L^{(p)}} C_q(X_T).$$

A straightforward argument shows that (E.2) is exact.

As explained above, there are two spectral sequences associated to the double complex. The second spectral sequence has $E^1_{pq} = H_q(C_{*,p})$. Since (E.2) is exact, $H_q(C_{*,p})$ is concentrated along the line $q = 0$, where $E^1_{p,0} = C_p(X)$. Taking horizontal homology, we get that $E^2$ is also concentrated on $q = 0$ and that $E^2_{p,0} = H_p(X)$. So the spectral sequence collapses and we conclude: $H_*(TC) \cong H_*(X)$.

The first spectral sequence has

$$E^1_{pq} = H_q(C_{p,*}) = \bigoplus_{\sigma_T \in L^{(p)}} H_q(X_T) := C_p(L; \mathfrak{h}_q),$$

where $\mathfrak{h}_q$ is the contravariant coefficient system on $L$ defined in Example E.3.1. Taking horizontal homology, gives $E^2_{pq} = H_p(L; \mathfrak{h}_q)$, a spectral sequence also converging to $H_*(TC)$ ($\cong H_*(X)$). This establishes the existence of the Mayer-Vietoris spectral sequence, which we state as the following.

**THEOREM E.3.2.** *Suppose, as above, that $\{X_s\}_{s\in S}$ is a cover of the CW complex $X$ by subcomplexes and that the nerve of the cover is $L$. There is a spectral sequence converging to $H_*(X)$ with $E^2$-term:*

$$E^2_{pq} = H_p(L; \mathfrak{h}_q),$$

*where $\mathfrak{h}_q$ denotes the coefficient system, $T \to H_q(X_T)$.*

### The Acyclic Covering Lemma

As above, $\{X_s\}_{s\in S}$ is a cover of a CW complex $X$ by subcomplexes and $L$ is the nerve of the cover.

**LEMMA E.3.3.** (The Acyclic Covering Lemma, see [42, p. 168].) *Suppose $X_T$ is acyclic for each $T \in \mathcal{S}(L)_{>\emptyset}$. Then $H_*(X) \cong H_*(L)$.*

*Proof.* The argument uses the Mayer-Vietoris spectral sequence. Since each $X_T$ is acyclic, $\mathfrak{h}_q := H_q(X_T)$ is 0 for $q > 0$ and for $q = 0$, it is the system of constant coefficients $\mathbb{Z}$. So the $E^2$-term is concentrated on the $p$-axis, where $E^2_{p,0} = H_p(L)$. The lemma follows. □

In Section 8.2, in the proof of Lemma 8.2.12, we needed the following more precise version of the Acyclic Covering Lemma.

**LEMMA E.3.4.** *Let $m$ be a nonnegative integer and suppose that, for each simplex $\sigma_T$ of $L$, $X_T$ is $(m - \dim \sigma_T)$-acyclic. Then $H_i(X) \cong H_i(L)$ for all $i \leqslant m$.*

*Proof.* By hypothesis, $\overline{H}_q(X_T) = 0$ for $q \leqslant m - \dim \sigma_T$. So, $E^1_{pq} := C_p(L; \mathfrak{h}_q) = 0$ if $p + q \leqslant m$ and $q \neq 0$. Moreover, since $X_T$ is 0-acyclic for $\dim \sigma_T \leqslant m$, $E^1_{p,0} = C_p(L; \mathbb{Z})$ for $p \leqslant m$. Also, $E^1_{m+1,0}$ maps onto $C_{m+1}(L; \mathbb{Z})$. Therefore, $E^2_{pq} = 0$ in the same range ($p + q \leqslant m$ and $q \neq 0$) and $E^2_{p,0} = H_p(L)$ for $p \leqslant m$. When $p \leqslant m$ all higher differentials mapping from the $p$-axis have image in groups which are 0. So, when the total degree $p + q$ is $\leqslant m$, the spectral sequence collapses and $E^2 = E^\infty$. The lemma follows. □

**LEMMA E.3.5.** *Let $m$ be an integer $\geqslant \mathrm{Card}(S) - 2$. Suppose that the nerve $L$ contains $\partial \Delta$, where $\Delta$ is the full simplex on $S$. (In other words, each proper intersection is nonempty.) Further assume for each proper subset $T$ of $S$, that $X_T$ is $(m - \dim \sigma_T)$-acyclic. Put $q := \min\{i \mid \overline{H}_i(X_S) \neq 0\}$. If $q \leqslant m - \dim \Delta$, then $\overline{H}_{q+\dim \Delta}(X) \neq 0$ (in particular, $X$ is not $m$-acyclic).*

*Proof.* If $q = -1$, then $L = \partial \Delta$ and $\dim \partial \Delta \leqslant m$. So, the previous lemma gives

$$\overline{H}_{-1+\dim \Delta}(X) = \overline{H}_{\dim \partial \Delta}(\partial \Delta) \cong \mathbb{Z} = \overline{H}_{-1}(\emptyset).$$

Suppose $q > -1$. Then $L = \Delta$. Consider the $E^2$-terms. The term of lowest total degree is $E^2_{0,0}$. If $q > 0$, this is $H_0(\Delta) = \mathbb{Z}$ (which accounts for $H_0(X)$). (When $q = 0$, we need to modify the argument slightly; the details are left to the reader.) The nonzero term of next lowest total degree is $E^2_{pq}$, where $p := \dim \Delta$ and $E^2_{pq} = H_p(\Delta; \mathfrak{h}_q) = H_q(X_S)$. Since this term cannot be affected by any higher differential, it survives to $E^\infty$ and therefore $\overline{H}_{p+q}(X) \neq 0$. □

## NOTES

**E.1.** Bass took the notes for the original French version of Serre's book [256] and wrote the appendix on the combinatorial definition of the fundamental group of a graph of groups (Definition E.1.4). A treatment of this material from a more topological viewpoint can be found in the paper of Scott and Wall [252].

"HNN" stands for G. Higman, B. Neumann, and H. Neumann.

Different proofs of Whitehead's Theorem E.1.15 can be found in [42, pp. 50–51] and [153].

Barnhill [17] has generalized Definition E.1.11 by defining a group $G$ to have *property* $\mathrm{FA}_n$ if any $G$-action on any CAT(0)-space of dimension $\leqslant n$ has a fixed point.

(CAT(0)-spaces are defined Appendix I.2.) She also generalized Example E.1.12: if a Coxeter group has the property that every special subgroup of rank $\leqslant n+1$ is spherical, then it has property $FA_n$.

**E.2.** The basic references for scwols and complexes of groups are [37] and two papers by Haefliger, [150, 151]. The construction of $\mathcal{U}(\mathcal{G})$ in (E.1) is also called the "basic construction" in [37].

In [151] it is shown that there is a category $C\mathcal{G}$ associated to a complex of groups $\mathcal{G}$ over a scwol $\mathcal{Y}$. The objects of $\mathcal{G}$ are the same as those of $\mathcal{Y}$. The set of morphisms from an object $u$ to itself is enlarged from the identity element to $G_u$. If $a$ is an edge in $\mathcal{Y}$, then the morphism set from $i(a)$ to $t(a)$ is $G_{t(a)}$.

One can define the "geometric realization" of any small category $C$ (not just a scwol). This is usually called the *classifying space* or the *nerve* of $C$ and denoted $BC$. (When $C$ is a group, i.e., when it has only one object and all morphisms are isomorphisms, then $BC$ is the usual classifying space.) In fact, $BC\mathcal{G}$ is a complex of spaces over $\mathcal{Y}$ and a cellular aspherical realization of $\mathcal{G}$ ([151]).

**E.3.** The type of coefficient system discussed in this section can be generalized to coefficients on (the nerve of) any small category $C$. One is interested in functors (either covariant or contravariant) from $C$ to the category of abelian groups. In the case of the category $C\mathcal{G}$ associated to $\mathcal{G}$, such functors are called $\mathcal{G}$-modules (either left or right) in [151].

A good general reference for spectral sequences is [199]. The discussion of the Mayer-Vietoris spectral sequences follows Brown's book [42, pp. 166–168]. Brown attributes the Acyclic Covering Lemma to Leray.

# Appendix F

# HOMOLOGY AND COHOMOLOGY OF GROUPS

## F.1. SOME BASIC DEFINITIONS

### Local Coefficients

Suppose $X$ is a CW complex with fundamental group $\pi$ and with universal cover $\widetilde{X}$. The cell structure on $X$ lifts to a cell structure on $\widetilde{X}$. A *local coefficient system* on $X$ means a $\mathbb{Z}\pi$-module $A$, for $\pi = \pi_1(X)$. $A$ is called a *constant coefficient system* if $\pi$ acts trivially on $A$.

Let $C_*(\widetilde{X})$ denote the cellular chain complex of $\widetilde{X}$. Since the action of $\pi$ on $\widetilde{X}$ by deck transformations freely permutes the cells, $C_k(\widetilde{X})$ is the free $\mathbb{Z}\pi$-module on the set of $k$-cells in $X$. Next we are going to define chains and cochains with "local coefficients" in a $\mathbb{Z}\pi$-module. We will try to stick to the convention that it is a right $\mathbb{Z}\pi$ module when we are dealing with chains or homology and a left $\mathbb{Z}\pi$ module when dealing with cochains or cohomology. Given a $\mathbb{Z}\pi$-module $A$, the cellular chain complex and the cellular cochain complex on $X$ *with local coefficients* in $A$ are defined, respectively, by

$$C_*(X;A) := A \otimes_\pi C_*(\widetilde{X}) \tag{F.1}$$

and

$$C^*(X;A) := \mathrm{Hom}_\pi(C_*(\widetilde{X}), A). \tag{F.2}$$

(Here and elsewhere $\otimes_\pi$ means $\otimes_{\mathbb{Z}\pi}$, the tensor product of a (right) $\mathbb{Z}\pi$-module and a (left) $\mathbb{Z}\pi$-module. Similarly, $\mathrm{Hom}_\pi(\,,\,)$ (or $\mathrm{Hom}_{\mathbb{Z}\pi}(\,,\,)$) means the set of $\pi$-equivariant homomorphisms between two $\mathbb{Z}\pi$-modules.) Taking homology of these chain complexes we get $H_*(X;A)$ and $H^*(X;A)$ which are, respectively, the homology and cohomology groups of $X$ with *local coefficients in* $A$.

**Example F.1.1**

(i) (*Constant coefficients.*) Suppose $A$ is the constant coefficient system $\mathbb{Z}$. Then $H_*(X;\mathbb{Z})$ and $H^*(X;\mathbb{Z})$ are the ordinary homology and cohomology groups of $X$.

(ii) (*Group ring coefficients.*) If $A = \mathbb{Z}\pi$, then $H_*(X;\mathbb{Z}\pi) = H_*(\widetilde{X})$ and if, in addition, $X$ is a finite complex, $H^*(X;\mathbb{Z}\pi)$ is $H^*_c(\widetilde{X})$, the cohomology of $\widetilde{X}$ with compact support. The point is that $\mathbb{Z}\pi$ is the ring of *finitely supported* functions $\pi \to \mathbb{Z}$ and for any $\mathbb{Z}\pi$-module $M$, $\mathrm{Hom}_\pi(M, \mathbb{Z}\pi)$ can be identified with $\mathrm{Hom}_c(M, \mathbb{Z})$, which, by definition is the set of $\mathbb{Z}$-module homomorphisms $f : M \to \mathbb{Z}$ such that for any $x \in M, f(gx) = 0$ for all but finitely many $g \in \pi$. (See [42, p. 208].) It follows that $C_*(X; \mathbb{Z}\pi)$ is the complex of ordinary (finitely supported) cellular chains on $\widetilde{X}$ while $C^*(X; \mathbb{Z}\pi)$ is the complex of finitely supported (i.e., compactly supported) cellular cochains on $\widetilde{X}$. See Section 8.5 and [42, Prop. 7.5, p. 209].)

(iii) More generally, suppose $X' \to X$ is a covering space with $\pi_1(X') = \pi'$. The fiber of $X' \to X$ can be identified with the $\pi$-set $\pi/\pi'$. Let

$$\mathbb{Z}(\pi/\pi') := \mathbb{Z}\pi \otimes_{\pi'} \mathbb{Z}$$

denote the permutation module on $\pi/\pi'$. Then $H_*(X; \mathbb{Z}(\pi/\pi')) = H_*(X')$ and if $X$ is a finite complex, then $H^*(X; \mathbb{Z}(\pi/\pi')) = H^*_c(X')$.

## The Topologist's Definitions of Group (Co)homology

Suppose $\pi$ is a group and that $B\pi$ is its classifying space (defined in Section 2.3). As is usual, the universal cover of $B\pi$ is denoted $E\pi$. ($E\pi$ is called the *universal space for* $\pi$.) Given a $\mathbb{Z}\pi$-module $A$, the homology and cohomology groups of $\pi$ with coefficients in $A$ are defined by

$$H_*(\pi; A) := H_*(B\pi; A) \quad \text{and} \quad H^*(\pi; A) := H^*(B\pi; A). \tag{F.3}$$

Let $\varepsilon : C_0(E\pi) \to \mathbb{Z}$ be the augmentation map. Since $E\pi$ is acyclic (it is contractible), the sequence

$$\longrightarrow C_k(E\pi) \longrightarrow \cdots \longrightarrow C_0(E\pi) \xrightarrow{\varepsilon} \mathbb{Z} \longrightarrow 0 \tag{F.4}$$

is exact. In other words, (F.4) is a resolution of $\mathbb{Z}$ by free $\mathbb{Z}\pi$-modules. (Here and throughout the group $\pi$ acts trivially on $\mathbb{Z}$.)

## The Algebraist's Definitions

A *projective resolution* (resp. a *free resolution*), $F_* = (F_k)_{k \geqslant 0}$, of $\mathbb{Z}$ is an exact sequence of projective $\mathbb{Z}\pi$-modules (resp. free $\mathbb{Z}\pi$-modules),

$$\longrightarrow F_k \longrightarrow \cdots \longrightarrow F_0 \longrightarrow \mathbb{Z} \longrightarrow 0. \tag{F.5}$$

Given a projective resolution $F_*$ and a $\pi$-module $A$, we have a chain complex $F_* \otimes_\pi A$ as well as a cochain complex $\mathrm{Hom}_\pi(F_*, A)$. The algebraist's definitions of group (co)homology are the following:

$$H_*(\pi;A) := H_*(F_* \otimes_\pi A) \quad \text{and} \quad H^*(\pi;A) := H^*(\mathrm{Hom}_\pi(F_*,A)). \qquad \text{(F.6)}$$

A standard argument (cf. [42, pp. 21–24]) shows any two projective resolutions are chain homotopic and hence, that the (co)homology groups defined by (F.6) are independent of the choice of projective resolution.

If $\pi$ acts freely and cellularly on an acyclic CW complex $Y$, then $C_*(Y)$ provides a free resolution of $\mathbb{Z}$ over $\mathbb{Z}\pi$. Hence, we have the following.

**PROPOSITION F.1.2.** *Suppose $\pi$ acts freely and cellularly on an acyclic CW complex $Y$ and that $A$ is a $\mathbb{Z}\pi$-module. Then*

$$H_*(\pi;A) \cong H_*(Y/\pi;A) \quad \textit{and} \quad H^*(\pi;A) \cong H^*(Y/\pi;A).$$

A similar result to this holds when the $\pi$-action on $Y$ is only required to be proper (rather than free) provided we replace $\mathbb{Z}$ by the rational numbers $\mathbb{Q}$. The reason is that for any finite subgroup $H$ of $\pi$, $\mathbb{Q}(\pi/H)$ is a projective $\mathbb{Q}\pi$-module (because $\mathbb{Q}(\pi/H)$ can be embeddded as a direct summand of $\mathbb{Q}\pi$). Hence, we have the following.

**PROPOSITION F.1.3.** *Suppose $\pi$ acts properly and cellularly on an acyclic CW complex $Y$ and that $A$ is a $\mathbb{Q}\pi$-module. Then $H_*(\pi;A)$ and $H^*(\pi;A)$ are the homology of the complexes*

$$A \otimes_{\mathbb{Q}\pi} C_*(Y) \quad \textit{and} \quad \mathrm{Hom}_{\mathbb{Q}\pi}(C_*(Y),A),$$

*respectively. In particular, $H_*(\pi;\mathbb{Q}) \cong H_*(Y/\pi;\mathbb{Q})$.*

One of the first theorems using the homology of groups (and still one of the best) is the following theorem of H. Hopf [156] concerning the cokernel of the Hurewicz map $h : \pi_2(X) \to H_2(X)$.

**THEOREM F.1.4.** (Hopf's Theorem.) *Let $X$ be a connected CW complex. Put $\pi_1(X) = \pi$. Then the sequence*

$$\pi_2(X) \xrightarrow{h} H_2(X) \xrightarrow{\psi} H_2(\pi) \longrightarrow 0$$

*is exact (where $\psi : H_2(X) \to H_2(\pi)$ is induced by the canonical map $X \to B\pi$).*

## F.2. EQUIVARIANT (CO)HOMOLOGY WITH GROUP RING COEFFICIENTS

Suppose $G$ is a discrete group acting cellularly on a CW complex $Y$. Given a $G$-module $A$, define the *equivariant (co)chains* of $Y$ with coefficients in $A$ by formulas similar to (F.1) and (F.2), namely, by

$$C_*^G(Y;A) := A \otimes_G C_*(Y) \tag{F.7}$$

and

$$C_G^*(Y;A) := \mathrm{Hom}_G(C_*(Y),A). \tag{F.8}$$

As before, $A$ is a right $G$-module for chains or homology and a left $G$-module for cochains or cohomology.

*Remark.* Some algebraic topologists believe that "equivariant (co)homology" only refers to the (co)homology of the Borel construction, $Y \times_G EG$. However, there are many other equivariant theories, e.g., the one defined by (F.7) and (F.8).

When $G$ acts freely, $Y \to Y/G$ is a covering projection, $\pi_1(Y/G)$ maps onto $G$ and the equivariant (co)chains reduce to the (co)chains on $Y/G$ with local coefficients in $A$. (Even when the action is not free, we can regard the equivariant (co)homology as computing the (co)homology of $Y/G$ with coefficients in a "system of coefficients" on $Y/G$, defined, for example, as in [151].)

**LEMMA F.2.1**

*(i)* $C_*^G(Y;\mathbb{Z}G) = C_*(Y)$.

*(ii) If $G$ acts properly and cocompactly on $Y$, then* $C_G^*(Y;\mathbb{Z}G) \cong C_c^*(Y)$.

*Proof.* (i) $C_i^G(Y;\mathbb{Z}G) = \mathbb{Z}G \otimes_{\mathbb{Z}G} C_i(Y) = C_i(Y)$.

(ii) For any $G$-module $A$, by [42, Lemma 7.4, p. 208], we have $\mathrm{Hom}_G(A,\mathbb{Z}G) \cong \mathrm{Hom}_c(A,\mathbb{Z})$, where $\mathrm{Hom}_c(A,\mathbb{Z})$ denotes the set of $\mathbb{Z}$-module maps $f : A \to \mathbb{Z}$ such that for each $a \in A$, $f(ga) = 0$ for all but finitely many $g \in G$. Hence, $C_G^i(Y;\mathbb{Z}G) = \mathrm{Hom}_G(C_i(Y),\mathbb{Z}G) = \mathrm{Hom}_c(C_i(Y),\mathbb{Z})$. $C_i(Y)$ is a direct sum of modules of the form $\mathbb{Z}(G/G_\sigma)$ where $\sigma$ ranges over a set of representatives for the orbits of $i$-cells. If $G_\sigma$ is finite, then $\mathrm{Hom}_c(\mathbb{Z}(G/G_\sigma),\mathbb{Z})$ can be identified with the $\mathbb{Z}$-module of finitely supported functions on $G/G_\sigma$. If the action is cocompact, then there are only finitely many orbits of cells and hence, only finitely many summands. So, when the action is both proper and cocompact, $\mathrm{Hom}_c(C_i(Y),\mathbb{Z}) = C_c^i(Y)$. □

So equivariant homology with group ring coefficients is the same as ordinary homology, while equivariant cohomology with group ring coefficients is the same as cohomology with compact supports.

**LEMMA F.2.2.** ([42, Proposition 7.5 and Exercise 4, p. 209].) *Suppose a group $G$ acts properly and cocompactly on an acyclic complex $Y$. Then $H^*(G;\mathbb{Z}G) \cong H_c^*(Y)$.*

(Some readers might think that this lemma has been stated incorrectly and that we should have assumed $\pi$ to be torsion-free. However, the statement is correct. The basic reason is that for any finite group $G$, $H^*(G;\mathbb{Z}G) = 0$ in positive dimensions.)

*Proof.* Let $C_* := C_*(Y)$ be the cellular chain complex of $Y$ and $F_*$ a free resolution of $\mathbb{Z}$ over $\mathbb{Z}G$. Since $C_*$ is acylic, $C_* \otimes F_*$ is also a free resolution of $\mathbb{Z}$ over $\mathbb{Z}G$; so, the cohomology of cochain complex $\mathrm{Hom}_G(C_* \otimes F_*, \mathbb{Z}G)$ is $H^*(G;\mathbb{Z}G)$. We can also compute $\mathrm{Hom}_G(C_* \otimes F_*, \mathbb{Z}G) = \mathrm{Hom}_G(C_*, \mathrm{Hom}(F_*, \mathbb{Z}G))$ via a spectral sequence, as follows. $C_p$ is a direct sum of modules of the form $\mathbb{Z}(G/G_\sigma)$ where $\sigma$ ranges over a set of representatives for the orbits of $p$-cells. For each such summand we have

$$\mathrm{Hom}_G(\mathbb{Z}(G/G_\sigma), \mathrm{Hom}(F_*, \mathbb{Z}G)) \cong \mathrm{Hom}_{G_\sigma}(F_*, \mathbb{Z}G).$$

The cohomology of this last complex is just $H^q(G_\sigma;\mathbb{Z}G)$. Taking cohomology first with respect to $q$ we get a spectral sequence whose $E_1^{pq}$ term is a sum of terms of the form $H^q(G_\sigma;\mathbb{Z}G)$. Since $\mathbb{Z}G$ is a free $G_\sigma$-module and since $G_\sigma$ is finite (because the action is proper), these groups vanish for $q > 0$. For $q = 0$ they are the invariants, $(\mathbb{Z}G)^{G_\sigma}$. All that remains is

$$E_1^{p,0} = \bigoplus (\mathbb{Z}G)^{G_\sigma} \cong \mathrm{Hom}_G(C_p, \mathbb{Z}G) = C_G^p(Y).$$

So $E_2^{p,0} \cong H_G^p(Y;\mathbb{Z}G)$. Since the $G$-action on $Y$ is cocompact, Lemma F.2.1 (i) gives $H_G^p(Y;\mathbb{Z}G) = H_c^p(Y)$, completing the proof. □

*Remark.* When $G$ contains a torsion-free subgroup $G'$ of finite index, one can give a different argument for this lemma. Then, as in Examples F.1.1, $H^*(G';\mathbb{Z}G') = H_c^*(Y)$. By Shapiro's Lemma ([42, p. 73]), $H^*(G;\mathrm{Coind}_{G'}^G A) \cong H^*(G;A)$ for any $G'$-module $A$. Since $G'$ is finite index in $G$ a coinduced module is isomorphic to the induced module (cf. [42, Prop. 5.9, p. 70]). In particular, when $A = \mathbb{Z}G'$, the induced module is

$$\mathbb{Z}G \otimes_{G'} \mathbb{Z}G' \cong \mathbb{Z}G.$$

So $H^*(G;\mathbb{Z}G) \cong H^*(G';\mathbb{Z}G') \cong H_c^*(Y)$.

## F.3. COHOMOLOGICAL DIMENSION AND GEOMETRIC DIMENSION

The *geometric dimension* of a group $\pi$, denoted $\operatorname{gd}\pi$, is the smallest integer $n$ so that there is an $n$-dimensional CW model for $B\pi$.

The cohomological version of this notion goes as follows. The *cohomological dimension* of $\pi$, denoted $\operatorname{cd}\pi$, is the projective dimension of $\mathbb{Z}$ over $\mathbb{Z}\pi$. In other words, it is the smallest integer $n$ such that $\mathbb{Z}$ admits a projective resolution

$$0 \longrightarrow P_n \longrightarrow \cdots \longrightarrow P_0 \longrightarrow 0 \tag{F.9}$$

(or $\infty$ if there is no such integer). The following basic lemma has been well-known since the beginning of the study of group cohomology.

**LEMMA F.3.1.** *If* $\operatorname{cd}\pi < \infty$, *then* $\pi$ *is torsion-free.*

*Proof.* Suppose that $\mathbb{Z}$ has a projective resolution over $\mathbb{Z}\pi$ as in (F.9) of length $n$. Let $\mathbf{C}_m$ be any cyclic sugroup of $\pi$. Then (F.9) is also a projective resolution for $\mathbf{C}_m$, so $\operatorname{cd}(\mathbf{C}_m) \leqslant n$. But $\operatorname{cd}(\mathbf{C}_m) = \infty$ whenever $m > 1$ (because $B\mathbf{C}_m$ is an infinite lens space which has cohomology in arbitrarily high dimensions). So $m = 1$ and $\pi$ is torsion-free. □

It is not hard to show ([42, p. 185]) that this definition is equivalent to the following:

$$\operatorname{cd}\pi = \sup\{n \mid H^n(\pi; A) \neq 0 \text{ for some } \pi\text{-module } A\}. \tag{F.10}$$

In fact, since any module $A$ is a quotient of some free module we can rewrite this as follows (cf. [42, Prop. 2.3, p. 186]:

$$\operatorname{cd}\pi = \sup\{n \mid H^n(\pi; F) \neq 0 \text{ for some free } \mathbb{Z}\pi\text{-module } F\}. \tag{F.11}$$

The chain complex (F.4) of an $n$-dimensional model for $E\pi$ gives a free resolution of $\mathbb{Z}$. Hence, $\operatorname{cd}\pi \leqslant \operatorname{gd}\pi$. When $\operatorname{cd}\pi \neq 2$, it is known that this inequality is an equality. 1n 1957 Eilenberg and Ganea [117] published the following theorem.

**THEOREM F.3.2.** (Eilenberg-Ganea.) *Suppose* $\pi$ *is a group.*

(i) *If* $\operatorname{cd}\pi > 2$, *then* $\operatorname{cd}\pi = \operatorname{gd}\pi$.

(ii) *If* $\operatorname{cd}\pi = 2$, *then* $\operatorname{gd}\pi \leqslant 3$.

A nontrivial free group $F$ obviously satisfies $\operatorname{cd} F = \operatorname{gd} F = 1$ (since we can take $BF$ to be a wedge of circles). A decade after the Eilenberg-Ganea paper, the case $\operatorname{cd}\pi = 1$ was taken care of by the following theorem.

**THEOREM F.3.3.** (Stallings [266] and Swan [274].) *If* $\text{cd}\,\pi = 1$*, then* $\pi$ *is a free group.*

Stallings [266] proved this theorem under the hypothesis that $\pi$ was finitely generated; Swan removed that hypothesis in [274].

The question of whether or not there exist groups of cohomological dimension 2 and geometric dimension 3 is known as the *Eilenberg-Ganea Problem.* (This is Problem D4 in Wall's Problem List [293, p. 381].) Possible counterexamples to this problem were discussed in Section 8.5.

**THEOREM F.3.4.** (Serre [255].) *Suppose* $\pi$ *is a torsion-free group and that* $\pi'$ *is a subgroup of finite index. Then* $\text{cd}\,\pi' = \text{cd}\,\pi$.

Recall that a group $\pi$ *virtually* possesses some property if it has a subgroup $\pi'$ of finite index which possesses that property. For example, $\pi$ is *virtually torsion-free* if it has a torsion-free subgroup of finite index. (This will be our main use of the term "virtually.") Given a virtually torsion-free group $\pi$, its *virtual cohomological dimension*, denoted $\text{vcd}\,\pi$, is the cohomological dimension of any torsion-free subgroup of finite index. In view of Serre's Theorem F.3.4 and the observation that the intersection of two finite index subgroups is also a finite index subgroup, we see that $\text{vcd}\,\pi$ is well defined.

## F.4. FINITENESS CONDITIONS

A CW complex is *finite type* if its skeleta are finite complexes. (In other words, $X$ is finite type if its $i$-skeleton, $X^i$, is a finite complex for each $i$.) A resolution $(P_i)_{i\geqslant 0}$ is *finite type* if each $P_i$ is a finitely generated module.

Previously, various "finiteness conditions" on a group $\pi$ have been considered. Here is the standard list. The group $\pi$ is *type*

- $F_n$ if the $n$-skeleton of $B\pi$ is a finite complex,
- $F_\infty$ if $B\pi$ is finite type,
- $F$ if $B\pi$ is a finite complex,
- $FL_n$ if $\mathbb{Z}$ has a partial free resolution of finite type and length $n$,
- $FL_\infty$ if $\mathbb{Z}$ has a free resolution of finite type,
- $FL$ if $\mathbb{Z}$ has a free resolution of finite type and finite length,
- $FP_n$ if $\mathbb{Z}$ has a partial projective resolution of finite type and length $n$,
- $FP_\infty$ if $\mathbb{Z}$ has a projective resolution of finite type,
- $FP$ if $\mathbb{Z}$ has a projective resolution of finite type and finite length.

Of course, in the above list, a phrase such as "$B\pi$ is a finite complex" is an abbreviation for "$B\pi$ has a model which is a finite complex" and a phrase such as "free resolution" is an abbreviation for "resolution by free $\mathbb{Z}\pi$-modules." The "$F$" stands for "finite," the "$L$" for "libre" (= "free") and the "$P$" for "projective." Standard references on such finite conditions are [41], [42] as well as [255].

Three other finite conditions are suggested in [24], $\pi$ is type

$FH_n$ if $\pi$ acts freely and cocompactly on an $(n-1)$-acyclic complex,

$FH_\infty$ if $\pi$ is type $FH_n$ for all $n$,

$FH$ if $\pi$ acts freely and cocompactly on an acyclic complex.

Obviously, $F \implies FH \implies FL \implies FP$, and the same implications are valid with the subscripts "$n$" or "$\infty$" appended. In fact, $FL_n \iff FP_n$ (cf. [42, Prop. 4.1, p. 192]). It is also obvious that $F \implies F_\infty \implies F_n$ and similarly, for $FH$, $FL$ and $FP$.

Types $F_1$, $FH_1$, $FL_1$ and $FP_1$ are each equivalent to the condition that $\pi$ is finitely generated. Type $F_2$ is equivalent to $\pi$ being finitely presented; however, types $FH_2$, $FL_2$ and $FP_2$ are weaker homological versions of finite presentation. (As we explained in Section 11.6, Bestvina–Brady [24] constructed examples of groups of type $FH$ which are not finitely presented.) On the other hand, Wall [292] proved that if $\pi$ is assumed to be finitely presented, then $FL_n \implies F_n$ and $FL \implies F$.

It is immediate from the definitions that if $\pi$ is type $FP$, then $\operatorname{cd}\pi < \infty$ and hence, that $\pi$ is torsion-free.

If $\pi$ is type $FP$, then the functor on $\pi$-modules $A \to H^n(\pi; A)$ commutes with direct limits. So, the definition of cohomological dimension in (F.11) can be simplified as follows.

**PROPOSITION F.4.1.** ([42, Prop. 6.7, p. 202].) *If $\pi$ is type FP, then* $\operatorname{cd}\pi = \sup\{n \mid H^n(\pi; \mathbb{Z}\pi) \neq 0\}$.

If any one of the above finiteness conditions holds for a group $\pi$, then it also holds for any subgroup of finite index in $\pi$. Since groups of type $FP$ are torsion-free, it is useful to define the corresponding virtual notions for a virtually torsion-free group $\pi$. Thus, $\pi$ is type $VF, VFH, VFL$ or $VFP$ if it contains a torsion-free subgroup of finite index which is type $F, FH, FL$ or $FP$, respectively.

*Remark F.4.2.* Given a commutative ring $R$, there are obvious definitions of what it means for a group $\pi$ to be type $FH_R$, $FL_R$, or $FP_R$ (namely, in the previous definitions replace $\mathbb{Z}$ by $R$ and $\mathbb{Z}\pi$ by $R\pi$). As before, it is obvious that $FH_R \implies FL_R \implies FP_R$. Similarly, we can define the cohomological dimension and virtual cohomological dimension *over* $R$, denoted by $\mathrm{cd}_R$ and $\mathrm{vcd}_R$, respectively.

### Finite Domination and Wall's Finiteness Obstruction

**DEFINITION F.4.3.** A space $X$ is *finitely dominated* (or is *dominated by a finite complex*) if there is a finite CW complex $Y$ together with maps $r : Y \to X$ and $i : X \to Y$ such that $r \circ i$ is homotopic to $id_X$.

Suppose $X$ is dominated by a finite complex $Y$. Then $\pi = \pi_1(X)$ is finitely presented (since $\pi_1(X)$ is a retract of $\pi_1(Y)$ and $\pi_1(Y)$ is finitely presented.) Let $\widetilde{X} \to X$ be the universal covering space and let $\widetilde{Y} \to Y$ be the covering space with group of deck transformations $i_*(\pi_1(X))$. Each cellular chain group $C_k(\widetilde{Y})$ is a finitely generated free $\mathbb{Z}\pi$-module. If $X$ is dominated by $Y$, then $C_*(\widetilde{X})$ is chain homotopy equivalent to a chain complex $C_*$, where $C_k$ is isomorphic to a direct summand of $C_k(\widetilde{Y})$, in other words, each $C_k$ is a finitely generated projective $\mathbb{Z}\pi$-module. Define *Wall's finiteness obstuction* $\sigma(X)$ to be the (reduced) Euler characteristic of this chain complex in the reduced projective class group $\widetilde{K}_0(\mathbb{Z}\pi)$. In other words, $\sigma(X)$ is the element of $\widetilde{K}_0(\mathbb{Z}\pi)$ defined by the formula:

$$\sigma(X) := \sum(-1)^k[C_k]. \tag{F.12}$$

(The *projective class group* $K_0(\mathbb{Z}\pi)$ is the Grothendieck group of finitely generated projective $\mathbb{Z}\pi$-modules. $K_0(\mathbb{Z}\pi) \cong \widetilde{K}_0(\mathbb{Z}\pi) \oplus \mathbb{Z}$, where the *reduced* projective class group $\widetilde{K}_0(\mathbb{Z}\pi)$ is the kernel of the homomorphism $K_0(\mathbb{Z}\pi) \to \mathbb{Z}$ which takes the rank of a projective module. Given a finitely generated projective $\mathbb{Z}\pi$-module $E$, $[E]$ denotes its class in $\widetilde{K}_0(\mathbb{Z}\pi)$.)

If $X$ is homotopy equivalent to a finite CW complex, then $C_*$ is chain homotopy equivalent to a finite chain complex of finitely generated free $\mathbb{Z}\pi$-modules and hence, $\sigma(X) = 0$. Conversely, in [292] Wall proved that $\sigma(X)$ is well-defined and that $X$ is finitely dominated if and only if $\pi_1(X)$ is finitely presented and $\sigma(X) = 0$. (See [218] for further discussion of Wall's finiteness obstruction.)

This gives us the topologist's interpretation of what it means for a finitely presented group $\pi$ to be type *FP*: it is *FP* if and only if $B\pi$ is finitely dominated.

There is no known example of a torsion-free group $\pi$ with $\widetilde{K}_0(\mathbb{Z}\pi) \neq 0$ and conjecturally, any such reduced projective class group is 0. Thus, conjecturally, a group of type *FP* is type *F* if and only if it is finitely presented. (See Problem F8 in Wall's Problem List [293, p. 387].)

## F.5. POINCARÉ DUALITY GROUPS AND DUALITY GROUPS

**DEFINITION F.5.1.** A group $\pi$ is an $n$-dimensional *Poincaré duality group* (or a $\mathrm{PD}^n$*-group* for short) if it is type *FP* and

$$H^i(\pi;\mathbb{Z}\pi) \cong \begin{cases} 0 & \text{if } i \neq n, \\ \mathbb{Z} & \text{if } i = n. \end{cases}$$

The *dualizing module* $D$ is the $\pi$-module $H^n(\pi;\mathbb{Z}\pi)$. (As an abelian group $D \cong \mathbb{Z}$. Necessarily, the action of $\pi$ on $D$ is via some homorphism $w_1 : \pi \to \{\pm 1\}$, called the *orientation character*. The $\mathrm{PD}^n$-group $\pi$ is *orientable* if its orientation character is trivial.

The fundamental group $\pi$ of a closed aspherical $n$-manifold $M^n$ is a $\mathrm{PD}^n$-group. The reason is that $H^*(\pi;\mathbb{Z}\pi)$ is isomorphic to the cohomology with compact supports of the universal cover $\widetilde{M}^n$ and since $\widetilde{M}^n$ is a contractible $n$-manifold, $H^*_c(\widetilde{M}^n) \cong H^*_c(\mathbb{R}^n)$ and furthermore, $H^*_c(\mathbb{R}^n)$ is as in the formula in Definition F.5.1. (Of course, this is the geometric intuition underlying the definition.)

Farrell proved the following.

**THEOREM F.5.2.** (Farrell [119].) *Suppose $\pi$ is a finitely presented group of type FP. Let $n$ be the smallest integer such that $H^n(\pi;\mathbb{Z}\pi) \neq 0$. If $H^n(\pi;\mathbb{Z}\pi)$ is finitely generated, then $\pi$ is a* $\mathrm{PD}^n$*-group.*

**LEMMA F.5.3.** *Suppose a virtually torsion-free group $G$ acts properly and cocompactly on an acyclic complex $Y$ whose cohomology with compact supports is given by*

$$H^i_c(Y) \cong \begin{cases} 0 & \text{if } i \neq n, \\ \mathbb{Z} & \text{if } i = n. \end{cases}$$

*Then $G$ is a virtual* $\mathrm{PD}^n$*-group.*

*Proof.* Since $Y/G$ is compact, $G$ is type *VFL*. By Lemma F.2.2,

$$H^i_c(G;\mathbb{Z}G) \cong H^i_c(Y) \cong \begin{cases} 0 & \text{if } i \neq n, \\ \mathbb{Z} & \text{if } i = n. \end{cases}$$

and the same formula holds for any torsion-free subgroup $\pi$ of finite index in $G$. □

Here is an equivalent definition ([42, p. 222]): $\pi$ is a $\mathrm{PD}^n$-group if there exists a $\pi$-module $D$ which is isomorphic to $\mathbb{Z}$ as an abelian group and if there is a *fundamental class* $\mu \in H_n(\pi;D)$ such that cap product with $\mu$ induces an isomorphism: $H^k(\pi;A) \cong H_{n-k}(\pi;A \otimes D)$. That is to say, $\pi$ is a $\mathrm{PD}^n$-group if $B\pi$ satisfies Poincaré duality with local coefficients. (We note that if $\pi$ is orientable, then $A$ and $A \otimes D$ are isomorphic $\pi$-modules.) In this version, it is not necessary to assume that $\pi$ is type *FP*, it is a consequence of the definition. If we assume that, in addition, $\pi$ is type $F$, then either version of the definition is equivalent to $B\pi$ being a Poincaré complex in the usual sense.

A virtually torsion-free group $\pi$ is said to be a *virtual Poincaré duality group* (a *$VPD^n$-group* for short) if it contains a finite index subgroup which is a $PD^n$-group. (This is in line with the definitions in the previous section.)

**DEFINITION F.5.4.** A group $\pi$ is an $n$-dimensional *duality group* (or a *$D^n$-group* for short) if it is type *FP* and if $H^i(\pi;\mathbb{Z}\pi) = 0$ for all $i \neq n$. The $\pi$-module $D := H^n(\pi;\mathbb{Z}\pi)$ is the *dualizing module*.

*Remark F.5.5.* Just as in Remark F.4.2, given a commutative ring $R$, there are obvious definitions of what it means for a group $\pi$ to be a $PD^n_R$-group or a $D^n_R$-group (it must be type $FP_R$ and its cohomology with $R\pi$ coefficients must be concentrated in degree $n$).

# Appendix G

## ALGEBRAIC TOPOLOGY AT INFINITY

### G.1. SOME ALGEBRA

A poset $\mathcal{A}$ is a *directed set* if for all $\alpha, \beta \in \mathcal{A}$, there exists an element $\gamma \in \mathcal{A}$ with $\gamma \geqslant \alpha, \gamma \geqslant \beta$. An *inverse system* in a category $\mathcal{C}$ is a contravariant functor from a directed set $\mathcal{A}$ to $\mathcal{C}$. In other words, an inverse system $\{C_\alpha, f_\beta^\alpha, \mathcal{A}\}$ is a collection of objects $C_\alpha$ indexed by $\mathcal{A}$ together with morphisms $f_\beta^\alpha : C_\alpha \to C_\beta$ for each pair $\beta, \alpha$ with $\beta \leqslant \alpha$ so that the following two conditions are satisfied: (1) for all $\alpha \in \mathcal{A}$, $f_\alpha^\alpha = id$ and (2) whenever $\gamma \leqslant \beta \leqslant \alpha$, $f_\gamma^\alpha = f_\beta^\alpha \circ f_\gamma^\beta$. The $f_\beta^\alpha$ are called the *bonds* of the inverse system. Let $\mathbb{N} = \{1, 2, \dots\}$ denote the natural numbers with their usual ordering. When $\mathcal{A} = \mathbb{N}$, the inverse system is called an *inverse sequence*. In this case the bonds are determined by specifying the morphisms $f_n : C_{n+1} \to C_n$.

A *morphism* between two inverse systems $\{C_\alpha, f_{\alpha'}^\alpha, \mathcal{A}\}$ and $\{D_\beta, g_{\beta'}^\beta, \mathcal{B}\}$ in $\mathcal{C}$ is a function $\phi : \mathcal{B} \to \mathcal{A}$ and for each $\beta \in \mathcal{B}$, a morphism $p_\beta : C_{\phi(\beta)} \to D_\beta$ of $\mathcal{C}$ so that whenever $\beta \leqslant \beta' \in \mathcal{B}$ there exists an $\alpha \in \mathcal{A}$ such that the following diagram commutes:

$$\begin{array}{ccccc} C_{\phi(\beta)} & \longleftarrow & C_\alpha & \longrightarrow & C_{\phi(\beta')} \\ {\scriptstyle p_\beta}\downarrow & & & & \downarrow{\scriptstyle p_{\beta'}} \\ D_\beta & & \longleftarrow & & D_{\beta'}. \end{array}$$

(The horizontal arrows are bonds.) Two morphisms $(\phi, \{p_\beta\})$ and $(\phi', \{p_{\beta'}\})$ are *equivalent* if for each $\beta \in \mathcal{B}$ there exists $\alpha \in \mathcal{A}$ with $\alpha \geqslant \phi(\beta), \alpha \geqslant \phi'(\beta)$ such that the following diagram commutes:

$$\begin{array}{ccccc} C_{\phi(\beta)} & \longleftarrow & C_\alpha & \longrightarrow & C_{\phi'(\beta)} \\ {\scriptstyle p_\beta}\downarrow & & & & \downarrow{\scriptstyle p'_\beta} \\ D_\beta & & \longleftarrow & & D_\beta. \end{array}$$

Given a category $\mathcal{C}$, we define another category, pro-$\mathcal{C}$. Its objects are inverse systems in $\mathcal{C}$ and its morphisms are equivalence classes of morphisms of the above type. A *pro-isomorphism* is an isomorphism in pro-$\mathcal{C}$.

Suppose $\mathcal{C}$ is a category in which arbitrary products exist. (Such categories include the categories of sets, groups, $R$-modules ($R$ a ring) and topological spaces.) If $\{C_\alpha, f_\beta^\alpha, \mathcal{A}\}$ is an inverse system, then its *inverse limit*, denoted $\varprojlim C_\alpha$, is the subset of the direct product $\prod_{\alpha \in \mathcal{A}} C_\alpha$ consisting of all $\mathcal{A}$-tuples $(c_\alpha)$ such that $f_\beta^\alpha(c_\alpha) = c_\beta$ for all $\beta \leqslant \alpha \in \mathcal{A}$.

Similarly, a *direct system* in $\mathcal{C}$ is a covariant functor from a directed set $\mathcal{A}$ to $\mathcal{C}$. It also has the form $\{C_\alpha, f_\beta^\alpha, \mathcal{A}\}$ except that this time there are bonds $f_\alpha^\beta : C_\beta \to C_\alpha$ whenever $\beta \leqslant \alpha$. If coproducts exist in $\mathcal{C}$, then the *direct limit* of a direct system of sets, $\{C_\alpha, f_\beta^\alpha, \mathcal{A}\}$, is denoted $\varinjlim C_\alpha$, and defined to be the quotient of the coproduct $\coprod_{\alpha \in \mathcal{A}} C_\alpha$ under the equivalence relation which identifies $c_\alpha \in C_\alpha$ with $c_\beta \in C_\beta$ whenever there is a $\gamma \in \mathcal{A}$ with $\gamma \geqslant \alpha, \beta$ such that $f_\gamma^\alpha(c_\alpha) = f_\gamma^\beta(c_\beta)$. (If $\mathcal{C}$ is the category of sets, coproduct means disjoint union; if $\mathcal{C}$ is the category of $R$-modules or the category of abelian groups, coproduct means direct sum; if it is the category of groups then coproduct means free product.)

It is clear that the inverse limit and the direct limit are characterized, up to canonical isomorphism, by a certain universal property. We leave the precise formulation of this property as an exercise for the reader.

In what follows we will primarily be interested in inverse systems which are inverse sequences.

## Semistability

Suppose

$$G_1 \longleftarrow \cdots G_{n-1} \xleftarrow{f_{n-1}} G_n \xleftarrow{f_n} G_{n+1} \longleftarrow \cdots$$

is an inverse sequence of groups. For $m > n$, $f_n^m : G_m \to G_n$ denotes the composition $f_n \circ \cdots \circ f_{m-1}$.

**DEFINITION G.1.1.** An inverse sequence of groups $(G_n, f_n)$ is *semistable* (or *Mittag-Leffler* or *pro-epimorphic*) if for any $n \in \mathbb{N}$, there is an integer $m > n$ so that for all $k \geqslant m$, $\operatorname{Im} f_n^k = \operatorname{Im} f_n^m$.

*Remark G.1.2.* An inverse sequence is pro-isomorphic to a sequence of epimorphisms if and only if it is semistable.

## The lim[1] Term

Given an inverse sequence $\{M_n\}$ of $R$-modules, define the *shift homomorphism*

$$s : \prod_{n=1}^{\infty} M_n \to \prod_{n=1}^{\infty} M_n$$

by $(x_1, x_2, \dots) \to (x_1 - f_1(x_2), x_2 - f_2(x_3), \dots)$. The kernel of $s$ is $\varprojlim M_n$. The cokernel is denoted $\lim^1 M_n$ and called the *derived limit*.

This definition can be extended to the category of groups. Given an inverse sequence of groups $\{G_n\}$ the kernel of the shift homomorphism is again the inverse limit. However, $\lim^1 G_n$ is defined to be the quotient set of $\prod_{n=1}^{\infty} G_n$ by the equivalence relation generated by putting $(x_n) \sim (y_n)$ if there exists $(z_n) \in \prod G_n$ such that $y_n = z_n x_n f_n(z_{n+1}^{-1})$ for all $n$. $\lim^1 G_n$ has the structure of a pointed set, where the base point $[1]$ is the equivalence class of $(1, 1 \dots)$.

The next result can be found in [135, Prop. 3.5.2].

**PROPOSITION G.1.3.** *Suppose $\{G_n\}$ is an inverse sequence of groups.*

*(i) If the sequence is semistable, then* $\lim^1 G_n = [1]$.

*(ii) Conversely, if each $G_n$ is countable and if* $\lim^1 G_n = [1]$, *then the sequence is semistable.*

The first assertion in the proposition follows from Remark G.1.2 and the fact that the derived limit depends only on the pro-isomorphism type of the inverse system.

## G.2. HOMOLOGY AND COHOMOLOGY AT INFINITY

Suppose $Y$ is a CW complex. Define the *chain groups at infinity* and the *homology groups at infinity* by

$$C^e_*(Y) := C^{lf}_{*+1}(Y)/C_{*+1}(Y), \tag{G.1}$$

$$H^e_k(Y) := H_k(C^e_*(Y)), \tag{G.2}$$

respectively. (The reasons for the shift in indexing in (G.1) will become clear below.) Define the *cochain groups at infinity* and the *cohomology groups at infinity* by

$$C^*_e(Y) := C^*(Y)/C^*_c(Y), \tag{G.3}$$

$$H^k_e(Y) := H_k(C^*_e(Y)). \tag{G.4}$$

The above notation is the same as in [135]; the subscript or superscript $e$ stands for "end." The groups defined by (G.2) and (G.4) are sometimes called the *end (co)homology* groups of $Y$.

Since the short exact sequences

$$\begin{aligned} 0 \longrightarrow C_*(Y) \longrightarrow C^{lf}_*(Y) \longrightarrow C^e_{*-1}(Y) \longrightarrow 0, \\ 0 \longrightarrow C^*_c(Y) \longrightarrow C^*(Y) \longrightarrow C^*_e(Y) \longrightarrow 0 \end{aligned} \tag{G.5}$$

induce long exact sequences in homology and cohomology, we have the following.

**PROPOSITION G.2.1.** *There are exact sequences*

$$\longrightarrow H_n^e(Y) \longrightarrow H_n(Y) \longrightarrow H_n^{lf}(Y) \longrightarrow H_{n-1}^e(Y) \longrightarrow,$$

$$\longrightarrow H_c^n(Y) \longrightarrow H^n(Y) \longrightarrow H_e^n(Y) \longrightarrow H_c^{n+1}(Y) \longrightarrow.$$

Suppose we have an increasing filtration of $Y$ by finite subcomplexes

$$C_1 \subset \cdots C_i \subset \cdots .$$

(In particular, this means that $Y$ is the union $C_i$.) This gives an inverse sequence:

$$Y - C_1 \supset \cdots Y - C_i \supset \cdots .$$

We want to compute the end (co)homology groups of $Y$ from the inverse sequence of chain groups $(C_*(Y, Y - C_i))$. The problem is that if we wish to remain within the context of cellular chains, $Y - C_i$ is, in general, not a subcomplex and neither is its closure $\overline{Y - C_i}$.

This can be remedied as in [135, Section 3.6]. A subcomplex $X$ of a CW complex $Y$ is a *full* subcomplex if it is the largest subcomplex of $Y$ having $X^0$ as its 0-skeleton. Given a subcomplex $A \subset Y$, its *CW complement* is the full subcomplex of $Y$ generated by the vertices of $Y^0 - A^0$. It is denoted $Y \setminus A$.

Given the increasing filtration $(C_i)$, the inverse sequence $(Y \setminus C_i)$ gives a direct sequence of cochain groups $(C^*(Y, Y \setminus C_i))$. It follows immediately from the definitions that

$$C_c^*(Y) \cong \varinjlim C^*(Y, Y \setminus C_i). \tag{G.6}$$

Comparing this with the short exact sequence (G.5), we also see that

$$C_e^*(Y) \cong \varinjlim C^*(Y \setminus C_i). \tag{G.7}$$

Since taking homology commutes with taking direct limits, we get the following.

**THEOREM G.2.2.** (Compare [135, Theorems 3.8.1, 3.8.2].) *The natural maps*

$$\varinjlim H^*(Y, Y \setminus C_i) \to H_c^*(Y) \quad \textit{and} \quad \varinjlim H^*(Y \setminus C_i) \to H_e^*(Y)$$

*are isomorphisms.*

The homology version is more problematic. The reason is that taking homology does not commute with taking inverse limits. The $\lim^1$ term intervenes. The homology version of Theorem G.2.2 is the following.

**THEOREM G.2.3.** (Compare [135, Theorems 3.6.12 and 3.6.13].) *The following sequences are short exact:*

$$0 \to \lim^1 H_{k+1}(Y, Y \setminus C_i) \to H_k^{lf}(Y) \to \varprojlim H_k(Y, Y \setminus C_i) \to 0,$$
$$0 \to \lim^1 H_{k+1}(Y \setminus C_i) \to H_k^e(Y) \to \varprojlim H_k(Y \setminus C_i) \to 0.$$

Under the condition of "homological semistability," defined in the next subsection, the $\lim^1$ terms in the above theorem vanish and Theorem G.2.3 becomes the direct analog of Theorem G.2.2.

### Homological Semistability

**DEFINITION G.2.4.** A CW complex $Y$ is *homologically semistable* if there is a filtration, $C_1 \subset \cdots C_i \subset \cdots$, by finite subcomplexes so that the inverse sequence $(H_*(Y \setminus C_i))$ is semistable.

One can show that if $(H_*(Y \setminus C_i))$ is semistable for one filtration $C_i$, then it is semistable for any such filtration. A version of Definition G.2.4 for the fundamental group will be discussed in Appendix G.4 below.

Combining Theorem G.2.3 and Proposition G.1.3 we get the following.

**PROPOSITION G.2.5.** *If $Y$ is homologically semistable, then*

$$H_*^e(Y) \cong \varprojlim H_*(Y \setminus C_i).$$

### Singular Homology and Cohomology

The above definitions can be extended to the singular theories in a straightforward fashion. So, for $Y$ an arbitrary space, $C_*(Y)$ and $C^*(Y)$ denote, respectively, the complexes of singular chains and singular cochains on $Y$, and $H_*(Y)$ and $H^*(Y)$ their respective (co)homology groups.

In the case of cohomology with compact supports one can proceed as in [142] as follows. Given compact subspaces $C$ and $D$ of $Y$ with $C \subset D$, we have $Y - D \subset Y - C$ and hence, a natural inclusion $C^*(Y, Y - C) \hookrightarrow C^*(Y, Y - D)$ and an induced homomorphism in cohomology $H^*(Y, Y - C) \to H^*(Y, Y - D)$. *Singular cohomology with compact supports* of $Y$ is defined by

$$H_c^*(Y) = \varinjlim H^*(Y, Y - C) \tag{G.8}$$

where the direct limit is over the directed system of all compact subspaces $C \subset Y$.

Locally finite singular homology is defined as follows. A *locally finite chain* $\alpha$ is a (possibly infinite) formal linear combination, $c = \sum m_\lambda \lambda$, of singular simplices $\lambda$ such that any $y \in Y$ has a neighborhood $U$ with $\operatorname{Im} \lambda \cap U \neq \emptyset$ for only finitely many $\lambda$ in the support of $c$. $C_*^{lf}(Y)$ denotes the chain complex of locally finite singular chains and $H_*^{lf}(Y)$ its homology.

## G.3. ENDS OF A SPACE

A continuous map $f : Y \to X$ of topological spaces is *proper* if $f^{-1}(C)$ is compact for every compact subset $C$ of $X$.

A *ray* in a topological space $X$ is a map $r : [0, \infty) \to X$. A ray $r : [0, \infty) \to X$ is proper if and only if for any compact $C \subset X$ there is a positive integer $N$ so that $r([N, \infty)) \subset X - C$. In other words, $r(t)$ "goes to infinity" as $t \to \infty$. We are interested in proper rays. The next definition is due to Freudenthal [132].

**DEFINITION G.3.1.** Two proper rays $r_1, r_2 : [0, \infty) \to X$ *determine the same end* if for any compact subset $C \subset X$, there is a positive integer $N$ so that $r_1([N, \infty))$ and $r_2([N, \infty))$ are contained in the same path component of $X - C$. This is an equivalence relation on the set of proper rays. An equivalence class is an *end* of $X$. The equivalence class of $r$ is denoted end($r$). The set of all such equivalence classes is denoted Ends($X$).

The *number of ends* of $X$ is Card(Ends($X$)). $X$ is *one-ended* if it has exactly one end. Suppose $X$ is locally path connected and connected. Then Ends($X$) $= \emptyset$ if and only if $X$ is compact.

There is a natural topology on Ends($X$). A sequence $(\text{end}(r_n))_{n \in \mathbb{N}}$ *converges* to end($r$) if for each compact $C \subset X$ there is a sequence $(N_n)$ of positive integers so that $r_n([N_n, \infty))$ and $r([N_n, \infty))$ are contained in the same path component of $X - C$. This gives a topology on Ends($X$): a subset $A$ of Ends($X$) is *closed* if and only if, for any sequence (end($r_n$)) in $A$, $\lim_{n \to \infty} \text{end}(r_n) = \text{end}(r)$ implies end($r$) $\in A$.

The poset $\mathcal{C}$ of all compact subsets of $X$ gives an inverse system $(X - C)_{C \in \mathcal{C}}$. So, we can take path components and then take the inverse limit. Under reasonable conditions, this inverse limit is the space of ends, i.e.,

$$\text{Ends}(X) \cong \varprojlim \pi_0(X - C). \tag{G.9}$$

The definition of "reasonable conditions" is given in Proposition G.3.3 below.

A reader unfamiliar with these notions will benefit from doing the following exercise.

**Exercise G.3.2.** Suppose $X$ is the regular trivalent tree. Then Ends($X$) is a Cantor set.

**PROPOSITION G.3.3.** *Suppose $X$ satisfies reasonable conditions (i.e., suppose $X$ is a connected, locally path connected, locally compact, second* *countable, Hausdorff space). Let $e(X)$ be the number of ends of $X$ and let $k$ be any field. Then*

$$e(X) = \dim_k H_0^e(X;k).$$

*Remark G.3.4.* If $X$ is path connected and $H_1(X;k) = 0$, then

$$e(X) = \dim_k H_1^{lf}(X;k) - 1$$

(i.e., $H_1^{lf}(X;k)$ is isomorphic to a codimension one subspace of $H_0^e(X;k)$). If $e(X)$ is finite, then we also have $e(X) = \dim_k H_e^0(X;k)$. If $e(X)$ is infinite, then so is $\dim_k H_c^0(X;k)$; however, it is countably infinite while $e(X)$ is uncountable. Similarly, if $H^1(X;k) = 0$ and $\dim_k H_c^1(X;k)$ is infinite, then so is $e(X)$.

## G.4. SEMISTABILITY AND THE FUNDAMENTAL GROUP AT INFINITY

Suppose $C_1 \subset C_2 \subset \cdots$ is a filtration of a CW complex $Y$ by finite subcomplexes and $r : [0,\infty) \to Y$ is a proper ray with $r(i) \in Y \setminus C_i$. Let $e := \text{end}(r)$ denote the equivalence class of $r$ in $\text{Ends}(Y)$. The inclusions induce an inverse sequence

$$\pi_1(Y \setminus C_{n+1}, r(n+1)) \to \pi_1(Y \setminus C_n, r(n)).$$

Our first concern is the semistability (Definition G.1.1) of this inverse sequence. By Proposition G.1.3 (i), the sequence is semistable if the derived limit is trivial.

Suppose $C'_1 \subset C'_2 \subset \cdots$ is another filtration of $Y$ by finite subcomplexes and $r'$ is another proper ray with $\text{end}(r') = e$ and $r'(i) \in Y \setminus C'_i$. It is proved in [135] that the two inverse sequences $(\pi_1(Y \setminus C_n, r(n)))$ and $(\pi_1(Y \setminus C'_n, r'(n)))$ are pro-isomorphic. So, their derived limits are simultaneously trivial or not. In other words, the two inverse sequences are simultaneously semistable or not.

**DEFINITION G.4.1.** Suppose a CW complex $Y$ has an end $e$ represented by a ray $r$. $Y$ is *semistable* at $e$ if there is a filtration $(C_n)_{n\in\mathbb{N}}$ by finite subcomplexes such that the inverse sequence $(\pi_1(Y \setminus C_n, r(n)))$ is semistable.

The notions of semistability (and stability) go back to Siebenmann's thesis [258]. The definition of a space being *simply connected at infinity* is given at the beginning of 9.2. Further results on semistability can be found in [164, 205, 206, 208].

Suppose $\Gamma$ is a finitely presented group, that $B$ is a finite CW complex with $\pi_1(B) = \Gamma$ and that $\widetilde{B}$ is its universal cover. Then $\Gamma$ is *semistable* (resp. *simply connected at infinity*) if $\widetilde{B}$ is semistable (resp. simply connected at infinity).

One shows this defintion is independent of the choice of $B$. In particular, it only depends on the 2-skeleton of $B$. More generally, if $\Gamma$ acts properly, cellularly and cocompactly on a simply connected CW complex $X$, then $\Gamma$ is semistable (resp. simply connected at infinity) if and only if $X$ is. The main problem in this area is the following.

**CONJECTURE G.4.2.** (Geoghegan's Conjecture.) *Every one-ended, finitely presented group is semistable.*

Geoghegan's Conjecture is equivalent to the conjecture that the universal cover of any finite CW complex is semistable.

If $Y$ is semistable, the pro-isomorphism type of $(\pi_1(Y \setminus C_n, r(n)))$ is independent of the choice of ray $r$ and filtration $\{C_n\}$. Hence, the inverse limit gives a group, well defined up to isomorphism, called the *fundamental group at* $e$:

$$\pi_1^e(Y) := \varprojlim(\pi_1(Y \setminus C_n, r(n))). \tag{G.10}$$

($e$ plays the role of a basepoint in this group.) If $Y$ is one-ended, then this group is called the *fundamental group at infinity* and denoted $\pi_1^\infty(Y)$.

If $\Gamma$ is finitely presented, one ended, and semistable, then define $\pi_1^\infty(\Gamma)$ to be $\pi_1^\infty(X)$, where $X$ is any complex on which $\Gamma$ acts properly and cocompactly.

Suppose $X$ is a one-ended space satisfying reasonable conditions (cf. Proposition G.3.3). The definitions of semistability and of the fundamental group at infinity also work in this context. A *neighborhood of infinity* in $X$ means the complement of a compact set. (In other words, a neighborhood of infinity is a deleted neighborhood of $\infty$ in the one point compactification.) $X$ is *simply connected at infinity* if given any neighborhood of infinity, $X - C$, there is a smaller neighborhood of infinity, $X - D$, such that any loop in $X - D$ is null-homotopic in $X - C$. ("Smaller" means $C \subset D$.) The proof of the next proposition is straightforward and is omitted.

**PROPOSITION G.4.3.** ([164, 135].) *Suppose $X$ is a one-ended space satisfying reasonable conditions. Then $X$ is simply connected at infinity if and only if it is semistable and $\pi_1^\infty(X)$ is trivial.*

*Remark.* When $X$ has more than one end, one can also define what it means for $X$ to be "simply connected at an end $e$." A similar result to the above proposition holds: $X$ is simply connected at $e$ if and only if it is semistable at $e$ and $\pi_1^e(X) = 1$. The details are left to the reader.

One can also make sense of various other connectivity conditions at infinity. For example, $X$ is *$m$-connected at infinity* if given any neighborhood of infinity $X - C$ there is a smaller neighborhood of infinity $X - D$ such that any map $S^k \to X - D$ is null-homotopic in $X - C$ for all $k \leqslant m$.

## NOTES

**G.1.** Most of this section is taken from [135, Chapter 3]. Another discussion of the derived limit of an inverse sequence can be found in [153, p. 313].

In the literature a slightly weaker definition of directed set is often used: the ordering need not give a partial order. Directed sets are necessary in the discussion of the pro-isomorphism type of an inverse system. However, the notions of direct limit and inverse limit are more general and can be defined for any functorial family of bonds.

**G.3.** Proofs for most of the results discussed in this section can be found in [37] and [135].

**G.4.** A interesting result on the fundamental group at infinity is proved by Geoghegan and Mihalik in [136].

# Appendix H

# THE NOVIKOV AND BOREL CONJECTURES

## H.1. AROUND THE BOREL CONJECTURE

The most famous conjecture concerning aspherical manifolds is the following.

**CONJECTURE H.1.1.** (The Borel Conjecture.) *Suppose $f : M \to M'$ is a homotopy equivalence between closed aspherical manifolds. Then $f$ is homotopic to a homeomorphism.*

*Remarks*

(i) At one time people believed that Conjecture H.1.1 might hold for all manifolds without the asphericity hypothesis. (This was the Hurewicz Conjecture.) This was soon seen to be false: Reidemeister showed that homotopy equivalent lens spaces could be distinguished by the use of "Reidemeister torsion." With the advent of surgery theory in the 1960's, it became clear that, in dimensions $n \geqslant 5$, nothing like the Hurewicz Conjecture could be true for simply connected manifolds. For example, for any simply connected $M^n$, with $n \geqslant 6$, and with a nontrivial Betti number in some dimension divisible by 4 and $\neq 0, n$, one can find homotopy equivalent manifolds $N^n$ not homeomorphic to $M^n$.

(ii) Apparently, Borel's reason in the 1950s for making this conjecture (or rather for asking it as a question) was the analogy with the rigidity results of Bierberbach (for flat manifolds) and Mostow (for solvmanifolds). Nowadays, when discussing Borel's Conjecture, people usually also mention the analogy with the Mostow Rigidity Theorem [220] on the rigidity of locally symmetric manifolds with universal cover having no factor which is either (a) compact, (b) Euclidean or (c) the hyperbolic plane. (See paragraph 6.10 of the Notes to Chapter 6 for a further discussion of the Mostow Rigidity Theorem.)

(iii) Even if $M$ and $M'$ are assumed to be smooth or PL manifolds, it is not possible to replace the word " homeomorphism" by "diffeomorphism" or "PL homeomorphism" in the conclusion of the Borel Conjecture. Indeed, if $M^n$ is any flat manifold (e.g., $\mathbb{T}^n$) or any closed, stably parallelizable, hyperbolic manifold and $\Sigma^n$ is any exotic sphere, then the connected sum $M^n \# \Sigma^n$ is never diffeomorphic to $M^n$. (A result of Sullivan [273] implies that any closed hyperbolic manifold has a finite sheeted cover which is stably parallelizable.) In the PL case, Ontaneda [230, 231] has shown that there are examples where $M$ and $M'$ are not PL homeomorphic.

(iv) Over the last twenty years, Farrell and Jones have made substantial progress in proving the Borel Conjecture for nonpositively curved Riemannian manifolds of dimension $\geqslant 5$. In particular, they have proved it in the case of closed nonpositively curved manifolds. (See [121, 122, 123, 124].)

(v) Of course, the Borel Conjecture is true in dimensions 1 and 2. It now seems that it has also been proved in dimension 3. Indeed, it is implied by Thurston's Geometrization Conjecture, which has recently been proved by Perelman [237, 239, 238]. (See also [49, 179].)

Closely related to the Borel Conjecture is the following.

**CONJECTURE H.1.2.** (The PD$^n$-Group Conjecture.) *Suppose $\pi$ is a finitely presented Poincaré duality group. Then $B\pi$ is homotopy equivalent to a closed manifold.*

The definition of "PD$^n$-group" can be found in Appendix F.5. As we saw in Theorem 11.6.3, without the hypothesis of finite presentation, the PD$^n$-group Conjecture is false.

### Poincaré Pairs

Before stating the relative versions of the above conjectures, we need to discuss Poincaré pairs.

**DEFINITION H.1.3.** An *n-dimensional Poincaré pair* is a CW pair $(X, \partial X)$, with $X$ is finitely dominated, together with a homomorphism $w_1 : \pi_1(X) \to \{\pm 1\}$ such that there is a fundamental class $\mu \in H_n(X, \partial X; D)$ so that for any local coefficient system $A$ on $X$, cap product with $\mu$ induces an isomorphisms

$$H^k(X;A) \cong H_{n-k}(X, \partial X; D \otimes A)$$
$$H_k(X;A) \cong H^{n-k}(X, \partial X; D \otimes A).$$

Here $D$ is the local coefficient system defined as follows: as an abelian group $D \cong \mathbb{Z}$ and $\pi_1(X)$ acts on $D$ via the homomorphism $w_1$. (For the notion of "finitely dominated," see Definition F.4.3. As in Appendix F.1, by a "local coefficient system" we mean a $\mathbb{Z}\pi_1(X)$-module.) $(X, \partial X)$ is *orientable* if $w_1$ is the trivial homomorphism. If $\partial X = \emptyset$, then $X$ is a *Poincaré complex*. $D$ is the *orientation module* and $\mu$ the *fundamental class*.

In 11.5 we need the following generalization of a Poincaré pair.

**DEFINITION H.1.4.** Suppose $X$ is a finitely dominated CW complex, $\partial X$ a subcomplex, $\pi$ a group and $\psi : \pi_1(X) \to \pi$ an epimorphism. As in Appendix F.1, any $\mathbb{Z}\pi$ module gives a local coefficient system on $X$. If $w_1 : \pi \to \{\pm 1\}$ is a homomorphism, then we get the $\mathbb{Z}\pi$-module $D$ which is isomorphic to $\mathbb{Z}$ as an abelian group and on which $\pi$ acts via $w_1$. The pair $(X, \partial X)$ together with the homomorphism $w_1$ is a *Poincaré pair over* $\pi$ of *formal dimension* $n$ if there is a class $\mu \in H_n(X, \partial X; D)$ such that

$$\cap \mu : H^i(X; M) \to H_{n-i}(X, \partial X; D \otimes M)$$

is an isomorphism for all $i$ and for any $\mathbb{Z}\pi$-module $M$.

Of course, when $\pi = \pi_1(X)$ and $\psi$ is the identity map, this reduces to the notion of "Poincaré pair" in Definition H.1.3.

**LEMMA H.1.5.** *Suppose $X$ is a finitely dominated CW complex, $\partial X$ is a subcomplex, $\pi$ is a group and $\psi : \pi_1(X) \to \pi$ an epimorphism. Let $\widetilde{X}$ be the covering space associated to $\psi$ and let $\partial\widetilde{X}$ denote the inverse image of $\partial X$ in $\widetilde{X}$. Assume $\widetilde{X}$ is acyclic. Then $(X, \partial X)$ is a Poincaré pair over $\pi$ if and only if*

$$H^i_c(\widetilde{X}, \partial\widetilde{X}; \mathbb{Z}) = \begin{cases} \mathbb{Z} & \text{if } i = n, \\ 0 & \text{otherwise.} \end{cases}$$

A proof in the absolute case can be found in [42, pp. 220–221]. In fact, the argument given there works in the relative case and proves the above lemma.

## Relative Versions of These Conjectures

**CONJECTURE H.1.6.** (Relative version of the Borel Conjecture.) *Suppose $(M, \partial M)$ and $(M', \partial M')$ are aspherical manifolds with boundary and $f : (M, \partial M) \to (M', \partial M')$ is a homotopy equivalence of pairs with $f|_{\partial M}$ a homeomorphism. Then $f$ is homotopic rel $\partial M$ to a homeomorphism.*

**CONJECTURE H.1.7.** (Relative version of the PD$^n$-group Conjecture.) *Suppose $\pi$ is a finitely presented group of type FP and $(X, \partial X)$ is a Poincaré pair with $X$ homotopy equivalent to $B\pi$ and $\partial X$ a manifold. Then $(X, \partial X)$ is homotopy equivalent rel $\partial X$ to a compact manifold with boundary.*

(What it means for a group to be of type $F$ or $FP$ is defined in Appendix F.4.)

In Theorem 11.4.1, by using the reflection group trick, we showed that the relative version of each of these conjectures is implied by its absolute version.

**Conjectures in Algebraic $K$-Theory**

Kirby and Siebenmann [177, p. 744] proved that any compact topological manifold $X$ is homotopy equivalent to a finite CW complex. (Replace $X$ by the total space $\widehat{X}$ of a normal disk bundle to $X$ in some Euclidean space of dimension $\geqslant 5$. Since $\widehat{X}$ is parallelizable, the main result of [177] implies that it admits a PL triangulation; in particular, it is a finite complex.) So, Conjecture H.1.7 implies that any finitely presented group $\pi$ of type $FP$ is type $F$. More precisely, if $(X, \partial X)$ is a Poincaré pair with $X \sim B\pi$, then its finiteness obstruction $\sigma(X)$ (defined by (F.12)) must vanish. As explained in Appendix F.4, the following conjecture says that $\sigma(X)$ always lies in the trivial group.

**CONJECTURE H.1.8.** (The Reduced Projective Class Group Conjecture.) *For any torsion-free group $\pi$, the reduced projective class group $\widetilde{K}_0(\mathbb{Z}\pi)$ vanishes.*

For manifolds with nontrivial fundamental groups, the h-Cobordism Theorem need not be true. The correct result is the s-Cobordism Theorem, where the "s" refers to a "simple" in front of "homotopy equivalence." (For smooth 4-manifolds, Donaldson proved that the h-Cobordism Theorem fails even in the simply connected case.) Simple homotopy equivalence is the equivalence relation generated by elementary expansions and collapses of simplices. Algebraically such an operation corresponds to change of basis for a free $\mathbb{Z}G$-module by an elementary matrix.

For a given ring $R$, define $K_1(R)$ to be the quotient $GL(R)/[GL(R), GL(R)]$, where $GL(R)$ denotes the infinite general linear group over $R$. It is known that the commutator subgroup $[GL(R), GL(R)]$ is generated by elementary matrices. Let $\{\pm g \mid g \in G\}$ be the subgroup of $K_1(\mathbb{Z}G)$ generated by $1 \times 1$ matrices of the form $\pm g$. The *Whitehead group* of $G$, denoted $Wh(G)$, is the quotient $K_1(\mathbb{Z}G)/\{\pm g \mid g \in G\}$. Associated to a homotopy equivalence $f : P \to Q$ between finite polyhedra, there is an element $\alpha(f) \in Wh(G)$ called its *Whitehead torsion*. The homotopy equivalence $f$ is *simple* if $\alpha(f) = 0$. The *torsion* of an h-cobordism $W^n$ between manifolds $M_i^{n-1}$, $i = 0, 1$, is the Whitehead torsion of the inclusion $M_0^{n-1} \hookrightarrow W^n$. $W^n$ is an *s-cobordism* if its torsion is 0. The s-Cobordism Theorem asserts that if $n \geqslant 6$ and $W^n$ is an s-cobordism, then $W^n$ is isomorphic to the cylinder $M^{n-1} \times [0, 1]$. (See [127].) Given a manifold $M_0^{n-1}$ with fundamental group $\pi$ and an element $\alpha \in Wh(\pi)$, it is also known that one can construct an h-cobordism $W^n$ with torsion $\alpha$. If $W$ is an h-cobordism from $M_0$ to $M_1$ with nontrivial torsion, then either $M_0$ is not homeomorphic to $M_1$ or else the two ends of $W$ can be glued together via a homeomorphism $M_0 \to M_1$ which is homotopic to but not pseudoisotopic to the identity, thus, producing a manifold homotopy equivalent to but not homeomorphic to $M_0 \times S^1$. In either case, if $M_0$ were a closed aspherical manifold, we would have a counterexample to Conjecture H.1.1. Similar remarks apply in the case of an aspherical manifold with boundary. Thus, Conjecture H.1.6 implies that for any group $\pi$ of type $F$ we should have $Wh(\pi) = 0$. In fact, there is no counterexample known to the following.

**CONJECTURE H.1.9.** (The Whitehead Group Conjecture.) *For any torsion-free group $\pi$, $Wh(\pi) = 0$.*

## H.2. SMOOTHING THEORY

For each positive integer $n$ there are H-spaces:

| | |
|---|---|
| $G(n)$ | the space of self homotopy equivalences of $S^{n-1}$, |
| $TOP(n)$ | the space of self-homeomorphisms of $\mathbb{R}^n$, |
| $PL(n)$ | the space of $PL$ self-homeomorphisms of $\mathbb{R}^n$, and |
| $O(n)$ | the Lie group of orthogonal linear automorphisms of $\mathbb{R}^n$. |

We have obvious inclusions $PL(n) \hookrightarrow TOP(n) \hookrightarrow G(n)$ and $O(n) \hookrightarrow TOP(n)$. $PL(n)$ is homotopy equivalent to the group $PD(n)$ of "piecewise differentiable" homeomorphisms of $\mathbb{R}^n$ and $O(n) \hookrightarrow PD(n)$; so, in this way we also have an "inclusion" $O(n) \to PL(n)$. Their classifying spaces are denoted $BG(n)$, $BTOP(n)$, $BPL(n)$ and $BO(n)$, respectively. $BG(n)$ is the classifying space for spherical fibrations with fiber $S^{n-1}$, $BO(n)$ is the classifying space for $n$-dimensional vector bundles, while $BTOP(n)$ and $BPL(n)$ are, respectively, the classifying spaces for topological and PL microbundles with fiber $\mathbb{R}^n$.

If we take the direct limit of the $G(n)$, $TOP(n)$, $PL(n)$ or $O(n)$ we obtain H-spaces $G$, $TOP$, $PL$ and $O$ and their respective classifying spaces $BG$, $BTOP$, $BPL$ and $BO$. These are the classifying spaces for "stable bundles." Let us recall what this means. Suppose $\xi_1$ and $\xi_2$ are two vector bundles over a CW complex $X$. For $i = 1, 2$, let $c_i : X \to BO(n_i)$ classify $\xi_i$ and let $\tilde{c}_i : X \to BO$ be the composition of $c_i$ and the natural map $BO(n_i) \to BO$. Bundles $\xi_1$ and $\xi_2$ are *stably isomorphic* if and only if it is possible to find trivial vector bundles $\varepsilon_1$ and $\varepsilon_2$ so that the Whitney sums $\xi_1 \oplus \varepsilon_1$ and $\xi_2 \oplus \varepsilon_2$ are isomorphic. It follows that $\xi_1$ is stably isomorphic to $\xi_2$ if and only if $\tilde{c}_1$ is homotopic to $\tilde{c}_2$. When we speak of a "stable vector bundle" we mean its class under the equivalence relation of stable isomorphism (in other words, its class in $\widetilde{KO}(X)$) and similarly, for $BG$, $BTOP$ and $BPL$. In what follows we are interested in the stable tangent bundle or its inverse, the stable normal bundle, of a compact manifold $M$.

The inclusions $TOP \hookrightarrow G$, $PL \to TOP$ and $O \hookrightarrow TOP$ induce fibrations $BG \to BTOP$, $BPL \to BTOP$ and $BO \to BTOP$ with fibers $G/TOP$, $TOP/PL$ and $TOP/O$, respectively. The main result in smoothing theory is the following.

**THEOREM H.2.1.** (Munkres [224], Hirsch and Mazur [155], Kirby and Siebenmann [177, 178].) *In the following statements,* $\dim M = n$, *and* $n \geqslant 5$.

*(i) A topological manifold M admits a PL structure if and only if the classifying map for its stable tangent bundle* $M \to BTOP$ *lifts to BPL.*

*(ii) If M admits at least one PL structure, then the set isomorphism classes of PL structures on M is in bijective correspondence with the set of homotopy classes* $[M, TOP/PL]$. *(This is the set of isomorphism classes of reductions of the "structure group" from TOP to PL.)*

*(iii) A topological or PL manifold M admits a smooth structure if and only if the classifying map of its stable tangent bundle lifts to BO.*

*(iv) If M admits at least one smooth structure, then the set of isomorphism classes of smooth structures on M is bijective with* $[M, TOP/O]$. *Similarly, the set of isomorphism classes of smooth structures with the same underlying PL structure is bijective with* $[M, PL/O]$.

*Remark.* In Theorem 11.3.2 we use this to show there exist closed aspherical manifolds not homotopy equivalent to smooth manifolds.

Kirby and Siebenmann also showed that

$$\pi_i(TOP/PL) \cong \begin{cases} \mathbb{Z}/2 & \text{if } i = 3, \\ 0 & \text{otherwise.} \end{cases}$$

From this and standard obstruction theory we get the following corollary.

**COROLLARY H.2.2.** (Kirby–Siebenmann [177], [178].) *Let M be an n-manifold,* $n \geqslant 5$.

*(i) M admits a PL structure if and only if a certain obstruction in* $H^4(M; \mathbb{Z}/2)$ *vanishes.*

*(ii) If M admits at least one PL structure, then the set isomorphism classes of PL structures on M is bijective with* $H^3(M; \mathbb{Z}/2)$.

In his Ph.D. thesis M. Spivak [262] showed that associated to any Poincaré complex $X$ there is a spherical fibration $\nu$ over $X$ with total space $E(\nu)$ such that $E(\nu)$ is fiber homotopy equivalent to the sphere bundle of the stable normal bundle of $X$ whenever $X$ is a manifold. It is called the *Spivak normal fibration*. The construction of $E(\nu)$ can be found in [40]. It goes as follows. Since $X$ is homotopy equivalent to a polyhedron we may assume that it actually is a polyhedron. Piecewise linearly embed $X$ in Euclidean space $\mathbb{R}^N$ with $N \geqslant 2\dim X + 1$. Let $R$ be a closed regular neighborhood of $X$ in $\mathbb{R}^N$. Then $R$ collapses onto $X$ and, as Spivak proved, the restriction of the collapse to $\partial R$ is the spherical fibration. In more detail, by a well-known construction (see [153, p. 407]), any map is homotopic to a fibration. If we use this construction to replace $\partial R \to X$ by $E(\nu) \to X$, then the fiber of $E(\nu)$ is $S^{k-1}$, where $k = N - \dim X$. An obvious obstruction for $X$ to be homotopy equivalent to a manifold is that the classifying map $c_\nu : X \to BG$ for the stable class of $\nu$ lifts to a map into $BTOP$ (or $BPL$ or $BO$ in the cases of PL or smooth manifolds, respectively). In analogy with Theorem H.2.1 one might expect this to be the complete obstruction (in dimensions $\geqslant 5$) and that manifold structures on $X$ would be bijective with $[X, G/TOP]$. We shall see in the next section that this is not, in fact, the case. Wall's surgery obstruction groups must be brought into the picture.

## H.3. THE SURGERY EXACT SEQUENCE AND THE ASSEMBLY MAP CONJECTURE

Suppose $(X, \partial X)$ is an $n$-dimensional Poincaré pair. A *structure* on $X$ is an $n$-dimensional manifold with boundary $(M, \partial M)$ and a homotopy equivalence of pairs $f : (M, \partial M) \to (X, \partial X)$ such that $f|_{\partial M}$ is a homeomorphism (so, $\partial X$ must be a manifold). Two structures $f_0 : (M_0, \partial M_0) \to (X, \partial X)$ and $f_1 : (M_1, \partial M_1) \to (X, \partial X)$ are *equivalent* if there is an h-cobordism $W$ rel $\partial M_0$ from $M_0$ to $M_1$ and a map $F : (W, \partial M_0 \times [0, 1]) \to (X, \partial X)$ such that for $i = 0, 1$, $F|_{M_i} = f_i$. Let $\mathcal{S}(X)$ denote the set of equivalence classes of structures on $X$.

Associated to a group $\pi$ and a nonnegative integer $n$, there is a certain abelian group called the "Wall surgery group" and denoted $L_n(\pi)$. This is defined either as a Witt group of Hermitian forms on free $\mathbb{Z}\pi$-modules (for $n$ even) or as a group of automorphisms of such a form (for $n$ odd). It follows that the surgery groups are 4-periodic, i.e., $L_n(\pi) = L_{n+4}(\pi)$. In the case of the trivial group, these groups were computed by Kervaire–Milnor [175] to be:

$$L_n(1) = \begin{cases} \mathbb{Z} & \text{if } i \equiv 0 \pmod 4, \\ \mathbb{Z}/2 & \text{if } i \equiv 2 \pmod 4, \\ 0 & \text{otherwise.} \end{cases} \tag{H.1}$$

For $\pi = \pi_1(X)$, we have Sullivan's *surgery exact sequence*,

$$L_{n+1}(\pi) \to \mathcal{S}(X) \to [(X, \partial X), (G/TOP, *)] \to L_n(\pi).$$

(The naive analogy with smoothing theory would suggest that the surgery groups $L_*(\pi)$ should not occur in the above sequence.)

By a *relative structure* on $(X, \partial X)$ we mean, as before, a homotopy equivalence of pairs $f : (M, \partial M) \to (X, \partial X)$, only this time there is no restriction on $f|_{\partial M}$. $\mathcal{S}(X, \partial X)$ stands for the set of equivalence classes of relative structures on $(X, \partial X)$. The corresponding surgery exact sequence is

$$L_{n+1}(\pi) \to \mathcal{S}(X, \partial X) \to [X, G/TOP] \to L_n(\pi).$$

(Our conventions about the meaning of $\mathcal{S}(X)$ and $\mathcal{S}(X, \partial X)$ are intended; they should not be reversed. The reason for these conventions is to make notation in various versions of the surgery exact sequence consistent.)

Not only are the surgery groups 4-periodic, so are the homotopy groups of $G/TOP$. They are given by the formula

$$\pi_i(G/TOP) = \begin{cases} \mathbb{Z} & \text{if } i \equiv 0 \pmod 4, \\ \mathbb{Z}/2 & \text{if } i \equiv 2 \pmod 4, \\ 0 & \text{otherwise.} \end{cases} \tag{H.2}$$

(The similarity in (H.1) and (H.2) is not accidental.) Moreover, the 4-fold loop space, $\Omega_4(\mathbb{Z} \times G/TOP)$, is homotopy equivalent to $\mathbb{Z} \times G/TOP$. It follows that $\mathbb{Z} \times G/TOP$ defines a spectrum $\mathbb{L}$ and a generalized homology theory $H_*(X; \mathbb{L})$. More is known about the homotopy type of $G/TOP$: localized at 2 it is isomorphic to a product of Eilenberg-MacLane spectra, away from 2 it is isomorphic to $BO$. For example,

$$H_m(X; \mathbb{L}) \otimes \mathbb{Q} \cong \sum H_{m-4k}(X; \mathbb{Q}).$$

The generalized cohomology theory asssociated to $\mathbb{L}$ is represented by homotopy classes of maps into $\mathbb{Z} \times G/TOP$, i.e., $H^0(X; \mathbb{L}) := [X, \mathbb{Z} \times G/TOP]$. For $n = \dim X$, Poincaré duality defines isomorphisms

$$[(X, \partial X), (G/TOP, *)] \cong H_n(X; \mathbb{L})$$

and

$$[X, G/TOP] \cong H_n(X, \partial X; \mathbb{L}),$$

where $* \in G/TOP$ is a base point.

The structure sets $\mathcal{S}(X)$ and $\mathcal{S}(X, \partial X)$ are also 4-periodic. (Strictly speaking, for this to be true, in the definition of "structure" it is necessary to replace the condition that $(M, \partial M)$ is a manifold with boundary by the condition that it be an ANR homology manifold with boundary satisfying the disjoint disk property. This is explained in [44, 296].) Then $\mathcal{S}(X) \cong \mathcal{S}(X \times D^4)$; moreover, $\mathcal{S}(X \times D^k)$ has the structure of an abelian group whenever $k \geqslant 2$.

So, if $\dim X = n$, define $\mathcal{S}_{n+k}(X) := \mathcal{S}(X \times D^k)$ and $\mathcal{S}_{n+k}(X, \partial X) := \mathcal{S}(X \times D^k, \partial(X \times D^k))$. By enforcing periodicity, we can then put an abelian group structure on $\mathcal{S}_{n+k}$ for $k < 2$ (indeed, even for $k$ negative). Using this, we get the modern versions of the above two surgery exact sequences:

$$\to L_{n+1}(\pi) \to \mathcal{S}_n(X) \to H_n(X; \mathbb{L}) \xrightarrow{A} L_n(\pi) \to,$$

$$\to L_{n+1}(\pi) \to \mathcal{S}_n(X, \partial X) \to H_n(X, \partial X; \mathbb{L}) \longrightarrow L_n(\pi) \to .$$

Both sequences are exact sequences of abelian groups. The map $A : H_n(X; \mathbb{L}) \to L_n(\pi)$ was first defined by Quinn. It is called the *assembly map*. For a more explanation of the surgery exact sequence, see [295].

**CONJECTURE H.3.1.** (The Assembly Map Conjecture.) *Suppose $M$ is a closed aspherical manifold or an aspherical Poincaré complex with fundamental group $\pi$. Then $A : H_*(M; \mathbb{L}) \to L_*(\pi)$ is an isomorphism.*

**CONJECTURE H.3.2.** (Relative version of the Assembly Map Conjecture.) *Suppose $\pi$ is a group of type FP. Then $A : H_*(B\pi; \mathbb{L}) \to L_*(\pi)$ is an isomorphism.*

Conceivably, this conjecture could be true for any torsion-free group $\pi$.

*Remark.* There are versions of the structure sets, Wall groups and the surgery exact sequences incorporating simple homotopy equivalences. One modifies the notation by adding a superscript "s" on $\mathcal{S}_*(\ )$ and $L_*(\ )$. $\mathcal{S}^s_*(X)$ is defined by considering simple homotopy equivalences from manifolds to $X$ and calling two such equivalent if they differ by an s-cobordism. Similarly, the Wall group $L^s_*(\mathbb{Z}\pi)$ contains $Wh(\pi)$. In this version the statement that the assembly map $H_*(B\pi; \mathbb{L}) \to L^s_*(\mathbb{Z}\pi)$ is an isomorphism implies $Wh(\pi) = 0$.

**THEOREM H.3.3.** *For a given group $\pi$ of type F and an integer $n \geqslant 5$, the relative version of the Borel Conjecture (H.1.6) holds whenever $X$ is a compact aspherical n-manifold with boundary with $\pi_1(X) = \pi$ if and only if the following two conditions hold:*

- *$Wh(\pi) = 0$ and*
- *$A : H_*(B\pi; \mathbb{L}) \to L_*(\pi)$ is an isomorphism.*

*Proof.* If $0 \neq a \in Wh(\pi)$ and $n \geqslant 5$, then we can find a manifold $V^{n+1}$ which is an h-cobordism rel boundary with one end $X$ and with Whitehead torsion $a$. If the other end $X'$ is not homeorphic to $X$, then it is a counterexample to Conjecture H.1.6. If $X'$ is homeorphic to $X$, then $V$ is a counterexample to the conjecture for $X \times [0, 1]$. If $A : H_n(B\pi; \mathbb{L}) \to L_n(\pi)$ is not injective, then $\mathcal{S}_n(B\pi) \neq 0$; hence, we can find $(X, \partial X)$, a $n + 4k$-manifold with boundary, such that $\mathcal{S}_n(X) \neq 0$ and Conjecture H.1.6 is false for $X$. Similarly, if $A : H_n(B\pi; \mathbb{L}) \to L_n(\pi)$ is not onto, then $\mathcal{S}_{n-1}(B\pi) \neq 0$ and we again get a counterexample.

Conversely, suppose both conditions hold. Since $A : H_*(B\pi; \mathbb{L}) \to L_*(\pi)$ is an isomorphism, $\mathcal{S}(X) = 0$; hence, up to an h-cobordism there is only one structure on $X$. Since $Wh(\pi) = 0$ and $n \geqslant 5$, the h-cobordism must be $X \times [0, 1]$. □

Similar considerations prove the following.

**THEOREM H.3.4.** *For a given group $\pi$ of type FP and an integer $n \geqslant 5$, the relative version of the* PD$^n$*-group Conjecture (Conjecture H.1.7) holds whenever $(X, \partial X)$ is an n-dimensional Poincaré pair with $X \sim B\pi$ if and only if the following two conditions hold:*

- *$\sigma(X) = 0$ (where $\sigma(X) \in \widetilde{K}_0(\mathbb{Z}\pi)$ is Wall's finiteness obstruction) and*
- *$A : H_*(B\pi; \mathbb{L}) \to L_*(\pi)$ is an isomorphism.*

## H.4. THE NOVIKOV CONJECTURE

The Hirzebruch *L-genus* of a closed manifold $M$ is a certain power series $\mathcal{L}_*(M)$ in the Pontrjagin classes (of the tangent bundle) of $M$. The part of $\mathcal{L}_*(M)$ which lies in $H^{4k}(M; \mathbb{Q})$ is denoted $\mathcal{L}_k(M)$; it is a polynomial in the Pontrjagin classes. The Hirzebruch Signature Theorem asserts that when $\dim M = 4k$, the signature of $M$ is $\langle \mathcal{L}_k(M), [M] \rangle$ (i.e., the cohomology class $\mathcal{L}_k(M)$ evaluated on the fundamental class $[M]$). For the definition of the $L$-genus and a proof of the Signature Theorem, see [216].

Suppose $\pi_1(M) = \pi$ and that $c : M \to B\pi$ is the canonical map. As $\alpha$ ranges over $H^*(B\pi; \mathbb{Q})$ we get numbers $\langle c^*(\alpha) \cup \mathcal{L}_*(M), [M] \rangle$, called the *higher signatures* of $M$. Novikov originally conjectured that these higher signatures were homotopy invariants, i.e., if $M$ and $M'$ are homotopy equivalent closed manifolds, then they have the same higher signatures. (Of course, it was well-known that, even in the simply connected case, the rational Pontrjagin classes of $M$ and $M'$ can be different.) It is also well-known that the cohomology classes $\mathcal{L}_*(M)$ determine the Pontrjagin classes (since $p_k(M)$ occurs with nonzero coefficient in $\mathcal{L}_k(M)$). It follows that in the case where $M = B\pi$ is an aspherical manifold, the higher signatures determine the Pontrjagin classes of $M$. So in this case, the Novikov Conjecture is the following weak form of the Borel Conjecture: if $M$ is aspherical, then its rational Pontrjagin classes are homotopy invariants, i.e., if $f : M \to M'$ is a homotopy equivalence, then $f^*(p_i(M')) = p_i(M)$. (This is implied by the Borel

Conjecture, since, as Novikov previously had shown, the rational Pontrjagin classes are homeomorphism invariants.)

With regard to the surgery exact sequence, if a homotopy equivalence $f : M' \to M$ represents an element in $\mathcal{S}(M)$, then its image in $H_n(M; \mathbb{L}) \otimes \mathbb{Q}$ can be computed from the difference of total Pontrjagin classes, $p_*(M') - f^*(p_*(M))$. So, for $M = B\pi$, Novikov's original conjecture is equivalent to the conjecture that assembly map $A : H_n(B\pi; \mathbb{L}) \otimes \mathbb{Q} \to L_n(\mathbb{Z}\pi) \otimes \mathbb{Q}$ is, rationally, a monomorphism. In fact, it turns out that this version is equivalent to the homotopy invariance of the the higher signatures. That is to say, the modern formulation of the Novikov Conjecture is the following.

**CONJECTURE H.4.1.** (The Novikov Conjecture.) *For any finitely presented group $\pi$, after tensoring with $\mathbb{Q}$, the assembly map, $A : H_n(B\pi; \mathbb{L}) \otimes \mathbb{Q} \to L_n(\pi) \otimes \mathbb{Q}$, is a monomorphism.*

## NOTES

References for this material include [40, 122, 183, 295].

# *Appendix I*

## NONPOSITIVE CURVATURE

### I.1. GEODESIC METRIC SPACES

Let $(X, d)$ be a metric space. A path $\gamma : [a, b] \to X$ is a *geodesic* (or a *geodesic segment*) if it is an isometric embedding, i.e., if $d(\gamma(s), \gamma(t)) = |s - t|$ for all $s, t \in [a, b]$. Similarly, an isometric embedding $[a, \infty) \to X$, $\mathbb{R} \to X$ or $\mathbb{S}^1 \to X$ is called, respectively, a *geodesic ray*, a *geodesic line* or a *closed geodesic*. ($\mathbb{S}^1$ denotes the circle with its standard metric, possibly rescaled so that its length can be arbitrary.) $(X, d)$ is a *geodesic space* if any two points can be connected by a geodesic segment.

Given a path $\gamma : [a, b] \to X$ in a metric space $X$, its *length* $l(\gamma)$ is defined by

$$l(\gamma) := \sup \left\{ \sum_{i=1}^{n} d(\gamma(t_{i-1}), \gamma(t_i)) \right\},$$

where $a = t_0 < t_1 < \cdots t_n = b$ runs over all possible subdivisions of $[a, b]$. $(X, d)$ is a *length space* if

$$d(x, y) = \inf\{l(\gamma) \mid \gamma \text{ is a path from } x \text{ to } y\}.$$

(Here we allow $\infty$ as a possible value of $d$.) Thus, a length space is a geodesic space if the above infimum is always realized and is $\neq \infty$.

### I.2. THE CAT($\kappa$)-INEQUALITY

As in Chapter 6, for each real number $\kappa$, $\mathbb{X}^2_\kappa$ denotes the simply connected, complete, Riemannian 2-manifold of constant curvature $\kappa$:

- $\mathbb{X}^2_0$ is the Euclidean plane $\mathbb{E}^2$.
- If $\kappa > 0$, then $\mathbb{X}^2_\kappa = \mathbb{S}^2$ with its metric rescaled so that its curvature is $\kappa$ (i.e., it is the sphere of radius $1/\sqrt{\kappa}$).
- If $\kappa < 0$, then $\mathbb{X}^2_\kappa = \mathbb{H}^2$, the hyperbolic plane, with its metric rescaled.

(The spaces $\mathbb{E}^2$, $\mathbb{S}^2$ and $\mathbb{H}^2$ are defined and explained in Section 6.2.)

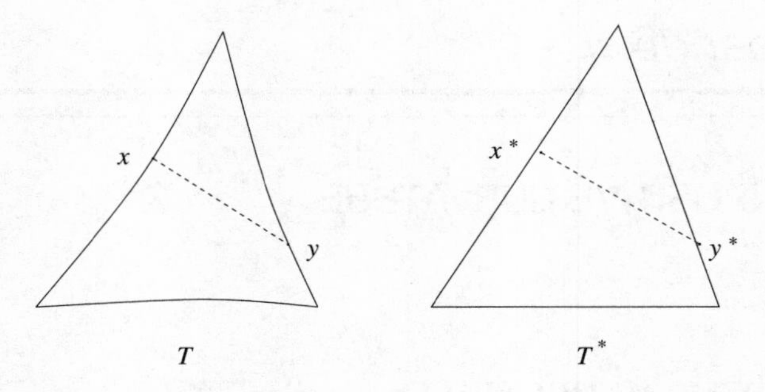


Figure I.1. The CAT($\kappa$)-inequality.

A *triangle* $\Delta$ in a metric space $X$ is a configuration of three geodesic segments ("edges") connecting three points ("vertices") in pairs. A *comparison triangle* for $\Delta$ is a triangle $\Delta^*$ in $\mathbb{X}^2_\kappa$ with the same edge lengths. When $\kappa \leqslant 0$ such comparison triangle always exists. When $\kappa > 0$ a necessary and sufficient condition for a comparison triangle to exist is that $l(\Delta) \leqslant 2\pi/\sqrt{\kappa}$, where $l(\Delta)$ denotes the sum of the lengths of the three edges of $\Delta$. ($2\pi/\sqrt{\kappa}$ is the length of the equator in a 2-sphere of curvature $\kappa$.) When $\kappa > 0$, we shall always make this assumption on $l(\Delta)$.

If $\Delta^*$ is a comparison triangle for $\Delta$, then for each edge of $\Delta$ there is a well-defined isometry, denoted by $x \to x^*$, which takes the given edge of $\Delta$ onto the corresponding edge of $\Delta^*$.

**DEFINITION I.2.1.** A metric space $X$ *satisfies* CAT($\kappa$) (or is a CAT($\kappa$)*-space*) if the following two conditions hold:

(a) If $\kappa \leqslant 0$, then $X$ is a geodesic space, while if $\kappa > 0$, it is required that there is a geodesic segment between any two points $< \pi/\sqrt{\kappa}$ apart.

(b) (*The* CAT($\kappa$)*-inequality*). For any triangle $\Delta$ (with $l(\Delta) < 2\pi/\sqrt{\kappa}$ if $\kappa > 0$) and any two points $x, y \in \Delta$, we have

$$d(x, y) \leqslant d^*(x^*, y^*),$$

where $x^*, y^*$ are the corresponding points in the comparison triangle $\Delta^*$ and $d^*$ is distance in $\mathbb{X}^2_\kappa$. (See Figure I.1.)

For $\kappa \leqslant 0$, condition (a) implies that $X$ is connected; however, if $\kappa > 0$, $X$ need not be connected.

**Exercise I.2.2.** Suppose $\kappa > 0$. Show that a circle of length $l$ is CAT($\kappa$) if and only if $l \geqslant 2\pi/\sqrt{\kappa}$.

**DEFINITION I.2.3.** A metric space $X$ *has curvature* $\leqslant \kappa$ if the CAT($\kappa$)-inequality holds locally in $X$.

*Remarks I.2.4*

(i) Suppose $\gamma_1, \gamma_2$ are two geodesic segments with the same initial point $x = \gamma_1(0) = \gamma_2(0)$. The *Aleksandrov angle* between $\gamma_1$ and $\gamma_2$ at $x$ is defined as

$$\limsup_{s,t \to 0} \alpha(s,t),$$

where $\alpha(s,t)$ is the angle at the vertex corresponding to $x$ in a comparison triangle in $\mathbb{E}^2$ for the triangle with vertices $x$, $\gamma_1(s)$, $\gamma_2(t)$. Aleksandrov proved that a metric space $X$ is CAT($\kappa$) if and only if it satisfies condition (a) above and if for any triangle $\Delta$ in $X$ (with $l(\Delta) < 2\pi/\sqrt{\kappa}$ when $\kappa > 0$) the Aleksandrov angle at any vertex is $\leqslant$ the corresponding angle in a comparison triangle $\Delta^* \subset \mathbb{X}^2_\kappa$. (For details, see [37, p. 161] or Troyanov's article in [138, p. 49].)

(ii) $\mathbb{X}^n_\kappa$ satisfies CAT($\kappa'$) for any $\kappa' \geqslant \kappa$ and any $n \geqslant 2$. It follows that if $X$ is CAT($\kappa$), then it is also CAT($\kappa'$) for every $\kappa' > \kappa$. (See [37, p. 165].)

(iii) The Comparison Theorem of Aleksandrov and Toponogov asserts that a Riemannian manifold has sectional curvature $\leqslant \kappa$ if and only if the CAT($\kappa$)-inequality holds locally. (Of course, this is the justification for the terminology in Definition I.2.3.) A proof can be found in Troyanov's article in [138, pp. 52–59] or in [1].

**THEOREM I.2.5.** (Uniqueness of geodesics.) *Suppose $X$ is* CAT($\kappa$). *If $\kappa \leqslant 0$, then there is a unique geodesic between any two points of $X$. If $\kappa > 0$, the geodesic segment between any two points of distance $< \pi/\sqrt{\kappa}$ is unique.*

In other words, there are no "digons" in $X$ of edge length $< \pi/\sqrt{\kappa}$ when $\kappa > 0$ and no digons at all when $\kappa \leqslant 0$.

*Proof.* Suppose we have two distinct geodesic segments between points $x$ and $y$ (of distance $< \pi/\kappa$ if $\kappa > 0$). Introduce a third vertex $z$ in a portion of one of the segments where they do not coincide. This gives a triangle in $X$. Its comparison triangle in $\mathbb{X}^2_\kappa$ degenerates to a single geodesic segment. So, the CAT($\kappa$)-inequality implies that the original two segments must be equal. □

Suppose $X$ is CAT(0). Choose a basepoint $x_0 \in X$. By the previous result, for each $x \in X$, there is a geodesic $\gamma_x : [0,d] \to X$ from $x_0$ to $x$, where $d := d(x_0, x)$. Define $h : X \times [0,1] \to X$ by $h(x,t) := \gamma_x(ts)$. It follows from the Arzelà–Ascoli Theorem that $h$ is continuous. So, $h$ is a homotopy from the constant map to $\mathrm{id}_X$. $h$ is called *geodesic contraction*. Hence, we have the following result.

**THEOREM I.2.6.** *A complete* CAT(0)*-space is contractible.*

The Cartan-Hadamard Theorem states that a simply connected, complete, nonpositively curved, Riemannian manifold is contractible. Gromov's version of this is the following.

**THEOREM I.2.7.** (Gromov's version of the Cartan-Hadamard Theorem.) *Suppose a geodesic space is complete, simply connected and has curvature $\leqslant \kappa$ for some $\kappa \leqslant 0$. Then it is* CAT($\kappa$).

The analog of this result for $\kappa > 0$ is the following.

**THEOREM I.2.8.** *Suppose a complete length space $X$ has curvature $\leqslant \kappa$, for some $\kappa > 0$. Then the following conditions are equivalent.*

(a) *Given any two points of distance $< \pi/\sqrt{\kappa}$, the geodesic between them is unique.*

(b) *There is no closed geodesic of length $< 2\pi/\sqrt{\kappa}$.*

(c) *$X$ is* CAT($\kappa$).

A corollary of Theorems I.2.6 and I.2.7 is the following result (mentioned previously in Section 2.3).

**COROLLARY I.2.9.** *If $X$ is complete and nonpositively curved, then it is aspherical.*

*Proof.* Its universal cover is CAT(0); hence, contractible. □

### The Center of a Bounded Set

The *radius* of a bounded subset $Y$ in a metric space $X$ is the real number

$$r_Y := \inf\{r \mid Y \subset B(x,r) \text{ for some } x \in X\}.$$

Here $B(x,r)$ denotes the open ball of radius $r$ centered at $x$. ($\overline{B}(x,r)$ denotes the closed ball.)

**PROPOSITION I.2.10.** (Compare [37, pp. 178–179] or [43, p. 188].) *Suppose $Y$ is a bounded subset of a complete* CAT(0)*-space $X$. Then there is a unique point $c_Y \in X$ (called the* center *of $Y$) such that $Y \subset \overline{B}(c_Y, r_Y)$.*

*Proof.* Let $(x_n)$ be a sequence in $X$ such that $Y \subset B(x_n, r_n)$ where $r_n \to r_Y$.

*Claim.* $(x_n)$ *is a Cauchy sequence.*

If we assume this claim there is nothing left to prove. Indeed, $(x_n)$ has a limit since $X$ is complete and $c_Y := \lim x_n$ has the required property. The fact that the sequence is Cauchy also implies uniqueness.

*Proof of Claim.* Given $\varepsilon > 0$, choose numbers $R > r_Y$ and $R' < r_Y$ so that the annulus $A := B(\mathbf{0}, R) - B(\mathbf{0}, R')$ in $\mathbb{R}^2$ contains no line segment of length $\geqslant \varepsilon$. (Here $\mathbf{0}$ denotes the origin in $\mathbb{R}^2$.) For sufficiently large $n, n'$, we have $r_n, r_{n'} < R$. For each $y \in Y$, consider the triangle $\Delta(y, x_n, x_{n'})$ in $X$ and its comparison triangle $\Delta^*_y := \Delta(\mathbf{0}, x^*_n, x^*_{n'})$ in $\mathbb{R}^2$. Let $m$ be the midpoint of the geodesic segment from $x_n$ to $x_{n'}$ and $m^*$ the midpoint of $[x_n, x_{n'}]$. If for every $y \in Y$ we had $m^* \in B(\mathbf{0}, R')$, then, by the CAT(0)-inequality, we would have $Y \subset B(m, R')$, contradicting $R' < r_Y$. Hence, there exists a point $y \in Y$ such that $m^*$ lies in $A$. Since this means that at least half of $[x_n, x_{n'}]$ lies in $A$, the length of $[x_n, x_{n'}]$ is $< 2\varepsilon$. □

### The Bruhat-Tits Fixed Point Theorem

A famous theorem of E. Cartan asserts the existence of a fixed point for any compact group of isometries on a simply connected, complete, nonpositively curved Riemannian manifold. This was generalized by Bruhat and Tits to the following.

**THEOREM I.2.11.** (The Bruhat-Tits Fixed Point Theorem.) *Let $G$ be a group of isometries of a* CAT(0)*-space $X$. If $G$ is compact (or more generally if $G$ has a bounded orbit on $X$), then $G$ has a fixed point on $X$.*

*Proof.* Suppose $X$ has a bounded orbit $Gx$. By Proposition I.2.10, this orbit has a center $c$. Since $Gx$ is $G$-stable so is $c$. □

**PROPOSITION I.2.12.** *Let $G$ be a group of isometries of a* CAT(0)*-space $X$. Then its fixed point set,* $\mathrm{Fix}(G, X)$*, is convex and hence,* CAT(0). *In particular, if* $\mathrm{Fix}(G, X)$ *is nonempty, it is contractible.*

*Proof.* Suppose $x, y \in \mathrm{Fix}(G, X)$ and let $\gamma$ be the geodesic segment from $x$ to $y$. For any $g \in G$, $g \cdot \gamma$ is another geodesic segment with the same endpoints. By Theorem I.2.5, $g \cdot \gamma = \gamma$, i.e., the geodesic segment from $x$ to $y$ lies in $\mathrm{Fix}(G, X)$. □

**COROLLARY I.2.13.** *Suppose $G$ is a discrete group of isometries acting properly on a complete* CAT(0)*-space $X$. Then $X$ is a model for $\underline{E}G$, the universal space for proper $G$-actions.*

(See Definition 2.3.1 in Section 2.3 for more about $\underline{E}G$.)

*Proof.* Let $F$ be a finite subgroup of $G$. By the Bruhat-Tits Fixed Point Theorem, $\mathrm{Fix}(F, X) \neq \emptyset$ and by Proposition I.2.12, it is contractible. □

**COROLLARY I.2.14.** *Suppose $G$ is a discrete group of isometries acting properly on a complete* CAT(0)*-space $X$. Further suppose $X/G$ is compact. Then there is a simplicial complex $\Omega$ on which $G$ acts properly, simplicially and with a finite number of orbits of simplices such that $\Omega$ is $G$-equivariantly homotopy equivalent to $X$. In particular, $\Omega$ is a CW model for $\underline{E}G$ with a finite number of $G$-orbits of cells.*

*Proof.* We shall construct $G$-equivariant covering of $X$ by open balls with the property that if $B(x,r)$ is one of these balls, then $gB(x,r) \cap B(x,r) = \emptyset$ for all $g \in G - G_x$. Since $X/G$ is compact we can cover it by a finite number of images of such balls. Lift this to a covering of $X$ and let $\Omega$ be the nerve of the covering. (In other words, $\mathrm{Vert}(\Omega)$ is the set of balls in the covering, a finite nonempty subset of $\mathrm{Vert}(\Omega)$ spans a simplex if and only if the corresponding intersection of balls is nonempty.) Since balls in $\mathbb{E}^2$ are convex, so are balls in any CAT(0) space (hence, they are contractible). Since the intersection of convex sets is convex, it follows that the natural map $X \to \Omega$ is an equivariant homotopy equivalence. □

## A Condition for Verifying the CAT(0)-Inequality

Suppose $x, y, z$ are vertices of a triangle in a geodesic space $X$ and that, for $t \in [0,1]$, $p_t$ is the point on the geodesic from $y$ to $x$ such that $d(y, p_t) = td(y,x)$ and $d(x, p_t) = (1-t)d(y,x)$. If $x, y, z$ are points in $\mathbb{E}^2$, then $p_t = ty + (1-t)x$. A simple argument ([43, p. 153]) shows

$$d^2(z, p_t) = (1-t)d^2(z,x) + td^2(z,y) - t(1-t)d^2(x,y). \tag{I.1}$$

Here $d$ denotes Euclidean distance and $d^2(x,y) := d(x,y)^2$.

We return to the situation where $\Delta$ is a triangle in $X$ with vertices $x, y, z$. Since the edge lengths of $\Delta$ are the same as those in the Euclidean comparison triangle, we have the following well-known lemma, which we needed in Section 18.3.

**LEMMA I.2.15.** *With notation as above, the* CAT(0)*-inequality for the triangle $\Delta$ in $X$ is equivalent to the inequality*

$$d^2(z, p_t) \leqslant (1-t)d^2(z,x) + td^2(z,y) - t(1-t)d^2(x,y),$$

*for all $t \in [0,1]$.*

### Local Geodesics

Suppose $c : [a, b] \to X$ is a path and $t_0 \in (a, b)$. Then $c$ is a *local geodesic* at $c(t_0)$ if there is an open interval $U$ containing $t_0$ such that $c|_U$ is a geodesic. A proof of the next lemma is can be found in [37, p. 160].

**LEMMA I.2.16.** *Suppose $X$ is a* CAT($\kappa$)*-space and $c : [a, b] \to X$ is a path. If $\kappa > 0$ also suppose $l(c) \leqslant \pi/\kappa$. Then $c$ is a geodesic if and only if it is a local geodesic.*

### The Cone on a CAT(1)-Space

We gave a definition of the "cone" on a topological space $X$ in Definition A.4.5. We slightly modify it here. The *cone on* $X$, denoted Cone($X$), is the quotient space of $X \times [0, \infty)$ by the equivalence relation $\sim$ defined by $(x, s) \sim (y, t)$ if and only if $(x, s) = (y, t)$ or $s = t = 0$. The image of $(x, s)$ in Cone($X$) is denoted $[x, s]$. The *cone of radius* $r$, denoted Cone($X, r$), is the image of $X \times [0, r]$ in Cone($X$). (In A.4.5 only the definition of the cone of radius 1 was given.)

Given a metric space $X$ and a real number $\kappa$, we shall define a metric $d_\kappa$ on Cone($X$). (When $\kappa > 0$, the definition only makes sense on the open cone of radius $\pi/\sqrt{\kappa}$.) The idea underlying the definition is that when $X = \mathbb{S}^{n-1}$, by using "polar coordinates" and the exponential map, Cone($X$) can be identified with $\mathbb{X}^n_\kappa$ (when $\kappa > 0$, provided the radius is $< \pi/\sqrt{\kappa}$). Transporting the constant curvature metric on $\mathbb{X}^n_\kappa$ to Cone($\mathbb{S}^{n-1}$), we obtain a formula for $d_\kappa$ on Cone($\mathbb{S}^{n-1}$). The same formula then defines a metric on Cone($X$) for any metric space $X$.

To write this formula we first need to recall the Law of Cosines in $\mathbb{S}^2_\kappa$, $\mathbb{E}^2$ or $\mathbb{H}^2_\kappa$. Suppose we have a triangle in $\mathbb{X}^2_\kappa$ with edge lengths $s$, $t$ and $d$ and angle $\theta$ between the first two sides. As in [37, p. 24], there is a *Law of Cosines* in each space:

- in $\mathbb{E}^2$:

$$d^2 = s^2 + t^2 - 2st \cos\theta$$

- in $\mathbb{S}^2_\kappa$:

$$\cos\sqrt{\kappa}d = \cos\sqrt{\kappa}s \cos\sqrt{\kappa}t + \sin\sqrt{\kappa}s \sin\sqrt{\kappa}t \cos\theta$$

- in $\mathbb{H}^2_\kappa$:

$$\cosh\sqrt{-\kappa}d = \cosh\sqrt{-\kappa}s \cosh\sqrt{-\kappa}t + \sinh\sqrt{-\kappa}s \sinh\sqrt{-\kappa}t \cos\theta$$

Given $x, y \in X$, put $\theta(x, y) := \min\{\pi, d(x, y)\}$. Define the metric $d_0$ on Cone($X$) by

$$d_0([x, s], [y, t]) := (s^2 + t^2 - 2st \cos\theta(x, y))^{1/2}.$$

Similarly, the metrics $d_\kappa$, $\kappa \neq 0$ are defined so that the appropriate Law of Cosines holds with $d = d_\kappa$ and $\theta = \theta(x, y)$. Cone$(X)$ equipped with the metric $d_\kappa$ is denoted by $\text{Cone}_\kappa(X)$.

*Remark.* If $X$ is a $(n-1)$-dimensional spherical polytope, then $\text{Cone}_\kappa(X)$ is isometric to a convex polyhedral cone in $\mathbb{X}_\kappa$ (see Section 6.2).

**PROPOSITION I.2.17.** *Suppose $X$ is a complete and any two points of distance $\leqslant \pi$ can be joined by a geodesic. Then*

*(i)* ([37, pp. 62–63].) $\text{Cone}_\kappa(X)$ *is a complete geodesic space.*

*(ii)* (Berestovskii [19] or [37, pp. 188–190].) $\text{Cone}_\kappa(X)$ *is* CAT$(\kappa)$ *if and only if $X$ is* CAT(1).

## Spherical Joins

In Definition A.4.5 we defined the join $X_1 * X_2$ of two nonempty topological spaces: it is the quotient space of $X_1 \times X_2 \times [0, 1]$ by an equivalence relation $\sim$. Now suppose $d^i$ is a metric on $X_i$. We are going to slightly modify this definition of $X_1 * X_2$ by replacing $[0, 1]$ by $[0, \pi/2]$. Then we will define a metric $d$ on the join. Thus,

$$X_1 * X_2 := (X_1 \times X_2 \times [0, \pi/2])/\sim,$$

where $\sim$ is defined by: $(x_1, x_2, \theta) \sim (x_1', x_2', \theta')$ whenever $\theta = \theta' = 0$ and $x_1 = x_1'$ or $\theta = \theta' = \pi/2$ and $x_2 = x_2'$ or $(x_1, x_2, \theta) = (x_1', x_2', \theta')$. We denote the image of $(x_1, x_2, \theta)$ in $X_1 * X_2$ by $(\cos\theta)x_1 + (\sin\theta)x_2$. For $i = 1, 2$, define a metric $d_\pi^i$ on $X_i$ by $d_\pi^i(x, y) := \min\{\pi, d^i(x, y)\}$. Define the distance $d$ between points $x = (\cos\theta)x_1 + (\sin\theta)x_2$ and $x' = (\cos\theta')x_1' + (\sin\theta')x_2'$ by requiring that it be at most $\pi$ and that

$$\cos(d(x, x')) = \cos\theta\cos\theta' \cos(d_\pi^1(x_1, x_1')) + \sin\theta\sin\theta' \cos(d_\pi^2(x_2, x_2')).$$

$(X_1 * X_2, d)$ is the *spherical join* of $X_1$ and $X_2$. If $X_1$ is the 0-sphere $\mathbb{S}^0$ and $X_2 = X$, then $\mathbb{S}^0 * X$ is called the *spherical suspension* and denoted by $SX$.)

Given two metric space $(Y, \rho)$ and $(Y' \rho')$, the *product metric $d$* on $Y \times Y'$ is defined by

$$d((y, y'), (z, z')) := \sqrt{\rho(y, z)^2 + \rho'(y', z')^2}. \tag{I.2}$$

The importance of the spherical join lies in the following lemma (the proof of which is left as an exercise for the reader).

**LEMMA I.2.18.** *Let $(X_1, d^1)$, $(X_2, d^2)$ be metric spaces. Then* $\text{Cone}_0(X_1) \times \text{Cone}_0(X_2)$ *endowed with the product metric is isometric with* $\text{Cone}_0(X_1 * X_2)$.

We give three references for the proof of the following important lemma.

**LEMMA I.2.19.** ([37, p. 190], [221] or [54, Appendix]). *Suppose $X_1$ and $X_2$ are metric spaces. The spherical join $X_1 * X_2$ is* CAT(1) *if and only if both $X_1$ and $X_2$ are* CAT(1).

## I.3. POLYHEDRA OF PIECEWISE CONSTANT CURVATURE

In 6.2 we defined the notion of a convex polytope in $\mathbb{X}^n_\kappa$. (As in 6.2 we are primarily interested in $\kappa \in \{+1, 0, -1\}$.) Call such a convex polytope an $\mathbb{X}_\kappa$-*polytope* when we do not want to specify $n$. In Definition A.2.12, we explained what it means for a poset $\mathcal{P}$ to be an abstract convex cell complex.

**DEFINITION I.3.1.** An $\mathbb{X}_\kappa$-*cell structure* on an abstract convex cell complex $\mathcal{P}$ is a family $(C_p)_{p \in \mathcal{P}}$ of $\mathbb{X}_\kappa$-polytopes so that if $p < p'$ and $F_p$ is the corresponding face of $C_{p'}$, then there is an isometry $C_p \cong F_p$ inducing the natural combinatorial identification of face posets. Thus, if $\Lambda$ is the geometric realization of $\mathcal{P}$, we have an identification, well-defined up to isometry, of each cell in $\Lambda$ with an $\mathbb{X}_\kappa$-polytope. We say that $\Lambda$ is a $\mathbb{X}_\kappa$-polyhedral complex.

**Examples I.3.2**

(i) (*Euclidean cell complexes.*) Suppose a collection of convex polytopes in $\mathbb{E}^n$ is a convex cell complex in the classical sense (Definition A.1.9). Then the union $\Lambda$ of these polytopes is a $\mathbb{X}_0$-polyhedral complex.

(ii) (*Regular cells.*) As explained in Appendix 1, there are three families of regular polytopes which occur in each dimension $n$: the $n$-simplex, the $n$-cube, and the $n$-octahedron. Each face of a simplex is a lower-dimensional simplex and each face of a cube is a lower-dimensional cube. On the other hand, the faces of an octahedron are simplices, not lower dimensional octahedra. By requiring each cell of a complex to be one of the above two types (simplices or cubes) we get the corresponding notion of a simplicial complex (Definition A.2.6) or cubical complex (Definition A.2.12). In each case we can be realize the polytope as a regular polytope in $\mathbb{X}^n_\kappa$ and this polytope is determined, up to congruence, by its edge length. (When $\kappa > 0$, there is an upper bound on this edge length, namely, $\frac{2}{\sqrt{\kappa}} \arctan \frac{d}{2}$, where $d$ is the edge length of the $n$-dimensional regular polytope in $\mathbb{R}^n$ centered at the origin with vertices on $\mathbb{S}^{n-1}$.) Hence, we can define an $\mathbb{X}_\kappa$-structure on a simplicial complex or a cubical complex simply by specifying an edge length.

Suppose $\Lambda$ is a $\mathbb{X}_\kappa$-polyhedral complex. A path $\gamma : [a,b] \to \Lambda$ is *piecewise geodesic* if there is a subdivision $a = t_0 < t_1 \cdots < t_k = b$ so that for $1 \leqslant i \leqslant k$ $\gamma([t_{i-1}, t_i])$ is contained in a single (closed) cell of $\Lambda$ and so that the restriction of $\gamma$ to $[t_{i-1}, t_i]$ is a geodesic segment in that cell. The *length* of the piecewise geodesic $\gamma$ is defined by $l(\gamma) := \sum_{i=1}^{n} d(\gamma(t_i), \gamma(t_{i-1}))$, i.e., $l(\gamma) := b - a$. $\Lambda$ has a natural length metric

$$d(x,y) := \inf\{l(\gamma) \mid \gamma \text{ is a piecewise geodesic from } x \text{ to } y\}.$$

(As usual we allow $\infty$ as a possible value of $d$.) The length space $\Lambda$ is called a *piecewise constant curvature polyhedron*. As $\kappa = +1, 0, -1$, we say that it is, respectively, *piecewise spherical*, *piecewise Euclidean*, or *piecewise hyperbolic*.

**DEFINITION I.3.3.** An $\mathbb{X}_\kappa$-polyhedral complex $\Lambda$ *has finitely many shapes of cells* if its cells represent only finitely many isometry classes.

For a proof of the following basic result see [37].

**PROPOSITION I.3.4.** (See [37, pp. 105–111] as well as [221]). *Let $\Lambda$ be a connected $\mathbb{X}_\kappa$-polyhedral complex with its length metric. Suppose that either (a) $\Lambda$ is locally finite or (b) $\Lambda$ has finitely many shapes of cells. Then $\Lambda$ is a complete geodesic space.*

### Geometric Links

Suppose $x$ is a point in an $n$-dimensional $\mathbb{X}_\kappa$-polytope $P$. The *geometric link*, $\mathrm{Lk}(x,P)$, of $x$ in $P$ is the set of all inward-pointing unit tangent vectors at $x$. It is (isometric to) an intersection of a finite number of half-spaces in $\mathbb{S}^{n-1}$. For example, if $x$ lies in the interior of $P$, then $\mathrm{Lk}(x,P) \cong \mathbb{S}^{n-1}$, while if $x$ is a vertex of $P$, then $\mathrm{Lk}(x,P)$ is a spherical polytope (i.e., it contains no pair of antipodal points). Similarly, if $\Lambda$ is an $\mathbb{X}_\kappa$-polyhedral complex and $x \in \Lambda$, define $\mathrm{Lk}(x,\Lambda)$ to be the union of all $\mathrm{Lk}(x,P)$ where $P$ is a cell of $\Lambda$ containing $x$ (i.e., if $x \in P \cap P'$, we glue together $\mathrm{Lk}(x,P)$ and $\mathrm{Lk}(x,P')$ along their common face $\mathrm{Lk}(x, P \cap P')$. Since each such $\mathrm{Lk}(x,P)$ can be subdivided into spherical polytopes, $\mathrm{Lk}(x,P)$ is a piecewise spherical length space. $\mathrm{Lk}(x,\Lambda)$ should be thought of as the space of directions at $x$.

It is well-known (and follows immediately from Proposition I.2.17 (ii)) that a surface of piecewise constant curvature $\kappa$ has curvature $\leqslant \kappa$ if and only if the sum of the angles at each vertex is $\geqslant 2\pi$. In view of Exercise I.2.2 this is the same as the requirement that the link of each vertex be CAT(1). In Theorem I.3.5 below, we state a similar result for an arbitrary $\mathbb{X}_\kappa$-polyhedral complex: it has curvature $\leqslant \kappa$ if and only if the link of each vertex is CAT(1). So, this CAT(1) condition on links of vertices should be viewed as a generalization of the condition in dimension 2 that the sum of the angles is $\geqslant 2\pi$.

As in 6.2, identify $\mathbb{X}^n_\kappa$ with a certain hypersurface in a $(n+1)$-dimensional vector space $E_\kappa$, equipped with a symmetric bilinear form. $E_\kappa$ is $\mathbb{R}^{n+1}$, $\mathbb{R}^{n+1}$ or $\mathbb{R}^{n,1}$ as $\kappa$ is, respectively, $> 0$, $= 0$ or $< 0$.) For any subset $P \subset \mathbb{X}^n_\kappa$, let $E(P) \subset E_\kappa$ denote the linear subspace spanned by $P$. Suppose $P$ is a convex polytope in $\mathbb{X}^n_\kappa$, that $F$ is a face of $P$ and $x \in F$. $E(F, P)$ denotes the orthogonal complement of $E(F)$ in $E(P)$. The restriction of the bilinear form to $E(F, P)$ is positive definite and $E(F, P)$ is naturally identified with the orthogonal complement of $T_xF$ in $T_xP$. (Here $T_xP$ denotes the tangent space at $x$.) In Appendix A.6, we defined the notion of the "link" of a face of a convex polytope in some affine space. We now define a similar notion for $\mathbb{X}_\kappa$-polytopes. Let $\mathrm{Cone}(F, P)$ denote the essential polyhedral cone in $E(F, P)$ consisting of all inward-pointing tangent vectors and set

$$\mathrm{Lk}(F, P) := \mathrm{Cone}(F, P) \cap \mathbb{S}(E(F, P)), \tag{I.3}$$

where $\mathbb{S}(E(F, P))$ denotes the unit sphere in $E(F, P)$. Thus, $\mathrm{Lk}(F, P)$ is spherical polytope. Its dimension is 1 less than the codimension of $F$ in $P$.

Suppose $\Lambda$ is an $\mathbb{X}_\kappa$-polyhedral complex and $F$ is one of its cells. As in Definition A.6.1, we have a cell complex $\mathrm{Lk}(F, \Lambda)$. By the previous paragraph it is piecewise spherical (i.e., it is an $\mathbb{X}_1$-polyhedral complex). When we wish to emphasize this piecewise spherical structure we will call $\mathrm{Lk}(F, P)$ the *geometric link*. The following theorem is main result of this section. Details of the proof can be found in [37, pp. 206–207].

**THEOREM I.3.5.** (The Link Condition.) *Suppose $\Lambda$ is an $\mathbb{X}_\kappa$-polyhedral complex with finitely many shapes of cells. The following conditions are equivalent.*

*(a)* *$\Lambda$ has curvature $\leqslant \kappa$.*

*(b)* *For each $x \in \Lambda$, $\mathrm{Lk}(x, \Lambda)$ is* CAT(1).

*(c)* *For each $v \in \mathrm{Vert}(\Lambda)$, the piecewise spherical complex $\mathrm{Lk}(v, \Lambda)$ is* CAT(1).

*(d)* *For each cell $F$ of $\Lambda$, $\mathrm{Lk}(F, L)$ has no closed geodesic of length $< 2\pi$.*

*(e)* *For each cell $F$ of $\Lambda$, $\mathrm{Lk}(F, L)$ is* CAT(1).

*Sketch of proof.* Since a neighborhood of $x$ in $\Lambda$ is isometric to a neighborhood of the cone point in $\mathrm{Cone}_\kappa(\mathrm{Lk}(x, \Lambda))$, the equivalence of (a) and (b) follows from Proposition I.2.17. If $x \in \mathrm{int}(F)$ for some cell $F$ of $\Lambda$, then $\mathrm{Lk}(x, \Lambda)$ is isometric to the spherical join $\mathbb{S}^{\dim F - 1} * \mathrm{Lk}(F, \Lambda)$. So, the equivalence of (b) and (e) follows from Lemma I.2.19. The equivalence of (d) and (e) follows from Theorem I.2.8 and induction on dimension. The equivalence of (c) and (e) is just the fact that if $v$ is a vertex of a cell $F$ and $\sigma_F$ is the cell of $\mathrm{Lk}(v, \Lambda)$ corresponding to $F$, then we have a canonical identification: $\mathrm{Lk}(\sigma_F, \mathrm{Lk}(v, \Lambda)) \cong \mathrm{Lk}(F, \Lambda)$. □

Given a geodesic space $X$ define its *systole*, $\mathrm{sys}(X)$, to be the infimum of the lengths of all closed geodesics. Call a piecewise spherical complex *large* if it is CAT(1). Theorem I.3.5 states that an $\mathbb{X}_\kappa$-polyhedral complex has curvature $\leqslant \kappa$ if and only if it has large links (or equivalently, if and only if the link of each of its cells has systole $\geqslant 2\pi$).

A piecewise spherical complex $L$ to be *extra large* if for each cell $\sigma$ of $L$ (including the empty cell) $\mathrm{sys}(\mathrm{Lk}(\sigma, L)) > 2\pi$.

**Infinitesimal Shadows**

Suppose $c : [a, b] \to \Lambda$ is a piecewise geodesic with $c(t_0) = x_0$. We then get two geodesic arcs, $c_{\text{out}} : [t_0, t_0 + \varepsilon) \to \Lambda$ and $c_{\text{in}} : [t_0, t_0 + \varepsilon) \to \Lambda$, both emanating from $x_0$, defined by $c_{\text{out}} := c|_{[t_0, t_0+\varepsilon]}$ and $c_{\text{in}}(t) := c(t_0 - t)$. Let $c'_{\text{out}}$ and $c'_{\text{in}}$ be their unit tangent vectors in $\mathrm{Lk}(x_0, \Lambda)$. The proof of next lemma is straightforward.

**LEMMA I.3.6.** (See [221, Lemmas 4.1 and 4.2].) *A piecewise geodesic path $c : [a, b] \to \Lambda$ is a local geodesic at $x_0$ if and only the distance from $c'_{\text{out}}$ to $c'_{\text{in}}$ in $\mathrm{Lk}(x_0, \Lambda)$ is $\geqslant \pi$.*

**Example I.3.7.** Let $\mathbb{S}^1(2\pi + \delta)$ denote the circle of circumference $2\pi + \delta$, $\delta \geqslant 0$. It is CAT(1). Let $X_\kappa := \mathrm{Cone}_\kappa(\mathbb{S}^1(2\pi + \delta))$. By Proposition I.2.17 (ii), it is CAT($\kappa$). Let $\theta_1, \theta_2 \in \mathbb{S}^1(2\pi + \delta)$ and define a path $c : [-a, a] \to X_\kappa$ by

$$c(t) := \begin{cases} [-t, \theta_1] & \text{if } t \leqslant 0, \\ [t, \theta_2] & \text{if } t \geqslant 0, \end{cases}$$

where, as usual, $[t, \theta]$ denotes the image of $(t, \theta)$ in the cone. By Lemmas I.2.16 and I.3.6, $c$ is a geodesic if and only if $d(\theta_1, \theta_2) \geqslant \pi$. So, if $\delta > 0$, there are many ways to extend the geodesic $c|_{[-a,0]}$ past the cone point. These extensions are parameterized by an arc of $\theta_2$'s, specifically, the arc of radius $\frac{1}{2}\delta$ centered at the point of distance $\pi + \frac{1}{2}\delta$ from $\theta_1$.

The above example illustrates a dramatic difference between a (singular) metric on piecewise constant curvature polyhedron and the metric on a Riemannian manifold: in the singular case extensions of geodesic segments need not be unique.

Suppose $\Lambda$ is a piecewise constant curvature cell complex, $x \in \Lambda$ and $u \in \mathrm{Lk}(x, \Lambda)$. The *infinitesimal shadow of $x$ with respect to $u$*, denoted $\mathrm{Shad}(x, u)$, is the set of all $v \in \mathrm{Lk}(x, \Lambda)$ such that there is a geodesic $c : [-a, a] \to \Lambda$ with $c(0) = x$ and with unit tangent vectors at $x$, $c_{\mathrm{in}} = u$ and $c_{\mathrm{out}} = v$. Thus, $\mathrm{Shad}(x, u)$ measures the outgoing directions of possible extensions of a geodesic coming into $x$ from the direction $u$. (In particular, $\Lambda$ has extendible geodesics if and only if $\mathrm{Shad}(x, u) \neq \emptyset$, for all $x \in \Lambda$ and $u \in \mathrm{Lk}(x, \Lambda)$.) For example, if $\mathrm{Lk}(x, \Lambda) = \mathbb{S}^n$ and $u \in \mathbb{S}^n$, then $\mathrm{Shad}(x, u) = \{-u\}$. As another example, if, as in I.3.7, $\mathrm{Lk}(x, \Lambda) = S^1(2\pi + \delta)$, then the infinitesimal shadow is an arc of radius $\frac{1}{2}\delta$.

The next lemma result follows immediately from Lemma I.3.6 and the definition of infinitesimal shadow.

**Lemma I.3.8.** *With notation as above,* $\mathrm{Shad}(x, u) = \mathrm{Lk}(x, u) - B(u, \pi)$, *where $B(u, \pi)$ is the open ball in* $\mathrm{Lk}(x, \Lambda)$ *of radius $\pi$ with center $u$.*

## I.4. PROPERTIES OF CAT(0) GROUPS

Call a group *CAT*(0) if it is isomorphic to a discrete group of isometries acting properly and cocompactly on some complete CAT(0) space.

**Theorem I.4.1.** *Suppose $G$ is a* CAT(0) *group. Then*

*(i) There is a model for $\underline{E}G$ (the universal space for proper $G$-actions) with $\underline{E}G/G$ compact.*

*(ii) $G$ is finitely presented.*

*(iii) There are only finitely many conjugacy classes of finite subgroups in $G$.*

*(iv)* $\mathrm{cd}_{\mathbb{Q}}(G) < \infty$.

*(v) $H^*(G; \mathbb{Q})$ is finite dimensional.*

*(vi) The Word and Conjugacy Problems for $G$ are solvable.*

*(vii) Any abelian subgroup of $G$ is finitely generated.*

*(viii) Any virtually solvable subgroup of $G$ is virtually abelian.*

Properties (i) through (v) are finiteness properties of the type considered in Appendix F.4. Property (i) is a consequence of the Bruhat-Tits Fixed Point Theorem (Corollary I.2.13). Given that (i) holds, Corollary I.2.14 implies that there is a CW model for $\underline{E}G$ with a finite number of orbits of cells. Properties (ii) through (v) follow from this.

Property (vi) gives two standard algorithmic consequences for a group which has such an action on a CAT(0) space. (The Word Problem is defined in

Section 3.4. Similarly, the Conjugacy Problem for a group $G$ asks if there is an algorithm for deciding when any two given elements are conjugate.)

Properties (vii) and (viii) are related to the Flat Torus Theorem, discussed below.

### The Flat Torus Theorem

Suppose $g$ is an isometry of a metric space $X$. Its *displacement function* $d_g : X \to [0, \infty)$ is defined by $d_g(x) := d(x, gx)$. Its *translation length* is the nonnegative real number $|g| := \inf\{d_g(x) \mid x \in X\}$. The set of points where this infimum is attained is denoted $\mathrm{Min}(g)$ (and called the *min set* of $g$). If $G$ is a subgroup of $\mathrm{Isom}(X)$, put

$$\mathrm{Min}(G) := \bigcap_{g \in G} \mathrm{Min}(g). \tag{I.4}$$

An isometry $g$ is *semisimple* if $\mathrm{Min}(g) \neq \emptyset$; otherwise it is *parabolic*. There are two types of semisimple isometries: if $|g| = 0$ (so that the fixed point set of $g$ is nonempty), $g$ is called *elliptic*, while if $|g| > 0$, $g$ is *hyperbolic*. If $g$ is an isometry of a CAT(0) space $X$, then $g$ is hyperbolic if and only if there exists a geodesic line $\mathbb{R} \hookrightarrow X$ which is translated nontrivially by $g$. It turns out that the magnitude of this translation is $|g|$. Such a geodesic line is called an *axis* for $g$. $\mathrm{Min}(g)$ is the union of all its axes. (See [37, pp. 231–232] for proofs of the statements in this paragraph.)

**THEOREM I.4.2.** (The Flat Torus Theorem, [37, pp. 244–245].) *Let $A$ be a free abelian group of rank $n$ acting by semisimple isometries on a* CAT(0) *space $X$. Then*

*(i)* $\mathrm{Min}(A) \neq \emptyset$ *and it decomposes as a product $Y \times \mathbb{E}^n$.*

*(ii) Every $a \in A$ stabilizes* $\mathrm{Min}(A)$ *and respects the product decomposition. Moreover, $a$ acts as the identity on the $Y$ factor and as a translation on the $\mathbb{E}^n$ factor.*

*(iii) For each $y \in Y$, $(y \times \mathbb{E}^n)/A$ is a flat $n$-torus.*

*(iv) If an element of* $\mathrm{Isom}(X)$ *normalizes $A$, then it stabilizes* $\mathrm{Min}(A)$ *and preserves the product decomposition.*

*(v) If a subgroup $G \subset$* $\mathrm{Isom}(X)$ *normalizes $A$, then a subgroup of finite index in $G$ centralizes $A$. Moreover, if $G$ is finitely generated, then $G$ has a subgroup of finite index which contains $A$ as a direct factor.*

One corollary of the Flat Torus Theorem is property (vii) of Theorem I.4.1. (See [37, pp. 247–248].) Another corollary is property (viii) which we state as the following result.

**THEOREM I.4.3.** (The Solvable Subgroup Theorem, [37, p. 249].) *Suppose $G$ is a* CAT(0) *group and that $H \subset G$ is a virtually solvable subgroup. Then $H$ is finitely generated and virtually abelian (i.e., $H$ contains an abelian subgroup of finite index).*

## I.5. PIECEWISE SPHERICAL POLYHEDRA

### The Polar Dual of a Spherical Polytope

Suppose $P \subset \mathbb{S}^n$ is an $n$-dimensional spherical polytope and $\{v_1, \dots, v_k\}$ is its vertex set. Define $k \times k$ symmetric matrices

$$l_{ij}(P) := d(v_i, v_j) = \cos^{-1}\langle v_i, v_j\rangle \tag{I.5}$$

and

$$c^*_{ij}(P) := \cos l_{ij} = \langle v_i, v_j\rangle. \tag{I.6}$$

where $d(v_i, v_j)$ is the length of the circular arc from $v_i$ to $v_j$. In other words, if $\{v_i, v_j\}$ is an edge of $P$ then $l_{ij}(P)$ is its length.

Similarly, if $u_1, \dots, u_l$ are the inward-pointing unit normal vectors to the codimension one faces of $P$, define $l \times l$ symmetric matrices:

$$\theta_{ij}(P) := \pi - \cos^{-1}\langle u_i, u_j\rangle \tag{I.7}$$

and

$$c_{ij}(P) := -\cos\theta_{ij} = \langle u_i, u_j\rangle. \tag{I.8}$$

By convention, $\theta_{ii} := \pi$. $(\theta_{ij}(P))$ is the *matrix of dihedral angles of $P$* and $(c_{ij}(P))$ is its *Gram matrix*. (See equations (6.12) and (6.13) in Section 6.8.)

Given a matrix $(l_{ij})_{0\leqslant i,j\leqslant n}$, with $l_{ij} \in (0, \pi)$ for $i \neq j$ and $l_{ij} = 0$ for $i = j$, define a matrix $(c^*_{ij})$ by

$$c^*_{ij} := \cos(l_{ij}). \tag{I.9}$$

The proof of the next lemma is essentially the same as that of Proposition 6.8.2.

**LEMMA I.5.1.** *Suppose $(l_{ij})_{0\leqslant i,j\leqslant n}$ is a symmetric matrix with $l_{ij} \in (0, \pi)$ whenever $i \neq j$ and $l_{ij} = 0$ for $i = j$. Then $(l_{ij})$ is the matrix of edge lengths of a spherical $n$-simplex $\sigma$ if and only if the matrix $(c^*_{ij})$ is positive definite.*

**DEFINITION I.5.2.** Suppose $P \subset \mathbb{S}^n$ is an $n$-dimensional spherical polytope with vertex set $\{v_1, \dots, v_k\}$. Its *polar dual* is the polytope $P^*$ given by the intersection of half-spaces in $\mathbb{S}^n$ defined by the inequalities $\langle v_i, x\rangle \geqslant 0$, $i = 1, \dots, k$. Equivalently, if $u_1, \dots, u_l$ are the unit inward-pointing normals to the codimension-one faces of $P$, then $\{u_1, \dots, u_l\}$ is the vertex set of $P^*$.

It is obvious that

- $P$ and $P^*$ are combinatorially dual. (Recall from Appendix A.2 that this means that their face posets are anti-isomorphic, i.e., $\widetilde{\mathcal{F}}(P^*) \cong \widetilde{\mathcal{F}}(P)^{op}$, where $\widetilde{\mathcal{F}}(P)$ is defined in Example A.2.3.)
- In particular, the dual of a spherical $n$-simplex $\sigma$ is a spherical $n$-simplex $\sigma^*$. The vertices $v_0, v_1, \dots, v_n$ of $\sigma$ are the inward-pointing normals to the codimension-one faces of $\sigma^*$ and vice versa.
- $P^{**} = P$.

From (I.5) and (I.7) we get the following lemma.

**LEMMA I.5.3.** *Suppose $P$ is a spherical polytope and $P^*$ is its dual. Then the edge lengths of $P$ are the exterior dihedral angles of $P^*$. In other words,*

$$l_{ij}(P) = \pi - \theta_{ij}(P^*) \quad \textit{and} \quad c^*_{ij}(P) = c_{ij}(P^*).$$

In Lemma 6.3.3 we proved that if all the dihedral angles of a spherical polytope are nonobtuse, then it is a simplex. Combining this with Lemma I.5.3, we get the following.

**COROLLARY I.5.4.** *Suppose each nontrivial edge of a spherical polytope $P$ has length $\geqslant \pi/2$. Then $P$ is a simplex.*

*Proof.* By Lemma 6.3.3, $P^*$ is a simplex. Therefore, so is $P$. □

**DEFINITION I.5.5.** A *face angle* of an $\mathbb{X}_\kappa$-polytope means an angle in some 2-dimensional (polygonal) face. An $\mathbb{X}_\kappa$-polytope $P$ *has nonacute face angles* if each of its face angles is $\geqslant \pi/2$.

**LEMMA I.5.6.** *If an $\mathbb{X}_\kappa$-polytope has nonacute face angles, then it is simple. (The definition of a "simple polytope" is given in Definition A.2.11.)*

*Proof.* Suppose $P$ is an $\mathbb{X}_\kappa$-polytope and $\sigma$ is the the link of a vertex $v$ of $P$. Then $\sigma$ is a spherical polytope. Moreover, the edge lengths of $\sigma$ are the face angles at $v$ of the 2-dimensional faces containing $v$. Hence, the edge lengths of $\sigma$ are all $\geqslant \pi/2$. By Corollary I.5.4, $\sigma$ is a simplex. Since this is true at each vertex, $P$ is simple. □

**DEFINITION I.5.7.** A spherical simplex is *all right* if each of its edge lengths is $\pi/2$. (It follows from (I.5) that a simplex is all right if and only if each of its dihedral angles is $\pi/2$.) Similarly, a piecewise spherical simplicial complex (i.e., an $\mathbb{X}_1$-simplicial complex) is *all right* if each of its simplices is all right.

**DEFINITION I.5.8.** A spherical simplex has *size* $\geqslant \pi/2$ if each of its edge lengths is $\geqslant \pi/2$. The phrase that a piecewise spherical simplicial complex *has simplices of size* $\geqslant \pi/2$ has a similar meaning.

**Examples I.5.9**

(i) Since each dihedral angle in a Euclidean cube is $\pi/2$, the link of each face of a Euclidean cube is an all right simplex. Hence, the link of any cell in a piecewise Euclidean cubical complex is a piecewise spherical, all right simplicial cell complex.

(ii) The link of a face of a simple polytope is a simplex. So, if all cells of a complex $\Lambda$ are simple polytopes, then the link of each cell in $\Lambda$ is a simplicial cell complex. The link of a vertex of a simple polytope $P$ with nonacute face angles is a simplex of size $\geqslant \pi/2$ and by Lemma I.5.11 below, the same is true for the link of any face of $P$. Hence, if each cell of an $\mathbb{X}_\kappa$-polyhedral complex is a simple polytope with nonacute face angles, then the link of each of its cells is a simplicial cell complex of size $\geqslant \pi/2$.

The next lemma is obvious and its proof is left to the reader.

**LEMMA I.5.10.** *Suppose L is a piecewise spherical, all right, simplicial complex. Then the link of any simplex in L is all right.*

However, the following generalization of the above lemma is not so obvious.

**LEMMA I.5.11.** (Moussong [221, Lemma 8.3].) *Suppose L is a piecewise spherical simplicial complex with simplices of size* $\geqslant \pi/2$. *Then the link of any simplex in L also has simplices of size* $\geqslant \pi/2$.

The proof depends on the next lemma.

**LEMMA I.5.12.** *Suppose* $\sigma \subset \mathbb{S}^n$ *is a spherical simplex of size* $\geqslant \pi/2$. *Let* $\{v_0, \dots, v_n\}$ *be its vertex set. Then the spherical simplex* $\mathrm{Lk}(v_0, \sigma)$ *has size* $\geqslant \pi/2$.

*Proof.* Let $p : \mathbb{R}^{n+1} \to \mathbb{R}^{n+1}$ be orthogonal projection onto the hyperplane $(v_0)^\perp$. Then $\mathrm{Lk}(v_0, \sigma)$ is the spherical $(n-1)$-simplex wtih vertex set $\{\hat{v}_1, \dots, \hat{v}_n\}$, where $\hat{v}_i := p(v_i)/|p(v_i)|$. (Here $|x| := \sqrt{\langle x, x\rangle}$.) We must show that for $i \neq j$, $\langle \hat{v}_i, \hat{v}_j\rangle \leqslant 0$, i.e., $\langle p(v_i), p(v_j)\rangle \leqslant 0$. This is the calculation:

$$\begin{aligned}\langle p(v_i), p(v_j)\rangle &= \langle v_i - \langle v_i, v_0\rangle v_0, v_j - \langle v_j, v_0\rangle v_0\rangle \\ &= \langle v_i, v_j\rangle - \langle v_i, v_0\rangle\langle v_j, v_0\rangle \leqslant 0.\end{aligned}$$

□

*Proof of Lemma I.5.11.* Suppose $\sigma$ is a simplex of $L$, $v \in \mathrm{Vert}(\sigma)$. Let $\sigma'$ (resp., $L'$) be the full subcomplex spanned by $\mathrm{Vert}(\sigma) - \{v\}$ (resp., by $\mathrm{Vert}(L) - \{v\}$). If $L$ has simplices of size $\geqslant \pi/2$, then so does $L'$. Also, $\mathrm{Lk}(\sigma, L)$ is naturally identified with $\mathrm{Lk}(\sigma', L')$. So, by induction on $\mathrm{Card}(\mathrm{Vert}(L))$, it suffices to do the case where $\sigma$ is a vertex. This case follows from Lemma I.5.12. □

## I.6. GROMOV'S LEMMA

Gromov proved the following lemma in [147, p. 122].

**LEMMA I.6.1.** (Gromov's Lemma.) *Suppose $L$ is an all right, piecewise spherical simplicial complex. Then $L$ is $CAT(1)$ if and only if it is a flag complex.*

(Flag complexes are discussed in Appendix A.3.)

**COROLLARY I.6.2.** (Berestovskii [19].) *Any polyhedron can be given a piecewise spherical structure which is* CAT(1).

*Proof.* As we pointed out in Remark A.3.10, since the barycentric subdivision of any cell complex is a flag complex, $L$ can be any polyhedron. □

Another important corollary of Gromov's Lemma is the following.

**COROLLARY I.6.3.** *A piecewise Euclidean cubical complex $X$ is nonpositively curved if and only if the link of each of its vertices is a flag complex.*

*Proof.* This follows from Theorem I.3.5 and Examples I.5.9 (i). □

The basic idea in the proof of Gromov's Lemma is the following.

**LEMMA I.6.4.** *If $L$ is an all right piecewise spherical flag complex, then the length of any closed geodesic in $L$ is $\geqslant 2\pi$.*

*Proof.* Given any vertex $v$ of $L$, its *open star*, denoted $O(v, L)$, is the union of all open simplices which have $v$ as a vertex. It is an open neighborhood of $v$ in $L$. The *closed star*, $\mathrm{Star}(v, L)$, is the union of all closed simplices containing $v$. Let $\gamma$ be the image of a closed geodesic in $L$ and $v$ a vertex of $L$. Suppose $\gamma$ has nonempty intersection with $O(v, L)$. Then $c := \gamma \cap \mathrm{Star}(v, L)$ is an arc of $\gamma$.

*Claim.* The length of $c$ is $\pi$.

*Proof of Claim.* Due to the all rightness condition, the space $\mathrm{Star}(v, L)$ is isometric to $\mathrm{Cone}_1(Lk(v, L))$ (the spherical cone on the link of $v$ in $L$) and we may identify $Lk(v, L)$ with the metric sphere of radius $\pi/2$ about $v$ in $L$. If $\gamma$ passes through $v$, the claim is obvious. So, suppose $\gamma$ misses $v$. Consider the surface $S \subset \mathrm{Star}(v, L)$ defined as the union of all geodesic rays of length $\pi/2$ emanating from $v$ and passing through points of $\gamma$. $S$ is a finite union of isosceles spherical triangles of height $\pi/2$ with common apex at $v$, matched together in succession along $\gamma$. Develop $S$ along $\gamma$ in a locally isometric fashion into $\mathbb{S}^2$ . Assume that the image of $v$ is the north pole. Under the developing map the curve $c$ must be mapped onto a local geodesic, therefore, onto a segment of a great circle of $\mathbb{S}^2$. Since the image of $c$ cuts inside the northern hemisphere and connects two points on the equator, it must have length $\pi$. □

It suffices to show that we can find two vertices, $v$, $w$, such that $O(v, L) \cap \gamma \neq \emptyset$, $O(w, L) \cap \gamma \neq \emptyset$ and $O(v, L) \cap O(w, L) = \emptyset$. Put $V = \{v \in \mathrm{Vert}(L) : O(v, L) \cap \gamma \neq \emptyset\}$, then $\gamma$ is contained in the full subcomplex of $L$ spanned by $V$. If $O(v, L) \cap O(w, L) \neq \emptyset$ held for all $v, w \in V$, then all pairs of vertices in $V$ would be connected by an edge in $L$. Since $L$ is a flag complex, $V$ would span a simplex in $L$. But a simplex cannot contain a closed geodesic. □

*Proof of Gromov's Lemma.* If $x$ is an interior point of a simplex $\sigma$ of $L$, then a neighborhood of $x$ is isometric to a neighborhood of the cone point in the spherical cone on $\mathbb{S}^{\dim \sigma - 1} * \mathrm{Lk}(\sigma, L)$. By Proposition I.2.17 (ii) and Lemma I.2.19, this neighborhood is CAT(1) if and only if $\mathrm{Lk}(\sigma, L)$ is CAT(1). (This is basically the argument for the proof of Theorem I.3.5.)

Suppose $L$ is a flag complex. The link of any simplex in a flag complex is (obviously) also a flag complex. By Lemma I.5.10, it is all right. So, by Lemma I.6.4, Theorem I.2.8 and induction on dimension, $L$ is CAT(1) if and only if for all simplices $\sigma$ in $L$ (including the empty simplex), $\mathrm{Lk}(\sigma, L)$ contains no closed geodesic of length $< 2\pi$. Since we proved that this holds in the previous lemma, $L$ is CAT(1).

Conversely, if $L$ is not a flag complex, then the link $L'$ of some simplex in $L$ must contain an "empty triangle," i.e., a circuit of 3 edges in its 1-skeleton which does not bound a 2-simplex. Since such a circuit is a closed geodesic in $L'$ of length $3\pi/2$, $L'$ is not CAT(1) and hence, neither is $L$. □

### The No □-Condition

Next we address the question of when an all right, piecewise spherical flag complex is extra large. In other words, when is its systole $> 2\pi$?

An *empty 4-circuit* in a simplicial complex $L$ is a circuit of 4 edges such that neither pair of opposite vertices is connected by an edge. In other words, the 4-circuit is a full subgraph of $L^1$. (If $L$ is a flag complex, a 4-circuit is empty if and only if it is not the boundary of two adjacent 2-simplices.) The simplicial complex $L$ satisfies the *no □-condition* if it has no empty 4-circuits.

(In Section 14.2, empty 4-circuits were called "full 4-cycles;" see Figure 14.2 for a picture of one.)

Gromov attributes the following result to Siebenmann. It follows from an argument similar to the proof of Lemma I.6.4.

**LEMMA I.6.5.** ([147, p. 123].) *Suppose L is an all right, piecewise spherical flag complex. Then L is extra large if and only if it satisfies the no □-condition.*

*Proof.* Note that if $L$ satisfies the no □-condition, then so does the link of each simplex of $L$. An empty 4-circuit in $L$ is a closed geodesic of length $2\pi$. So, extra largeness implies the no □-condition.

Conversely, suppose that in $L$ there is a closed geodesic $\gamma$ of length $2\pi$. As in the proof of Lemma I.6.4, we can find vertices $v$ and $w$ such that $\gamma$ has nonempty intersection with both open stars $O(v, L)$ and $O(w, L)$ and such that these open stars are disjoint. Moreover, the intersection of $\gamma$ with either closed star is an arc of length $\pi$. Since $l(\gamma) = 2\pi$, $\mathrm{Star}(v, L) \cap \mathrm{Star}(w, L)$ must contain at least 2 points of $\gamma$ of distance $\pi$. It follows that the intersection must contain at least 2 vertices not connected by an edge. Hence, we get an empty 4-circuit passing through $v$ and $w$. □

The next proposition follows from the arguments in Section 6.11 on the nonexistence of right-angled polytopes in hyperbolic spaces of dimension > 4.

**PROPOSITION I.6.6.** *Suppose L is an all right, piecewise spherical flag complex which is*

- *a* $\mathrm{GHS}^{n-1}$ *and*
- *extra large.*

*Then* $n \leqslant 4$.

(The notion of a $\mathrm{GHS}^{n-1}$ or "generalized homology $(n-1)$-sphere," was defined in 10.4.5.)

*Proof.* Suppose $L$ is a flag complex and a $\mathrm{GHS}^{n-1}$. We will show that if $n > 4$, then $L$ must contain an empty 4-circuit. The proof is basically the same as that of Corollary 6.11.6. (This corollary deals with simple polytopes. In order to translate it to our situation, one should think of $L$ as being the dual of the boundary complex of a simple polytope.)

Let $\sigma$ be a simplex of codimension 2 in $L$. Since $L$ is a generalized homology sphere, the link of $\sigma$ is a triangulation of $S^1$ (i.e., it is the boundary of a polygon). Let $m_\sigma$ be the number of edges in $\mathrm{Lk}(\sigma, L)$. As in Section 6.11, let $A_2$ be the average value of $m_\sigma$.

*Claim.* (Compare Lemma 6.11.5.) If $n > 4$, then $A_2 < 5$.

*Proof of Claim.* Define $f_i$ to be the number of $(n-i-1)$-simplices in $L$ (and $f_n = 1$) and then define the $h$-polynomial of $L$ by formula (6.24) of Section 6.11. It is known that Lemma 6.11.4 holds for generalized homology spheres, i.e., $h_i = h_{n-i}$ and $h_i \geqslant 0$. The estimate of $A_2$ then proceeds exactly as in the proof of Lemma 6.11.5. $\square$

Since $L$ is a flag complex, so is $\mathrm{Lk}(\sigma, L)$. Hence, each $m_\sigma > 3$. If $n > 4$, then, by the Claim, there must exist at least one $\sigma$ with $m_\sigma = 4$. Regarding $\mathrm{Lk}(\sigma, L)$ as a subcomplex of $L$, we see that it is, in fact, an empty 4-circuit. (Suppose to the contrary that there were an edge $e$ connecting opposite edges of this 4-circuit. Then $L$ contains the 1-skeleton of the join $e * \sigma$ and hence, by the flag condition, the entire simplex $e * \sigma$. But then $e$ represents another edge in $\mathrm{Lk}(\sigma, L)$, a contradiction.) $\square$

### Deformations of Piecewise Spherical Structures

Suppose $\sigma, \sigma'$ are spherical simplices. Let $\delta \geqslant 1$. Call $\sigma'$ a *$\delta$-change* of $\sigma$ if there exists a bi-Lipschitz homeomorphism $f : \sigma' \to \sigma$ with Lipschitz constant $\delta$, i.e.,

$$\frac{1}{\delta} d(x, y) \leqslant d(f(x), f(y)) \leqslant \delta d(x, y).$$

$f$ is called a *$\delta$-map*. Similarly, given two piecewise spherical simplicial complexes $L'$ and $L$, call $L'$ a *$\delta$-change* of $L$ if there exists a simplicial isomorphism $f : L' \to L$ which restricts to a $\delta$-map on each simplex. Moussong proved the following.

**Lemma I.6.7.** (Moussong [221, Lemma 5.11].) *Suppose L is a piecewise spherical simplicial complex. For any positive number $\alpha <$* $\mathrm{sys}(L)$*, there is a number $\delta > 1$ such that* $\mathrm{sys}(L') \geqslant \alpha$ *for any $\delta$-change $L'$ of L.*

*Remark.* In general, $\mathrm{sys}(L)$ is not an upper semicontinuous function of the spherical structure on $L$. (In other words, it is not an upper semicontinuous in the edge lengths of simplices of $L$.)

### CAT(−1)-Cubical Complexes

As we pointed out in Example I.3.2, every cubical complex has a canonical $\mathbb{X}_\kappa$ structure (once we specify the edge length of a cube) formed by declaring each cube to be a regular polytope in $\mathbb{X}_\kappa$. The most important case is when $\kappa = 0$; however, $\kappa < 0$ is also interesting. There are two equivalent ways to look at this. Either we can let the curvature $\kappa$ vary and hold the edge length fixed or we can vary the edge length $\varepsilon$ and hold the curvature $= -1$. In the first way, the dihedral angle between two codimension one faces of a cube increases to its Euclidean value of $\pi/2$ as $\kappa \to 0$. In the second way, the dihedral angle goes to $\pi/2$ as $\varepsilon \to 0$. We adopt the second point of view.

Suppose $\Lambda$ is a cubical cell complex. It admits a family of piecewise hyperbolic structures, parameterized by a positive real number $\varepsilon$, defined by declaring each cube to be isometric to a regular cube of edge length $\varepsilon$ in $\mathbb{H}^n$. Let $\Lambda_\varepsilon$ denote $\Lambda$ equipped with this structure. As we increase $\varepsilon$ away from 0 the dihedral angle of each cube will decrease from $\pi/2$. In other words, the piecewise spherical structure on the link of any vertex will be a $\delta$-change of the all right structure in which the edge lengths are $< \pi/2$. When will $\Lambda_\varepsilon$ have curvature $\leqslant -1$? By Theorem I.3.5 this will be the case if and only if the link of each vertex is CAT(1). By Lemma I.6.7 such a $\delta$-change of the all right structure on a link will still be CAT(1), for $\delta$ sufficiently close to 1 if and only if the original all right structure is extra large. If there are only a finite number of isomorphism types of links, then for $\varepsilon$ sufficiently small we will obtain a sufficiently small $\delta$-change of each link. Hence, we have the following.

**PROPOSITION I.6.8.** *Suppose $\Lambda$ is a locally finite cubical complex. Further suppose that there are only finitely many isomorphism types of links of vertices in $\Lambda$. Then $\Lambda_\varepsilon$ has curvature $\leqslant -1$ for some $\varepsilon > 0$ if and only if the link of each of its vertices is a flag complex satisfying the no $\square$-condition.*

Combining this with Proposition I.6.6 we get the following.

**COROLLARY I.6.9.** *Let $\Lambda$ be a cubical complex endowed with a piecewise hyperbolic structure so that each cube is regular. Suppose $\Lambda$*

- *is a homology n-manifold and*
- *has curvature $\leqslant -1$.*

*Then $n \leqslant 4$.*

## I.7. MOUSSONG'S LEMMA

Throughout this section, $L$ is a piecewise spherical cell complex and each cell is a simplex of size $\geqslant \pi/2$.

**DEFINITION I.7.1.** $L$ is a *metric flag complex* if it is a simplicial complex and if the following condition $(*)$ holds.

$(*)$ Suppose $\{v_0, \dots, v_k\}$ is a set of vertices of $L$ any two of which are connected by an edge. Put $c_{ij} = \cos(d(v_i, v_j))$. Then $\{v_0, \dots, v_k\}$ spans a simplex in $L$ if and only if $(c_{ij})$ is positive definite.

In other words, $L$ is a metric flag complex if it is "metrically determined by its 1-skeleton."

**Examples I.7.2**

(i) Suppose a 1-dimensional simplicial complex $L$ is a triangle with edge lengths $l_1$, $l_2$, and $l_3$. $L$ is the boundary of a spherical 2-simplex if and only if $l_1 + l_2 + l_3 < 2\pi$ and the three triangle inequalities hold. (See the discussion at the beginning of Section 6.5.) Hence, $L$ is a metric flag complex if and only if the sum of the three edge lengths is $\geqslant 2\pi$. If the sum is $< 2\pi$, it is necessary to fill in the spherical 2-simplex to get a metric flag complex.

(ii) Any flag complex with its all right, piecewise spherical structure is a metric flag complex.

**Lemma I.7.3.** *The link of a simplex in a metric flag complex is a metric flag complex.*

*Proof.* Suppose $L$ is a metric flag complex. Let $\mathbb{R}^S$ be the vector space on the vertex set of $L$. As in formula (I.10) below, we have a symmetric bilinear form $\langle\, ,\rangle$ on $\mathbb{R}^S$ defined by

$$\langle v_i, v_j\rangle := \begin{cases} 1 & \text{if } i = j, \\ \cos(d(v_i, v_j)) & \text{if } \{v_i, v_j\} \in \mathrm{Edge}(L), \\ -1 & \text{if } \{v_i, v_j\} \notin \mathrm{Edge}(L). \end{cases}$$

As in the proof of Lemma I.5.11, it suffices to prove the case of a link of a vertex $v_0$ in $L$. Let $T$ be the union of $\{v_0\}$ with the set of vertices which are connected to $v_0$ by an edge of $L$. Let $V$ $(:= (v_0)^{\perp})$ denote the orthogonal complement of $v_0$ in $\mathbb{R}^T$ and let $p : \mathbb{R}^T \to V$ be orthogonal projection. As in I.5.11, put $\hat{v}_i := p(v)/|p(v_i)|$. For $i \neq 0$, the $\hat{v}_i$ can be identified with the vertices of $\mathrm{Lk}(v_0, L)$. Suppose $\hat{v}_1, \dots, \hat{v}_k$ are pairwise connected by edges in $\mathrm{Lk}(v_0, L)$. To prove the lemma we must show that the restriction of the bilinear form to $\mathrm{Span}(\hat{v}_1, \dots \hat{v}_k)$ is positive definite if and only if it is positive definite on $\mathrm{Span}(v_0, v_1 \dots v_k)$. This is obvious. □

The following result of [221] generalizes Gromov's Lemma.

**Lemma I.7.4.** (Moussong's Lemma.) *Suppose L is a piecewise spherical cell complex in which all cells are simplices of size* $\geqslant \pi/2$. *Then L is* CAT(1) *if and only if it is a metric flag complex.*

We sketch a proof taken from [182] (or [90]). The argument is fairly simple if one is willing to assume the following lemma of Bowditch. It gives a characterization of CAT(1) spaces. Call a closed rectifiable curve in a metric space *shrinkable* if it can be homotoped through a family of closed rectifiable curves so that the lengths of the curves do not increase and do not remain constant during the homotopy. (Note that a shrinkable closed curve in a compact space is always shrinkable to a constant curve.) Bowditch [33] uses the Birkhoff curve-shortening process to prove the following.

**LEMMA I.7.5.** (Bowditch's Lemma.) *Let $X$ be a compact metric space of curvature $\leqslant 1$. Suppose that every nonconstant closed rectifiable curve of length $< 2\pi$ is shrinkable in $X$. Then $X$ is* CAT(1).

*Proof of Moussong's Lemma.* As in Gromov's Lemma I.6.1, the "only if" implication is easy; it is left as an exercise for the reader. Let $L$ be a metric flag complex in which all simplices have size $\geqslant \pi/2$. These properties are inherited by links (Lemmas I.5.11 and I.7.3). By Theorem I.3.5 and induction on dimension, we may assume that $L$ has curvature $\leqslant 1$. Suppose that $L$ is not CAT(1). Then by Bowditch's Lemma there exists a nonconstant, nonshrinkable, closed curve of length $<2\pi$ in $L$. Choose a shortest possible among such curves. This curve, denoted $\gamma$, is a nonshrinkable closed geodesic of length $<2\pi$. Since $L$ consists of simplices of size $\geqslant \pi/2$, for each vertex $v$ of $L$, the closed $\pi/2$-ball $B_v$ about $v$ is isometric to the spherical cone $\mathrm{Cone}_1(Lk(v, L))$ and is contained in the star $\mathrm{Star}(v, L)$. It is easy to see that $L$ is covered by the interiors of balls $B_v$, $v \in \mathrm{Vert}(L)$. As in the proof of Lemma I.6.4, $\gamma \cap B_v$ is an arc of length $\pi$ whenever $\gamma$ meets the interior of $B_v$.

Suppose a vertex $v$ is such that $\gamma$ meets the interior of $B_v$ and does not pass through $v$. Let us perform the following modification on $\gamma$: replace the segment $\gamma \cap B_v$ with the union of the two geodesic segments from $v$ to $\gamma \cap \partial B_v$. This modification can clearly be achieved via a homotopy through curves that have constant length. It follows that the modified closed curve continues to be a nonshrinkable closed geodesic of same length. Repeated application of such modifications results in a nonshrinkable closed geodesic that passes through the maximum number of vertices. Such a curve is necessarily contained in the 1-skeleton of $L$. Since all edges have length $\geqslant \pi/2$, the curve is a 3-circuit in the 1-skeleton. But in a metric flag complex every 3-circuit of perimeter $< 2\pi$ is the boundary of a 2-simplex. Therefore, it cannot be a closed geodesic, a contradiction. □

## The Nerve of an Almost Negative Matrix

Let $S$ be a finite set. Following [221], call a symmetric $S \times S$ matrix $(c_{st})$ *almost negative* if each entry on the diagonal is $> 0$ and each off-diagonal matrix is $\leqslant 0$. An almost negative matrix is *normalized* if each diagonal entry is 1. Any almost negative matrix $C$ can be normalized to $C' := DCD$ where $D := (d_{st})$ is the diagonal matrix defined by $d_{ss} := 1/\sqrt{c_{ss}}$. $C'$ is the *normalization* of $C$. Suppose $C = (c_{st})$ is a normalized almost negative matrix. Note that $C$ is positive definite if and only if it is the cosine matrix associated to the set of edge lengths of a spherical simplex (as in (I.9)). For each $T \subset S$ let $C_T$ be the $T \times T$ submatrix obtained by considering only the entries indexed by $T \times T$. We associate to a normalized almost negative matrix $C$ a piecewise spherical metric flag complex $L$ $(= L(C))$ as follows: $\mathrm{Vert}(L) := S$ and a nonempty subset $T$ of $S$ is the vertex set of a simplex $\sigma_T$ in $L$ if and only if $C_T$ is positive definite. $L(C)$ is called the *nerve* of $C$. The piecewise spherical structure is defined by identifying $\sigma_T$ with the spherical simplex with cosine matrix $C_T$. In particular, distinct vertices $s$ and $t$ are connected by an edge if and only if $0 \geqslant c_{st} > -1$ and if this is the case, the length $l_{st}$ of the edge is $\cos^{-1}(c_{st})$.

A symmetric $T \times T$ matrix $(c_{st})$ is *decomposable* if there is a nontrivial partition $T = T' \cup T''$ so that $c_{st} = 0$ for all $s \in T'$, $t \in T''$. If $C_T$ is decomposable, write $C_T = C_{T'} \oplus C_{T''}$. Note that if $C_T$ has such a decomposition, then $L(C_T)$ is the spherical join $L(C_{T'}) * L(C_{T''})$.

## Extra Largeness

Besides proving Lemma I.7.4, Moussong established the following generalization of the no $\square$-condition (Lemma I.6.5).

**LEMMA I.7.6.** (Moussong [221, Lemma 10.3].) *Suppose $C$ is an $S \times S$ normalized almost negative matrix. Then any closed geodesic of length $2\pi$ is contained in a subcomplex of the form $L(C_T)$, for some $T \subset S$, where either $C_T$*

*(a) is indecomposable and positive semidefinite of rank equal to* $\mathrm{Card}(T) - 1$, *with* $\mathrm{Card}(T) \geqslant 3$ *or*

*(b) decomposes as $C_T = C_{T'} \oplus C_{T''}$, with neither $C_{T'}$ nor $C_{T''}$ positive definite.*

Any metric flag complex with simplices of size $\geqslant \pi/2$ arises as the nerve of a normalized almost negative matrix. Indeed, given such an $L$, set

$$c_{st} := \begin{cases} 1 & \text{if } s = t, \\ \cos(l_{st}) & \text{if } \{s,t\} \in \mathrm{Edge}(L), \\ -1 & \text{if } \{s,t\} \notin \mathrm{Edge}(L), \end{cases} \tag{I.10}$$

where $l_{st}$ denotes the length of the edge spanned by $\{s.t\}$. The matrix $(c_{st})$ is almost negative (since $l_{st} \geqslant \pi/2$ whenever $\{s,t\} \in \mathrm{Edge}(L)$). Moreover, $L$ is the nerve of this matrix.

Given a subset $T$ of vertices of a piecewise spherical complex $L$, let $L_T$ denote the full subcomplex spanned by $T$.

**COROLLARY I.7.7.** *Suppose L is a piecewise spherical metric flag complex with simplices of size $\geqslant \pi/2$. Then L is extra large if and only if for all* $T \subset \mathrm{Vert}(L)$ *neither of the following conditions holds:*

*(a)* $L_T$ *is isometric to the polar dual of a Euclidean simplex of dimension* $\geqslant 2$.

*(b)* $T = T' \cup T''$ *and* $L_T$ *is isometric to the spherical join* $L_{T'} * L_{T''}$, *where neither* $T'$ *nor* $T''$ *is the vertex set of a simplex in L.*

*Proof.* We deduce this from Lemma I.7.6. Suppose $L$ is associated to the normalized almost negative matrix $C = (c_{st})$. Let $U \subset S$ be such that $C_U$ is indecomposable, positive semidefinite of corank 1 and $\mathrm{Card}(U) \geqslant 3$. Let $T \subset U$ be a minimal subset such that $C_T$ is not positive definite. Then $\mathrm{Card}(T) \neq 2$. (Otherwise, $C_T = \left(\begin{smallmatrix} 1 & -1 \\ -1 & 1 \end{smallmatrix}\right)$. By calculating a $3 \times 3$ determinant we see that any $3 \times 3$ positive semidefinite matrix containing this matrix as a $2 \times 2$ minor must have all other off-diagonal entries equal to 0. So, if $C_T$ if $2 \times 2$, $C_U$ is decomposable.) $T$ has the property that for every nonempty, proper subset $T'$ of $T$, $C_{T'}$ is positive definite. By Proposition 6.8.8, $C_T$ is the Gram matrix of a Euclidean simplex, i.e., $L_T$ is polar dual of this same simplex. Thus, condition (a) of the corollary is equivalent to condition (a) of Lemma I.7.6. □

## I.8. THE VISUAL BOUNDARY OF A CAT(0)-SPACE

Here we explain how to adjoin a space $\partial X$ of "ideal points" to a complete CAT(0)-space $X$ obtaining $\overline{X} = X \cup \partial X$. When $X$ is locally compact, $\overline{X}$ will be a compactification of $X$. We give several equivalent definitions of $\partial X$.

Fix a base point $x_0 \in X$. The rough idea is that $\overline{X}$ is formed by adding an "endpoint" $c(\infty)$ to each geodesic ray $c : [0, \infty) \to X$ beginning at $x_0$. $\partial X$ is the set of such endpoints. In other words, $\partial X$ is the set of geodesic rays emanating from $x_0$. The topology on $\overline{X}$ is the "inverse limit topology" (called the "cone topology" in [37]). It can be described as follows. Consider the system of closed balls centered at $x_0$, $\{\overline{B}(x_0, r)\}_{r \in [0,\infty)}$. For each $s > r$, there is a retraction $p_{s,r} : \overline{B}(x_0, s) \to \overline{B}(x_0, r)$, defined as the "nearest point projection." In other words, if $c$ is a geodesic segment starting from $x_0$, then $p_{s,r}$ takes $c(t)$ to itself when $0 \leqslant t \leqslant r$ and to $c(r)$ when $r < t \leqslant s$. This gives an inverse system of balls. (See Appendix G.1 for a discussion of inverse systems.) The inverse limit is denoted $\varprojlim \overline{B}(x_0, r)$. A point in $\varprojlim \overline{B}(x_0, r)$ can be identified with a map $c : [0, \infty) \to X$ such that $c(0) = x_0$ and $p_{s,r}(c(s)) = c(r)$ for all $s > r$. There are two types of such maps: either $c(s) \neq c(r)$ for all $s > r$ (in which case $c$ is a geodesic ray) or else there is a minimum value $r_0$ such that $c(s) = c(r_0)$ (in which case $c$ is a geodesic on $[0, r_0]$ followed by the constant map on $[r_0, \infty)$). There is an obvious bijection $\phi : \overline{X} \to \varprojlim \overline{B}(x_0, r)$ which associates to $x \in X$ the map $c_x : [0, \infty) \to X$ whose restriction to $[0, d(x_0, x)]$ is the geodesic segment from $x_0$ to $x$ and which is constant on $[d(x_0, x), \infty)$. The inverse limit has a natural topology (as a subspace of the direct product). It is compact if $X$ is locally compact. The map $\phi$ takes $X$ homeomorphically onto an open dense subset of the inverse limit. So, it makes sense to topologize $\overline{X}$ by declaring $\phi$ to be a homeomorphism.

**Examples I.8.1.** If $X = \mathbb{E}^n$ or $\mathbb{H}^n$, then $\overline{X}$ is homeomorphic to the $n$-disk $D^n$ and $\partial X$ to $S^{n-1}$ (the "sphere at infinity").

The problem with the previous definition is its dependence on the choice of base point $x_0$. This is remedied as follows. Call two geodesics $c : [0, \infty) \to X$ and $c' : [0, \infty) \to X$ with possibly different base points $x_0$ and $x'_0$, *parallel* if there exists a constant $C$ such that $d(c(t), c'(t)) \leqslant C$ for all $t$. This is an equivalence relation on the set of geodesic rays. The equivalence class of $c$ is denoted $c(\infty)$. For any choice of $x_0$, the equivalence class of any ray has a unique representative emanating from $x_0$. (This follows from the fact that $X$ is CAT(0), cf. [37, p. 261].) Moreover, different choices of base point yield the same topology on $\overline{X}$ ([37, p. 264]). It is often better to regard $\partial X$ as the space of parallel classes of geodesic rays in $X$. For example, from this point of view, it is clear that the action of Isom($X$) on $X$ extends to the boundary.

### Horofunctions

Let $C(X)$ be the vector space of real-valued, continuous functions on $X$, equipped with the topology of uniform convergence on bounded sets. $\mathcal{C}(X)$ denotes its quotient by the 1-dimensional subspace of constant functions. The equivalence class of a continuous function $f$ is denoted $\overline{f}$. Define an embedding $\iota : X \hookrightarrow \mathcal{C}(X)$ by $\iota(x) := \overline{d_x}$, where $d_x(y) := d(x, y)$. The closure of $\iota(X)$ in $\mathcal{C}(X)$ is denoted $\widehat{X}$. The following theorem is proved in [37, pp. 267–271].

**THEOREM I.8.2.** *The map $\iota : X \to \widehat{X}$ extends to a homeomorphism $\overline{X} \to \widehat{X}$.*

A function $h \in C(X)$ is a *horofunction* if its image in $\mathcal{C}(X)$ lies in $\widehat{X} - \iota(X)$. Suppose $c : [0, \infty) \to X$ is a geodesic ray. The *Busemann function* associated to $c$ is the function $b_c \in C(X)$ defined by

$$b_c(x) := \lim_{t \to \infty} (d(x, c(t)) - t).$$

Its image $\overline{b_c}$ in $\mathcal{C}(X)$ depends only on the parallel class of $c$. The content of Theorem I.8.2 is that every horofunction is the Busemann function of a unique parallel class of geodesic ray.

## Z-Set Compactications

A compact Hausdorff space $Y$ is a *Euclidean neighborhood retract* (an "ENR" for short) if it is finite dimensional and locally contractible. A closed subset $Z$ of an ENR $Y$ is a *Z-set* if for every open subset $U \subset Y$, the inclusion $U - Z \hookrightarrow U$ is a homotopy equivalence. $Y$ is called a *Z-set compactification* of $Y - Z$. The standard example of this is where $Y$ is a compact manifold with boundary and $Z$ is a closed subset of its boundary.

**THEOREM I.8.3.** *Suppose $X$ is a finite-dimensional, complete, locally compact,* CAT(0)*-space. Then*

*(i) $\partial X$ is a Z-set in $\overline{X}$.*

*(ii) $H^*_c(X) \cong H^*(\overline{X}, \partial X) \cong \check{H}^{*-1}(\partial X)$, where $\check{H}^*(\ )$ means reduced Cěch-cohomology.*

*Proof.* (i) Given $t \in (0, \infty)$, let $R_t : [0, \infty] \to [0, t]$ be the obvious retraction. Given a continuous function $f : \partial X \to (0, \infty)$, let $r_f : \overline{X} \to X$ be the map which retracts each geodesic ray $c$ starting at $x_0$ onto the restriction of the ray to $[0, f(c(\infty))]$. This is a deformation retraction onto $r_f(\overline{X})$. Given an open set $U \subset \overline{X}$, we can choose $f$ so that $r_f : U \to U - \partial X$ is a homotopy equivalence. It follows that $\overline{X}$ is locally contractible (since any CAT(0)-space is locally contractible) and that $\partial X$ is a $Z$-set.

(ii) Since $\overline{X}$ is compact and $\partial X$ is a $Z$-set, it follows from the definitions in Appendix G.2 that $H^*_c(X) = H^*(\overline{X}, \partial X)$ and $H^*_e(X) = \check{H}^*(\partial X)$, where $H^*_e(X)$ denotes the end cohomology of $X$. (Compare Theorem G.2.2.) Since $\overline{X}$ is contractible, the long exact sequence of the pair gives $H^*(\overline{X}, \partial X) \cong \check{H}^{*-1}(\partial X)$. □

## Polyhedra with Isolated PL Singularities

In this paragraph $\Lambda$ is an $\mathbb{X}_\kappa$-polyhedral complex and a complete CAT($\kappa$)-space. Further suppose that the link of each vertex is a PL $n$-manifold. To simplify notation, for each $v \in \mathrm{Vert}(\Lambda)$, write $L(v)$ instead of $\mathrm{Lk}(v, \Lambda)$. Choose a base point $x_0$ not in $\mathrm{Vert}(\Lambda)$. In the next theorem we consider the inverse system of metric spheres $\partial B(x_0, r)$ centered at $x_0$.

**THEOREM I.8.4.** ([83].) *With hypotheses as above, suppose $v_1, \dots, v_m$ are the vertices of $\Lambda$ which lie in $B(x_0, r)$. Then*

*(i) $\partial B(x_0, r) = L(v_1)\sharp \cdots \sharp L(v_m)$, where the symbol $\sharp$ means connected sum.*

*(ii) The inverse system of metric spheres is pro-isomorphic to the inverse system $\{L(v_1)\sharp \cdots \sharp L(v_m)\}$, where the bonding maps are the obvious*

*ones which collapse those factors in the domain which are not in the range. Hence,* $\partial\Lambda = \varprojlim(L(v_1)\sharp\cdots\sharp L(v_m))$.

(iii) *Suppose* $\Lambda$ *is a* $(n+1)$*-dimensional PL manifold (i.e., suppose the link of each of its vertices is PL homeomorphic to* $S^n$*). Then its compactification* $\overline{\Lambda}$ *is homeomorphic to* $D^{n+1}$ *and* $\partial\Lambda$ *is homeomorphic to* $S^n$.

*Sketch of proof.* A corollary to Theorem I.2.5 is that in a CAT(1)-space, any open ball of radius $\leqslant \pi$ is contractible. In the case at hand, the CAT(1)-space is a link of the form $L(v)$ and hence, is a PL $n$-manifold. So, if we extend the statement of the theorem to $\kappa > 0$ by making the obvious modification for open balls of radius $\leqslant \pi/\sqrt{\kappa}$, then, by induction on dimension, we can assume that for each $v \in \mathrm{Vert}(\Lambda)$, $u \in L(v)$ and $r < \pi$, we have proved $\overline{B}(u,r) \cong D^n$ (where $\overline{B}(u,r)$ means the closed ball in $L(v)$). Moreover, $B(u,\pi) \cong \mathbb{R}^n$.

Define $\rho : \Lambda \to [0,\infty)$ by $\rho(x) := d(x_0, x)$. The level set $\rho^{-1}(r)$ is the metric sphere $\partial B(x_0, r)$. This is analogous to the usual picture in Morse theory. The vertices of $\Lambda$ play the role of the critical points of $\rho$. The topological type of a level set does not change if there is no vertex in the region between $\rho^{-1}(r)$ and $\rho^{-1}(r+\varepsilon)$; however, when the level set crosses a vertex $v$, the effect is to remove an $n$-disk of the form $\overline{B}(u,r)$ for some $u \in L(v)$ and $r < \pi$ and replace it with $L(v) - B(u,r)$. In other words, the level set changes by taking connected sum with $L(v)$. This is the argument for (i). Statement (ii) follows. As for (iii), if $\Lambda$ is a PL $(n+1)$-manifold, then each $L(v)$ is PL homeomorphic to $S^n$. By (i) and (ii) each metric sphere is homeomorphic to $S^n$ and $\partial\Lambda \cong S^n$. Similarly, one can see directly that $\overline{\Lambda} \cong D^{n+1}$. □

*Remark I.8.5.* The notion of a "word hyperbolic group" was defined in 12.5. Given a word hyperbolic group $\Gamma$, Gromov [147] showed how to associate a compact metric space $\partial\Gamma$. If $\Gamma$ acts isometrically and cocompactly on a CAT(0)-space $X$, then $\partial\Gamma$ is homeomorphic in $\partial X$. (In particular, if $\Gamma$ acts this way on two different CAT(0)-spaces, their boundaries are homeomorphic.) More generally, it is a result of Bestvina and Mess [25] that the Rips complex of $\Gamma$ has an equivariant $Z$-set compactification formed by adjoining $\partial\Gamma$.

## NOTES

The basic reference for the material in this appendix is the book [37] of Bridson and Haefliger. Their notation is somewhat different from ours. For example, they use $M_\kappa$ instead of $\mathbb{X}_\kappa$ to denote a space of constant curvature $\kappa$. The organization of this appendix is similar to that in my expository paper [90] written with G. Moussong.

**I.2.** In the 1940s and 1950s, A. D. Aleksandrov introduced the notion of a length space and the idea of curvature bounds on length spaces. He was primarily interested in lower curvature bounds (defined by reversing the CAT($\kappa$)-inequality). Aleksandrov's motivation was the following classical problem of H. Weyl: is every positvely curved Riemannian metric on $S^2$ isometric to the boundary of a convex body in $\mathbb{E}^3$? This was answered affirmatively for analytic metrics by H. Lewy. The answer was later extended to $C^k$ metrics for $k \geqslant 4$ by Nirenberg and eventually for $k \geqslant 2$ by Heinz. It was proved for $C^2$ metrics in 1942 by Aleksandrov using polyhedral techniques. Somewhat later Pogorelov proved that a geodesic metric on the 2-sphere has nonnegative curvature if and only if it is isometric to the boundary of a convex body. Pogorelov first showed that any nonnegatively curved piecewise flat metric on $S^2$ was isometric to the boundary of a convex polytope. He deduced the general result from this by using approximation techniques. (For a nice discussion of Weyl's problem including a list of further references, see [263, pp. 226–227].)

Call a smooth convex hypersurface in Minkowski space $\mathbb{R}^{n,1}$ *spacelike* if it has a spacelike normal vector at each point. (See 6.2 for the definitions of $\mathbb{R}^{n,1}$ and of a "spacelike" vector.) It follows from Gauss' equation that a spacelike, convex hypersurface in $\mathbb{R}^{n,1}$ is a nonpositively curved Riemannian manifold (e.g., $\mathbb{H}^n$ is nonpositively curved.) More generally, any piecewise Euclidean, spacelike, convex hypersurface in $\mathbb{R}^{n,1}$ is CAT(0), cf. [61].

Another early paper on nonpositively curved spaces was Busemann's [46].

Gromov's version of the Cartan–Hadamard Theorem (Theorem I.2.7) is stated in [147]. A proof can be found in Ballman's article in [138, pp. 195–196] under the additional hypothesis that $X$ is locally compact. A proof which doesn't use this hypothesis can be found in [37, Chapter II.4]; the argument there follows the one given in [3].

The CAT($\kappa$) terminology was introduced by Gromov [147]. When asked what it abbreviates, he sometimes answers "Cartan, Aleksandrov, Toponogov;" other times Cartan is replaced by "Comparison."

The proof of Proposition I.2.10 (asserting the existence of a center of a bounded set) is taken from [37, p. 179].

**I.3.** The first person to focus attention on nonpositively curved polyhedral metrics was Gromov in his seminal paper [147]. Soon afterwards, proofs of many of the basic facts about piecewise constant curvature polyhedra were given by Moussong in his thesis [221].

**I.4.** In the context of Riemannian manifolds, the Flat Torus Theorem and the Solvable Subgroup Theorem were proved independently by Gromoll and Wolf [143] and by Lawson and Yau [186]. In [147] Gromov states that both results hold for CAT(0) spaces. As indicated in our exposition, proofs were provided by Bridson and Haefliger in [37].

**I.5.** The boundary of an $n$-dimensional spherical polytope $P$ is a piecewise spherical structure on $S^{n-1}$ and it satisfies the "reverse CAT(1)-inequality" (as does the boundary of any convex body in $\mathbb{S}^n$). In particular, the boundary of the polar dual of $P$ satisfies the reverse CAT(1)-inequality. Given an $n$-dimensional Euclidean polytope $P \subset \mathbb{E}^n$, its *polar dual* is the cellulation of $\mathbb{S}^{n-1}$ obtained by taking all normal vectors to $\partial P$. (The $(n-1)$-cells of this cellulation are the polar duals of the spherical polytopes of the form $\mathrm{Lk}(v, P)$, $v \in \mathrm{Vert}(P)$.) Similarly, the polar dual of a polytope $P$ in $\mathbb{H}^n$ is a piecewise spherical structure on $S^{n-1}$. It is proved in [58] that it is CAT(1). (In fact, it is "extra large" in the sense of the definition following Theorem I.3.5.) Conversely, Rivin and Hodgson [246] proved that a piecewise spherical structure on $S^2$ was CAT(1) and extra large if and only if it was the polar dual of a polytope in $\mathbb{H}^3$.

**I.6.** During the last fifteen years nonpositively curved cubical complexes have become ubiquitous in geometric group theory e.g., see [24, 59, 88, 90, 91, 187, 228, 251]. There are two reasons for this: first, along with simplicial complexes, cubical complexes are the most likely to occur in nature and second, Gromov's Lemma I.6.1 (or more precisely, Corollary I.6.3) gives a combinatorial condition for nonpositive curvature (links are flag complexes) which is easy to check.

Corollary I.6.3 has many further applications. For example, as explained in [83] and [236], it is the reason why Gromov's "hyperbolization procedures" produce nonpositively curved spaces. It is also the basic mechanism behind the proofs in [84, 85] that various blowups of real projective space are nonpositively curved.

**I.7.** The basic reference for Moussong's Lemma is his Ph.D. thesis, [221]. Our proof here closely follows the exposition in [90, pp. 64–65]. The best example of a metric flag complex is the nerve of a Coxeter system $(W, S)$ with its natural piecewise spherical structure and the best application of Moussong's Lemma is Moussong's Theorem 12.3.3, which states that $\Sigma(W, S)$ is CAT(0). Similarly, the most important application of Lemma I.7.6 is Moussong's result (Corollary 12.6.3) on the word hyperbolicity of $W$. These results are explained in Chapter 12.

**I.8.** A good reference for this material is [37, Chapter II.8]. Busemann functions and horofunctions are discussed in [15]. Other references include [23, 83, 104]. Most of Theorem I.8.4 comes from [83]. Case (iii) of this theorem is similar to a result proved earlier by D. Stone [271]: he showed that when $\Lambda$ is a PL $(n+1)$-manifold, it is PL homeomorphic to $\mathbb{R}^{n+1}$.

In [23] Bestvina discusses the notion of a "Z-set compactification" of a group $\Gamma$. This means a proper action of $\Gamma$ on a contractible ENR $X$ (i.e., an "ER") together with a $Z$-set compactification to $\overline{X}$.

# Appendix J

# $L^2$-(CO)HOMOLOGY

## J.1. BACKGROUND ON VON NEUMANN ALGEBRAS

Suppose $H$ is a Hilbert space with inner product $\langle\, ,\rangle$. Let $\mathcal{L}(H)$ be the algebra of bounded linear operators on $H$, i.e., all continuous linear endomorphisms $\varphi : H \to H$. The *operator norm* of $\varphi$, denoted $\|\varphi\|$, is the supremum over all $x$ in the unit ball of $H$ of $\|\varphi(x)\|$. The topology on $\mathcal{L}(H)$ defined by this norm is called the *uniform topology*, since with this topology a sequence converges if and only if it converges uniformly on all bounded subsets of $H$. The *weak topology* on $\mathcal{L}(H)$ is defined as follows. Suppose $\mathcal{E} = \{(x_1, y_1), \dots (x_n, y_n)\}$ is a finite collection of pairs in $H$. Let $U_{\mathcal{E}}$ denote the set of $\varphi \in \mathcal{L}(H)$ such that $\langle\varphi(x_i), y_i\rangle < 1$, for $i = 1, \dots, n$. The weak topology is defined so that $\{U_{\mathcal{E}}\}$ is a neighborhood basis of 0 (and so that translation is a homeomorphism). It is the weakest topology in which the maps $\mathcal{L}(H) \to \mathbb{C}$ defined by $\varphi \to \langle\varphi(x), y\rangle$ are continuous for all $x, y \in H$.

A unital subalgebra $\mathcal{A} \subset \mathcal{L}(H)$ is a $*$-*algebra* if it is closed under taking adjoints. (*Unital* means that it contains the identity operator.) It is a $C^*$-*algebra* if, in addition, it is closed in the uniform topology. It is a *von Neumann algebra* if it is also closed in the weak topology. For any $*$-algebra $\mathcal{A} \subset \mathcal{L}(H)$, its *commutant* $\mathcal{A}'$ is the set of all $\varphi \in \mathcal{L}(H)$ which commute with $\mathcal{A}$; its *bicommutant* is $\mathcal{A}''$. Obviously, $\mathcal{A} \subset \mathcal{A}''$. Basic facts are that $\mathcal{A}''$ is a von Neumann algebra and that $\mathcal{A}$ is a von Neumann algebra if and only if $\mathcal{A} = \mathcal{A}''$.

The center of $\mathcal{A}$ always contains the scalar multiples of the identity; if there is nothing else, we say it has *trivial center*. A von Neumann algebra is a *factor* if it has trivial center. Every von Neumann algebra is a direct integral of factors.

## J.2. THE REGULAR REPRESENTATION

Suppose $\Gamma$ is a countable discrete group. $L^2(\Gamma)$ denotes the vector space of square-summable, real-valued functions on $\Gamma$, i.e.,

$$L^2(\Gamma) := \left\{ f : \Gamma \to \mathbb{R} \;\middle|\; \sum f(\gamma)^2 < \infty \right\}.$$

It is a Hilbert space with inner product $\langle\ ,\ \rangle$ defined by

$$\langle f, f' \rangle := \sum_{\gamma \in \Gamma} f(\gamma) f'(\gamma). \tag{J.1}$$

The group algebra $\mathbb{R}\Gamma$ can be identified with the dense subspace of $L^2(\Gamma)$ consisting of the functions with finite support. For each $\gamma \in \Gamma$, $e_\gamma$ denotes the characteristic function of $\{\gamma\}$, i.e.,

$$e_\gamma(\gamma') := \begin{cases} 1 & \text{if } \gamma = \gamma', \\ 0 & \text{otherwise.} \end{cases}$$

So $\{e_\gamma\}_{\gamma \in \Gamma}$ is an orthonormal basis for $L^2(\Gamma)$ in the sense that it spans a dense subspace. There is a left action of $\Gamma$ on $L^2(\Gamma)$ defined by right translation:

$$(\gamma \cdot f)(\gamma') := f(\gamma' \gamma).$$

This action is called the left *regular representation* of $\Gamma$. Similarly, left translation defines the right regular representation. Both actions are orthogonal with respect to the inner product (J.1). The left $\Gamma$-action gives $L^2(\Gamma)$ the structure of a left $\mathbb{R}\Gamma$-module and similarly, for the right action. Given $x = \sum x_\gamma e_\gamma \in \mathbb{R}\Gamma$, put $|x| := \sum |x_\gamma|$. Then, for any $f \in L^2(\Gamma)$,

$$\|x \cdot f\| \leqslant \sum_{\gamma \in \mathbb{R}\Gamma} |x_\gamma| \, \|\gamma \cdot f\| = |x| \, \|f\|, \tag{J.2}$$

where $\|f\| := \langle f, f \rangle$ is the $L^2$ norm. In other words, the left action of $x$ on $L^2(\Gamma)$ is a bounded operator and similarly, for the right action. The adjoint of translation by $e_\gamma$ is translation by $e_{\gamma^{-1}}$. So we have proved the following.

**Lemma J.2.1.** *Both the left and right actions make* $\mathbb{R}\Gamma$ *into a* $*$*-algebra of bounded linear operators on* $L^2(\Gamma)$.

Let $(\mathbb{R}\Gamma)^n$ (resp. $((L^2(\Gamma))^n)$ denote the direct sum of $n$ copies of $\mathbb{R}\Gamma$ (resp. of $L^2(\Gamma)$). Using (J.2), we get the following.

**Lemma J.2.2.** ([113, Lemma 2.2.1, p. 190].) *Let* $\varphi : (\mathbb{R}\Gamma)^n \to (\mathbb{R}\Gamma)^m$ *be a morphism of free left* $\mathbb{R}\Gamma$*-modules (i.e., a* $\Gamma$*-equivariant linear map). Then the induced map* $\Phi := L^2(\Gamma) \otimes_{\mathbb{R}\Gamma} \varphi : (L^2(\Gamma))^n \to (L^2(\Gamma))^m$ *is bounded.*

*Proof.* We can represent $\varphi$ by right multiplication by the matrix $X = (x_{ij})$, with $x_{ij} \in \mathbb{R}\Gamma$. In other words, $\varphi$ takes the row vector $\mathbf{a} = (a_1, \ldots, a_n)$ to $\mathbf{a} \cdot X$.

Then $\Phi$ takes the row vector $\mathbf{f} = (f_1, \ldots, f_n) \in (L^2(\Gamma))^n$ to $\mathbf{f} \cdot X$. So,

$$\|\mathbf{f} \cdot X\|^2 = \sum_j \left\| \sum_i f_i x_{ij} \right\|^2 \leqslant \sum_{i,j} |x_{ij}|^2 \|f_i\|^2$$
$$\leqslant \left( \sum_{i,j} |x_{ij}|^2 \right) \|\mathbf{f}\|^2,$$

which means $\Phi$ is bounded. □

Suppose $V$ and $V'$ are Hilbert spaces with orthogonal $\Gamma$-actions. By a *map* from $V$ to $V'$ we will always mean a $\Gamma$-equivariant, bounded linear map $f : V \to V'$. The kernel of a map is a closed subspace; however, the image need not be. The map $f$ is a *weak surjection* if its image is dense in $V'$; it is a *weak isomorphism* if, in addition, it is injective. A sequence of maps

$$\cdots \longrightarrow V \xrightarrow{f} V' \xrightarrow{g} V'' \cdots$$

is *weakly exact* at $V'$ if $\overline{\operatorname{Im} f} = \operatorname{Ker} g$.

### Hilbert $\Gamma$-Modules

A Hilbert space with orthogonal $\Gamma$-action is a *Hilbert $\Gamma$-module* if it is isomorphic to a closed $\Gamma$-stable subspace of a finite (orthogonal) direct sum of copies of $L^2(\Gamma)$ with the diagonal $\Gamma$-action. (In the literature such a Hilbert space with orthogonal $\Gamma$-action is sometimes called a "finitely generated" Hilbert $\Gamma$-module.)

**Example J.2.3.** If $F$ is a finite subgroup of $\Gamma$, then $L^2(\Gamma/F)$, the space of square summable functions on $\Gamma/F$, can be identified with the subspace of $L^2(\Gamma)$ consisting of the square summable functions on $\Gamma$ which are constant on each coset. This subspace is clearly closed and $\Gamma$-stable; hence, $L^2(\Gamma/F)$ is a Hilbert $\Gamma$-module.

**Example J.2.4.** (*The completed tensor product*) Suppose $\Gamma = \Gamma_1 \times \Gamma_2$ and for $j = 1, 2$, that $V_j$ is a Hilbert $\Gamma_j$-module. The $L^2$-completion of the tensor product is denoted $V_1 \widehat{\otimes} V_2$. It is a Hilbert $\Gamma$-module.

Next we want to state a mild generalization of Lemma J.2.2 which will be needed in the next section. Suppose

$$E = \bigoplus_{i=1}^{n} \mathbb{R}(\Gamma/F_i) \quad \text{and} \quad E' = \bigoplus_{j=1}^{m} \mathbb{R}(\Gamma/F'_j),$$

where $F_i$ and $F'_j$ are finite subgroups of $\Gamma$. Then $V = L^2(\Gamma) \otimes_{\mathbb{R}\Gamma} E$ and $V' = L^2(\Gamma) \otimes_{\mathbb{R}\Gamma} E'$ are Hilbert $\Gamma$-modules.

**LEMMA J.2.5.** *With notation as above, let $\varphi : E \to E'$ be a morphism left $\mathbb{R}\Gamma$-modules. Then the induced map $\Phi := L^2(\Gamma) \otimes_{\mathbb{R}\Gamma} \varphi : V \to V'$ is bounded and hence is a map of Hilbert $\Gamma$-modules.*

*Proof.* Let $p : (\mathbb{R}\Gamma)^n \to E$ be orthogonal projection and $\iota : E' \to (\mathbb{R}\Gamma)^m$ inclusion. Then $\iota \circ \varphi \circ p : (\mathbb{R}\Gamma)^n \to (\mathbb{R}\Gamma)^m$ is $\Gamma$-equivariant and hence can be represented by an $(n \times m)$-matrix with coefficients in $\mathbb{R}\Gamma$ as in the proof Lemma J.2.2. As before, this same matrix represents $\Phi$; so, $\Phi$ is bounded. □

An isomorphism of finite-dimensional Hilbert spaces can be factored as the composition of a self-adjoint map and an isometry ("polar decomposition"). This fact can be promoted to the following.

**LEMMA J.2.6.** ([113, Lemma 2.5.3, p. 194].) *If two Hilbert $\Gamma$-modules are weakly isomorphic, then they are $\Gamma$-isometric.*

*Proof.* Suppose $f : V_1 \to V_2$ is a weak isomorphism. Then $f^* \circ f : V_1 \to V_1$ is self-adjoint, positive definite, and has dense image. Put

$$g = \sqrt{f^* \circ f}.$$

Then $g$ is self-adjoint, positive definite, and $\mathrm{Im}(g) \supset \mathrm{Im}(f^* \circ f)$. Put $h := f \circ g^{-1} : \mathrm{Im}(g) \to V_2$. We compute

$$\begin{aligned}\langle h(x), h(y)\rangle &= \langle f^* \circ (f \circ g^{-1})(x), g^{-1}(y)\rangle = \langle (g^2 \circ g^{-1})(x), g^{-1}(y)\rangle \\ &= \langle (g \circ g^{-1})(x), (g^* \circ g^{-1})(y)\rangle = \langle x, y\rangle;\end{aligned}$$

so $h : \mathrm{Im}(g) \to \mathrm{Im}(f)$ is an isometry. Hence, it extends to an isometry $V_1 \to V_2$, which we also denote by $h$. We easily check that since $f$ and $f^*$ are $\Gamma$-equivariant so are $g$ and $h$. So $h$ is the required isometry of Hilbert $\Gamma$-modules. □

### Induced Representations

Let $H$ be a subgroup of $\Gamma$ and $V$ a Hilbert $H$-module. The *induced representation* $\mathrm{Ind}_H^\Gamma(V)$ is the $L^2$-completion of $\mathbb{R}\Gamma \otimes_{\mathbb{R}H} V$. Alternatively, it is the vector space of all square summable sections of the vector bundle $\Gamma \times_H V \to \Gamma/H$. ($\Gamma/H$ is discrete.) The induced representation is obviously a Hilbert space with orthogonal $\Gamma$-action. Denote by $L^2(H)^n$ the direct sum of $n$ copies of $L^2(H)$. If $V$ is a closed subspace of $L^2(H)^n$, then $\mathrm{Ind}_H^\Gamma(V)$ is a closed subspace of $L^2(\Gamma)^n$. (This follows from the observation that $\mathrm{Ind}_H^\Gamma(L^2(H))$ can be identified with $L^2(\Gamma)$.) Thus, $\mathrm{Ind}_H^\Gamma(V)$ is a Hilbert $\Gamma$-module. For example, if $F$ is a finite subgroup of $\Gamma$ and $\mathbb{R}$ denotes the trivial 1-dimensional representation of $F$, then $\mathrm{Ind}_F^\Gamma(\mathbb{R})$ can be identified with $L^2(\Gamma/F)$.

### The Von Neumann Algebra $\mathcal{N}(\Gamma)$

$L^2(\Gamma)$ is an $\mathbb{R}\Gamma$-bimodule. Here are three equivalent definitions of the von Neumann algebra $\mathcal{N}(\Gamma)$ associated to the group algebra $\mathbb{R}\Gamma$.

- $\mathcal{N}(\Gamma)$ is the algebra of all maps from $L^2(\Gamma)$ to itself. (Recall that "map" means a bounded linear operator which is equivariant with respect to the left $\Gamma$-action.)
- $\mathcal{N}(\Gamma)$ is the double commutant of the right $\mathbb{R}\Gamma$-action on $L^2(\Gamma)$.
- $\mathcal{N}(\Gamma)$ is the weak closure of the algebra of operators $\mathbb{R}\Gamma$ acting from the right on $L^2(\Gamma)$.

There is also a version of this von Neumann algebra acting from the left. We also denote it $\mathcal{N}(\Gamma)$.

### The $\Gamma$-Trace

The $\Gamma$-*trace* of an element $\varphi \in \mathcal{N}(\Gamma)$ is defined by

$$\mathrm{tr}_\Gamma(\varphi) := \langle \varphi(e_1), e_1 \rangle \tag{J.3}$$

(where $e_1 \in L^2(\Gamma)$ is the basis element corresponding to the identity element of $\Gamma$). The restriction of $\mathrm{tr}_\Gamma$ to the subalgebra $\mathbb{R}\Gamma \subset \mathcal{N}(\Gamma)$ is the classical *Kaplansky trace*.

Let $M_n(\mathcal{N}(\Gamma))$ denote the set of $n \times n$ matrices with coefficients in $\mathcal{N}(\Gamma)$. Given $\Phi = (\varphi_{ij}) \in M_n(\mathcal{N}(\Gamma))$, define

$$\mathrm{tr}_\Gamma(\Phi) := \sum_{i=1}^{n} \mathrm{tr}_\Gamma(\varphi_{ii}). \tag{J.4}$$

The next lemma states that $\mathrm{tr}_\Gamma(\ )$ has the expected properties. It is proved by standard arguments.

**Lemma J.2.7.** (Compare [113, pp. 199–200].) *Suppose* $\Phi, \Psi \in M_n(\mathcal{N}(\Gamma))$. *Then*

*(i)* $\mathrm{tr}_\Gamma(\Phi \circ \Psi) = \mathrm{tr}_\Gamma(\Psi \circ \Phi)$.

*(ii)* $\mathrm{tr}_\Gamma(\Phi) = \mathrm{tr}_\Gamma(\Phi^*)$ *(where* $\Phi^*$ *is the adjoint of* $\Phi$ *with respect to the inner product* $\langle\, , \rangle$*).*

*(iii) Suppose* $\Phi$ *is self-adjoint and idempotent. Then* $\mathrm{tr}_\Gamma(\Phi) \geqslant 0$ *with equality if and only if* $\Phi = 0$.

Similarly, given a Hilbert $\Gamma$-module $V$ isomorphic to $L^2(\Gamma)^n$ and a self-map $\Phi$ of $V$, by choosing an identification $V \cong L^2(\Gamma)^n$, one can define $\mathrm{tr}_\Gamma(\Phi)$ by (J.4). The resulting number is independent of the choice of isomorphism $V \cong L^2(\Gamma)^n$.

**Von Neumann Dimension**

Let $V$ be a Hilbert $\Gamma$-module. Choose an embedding of $V$ as a closed, $\Gamma$-stable subspace of $L^2(\Gamma)^n$ for some $n \in \mathbb{N}$ and let $p_V : L^2(\Gamma)^n \to L^2(\Gamma)^n$ be orthogonal projection onto $V$. Then $\dim_\Gamma(V)$, the *von Neumann dimension* of $V$ (also called its $\Gamma$*-dimension*), is defined by

$$\dim_\Gamma(V) := \mathrm{tr}_\Gamma(p_V). \tag{J.5}$$

If $E \subset L^2(\Gamma)^n$ is a not necessarily closed, $\Gamma$-stable subspace, put $\dim_\Gamma(E) := \dim_\Gamma(\overline{E})$. We list some properties of $\dim_\Gamma(V)$. Proofs can be found in [113].

**PROPOSITION J.2.8.** *Suppose $V, W$ are Hilbert $\Gamma$-modules.*

*(i)* $\dim_\Gamma(V) \in [0, \infty)$.

*(ii)* $\dim_\Gamma(V) = 0$ *if and only if* $V = 0$.

*(iii) If $\Gamma$ is the trivial group (so that $V$ is finite dimensional), then* $\dim_\Gamma(V) = \dim(V)$.

*(iv)* $\dim_\Gamma(L^2(\Gamma)) = 1$.

*(v)* $\dim_\Gamma(V \oplus W) = \dim_\Gamma(V) + \dim_\Gamma(W)$.

*(vi) If $f : V \to W$ is a map of Hilbert $\Gamma$-modules, then*

$$\dim_\Gamma(V) = \dim_\Gamma(\mathrm{Ker} f) + \dim_\Gamma(\mathrm{Im} f).$$

*(vii) If $f : V \to W$ is a map of Hilbert $\Gamma$-modules and $f^* : W \to V$ its adjoint, then* $\mathrm{Ker} f$ *and* $\mathrm{Im} f^*$ *are orthogonal complements in $V$. Hence,*

$$\dim_\Gamma(V) = \dim_\Gamma(\mathrm{Ker} f) + \dim_\Gamma(\mathrm{Im} f^*).$$

*So by* (vi), $\dim_\Gamma(\mathrm{Im} f) = \dim_\Gamma(\mathrm{Im} f^*)$.

*(viii) By* (v) *and* (vi), *if* $0 \to V_n \to \cdots \to V_0 \to 0$ *is a weak exact sequence of Hilbert $\Gamma$-modules, then*

$$\sum_{i=0}^{n} (-1)^i \dim_\Gamma(V_i) = 0.$$

*(ix) If $H$ is a subgroup of finite index $m$ in $\Gamma$, then*

$$\dim_H(V) = m \dim_\Gamma(V).$$

*(x) If $\Gamma$ is finite, then*

$$\dim_\Gamma(V) = \frac{1}{|\Gamma|} \dim(V).$$

*(xi) If $H$ is a subgroup of $\Gamma$ and $W$ is a Hilbert $H$-module, then*

$$\dim_\Gamma(\mathrm{Ind}_H^\Gamma(W)) = \dim_H(W).$$

*(xii) If $F$ is a finite subgroup of $\Gamma$, then by* (x) *and the fact that $L^2(\Gamma/F)$ is identified with* $\mathrm{Ind}_F^\Gamma(\mathbb{R})$,

$$\dim_\Gamma(L^2(\Gamma/F)) = \frac{1}{|F|}.$$

*(xiii) As in Example J.2.4, suppose $\Gamma = \Gamma_1 \times \Gamma_2$ and $V_j$ is a Hilbert $\Gamma_j$-module, for $j = 1, 2$. Then*

$$\dim_\Gamma(V_1 \widehat{\otimes} V_2) = \dim_{\Gamma_1}(V_1) \dim_{\Gamma_2}(V_2).$$

**LEMMA J.2.9.** *Let $V$, $V'$ be Hilbert $\Gamma$-modules with* $\dim_\Gamma(V) = \dim_\Gamma(V')$. *Then the following statements are equivalent:*

*(a) $f$ is injective;*

*(b) $f$ is weakly surjective;*

*(c) $f$ is a weak isomorphism.*

*Proof.* Statement (c) is equivalent to the conjunction of (a) and (b). So, it suffices to show (a) and (b) are equivalent. By Proposition J.2.8 (ii), $\dim_\Gamma(\mathrm{Im}(f)) = \dim_\Gamma(V) - \dim_\Gamma(\mathrm{Ker}(f))$. So, $\dim_\Gamma(\mathrm{Ker}(f)) = 0$ if and only if $\dim_\Gamma(\mathrm{Im}(f)) = \dim_\Gamma(V) = \dim_\Gamma(V')$. □

**Example J.2.10.** *(The infinite cyclic group and Fourier series.)* In this example $L^2(\Gamma)$ will denote the Hilbert space of square summable, complex-valued (rather than real-valued) functions on $\Gamma$. When $\Gamma = \mathbf{C}_\infty$, the infinite cyclic group, the von Neumann dimension can be understood analytically. Let $L^2(S^1)$ be the Hilbert space of equivalence classes of square integrable, complex-valued functions on the circle (where two functions are *equivalent* if they differ only on a set of measure 0). An action of $\mathbf{C}_\infty$ on $L^2(S^1)$ is defined by letting $k$ times the generator act by pointwise multiplication with the function $z \to z^k$. Fourier transform gives a $\mathbf{C}_\infty$-equivariant isometry $L^2(\mathbf{C}_\infty) \cong L^2(S^1)$. The von Neumann algebra $\mathcal{N}(\mathbf{C}_\infty)$ is $\mathcal{B}(L^2(S^1))^{\mathbf{C}_\infty}$, the algebra of $\mathbf{C}_\infty$-equivariant, bounded linear operators on $L^2(S^1)$. Let $L^\infty(S^1)$ be the Banach space of equivalence classes of essentially bounded functions on $S^1$ (*essentially bounded* means bounded on the complement of a set of measure 0). Given $f \in L^2(S^1)$ we get an operator $M_f \in \mathcal{B}(L^2(S^1)^{\mathbf{C}_\infty}$ which sends $g \in L^2(S^1)$ to $g \cdot f$ (pointwise multiplication by $f$). In this way, $\mathcal{N}(\mathbf{C}_\infty)$ is identified with $L^\infty(S^1)$ and the $\Gamma$-trace, $L^\infty(S^1) \to \mathbb{C}$ (where $\Gamma = \mathbf{C}_\infty$) with

$$f \to \int_{S^1} f \, d\mu,$$

where $d\mu$ is Lebesgue measure, normalized so that the circle has length 1. A function $f$ corresponds to an idempotent in $\mathcal{N}(\mathbf{C}_\infty)$ if and only if it is the characteristic function $\chi_E$ of some measurable set $E \subset S^1$. The corresponding idempotent is projection onto the closed subspace of $L^2(S^1)$ consisting of those functions which are supported on $E$. If $V_E$ is the corresponding subspace of $L^2(\mathbf{C}_\infty)$, then $\dim_\Gamma V_E = \mu(E)$. This can be any number $\lambda$ in $[0, 1]$ (take $E$ to be an arc of length $\lambda$). It follows that there are Hilbert $C_\infty$-modules with von Neumann dimension any nonnegative real number. By Proposition J.2.8 (xi) the same holds for an arbitrary group $\Gamma$, provided it contains at least one element of infinite order.

## J.3. $L^2$-(CO)HOMOLOGY

Suppose $\Gamma$ acts properly, cellularly and cocompactly on a CW complex $X$. Cocompactness implies that there are only finitely many $\Gamma$-orbits of cells in $X$. Let $C_*(X)$ denote the usual cellular chain complex on $X$ with coefficients in $\mathbb{R}$. An element of $C_*(X)$ is a finitely supported function $\varphi$ from the set of oriented $k$-cells in $X$ to $\mathbb{R}$ satisfying $\varphi(\overline{c}) = -\varphi(c)$ (where $c$ and $\overline{c}$ denote the same cell but with opposite orientations). Define $L^2C_i(X)$ to be the space of $\Gamma$-equivariant $i$-chains with coefficients in $L^2(\Gamma)$: $L^2C_i(X) := C_i^\Gamma(X; L^2(\Gamma))$ (see (F.7) in Appendix F.2.) In other words,

$$L^2C_i(X) := L^2(\Gamma) \otimes_{\mathbb{R}\Gamma} C_i(X). \tag{J.6}$$

An element of $L^2C_i(X)$ is an $L^2$-*chain*. It can be regarded as an infinite chain with square summable coefficients. If $X$ is connected and $\Gamma$ acts freely on it, then $\Gamma$ is a quotient of $\pi_1(X/\Gamma)$, $L^2(\Gamma)$ is a $\pi_1(X/\Gamma)$-module and $L^2C_i(X)$ is just the space of $i$-chains on $X/\Gamma$ with local coefficients in $L^2(\Gamma)$ (as defined by formula (F.1) of Appendix F.1).

The definition of the space of $L^2$-cochains on $X$ is the same, i.e., $L^2C^i(X) := L^2C_i(X)$. (Since $L^2(\Gamma)$ can be identified with its dual, $L^2C_i(X)$ is naturally identified with $C^i_\Gamma(X; L^2(\Gamma)) := \mathrm{Hom}_\Gamma(C_i(X), L^2(\Gamma))$, the equivariant cochains with coefficients in $L^2(\Gamma)$.) If $c$ is an $i$-cell of $X$, then the space of $L^2$-chains which are supported on the $\Gamma$-orbit of $c$ can be identified with $L^2(\Gamma/\Gamma_c)$. Since $\Gamma_c$ is finite, $L^2(\Gamma/\Gamma_c)$ is a Hilbert $\Gamma$-module. (Here we are assuming that the stabilizer of $\Gamma_c$ fixes $c$ pointwise. In the general case, it is induced from the one-dimensional represention of $\Gamma_c$ on $H_*(c, \partial c; \mathbb{R})$.) Since there are a finite number of $\Gamma$-orbits of $i$-cells, $L^2C_i(X)$ is the direct sum of a finite number of such subspaces; hence, it is a Hilbert $\Gamma$-module.

## Unreduced and Reduced $L^2$-Homology

The boundary maps $\partial_i : L^2C_i(X) \to L^2C_{i-1}(X)$ and coboundary maps $\delta^i : L^2C^i(X) \to L^2C^{i+1}(X)$ are defined by the usual formulas.

**LEMMA J.3.1.** *The boundary and the coboundary operators are maps of Hilbert $\Gamma$-modules, i.e., they are $\Gamma$-equivariant, bounded linear operators.*

*Proof.* The linear operator $\partial_i : L^2C_i(X) \to L^2C_{i-1}(X)$ is $\Gamma$-equivariant since the action on $X$ is cellular and it is bounded (by Lemma J.2.5) since it is induced from the boundary operator $C_i(X) \to C_{i-1}(X)$. The same holds for $\delta^i$ since it is the adjoint of $\partial_{i+1}$. □

Define subspaces of $L^2C_i(X)$:

$$Z_i(X) := \operatorname{Ker} \partial_i, \qquad Z^i(X) := \operatorname{Ker} \delta^i,$$

$$B_i(X) := \operatorname{Im} \partial_{i+1}, \qquad B^i(X) := \operatorname{Im} \delta^{i-1},$$

the $L^2$*-cycles, -cocycles, -boundaries*, and *-coboundaries*, respectively. The corresponding quotient spaces

$$L^2H_i(X) := Z_i(X)/B_i(X),$$

$$L^2H^i(X) := Z^i(X)/B^i(X).$$

are the *unreduced $L^2$-homology* and *-cohomology groups*, respectively. (In other words, $L^2H_i(X)$ is the equivariant homology of $X$ with coefficients in $L^2(\Gamma)$, that is, $L^2H_i(X) := H_i^\Gamma(X; L^2(\Gamma))$.) Since the subspaces $B_i(X)$ and $B^i(X)$ need not be closed, these quotient spaces need not be Hilbert spaces.

Let $\overline{B_i}(X)$ (resp. $\overline{B^i}(X)$) denote the closure of $B_i(X)$ (resp. $B^i(X)$). The *reduced $L^2$-homology* and *-cohomology groups* are defined by

$$L^2\mathcal{H}_i(X) := Z_i(X)/\overline{B_i}(X),$$

$$L^2\mathcal{H}^i(X) := Z^i(X)/\overline{B^i}(X).$$

They are Hilbert $\Gamma$-modules (since each can be identified with the orthogonal complement of a $\Gamma$-stable subspace in a closed $\Gamma$-stable subspace of $L^2C_i(X)$).

**Hodge Decomposition**

Since $\langle \delta^{i-1}(x), y\rangle = \langle x, \partial_i(y)\rangle$ for all $x \in L^2C_{i-1}(X)$, $y \in L^2C_i(X)$, we have orthogonal direct sum decompositions

$$L^2C_i(X) = \overline{B_i}(X) \oplus Z^i(X),$$
$$L^2C_i(X) = \overline{B^i}(X) \oplus Z_i(X).$$

Since $\langle \delta^{i-1}(x), \partial_{i+1}(y)\rangle = \langle x, \partial_i\partial_{i+1}(y)\rangle = 0$, the subspaces $\overline{B_i}(X)$ and $\overline{B^i}(X)$ are orthogonal. This gives the *Hodge decomposition*

$$L^2C_i(X) = \overline{B_i}(X) \oplus \overline{B^i}(X) \oplus (Z_i(X) \cap Z^i(X)).$$

It follows that the reduced $L^2$-homology and $L^2$-cohomology groups can both be identified with the subspace $Z_i(X) \cap Z^i(X)$. We also denote this intersection $\mathcal{H}_i(X)$ and call it the subspace of *harmonic* $i$-cycles. Thus, an $i$-chain is harmonic if and only if it is simultaneously a cycle and a cocycle. The *combinatorial Laplacian* $\Delta : L^2C_i(X) \to L^2C_i(X)$ is defined by

$$\Delta = \delta^{i-1}\partial_i + \partial_{i+1}\delta^i.$$

One can check that $\mathcal{H}_i(X) = \operatorname{Ker}\Delta$.

**LEMMA J.3.2.** *Suppose $C^*$ and $D^*$ are cochain complexes of Hilbert $\Gamma$-modules and $f : C^* \to D^*$ is a weak isomorphism of cochain complexes (i.e., $f$ is a cochain map and for each $i$, $f_i : C^i \to D^i$ is a weak isomorphism). Then the induced map $\mathcal{H}(f) : \mathcal{H}^i(C^*) \to \mathcal{H}^i(D^*)$ is also a weak isomorphism. In particular (by Lemma J.2.6), $\mathcal{H}^i(C^*)$ and $\mathcal{H}^i(D^*)$ are isometric Hilbert $\Gamma$-modules.*

*Proof.* Let $Z^i(C)$ and $B^i(C)$ denote the cocycles and coboundaries in $C^i$ (and similarly for $D^*$). Since $f$ is a cochain map, $f(Z^i(C)) \subset Z^i(D)$ and $f(B^i(C)) \subset B^i(D)$. Since $f$ is continuous, $f(\overline{B^i(C)}) \subset \overline{B^i(D)}$. Since $f$ is injective,

$$\dim_\Gamma(Z^i(D)) \geqslant \dim_\Gamma(Z^i(C)) \quad \text{and} \quad \dim_\Gamma(B^i(D)) \geqslant \dim_\Gamma(B^i(C)). \tag{J.7}$$

From the short exact sequences

$$0 \to Z^i(C) \to C^i \to B^{i+1}(C) \to 0,$$
$$0 \to Z^i(D) \to D^i \to B^{i+1}(D) \to 0,$$

one obtains

$$\dim_\Gamma(C^i) = \dim_\Gamma(Z^i(C)) + \dim_\Gamma(B^{i+1}(C))$$

and

$$\dim_\Gamma(D^i) = \dim_\Gamma(Z^i(D)) + \dim_\Gamma(B^{i+1}(D)).$$

Since $f$ is a weak isomorphism, one has that $\dim_\Gamma(D^i) = \dim_\Gamma(C^i)$. Hence,

$$\begin{aligned}\dim_\Gamma(C^i) = \dim_\Gamma(D^i) &= \dim_\Gamma(Z^i(D)) + \dim_\Gamma(B^{i+1}(D))\\ &\geqslant \dim_\Gamma(Z^i(C)) + \dim_\Gamma(B^{i+1}(C)) = \dim_\Gamma(C^i),\end{aligned}$$

which together with (J.7) implies

$$\dim_\Gamma(Z^i(D)) = \dim_\Gamma(Z^i(C)) \quad \text{and} \quad \dim_\Gamma(B^i(D)) = \dim_\Gamma(B^i(C)). \tag{J.8}$$

It follows from Lemma J.2.9 that the maps $f : Z^i(C) \to Z^i(C)$ and $f : B^i(C) \to B^i(D)$ are weak isomorphisms. This implies that $\mathcal{H}(f) : \mathcal{H}^i(C^*) \to \mathcal{H}^i(D^*)$ is weakly surjective.

To show $\mathcal{H}(f)$ is injective, let $H$ be the orthogonal complement to $B^i(C)$ in $Z^i(C)$. Note that $H$ is closed and $\Gamma$-stable. We have

$$\begin{aligned}\dim_\Gamma(Z^i(D)) &= \dim_\Gamma\left(f\left(H + \overline{B^i(C)}\right)\right)\\ &= \dim_\Gamma(f(H)) + \dim_\Gamma\left(B^i(D)\right) - \dim_\Gamma\left(f(H) \cap B^i(D)\right)\\ &= \dim_\Gamma(Z^i(D)) - \dim_\Gamma\left(f(H) \cap B^i(D)\right),\end{aligned}$$

where the third equality follows from (J.8). Hence, $\dim_\Gamma\left(\overline{f(H)} \cap \overline{B^i(D)}\right) = 0$ and therefore $\overline{f(H)}$ is a complementary subspace for $\overline{B^i(D)}$. Since $f$ is injective when restricted to $H$, so is $\mathcal{H}(f)$. □

## J.4. BASIC $L^2$ ALGEBRAIC TOPOLOGY

Suppose $\Gamma$ acts cellularly, properly and cocompactly on a CW complex $X$ and that $Y$ is a $\Gamma$-stable subcomplex. The reduced $L^2$-(co)homology groups $L^2\mathcal{H}_i(X, Y)$ are defined similarly to before. Versions of most of the Eilenberg-Steenrod Axioms hold for $L^2\mathcal{H}_*(X, Y)$. We list some standard properties below. (Of course, similar results hold for the contravariant $L^2$-cohomology functor.)

### Functoriality

For $i = 1, 2$, suppose $(X_i, Y_i)$ is a pair of $\Gamma$-complexes and $f : (X_1, Y_1) \to (X_2, Y_2)$ is a $\Gamma$-equivariant map (a $\Gamma$-*map* for short). Then there is an induced map $f_* : L^2\mathcal{H}_i(X_1, Y_1) \to L^2\mathcal{H}_i(X_2, Y_2)$. This gives a functor from pairs of $\Gamma$-complexes to Hilbert $\Gamma$-modules. Moreover, if $f' : (X_1, Y_1) \to (X_2, Y_2)$ is another $\Gamma$-map which is $\Gamma$-homotopic to $f$, then $f_* = f'_*$.

**Exact Sequence of a Pair**

The sequence of a pair $(X, Y)$,

$$\to L^2\mathcal{H}_i(Y) \to L^2\mathcal{H}_i(X) \to L^2\mathcal{H}_i(X, Y) \to,$$

is weakly exact.

**Excision**

Suppose $(X, Y)$ is a pair of $\Gamma$-complexes and $U$ is a $\Gamma$-stable subset of $Y$ such that $Y - U$ is a subcomplex. Then the inclusion $(X - U, Y - U) \to (X, Y)$ induces an isomorphism

$$L^2\mathcal{H}_i(X - U, Y - U) \cong L^2\mathcal{H}_i(X, Y).$$

A standard consequence of the last two properties is the following.

**Mayer-Vietoris Sequences**

Suppose $X = X_1 \cup X_2$, where $X_1$ and $X_2$ are $\Gamma$-stable subcomplexes of $X$. Then $X_1 \cap X_2$ is also $\Gamma$-stable and the Mayer-Vietoris sequence,

$$\to L^2\mathcal{H}_i(X_1 \cap X_2) \to L^2\mathcal{H}_i(X_1) \oplus L^2\mathcal{H}_i(X_2) \to L^2\mathcal{H}_i(X) \to,$$

is weakly exact.

**Twisted Products and the Induced Representation**

Suppose that $H$ is a subgroup of $\Gamma$ and that $Y$ is a space on which $H$ acts. The twisted product, $\Gamma \times_H Y$, is a left $\Gamma$-space and a $\Gamma$-bundle over $\Gamma/H$. Since $\Gamma/H$ is discrete, $\Gamma \times_H Y$ is a disjoint union of copies of $Y$, one for each element of $\Gamma/H$. If $Y$ is an $H$ complex, then $\Gamma \times_H Y$ is a $\Gamma$ complex and the following formula holds:

$$L^2\mathcal{H}_i(\Gamma \times_H Y) \cong \mathrm{Ind}_H^\Gamma(L^2\mathcal{H}_i(Y)).$$

**Künneth Formula**

Suppose $\Gamma = \Gamma_1 \times \Gamma_2$ and that $X_j$ is a CW complex with $\Gamma_j$-action for $j = 1, 2$. Then $X_1 \times X_2$ is a $\Gamma$-complex and

$$L^2\mathcal{H}_k(X_1 \times X_2) \cong \sum_{i+j=k} L^2\mathcal{H}_i(X_1)\widehat{\otimes} L^2\mathcal{H}_j(X_2).$$

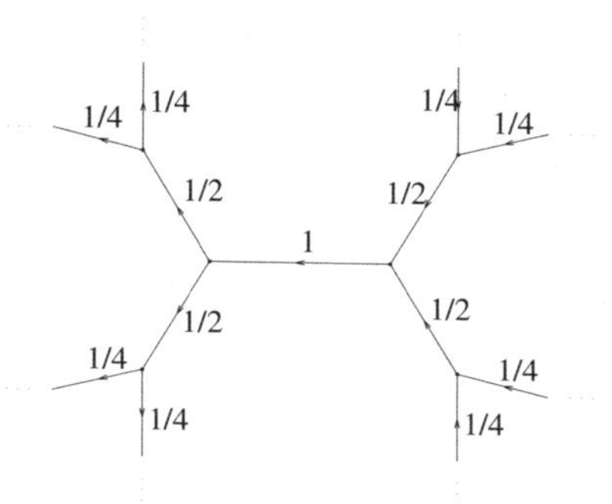


Figure J.1. An $L^2$ 1-cycle.

## (Co)homology in Dimension 0

An element of $C^0(X)$ is an $L^2$ function on the set of vertices of $X$; it is a 0-cocycle if and only if it takes the same value on the endpoints of each edge. Hence, if $X$ is connected, any 0-cocycle is constant. If, in addition, $\Gamma$ is infinite (so that the 1-skeleton of $X$ is infinite), then this constant must be 0. So, when $X$ is connected and $\Gamma$ is infinite, $L^2H^0(X) = L^2\mathcal{H}^0(X) = 0$. Hence,

$$L^2\mathcal{H}_0(X) = 0. \tag{J.9}$$

On the other hand, the unreduced homology $L^2H_0(X)$ need not be 0. For example, if $X = \mathbb{R}$, cellulated as the union of intervals $[n, n+1]$, and $\Gamma = \mathbf{C}_\infty$, then any vertex of $\mathbb{R}$ is an $L^2$-0-cycle which is not an $L^2$-boundary. (A vertex bounds a half-line which can be thought of as an infinite 1-chain, however, this 1-chain is not square summable.) In fact, if $\Gamma$ is infinite, then $L^2H_0(X) = 0$ if and only if $\Gamma$ is not amenable.

**Example J.4.1.** Suppose $X$ is a regular trivalent tree. The previous paragraph shows $L^2H^0(X) = 0$ (since $X$ is infinite). By an easy computation, the 1-cycle pictured in Figure J.1 is square summable. Since any 1-chain on a graph is automatically a cocycle, this cycle is harmonic and hence represents a nonzero element in $L^2\mathcal{H}_1(X)$ (Hodge decomposition).

## The Top-Dimensional Homology of a Pseudomanifold

Suppose a regular $\Gamma$ cell complex $X$ is an $n$-dimensional, pseudomanifold. This means that each $(n-1)$-cell is contained in precisely two $n$-cells. If a component of the complement of the $(n-2)$-skeleton is not orientable, then it does not support a nonzero $n$-cycle (with coefficients in $\mathbb{R}$). If such a component is orientable, then any $n$-cycle supported on it is a constant multiple of the $n$-cycle with all coefficients $+1$. If the component has an infinite number of $n$-cells, then this $n$-cycle does not have square summable coefficients. Hence, if each component of the complement of the $(n-2)$-skeleton is either infinite or nonorientable, then $L^2H_n(X) = 0$. In particular, if the complement of the $(n-2)$-skeleton is connected and $\Gamma$ is infinite, then $L^2H_n(X) = 0$.

**Example J.4.2.** It follows from the previous two paragraphs that $L^2\mathcal{H}(\mathbb{E}^1)$ vanishes in all dimensions (where $\mathbb{E}^1$ is cellulated by unit intervals). So, by the Künneth Formula, $L^2\mathcal{H}(\mathbb{E}^n)$ also vanishes in all dimensions.

## J.5. $L^2$-BETTI NUMBERS AND EULER CHARACTERISTICS

The $i^{\text{th}}$ *$L^2$-Betti number* of $X$ is defined by

$$L^2b^i(X;\Gamma) := \dim_\Gamma L^2\mathcal{H}^i(X), \tag{J.10}$$

where $\dim_\Gamma(\ )$ is the $\Gamma$-dimension (or von Neumann dimension) defined by (J.5).

If $X$ and $X'$ are contractible (proper and cocompact) $\Gamma$ complexes, their $L^2$-Betti numbers are equal. (The reason is that the chain complex $C_*(X;\mathbb{R})$ gives a projective resolution of $\mathbb{R}$ by projective $\mathbb{R}\Gamma$-modules. It follows that the unreduced $L^2$-cohomology, $L^2H^i(X)$ is equal to $H^i(\Gamma; L^2(\Gamma)$. Hence, when we take reduced cohomology, the result again depends only on $\Gamma$.) So, when $X$ is contractible (or just acyclic), we will sometimes write $L^2b^i(\Gamma)$ instead of $L^2b^i(X;\Gamma)$. We summarize some of basic properties of $L^2$-Betti numbers in the next proposition.

**PROPOSITION J.5.1**

*(i)* $L^2b^i(X;\Gamma) = 0$ *if and only if* $L^2\mathcal{H}^i(X) = 0$.

*(ii) If $H$ is a subgroup of finite index $m$ in $\Gamma$, then* $L^2b^i(X;H) = mL^2b^i(X;\Gamma)$.

*(iii) If $Y$ is a CW complex with $H$-action, $H$ a subgroup of $\Gamma$, then*

$$L^2b^i(\Gamma \times_H Y;\Gamma) = L^2b^i(Y;H).$$

*(iv) If $\Gamma = \Gamma_1 \times \Gamma_2$ and for $j = 1, 2$, $X_j$ is a CW complex with $\Gamma_j$ action, then*

$$L^2b^k(X_1 \times X_2;\Gamma) = \sum_{i+j=k} L^2b^i(X_1;\Gamma)L^2b^j(X_2;\Gamma).$$

*Proof.* The first three statements follow from, respectively, parts (ii), (ix), and (xi) of Proposition J.2.8. Statement (iv) is the combination of part (xiii) of the same proposition with the Künneth Formula. □

The *$L^2$-Euler characteristic* of $X$ is defined by

$$L^2\chi(X;\Gamma) := \sum_{i=0}^{\infty}(-1)^i L^2 b^i(X;\Gamma). \tag{J.11}$$

Supposing, as always, that the $\Gamma$-complex $X$ is proper and cocompact, the *orbihedral Euler characteristic* $\chi^{\text{orb}}(X/\!/\Gamma)$ is defined by formula (16.4), which we recall here:

$$\chi^{\text{orb}}(X/\!/\Gamma) := \sum_{\substack{\text{orbits}\\ \text{of cells}}} \frac{(-1)^{\dim c}}{|\Gamma_c|},$$

where $|\Gamma_c|$ denotes the order of the stabilizer of the cell $c$. Note that if the $\Gamma$-action is free, then $\chi^{\text{orb}}(X/\!/\Gamma)$ is just the ordinary Euler characteristic $\chi(X/\Gamma)$.

## Atiyah's Formula

**LEMMA J.5.2.** *Suppose $C_*$ is a chain complex of (finitely generated) Hilbert $\Gamma$-modules and that $\mathcal{H}_*$ is its reduced homology. Set*

$$c_i := \dim_\Gamma C_i \quad \textit{and} \quad b_i := \dim_\Gamma \mathcal{H}_i.$$

*Further suppose that there is an integer $n$ such that $c_i = 0$ for all $i > n$. Then*

$$\sum_{i=0}^{n}(-1)^i c_i = \sum_{i=0}^{n}(-1)^i b_i$$

*Proof.* (From [113, Theorem 3.6.1]). Let $B^i := \operatorname{Im}\delta \subset C_i$ and $B_i := \operatorname{Im}\partial \subset C_i$ be the spaces of cocycles and cycles, respectively. We have the Hodge decomposition $C_i = \overline{B^i} \oplus \overline{B_i} \oplus \mathcal{H}_i$. So

$$c_i = \dim \overline{B^i} + \dim \overline{B_i} + b_i, \tag{J.12}$$

where, to simplify notation, we are writing dim( ) for $\dim_\Gamma$( ). Since the kernel of $\partial_i : C_i \to C_{i-1}$ is $\overline{B_i} \oplus \mathcal{H}_i$, the restriction of $\partial_i$ to $\overline{B^i}$ is a weak isomorphism to $\overline{B_{i-1}}$. So, $\dim \overline{B^i} = \dim \overline{B_{i-1}}$. Hence, if we take the alternating sum of both sides of (J.12), the terms on the right-hand side involving $\dim \overline{B^i}$ and $\dim \overline{B_i}$ cancel. □

**THEOREM J.5.3.** (Atiyah's Formula [11].) *Suppose the $\Gamma$-action on $X$ is proper and cocompact. Then*

$$\chi^{\text{orb}}(X/\!/\Gamma) = L^2\chi(X;\Gamma).$$

*Proof.* Given a cell $c$ in $X$, by Proposition J.2.8 (xii), we have $\dim_\Gamma L^2(\Gamma/\Gamma_c) = 1/|\Gamma_c|$. Since $L^2C_i(X)$ is a direct sum over orbits of $i$-cells $c$ of $L^2(\Gamma/\Gamma_c)$,

$$c_i := \dim_\Gamma L^2C_i(X) = \sum_{\substack{\text{orbits}\\ \text{of } i\text{-cells}}} \frac{1}{|\Gamma_c|}.$$

So, the formula follows from the previous lemma and the definition of $\chi^{\text{orb}}(\,)$. □

**Example J.5.4.** Suppose $\Gamma$ is the fundamental group of $M^2$, a closed surface of genus $g > 0$ and $X$ is its universal cover. By results in the previous section, $L^2\mathcal{H}^0(X) = 0$ and $L^2\mathcal{H}^2(X) = 0$; hence, only $L^2\mathcal{H}^1(X)$ can be nonzero. By Atiyah's Formula, $b^1(X;\Gamma) = -\chi(M^2) = 2g - 2$.

*Remark J.5.5.* Since $L^2\mathcal{H}_i(X)$ is a Hilbert $\Gamma$-submodule of $L^2C_i(X)$,

$$L^2b^i(X;\Gamma) \leqslant \dim_\Gamma L^2C_i(X).$$

In particular, if $X$ is the universal cover of a finite CW complex $Y$ and $\Gamma = \pi_1(Y)$, we get the estimate, $L^2b^i(X;\Gamma) \leqslant c_i(Y)$, where $c_i(Y)$ is the number of $i$-cells in $Y$.

## J.6. POINCARÉ DUALITY

Suppose $(X, \partial X)$ is a pair of $\Gamma$-complexes and that $X$ is an $n$-dimensional manifold with boundary. Then

$$L^2\mathcal{H}_i(X, \partial X) \cong L^2\mathcal{H}^{n-i}(X), \tag{J.13}$$

$$L^2\mathcal{H}_i(X) \cong L^2\mathcal{H}^{n-i}(X, \partial X).$$

When $X$ is cellulated as a *PL* manifold with boundary, these isomorphisms are induced by the bijective correspondence $\sigma \leftrightarrow D\sigma$ which associates to each $i$-cell $\sigma$ its dual $(n-i)$-cell $D\sigma$. A slight elaboration of this argument also works in the case where $(X, \partial X)$ is a polyhedral homology manifold with boundary; the only complication being that the "dual cells" need not actually be cells, rather they are "generalized homology disks" as in Definition 10.4.5.

In fact, as is shown in [113, Theorem 3.7.2], in order to have the Poincaré duality isomorphisms of J.13, all one need assume is that $(X, \partial X)$ is a "virtual $PD^n$-pair," in the sense that there is a subgroup $H$ of finite index in $\Gamma$ so that the chain complexes $K_*(X, \partial X)$ and ${}^nDK_*(X)$ are chain homotopy equivalent, where ${}^nDK_i(X)$ is defined by ${}^nDK_i(X) = \mathrm{Hom}_{\mathbb{Z}H}(K_{n-i}(X), \mathbb{Z}H)$.

**Example J.6.1.** (PD$^2$-*groups.*) Suppose $\Gamma$ is a PD$^2$-group (Definition F.5.1). Since this implies $\Gamma$ is infinite, $L^2b^0(\Gamma) = 0$ (by formula (J.9) in J.4). By Poincaré duality, $L^2b^2(\Gamma) = 0$. So, $\chi(\Gamma) = -L^2b^1(\Gamma) \leqslant 0$. Hence, if $\Gamma$ is orientable, its ordinary Betti numbers satisfy

$$b^1(\Gamma) - 2 = -b^0(\Gamma) + b^1(\Gamma) - b^2(\Gamma) \geqslant 0.$$

So $b^1(\Gamma) = \mathrm{rk}(\Gamma^{ab}) \geqslant 2$. Similarly, when $\Gamma$ is nonorientable, $b^1(\Gamma) \geqslant 1$. This fact was crucial in the proof by Eckmann and his collaborators that PD$^2$-groups are surface groups, see [112]. (The actual argument in [114] used the Hattori-Stallings rank instead of $L^2$-Betti numbers.)

## J.7. THE SINGER CONJECTURE

Shortly after Atiyah's Formula (Theorem J.5.3) became known, Dodziuk and Singer independently pointed out that the Euler Characteristic Conjecture of 16.2 follows if one can prove the reduced $L^2$-(co)homology of the universal cover of any even dimensional, closed, aspherical manifold vanishes except in the middle dimension. (This was first explained in the introduction of [102]. We give the argument below.) This led to the following.

**CONJECTURE J.7.1.** (The Singer Conjecture.) *If $M^n$ is a closed aspherical manifold with universal cover $\widetilde{M}^n$, then*

$$L^2\mathcal{H}_i(\widetilde{M}^n) = 0 \quad \textit{for all } i \neq \tfrac{n}{2}.$$

**PROPOSITION J.7.2.** *The Singer Conjecture implies the Euler Characteristic Conjecture (= Conjecture 16.2.1).*

*Proof.* Let $M^{2k}$ be a closed aspherical manifold with fundamental group $\pi$. The Euler Characteristic Conjecture asserts that $(-1)^k\chi(M^{2k}) \geqslant 0$. If the Singer Conjecture holds for $M^{2k}$, then $L^2b^i(\widetilde{M}^{2k};\pi) = 0$ for $i \neq k$. Atiyah's Formula gives: $\chi(M^{2k}) = \sum(-1)^iL^2b^i(\widetilde{M}^{2k};\pi) = (-1)^kL^2b^k(\widetilde{M}^{2k};\pi)$. So, $(-1)^k\chi(M^{2k}) = L^2b^k(\widetilde{M}^{2k};\pi) \geqslant 0$. □

The Singer Conjecture holds for elementary reasons in dimensions $\leqslant 2$ (Examples J.4.2 and J.5.4). In [189] Lott and Lück proved that it holds for those aspherical 3-manifolds for which Thurston's Geometrization Conjecture is true (hence, by [237], for all aspherical 3-manifolds). It is also known to hold for (a) locally symmetric spaces, (b) negatively curved Kähler manifolds (by [148]), (c) Riemannian manifolds of sufficiently pinched negative sectional curvature (by [103]), (d) closed aspherical manifolds with fundamental group containing an infinite amenable normal subgroup (by [62]), and (e) manifolds which fiber over $S^1$ (by [191]).

We note that the Euler Characteristic Conjecture and Singer's Conjecture both make sense for closed aspherical orbifolds or, for that matter, for virtual Poincaré duality groups (Definition F.5.1 in Appendix F.5).

## J.8. VANISHING THEOREMS

Here we state some basic results about the vanishing of $L^2$-cohomology, providing evidence for the Singer Conjecture.

### Mapping Tori

Suppose a closed manifold $M$ fibers over a circle. The product formula for Euler characteristics shows $\chi(M) = 0$. Gromov asked if the reduced $L^2$-homology of its universal cover vanishes in all dimensions. It does, as Lück showed in [191], by proving a more general result for "mapping tori."

Suppose $F$ is a CW complex and $f : F \to F$ is a self-map. The *mapping torus of* $f$ is the quotient space $T_f$ of $F \times [0, 1]$ by the equivalence relation which identifies $(x, 0)$ with $(f(x), 1)$. Projection of $F \times [0, 1]$ onto its second factor induces a natural projection $p : T_f \to S^1$ (we regard $S^1$ as the space formed by identifying the endpoints of $[0, 1]$). The homotopy type of $T_f$ depends only on the homotopy class of $f$.

*Remark.* If $f$ is a homeomorphism, then $T_f \to S^1$ is a fiber bundle. Conversely, any fiber bundle over $S^1$ is the mapping torus of a homeomorphism. Similarly, if $f$ is a homotopy equivalence, then $T_f \to S^1$ is a fibration and any fibration over $S^1$ has this form.

Next, suppose $f$ is cellular (i.e., $f$ takes the $i$-skeleton of $F$ to itself, for all $i \geqslant 0$). Then $T_f$ has a cell structure induced from that of $F \times [0, 1)$, i.e., each $i$-cell $c$ of $F$ contributes two cells to $T_f$, the $i$-cell $c \times 0$ and the $(i + 1)$-cell $c \times (0, 1)$. Put $\Gamma = \pi_1(T_f)$. At the level of fundamental groups the projection $p$ induces a surjection $p_* : \Gamma \to \mathbb{Z}$. Put $\Gamma_n := p_*^{-1}(n\mathbb{Z})$. It is a subgroup of index $n$ in $\Gamma$.

**THEOREM J.8.1.** (Lück [191]). *Suppose $f : F \to F$ is a self-map of a finite CW complex and that $T_f$ is its mapping torus. Let $\widetilde{T}_f$ be the universal cover of $T_f$ and $\Gamma := \pi_1(Y)$. Then $L^2b^i(\widetilde{T}_f; \Gamma) = 0$ for all $i$.*

*Proof.* Let $c_i(Y)$ denote the number of $i$-cells in a CW complex $Y$. By the Cellular Approximation Theorem ([153, p. 349]), we can assume $f$ is cellular. As mentioned above, $T_f$ has a cell structure with $c_i(T_f) = c_{i-1}(F) + c_i(F)$. Note that this formula is independent of the map $f$. We leave it as an exercise for the reader to show that the $n$-fold covering space $\widetilde{T}_f/\Gamma_n$ of $T_f$ is homotopy equivalent to the mapping torus $T_{f^n}$, where $f^n$ denotes the $n$-fold composition $f \circ \cdots \circ f$. By Remark J.5.5,

$$L^2b^i(\widetilde{T}_{f^n}; \Gamma_n) \leqslant c_i(T_{f^n}) = c_{i-1}(F) + c_i(F).$$

So, by Proposition J.5.1 (ii),

$$L^2b^i(\widetilde{T}_f; \Gamma) = \frac{L^2b^i(\widetilde{T}_f; \Gamma_n)}{n} \leqslant \frac{c_{i-1}(F) + c_i(F)}{n}.$$

Since the right-hand side goes to 0 as $n \to \infty$, we get $L^2b^i(\widetilde{T}_f; \Gamma) = 0$. □

### Actions of Amenable Groups

There is a canonical map $L^2H^i(X) \to H^i(X; \mathbb{R})$, which regards an $L^2$-cocycle as an ordinary cocycle. Restricting this map to the harmonic cocycles we get a map, can : $L^2\mathcal{H}^i(X) \to H^i(X; \mathbb{R})$.

**THEOREM J.8.2.** (Cheeger-Gromov [62].) *Suppose $\Gamma$ is an infinite amenable group. Then the canonical map $L^2\mathcal{H}^i(X) \to H^i(X; \mathbb{R})$ is injective.*

In the case when the action is proper and cocompact a simple proof of this theorem appears in [113]. We outline it below.

*Outline of Eckmann's proof of Theorem J.8.2.* Let $\mathcal{K}$ be the kernel of the canonical map $L^2\mathcal{H}^i(X) \to H^i(X; \mathbb{R})$. The idea is to use the Følner Condition to show that $\dim_\Gamma \mathcal{K} = 0$. (The Følner Condition for a finitely generated group to be amenable is explained at the beginning of 17.2.) Since $\Gamma$ is countable, the Følner Condition implies we can find an exhaustion of $\Gamma$ by finite sets, $F_1 \subset F_2 \subset \cdots$ so that

$$\bigcup_{j=1}^{\infty} F_j = \Gamma \quad \text{and} \quad \lim_{j\to\infty} \frac{|\partial F_j|}{|F_j|} = 0.$$

Choose a fundamental domain $D$ for $X$ as follows. Pick a representative cell for each $\Gamma$-orbit of cells and let $D$ be the union of the closures of these representatives. Put $X_j = \bigcup F_j D$ and let $\partial X_j$ be its topological boundary in $X$. Let $P : L^2C^i(X) \to L^2\mathcal{H}^i(X)$ be orthogonal projection onto the harmonic cocycles and let $\pi_j$ be the composition of $P$ with the inclusion $L^2C^i(X_j) \hookrightarrow L^2C^i(X)$. (Since $X_j$ is a finite complex, $L^2C^i(X_j) = C^i(X_j)$.) We have

$$\begin{aligned}\dim_{\mathbb{R}} \pi_j(\mathcal{K}) &= \sum_{\substack{\gamma, c \\ \gamma c \subset X_j}} \langle \pi_j(\gamma c), c\rangle = |F_j| \sum_{c \subset D} \langle P(c), c\rangle \\ &= |F_j| \dim_\Gamma \mathcal{K}.\end{aligned}$$

So

$$\dim_\Gamma \mathcal{K} = \frac{\dim_\mathbb{R} \pi_j(\mathcal{K})}{|F_j|}. \tag{J.14}$$

Using the local finiteness of $X$ (and some work), one can obtain the estimates: $\dim_\mathbb{R} \pi_j(\mathcal{K}) \leqslant \dim_\mathbb{R} C^i(\partial X_j) \leqslant |\partial F_j|\alpha_i$, where $\alpha_i$ is the number of $i$-cells in $D$. Combining this with (J.14), we get

$$\dim_\Gamma \mathcal{K} \leqslant \frac{|\partial F_j|}{|F_j|}.$$

By the Følner Conditon, the right-hand side goes to 0 as $j \to \infty$. □

**COROLLARY J.8.3.** *Suppose* $\Gamma$ *is an infinite amenable group and* $X$ *is contractible. Then for all* $i \geq 0$, $L^2\mathcal{H}^i(X) = 0$ *and hence,* $L^2 b^i(\Gamma) = 0$.

Cheeger and Gromov actually prove a stronger form of Theorem J.8.2. They do not need to assume the action is cocompact but only that $X$ has "bounded geometry." For example, they reach the same conclusion when $\Gamma$ is only assumed to contain an infinite, normal amenable subgroup $A$ and the $\Gamma$-action is cocompact. Then Theorem J.8.2 applies to the $A$-action on $X$ to give the following.

**COROLLARY J.8.4.** *Suppose a group* $\Gamma$ *of type VF contains an infinite, normal amenable subgroup. Then* $L^2 b^i(\Gamma) = 0$ *for all* $i \geqslant 0$*. Hence,* $\chi(\Gamma) = 0$.

**Example J.8.5.** (*Circle bundles.*) Suppose $B\Gamma \to B$ is a bundle with fiber $S^1$ and $B\Gamma$ homotopy equivalent to a finite CW complex. Then the image of $\pi_1(S^1)$ in $\pi_1(B\Gamma)$ is a normal, infinite amenable subgroup; hence, $L^2 b^i(\Gamma) = 0$ for all $i \geqslant 0$.

## Complements on Square Summable Forms and the DeRham Theorem

Suppose $M$ is a smooth closed manifold , $\Omega^p(M)$ is the vector space of closed $p$-forms on $M$, and $d : \Omega^p(M) \to \Omega^{p+1}(M)$ is the exterior differential. This gives us the *de Rham cochain complex*

$$\cdots \longrightarrow \Omega^p(M) \xrightarrow{d} \Omega^{p+1}(M) \longrightarrow \cdots,$$

and corresponding cohomology groups $H^*_{\mathrm{dR}}(M)$. The classical de Rham Theorem states that if $M$ has a smooth triangulation, then integration over $p$-simplices defines an isomorphism $H^p_{\mathrm{dR}}(M) \xrightarrow{\cong} H^p(M;\mathbb{R})$.

Let $\Lambda^p(V)$ denote the space of alternating multilinear $p$-forms on a vector space $V$. If $V$ has an inner product and $\dim V = n$, then there is a canonical isomorphism $\Lambda^p(V) \cong \Lambda^{n-p}(V)$. This isomophism takes the wedge product of $p$ elements in an orthonormal basis for $V^*$ and maps them to the wedge product of the complementary elements in the basis. Now suppose $M^n$ is a (not necessarily closed) $n$-dimensional, Riemannian manifold. If $M^n$ is not closed, then $\Omega^P(M^n)$ means the compactly supported smooth $p$-forms. If $T_x$ denotes the tangent space of $M^n$ at $x$, we get an isomorphism $\Lambda^p(T_x) \cong \bigwedge^{n-p}(T_x)$. This isomorphism induces the *Hodge star operator*, $* : \Omega^p(M^n) \to \Omega^{n-p}(M^n)$. (There is a choice of $\pm$ sign here, which we are ignoring.) Define an inner product on $\Omega^p(M^p)$ by

$$\langle \omega, \eta \rangle := \int_M \omega \wedge \eta$$

The Hilbert space completion of $\Omega^p(M^n)$ is denoted $L^2\Omega^p(M^n)$. If $M^n$ has bounded geometry (i.e., if it is the universal cover of a closed manifold), then $d$ is a bounded operator. Define the *Laplacian*, $\Delta := d \circ d^* + d^* \circ d : L^2\Omega^p(M^n) \to L^2\Omega^p(M^n)$. Its kernel $L^2\mathcal{H}^p_{\mathrm{dR}}(M^n)$ is the space of *square integrable hamonic forms*. It can be identified with the reduced $L^2$-cohomology of the cochain complex $\{L^2\Omega^*(M^n)\}$. The $L^2$ version of the deRham Theorem was proved by Dodziuk.

**THEOREM J.8.6.** ($L^2$ de Rham Theorem of [101].) *Suppose $M^n$ is a closed manifold with a Riemannian metric and a smooth triangulation. Give its universal cover $\widetilde{M}^n$ the induced metric and induced triangulation. Then integration of closed p-forms over p-simplices defines an isomorphism* $L^2\mathcal{H}^p_{dR}(\widetilde{M}^n) \xrightarrow{\cong} L^2\mathcal{H}^p(\widetilde{M}^n)$.

## Symmetric Spaces

A *symmetric space of noncompact type* (without Euclidean factor) means a manifold of the form $G/K$, where $G$ is a semisimple Lie group (without compact factors) and $K$ is a maximal compact subgroup. For example, hyperbolic $n$-space $\mathbb{H}^n$ is a symmetric space.

**THEOREM J.8.7.** (Dodziuk [102].) $L^2\mathcal{H}^i_{dR}(\mathbb{H}^n) = 0$, *for all* $i \neq \frac{n}{2}$.

*Outline of Dodziuk's proof of Theorem J.8.7.* The argument only uses the fact that we have a "rotationally symmetric" Riemannian metric on a manifold $M^n$ diffeomorphic to $\mathbb{R}^n$. This means that in polar coordinates the metric has the form $ds^2 = dr^2 + f(r)^2 d\theta^2$ where $d\theta$ is the standard round metric on $\mathbb{S}^{n-1}$, where $r$ is the distance to the origin and $f : [0, \infty) \to [0, \infty)$ is a smooth increasing function satisfying

$$f(0) = 0, \quad f'(0) = 1, \quad \lim_{r \to \infty} f(r) = \infty.$$

(When $M^n = \mathbb{H}^n, f(r) = e^r$.)

Let $\omega$ be a harmonic $p$-form. Using the fact that the Hodge star operator on $\mathbb{S}^{n-1}$ is conformally invariant, after some computations, Dodziuk shows that $\omega \neq 0$ implies $\int_1^\infty f^{n-2p-1}(r)dr < \infty$ (where $f^k(r) := f(r)^k$). Applying this in dimension $n-p$, we see that if $L^2\mathcal{H}^{n-p}(M^n) \neq 0$, then $\int_1^\infty f^{-n+2p-1}(r)dr < \infty$. By Poincaré duality, $L^2\mathcal{H}^p(M^n) \cong L^2\mathcal{H}^{n-p}(M^n)$; so, both exponents must give convergent integrals. If $n = 2p$, both exponents equal $-1$ and we get the condition $\int_1^\infty \frac{1}{f(r)}dr < \infty$. Notice that $(n-2p-1)(-n+2p-1) = 1-(n-2p)^2$. So, if $n-2p = \pm 1$, one exponent is 0 and the integral diverges. Otherwise, the exponents have different signs, so at least one of them diverges. □

In fact, the following theorem states that the same is true for any symmetric space of noncompact type ( for a proof, see [192, Cor. 5.16]).

**THEOREM J.8.8.** *Suppose $X^n$ is a symmetric space of noncompact type. Then $L^2\mathcal{H}^i_{dR}(X^n) = 0$, for all $i \neq \frac{n}{2}$.*

By the de Rham Theorem, we get the same conclusion about the vanishing of the cellular $L^2$-cohomology of $X^n$ whenever the cell structure on $X^n$ is pulled back from a cellulation of a locally symmetric manifold $M^n$ (of which $X^n$ is the universal cover). Thus, Theorem J.8.8 shows the Singer Conjecture holds for locally symmetric manifolds. We also see that $L^2\mathcal{H}^k(X^{2k}) \neq 0$ whenever $X^{2k}$ is the universal cover of a locally symmetric manifold of nonzero Euler characteristic.

## Complements on Kähler Manifolds and the Hard Lefschetz Theorem

Suppose $M$ is a complex manifold of complex dimension $n$ (hence, of real dimension $2n$). We can equip $M$ with a Hermitian metric (i.e., a Hermitian inner product on each tangent space). The imaginary part of the Hermitian metric is a nondegenerate 2-form $\omega$. $M$ is a *Kähler manifold* if $\omega$ is closed. For example, $\mathbb{C}P^n$ is a Káhler manifold, so is any smooth projective variety $M \subset \mathbb{C}P^n$.

Suppose $(M, \omega)$ is a Kähler manifold. Let $\mathcal{L} : \Omega^p(M) \to \Omega^{p+2}(M)$ be the linear operator $\lambda \to \lambda \wedge \omega$. The key fact is that $L$ preserves harmonic forms. Let $[\omega] \in \mathcal{H}^2_{\mathrm{dR}}(M)$ be the Kähler class. Put $\ell := \wedge [\omega] : \mathcal{H}^p_{\mathrm{dR}}(M) \to \mathcal{H}^{p+2}_{\mathrm{dR}}(M)$ and $\ell^k := \ell \circ \cdots \ell$, the $k$-fold composition. ($\mathcal{H}^p_{\mathrm{dR}}(\;)$ denotes the harmonic $p$-forms.)

**THEOREM J.8.9.** (The Hard Lefschetz Theorem.) *Suppose $(M, \omega)$ is a closed Kähler manifold of complex dimension n. Then $\ell^{n-p} : \mathcal{H}^p_{dR}(M) \to \mathcal{H}^{2n-p}_{dR}(M)$ is an isomorphism.*

*Sketch of proof.* Wedging $(n-p)$ times with a nondegenerate 2-form on $T_x$ gives an isomorphism $\Lambda^p(T_x) \to \Lambda^{2n-p}(T_x)$. Hence, $\mathcal{L}^{n-p} : \Omega^p(M) \to \Omega^{2n-p}(M)$ is an isomorphism. Since $\mathcal{L}^{n-p}$ takes harmonic forms to harmonic forms and restricts to $\ell^{n-p}$ on $\mathcal{H}^p_{\mathrm{dR}}(M)$, the theorem follows. □

The same argument proves the following.

**THEOREM J.8.10.** ($L^2$ version of Hard Lefschetz Theorem.) *Suppose $(M, \omega)$ is a (not necessarily compact) Kähler manifold of complex dimension $n$. Then $\ell^{n-p} : L^2\mathcal{H}^p_{dR}(M) \to L^2\mathcal{H}^{2n-p}_{dR}(M)$ is an isomorphism.*

### Kähler Hyperbolicity

The ideas in this paragraph are from Gromov [148].

**DEFINITION J.8.11.** Suppose $(M, \omega)$ is a closed Kähler manifold, that $\widetilde{M}$ is its universal cover and $\tilde{\omega}$ is the pullback of $\omega$ to $\widetilde{M}$. The n$(M, \omega)$ (or $(\widetilde{M}, \tilde{\omega})$) is *Kähler hyperbolic* if there exists a bounded 1-form $\eta$ on $\widetilde{M}$ such that $\tilde{\omega} = d\eta$ (i.e., $\tilde{\omega}$ is $d$(bounded)).

Gromov proved the following two theorems.

**THEOREM J.8.12.** (Gromov [148].) *A closed Kähler manifold of negative sectional curvature is Kähler hyperbolic.*

**THEOREM J.8.13.** (Gromov [148].) *Suppose $(M, \omega)$ is Kähler hyperbolic of complex dimension $n$ and that $\pi = \pi_1(M)$. Then*

(i) *$L^2\mathcal{H}^p(\widetilde{M}) = 0$ for all $p \neq n$ (and hence, $L^2b^p(\widetilde{M}) = 0$ for all $p \neq n$.*

(ii) *$L^2b^n(\widetilde{M}) \neq 0$ (and hence, $(-1)^n\chi(M) > 0$).*

*Proof of part (i) of Theorem J.8.13.* Suppose $\lambda \in L^2\Omega^p(\widetilde{M})$ is a closed $p$-form. By Theorem J.8.10 it suffices to show $\mathcal{L}^{n-p}(\lambda)$ represents 0 in cohomology. We will show, in fact, that $\mathcal{L}(\lambda)$ is 0 in cohomology. Note that if $\eta$ is any bounded form then $\lambda \wedge \eta$ is square summable. In particular, by Kähler hyperbolicity, $\tilde{\omega} = d\eta$, where $\eta$ is bounded. Hence,

$$d(\lambda \wedge \eta) = (d\lambda \wedge \eta) - (\lambda \wedge \tilde{\omega}) = -\mathcal{L}(\lambda).$$

So $\mathcal{L}(\lambda)$ is 0 in cohomology. (The proof of (ii) is more difficult.) □

*Remark.* Lück [192] points out the following are examples of Kähler manifolds $(M, \omega)$ which are Kähler hyperbolic:

- $\pi_1(M)$ is word hyperbolic and $\pi_2(M) = 0$.
- $M$ is a submanifold of a Kähler hyperbolic manifold.
- $\widetilde{M}$ is a Hermitian symmetric space of noncompact type (and without Euclidean factor).

## NOTES

A good introduction to $L^2$-cohomology along the same lines as here can be found in Eckmann's paper [113]. Much more can be found in Lück's book [192].

Farber defines an "extended $L^2$-(co)homology" theory in [118] and demonstrates that this is the correct categorical framework for $L^2$-(co)homology. An extended $L^2$-(co)homology object is isomorphic to the sum of its "projective part" and its "torsion part." The projective part is essentially the reduced $L^2$-(co)homology group while its torsion part contains information such as Novikov-Shubin invariants.

In [193, 192] Lück introduces a different approach to $L^2$-Betti numbers. He considers the ordinary equivariant homology $H^*_\Gamma(X; \mathcal{N}(\Gamma))$ with coefficients in the $\Gamma$-module $\mathcal{N}(\Gamma)$ defined as in Appendix F.2. Since $\mathcal{N}(\Gamma)$ is an $\mathcal{N}(\Gamma)$-bimodule, $H^*_\Gamma(X; \mathcal{N}(\Gamma))$ is a right $\mathcal{N}(\Gamma)$-module. (The situation is analogous to equivariant (co)homology with group ring coefficients discussed in Lemma F.2.1 and in the introduction to Chapter 15.) Lück shows that one can associate a "dimension" to a $\mathcal{N}(\Gamma)$-module in a way which is compatible with the $\Gamma$-dimension and $L^2$-Betti numbers. More exactly, this dimension of $H_k(X; \mathcal{N}(\Gamma))$ is equal to $L^2b^k(X; \Gamma)$. There are several advantages to Lück's approach. First, $L^2$-Betti numbers can be defined even when the action is not proper and not cocompact (provided we allow $\infty$ as a possible value). Second, Lück's definition makes sense equally well for singular (co)homology and other (co)homology theories.

# Bibliography

[1] A. D. Aleksandrov, V. N. Berestovskii, and I. Nikolaev. Generalized Riemannian spaces. *Russian Math. Surveys*, 41:1–54, 1986.

[2] D. V. Alekseevskij and E. B. Vinberg. Geometry of spaces of constant curvature. In E. B. Vinberg, editor, *Geometry II: Spaces of Constant Curvature*, volume 29 of *Encyclopaedia of Mathematical Sciences*, pages 1–138. Springer-Verlag, Berlin, 1993.

[3] S. B. Alexander and R. L. Bishop. The Hadamard–Cartan Theorem in locally convex spaces. *Enseign. Math.*, 36:309–320, 1990.

[4] D. Allcock. Infinitely many hyperbolic Coxeter groups through dimension 19. *Geometry & Topology*, 10:737–758, 2006.

[5] F. D. Ancel, M. W. Davis, and C. R. Guilbault. CAT(0) reflection manifolds. In *Geometric Topology (Athens, GA 1993)*, volume 2 of *AMS/IP Studies in Advanced Mathematics*, pages 441–445. Amer. Math. Soc., Providence, R.I., 1997.

[6] F. D. Ancel and L. Siebenmann. The construction of homogeneous homology manifolds. *Abstracts Amer. Math. Soc.*, 6, 1985. Abstract number 816-57-72.

[7] E. M. Andreev. On convex polyhedra in Lobacevskii spaces. *Math. USSR Sbornik*, 10(5):413–440, 1970.

[8] E. M. Andreev. On convex polyhedra of finite volume in Lobacevskii spaces. *Math. USSR Sbornik*, 12:255–259, 1970.

[9] E. M. Andreev. On the intersections of plane boundaries of acute-angled polyhedra. *Math. Notes*, 8:761–764, 1971.

[10] M. A. Armstrong. *Basic Topology. Undergraduate Texts in Mathematics*. Springer–Verlag, Berlin, 1983.

[11] M. F. Atiyah. Elliptic operators, discrete groups and von Neumann algebras. volume 32-33 of *Astérisque*, pages 43–72. Soc. Math. France, Paris, 1976.

[12] P. Bahls. *The Isomorphism Problem in Coxeter Groups*. Imperial College Press, London, 2005.

[13] P. Bahls and M. Mihalik. Reflection independence in even Coxeter groups. *Geom. Dedicata*, 110:63–80, 2005.

[14] W. Ballman and M. Brin. Polygonal complexes and combinatorial group theory. *Geom. Dedicata*, 50:165–191, 1994.

[15] W. Ballman, M. Gromov, and V. Schroeder. *Manifolds of Nonpositive Curvature*, volume 61 of *Progress in Mathematics* Birkhäuser, Stuttgart, 1985.

[16] D. Barnette. A proof of the lower bound conjecture for convex polytopes. *Pacific J. Math.*, 46:349–354, 1973.

[17] A. Barnhill. The $FA_n$ Conjecture for Coxeter groups. *Algebraic Geometric Topology*, 6:2117–2150, 2006.

[18] N. Benakli. Polyèdre à géométrie locale donée. *C. R. Acad. Sci. Paris*, 313: 561–564, 1991.

[19] V. N. Berestovskii. Borsuk's problem on the metrization of a polyhedron. *Soviet Math. Doklady*, 27:56–59, 1983.

[20] M. Berger. *Geometry I and II*. Springer–Verlag, Berlin, 1987.

[21] M. Bestvina. Questions in geometric group theory. http://www.math.utah.edu/ bestvina.

[22] M. Bestvina. The virtual cohomological dimension of Coxeter groups. In I. G. Niblo and M. Roller, editors, *Geometric Group Theory, Volume 1*, volume 181 of *London Mathematical Society Lecture Notes*, pages 19–23. Cambrige Univ. Press, Cambridge, 1993.

[23] M. Bestvina. The local homology properties of boundaries of groups. *Michigan Math. J.*, 43:123–139, 1996.

[24] M. Bestvina and N. Brady. Morse theory and finiteness properties of groups. *Invent. Math.*, 129:445–470, 1997.

[25] M. Bestvina and G. Mess. The boundary of negatively curved groups. *J. Amer. Math. Soc.*, 4:469–481, 1991.

[26] A. Björner and F. Brenti. *Combinatorics of Coxeter Groups*, volume 231 of *Graduate Texts in Mathematics*. Springer-Verlag, Berlin, 2005.

[27] A. Borel. *Seminar on Transformation Groups*, volume 46 of *Annals of Mathematical Studies*. Princeton University Press, Princeton, N.J., 1960.

[28] R. Bott. An application of Morse theory to the topology of Lie groups. *Bull. Soc. Math. France*, 84:251–281, 1956.

[29] N. Bourbaki. *Lie Groups and Lie Algebras, Chapters 4–6*. Springer-Verlag, Berlin, 2002.

[30] M. Bourdon. Immeubles hyperboliques, dimension conforme et rigidité de Mostow. *Geom. Funct. Anal.*, 7:245–268, 1997.

[31] M. Bourdon. Sur les immeubles fuchsiens et leur type de quasi-isométrie. *Ergodic Theory and Dynamical Systems*, 20:343–364, 2000.

[32] M. Bourdon and H. Pajot. Rigidity of quasi-isometries for some hyperbolic buildings. *Comment. Math. Helv.*, 75:701–736, 2000.

[33] B. H. Bowditch. Notes on locally $CAT(1)$-spaces. In R. Charney, M. Davis, and M. Shapiro, editors, *Geometric Group Theory*, volume 3 of *Ohio State University Mathematical Research Institute Publication*, pages 1–48. de Gruyter, Berlin, 1995.

[34] N. Brady, I. Leary, and B. E. A. Nucinkis. On algebraic and geometric dimensions for groups with torsion. *J. London Math. Soc.*, 64:489–500, 2001.

[35] N. Brady, J. McCammond, B. Mühlherr, and W. Neumann. Rigidity of Coxeter groups and Artin groups. *Geom. Dedicata*, 94:91–109, 2002.

[36] G. Bredon. *Introduction to Compact Transformation Groups*, volume 46 of *Pure and Applied Mathematics*. Academic Press, New York, 1972.

[37] M. Bridson and A. Haefliger. *Metric Spaces of Non-Positive Curvature*. Springer-Verlag, Berlin, 1999.

[38] B. Brink and R. Howlett. Normalizers of parabolic subgroups in Coxeter groups. *Invent. Math.*, 136:323–351, 1999.

[39] A. Brønsted. *An Introduction to Convex Polytopes*, volume 90 of *Graduate Texts in Mathematics*. Springer-Verlag, Berlin, 1983.

[40] W. Browder. *Surgery on simply-connected manifolds*. Ergebnisse der Mathematik und ihrer Grenzgebiete, Band 65. Springer-Verlag, Berlin, 1972.

[41] K. S. Brown. Homological criteria for finiteness. *Comment. Math. Helv.*, 50: 129–135, 1975.

[42] K. S. Brown. *Cohomology of Groups*, volume 87 of *Graduate Texts in Mathematics*. Springer-Verlag, Berlin, 1982.

[43] K. S. Brown. *Buildings*. Springer-Verlag, Berlin, 1989.

[44] J. Bryant, S. C. Ferry, W. Mio, and S. Weinberger. Topology of homology manifolds. *Ann. Math.*, 143:435–467, 1996.

[45] D. Burago, Y. Burago, and S. Ivanov. *A Course in Metric Geometry*, volume 33 of *Graduate Studies in Mathematics*. American Math. Society, Providence, 2001.

[46] H. Busemann. Spaces with non-positive curvature. *Acta Math.*, 80:259–310, 1948.

[47] J. W. Cannon. Shrinking cell-like decompositions of manifolds: codimension three. *Ann. Math.*, 110:83–112, 1979.

[48] J. W. Cannon. The combinatorial structure of cocompact discrete hyperbolic groups. *Geom. Dedicata*, 16:123–148, 1984.

[49] H.-D. Cao and X.-P. Zhu. A complete proof of the Poincaré and Geometrization Conjectures—Applications of the Hamilton-Perelman Theory of the Ricci flow. *Asian J. Math.*, 10:145–492, 2006.

[50] P.-E. Caprace. Conjugacy of 2-spherical subgroups of Coxeter groups and parallel walls. *Algebraic & Geometric Topology*, 61:1987–2029, 2006.

[51] P.-E. Caprace and B. Mühlherr. Reflection rigidity of 2-spherical Coxeter groups. *Proc. London Math. Soc.*, to appear.

[52] P.-E. Caprace and B. Rémy. Simplicité abstraite des groupes de Kac–Moody non affines. *C. R. Acad. Sc. Paris*, 342:539–544, 2006.

[53] R. Charney and M. W. Davis. Reciprocity of growth functions of Coxeter groups. *Geom. Dedicata*, 39:373–378, 1991.

[54] R. Charney and M. W. Davis. Singular metrics of nonpositive curvature on branched covers of Riemannian manifolds. *Amer. J. Math.*, 115:929–1009, 1993.

[55] R. Charney and M. W. Davis. The Euler characteristic of a nonpositively curved, piecewise Euclidean manifold. *Pac. J. Math*, 171:117–137, 1995.

[56] R. Charney and M. W. Davis. Finite $K(\pi, 1)$s for Artin groups. In F. Quinn, editor, *Prospects in Topology*, volume 138 of *Annals of Mathematical Studies*, pages 110–124. Princeton University Press, Princeton, 1995.

[57] R. Charney and M. W. Davis. The $K(\pi, 1)$-problem for hyperplane complements associated to infinite reflection groups. *J. Amer. Math.*, 8:597–627, 1995.

[58] R. Charney and M. W. Davis. The polar dual of a convex polyhedral set in hyperbolic space. *Michigan Math. J.*, 42:479–510, 1995.

[59] R. Charney and M. W. Davis. Strict hyperbolization. *Topology*, 34:329–350, 1995.

[60] R. Charney and M. W. Davis. When is a Coxeter system determined by its Coxeter group? *J. London Math. Soc.*, 61:441–461, 2000.

[61] R. Charney, M. W. Davis, and G. Moussong. Nonpositively curved, piecewise euclidean structures on hyperbolic manifolds. *Michigan Math. J.*, 44:201–208, 1997.

[62] J. Cheeger and M. Gromov. $L_2$-cohomology and group cohomology. *Topology*, 25:189–215, 1986.

[63] J. Cheeger, W. Müller, and R. Schrader. On the curvature of piecewise flat spaces. *Commun. Math. Phys.*, 92:405–454, 1984.

[64] S. S. Chern. On the curvature and characteristic classes of a Riemannian manifold. *Abh. Math. Sem. Hamburg*, 20:117–126, 1956.

[65] D. Cooper, D. Long, and A. Reid. Infinite Coxeter groups are virtually indicable. *Proc. Edinburgh Math. Soc.*, 41:303–313, 1998.

[66] J. M. Corson. Complexes of groups. *Proc. London Math. Soc.*, 65:199–224, 1992.

[67] H. S. M. Coxeter. Discrete groups generated by reflections. *Ann. Math.*, 35: 588–621, 1934.

[68] H. S. M. Coxeter. Regular honeycombs in hyperbolic space. In *Proceedings of the International Congress of Mathematicians, Amsterdam 1954, vol. 3*, pages 155–169. 1954.

[69] H. S. M. Coxeter. *Regular Polytopes*. Macmillan, New York, second edition, 1963.

[70] J. S. Crisp. Symmetrical subgroups of Artin groups. *Adv. Math.*, 152:159–177, 2000.

[71] M. W. Davis. Groups generated by reflections and aspherical manifolds not covered by Euclidean space. *Ann. Math.*, 117:293–325, 1983.

[72] M. W. Davis. Coxeter groups and aspherical manifolds. In I. Madsen and R. Oliver, editors, *Algebraic Topology Aarhus 1982*, volume 1051 of *Lecture Notes in Mathematics*, pages 197–221. Springer-Verlag, Berlin, 1984.

[73] M. W. Davis. The homology of a space on which a reflection group acts. *Duke Math. J.*, 55:97–104, 1987.

[74] M. W. Davis. Regular convex cell complexes. In C. McCrory and T. Shifrin, editors, *Geometry and Topology: Manifolds, Varieties, and Knots*, pages 53–84. Marcel Dekker, New York, 1987.

[75] M. W. Davis. Some aspherical manifolds. *Duke Math. J.*, 55:105–139, 1987.

[76] M. W. Davis. Buildings are *CAT*(0). In P. Kropholler, G. Niblo, and R. Stöhr, editors, *Geometry and Cohomology in Group Theory*, volume 252 of *London Mathematical Society Lecture Notes*, pages 108–123. Cambridge Univ. Press, Cambridge, 1998.

[77] M. W. Davis. The cohomology of a Coxeter group with group ring coefficients. *Duke Math. J.*, 91:297–314, 1998.

[78] M. W. Davis. Nonpositive curvature and reflection groups. In R. J. Daverman and R. B. Sher, editors, *Handbook of Geometric Topology*, pages 373–422. Elsevier, Amsterdam, 2002.

[79] M. W. Davis, J. Dymara, T. Januszkiewicz, and B. Okun. Weighted $L^2$-cohomology of Coxeter groups, *Geometry & Topology*, 11:47–138, 2007.

[80] M. W. Davis, J. Dymara, T. Januszkiewicz, and B. Okun. Cohomology of Coxeter groups with group ring coefficients: II. *Algebraic & Geometric Topology*, 6:1289–1318, 2006.

[81] M. W. Davis and J.-C. Hausmann. Aspherical manifolds without smooth or PL structure. In G. Carlsson, R. Cohen, H. Miller, and D. Ravenel, editors,

*Algebraic Topology*, volume 1370 of *Lecture Notes in Mathematics*, pages 135–142. Springer-Verlag, Berlin, 1989.

[82] M. W. Davis and T. Januszkiewicz. Convex polytopes, Coxeter orbifolds and torus actions. *Duke Math. J.*, 62:417–451, 1991.

[83] M. W. Davis and T. Januszkiewicz. Hyperbolization of polyhedra. *J. Differential Geometry*, 34:347–388, 1991.

[84] M. W. Davis, T. Januszkiewicz, and R. Scott. Nonpositive curvature of blow-ups. *Selecta math., New Series.*, 4:491–547, 1998.

[85] M. W. Davis, T. Januszkiewicz, and R. Scott. Fundamental groups of blow-ups. *Adv. Math.*, 177:115–179, 2003.

[86] M. W. Davis, T. Januszkiewicz, and S. Weinberger. Relative hyperbolization and aspherical bordism: an addendum to hyperbolization of polyhedra. *J. Differential Geometry*, 58:535–541, 2001.

[87] M. W. Davis and I. J. Leary. $\ell^2$-cohomology of Artin groups. *J. London Math. Soc.*, 68:493–510, 2003.

[88] M. W. Davis and I. J. Leary. Some examples of discrete group actions on aspherical manifolds. In F. T. Farrell and W. Lück, editors, *High-dimensional Manifold Topology*, pages 139–150, World Scientific, Englewood Clifts, N.J., 2003.

[89] M. W. Davis and J. Meier. The topology at infinity of Coxeter groups and buildings. *Comment. Math. Helv.*, 77:746–766, 2002.

[90] M. W. Davis and G. Moussong. Notes on nonpositively curved polyhedra. In W. Neumann K. Böröczky and A. Stipsicicz, editors, *Low Dimensional Topology*, volume 8 of *Bolyai Society Mathematical Studies*, pages 11–94. Janos Bolyai Society, Budapest, 1998.

[91] M. W. Davis and B. Okun. Vanishing theorems and conjectures for the $\ell^2$-homology of right-angled Coxeter groups. *Geometry & Topology*, 5:7–74, 2001.

[92] P. Deligne. Les immeubles des groupes de tresses généralisés. *Invent. Math.*, 17:273–302, 1972.

[93] V. V. Deodhar. On the root system of a Coxeter group. *Comm. Algebra*, 10:611–630, 1982.

[94] V. V. Deodhar. A note on subgroups generated by reflections in Coxeter groups. *Arch. Math.*, 53:543–546, 1989.

[95] W. Dicks and M. J. Dunwoody. *Groups Acting on Graphs*. Cambridge Univ. Press, Cambridge, 1989.

[96] W. Dicks and I. J. Leary. Presentations for subgroups of Artin groups. *Proc. Amer. Math. Soc.*, 127:343–348, 1999.

[97] W. Dicks and I. J. Leary. On subgroups of Coxeter groups. In G. Niblo P. Kropholler and R. Stöhr, editors, *Geometry and Cohomology in Group Theory*, volume 252 of *London Mathematical Society Lecture Notes*, pages 124–160. Cambridge Univ. Press, Cambridge, 1998.

[98] J. Dieudonne. *A History of Algebraic Topology, 1900–1960*. Birkhäuser, Boston, 1989.

[99] G. A. Dirac. On rigid circuit graphs. *Abh. Math. Sem. Univ. Hamburg*, 25:71–76, 1961.

[100] J. Dixmier. *Von Neumann Algebras*. North-Holland, Amsterdam, 1981.

[101] J. Dodziuk. de Rham–Hodge theory for $l^2$-cohomology of infinite coverings. *Topology*, 16:157–165, 1977.

[102] J. Dodziuk. $l^2$ harmonic forms on rotationally symmetric Riemannian manifolds. *Proc. Amer. Math. Soc.*, 77:395–400, 1979.

[103] H. Donnelly and F. Xavier. On the differential form spectrum of negatively curved Riemannian manifolds. *Amer. J. Math.*, 106:169–185, 1984.

[104] A. N. Dranishnikov. The virtual cohomological dimension of Coxeter groups. *Proc. Amer. Math. Soc.*, 125:1885–1891, 1997.

[105] M. J. Dunwoody. The accessibility of finitely presented groups. *Inventiones Math.*, 81:449–457, 1985.

[106] M. J. Dunwoody. An inacessible group. In G. A. Niblo and M. A. Roller, editors, *Geometric Group Theory, Volume 1*, volume 181 of *London Mathematical Society Lecture Notes*, pages 75–78. Cambridge Univ. Press, Cambridge, 1993.

[107] M. Dyer. Reflection subgroups of Coxeter systems. *J. Algebra*, 135:57–73, 1990.

[108] J. Dymara. $l^2$-cohomology of buildings with fundamental class. *Proc. Amer. Math. Soc.*, 132:1839–1843, 2004.

[109] J. Dymara. Thin buildings. *Geometry & Toplogy*, 10:667–694, 2006.

[110] J. Dymara and T. Januszkiewicz. Cohomology of buildings and their automorphism groups. *Invent. Math.*, 150:579–627, 2002.

[111] J. Dymara and D. Osajda. Boundaries of right-angled hyperbolic buildings. preprint, 2005.

[112] B. Eckmann. Poincaré duality groups of dimension two are surface groups. In S. Gersten and J. Stallings, editors, *Combinatorial Group Theory and Topology*, volume 111 of *Annals. of Mathematical Studies*, pages 35–51. Princeton University Press, Princeton, 1987.

[113] B. Eckmann. Introduction to $\ell_2$-methods in topology: reduced $\ell_2$-homology, harmonic chains, $\ell_2$-betti numbers. *Israel J. Math.*, 117:183–219, 2000.

[114] B. Eckmann and P. A. Linnell. Poincaré duality groups of dimension two. *Comment. Math. Helv.*, 58:111–114, 1983.

[115] A. L. Edmonds, J. H. Ewing, and R. S. Kulkarni. Regular tessellations of surfaces and $(p, q, r)$-triangle groups. *Ann. Math.*, 116:113–132, 1982.

[116] R. D. Edwards. The topology of manifolds and cell-like maps. In *Proceedings of the International Congress of Mathematicians, Helsinki 1978, Vol. 1*, pages 111–127. Academia Scientiarium Fennica, Hungary, 1980.

[117] S. Eilenberg and T. Ganea. On the Lusternik-Schnirelmann category of abstract groups. *Ann. Math.*, 65:517–518, 1957.

[118] M. Farber. Homological algebra of Novikov-Shubin invariants and morse inequalities. *Geom. Funct. Analysis*, 6:628–665, 1996.

[119] F. T. Farrell. Poincaré duality and groups of type (FP). *Comment. Math. Helv.*, 50:187–195, 1975.

[120] F. T. Farrell and W. C. Hsiang. On Novikov's Conjecture for nonpositively curved manifolds. *Ann. Math. (2)*, 113:199–209, 1981.

[121] F. T. Farrell and L. E. Jones. A topological analogue of Mostow's rigidity theorem. *J. Amer. Math. Soc.*, 2:257–369, 1989.

[122] F. T. Farrell and L. E. Jones. *Classical Aspherical Manifolds*, volume 75 of *CBMS Regional Conference Series in Mathematics*. Amer. Math. Soc., Providence, 1990.

[123] F. T. Farrell and L. E. Jones. Topological rigidity for compact non-positively curved manifolds. In *Differential Geometry, Los Angeles 1990*, volume 54 of *Proceedings and Symposia in Pure Mathematics*, pages 224–274. Amer. Math. Soc., Providence, 1993.

[124] F. T. Farrell and L. E. Jones. Rigidity for aspherical manifolds with $\pi_1 \subset GL_m(\mathbf{R})$. *Asian J. Math.*, 2:215–262, 1998.

[125] F. T. Farrell and J. Lafont. Finite automorphisms of negatively curved Poincaré duality groups. *Geom. Funct. Anal.*, 14:283–294, 2004.

[126] W. Feit and G. Higman. The nonexistence of certain generalized polygons. *J. Algebra*, 1:114–131, 1964.

[127] S. Ferry. Notes for polyhedral and TOP topology 1992–1993. Unpublished manuscript, 1993.

[128] H. Fischer. Boundaries of right-angled Coxeter groups with manifold nerves. *Topology*, 42:423–446, 2003.

[129] W. J. Floyd and S. P. Plotnick. Symmetries of planar growth functions. *Invent. Math.*, 93:501–543, 1988.

[130] W. N. Franszen and R. B. Howlett. Automorphisms of nearly finite Coxeter groups. *Advances in Geometry*, 3(3):301–338, 2003.

[131] M. Freedman. The topology of four–dimensional manifolds. *J. Differential Geometry*, 17:357–453, 1982.

[132] H. Freudenthal. Über die Enden topologischer Räume und Gruppen. *Math. Z.*, 33:692–713, 1931.

[133] D. Gaboriau and F. Paulin. Sur les immeubles hyperboliques. *Geom. Dedicata*, 88:153–197, 2001.

[134] S. Gal. Real root conjecture fails for five- and higher-dimensional spheres. *Discrete Comput. Geom*, 34:269–284, 2005.

[135] R. Geoghegan. *Topological Methods in Group Theory*. In *Graduate Texts in Mathematics*. Springer-Verlag, Berlin, to appear.

[136] R. Geoghegan and M. Mihalik. The fundamental group at infinity. *Topology*, 35:655–669, 1996.

[137] R. Geroch. Positive sectional curvature does not imply positive Gauss-Bonnet integrand. *Proc. Amer. Math. Soc.*, 54:267–270, 1976.

[138] E. Ghys and P. de la Harpe (eds.). *Sur les groupes hyperboliques d'aprés Mikhael Gromov*, volume 83 of *Progress in Mathematics*. Birkhäuser, Boston, 1990.

[139] C. Gonciulea. Infinite Coxeter groups virtually surject onto $\mathbb{Z}$. *Comment. Math. Helv.*, 72:257–265, 1997.

[140] C. Gonciulea. *Virtual epimorphisms of Coxeter groups onto free groups*. PhD thesis, The Ohio State University, 2000.

[141] C. McA. Gordon, D. D. Long, and A. Reid. Surface subgroups of Coxeter groups and Artin groups. *J. Pure and Applied Algebra*, 189:135–148, 2004.

[142] M. Greenberg and J. Harper. *Algebraic Topology: a First Course*. Benjamin/Cummings, London, 1981.

[143] D. Gromoll and J. Wolf. Some relations between the metric structure and algebraic structure of the fundamental group in manifolds of non-positive curvature. *Bull. Amer. Math. Soc.*, 77:545–552, 1971.

[144] M. Gromov. Hyperbolic manifolds, groups and actions. In *Riemann Surfaces and Related Topics*, volume 97 of *Annals of Mathematical Studies*, pages 183–215. Princeton University Press, Princeton, 1981.

[145] M. Gromov. Volume and bounded cohomology. *IHES Publ. Math.*, 56:213–307, 1983.

[146] M. Gromov. Infinite groups as geometric objects. In *Proceedings of the International Congress of Mathematicians, Warsaw 1983, vol. 1*, pages 385–392, Polish Scientific Publishers, Warsaw, 1984.

[147] M. Gromov. Hyperbolic groups. In S. Gersten, editor, *Essays in Group Theory*, volume 8 of *MSRI Publications*, pages 75–264. Springer-Verlag, Berlin, 1987.

[148] M. Gromov. Kähler hyperbolicity and $L_2$-Hodge theory. *J. Differential Geom.*, 33:1991, 1991.

[149] A. Haefliger. Groupoides d'holonomie et classifiants. In *Structure Transverse des Feuillatages, Toulouse 1982*, volume 116 of *Astérisque*, pages 70–97. 1984.

[150] A. Haefliger. Complexes of groups and orbihedra. In E. Ghys, A. Haefliger, and A. Verjovsky, editors, *Group Theory from Geometrical Viewpoint (Trieste, 1990)*, pages 504–540. World Scientific, Singapore, 1991.

[151] A. Haefliger. Extensions of complexes of groups. *Ann. Inst. Fourier (Grenoble)*, 42:275–311, 1992.

[152] F. Haglund. Polyèdre de Gromov. *C.R. Acad. Sci. Paris*, 313:603–606, 1991.

[153] A. Hatcher. *Algebraic Topology*. Cambridge University Press, Cambridge, 2002.

[154] J.-C. Hausmann and S. Weinberger. Caracteristiques d'Euler et groupes fondamentaux de dimension 4. *Comment. Math. Helv.*, 60:139–144, 1985.

[155] M. Hirsch and B. Mazur. *Smoothings of Piecewise Linear Manifolds*, volume 80 of *Ann. of Math. Studies*. Princeton University Press, Princeton, 1974.

[156] H. Hopf. Fundamentalgruppe und zweite Bettische Gruppe. *Comment. Math. Helv.*, 14:257–309, 1942.

[157] H. Hopf. Enden offener Räume und unendliche diskontinuierliche Gruppen. *Comment. Math. Helv.*, 16:81–100, 1943.

[158] R. Howlett. Normalizers of parabolic subgroups of reflection groups. *J. London Math. Soc.*, 21:62–80, 1980.

[159] W. C. Hsiang and W. Y. Hsiang. Differentiable actions of compact connected Lie groups I. *Amer. J. Math.*, 89:705–786, 1967.

[160] W. C. Hsiang and W. Y. Hsiang. Differentiable actions of compact connected Lie groups III. *Ann. Math.*, 99:220–256, 1974.

[161] B. Hu. Whitehead groups of finite polyhedra with nonpositive curvature. *J. Differential Geometry*, 38:501–517, 1993.

[162] B. Hu. Retractions of closed manifolds with nonpositive curvature. In R. Charney, M. W. Davis, and M. Shapiro, editors, *Geometric Group Theory*, volume 3 of *Ohio State University Mathematical Research Institute Publications*, pages 135–148. de Gruyter, Berlin, 1995.

[163] J. E. Humphreys. *Reflection Groups and Coxeter Groups*, volume 29 of *Cambridge Studies in Advanced Mathematics*. Cambridge Univ. Press, Cambridge, 1990.

[164] B. Jackson. End invariants of group extensions. *Topology*, 21:71–81, 1981.

[165] W. Jakobsche. Homogeneous cohomology manifolds which are inverse limits. *Fund. Math.*, 137:81–95, 1991.

[166] T. Januszkiewicz and J. Świątkowski. Hyperbolic Coxeter groups of large dimension. *Comment. Math. Helv.*, 78:555–583, 2003.

[167] F. E. A. Johnson. Manifolds of homotopy type of $K(\pi, 1)$: II. *Proc. Cambridge Phil. Soc.*, 75:165–173, 1974.

[168] E. R. van Kampen. On the connection between the fundamental groups of some related spaces. *Amer. J. Math.*, 55:261–267, 1933.

[169] I. M. Kaplinskaja. Discrete groups generated by reflections in the faces of simplicial prisms in Lobachevskian spaces. *Math. Notes*, 15:88–91, 1974.

[170] M. Kato. Some problems in topology. In *Manifolds—Tokyo 1973*, pages 421–431. Univ. of Tokyo Press, Tokyo, 1975.

[171] M. Kato. On combinatorial space forms. *Scientific Papers of the College of General Education, Univ. of Tokyo*, 30:107–146, 1980.

[172] A. Kaul. *Rigidity for a class of Coxeter groups*. PhD thesis, Oregon State University, 2000.

[173] D. Kazhdan and G. Lusztig. Representations of Coxeter groups and Hecke algebras. *Invent. Math.*, 53:165–184, 1979.

[174] M. A. Kervaire. Smooth homology spheres and their fundamental groups. *Trans. Amer. Math. Soc.*, 144:67–72, 1969.

[175] M. A. Kervaire and J. Milnor. Groups of homotopy spheres, I. *Ann. Math.*, 77:504–537, 1963.

[176] K. H. Kim and F. W. Roush. Homology of certain algebras defined by graphs. *J. Pure Appl. Algebra*, 17:179–186, 1980.

[177] R. Kirby and L. Siebenmann. On the triangulation of manifolds and the Hauptvermutung. *Bull. Amer. Math. Soc.*, 75:742–749, 1969.

[178] R. Kirby and L. Siebenmann. *Foundational Essays on Topological Manifolds, Smoothings, and Triangulations*, volume 88 of *Annals. of Mathematical Studies*. Princeton University Press, Princeton, 1977.

[179] B. Kleiner and J. Lott. Notes on Perelman's papers. arXiv:math.DG/0605667, 2006.

[180] J. L. Koszul. *Lectures on Groups of Transformations*, volume 32 of *Tata Institue of Fundamental Research Notes*. Tata Institute, Bombay, 1965.

[181] D. Krammer. *The conjugacy problem for Coxeter groups*. PhD thesis, Universiteit Utrecht, 1994.

[182] D. Krammer. A proof of the Moussong Lemma. Preprint, 1996.

[183] M. Kreck and W. Lück. *The Novikov Conjecture, Geometry and Algebra*. Birkhäuser, Basel, 2005.

[184] F. Lannér. On complexes with transitive groups of automorphisms. *Comm. Sem. Math. Univ. Lund*, 11:1–71, 1950.

[185] R. Lashof. Problems in differential and algebraic topology. Seattle Conference,1963. *Ann. Math.*, 81:565–591, 1965.

[186] H. B. Lawson and S. T. Yau. Compact manifolds of nonpositive curvature. *J. Differential Geom.*, 7:211–228, 1972.

[187] I. Leary and B. E. A. Nucinkis. Some groups of type *VF*. *Invent. Math.*, 151: 135–165, 2003.

[188] R. Lee and F. Raymond. Manifolds covered by Euclidean space. *Topology*, 14:49–57, 1975.

[189] J. Lott and W. Lück. $l^2$-topological invariants of 3-manifolds. *Invent. Math.*, 120:15–60, 1995.

[190] A. Lubotzky. Free quotients and first Betti numbers of some hyperbolic manifolds. *Transformation Groups*, 1:71–82, 1996.

[191] W. Lück. $l^2$-Betti numbers of mapping tori and groups. *Topology*, 33:203–214, 1994.

[192] W. Lück. *$L^2$-invariants and K-theory*. Springer-Verlag, Berlin, 2002.

[193] W. Lück. $l^2$-invariants of regular coverings of compact manifolds and CW-complexes. In R. Daverman and R. Sher, editors, *Handbook of Geometric Topology*. Elsevier, Amsterdam, 2002.

[194] G. Lusztig. Left cells in Weyl groups. In R.L.R. Herb and J. Rosenberg, editors, *Lie Group Representations*, volume 1024, pages 99–111. Springer-Verlag, Berlin, 1983.

[195] R. Lyndon and P. Schupp. *Combinatorial Group Theory*. Springer–Verlag, Berlin, 1978.

[196] G. Margulis and E. B. Vinberg. Some linear groups virtually having a free quotient. *J. Lie Theory*, 10:171–180, 2000.

[197] W. Massey. *Algebraic Topology: an Introduction*, volume 56 of *Graduate Texts in Mathematics*. Springer-Verlag, Berlin, 1967.

[198] B. Mazur. A note on some contractible 4-manifolds. *Ann. Math.*, 73:221–228, 1961.

[199] J. McCleary. *User's Guide to Spectral Sequences*. Publish or Perish, Wilmington, Dela., 1985.

[200] C. McMullen. Coxeter groups, Salem numbers and the Hilbert metric. *Publ. Math. Inst. Hautes Études Sci.*, 95:151–183, 2002.

[201] P. McMullen. Combinatorially regular polytopes. *Mathematika*, 14:142–150, 1967.

[202] J. Meier. When is the graph product of hyperbolic groups hyperbolic? *Geom. Dedicata*, 61:29–41, 1996.

[203] D. Meintrup and T. Schick. A model for the universal space for proper actions of a hyperbolic group. *New York J. Math.*, 8:1–7.

[204] G. Mess. Examples of Poincaré duality groups. *Proc. Amer. Math. Soc.*, 110:1145–1146, 1990.

[205] M. Mihalik. Semi-stability at the end of a group extension. *Trans. Amer. Math. Soc.*, pages 307–321.

[206] M. Mihalik. Semistability at $\infty$, $\infty$-ended groups and group cohomology. *Trans. Amer. Math. Soc.*, 303:479–485, 1987.

[207] M. Mihalik. Semistability of Artin groups and Coxeter groups. *J. Pure and Appl. Algebra*, 111:205–211, 1996.

[208] M. Mihalik and S. Tschantz. Ends of amalgamated products and HNN extensions. *Mem. Amer. Math. Soc.*, 98, 1992.

[209] C. F. Miller. Decision problems for groups — survey and reflections. In G. Baumslag and C. F. Miller, editors, *Algorithms and Classification in Combinatorial Group Theory*, volume 23 of *MSRI Publications*, pages 1–60. Springer–Verlag, Berlin, 1991.

[210] J. Millson. On the first Betti number of a constant negatively curved manifold. *Annals Math.*, 104:235–247, 1976.

[211] J. Milnor. The Poincaré Conjecture 99 years later: a progress report. *The Clay Mathematics Institute 2002 Annual Report*, pages 6–7, 16–21.

[212] J. Milnor. A note on curvature and the fundamental group. *J. Differential Geom.*, 2:1–7, 1968.

[213] J. Milnor. *Singular Points of Complex Hypersurfaces*, volume 61 of *Annals of Mathematical Studies*. Princeton University Press, Princeton, 1968.

[214] J. Milnor. Towards the Poincaré Conjecture and the classification of 3-manifolds. *Notices of the AMS*, 50(10):1226–1233, 2003.

[215] J. Milnor. The Poincaré Conjecture. In J. Carlson, A. Jaffe, and A. Wiles, editors, *The Millenium Prize Problems*, pages 71–86. The Clay Math. Institute and the Amer. Math. Soc., Cambridge, MA and Providence, RI, 2006.

[216] J. Milnor and J. Stasheff. *Characteristic Classes*, volume 76 of *Annals of Mathematical Studies*. Princeton University Press, Princeton, 1974.

[217] H. Minkowski. Zur Theorie der positiven quadratischen Formen. *J. Crelle*, 101:196–202, 1887.

[218] G. Mislin. Wall's finiteness obstruction. In *Handbook of Algebraic Topology*, pages 1259–1291. Elsevier, Amsterdam, 1995.

[219] J. Morgan and G. Tian. *Ricci Flow and the Poincaré Conjecture*. http://arXiv.org/abs/math/0607607.

[220] G. D. Mostow. *Strong Rigidity of locally Symmetric Spaces*, volume 78 of *Annals of Mathematical Studies*. Princeton University Press, Princeton, 1973.

[221] G. Moussong. *Hyperbolic Coxeter groups*. PhD thesis, The Ohio State University, 1988.

[222] B. Mühlherr. On isomorphisms between Coxeter groups. *Designs, Codes and Cryptography*, 21:189, 2000.

[223] B. Mühlherr. The Isomorphism Problem for Coxeter groups. In *The Coxeter Legacy*, pages 1–15. Amer. Math. Soc., Providence, RI, 2006.

[224] J. Munkres. Obstruction to imposing differentiable structures. *Ill. J. Math.*, 12:610–615, 1968.

[225] M. H. A. Newman. A theorem on periodic transformations of spaces. *Quart. J. Math. Oxford Ser.*, 2:1–9, 1931.

[226] M. H. A. Newman. Boundaries of ULC sets in Euclidean $n$-space. *Proc. N.A.S.*, 34:193–196, 1948.

[227] M. H. A. Newman. The engulfing theorem for topological manifolds. *Ann. of Math.*, 84:555–571, 1966.

[228] G. Niblo and L. Reeves. Coxeter groups act on CAT(0) cube complexes. *J. Group Theory*, 6:399–413, 2003.

[229] G. A. Noskov and È. B. Vinberg. Strong Tits alternative for subgroups of Coxeter groups. *J. Lie Theory*, 12:259–264, 2002.

[230] P. Ontaneda. Hyperbolic manifolds with negatively curved exotic triangulations in dimension six. *J. Differential Geom.*, 40:7–22, 1994.

[231] P. Ontaneda. The double of a hyperbolic manifold and non-positively curved exotic *PL* structures. *Trans. Amer. Math. Soc.*, 355:935–965, 2003.

[232] C. D. Papakyriakopoulos. On Dehn's Lemma and the asphericity of knots. *Ann. of Math.*, 66:1–26, 1957.

[233] L. Paris. Irreducible Coxeter groups. *Internat. J. Algebra Comput.*, 193:79–93, 2007.

[234] L. Paris. Growth series of Coxeter groups. In E. Ghys, A. Haefliger, and A. Verjovsky, editors, *Group Theory from Geometrical Viewpoint (Trieste, 1990)*. World Scientific, Singapore, 1991.

[235] W. Parry. Growth series of Coxeter groups and Salem numbers. *J. Algebra*, 154:406–415, 1993.

[236] F. Paulin. Hyperbolization of polyhedra. In E. Ghys, A. Haefliger, and A. Verjovsky, editors, *Group Theory from Geometrical Viewpoint (Trieste, 1990)*. World Scientific, Singapore, 1991.

[237] G. Perelman. The entropy formula for the Ricci flow and its geometric applications. http://arXiv.org/abs/math.DG/0211159, 2002.

[238] G. Perelman. Finite extinction time for the solutions to the Ricci flow. http://arXiv.org/abs/math.DG/0307245, 2003.

[239] G. Perelman. Ricci flow with surgery on three-manifolds. http://arXiv.org/abs/math.DG/0303109, 2003.

[240] S. Prassidis and B. Spieler. Rigidity of Coxeter groups. *Trans. Amer. Math. Soc.*, 352:2619–2642, 2000.

[241] M. Prokhorov. Absence of discrete reflection groups with a non-compact polyhedron of finite volume in Lobachevsky spaces of large dimension. *Math. USSR Izv.*, 28:401–411, 1987.

[242] D. Qi. On irreducible, infinite, non-affine Coxeter groups. *Fundamenta Mathematicae*, 193:79–93, 2007.

[243] D. Quillen. Higher algebraic K-theory: I. In *Battelle Institute Conf., 1972*, volume 341 of *Lecture Notes in Mathematics*, pages 77–139. Springer-Verlag, Berlin, 1973.

[244] F. Raymond. Separation and union theorems for generalized manifolds with boundary. *Mich. Math. J.*, 7:7–21, 1960.

[245] B. Rémy. Immeubles de Kac–Moody hyperboliques. Isomorphismes abstraits entre groupes de même immeuble. *Geometriae Dedicata*, 90:29–44, 2002.

[246] I. Rivin and C. Hodgson. A characterization of compact convex polyhedra in hyperbolic 3-space. *Invent. Math.*, 111:77–111, 1993.

[247] V. A. Rohlin. A new result in the theory of 4-dimensional manifolds. *Soviet Math. Doklady*, 8:221–224, 1952.

[248] M. Ronan. *Lectures on Buildings*. Academic Press, San Diego, 1989.

[249] E. Rosas. Rigidity theorems for right-angled reflection groups. *Trans. Amer. Math. Soc.*, 308:837–848, 1988.

[250] C. Rourke and B. Sanderson. Block bundles I. *Ann. Math.*, 87:1–28, 1968.

[251] M. Sageev. Ends of group pairs and non-positively curved cube complexes. *Proc. London Math. Soc.*, 17:585–617, 1995.

[252] P. Scott and C.T.C. Wall. Topological methods in group theory. In C.T.C. Wall, editor, *Homological Group Theory*, volume 36 of *London Mathematical Society Lecture Notes*, pages 137–204. Cambridge Univ. Press, Cambridge, 1979.

[253] H. Seifert. Konstruktion dreidimensionaler geschlossener Räume. *Berlin Verb. Sachs. Akad. Leipzig, Math.-Phys. Kl.*, 83:26–66, 1931.

[254] A. Selberg. On discontinuous groups in higher dimensional symmetric spaces. In *Int. Colloquium on Function Theory*, pages 147–160. Tata Institute, Bombay, 1960.

[255] J.-P. Serre. Cohomologie des groupes discrets. In *Prospects in Mathematics*, volume 70 of *Annals. of Mathematical Studies*, pages 77–169. Princeton University Press, Princeton, 1971.

[256] J.-P. Serre. *Trees*. Springer-Verlag, Berlin, 1980.

[257] O. V. Shvartsman and È. B. Vinberg. Discrete groups of motion of spaces of constant curvature. In È. B. Vinberg, editor, *Geometry II: Spaces of Constant Curvature*, volume 29 of *Encyclopaedia of Mathematical Sciences*, pages 139–248. Springer-Verlag, Berlin, 1993.

[258] L. Siebenmann. *The obstruction to finding a boundary for an open manifold of dimension* $\geqslant 5$. Ph.D. thesis, Princeton University, 1965.

[259] S. Smale. Generalized Poincaré's Conjecture in dimensions greater than four. *Ann. Math.*, 74:391–406, 1961.

[260] P. A. Smith. Transformations of finite period II. *Ann. Math.*, 40:690–711, 1939.

[261] L. Solomon. A decomposition of the group algebra of a finite Coxeter group. *J. Algebra*, 9:220–239, 1968.

[262] M. Spivak. Spaces satisfying Poincaré duality. *Topology*, 6:77–102, 1967.

[263] M. Spivak. *A Comprehensive Introduction to Differential Geometry, Volume 5*. Publish or Perish, Boston, 1975.

[264] J. R. Stallings. Polyhedral homotopy spheres. *Bull. Amer. Math. Soc.*, 66:485–488, 1960.

[265] J. R. Stallings. The piecewise–linear structure of Euclidean space. *Proc. Cambridge Phil. Soc.*, 58:481–488, 1962.

[266] J. R. Stallings. On torsion-free groups with infinitely many ends. *Ann. of Math.*, 88:312–334, 1968.

[267] J. R. Stallings. *Group Theory and Three-dimensional Manifolds*. Yale Univ. Press, New Haven, 1971.

[268] R. Stanley. *Combinatorics and Commutative Algebra*, volume 41 of *Progress in Mathematics*. Birkhäuser, Boston, 1983.

[269] R. Stanley. Flag $f$-vectors and the cd-index. *Math. Z.*, 216:483–499, 1994.

[270] R. Steinberg. Endomorphisms of linear algebraic groups. *Mem. Amer. Math. Soc.*, 80, 1968.

[271] D. Stone. Geodesics in piecewise linear manifolds. *Trans. Amer. Math. Soc.*, 215:1–44, 1976.

[272] R. Strebel. A remark on subgroups of infinite index in Poincaré duality groups. *Comment. Math. Helv.*, 52:317–324, 1977.

[273] D. Sullivan. Hyperbolic geometry and homeomorphisms. In *Proc. Georgia Topology Conf., Athens, Ga., 1977*, pages 543–555. Academic Press, New York, 1979.

[274] R. G. Swan. Groups of cohomological dimension one. *J. Algebra*, 12:588–610, 1969.

[275] J. Świątkowski. Trivalent polygonal complexes of nonpositive curvature and Platonic symmetry. *Geom. Dedicata*, 70:87–110, 1998.

[276] W. Thurston. Three dimensional manifolds, Kleinian groups and hyperbolic geometry. *Bull. Amer. Math. Soc.*, 6:357–381, 1982.

[277] W. Thurston. Hyperbolic structures on 3-manifolds I: Deformation of acylindrical manifolds. *Ann. Math.*, 124:203–246, 1986.

[278] W. Thurston. Three-dimensional geometry and topology. Unpublished manuscript continuing book of same name, 1992.

[279] W. Thurston. *Three-Dimensional Geometry and Topology*. Princeton University Press, Princeton, 1997.

[280] W. Thurston. Shapes of polyhedra. In C. Rourke I. Rivin and C. Series, editors, *The Epstein Birthday Schrift*, volume 1 of *Geometry & Topology Monographs*, pages 511–549. International Press, Cambridge, MA, 1998.

[281] J. Tits. Groupes et géométries de Coxeter. Unpublished manuscript, 1961.

[282] J. Tits. Structures et groupes de Weyl. In *Séminaire Bourbaki 1964–65, exposé no. 288*. February 1965.

[283] J. Tits. Le problém des mots dans les groupes de Coxeter. In *Symposia Mathematica (INDAM, Rome, 1967/68)*, volume 1, pages 175–185. Academic Press, London, 1969.

[284] J. Tits. *Buildings of spherical type and finite BN-pairs*, volume 386 of *Lecture Notes in Mathematics*. Springer-Verlag, Berlin, 1974.

[285] J. Tits. On buildings and their applications. In *Proceedings of the International Congress of Mathematicians, Vancouver 1974, vol. 1*, pages 209–220. Canad. Math. Congress, Montreal, 1975.

[286] J. Tits. A local approach to buildings. In C. Davis, B. Grunbaum, and F. A. Sherk, editors, *The Geometric Vein: Coxeter Festschrift*, pages 519–547. Springer-Verlag, Berlin, 1982.

[287] J. Tits. Buildings and group amalgamations. In *Proceedings of Groups, St. Andrews 1985*, volume 121 of *LMS Lecture Notes Series*, pages 110–127. Cambridge Univ. Press, Cambridge, 1986.

[288] T. tom Dieck. Orbittypen und äquivariante Homologie I. *Arch. Math.*, 23: 307–317, 1972.

[289] H. van der Lek. *The homotopy type of complex hyperplane complements*. Ph.D. thesis, University of Nijmegan, 1983.

[290] È. B. Vinberg. Discrete linear groups generated by reflections. *Math. USSR Izvestija*, 5(5):1083–1119, 1971.

[291] È. B. Vinberg. Hyperbolic reflection groups. *Russian Math. Surveys*, 40:31–75, 1985.

[292] C. T. C. Wall. Finiteness conditions for CW-complexes. *Ann. Math.*, 81:56–59, 1965.

[293] C. T. C. Wall. List of problems. In C. T. C. Wall, editor, *Homological Group Theory*, volume 36 of *London Mathematical Society Lecture Notes*, pages 369–394. Cambridge Univ. Press, Cambridge, 1979.

[294] C. A. Weibel. *An Introduction to Homological Algebra*, volume 38 of *Cambridge Studies in Advanced Mathematics*. Cambridge Univ. Press, Cambridge, 1994.

[295] S. Weinberger. *The Topological Classification of Stratified Spaces*. *Chicago Lectures in Mathematics*. Univ. of Chicago Press, Chicago, 1994.

[296] S. Weinberger. Homology manifolds. In R. J. Daverman and R. B. Sher, editors, *Handbook of Geometric Topology*, pages 1085–1102. Elsevier, Amsterdam, 2002.

[297] S. Weinberger. *Computers, Rigidity, and Moduli*. Princeton Univ. Press, Princeton, 2005.

[298] R. Weiss. *The Structure of Spherical Buildings*. Princeton Univ. Press, Princeton, 2004.

[299] J. H. C. Whitehead. Certain theorems about three-dimensional manifolds. *Quart. Jour. Math*, 5:308–320, 1934.

[300] J. H. C. Whitehead. A certain open manifold whose group is unity. *Quart. Jour. Math.*, 6:268–279, 1935.

[301] J. H. C. Whitehead. Combinatorial homotopy I. *Bull. Amer. Math. Soc.*, 55:213–245, 1949.

[302] B. Williams. *Two topics in geometric group theory*. Ph.D. thesis, University of Southampton, 1999.

[303] E. Witt. Spiegelungsgruppen und Aufzählung halbeinfacher Liescher Ringe. *Abh. Math. Sem, Univ. Hamburg*, 14:289–322, 1941.

[304] J. Wolf. *Spaces of Constant Curvature*. 4th ed. Publish or Perish, Berkeley, 1972.

# Index